AF527783

Perspective of Extension Education

The Author

Professor Manas Mohan Adhikary (b.1951), a retired Professor of Agricultural Extension, Faculty of Agriculture, Bidhan Chandra Krishi Viswavidyalaya, West Bengal was attached with the profession of teaching, research and extension for about thirty-seven years. His immense potentiality of academic activities and capability to work with the people, sense of research and extension works helped him to serve as Dean, Faculty of Agriculture, Registrar, Bidhan Chandra Krishi Viswavidyalaya,

Vice-Chancellor, Bidhan Chandra Krishi Viswavidyalaya, and Director (In charge), Directorate of Extension Education in the same University as well as Head of the Department of Agricultural Extension. To his credit there are 98 numbers of scientific research papers published in National & International Journals of repute, a good score of chapters and 24 numbers of books .

Perspective of Extension Education

Author

***Professor* Manas Mohan Adhikary**

2019

Daya Publishing House®

A Division of

Astral International Pvt. Ltd.

New Delhi–110 002

ISBN: 978-93-88173-79-7 (HB)

Publisher's Note:

Every possible effort has been made to ensure that the information contained in this book is accurate at the time of going to press, and the publisher and author cannot accept responsibility for any errors or omissions, however caused. No responsibility for loss or damage occasioned to any person acting, or refraining from action, as a result of the material in this publication can be accepted by the editor, the publisher or the author. The Publisher is not associated with any product or vendor mentioned in the book. The contents of this work are intended to further general scientific research, understanding and discussion only. Readers should consult with a specialist where appropriate.

Every effort has been made to trace the owners of copyright material used in this book, if any. The author and the publisher will be grateful for any omission brought to their notice for acknowledgement in the future editions of the book.

Published by : **Daya Publishing House®**
A Division of
Astral International Pvt. Ltd.
– ISO 9001:2015 Certified Company –
4736/23, Ansari Road, Darya Ganj
New Delhi-110 002
Ph. 011-43549197, 23278134
E-mail: info@astralint.com
Website: www.astralint.com

Digitally Printed at : **Replika Press Pvt. Ltd.**

This copy is for sale in Indian Sub-Continent.

CENTRAL AGRICULTURAL UNIVERSITY
LAMPHELPAT, IMPHAL-795004, MANIPUR (INDIA)

Prof. M. Premjit Singh,
Vice-Chancellor

Tel : (0385) 2415933(O)
Gram : AGRIVARSITY
Fax : 2410414

Foreword

The coming turn of the century provokes various creative ideas looking back and forth and speculating on the future of agricultural education and extension. With the touch of globalization & worldwide market concept agriculture seems to be conglomeration of all types of rural livelihood aspects. In perspective of our country, India is the world's largest producer of many fresh fruits like banana, mango, guava, papaya, lemon and vegetables like chickpea, okra and milk, major spices like chilli, pepper, ginger, fibrous crops such as jute, staples such as millets and castor oil seed *etc.* India is the second largest producer of wheat and rice, the world's major food staples. India has shown a steady average nationwide annual increase in the kilograms produced per hectare for some agricultural items over the last 60 years. Before the era of mid-sixties, India relied on imports and food aid which has been cured by applying proper policy reforms & steered by Green revolution. No doubt, extension education plays a pivotal role to accelerate the growth and to achieve self-sufficiency and self-dependency in food grain production. Despite of that with time and growing population, the per capita land holding of farmers is decreasing whereas demands of food grain for survival are increasing. Recent accomplishments, agriculture has the potential for major productivity and total output gains, because crop yields in India are still just 30 per cent to 60 per cent of the best sustainable crop yields achievable in the farms of developed and other developing countries. Additionally, losses after harvest due to poor infrastructure and unorganized retail cause India to experience some of the highest food losses in the world. Globalization affects on daily life style as well as status of living. This results a complex situation to cope up with the reality. Obviously to combat the complex situation, we had to modernize, reschedule and renovate not only the theoretical analysis but also the practical implications of extension education as an instrument.

Now, if we consider the basic origin of extension education, we will find that the first agricultural extension service of a modem kind came into existence as the result of a crisis and the initiative of the occupant of a high office of authority. The crisis was the outbreak of potato blight in Europe in 1845. In Ireland, its effects were particularly severe because the predominantly peasant population relied on potatoes in their diet, and "the potato famine" persisted until 1851. The growth of agricultural education and extension work in continental Europe was to have a strong impact on the emergence of comparable activity in the United Kingdom.

If we consider Indian economy, still its backbone is Agriculture. It sustains over 70 per cent of the populations contributes 35.6 per cent to gross national products and adds 39 per cent as it's share towards exports. The growth in population and migration of the people to the city has led to an increasing demand for agricultural products. Side by side this migration causes a crisis of young blood, young generation in agricultural sector, which affects drastically in adoption a good technology. Agricultural extension efforts in India have made significant impact towards development of the agricultural sector. The report of the National Commission on Agriculture has emphasized the need for massive extension efforts to modernize the outlook of the farmers and to make them more enterprising and willing to adapt readily to innovations so that agricultural production could be increased. Agricultural extension is favourite whipping boy in the agricultural system. It is an applied behavioural science, the knowledge of which is applied to bring farmers or people through various strategies and programme of change by applying the latest scientific and technological innovations. Agricultural extension education, its principles, methods and techniques are applicable not only to agriculture but also to other sciences like veterinary, animal husbandry, dairying, health and family planning *etc.* To summarize, agricultural extension viewed as an educational programme to be undertaken by public agencies to activate the process of transferring knowledge, science and technology from laboratories to people or farmers and to help them in farm planning, decision making, record keeping, use of inputs, storage, processing and marketing, ensure supplies and services, increase their production, develop people and their leaders, improve their occupation, family and community life. The field of extension now encompasses a wider range of communication and learning activities organized for rural people by professionals from different disciplines, including agriculture, health and business studies.

I hope that the book entitled *"Perspectives of Extension Education"* has correctly caught the demand of time and come out as an instrument to encourage a clear focus on improving income, livelihood, nutrition and environmental management with the core areas of extension like concept, dimensions, objectives, philosophies as well as newer concepts of extension education. The book will cater to the needs of both undergraduate and post graduate students of agriculture and allied sciences as well as the Doctoral level, delineated by the Indian Council of Agricultural Research, New Delhi, the apex body of agricultural research, agricultural education and application of technology.

The book will certainly provide a strong base in Perspective of Extension Education for the UG and PG students, research scholars, academicians and the teachers of all the Universities/research institutes across the country and abroad. The context of this book is concentrated in a way that it serves as the base line guide for those who are dealing with the issues related to the present base needs and directions of extension education. Side by side, it also serves as the base in a multilevel stakeholders environment requiring delineation of roles, duties and responsibilities. The role of extension in micro level organization including macro level nationalized development has been implied in this book with a commendable balance. The text has been dealt with the vast domain of agriculture, prioritization of many of the issues of developmental studies, methodological researches, policy formulation and management applications.

Prof. Manas Mohan Adhikary with his almost four decades of professional experience in the field of extension education as teacher, planner, administrator and policy maker in various capacities including Vice Chancellor, Registrar, Dean, Director of Extension Education and Head of the Department; and further exposure in the process of working in many apex professional bodies and participation in various discussions held overseas, is definitely a person of right attitude and aptitude to deal with a complex subject of Extension Education in totality including its basis, sociology and psychology, rural development, participatory development, training, communication and diffusion of innovation, audio-visual aids including research, teaching and extension methods in the context of not only organizational but also country wide perspectives in connection to the integration of extension education.

I sturdily believe that this book will be applauded by the scholars, teachers, researchers, planners and policy makers, striving hard and persistent to ensure that Indian agriculture be a major player in the global market of twenty first century and Agricultural Extension will emerge as a key factor to fulfill this goal. Keep generating valuable feedback for its continuous perfection.

Professor Moirangthem Premjit Singh

Preface

Agriculture is the mainstay of Indian economy. It sustains over 70 per cent of the populations contributes 35.6 per cent to gross national products and adds 39 per cent as it's share towards exports. The growth in population and migration of the people to the city has led to an increasing demand for agricultural products. Agricultural extension efforts in India have made significant strides towards development of the agricultural sector. In the context of sustainable agricultural development, agricultural extension has a very crucial role to play. The tasks and responsibilities of extension service will need to be broad-based and holistic in contents and scope, thus beyond agricultural technology transfer. Its normal task of transferring and disseminating to farmers appropriate agricultural technologies and good farm practices would not be sufficient. Extension agencies, services, and workers will need to exercise a more proactive and participatory role and serve as knowledge/information "brokers", initiating and facilitating mutually meaningful and equitable knowledge-based transactions among agricultural researchers, trainers, and primary producers. To improve its cost-effectiveness, proper strategies to advocate favourable and explicit agricultural extension policies are needed. Modern strategic planning and quality management tools and approaches should be utilized in developing or restructuring extension organizations or institutions. More innovative methods must be developed to identify systematically farmer's problems and felt needs, and to help formulate and set agricultural research agenda based on such needs and problems. In short, there is a need to develop and improve the conceptual, technical, and operational methods and tools in order to strategically plan, efficiently manage, and scientifically evaluate a problem-solving, demand-driven and needs-based agricultural extension programmes. To summarize, agriculture extension viewed as an educational programme to be undertaken by public agencies to activate the process of transferring knowledge, science and technology from laboratories to people or farmer and to help them in farm planning, decision making, record keeping, use of inputs, storage, processing and marketing, ensure supplies and services, increase their production, develop people and their leaders, improve their occupation, family, and community life.

The book "Perspective of Extension Education" will help in gathering knowledge about the upcoming decades choosing communication intervention out of huge choices of communication. The modern civilization has up toned the textural configuration of rural people; so the role of prompt, precise, palpable information and its proper communication would write a new biography for million of farmers. The birth of the modern extension service has been attributed to events that took place in Ireland in the middle of the 19th century. In-between 1845 – 1851, the Irish potato crop was destroyed by fungal diseases and a severe famine occurred. The British Government arranged for "practical instructors" to travel to rural areas and teach small farmers how to cultivate alternative crops. This scheme attracted the attention of government officials in Germany, who organized their own system of travelling instructors. By the end of the 19th century, the idea had spread to Denmark, Netherlands, Italy, and France. An agricultural extension service offers technical advice on agriculture to farmers, and also supplies them with the necessary inputs and services to support their agricultural production. It provides information to farmers and passes to the farmer's new ideas developed by agricultural research stations. Agricultural extension programmes cover a broad area including improved crop varieties, better livestock control, improved water management, and the control of weeds, pests or plant diseases. Where appropriate, agricultural extension may also help to build up local farmers' groups and organizations so that they can benefit from extension programmes. This book may play a pivotal role to reveal the philosophy of extension to communicate with the target people and it may be utilized as a light house to explore the new era of the subject

I am grateful to my wife Shibani, son Mayukh and daughter Manjeer who acts as a source of inspiration to me during writing of this book. I also give my profuse thanks to the Publisher, M/S Daya Publishing House a division of Astral International Private Limited specially, Sri Anil Mittal Ji, for their help in bringing out the book in time.

Manas Mohan Adhikary

Contents

Part II: Society with Sociology and Psychology

Part III: Rural Development

Part IV: Participatory Development

Part V: Training and Development

Part VI: Communication and Diffusion of Innovations

Part VII: Audio-Visual Aids

Part VIII: Research, Teaching and Extension

Part 1

Basis of Extension Education

1

Fundamental of Extension Education

The term extension was first used in the United States of America in the first decade of this century to con note the extension of knowledge from land grant colleges to the farmers through the process of informal education. In India, extension work was primarily started by F.L. Brayne (1920) in Punjab the term community development and extension education became more popular with the launching of community development projects in 1952 and with the establishment of the national extension service in 1953, Since then, community Development has been regarded as a programme for all-round development of the rural people and extension education as the means to achieve this objective.

Extension Terminology

Extension began in Dublin, Ireland in 1847 with Lord Clarendon's itinerant instructors during the great famine. It expanded in Germany in the 1850s, through the itinerant agricultural teachers *Wanderlehrer* and later in the USA via the cooperative extension system authorized by the Smith-Lever Act in 1914. The term was later adopted in the United States of America, while in Britain it was replaced with "advisory service" in the 20th century. A number of other terms are used in different parts of the world to describe the same or similar concept:

- ☆ Arabic: *Al-Ershad* ("guidance")
- ☆ Dutch: *Voorlichting* ("lighting the path")
- ☆ German: *Beratung* ("advisory work")
- ☆ French: *Vulgarisation* ("popularization")
- ☆ Spanish: *Capacitación* ("training" "capacity building")

- ☆ Thai, Lao: *Song-Suem* ("to promote")
- ☆ Persian: *Tarvij* and *Gostaresh* ("to promote and to extend")-

In the US, an extension agent is a university employee who develops and delivers educational programs to assist people in economic and community development, leadership, family issues, agriculture and environment. Another program area provided by extension agents is 4-H and youth activities. Many extension agents work for cooperative extension service programs at land-grant universities. They are sometimes referred to as county agents, or extension educators. Often confused with Extension agents, Extension specialists are subject matter experts usually employed as scientists and university professors in various departments in the land-grant university system. Subjects ranges from agriculture, life sciences, economics, engineering, food safety, pest management, veterinary medicine, and various other allied disciplines. These subject matter specialists work with agents (usually in a state wide or regional team environment) to support programs within the cooperative extension system. Extension is constantly evolving and the availability of rapid sharing has seen new platforms like Digital Green and Plant Village.

Definitions of Extension

There is no widely accepted definition of agricultural extension. The examples given below are taken from a number of books on extension published over a period of more than 50 years:

- ☆ 1949: The central task of extension is to help rural families help themselves by applying science, whether physical or social, to the daily routines of farming, homemaking, and family and community living.
- ☆ 1965: Agricultural extension has been described as a system of out-of-school education for rural people.
- ☆ 1966: Extension personnel have the task of bringing scientific knowledge to farm families in the farms and homes. The object of the task is to improve the efficiency of agriculture.
- ☆ 1973: Extension is a service or system which assists farm people, through educational procedures, in improving farming methods and techniques, increasing production efficiency and income, bettering their standard of living and lifting social and educational standards.
- ☆ 1974: Extension involves the conscious use of communication of information to help people form sound opinions and make good decisions.
- ☆ 1982: Agricultural Extension: Assistance to farmers to help them identify and analyze their production problems and become aware of the opportunities for improvement.
- ☆ 1988: Extension is a professional communication intervention deployed by an institution to induce change in voluntary behaviours with a presumed public or collective utility.

- ☆ 1997: Extension is the organized exchange of information and the deliberate transfer of skills.
- ☆ 1999: The essence of agricultural extension is to facilitate interplay and nurture synergies within a total information system involving agricultural research, agricultural education and a vast complex of information-providing businesses.
- ☆ 2004: Extension is a series of embedded communicative interventions that are meant, among other goals, to develop and/or induce innovations which help to resolve (usually multi-actor) problematic situations.
- ☆ 2006: Extension is the process of enabling change in individuals, communities and industries involved in the primary industry sector and in natural resource management.

The Term "Extension"

The use of the word "extension" derives from an educational development in England during the second half of the nineteenth century. Around 1850, discussions began in the two ancient universities of Oxford and Cambridge about how they could serve the educational needs, near to their homes, of the rapidly growing populations in the industrial, urban area. It was not until 1867 that a first practical attempt was made in what was designated "university extension," but the activity developed quickly to become a well-established movement before the end of the century. Initially, most of the lectures given were on literary and social topics, but by the 1890s agricultural subjects were being covered by peripatetic lecturers in rural areas (Jones, 1994). The growth and success of this work in Britain influenced the initiation of similar activity elsewhere, especially in the United States. There, in many states, comparable out-of-college lectures were becoming established by the 1890s (True, 1900, 1928). During the first two decades of this century, the extramural work of the land-grant colleges, concerned with serving the needs of farm families, was to expand dramatically and become formally organized; but the use of the term "extension" continued and has persisted as the designation for the work.

The overt use of the notion of "extending" relevant and useful information to the adult population at large, however, predates the university extension movement. Earlier in the nineteenth century, a British politician, Lord Henry Brougham, an influential advocate of formal education for the poor and of mass adult education, founded the Society for the Diffusion of Useful Knowledge in 1826. Its objective was "imparting useful information to all classes of the community, particularly to such as are unable to avail themselves of experienced teachers, or may prefer learning by themselves." The society sought to do this largely through producing low-priced publications and establishing local committees throughout the country "for extending the object of the Society" (Society for the Diffusion of Useful Knowledge 1827). During its twenty years' existence, agricultural topics were well covered in the society's publications. Similar, albeit short-lived, societies were also established before 1840 in several other European countries, India, China, Malaysia, and the United States (in Virginia) (Grobel, 1933; Smith, 1972).

The Origins

The dissemination of relevant information and advice to farmers, however, has a long if chequered history prior to the emergence of modem forms of agricultural extension in the nineteenth century.

The first known example was in Mesopotamia (roughly, present-day Iraq) around 1800 B.C. Archaeologists have unearthed clay tablets of the time on which were inscribed advice on watering crops and getting rid of rats - important for mitigating any potential loss of taxation revenue from farmers (Ahmed, 1982, as quoted in Bne Saad, 1990). Some hieroglyphs on Egyptian columns also gave advice on avoiding crop damage and loss of life from the Nile's floods. An important advance was the beginning of agricultural writings. Though few have survived, the earliest were written during the ancient Greek and Phoenician civilizations, but some of them were adapted by Roman writers. From the second century B.C. to the fourth century A.D., several important Latin texts were written, frequently drawing on practical farming experience, which aimed to help Roman landowners to maintain and improve then-estates and their revenues (White, 1970, 1977).

At around the same period in imperial China, early forms of advancing and disseminating agricultural information also began. That landowners and their tenants should improve their production was a matter of concern to the state since, from the sixth century B.C. onwards, it relied heavily on land taxes for its revenues. The support of relevant agricultural research and the dissemination of information and advice had certainly begun by the late Han Dynasty (25-220 A.D.). The oldest fully surviving Chinese agricultural treatise, *Essential Techniques/or the Peasantry,* dating from 535 A.D., aimed to show landowners how to improve their estate management through the advice they gave to their tenants. The Sung and Yuan Dynasties (960-1368) with their firm local government administrations were notable in organizing and promoting agricultural research, extension work, and the teaching of agriculture and sericulture, much facilitated by the invention of woodblock printing, which allowed agricultural treatises and practical handbooks to be widely distributed. Similar activities continued during the succeeding Ming (1368-1644) and Chi'ing (1644-1912) Dynasties, driven not only by the growing population and periodic threats of famine, but also by the state's recognition of the importance of well-coordinated extension work on agricultural recommendations if the most benefit was to be achieved (Perkins, 1969; Elvin, 1973; Bray, 1984; Delman, 1991).

Necessary Conditions for Agricultural Extension to Evolve

Apart from the importance of farmers and agriculture in the society and economy concerned, several conditions appear to be necessary for the initiation and organized development of agricultural extension work.

The prime condition is that information has been assembled, systematized, and made available on good or progressive or new agricultural practices suited to a particular environment, and is based on either (or both) the accumulation of experience or findings from research (however rudimentary).

Second, this information is used, among other things, to educate professional agriculturists who may further enlarge or refine this body of knowledge or become active promoters and disseminators of it.

Third, an appropriate administrative or organizational structure exists by and within which the dissemination activities may be established and conducted.

Fourth, there is a legislative or some other official mandate or influential proponent which prescribes or enables that agricultural extension work is desirable and must occur.

Fifth, there are invariably a variety of antecedents which have attempted protoforms of agricultural information and advice dissemination.

In addition, the incidence of critical situations, such as famine, crop failure, soil exhaustion, or altered economic conditions or relationships, may create an immediate cause for initiating the organization of extension work. All or several of these conditions have been present in the evolution of modem forms of agricultural extension.

Towards the Modern Era

The direct antecedents of organized agricultural research and dissemination of its results which occurred in nineteenth century Europe and North America can be traced back to the "renaissance" which began in the fourteenth century. Between 1300 and 1700, European society became transformed from its medieval feudal forms into recognizably modem social systems. It was a period of complex, multi stranded development. Along with the growth of national states and European exploration and "discovery" of the rest of the world was the "new learning." This involved not only a fresh appreciation of rediscovered classical writings and art forms, but also many novel ideas and activities, a spirit of humanism, and rational enquiry. All of this was considerably facilitated by the invention of printing using movable type, usually attributed to Gutenberg around 1450, and the rapid diffusion over Europe of the printing press, for whose output there existed a ready market.

The earliest known renaissance agricultural text was written in Latin by Pietro de Crescenzi in 1304 and was translated into Italian and French. This became the first book on agriculture to be printed in the mid-fifteenth century. Others soon followed, often based on the old Latin texts or on the collected wisdom of farmers and their families. A well-known example, a compendium of helpful advice in simple verse and a bestseller in Tudor England, was Thomas Tusser's *A hundredth goode pointes of husbandrie*, published in 1557 and expanded in 1573 to five hundred good points with as many on "goode house wiferie" (Tusser, 1580). Less popular, but of greater significance, were Francis Bacon's writings early in the next century based on his observations and scientific experiments on his estate north of London - the beginnings of the application of science and scientific method to agriculture (Russell, 1966).

By the mid-eighteenth century, throughout much of Europe, progressive landowners (frequently aristocrats) and their agents and a few similarly minded farmers were being known as "improvers." These, along with some "men of science," were the main proponents of agricultural clubs or societies. At their regular meetings and demonstrations, locally and regionally, landowners and leading farmers exchanged ideas and information and discussed farming improvements. Two main forces underlay the movement. First, many landowners were eager to learn of ways to improve their estates and the production capabilities of their tenants so as to increase the value of their estates and their rental incomes. Secondly, progress was being made towards modern science and its application to agriculture, especially in agricultural chemistry and plant physiology (Russell, 1966). These societies sought to alter radically the traditional modes of farming by initiating experiments, arranging demonstrations, disseminating information, and advocating the adoption of innovations. It was considered almost a duty by their elite membership to make their initiatives and activities known to "the generality" of farmers through publishing their proceedings and reporting their meetings in newspapers (Hudson, 1972). Although such agricultural societies initially spread slowly - the first had been formed at Rezzato near Milan in 1548 (Coletti, 1900) - they had become common throughout much of Europe by 1800, and a small number had been established by that year in the young United States and eastern Canada.

It is not possible, here, to enter into detail on the interactions between a growing scientific knowledge of agriculture and its application in practice, the many examples of increasingly widespread agricultural improvement, and the numerous personalities involved in Europe and North America during the century or so after 1750. Reference must, however, be made to one figure whose ideas and activities were of pivotal significance to the developments of the time, and later. This was Philipp Emanuel von Fellenberg (1771-1844), who in 1799 purchased the estate of Wylhof, which he renamed Hofwyl, near Bern in Switzerland (Gray, 1952; Guggisberg, 1953). Over the next decade or so, he established agricultural schools at Hofwyl for the children of peasants and of the poor and for the aristocracy and their agents.

By the 1820s, most of the elements for creating modem forms of agricultural extension were in being, although each was to develop considerably during the nineteenth century. A crucial missing element, however, was an effective means by which the "generality" of farmers could be directly given information, advice, and encouragement. This required itinerant agriculturists who could meet farmers in their home localities, give instructional talks and demonstrations, advocate superior or new practices, and have discussions with the farmers. The notion of "itinerancy" was not new: since late medieval times, tradesmen and proto-professional men had travelled through rural areas to serve their clients. The first examples of itinerant agricultural lecturers-cum-instructors were in parts of New England and New York in the 1820s (True, 1928) and in France, where a first migratory agricultural teacher was appointed in the Gironde in 1837, followed by nine more in various areas of the country in succeeding years (Boulet n.d.). During the 1840s, further sporadic developments also occurred in the United States, particularly in New York, Ohio, and Maryland (True, 1928), while in Württemberg, in southwest Germany, a pasture specialist (*Wiesenbaumeister*) together with a staff of eighteen technicians was employed by the state agricultural society to advise farmers, landowners, and town administrations on land drainage, irrigation, and improved pasture management (CLVS, 1845).

In Europe, agricultural science was evolving rapidly by the 1840s, with notable strides being made in Germany by Justus von Liebig at Giessen, and with the establishment of agricultural experiments at Rothamsted in England in 1843 by John Bennet Lawes and Henry Gilbert. Agricultural societies and their shows were flourishing. Numerous publications and periodicals were aimed at farmers. Agricultural schools, if not commonplace, had been established in most European countries. Thus a small minority of younger landowners and farmers had received a formal education in their calling, while purposely trained agriculturists were available to be engaged as estate agents or teachers. Many of the more progressive landowners employed agents to travel around their estates to urge improved methods on their tenants. The main element necessary to create modern agricultural extension services was for legitimate authorities to establish the necessary organizations - and the germ of this had already been present in France, Germany, and the United States.

The Birth of Modern Agricultural Extension Services

The first agricultural extension service of a modem kind came into existence as the result of a crisis and the initiative of the occupant of a high office of authority. The crisis was the outbreak of potato blight in Europe in 1845. In Ireland its effects were particularly severe because the predominantly peasant population relied on potatoes in their diet, and "the potato famine" persisted until 1851. The new British viceroy appointed to Ireland in 1847, the Earl of Clarendon, soon after his arrival in Dublin wrote a letter (Jones, 1982) to the president of the Royal Agricultural Improvement Society of Ireland (founded in 1841), which acted as the central society for numerous local agricultural societies. This letter, no less than an official directive, urged the society to appoint itinerant lecturers to travel around the most distressed districts to inform and show small farmers, in simple terms, how to improve their cultivation and how to grow nutritious root crops other than potatoes. "Lord Clarendon's practical instructors in husbandry," as they became known, were centrally appointed, deployed, and paid and reported weekly to the society in Dublin, with some local control of their activities being exercised by the major landowners in their areas. Over the four years of its existence, the scheme was funded to about half its total cost by landowners and charitable donations, with the remainder coming from government-controlled funds (Jones, 1979, 1981).

The potato famine also led to consideration being given in Württemberg to employing itinerant farm advisers, but the proposal failed to gain approval (R. Bühler, personal communication). However, from the mid-1850s, first in Württemberg, Hesse, and western Prussia, itinerant agricultural teachers (*Wanderlehrer*) began to be appointed under the auspices of central agricultural societies. After some ten years, the system grew rapidly, influenced in part by the crisis among vine growers resulting from the devastation caused by *phylloxera* aphid infestations, and became formalized (Jones, 1981). Normally, the *Wanderlehrer* spent the summer half of the year travelling around their districts giving talks, demonstrations, and advice to farmers; during the remainder of the year they taught farmers' sons at winter agricultural schools. Although officially they were part of the activities of the agricultural associations, their work was in all cases supported heavily by state funds, and their advice was free to farmers. When the system was adopted in the kingdom of Bavaria in 1896, it was as an integral part of the state civil service; the extension workers were grandly titled Royal Agricultural Teachers (*Königliche Landwirtschafts-lehrer*) (Maier-Bode, 1910).

By the close of the nineteenth century, agricultural extension systems modelled to a considerable extent on the German *Wanderlehrer* had spread: to Denmark from 1870 onwards; to the Netherlands, where a few extension workers (*wandelleraren*) had been appointed by agricultural societies in the late 1840s and 1850s, but had then disappeared before being revived as a government system in the 1890s; to Italy, where the first itinerant agricultural teacher (*cattedra ambulante di agricoltura*) was appointed in 1886 at Rovigo, near the estuary of the River Po, with many others following in the next decade and funded largely by public donations, the church, and the banks; to Switzerland; to much of the Austro-Hungarian Empire; and to Russia.

The growth of agricultural education and extension work in continental Europe was to have a strong impact on the emergence of comparable activity in the United Kingdom. An official commission on technical education in the early 1880s included a detailed review of the European developments (Jenkins, 1884). At the end of the decade, a cluster of enactment's, which established county-based local government, created a board of agriculture, promoted technical (including agricultural) education, and allocated funds for the purpose, enabled agricultural extension work to be initiated. It was to be part of the services provided by the local government authorities. They either employed their own agricultural officer or more commonly sponsored lectures and travelling schools on agriculture (especially dairying) as part of the university extension system. This meant drawing on the staffs of the agricultural departments which were being created in new institutions of higher education. Government funds were available to support these activities, but funding also had to be provided by the local county authorities (Jones, 1994). By the turn of the century, such work existed throughout Great Britain.

This system and its underlying legislation, however, did not apply to Ireland (then entirely a part of the United Kingdom). There, agricultural extension work became established in 1900 as a result of the initiative of Horace Plunkett, well known for his advocacy of agricultural cooperation. An official committee in 1896, chaired by Plunkett, reviewed the developments in Europe and North America (Report, Recess Committee, 1896) and set out to adapt the various systems to suit Irish conditions. In 1900, a Department of Agriculture and Technical Instruction was established in Dublin, governed by a board of representative Irishmen. This initiated itinerant agricultural instruction, organized within each county as in Britain and similarly resourced partly from local and partly from central funds. A vague recollection existed of Lord Clarendon's "practical instructors" half a century earlier, and the title "itinerant instructors" was applied to the new extension workers, who were expected to provide information and advice, each to be "the guide, philosopher and friend of the existing farmers" (Plunkett, 1901-02).

Many visitors and several official delegations from North America to Europe, particularly from the mid-nineteenth century onwards, reported back on the progress in agricultural research and education, including the itinerant teachers. In the United States and eastern Canada, agricultural societies had become common during the first half of the century and, usually supported by their state or provincial legislatures, some had at times sponsored itinerant lecturers in agriculture. However, two other developments after 1850 were of more significance to the evolution of agricultural extension in the United States. First was the Morrill Act of 1862, signed by President Lincoln during the Civil War, which was seminal in the creation of state colleges "of agriculture and the mechanic arts" in the northern United States; its land-grant provisions enabled the states to establish and fund their colleges. Second was the beginning at about the same time of the farmers' institute movement. These institutes organized one or two-day (and later longer) meetings, which became popular after 1860, arranged by and for farmers.

Both developments had been widely discussed during the previous decade, and their growth over the next half century was closely interwoven. The visiting speakers at the institutes were largely professors at the state colleges of agriculture, and both depended on the formal support of their state legislatures and of farmers, especially through their agricultural societies (True, 1895, 1928; Kile, 1921). Over the next forty years, these activities were influenced also by the university extension movement in Britain and the growing interest in adult self-improvement (inspired, for example, by the Chautauqua adult education institution in New York State). By 1890, when the second Morrill Act granted federal funds for the establishment of agricultural colleges in the remainder of the United States, the farmers' institutes had spread throughout and become a national institution with federal support and supervision, further stimulated by the formal establishment of experimental work at the state colleges of agriculture under the 1887 Hatch Act. A comparable development of farmers' institutes began in Ontario, Canada, in 1885. These were financially supported by the provincial legislature and spread rapidly with lecturers mainly from the Ontario Agricultural College at Guelph (founded in 1874). A somewhat similar system began in Prince Edward Island (Province of Ontario, 1900; Blackburn and Vist, 1984).

Thus, by the end of the last century, a system of agricultural extension work had become well established in a large part of North America. In the United States, the colleges and their leading professors, including several notable proponents of more practical extension work, progressively took over the initiation and organization of the activity. This culminated in 1914 with the passage of the Smith-Lever Act, establishing the Cooperative Extension Service - a tripartite cooperation of federal, state, and local county governments, with the state college as the extension agency - "in order to aid in diffusing among the people of the United States useful and practical information on subjects relating to agriculture and home economics, and to encourage the application of the same."

In the Southern Hemisphere, extension work also became established along the wide coastal belts of southern and eastern Australia. Several agricultural ("show") societies were formed in the second half of the nineteenth century, although their effect was slight, but as the state administrations became more organized, departments of agriculture were established in the 1870s and 1880s with the aim of developing the potential of their territories. They recognized the importance of agricultural education, influenced by British, Irish, and some American examples whose activities were widely reported in the Australian press. Before the end of the century, under specific state legislation, the departments of agriculture had established agricultural colleges and experimental work in Victoria, South Australia, New South Wales, and Queensland (Black, 1976).

Associated with this development was the official appointment in these states of the first itinerant agricultural instructors in the late 1880s. At the same time, because of the potential importance of milk products, travelling dairy schools were begun, while state exhibitions, especially the Centennial Exhibition in Melbourne in 1888, showed what was possible and gave considerable impetus to farming improvements. The few "government experts," some from the United Kingdom or the United States, grew in number during the 1890s and the first decade of this century, developing the range of the extension work. Its impact and that of the agricultural colleges in their early years was probably slight, but the basis had been laid for further development (Logan, 1984).

Agricultural extension work had also started before 1900 in Japan. Following the Meiji Restoration in 1868, new administrative structures and various modernizing policies were adopted. Two agricultural colleges were established in the mid-1870s, staffed by Western (mainly European) teachers. At these colleges and government farms, experimental work was conducted and new practices were tested and developed. At the same time, agricultural fairs and exhibitions were begun, and progressive Japanese farmers gave talks and demonstrations at them. These led to the development of many agricultural societies from 1881 onwards, a "movement" formalized by legislation in 1899. In 1885, the government also initiated, at national and prefectural levels, a system of appointing experienced farmers as itinerant agricultural lecturers (because the Western "experts" knew little about rice husbandry). Supported by the work at government experiment stations, established from 1893, these farmers formed the basis of agricultural extension work. This activity, including the establishment of demonstration farms, was allocated in 1903 to the numerous agricultural societies which, with state funds, appointed agricultural technicians. In 1910, the 1899 law was strengthened; thereafter, farmers were required to belong to a village agricultural society which was linked to a national network and hierarchy of societies, and farmers were compelled to adopt the technical guidance and recommendations of the societies' extension workers - what became known as "forced extension" (Tajima, 1991; Ministry of Agriculture, 1993).

The development and organization of agricultural extension work was not entirely confined to temperate countries. In a variety of ways, it had also begun in tropical areas, especially in colonial territories. The European colonial powers looked to their overseas territories as a source of tropical agricultural products.

Despite a long connection with some of the colonial areas, the Europeans remained largely ignorant of many tropical agricultural plants. The solution was to establish experimental and demonstration "botanical gardens." The earliest was opened in 1821 at Peradeniya, Sri Lanka (Ceylon), and two others were established in the country later in the nineteenth century. Smaller ones were also created in several Caribbean islands and some West African territories. During the early years of this century, some of these developed considerably, although others were short-lived. Those which succeeded provided important sources of agricultural knowledge and innovation and formed the basis for an interest in agricultural societies and agricultural instruction. Some attempts were also undertaken to improve "native agriculture." This was often associated with the creation, as part of the administration, of departments of agriculture and the appointment of professional agriculturists as directors of agriculture.

A central department of agriculture was established in India after the 1866 Orissa famine, and the government of India soon after resolved to establish departments in each province. However, it was 1905 before a central government directive ordered every province to appoint a full time director of agriculture who should organize agricultural research and demonstration farms with staff who could advise farmers (Mook, 1982). The first British colony to appoint a director of agriculture was Zanzibar in 1896. Of more significance, however, was the creation in 1898 of the Imperial Department of Agriculture for the West Indies, with headquarters in Barbados. Before 1914, such departments of agriculture had been created in several African and Southeast Asian territories, as well as in several Caribbean islands (Masefield, 1950). In Sri Lanka, a few agricultural instructors had been appointed about 1880 to work alongside government agents. When in 1904 the Ceylon Agricultural Society was formed to promote experimental work, it also began an agricultural extension service with the objective of reaching native cultivators (Arasasingham, 1981).Along with school gardens (Willis, 1922), the extension workers were considered an effective way of demonstrating improved cultivation practices to villagers. Similar developments also occurred in the Caribbean.

What is Agricultural Extension?

Agriculture is the mainstay of Indian economy. It sustains over 70 per cent of the populations contributes 35.6 per cent to gross national products and adds 39 per cent as it's share towards exports. The growth in population and migration of the people to the city has led to an increasing demand for agricultural products. Agricultural extension efforts in India have made significant strides towards development of the agricultural sector. The report of the national commission on Agriculture (1976) has emphasized the need for massive extension efforts to modernize the outlook of the farmers and to make them more enterprising and willing to adapt readily to innovations so that agricultural production could be increased.

Agriculture extension is favorite whipping boy in the agricultural system. It is an applied behavioural science, the knowledge of which is applied to bring farmers or people through various strategies and programme of change by applying the latest scientific and technological innovations. Agriculture extension education, its principles, methods and techniques are applicable not only to agriculture but also to other sciences like veterinary, animal husbandry, dairying, health and family planning *etc.*

To summarize, agriculture extension viewed as an educational programme to be undertaken by public agencies to activate the process of transferring knowledge, science and technology from laboratories to people or farmer and to help them in farm planning, decision making, record keeping, use of inputs, storage, processing and marketing, ensure supplies and services, increase their production, develop people and their leaders, improve their occupation, family, and community life.

Agricultural extension is known as the application of scientific research and new knowledge to agricultural practices through farmer education. The field of extension now encompasses a wider range of the communication and learning activities organized for rural people by professionals from different disciplines, including agriculture, health and business studies.

Thrust Areas of Agricultural Extension

a) Assessment refinement and demonstration of technology/products.

b) Training of farmers.

c) Training of extension personnel areas of technology.

d) Single window delivery system of technology products, diagnostic service and information through Agricultural Technology Information Centers.

e) Development of gender-specific technologies.

f) Creating awareness of improved agricultural technologies among the farmers

Agricultural Extension aims at improving technology dissemination of farm technologies to the farmers. The Extension system has undergone several changes over a period, in terms of content and delivery approaches.

Origins of Agricultural Extension

It is not known where or when the first extension activities took place. It is known, however, that Chinese officials were creating agricultural policies, documenting practical knowledge, and disseminating advice to farmers at least 2,000 years ago. For example, in approximately 800 BC, the minister responsible for agriculture under one of the Zhou dynasty emperors organized the teaching of crop rotation and drainage to farmers. The minister also leased equipment to farmers, built grain stores and supplied free food during times of famine.

The birth of the modern extension service has been attributed to events that took place in Ireland in the middle of the 19th century. In-between 1845 – 1851, the Irish potato crop was destroyed by fungal diseases and a severe famine occurred. The British Government arranged for "practical instructors" to travel to rural areas and teach small farmers how to cultivate alternative crops. This scheme attracted the attention of government officials in Germany, who organized their own system of travelling instructors. By the end of the 19th century, the idea had spread to Denmark, Netherlands, Italy, and France.

The term "university extension" was first used by the Universities of Cambridge and Oxford in 1867 to describe teaching activities that extended the work of the institution beyond the campus. Most of these early activities were not, however, related to agriculture. It was not until the beginning of the 20th century, when colleges in the United States started conducting demonstrations at agricultural shows and giving lectures to farmer's clubs, that the term "extension service" was applied to the type of work that we now recognize by that name.

In the United States, the Hatch Act of 1887 established a system of agricultural experiment stations in conjunction with each state's land-grant university, and the Smith-Lever Act of 1914 created a system of cooperative extension to be operated by those universities in order to inform people about current developments in agriculture, home economics, and related subjects.

Four Generations of Extension in Asia

The development of extension services in modern Asia has differed from country to country. Despite the variations, it is possible to identify a general sequence of four periods or "generations":

- **Colonial agriculture:** Experimental stations were established in many Asian countries by the colonial powers. The focus of attention was usually on export crops such as rubber, tea, cotton, and sugar. Technical advice was provided to plantation managers and large landowners. Assistance to small farmers who grew subsistence crops was rare, except in times of crisis.
- **Diverse top-down extension:** After independence, commodity-based extension services emerged from the remnants of the colonial system, with production targets established as part of five-year development plans. In addition, various schemes were initiated to meet the needs of small farmers, with support from foreign donors.
- **Unified top-down extension:** During the 1970s and 1980s, the Training and Visit system (T&V) was introduced by the World Bank. Existing organizations were merged into a single national service. Regular messages were delivered to groups of farmers, promoting the adoption of "Green Revolution" technologies.
- **Diverse bottom-up extension:** When World Bank funding came to an end, the T&V system collapsed in many countries, leaving behind a patchwork of programs and projects funded from various other sources. The decline of central planning, combined with a growing concern for sustainability and equity, has resulted in participatory methods gradually replacing top-down approaches.

The fourth generation is well established in some countries, while it has only just begun in other places. While it seems likely that participatory approaches will continue to spread in the next few years, it is impossible to predict the long-term future of extension. Compared to 20 years ago, agricultural extension now receives considerably less support from donor agencies. Among academics working in this field, some have recently argued that agricultural extension needs to be reinvented as a professional practice. Other authors have abandoned the idea of extension as a distinct concept and prefer to think in terms of "knowledge systems" in which farmers are seen as experts rather than adopters.

Aspects of Future Extension Education

a) Evolution of extension system and operationalisation of approaches
b) Future extension education initiatives
c) Collegiate participation of farmers
d) Web enabled technology dissemination
e) Developing cases as tools for technology dissemination
f) Agriculture as a profitable venture
g) Scaling up of group mobilization
h) Micro-enterprises promotion

Several of the institutional innovations that have come up in response to the weaknesses in public research and extension system have given enough indications of the emergence of an agricultural innovation system in India. This has resulted in the blurring of the clearly demarcated institutional boundaries between research, extension, farmers, farmers' groups, NGOs and private enterprises. Extension should play the role of facilitating the access to and transfer of knowledge among the different entities involved in the innovation system and creates competent institutional modes to improve the overall performance of the innovation system. Inability to play this important role would further marginalize extension efforts.

Communication Processes

The term "extension" has been used to cover widely differing communication systems. Two particular issues help to define the type of extension: how communication takes place, and why it takes place.

The related but separate field of agricultural communication has emerged to contribute to in-depth examinations of the communication processes among various actors within and external to the agricultural system. This field refers to the participatory extension model as a form of public relations-rooted two-way symmetric communication based on mutual respect, understanding, and influence between an organization and its stakeholders.

Agricultural communication can take three modes face-to-face training, training "products" such as manuals and videos, or information and communication technologies (ICTs), such as radio and short message system (SMS). The most effective systems facilitate two-way communication and often combine different modes.

Four Paradigms of Agricultural Extension

Any particular extension system can be described in terms of both how communication takes place and why it takes place. It is not the case that paternalistic systems are always persuasive, nor is it the case that participatory projects are necessarily educational. Instead there are four possible combinations, each of which represents a different extension paradigm, as follows:

- **Technology transfer (persuasive + paternalistic):** This paradigm was prevalent in colonial times and reappeared in the 1970s and 1980s when the "Training and Visit" system was established across Asia. Technology transfer involves a top-down approach that delivers specific recommendations to farmers about the practices they should adopt.
- **Advisory work (persuasive + participatory):** This paradigm can be seen today where government organizations or private consulting companies respond to farmers' inquiries with technical prescriptions. It also takes the form of projects managed by donor agencies and NGOs that use participatory approaches to promote predetermined packages of technology.
- **Human resource development (educational + paternalistic):** This paradigm dominated the earliest days of extension in Europe and North America, when universities gave training to rural people who were too poor to attend full-time courses. It continues today in the outreach activities of colleges around the world. Top-down teaching methods are employed, but students are expected to make their own decisions about how to use the knowledge they acquire.
- **Facilitation for empowerment (educational + participatory):** This paradigm involves methods such as experiential learning and farmer-to-farmer exchanges. Knowledge is gained through interactive processes and the participants are encouraged to make their own decisions. The best known examples in Asia are projects that use Farmer Field Schools (FFS) or participatory technology development (PTD).

There is some disagreement about whether or not the concept and name of 'extension' really encompasses all four paradigms. Some experts believe that the term should be restricted to persuasive approaches, while others believe it should only be used for educational activities. Paulo Freire has argued that the terms 'extension' and 'participation' are contradictory. There are philosophical reasons behind these disagreements. From a practical point of view, however, communication processes that conform to each of these four paradigms are currently being organized under the name of extension in one part of the world or another. Pragmatically, if not ideologically, all of these activities are considered to be represented in agricultural extension.

Types of Extension

There is no one universal type of extension but a variety of activities and approaches which can be called extension. It has already been stated that since agriculture is the basis of a rural economy, agricultural extension is the most common type of extension to be found in rural areas. But the areas of knowledge and new ideas that farmers and their families require are not restricted to agriculture. There are other aspects of family life in which new knowledge and practices can lead to improvement. Extension is any activity that works with farmers and their families in order to improve the economic and social conditions of their lives and to develop their ability to take responsibility for their own future development. This extension, however, can take different forms and it would be useful to review the two principal ones.

Agricultural Extension

There is probably more extension agents involved in agricultural activities than in any other aspect of rural life. Given the importance of agriculture and the need to produce food both for the farm family and for the nation as a whole, this emphasis upon agricultural extension is understandable. Some agricultural extension services are based upon a single crop, while others adopt more of a "whole farm" approach. The choice is very much dependent upon the local agricultural system and the national crop requirements. In regions where cash crops such as cotton, cocoa or sugar grow, the single crop extension approach is more common.

An agricultural extension service offers technical advice on agriculture to farmers, and also supplies them with the necessary inputs and services to support their agricultural production. It provides information to farmers and passes to the farmers new ideas developed by agricultural research stations. Agricultural extension programmes cover a broad area including improved crop varieties, better livestock control, improved water management, and the control of weeds, pests or plant diseases. Where appropriate, agricultural extension may also help to build up local farmers' groups and organizations so that they can benefit from extension programmes. Agricultural extension, therefore, provides the indispensable elements that farmers need to improve their agricultural productivity.

Non-Agricultural Extension

In the absence of a collective term to cover the other types of extension, it is convenient to refer to them all as non-agricultural extension. This term includes all activities and efforts not directly related to agriculture or livestock production, but which are important to the farm families. Home economics, family health and nutrition, population education and community development are all non-agricultural extension activities.

Rural Extension Covers many Aspects of Rural Life

When talking of extension and extension agents, therefore, all activities of the above type are included. These activities also involve the basic elements and principles of extension outlined earlier in this chapter, such as knowledge, learning and practice. Home economists and community development workers, therefore, are extension agents who deal with farm families in the same way as agricultural extension agents. The only difference is their areas of concern.

In fact, it is becoming increasingly common to talk of rural extension as a collective term which brings together all agricultural and non-agricultural extension activities. The feature common to both types of extension is that they work with families in rural areas and deal with problems in a rural environment. Their different programmes and approaches have a common aim, which is the improvement of the lives of the rural people, and they are both guided by common principles and ideals.

This guide, therefore, is a guide to rural extension and is relevant to both agricultural and non-agricultural extension agents. Given the predominance of agriculture in the rural economy, however, there will be some emphasis on agricultural extension within the guide. The understanding of extension, the methods used by extension agents, the planning processes involved and the qualities and skills required by agents are factors relevant to all forms of rural extension. The content and subject-matter may be different, but the same general principles apply in both types of extension.

Dimensions of Agricultural Extension

What is Education?

What is the meaning of education? Why are you a teacher? What is education any way? What does it mean to educate another human being especially a child? All these are important questions. The word "educate" comes from the Latin 'educere', meaning "to lead out", "to bring out" To educate means to bring out of the child, the student, that sprit of learning and wonder, the desire to know, that thirsts for knowledge.

True education is the harmonious development of the physical, mental, moral and social faculties, the four dimensions of life. The illiterate of 21st century will not be those who cannot read and write, but those who cannot learn, unlearn and relearn. The world may not be dividing into the weak and the strong or the successes and failures. But it may be dividing into learners and non-learners. Education is the cheapest, most read and most reliable path to economic advancement under present conditions. If education is the door to economic, social, and personal advancement, then literacy-the ability to read-may very well be the key to that door and teachers stand as the gatekeepers to the doors.

Education can be defined as production of desirable changes in the behavior of people or individual. The changes which are not desirable are not considered as a education. What is desirable and what is undesirable is decided by social values. Education is continuous process in the life of an individual. This process begins immediately after birth and continuous till death. He may be conscious of it or he may not be. In everyday life a man learned so many things but he is not conscious (aware) of this type of education.

Education is important for all. If village people are properly educated, they shall be able to know, what is expected of them and what way, they can be useful members of the society.

Definitions of Education

Webster defines education as the process of education or teaching to develop the knowledge, skill, or character of the student

- ✰ Education is the process of bringing desirable change into the behavior of human beings.

- ☆ It can also be defined as the process of imparting or acquiring knowledge and habits through instruction or study.
- ☆ Sociologist Rodney Stark declares that, "education is the cheapest, most rapid, and most reliable path to economic advancement under present conditions."
- ☆ Education is the harmonious development of the physical, mental, moral (spiritual), and social faculties, the four dimensions of life, for a life of dedicated service.

Types of Education

With the development of society, education has taken many shapes, such as

a) Child education
b) Adult education
c) Technical education
d) Education in the arts and crafts.
e) Physical education
f) Health education

Education is the Humanities and social sciences and several others.

It is Important for the Rural People because of the Following Factors:

a) Backwardness of the village life and the agriculture.
b) For village progress and reconstruction.
c) For proper training of the 70 per cent of the population of the country which depends on agriculture.
d) To remove their superstitions.
e) To educate them about rural indebtedness which is responsible for various drawbacks of the village society?

Difference between Formal and Informal Education

Sl.No.	*Formal*	*Informal*
1	There are definite rules and regulations and generally they are in written form. Disobeying of this result to severe consequences.	There are no rules and regulations and hence the question of disobeying does not arise.
2	There is definite time and fixed place for teaching. The procedure of teaching is also fixed.	It can be given at any time. No fix procedure for teaching.
3	There is fix syllabus about the subject matter.	There is no fix syllabus.
4	The level of educational qualifications, experience and age of learner is almost same	There is considerable difference in the level of age, experience and educational qualification.

In a village generally primary school acts as an educational institution. The basic knowledge like reading, writing, arithmetic *etc.* is imparted to rural children in school.

At present primary education in the villages is looked after by the Zilla Parishad.

Different Problems Faced by the Village Primary School

a) Lack of proper school building.

b) Lack of adequate and trained teaching staff.

c) Lack of adequate teaching and sport material.

d) Lack of interest on the part of parents.

e) Lack of interest on the part of students.

Present System of Education and its Utility for Indian Rural Society

The present system of education is more or less futile for the Indian villages. It has no relevance to Indian life, much or less to the rural life. If it has to suit the needs of the rural society, it shall have to be-organized and change in many respects. The defects and drawbacks of the present system of education are discussed below:

a) Impractical and Expensive: The present system of education is impractical and too is based on bookish knowledge. It has no relationship and with the objectives of the society and the life. It does not teach the young boys and girls, the ways through which they can earn their livelihood.

b) Lacks of practical utility, it is very expensive; it is beyond the means of poor people to acquire education. The education facilities are more expensive and it is not possible for ordinary people to afford it. Villagers are very poor and the present day education is much beyond their reach.

c) No Relevance to the Interest of the Students: The present system of education is neither interested nor aptitude oriented. Only that education can be effective, useful and helpful which is based on the interest and the aptitude of the children. From this point of view, the Indian education is not at all useful for the needs of the Indian rural society.

d) Low Standard and Defective Examination System: The education system today is of very low standard. It means that the education Institution side not has all the equipments that are required for proper education. Therefore, the boys are not able to take full advantage of the education facilities that exists.

e) The examination system which is most important aspect of the present day education is also defective. It is neither an effective method of evaluating the competence of the education nor does it provide an opportunity to the child to go through his studies in scientific manner. Rural society needs a system of education which can be indigenous and which should provide for the need of the rural people. All these things are lacking form the Indian system of education.

f) **Lack of Facilities for Education:** In rural society, there is dearth of facilities for education. There are no institutions nor there any arrangement through which cans the people be educated. Whatever education facilities are available is less in preparation to the requirements of the people.

Classification of Education

A) Formal education

B) Non-formal education

C) Informal education

A) Formal Education

Formal education is basically an institutional activity, uniform and subject oriented, full time, Sequential, hierarchically structured, leading to certificates degrees and diplomas.

Characteristics of Formal Education

- ☆ Hierarchically structured
- ☆ Full time education.
- ☆ Technical and professional training.
- ☆ A variety of specialized programmes.
- ☆ Running from primary school through the university.
- ☆ Chronologically graded education system.

B) Non-formal Education

Any organized education activity outside the established formal system whether operation separately or as an important feature of some broader activity that is intended to serve Identifiable learning clienteles and learning objectives.

Characteristics of Non Formal Education

- ☆ It is flexible.
- ☆ It is life, environment and learner oriented.
- ☆ It is diversified in content and method.
- ☆ It is non-authoritarian
- ☆ It is built on learner-participation
- ☆ It organizer human and environmental potential.
- ☆ It enhances human and environmental potential.

Provide selected types of learning to particular subgroups in the population, adults as well as children.

C) Informal Education

The truly lifelong process whereby every individual acquires attitude, values, skills and knowledge from daily experience and the educative inferences and resources in his or her environment from family and the educative inferences and resources in his or her environment from family and neighbours, form work and play, from the market place, the library and the mass media.

Characteristics of Informal Education

- ✰ Informal education is the least controlled, that's why this type of education cannot be excluded of somebody's life.
- ✰ Informal education consists of accidental, unclear, quantitative information.
- ✰ It usually has a quantitative aspect that a qualitative one
- ✰ Informal education refers even to emotions, feelings, beliefs, superstitions *etc.*
- ✰ Informal education offers responsiveness ready response when interact with environment.
- ✰ Informal education offers possibility to freely act in unknown situation.
- ✰ Informal education offers freedom of self-formation.

Difference between Formal Education and Non-Formal Education

	Formal Education	*Non Formal Education*
Target Group	Mainly young, Universal, Compulsory, Selective	Mainly adults, those interested, voluntary and open
Time Scale	Full time and Primary activity	Part time and Secondary activity of participants
Relevance	Separate form life, In special institution, In sole purpose buildings	Integrated with life, In the community, In all kinds of settings
Programme	Run by professionals, Excludes large parts of life	It is participatory, Includes large parts of life.
Curriculum	One kind of education for all	Education to meet learner
Methods	Teacher centered, Mainly written	Learner centered, Much is Oral
Objectives	Conformist	Promotes
Independence	Set by teachers, Competitive	Set by learners and Controlled by Learners
Orientation	Future	Present
Relationship	Hierarchical	Egalitarian believing in Equal Right
Validation	Terminal at each stage, Validated by education Professional	Continuing validated by learners

Drawbacks of Non Formal Education

There are drawbacks to this approach to defining non formal education, drawbacks which immediately leap to the eyes of distance educators, since most of the features listed under non formal education are also virtues claimed by distance educators. It is clearly the case that the positive virtues attributed here to NFE are idealized. Rather than characteristics of NFE, they are goals towards which NFE are idealized much formal education is striving. Although this gives such lists limited utility in making clear distinctions between formal and non-formal education, it is nonetheless possible to identify some important differences between the two sectors, NFE courses do tend to be geared more to the immediate needs of learner; they tend to give learners more in influence in shaping their experience of teaching and learning by not holding the hammer of external testing and validation over their heads.

Extension Education

The term extension was first used in the United States of America in the first decade of this century to con note the extension of knowledge from land grant colleges to the farmers through the process of informal education. In India, extension work was primarily started by F.L. Brayne (1920) in Punjab the term community development and extension education became more popular with the launching of community development projects in 1952 and with the establishment of the national extension service in 1953, Since then, community Development has been regarded as a programme for all-round development of the rural people and extension education as the means to achieve this objective.

Difference between Formal Education and Extension Education

Formal Education	*Extension Education*
The teacher starts with theory and works up to practical	The teacher (Extension worker) starts with practical and may take up theory later on
Students study subjects	Farmers study problems
Students must adapt themselves to the fixed curriculum offered	It has no fixed curriculum or course study and the farmers help to formulate curriculum
Authority rests with the teacher	Authority tests with the farmers
Class attendance is compulsory	Participation is voluntary
Teacher instructs the student	Teaching is also through local leaders
Teaching is only through instructors	Teaching is also through local leaders
Teaching is mainly vertical	Teaching is mainly horizontal
The teacher has more or less homogeneous audience	The teacher has a large and heterogeneous audience
It is rigid	It is flexible
It has all pre-planned programme	It has freedom to develop programme
It is more theoretical	It is more practical and intended for immediate application in the solution of problem

Basic of Extension Education

On the eve of twentieth century, extension education has emerged distinctly as discipline and profession responsible for initiating the technological changes. The ingredients have been represented through the technologies to bring the behavioural changes of the stakeholders. The behavioural changes through extension education have been exemplified by using communication as a tool. The intervention through communication motivates the people to indulge themselves directly in programme of changes. The more interactive participatory research and extension has been incorporated in extension education. The field has grown to its full bloom with varied international experience. The unique contextual and relevant knowledge bank has been developed today in this discipline. The wide variation of perception regarding the concept of extension education produces a holistic approach of conceptualization. The inherent generic nature of extension education leads to its dynamic and flexible concept for the development of the individual as well as all categories of constituents of society, concerned directly or indirectly with it. The role of extension education is developing human resource at the grass-root level by mobilizing social conscience, entrepreneurship and rural leadership.

Sohal has remarked: "On account of its simple but generic nature, extension education is taken for granted to the thoroughly and completely understood in its entirety by each and every person concerned directly or indirectly with it. But whereas it can be admitted that one or a few aspects of this subject can be expected to be known extensively, yet there are only a few individuals with professional insight into the subject who really known all of its varied facets."

Analysis of Definitions

There are as many definitions as those who define. Shukla (1972) collected and studied several prevailing concepts of extension education using psycho-physical methods to find out the comparative attitudinal positions for these concepts. The following definitions of extension education in the light of current development help to conceptualise the discipline in a proper manner.

- ✰ Extension education is an applied science consisting of contents derived from researches, accumulated field experiences and relevant principles drawn from the behavioural sciences, synthesised with useful technology, in a body of philosophy, principles, contents and methods focussed on the problems of out of school education for adults and youths (**Leagans, J.P.**).
- ✰ Extension is an education and its purpose is to change the attitude and practices of the people with whom the work is done (**Ensminger.D.**).
- ✰ Extension education is defined as an educational process to provide knowledge to the rural people about the improved practices in a convincing manner and to help them to take decisions within their specific local conditions (**Dahama, O.P.**).
- ✰ Extension education is a science which deals with creation, transmission and application of knowledge design to bring about planned changes in the behaviour complex of people with a view to help them live better by learning the ways of improving their vocation, enterprises and institutions (**Reddy, A. Adivi.**).

- It is the art and science of extending knowledge, attitude and skill through the process of teaching and learning to young and adult people in informal ways mainly through learning by doing (**Choubey, B. K.**).
- Extension education is an out of school process which aims at helping people to help themselves in solving their problems. (**Waghmare, S. K. and Vanshree Waghmare)**
- Extension education is a Behavioural Science following a continuous, persuasive and discriminating educational process. It aims at affecting the behavioural components of people in a desirable direction, through conviction, communication and diffusion, by its proven methods, principles and philosophies resulting in learning involvement of both client and change-agent systems(**Dahama and Bhatnagar**).
- Agricultural extension is concerned with agricultural education aimed at assisting rural people to bring about continuous improvement in their physical, economic and social well being, through individual and cooperative efforts. It makes available to the villages, scientific and other factual information and training and guidance for the solution of problems of agriculture and rural life.
- Extension education is the act of putting across to the people, in an understandable manner, new ideas and improved technology of practical utility and to enable them to put them into practice so as to improve their general standard of living through their own realisation and efforts.
- Extension education is a science that deals with various strategies of change in the behavioural patterns of human beings through technological and scientific innovations for the improvement of their standard of living.
- Extension is a continuous process designed to make the rural people aware of their problems and indicating to them the ways and means by which they can solve them. It involves not only educating rural people in determining their problems and methods of solving them but also inspiring them towards positive action in achieving them.
- Agricultural extension is a bridge that fills the gap between agricultural research stations on the one hand and the farming population on the other by establishing a suitable teaching organisation at various levels of administration.
- Extension education is an applied behavioural science, the knowledge of which is to be applied for desirable change in the behavioural complex of the people.
- Extension is defined primarily as an educational process aiming at the development of individuals; through this process the villagers are helped to become discontented with the present conditions and are helped by extension workers to improve their conditions of living.
- Extension is to teach a person how to think, not what to think, and to teach people to determine accurately their own needs to find solution to

their own problems and to help them acquire knowledge and develop convictions in that direction.

- Extension is an out-of-school system of education in which adults and young people learn by doing. It is a partnership between Government, the Land Grant Colleges and the people, which provides services and education designed to meet the needs of the people.
- Extension or agricultural extension is a method, or a series of methods, by which the technical know-how of science is carried to and included in the practices of the cultivators.
- Extension education is the education of the people as to what more to want as well as how to work out ways of satisfying them. Informing people not to remain content with their present lot and inspiring them to work vigorously towards fulfilment of their self-created, increased wants or desires.
- Extension is the education of the rural adults outside the school in matters of their choice and interest. It is education for freedom, which seeks to help persons to use the liberty of action with which democratic society is constructed.
- Extension Education deals with practical items of information which is useful for rural people which solve their daily problems, specially those relations to agricultural production.
- Extension Education is an integral behavioral science which contributes towards the understanding and formulation of methods and procedures for bringing planned change in human behavior.
- Extension education is education for the betterment of people and for changing their behavior *i.e.* knowledge, skill and attitude.
- Extension education is the dissemination of useful research findings and ideas among rural people to bring out desirable changes in their social and cultural behavior.
- Extension education in an applied behavioral science, the knowledge of which is to be applied for desirable changes in the behavioral complex of the people.
- Extension is an education and its purpose is to change the attitude and its purpose is to change the attitude and practices of the people with whom the work is done.
- Extension education is a science which deals with various strategies of change in the behavioral patterns of human beings through technological and scientific innovation for the improvement of their standard of living.
- Extension is to teach a person how to think, not what to think, and to teach people, to determine accurately their own needs to find solution to their own problems and to help them acquire knowledge and develop convictions in that direction.

- ☆ Extension is an out-of school system of education in which adults and young people learn by doing. It is a partnership between government, the land grant colleges and the people, which provider services and education designed to meet the needs of the people

Keeping all the accepted views regarding extension education discipline in the present context according to the author, it is very much appropriate to define extension education as **"Extension education is an applied behavioural science, mostly informal in nature which deals with the theory, procedures and methods for disseminating the information and knowledge based on research and experiences of people through several teaching methods so as to they can perceive, accept, use it and change their behaviour to solve the existing problem faced by them".**

Agricultural extension is the application of scientific research and new knowledge to agricultural practices through farmer education. The field of 'extension' now encompasses a wider range of communication and learning activities organized for rural people by educators from different disciplines, including agriculture, agricultural marketing, health, and business studies. Extension practitioners can be found throughout the world, usually working for government agencies. They are represented by several professional organizations, networks and extension journals. Agricultural extension agencies in developing countries receive large amounts of support from international development organizations such as the World Bank and the Food and Agriculture Organization of the United Nations.

Agricultural extension work has a venerable, albeit largely unrecorded, history. It is a significant social innovation, an important force in agricultural change, which has been created and recreated, adapted and developed over the centuries. Its evolution extends over nearly four thousand years, although its modem forms are largely a product of the past two centuries. Today, the organizations and personnel engaged in agricultural extension encompass a diverse range of socially sanctioned and legitimate activities which seek to enlarge and improve the abilities of farm people to adopt more appropriate and often new practices and to adjust to changing conditions and societal needs.

Importance of Extension Education

- ☆ Extension uses democratic methods in educating the farmers.
- ☆ Extension helps in adoption of innovations.
- ☆ Extension helps in studying and solving the rural problems.
- ☆ Extension increases farm yields and improve the standard of living of farmers.
- ☆ Extension makes good communities better and progressive.
- ☆ Extension contributes to national development programmers.

Scope of Extension Education

It includes all activities of rural development. So extension programmers should be dynamic and flexible. The areas indicating scope of Extension are listed below:

a) Increasing efficiency in agricultural production.

b) Increasing efficiency in marketing, distribution and utilization of agricultural inputs and outputs.

c) Conservation, development and use of natural resources.

d) Proper farm and home management.

e) Better family living.

f) Youth development

g) Leadership development.

h) Community and rural development.

i) Improving public affairs for all round development.

Important Concepts of Extension Education

"A concept is an abstraction from observed events or a short hand representation of variety of facts." The purpose of concept is to simplify thinking by sub summing a number of events under general one heading. "Concept is an idea, general notion or way in which one can see a thing in his mind. In more general, concept means assigning meaning to the words."

Some important concepts of extension education are presented below:

☆ **Extension** – The word extension is mostly used for Extension Education. The basic concept of Extension is that it is Education – Means production of desirable change in human behaviour which includes knowledge, attitude and skill.

☆ **Extension Work** – It means the whole structure of Extension work. It includes the process of Extension Education *i.e.* the process of teaching and learning. Besides the process, in extension work are included organizations, administration, supervision, finances as well as the programs for the overall development.

☆ **Extension Service** –It means an organization and/or a programme for the welfare and development, which employs the extension educational process for the implementation of programme.

☆ **Extension Job** – The job of extension in agriculture and home science is to assist people engaged in farming and home-making to utilize their own resources more effectively and those that are available to them, in solving the current problems and in meeting the changing economic and social conditions.

☆ **Extension Educational Process** – The extension process is working with the people with their immediate needs and interests which can make available additional occupation, make improvement in the socio-economic status, better home-management and expedite welfare of the rural people.

Philosophy of Extension Education

Philosophy, in the original and wider sense, is the pursuit of wisdom, or knowledge of things and their causes, both theoretical and practical. It is also defined as moral wisdom. Philosophy is interwoven in daily life. Then it can become a working tool for improving life. Extension education is becoming an important intervention in most of the parts of the world in rural areas. Due to dynamic changes in the socio-economic perspective of the developing countries, technologies and political situations there is a need to reorient the philosophy of extension in several perspectives. The philosophy of extension must have been influenced by the values, knowledge and situation existing during the period. The practical implication of it is that philosophy of a particular field would frame the principles and the guidelines with which to shape or mould the activities related to that particular field.

The philosophy of extension education has been described and interpreted in different ways by different authors and a clear picture cannot be drawn due to the very complexity of its nature. All one can do is try to gain a comprehensive idea by examining the viewpoints of various authors.

Strategies of aquaculture extension are as diverse as national political philosophies and goals. The following chapter is a discussion of several of the predominant approaches to agriculture extension programmes. The discussion will focus on (a) centralized versus decentralized control over extension activities, (b) selection of extension agents, (c) direction of information flow, and (d) specific techniques of extension appropriate for different target groups for different levels of extension involvement.

Axinn (1988) identified 8 different approaches to extension work. These are briefly summarized below:

a) The General Agricultural Extension Approach

The purpose is to help farmers increase their production. Planning is done on a national basis by the central government "which knows better than farmers". This is a typical case of top-down planning. Field personnel tend to be large in number and high in cost, with the central government bearing most of the cost. The rate of adoption of important recommendations and increases in national production are the measures of success. A survey of agricultural extension programmes indicated that agricultural extension generally was part of the Ministry of Agriculture, with field extension officers at the bottom of the hierarchy and a minister at the top (FAO, 1971). This approach lacks a two-way flow of information. It fails to adjust messages for each different locality. Only farmers who seek advice benefit, and these tend to be large-scale wealthier farmers. This approach does provide farmers with information on a number of production alternatives from one single source.

b) The Commodity Specialized Approach

All functions related to a particular commodity are grouped together, including extension, research, input supply, output marketing, and prices. Planning is controlled by a commodity organization for the purpose of increasing production of a particular commodity. Highly trained scientific personnel equipped with expensive vehicles and field scientific apparatus are employed. Techniques recommended must produce financial benefits for farmers, and be demonstrable on a farmer's own field. New inputs must be accessible, a credit scheme established, and the ratio between farm-gate inputs and commodity prices considered. Technology tends to be appropriate and distributed in a timely manner because it focuses on a narrow range of technical concerns. Interests of farmers, however, may have less priority than those of commodity production organizations.

c) The Training and Visit Approach

The purpose of the training and visit approach (often called T&V) is to induce farmers to increase production of specified crops. Planning is controlled centrally and field personnel tend to be numerous and dependent on central resources. There is a rigid pattern of visits to farmers and in-service training of field staff. Success is measured in terms of production increases of the particular crops covered by the programme. The training and visit approach is another top-down approach. The emphasis is on disseminating unsophisticated, low-cost improved practices, and teaching farmers to make best use of available resources. There is pressure on the government to reorganize into a more integrated service, and to send extension officers into the field to meet with farmers. It provides closer technical supervision and logistic support, but at a high cost. Actual two-way communication is lacking and there is little flexibility.

d) The Agricultural Extension Participatory Approach

This approach assumes that farmers are skilled in food production from their land, but their levels of living could be improved by additional knowledge. Active participation by farmers themselves is necessary and produces a reinforcing effect in group learning and group action. Much of the work is through group meetings, demonstrations, individual and group travel, and local sharing of appropriate technologies. Success is measured through numbers of farmers actively participating, and the continuity of the programme. There is much to be gained by combining indigenous knowledge with science. Expressed needs of farmers are targeted. The system requires that extension workers, who are also animators and catalysts, stimulate farmers to organize for group efforts. Local people evaluate their own programmes and play a role in establishing research agendas.The agricultural extension participatory approach costs less, fits needs well, and is more efficient. However, it is more work for extension agents to organize and motivate farmers. It requires agents to live and to socialize with farmers. Where a government job is seen as a reward for good friends, the "hardship" implied by this approach dooms it to failure. The agent will be there only "part time" and have no personal stake in the outcome.

e) The Project Approach

This approach uses large infusions of outside resources for a few years to demonstrate the potential of new technologies. Control is at the central government level and there is an often considerable financial and technical input from an international development agency. Short-term change is the measure of success. In the aquaculture project in Nepal, for example, a loan from the Asian Development Bank was used by the Ministry of Agriculture to support extension work by fisheries officers in many different locations throughout the country. They were able to introduce pond fisheries through an effort which combined the project approach with the specialized commodity approach. One problem with this approach, however, is that a flow of ideas outside the project rarely occurs.

f) The Farming Systems Development Approach

This approach assumes that technology which fits the needs of farmers, particularly small-scale farmers, is not available and needs to be generated locally. Planning evolves slowly and may be different for each agro-climatic farm ecosystem. This approach is implemented through a partnership of research and extension personnel using a systems approach. Analyses and field trials are carried out on farmers' fields and in homes. The measure of success is the extent to which farm people adopt technologies developed by the programme and continue to use them over time. Control of the programme is shared jointly by local farm families, extension officers, and researchers. Advantages of this system include strong linkages between extension and research personnel, and the commitment of farmers to using technologies they helped to develop. Costs can be high, and results can be slow in coming.

g) The Cost Sharing Approach

This approach is based on local people sharing part of the cost of the extension programme. Control and planning is shared by various entities and is responsive to local interests. Success is measured by farmers' willingness and ability to provide some share of the cost, be it individually or through local government units. Problems may arise if local farmers are pressured into investing in unproven enterprises.

h) The Educational Institution Approach

In the educational institution approach, planning is controlled by those determining the curriculum of the educational institution. Implementation is through non formal instruction in groups or individuals through a college or university. Attendance and the extent of participation by farmers in agricultural extension activities are the measures of success. Ideally researchers learn from extension personnel who, in turn, learn from farmers. However, this rarely occurs in practice. The advantage of this approach is the relationship of specialized scientists to the field extension personnel.

According to Kelsey and Hearne

The philosophy of extension work is based on the importance of the individual in the promotion of progress for rural people and for the nation. Extension educators work with the people to help them to develop themselves and achieve superior personal well-being. Together they establish specific objectives, expressed in terms of everyday life, which lead them in the direction of overall objectives. Some will make progress in one direction while others will do so in another direction. Progress varies with individual needs, interests and abilities. Through this process the whole community improves, as a result of cooperative participation and leadership development.

According to Ensminger

The philosophy of extension can be expressed in the following lines:

a) It is an educational process. Extension is changing the attitudes, knowledge and skills of the people;
b) Extension is working with men and women, young people, boys and girls to answer their needs and their wants. Extension is teaching people what to want and ways to satisfy their wants.
c) Extension is "helping people to help themselves";
d) Extension is " learning by doing" and "seeing is believing";
e) Extension is development of individuals, their leaders, their society and their world as a whole;
f) Extension is working together to expand the welfare and happiness of people;
g) Extension is working in harmony with the culture of the people;
h) Extension is a living relationship, respect and trust for each other;
i) Extension is a two-way channel; and
j) Extension is a continuous, educational process;

According to Dahama

The following points are the "Philosophy of Extension".

a) Self-help;
b) People are the greatest resources;
c) It is a cooperative effort;
d) It has its foundation in democracy;
e) It involves a two-way channel of knowledge and experience;
f) It is based on creating interest by seeing and doing;
g) Voluntary, cooperative participation in programmes;
h) Persuasion and education of the people;
i) The programme is based on the attitudes and values of the people; and
j) It is a never-ending process.

According to Mildred Horton

The philosophy of extension education is the combination of the followings:

a) The individual is supreme in a democracy.
b) The home is the fundamental unit in a civilisation.
c) The family is the first training group of the human race.
d) The foundation of any permanent civilisation must rest on the partnership of man and land.

According to Shukla

"Extension programme revolves around the individual, the cultivator, and we have to bring change in his attitude, knowledge, skill, understanding, capacity and ability through persuasion by educational means."

According to Rudramurthy

The philosophy of extension work links with the Vedas, the Upanishads, the Gita as well as the orthodox and unorthodox schools of philosophy. This is based on the concept of man and the values are worthy of pursuit.

According to Bhatnagar

Extension is combination of the activities of the State Governments (with, or without, the help of Central Government, or other agencies) which provide the farmers with technical know-how as a guide to improved methods, in order to bring desirable changes in their behaviour with the aim of attaining higher production. In this sense, the extension activities relate closely to the activities in research and education.

According to Supe

The basic philosophy of extension is directed towards changing the outlook of man by educating him. Compulsion does not persuade the people to act in a proper way. The only way to secure wholehearted cooperation of a person is to educate him. Education is not mere transfer of information. It is more than that. Its primary aim is to transform the people by bringing about desired changes in their knowledge, attitude and skills. If this has not happened then the activities chosen for extension are not educational and its effect cannot be long lasting.

According to Waghmare

The ultimate aim of extension education is the development of people. Unless people are changed, no permanent change can be achieved. Physical and economic developments become sterile without the development of people.

Basic Aspects of Philosophy of Extension Education

Based on the views of various outstanding scholars, it can be inferred that the philosophy of extension education includes certain basic aspects. These are:

- ☆ Extension education is based on the importance of individual.

- Self-help (help them develop themselves) is important part of the philosophy of extension education.
- Participation in extension education is voluntary and cooperative.
- Extension education is democratic in nature and procedure (programme by the people and for the people based on their needs).
- Extension education philosophy does not believe in giving orders but depends on persuasion and education.
- Extension education works on the assumption that programme is people's programme and is not that of the government programme.
- People should be given full responsibility.
- Development of family in relation to village and rest of world is the guiding Philosophy of extension education.
- Changes in terms of attitudes besides economic and social change are an important part of extension philosophy.

Extension Education as a Profession/Discipline

It has been observed that there are some faculty members as teachers are found in the discipline of Extension Education who does not possess even the minimum requisite desired temperamental, personal and academic qualifications for admission in to the discipline/profession. Barring this highly serious lacuna, the extension education both as a discipline as well as a profession is developing rapidly. Extension Education as a branch of education is fully dedicated and responsible to teach its clientele. Such types of individual tasks are the tasks for professionals. Extension Education is a well-developed discipline having its own swiftly growing body of knowledge with its own well-developed tools and techniques.

Discipline of Extension Education

Discipline refers to a branch of instruction/education, may it be a branch of art, science as vocation. Extension education has established itself as a discipline by virtue of systematic body of knowledge and well distinguished content of practical relevance, as detailed below.

(i) Systematic Body of Knowledge

Extension education has its own principles, philosophy, methods and contents. The contents of extension education have been borrowed from various social/ behavioural sciences like Sociology, Psychology, Rural Sociology, Social Psychology, Communication, Journalism, Management, Administration, *etc.* Such contents have been focussed on the problems of rural areas. The contents of extension education have certain distinguishing features as given below:

- Extension education did not begin as subject first. The extension works were initiated first and the rich experiences, wisdom and experiments led to evolution of discipline much later.

- It belongs to social science but has bias towards technology be it Agriculture, Home Science, Forestry, Fisheries, Family Welfare, Public Health Energy.
- It is largely rurally oriented though much of extension work in developed nations now focussed on urban audience too. It is problem-solving in nature, requiring professionals to diagnose problems and provide strategic solutions.
- It is inter-disciplinary in nature calling for use of other sister disciplines to solve problems of complex nature.

Extension education has established itself as well developed discipline. In India, it has acquired a status of its own within the fold of higher agricultural education.

(ii) Growth of Academic Discipline in India

Though provisions had been made to specialise in agricultural extension within Master's programme (Associateship of ICAR) of Agronomy division, creation of separate department of Agricultural Extension with assistance from Ford Foundation under expert advice of legendary J. Paul Leagans of Cornell (USA) was a landmark event in the history of agricultural extension. Well-founded Ph.D. programme initiated in 1961 at IARI attracted many interested students, most of whom in service students.

(iii) Researches in Extension education

Research is an important requirement for continuous growth. It is important for a professional practice to be guided by science and scholarship rather than trial and error. With changing technologies and social situations, new techniques of working with people are needed. Extension researches are needed to assess effectiveness of methods and find answer to problems faced by practitioners in the field. The quest for scientific enquiry is evident among academicians as empirical researches are being carried out right since the inception of the discipline in India.

(iv) Practical Utility

The discipline of extension education has immediate practical relevance in enhancing production of food and development of human resource in the rural areas. It has proved beyond doubt that people with training and education in the discipline have been able to develop villages, encourage adoption of new technology, organise village communities and mobilise leadership.

Extension Education as Profession

A profession refers to a particular field of vocation or occupational organisation. Webster defined profession as a calling in which one professes to have acquired some special knowledge used by way of instructing, guiding or advising others or of serving them in some act. There are a number of criteria that must be fulfilled to call a calling a profession (Brown, 1965 and Clark, 1965).

- ☆ It should have a body of knowledge based on scientific investigation, which is a member must master in order to enter profession.
- ☆ It must have continued research and exchange of information among members.
- ☆ It has established requirements of training and competency as a basis for entry into profession.
- ☆ It requires personal responsibility on the part of members for self-direction and self-evaluation.
- ☆ It must exhibit high standard of ethical conduct.
- ☆ The profession must show monopoly over the body of knowledge which it calls its own. Because of the complex and difficult nature of knowledge required, only members of a profession are in a position to pass intelligent judgement.
- ☆ The body of knowledge which the profession possessed is necessary for the good of the society.
- ☆ Its members organise association through which they act collectively to maintain and improve the service.
- ☆ The occupational group and its members have an esteemed position in the society to have earned the right to voice in public in areas concerned with professional concerns and competence.

Extension education has established itself as a well developed discipline. Master's and doctoral degrees are awarded to those aspiring to specialize in it. Job-opportunities are now open to those trained post-graduates in training, rural development, sales, teaching, research and administrative jobs. There are positions of Scientists or Subject-Matter Specialists in Extension Education as well in most teaching, research and extension organisations. Besides, voluntary organisations and input organisations also advertise for positions to be filled by post-graduates in extension education. Thus, the discipline has potentialities to provide employment on the basis of knowledge and competencies. There are good reasons to believe that extension education fulfils the criteria for being called as profession, as discussed below.

Professional Society

There is well developed Indian Society of Extension Education with members from all over the country and abroad. The society has branched into many chapters as well. It has a research-based journal, Indian Journal of Extension education.

Continuing Research

A profession is not dynamic if it does not have continuous research. The founding fathers of extension education have paved ways for research ever since co-operative extension service era in USA. In India it began with post-graduate research. Most post-graduate departments have faculty lead projects.

Significant Service

One of the requirements of profession is its significant contribution to the society. In India, Extension Education has contributed in bringing about technological awareness among farmers. Over the years, farming on scientific line has been adopted. It has helped in enhancing food production. It has enabled agricultural universities and research institutions to extend its frontiers beyond their four walls. It has given orientation to agricultural research.

Objectives in Extension Education

1. To raise the standard of living of the rural people by helping them in right use of their resources.
2. To help in planning and implementing the family and village plans for increasing production in various occupations.
3. To provide facilities for better family living.

Specific Objectives

1. To provide knowledge and help for better management of farms and increase incomes.
2. To encourage the farmers to grow his own food, eat well and live well.
3. To promote better social, natural recreational intellectual and spiritual file among the people.
4. To help rural families in better appreciation of SWOT in the village.
5. To open new opportunities for developing talents and leadership of rural people.
6. To build rural citizens who are:
   a) Proud of their occupation
   b) Independent in thinking.
   c) Constructive in outlook.
   d) capable, efficient and self-reliant in character
   e) having love of home and country m their heart

Objectives

"Objectives are expression of the ends towards which our efforts are directed" In other words, an objective means a direction of movement before, starting any programme, its objectives must be clearly stated, so that one knows where to go and what is to be achieved. The objectives should be such which provide night direction to the large number of people to set a direction and travel the distance between theory and practice. Again an objective "is a statement of change in knowledge, feeling or reaction we want to bring about in the people."

An objective is that which a person, group or agency sets before itself as an object or condition to be attained.

Objectives and Goals

"Objectives are directions of movement" where as a goal is "the distance in any given direction one expects to go during a given period of time."

Different Levels of Objectives

The fundamental objective of extension education is to raise the standard of lining of the rural people by helping them in wing their natural resources in the night way. It should also help in providing minimum health, recreational, educational and haring facilities for improving family living conditions in the village.

Extension education in our country is primarily concerned with the following main objectives:

a) The basic objectives of the extension education are the overall development of the rural people.

b) To bring about desirable changes in the human behavior, which includes change in knowledge, skill and attitude?

c) The dissemination of useful and practical information relating to agriculture, including improved seeds, fertilizers, implements, pesticides, improved cultural practices, dairying, poultry nutrition *etc.*

d) To make the people aware that agriculture is a profit table profession.

e) To create an environment for rural people so that they can show their talent, leadership and efficiency.

f) To provide appropriate solution of the farmer's problems.

g) To bring the scientist closer to the farmers.

Fundamental Objectives

The central objective of Extension Education is "to bring about all round improvement in the life clientele. This will include material, moral, civic and social development. It is to help each individual, each family and each community achieve the highest level of living that is capable of economically, socially and culturally by means of added self help through education".

General Objectives

These are more definite than fundamental objectives which are directly associated with the extension service.

General Objectives of Extension Education are:

a) To assist people to discover and analyze their problems, their felt and unfelt needs.

b) To develop leadership among people and help them in organizing groups to solve their problems.

c) To disseminate information based on research and or practical experience, in such a manner that the people would accept it and put it into actual practice.

d) To keep the research workers informed of the people's problem from time to time, so that they may offer solutions based on necessary research.

Working Objectives

These are specific subject matter approaches for example, the Urd growers of village A to increase the yield of black gram/ha by using H.Y.V. RBU-38.

Working objective has three parts:

a) Audience: The people to be changed.

b) Type of Change: Whether the change is desired in knowledge, attitude or skill.

c) Content: What actually is needed to know, feel or do by the people?

Role of Extension Education in Agricultural Development

a) To provide exposure and information to the people about the new technology.

b) To teach the people – about new technology.

c) To convince the people and to put the new technology in practice.

d) To help the people to put the new technology in practice.

e) Evaluation and follow-up continuously.

f) To communicate the problems of people to the research for their suitable solution.

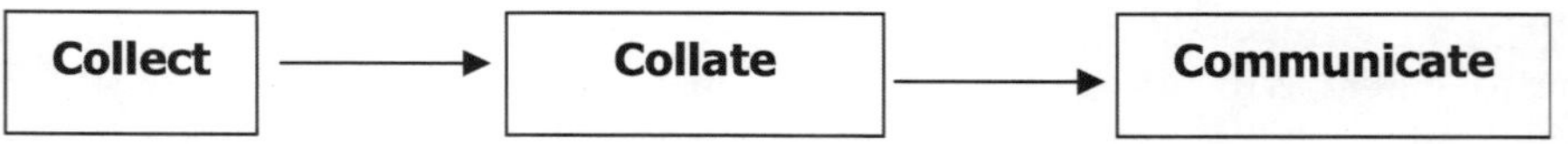

Role of Extension Education in Agricultural Development.

Principles of Extension Education

Principles are the fundamental laws and rules discovered by educational researches and proven by years of experiences that now serve as guide lines for educational endeavors which necessary for an extension worker. Some of this related to agricultural extension is given here.

1. Principle of Interest and Needs

The rural people should voluntarily participate in the extension work. Extension work must be based on the needs and interests of the people. These interest and needs differ from individual to individual, from village to village, form block to block and form state to state and therefore there cannot be one programmes for all people. To be effective, extension work must begin with the interested and needs of the people. Many times the interests of the rural people are not the interest of the extension worker. Even though the sees the needs of the people better than they do themselves, he must begin with the beds and interests as they (the people) see them.

2. Principle of Cultural Difference

The educational methods should be in line with the culture of the people in order to make extension education effective. Extension work is based of the cultural background of the people with whom the work is done. Improvement can only begin form the level of the extension worker has to know the level of the knowledge and the skills of the people methods and tools used by them, their customs, traditions, beliefs, values *etc.* before starting the extension programme.

In' a vast country, like India different extension methods need to be used for different states, as people in these states differ in their thinking living and culture. The blueprint of the extension programme developed for one area may not be applicable as such in another area but it can serve as a guide in similar cultural areas.

3. Principle of Cultural Change

To change the behaviours of the people through extension education, the extension worker should gain the confidence of the rural people. It is obvious that the change agent who works personally with the villagers must know what the villagers know and what they think. They should believe that what the extension worker says has relevance to their daily life. For example, extension worker shows same demonstrations or trials on the farmer's field. He gave the information about positive and negative point of that trial and also discuss the about their difficulties with the farmers. Then the farmers that, he is doing for their welfare only. Taking the changes in the needs of the rural people into account the extension worker has to change his area of work. The situation prevailing twenty years ago when the extension service started and today is quite different, and therefore, with growth and development the extension work has to be changed to meet the cultural changes among the people.

4. Principle of Participation

The participation of the people is of fundamental importance of the success of any educational endeavour. People must share in the development of a programme and must feel that it is their own programme. Good extension work helps the rural people identifying their problems and then helping them in solving these problems, people will not feel attached to the work if they are given ready-made things. Extension helps people to help themselves. Actual participation and experience of people in these programmes creates self-confidence in them and also they learn more by doing.

5. Principle of Adaptability in the Use of Teaching Methods

People differ from each other, one group differs from another group and conditions also differ from place to place. These people differ in their level of understanding and knowledge and therefore, only one extension method will not be of use in providing information to all, an extension programme should be flexible, so that necessary changes can be made whenever needed to meet the varying conditions. No single extension teaching method is effective under all situations. Written material will be of use for those who can read it, radio programmes for those who have radio, meetings for those who can attend. Like these different situations requires difference teaching methods. New situations also arise where a special combination of method is necessary. Thus the extension methods must have flexibility to be adapted to members of a community who differ in age, education, economic and social status.

6. The Grassroots Principle of Organization

A group of rural people in local community should sponsor extension work. The programme should fit in with the local conditions. The aim of organizing the local group is to demonstrate the value of the new practices or programs so that more people would participate.

Different groups work in rural communities. The extension worker should pay attention to the needs and interests of these groups while planning extension programmes. In same places the extension programmes are thrust on the people and the imposed innovations; many times have no relevance to the interests of these small groups. Unless the farmers feel that the ideas coming to them are useful they will not come ahead to adopt them. As the buildings based on should foundation last long similarly the programmes based on the needs and interests of these small groups would give better results in extension work.

7. Principle of Leadership

It is said that there is one leader in ten persons. Local leaders are the guardians of local thoughts and action and can be trained and developed to best serve as interpreters of new ideas to the villagers. They are the representatives of the local situation. Extension work is based of the full utilization of local leadership. The selection and training of local leaders to enable them to help to carry our extension work is essential to the success of the programme. People have more faith in local leaders and they should be used to put across a new idea so that it is accepted with the last resistance. All communities have leaders or potential leaders Extension workers and specialists should train and orient these voluntary leaders for good extension work.

8. The Whole-family Principle

The family is the unit of any society. All the members have equal importance and they should be developing equally. Extension work is for the whole family and it should reach all the members of the family. Extension work will have a better chance of success if the extension workers have whole-family approach, instead of separate and un-integrated approach. There should be place for the farer, his wife and children in the programmes. Such programme will be useful and popular. Any innovation liked by male member of the family but did not like by other, then decision of adoption may affected.

9. Principle of Satisfaction

The end product of the efforts of extension teaching is the satisfaction that comes to the farmer his wife or youngsters as the result of solving a problem, meeting a need, acquiring a new skill or some other changes in behaviour. If the peoples are not satisfied by participating in democratic society's people cannot be made to move like machines. They continue the things only when they get satisfaction through these innovations. Satisfaction is the key to success in extension work. "A satisfied customer is the best advertisement".

10. Principle of Evaluation

Extension is based upon the methods of science and it needs constant evaluation. It is necessary to determine the teaching results in an unbiased way. For this it is necessary to review the development made so far and see whether the extension work is proceeding in the right direction. If it is not, then it is necessary to take corrective measures. Extension work is educational in nature and therefore, its effectiveness should be measured by measuring the changes in people resulting from the teaching process. It is no longer enough to have only plans and methods but it is necessary to determine the teaching results by scientific way. The results of such evolutions would help the extension workers in improving the quality of the programmes in the future.

11. Principle of Applied Science and Democracy

Applied agricultural science is not a one-way process. The problem of the people is taken to the scientists who do the experimentation necessary to find out the solution. The extension worker transmits the findings of the laboratories to the farmers, but the farmer has the freedom to decide the adoption or rejection of the innovation. In democracy, freedom of thought and the unbiased and objective approach of the scientist, establish facts used in the solution of problem. But farm families can voluntarily adopt these facts to satisfy their own needs.

12. Principle of Trained Specialists

Extension is the bridge between scientists and farmers. It is very difficult for a multipurpose extension worker to keep himself abreast with all the latest findings of research in all the branches of science. If there are no trained cannot thrive. These specialists have the responsibility of solving the problems of the extension workers in their subject. The subject matter specialist should have a broad outlook and he should know other subject matter fields related to family problems thereby concentrating on the welfare of the whole family and making his special.

13. Principle of Co-operation

Extension programmes are for the people by the people. The idea behind the extension work is the coming together of the rural people and the extension workers for social up-liftmen. The extension programmes should be the people's programme with government aid. Most members of the village community will willingly co-operative in carrying out a project. All should co-operate and help each other for this task of social up-liftmen.

14. Principle of Learning by doing

In extension work, farmers should be encouraged to learn new things by doing and buy to believe on innovations or new ideas, until they see the results with their own eyes. So that extension workers have to develop their faith. The motive for improvement must come from the people, and they must practice the new ideas by actually doing them. It is learning by doing, which is most effective in changing people's behaviour and developing the confidence to use the new methods in future.

2
Modern Agricultural Extension

In the early years of this century, extension services were in their formative stage; they were relatively small in scale and limited in the scope of their work and contact with farmers, and their organization was often somewhat haphazard even though based on legislation. They were organized predominantly either by central or local governments, or by agricultural colleges, usually in close association with experiment stations, or by farmers' organizations (agricultural societies, cooperatives, farmers' unions, or chambers of agriculture), or combinations of these parent bodies. As the century has progressed, the organizations have matured. Changes have often occurred to their parent affiliations, government funding has become relatively more important, their objectives have become broader, especially in "the North," and the extension workers have become better trained and more professional. In addition, several other kinds of organizations have developed comparable work: agriculture-related commercial companies; agricultural commodity marketing boards, concerned to assure the supply and quality of their specific product; agricultural development projects, many of considerable territorial scale; and a variety of nongovernmental organizations (especially religious and charitable) involved in agricultural and rural development.

As agricultural extension organizations have grown and changed, they have invariably become more bureaucratic with distinct hierarchical structures. The work of dispersed extension workers had to be administered and controlled so that one or more levels of intermediary structure (for example, district, region) have been created between the field-level agents and their headquarters. Thus the management of extension activities has become a major preoccupation, and many organizations have been open to the criticism of being too heavy and top-down in their approach.

However, with funding derived largely from national revenues (or international donors), senior managers have necessarily had to account for and justify their organization's activities. This has been equally pronounced in the North as in the South where, after colonial territories gained their independence, extension work has commonly been rein-vented and staffed by nationals under the aegis of their new administrations (usually ministries of agriculture).

During the past quarter century, the work of extension services has often become more diversified. In the less developed countries, the main focus remains on agricultural (mainly food) production, but there has been a growing recognition of the need to reach, influence, and benefit the multitudes of small, resource-poor farmers. Strong efforts have been made in this direction, notably through the training and visit system. Among the commercial farmers of the North, a major problem has become surplus production, with farmers facing economic and policy pressures to restrict it. Associated with intensive production methods, many issues and problems regarding environmental deterioration and livestock welfare have also arisen. Thus these have become important aspects of extension work, particularly socioeconomic guidance which focuses both on means by which farmers might maintain their income levels from their resources (for example, introduction of novel crops or livestock and involvement in various rural enterprises) and on the ways of assuring the longer term welfare of farmers and their families. Agricultural extension services are thus adding a strong social dimension to their activities.

Agricultural extension has now become recognised as an essential mechanism for delivering information and advice as an "input" into modem farming. Since commercial farmers can derive direct financial benefits from these inputs, there is a trend towards the privatization of the extension organizations, often as parastatal or quasigovernmental agencies, with farmers being required to pay for services which they had previously received free of charge. This trend is strong in the North, and there are examples of it beginning in the South.

The pace of change in the organization, functions, strategies, and approaches of agricultural extension is clearly accelerating.

The Future

The need for agricultural and rural information and advisory services is likely to intensify in the foreseeable future. In much of the world, agriculture faces the challenge of keeping pace with rapidly increasing population with few reserves of potentially cultivable land. Farmers will have to become more efficient and specialized.

From government perspectives, whatever priority is given to production, extension will remain a key policy tool for promoting ecologically and socially sustainable farming practices.

Some of the most promising recent developments in extension methodology have occurred where the key agenda is environmental or is concerned with equity, for example in the need for the joint management of forests by professionals and local forest users and in integrated pest management. A consistent theme running through the innovative approaches being used, such as participatory rural appraisal (Chambers, 1993), is a fundamental change in what are the respective roles of extension agent and clients. The agent is no longer seen as the expert who has all the useful information and technical solutions; the clients' own knowledge and ingenuity, individually and collectively, are recognized as a major resource; solutions to local problems are to be developed in partnership between agent and clients. Since the scale at which extension support is required is thus often larger than the individual farm, extension workers need new skills of negotiation, conflict resolution, and the nurturing of emerging community organizations (Garforth, 1993; Smith, 1994).

The future is also likely to witness a reversal of recent trends towards bureaucratization within hierarchical extension services and a reduction in their levels of public funding. Moreover, a rapid increase can be expected in the use of information technology in support of extension. The forces for change in these areas (see Rivera and Gustafson, 1991) will come from four main directions.

Economic and Policy Climate

With the collapse during the past decade of socialist forms of economic organization, the (dominant) role of the public sector in national economies has become questionable, with a strengthening trend to reduce levels of public spending. Thus government extension services and those which are largely publicly financed are, and will continue to be, under pressure to become more efficient, to reduce their expenditure and staff, and to pass on (some of) the costs of provision to their clients who directly benefit financially. This is particularly the case in countries where the farm population forms a small minority and agricultural production is in surplus. The case is weaker, but not absent, in less developed countries where farming households form a high proportion of the total population and where increasing food production is still important. Thus charging clients for services is likely to become more widespread, while governments will find it attractive to contract out the operation of services to the private or the voluntary sector.

Social Context in Rural Areas

In the future, rural populations will undoubtedly be progressively better educated, while their exposure to the mass media will continue to reduce their isolation and detachment from information, ideas, and an awareness of their situation within a national and international context. However, this exposure will not reduce the need for extension. Rather, given the changing demands on agricultural producers from population growth, increasing urbanization, legislative changes, and market requirements, the more knowledgeable farming population will require different kinds of extension services. Social and economic trends within rural areas will therefore necessitate more highly trained, specialized, and technically competent workers, who also know where to obtain relevant information and problem solutions and various provision and organizational forms (Moris, 1991; Hayward, 1990) to replace monolithic government extension agencies. These agencies will need to recognize and serve different types of clients defined not in terms of "adopter categories" but of access to markets, degree of commercialization, and relative dependence on agriculture for family income and welfare.

Systems Knowledge

Recognition of the locale-specific nature of farming systems and the agricultural information systems which support them is an important source of the pressure towards the de-bureaucratization and devolution of extension services. This recognition also implies that extension workers and farmers be jointly involved in the verification and adaptation of new technology, and thus that the extension workers respect farmers as experimenters, developers, and adapters of technology and devote more energy on communication within their local areas. The devolution of extension services to become local organizations is a reasonable corollary of this. Developments in mass media technology, already apparent over a decade ago (Garforth, 1986), will continue to support this localization of extension effort.

Information Technology

The continuing rapid development of telecommunications and computer-based information technology (IT) is probably the biggest factor for change in extension, one which will facilitate and reinforce other changes. There are many possibilities for the potential applications of the technology in agricultural extension (FAO, 1993; Zijp, 1994). IT will bring new information services to rural areas over which farmers, as users, will have much greater control than over current information channels. Even if every farmer does not have a computer terminal, these could become readily available at local information resource centres, with computers carrying expert systems to help farmers to make decisions. However, it will not make extension workers redundant. Rather, they will be able to concentrate on tasks and services where human interaction is essential - in helping farmers individually and in small groups to diagnose problems, to interpret data, and to apply their meaning (Leeuwis, 1993).

The future will call for more able, more independent, more client-oriented extension workers. The emphasis will be on the quality of interaction between agent and client rather than on the movement of "messages" through a hierarchical system.

Flexibility and adaptability will be seen as virtues rather than aberrations. Paradoxically, these trends will bring us full circle to the early manifestations of modem extension in Europe. The itinerant agricultural teachers, unencumbered by large bureaucracies and tall hierarchies, will find their modem counterparts in the computer-carrying extension workers who are at ease helping farmers to identify the information they need in order to realize the potential of their farming operations. Looking back, we can regard the period from 1970 to 1995 as a necessary but expensive stage in the evolution of extension systems, after which extension agents were able to settle down to their main task - bringing together the expertise of farmers and the best available scientific knowledge to develop farms and local agricultural economies.

Alternative Approaches to Organizing Extension

Extension, as the organized exchange of information and the purposive transfer of skills, is a rather recent phenomenon. Obviously, transfer of information and skills has existed since the emergence of permanent agriculture. Today's practice is different in that the process is dominated by organizations, and its scope has extended from disconnected local events to a complicated, large-scale, and even worldwide activity.

Here, extension approaches are presented in terms of their most important organizational forms and their respective goals. The goal system reflects the power positions of various groups of actors. Therefore, without an understanding of the historical development and of the interest groups involved, present achievements and shortcomings of extension approaches cannot be evaluated. It is assumed that different forms of organizing extension are per se neither "good" nor "bad." Rather, extension services must be judged against their proper goals. The one universal yardstick, however, is their service function to the rural communities. Extension which is not in touch with and does not significantly contribute to improving the life situation of its clientele has lost its legitimization.

Extension Goals

Goals lead the actions of individuals, groups, and organizations. While pointing towards a future state, they are influenced if not determined by past experiences. They reflect the interests of their stakeholders and differ, therefore, according to specific life situations, power positions, and development philosophies. The prominent features of a system, such as its organizational structure, the choice of clientele, its operational design, and the methods used, are directly influenced by its set of goals and must be evaluated in terms of their contribution to goal achievement.

Main actors within the extension system are the members of rural communities, extension and other development personnel, researchers, and staff of commercial or public service and support organizations. Empirical evidence shows a variety of forms in which interaction among these groups is institutionalized. The variety of forms suggests a similar variety of goals, and either could be used to classify extension approaches. In practice, however, one finds an almost inseparable mixture of goals inhibiting a clear-cut classification. It seems more appropriate, then, to use a broader category, namely, selectivity with regard to clientele, and treat the respective goals as a continuum. The two end points of this continuum would be marked as technology transfer and human resource development, suggesting either a rather narrow technical or a broader socioeconomic view of development.

Technology Transfer

Until the end of the eighteenth century, farming techniques developed gradually and steadily over centuries with few qualitative leaps. Colonialism and imperial expansion introduced innovations - the spread of maize, tobacco, potatoes are striking examples - but experimentation and dissemination of knowledge were basically at the local farm level. The rise of agricultural sciences has induced dramatic changes in this respect. Increasingly, new technology has been created outside the actual farming sector by public sector research organizations. More recently, private firms in industrialized economies find agricultural technology research and development a highly profitable business.

For decades the research-extension-farmer linkage, especially in developing countries, was based on a rather simple model. In order to achieve development, "modem" research results had to be transferred to the "traditional" farmer, and extension seemed to be the appropriate means to do so. The general faith in science and the commitment to modernization led to discrediting indigenous knowledge. Although this view is still held by many administrators, researchers, and extension agents, it is now being seriously questioned. Farming systems research and the "rediscovery" of farmers' knowledge (Chambers, Pacey, and Thrupp, 1989) have shown that "improved technology is a package of inputs and practices that usually comes from many sources" (UNDP, 1991, p. 2). The re-examination of the conventional view on agricultural knowledge cannot, however, result in questioning the important role of research as the source of new technology. For developing countries, one observes that the accelerated growth and spread of problems - such as the degradation of marginal land - surpass the problem-solving capacities of the local population. What is called for is a setting of new priorities and the building of knowledge systems based on problem solving rather than on information transfer.

Human Resource Development

The concept of human resource development is much broader than that of technology transfer, though both are closely interrelated. Increasing complexity not only of technology but also of the life situation of farmers even in remote areas demands new skills. With the help of these skills, rural women and men "acquire a better insight into the network of problems and recognize the alternative solutions available" (Albrecht *et al.*, 1989, p. 34). Traditionally, teaching the basic skills of literacy and numeracy has not been an extension activity. The limited success of literacy programmes in poor countries has drawn attention to nonformal education in which extension has an important part to play (Coombs and Ahmed, 1974). Whereas in most cases this would require a coordinated effort of different organizations of which extension is but one, human resource development may also be regarded as genuine extension content.

Extension may substitute over a certain period activities such as vocational education that are not yet in place, but more important will be the teaching of managerial and organizational skills that will enable farmers to increasingly solve their own problems. Human resource development thus aims at what may be called "critical competence." Extension clients know what to ask for, they can evaluate the appropriateness of technical information, they are responsible decision makers. Persons with this qualification exist in every rural community, and they will be the ones who actively seek further assistance. One important task of any extension system will therefore be to extend human resource development to underprivileged groups with less access to formal or vocational education - women farmers, rural youth, and generally small farmers in remote areas.

Alternative Ways of Organizing Eextension

The goals of extension may vary, as was shown, within the overall system as well as between different extension organizations. In addition, specific objectives may sometimes contradict each other. While smaller systems may come close to pursuing a consistent set of objectives or reconciling conflicting interests, large-scale organizations must work on a compromise basis. In this respect, Axinn's principal observation is of particular importance: "The success of an agricultural extension programme tends to be directly related to the extent to which its approach fits the programme goals for which it was established" (Axinn, 1988, p. 135).

The alternatives to organizing extension demand choices on various levels:

a) Public versus private

b) Government versus nongovernment

c) Top-down (bureaucratic) versus bottom-up (participatory)

d) Profit versus non-profit

e) Free versus cost-recovery

f) General versus sector

g) Multipurpose versus single purpose

h) Technology driven versus need oriented

In practice, extension organizations everywhere pursue the overall goals of technology transfer and human resource development, though the emphasis will differ. Within each organization there is a mix of objectives, and within countries there is often a mix of organizational patterns. When presenting an overview on the most important patterns, we will be using a well-established terminology (Axinn, 1988; UNDP, 1991), though the grouping is different. We will differentiate between approaches that, at least in principle, target all persons in rural areas engaged in farming and those that purposely select clientele according to specific criteria.

General Clientele Approaches

Ministry-Based General Extension shortly before or after independence, organizing agricultural extension work under the wings of the ministry of agriculture seemed to be an ideal solution for many African and Asian governments. All options for reaching large numbers of clients and serving their needs in terms of quality information and assistance appeared to be open. The original colonial model combined research and extension within the same organization. All important aspects of small holder agriculture plant production, animal husbandry, home economics could be attended to as the ministry established respective sections under its jurisdiction. The fact that the ministerial hierarchy followed the country's territorial subdivision allowed the systematic expansion of the system "down" to the village. The generalist nature of field extension staff functions corresponded to the set of problems faced by non-commercial growers. To cater to specific needs in terms of technology or in terms of target group's specialists could be employed. Thus clientele included in principle all persons engaged in agriculture. Commercial service and support organizations lacking, village-level extension staff could be expected to supplement information by rendering services necessary to apply it productively. A uniform and nationwide organizational pattern seemed to facilitate information flow - including the infusion of expatriate expertise - and corrective measures whenever weaknesses were identified. Public interest was to guide goal setting, programme formulation, and the implementation of fieldwork.

Ministry based extension has been unable to reach a majority of its potential clientele for economic, socio-psychological, and technical reasons. Even dramatic quantitative increases in personnel more staff closer to the farmer have not produced manageable client-to-agent ratios. In recent years, the trend has even been negative. Financial constraints have produced a strong pressure to reduce staff, and the field level has been hit hardest. Those remaining have little if any material resources left to maintain mobility.

Today's situation is aggravated by two additional aspects which refer to the internal structure of the service: management problems and lack of control from below. Ministry extension employs thousands of persons working under a wide variety of circumstances. Decision making and management are highly centralized and formalized. Extension fieldwork, on the other hand, demands location-specific, flexible, and often quick decisions and actions. Managing the "invisible" man or woman (Chambers, 1974) must be highly ineffective as long as he or she is expected to receive and execute orders.

Training and Visit Extension (T&V) in the strict sense of the word, T&V is not a separate approach but one way to organize ministry-based extension. The controversial debate on the merits of T&V tends to obscure the fact that it was originally meant to solve some very specific problems of conventional extension services.

Benor and Harrison's original paper - one of the most influential extension publications ever - critically evaluates the ministry-based extension system of the 1970s (Benor and Harrison, 1977). They found:

- ✰ An inadequate internal organizational structure
- ✰ Inefficiency of extension personnel
- ✰ Inappropriateness or irrelevance of extension content
- ✰ Dilution of extension impact

In addition, T&V operates under the assumption that its extension workers are exclusively engaged in educational activities and that a unified extension service exists. Agricultural research must not only be effective but also work in close collaboration with extension. Both external and internal evaluations are to be used to constantly modify and adapt the system to changing conditions.

Simple as the prescriptions seemed, implementation proved to be difficult. First, the contact farmer concept implying a two-step flow of information from the extension worker to the contact, farmer and from there to other farmers has frequently failed. Extension workers have been blamed for "wrong selection," but the root of the problem lies within the purely technical philosophy of T&V. Other aspects such as communication skills, leadership, and organizational capacities are neglected. In practice, T&V has been a top-down approach leaving little possibility for participation and initiative, both for farmers and village extension workers. Too little emphasis has been put on critical feedback based on self-evaluation. As a result, rigidity rather than flexibility characterizes local fieldwork.

Secondly, Benor's fear that extension services may "rapidly run out of anything to extend" (Benor and Harrison, 1977) characterizes many T&V field situations. The standardized messages passed on are often of little relevance to local conditions. Once T&V was extended to less favoured regions, it soon became clear that technology of the green revolution type showing quick and visible results is not available. Still, training sessions were held and visits made according to schedule, leaving behind disinterested farmers and de-motivated extension workers.

The limited success of T&V in its present form as a nationwide extension system should not discredit the quality and appropriateness of many of its elements. Applied less rigidly and combined with the tools of human resource development as well as with the concept of participation, these elements may constitute a valuable base for reforming extension organizations, large or small (Nagel *et al.*, 1992).

The Integrated (Project) Approach

Integrated approaches aim at influencing the entire rural development process. Extension is only one though often crucial element in this strategy which targets the entire population in a given area but emphasizes work with disadvantaged groups. Integrated approaches are generally implemented in the form of large-scale and foreign-funded projects aiming at alleviating mass poverty in rural areas on the basis of "a simultaneous improvement in the utilization of natural resources and of human potential" (Rauch, 1993). Measures to promote production are coupled with a strong emphasis on self-help. The underlying concept is typically multisectoral.

Evaluations of more than a decade of integrated rural development (IRD) projects have revealed serious shortcomings in reaching the goal of mass poverty alleviation (IBRD, 1987; BMZ, 1990). Sizeable numbers of the poor were not reached by project activities, nor were positive effects consolidated on a sustainable basis. Project deficiencies were in part management related and very often due to a serious underestimation of the great complexity of multisectoral programmes with ambitious goals. The disregard of the target group principle and of due consideration for framework conditions (economic and institutional) played an even more important role, as did the lack of compatible technical solutions.

University Based Extension

While the Cooperative Extension Service (CES) of the United States is still the only system in which the main extension function remains within the university, some developing countries, notably India, have integrated educational institutions into practical extension work. Within the United States of America, state universities have traditionally cooperated with local counties and the U.S. Department of Agriculture in doing extension besides education and research. Within the last

130 years, extension goals of the land-grant colleges have shifted from practical education to technology transfer and, more recently, to a much broader concept of human resource development.

With the emergence of strong private and other public sector research and development organizations and dramatic changes within the agricultural production sector, CES is facing new challenges with regard to coordination and cooperation. Apart from its traditional roles, networking will become a primary role (Bennet, 1990). In this model, industry as well as intermediate and end users of knowledge become part of the extension system.

Extension to Selected Clientele

Commodity Based Extension. Next to the ministry-operated general approach, commodity-based extension run by government, parastatals, or private firms is the most frequent extension organization. Clients may be dispersed over a large area or closely connected, as in the case of large, centrally operated irrigation projects. Commodity-based extension is the predominant feature in many francophone countries of Africa (Schulz, 1973), but is also strong in other countries with commercial or export crops.

The original rationale was the generation of revenue as well as the assured supply of tropical products for the colonial powers. Today, goals are still clearly and intentionally production and profit oriented. All aspects of producing and marketing a particular crop are vertically integrated, spanning the whole range from research, advice, and material support given to farmers, to organizing marketing and even exports. Proponents of the approach argue that, by infusing modern technologies and monetary incentives into traditional farming, a cumulative chain of effects are triggered, thus contributing to overall development.

Advantages in terms of organizing the extension function seem obvious. One generally works with well-tested technologies. Objectives and targets can be clearly defined and the organizational structure kept simple. The focus on only one or two crops facilitates training of extension workers who are agents of the society or board concerned. Control of agents and farmers is easy, because they are judged in terms of defined targets.

Extension as a Commercial Service

Commercial extension is a rather recent phenomenon and typical of either industrialized forms of agriculture or the most modem sector of an otherwise traditional agriculture. It may be either part of the sales strategy of input supply firms or a specialized consultancy service demanded by an agricultural producer. In both cases, the goal of the organization or the individual is profit earning, which in turn is tied very closely to customer satisfaction. Most directly this is the case for private consultants who will be retired only if their clients feel that expenses made have been profitable. Large input supply firms or rural banks that use their own extension workers as sales personnel must also have a long-term perspective with regard to the competitiveness of their products and services. Negative effects of incorrect application or use will be attributed to the product itself. The clients of commercial extension will also be profit oriented. Their objective is the optimal utilization of purchased inputs or contracted expertise.

The emergence of commercial extension has influenced the debate on who should bear the costs of extension. With escalating budget deficits, the idea of extension as a free public service is no longer being generally accepted. It is argued that those who can afford it should actually pay for advisory services. In the case of commercial input suppliers, the solution is very simple: the costs of extension are included in the product price, as are the costs for research or advertisement. Private consultancy, on the other hand, is costly and affordable only to either large-scale or highly specialized producers.

Client-Based and Client-Controlled Extension

One way of dealing with the shortcomings of large extension systems has been to localize extension and utilize the self-help potential of rural groups. Often organized by outsiders, these decentralized approaches are in a better position to serve the needs of specific target groups, notably those in disadvantaged positions. Close contact with their clients and intimate knowledge of their life situations are essential for the planning of problem-oriented extension activities. Local personalities are identified who take over leader functions once the external (nongovernmental) organization withdraws. The principles of these organizations (awareness, empowerment, participation, self-help) are close to the philosophy of Animation Rurale without the national dimension. The impact of client-based approaches must be seen on two levels. Directly, they provide benefits to their clients. The diversity and large number of small projects forbid a general statement on their effectiveness in terms of human resource development.

Present and Future Role of Extension Staff

Person-to-person communication has traditionally been the most important form of information transfer. Print media as well as radio and television were of a supplementary nature because they frequently lacked target group or location specificity and information was not up-to-date. Revolutionary changes in communication technology have dramatically increased the speed and quality of information transfer and changed the role of extension workers in industrialized countries. Electronic communications systems may in part replace personal visits, and one of the major tasks of any agent will be to link her or his clients with other suppliers of information.

Notwithstanding the fact that their use is not restricted to industrialized countries, the fascination with modem communication means tends to obscure the fact that most extension personnel - 90 per cent of which are located in developing countries (FAO, 1990) - are working under extremely difficult and disadvantageous conditions. In fact, little has changed during the last two decades to remedy the basic ills, and the field agent is still the weakest link within the system.

There is a wide discrepancy between organizational goals and the potential of even well-trained and dedicated field staff. Early Indian community development activities covered close to fifty areas (from reclamation of wasteland to improved rural housing on a self-help basis), all to be administered by the local village-level worker (Dube, 1958). Fieldwork in most developing countries is characterized by conditions that foster low morale: lack of mobility, virtually no equipment, and extremely low salaries. For many extension workers, tapping additional income sources is a question of physical survival. Quality performance is further impeded by low educational qualifications and lack of advancement possibilities (Swanson, Farner, and Bahal, 1990).

Agricultural extension, whether public or private, operates in a context or an environment that influences the organization, form, and content of transfer activities (Moris, 1991). The dominant characteristic of that context is change. Because the changes affect all aspects of extension, the context should be examined and understood so that extension can be better managed.

A Systems Perspective

The usefulness of a systems approach for understanding and analysing agricultural technology generation and dissemination is widely acknowledged (among others, Nagel, 1979; Swanson and Peterson, 1991; Roling, 1991; Elliott, 1994). Although they are sometimes criticized because they are so abstract, systems analyses offer holistic vantage points for understanding the factors that impede or enhance the two-way flow of technology and information between farmers and the public organizations that constitute the system.

The components are technology generation, technology transfer (knowledge and input transfer), technology utilization, and agricultural policy (Swanson, Sands, and Peterson, 1990). The organizations that constitute the components, as well as others in the system environment influence each other in complex ways. Such publicly funded systems are established by governments to improve the conditions of life and well-being of rural and urban populations and to increase agricultural productivity.

The functions and linkages related to the flow and feedback of technology and information in the system define the components. In the public sector, agricultural research organizations (technology generation) and extension (technology transfer) are major actors, although commercial companies and NGOs (the private sector) are commercial companies and NGOs (the private sector) are also involved. The way these are divided among organizations differs from country to country.

- ☆ **Technology generation** consists of planning, administration, and implementation of research activities that develop, assess, adapt, and test improved agricultural technology for farmers and other users. In the public sector, these tasks, as well as some dissemination work, are carried out by agricultural research organizations.
- ☆ **Technology transfer** further evaluates and adapts research outputs for users and then widely disseminates the knowledge and inputs to different target adopters - farmers of different categories, private companies, and so on. Two parts of the transfer component, namely, knowledge and inputs. In many countries, government extension does both types of

transfer, although the emphasis is on knowledge. Private organizations are also involved in both types of transfer, although farmer coverage is more limited and input transfer may play a bigger role.

- **The technology utilization** component encompasses the users of the agricultural technology, mainly farmers. User awareness, adaptation, and adoption of improved technology from various sources affect farm-level productivity and profitability and, ultimately, economic growth at the national level. Interaction and feedback between users and research and transfer organizations improve cooperation and the relevance of technology.
- **The policy component** relates to government development goals and strategies, market and price policies, and the levels of resource investments in the system. Various government bodies play a role in setting development policy. Technology development and transfer organizations are affected by the policy in fundamental ways.

The division of generation and transfer tasks organizations and sectors differs from country to country, as does the nature of the policy and utilization components. However, the flow of technology can be mapped as it moves between or is influenced by organizations constituting the agricultural technology system. Flow analysis shows the impediments to generation and transfer and the influences of other institutional actors and highlights contextual issues.

The agricultural technology system context in which government extension organizations operate can be described under two headings: macro-factors and institutional factors. The discussion that follows introduces the more important ones in each category that need to be understood by extension planners and managers. Any of these factors can impede or influence the flow of technology and information between farmers and extension.

Different Macro-factors

Agro-ecological

Because the natural environment strongly influences extension planning and operations, extension should respond to the technological needs of farmers in different agroecological zones. The variation represented by agroecological zones in a given country can be substantial. Differences in temperatures, rainfall, soil types, evapotranspiration, and so forth are reflected in the diversity of farming conditions and production systems found.

Extension planners face some difficult choices because of the need to respond to the diverse technology and information needs of farmers from many different zones and, at the same time, to satisfy a requirement for extensive countrywide coverage of the rural population. For example, extension resource investments can be determined by farmer population concentrations, potential productivity of selected agroecological zones, or a combination of both.

Political-Economic

The political and economic environment affects extension in many ways. One of the most significant factors is a country's stage of economic development. Another is the level of government investment in public sector extension. This is influenced by the presence or absence of a structural adjustment programme, the degree of economic dependence on agriculture, and the proportion of the population economically active in agriculture as opposed to industry. The percentage of resource-poor smallholder farmers influences the type of technology to be transferred, particularly if government is concerned with the equitable coverage of all categories of farmers.

Politics has another, infrequently identified impact on public extension organizations. Political shifts at the national level often result in changes in extension personnel, management, and programmes. High turnover of top managers undermines management experience and continuity in leadership. In addition, political unrest and war often make extension efforts impossible (Gustafson, 1994).

Socio-cultural

In many countries, socio-cultural factors are leading constraints to the effectiveness of extension. Language differences and illiteracy can impede the communication of improved technology unless they are taken into account. The division of labour between the sexes can differ along cultural lines and influences the nature of farming systems in different regions. In many countries, the men are employed off-farm, leaving the farm operations to women. In extension organizations, under representation of women on the extension force means that the production responsibilities and needs of women at the farm level may not be adequately addressed.

Cultural differences among farmers, as well as differences in their resource endowments, also need to be taken into account. In particular, these are reflected in land-use strategies. Pastoral herders, for example, will require different types of subject-matter expertise, and extension will need to use different strategies to transfer technology to them than to permanent field agriculturalists. The resource endowments of different categories of farmers also affect technology adoption levels. Subsistence farmers adopt mainly low-cost technologies. For this reason, extension work that focuses on cultural practices and affordable technologies may be more appropriate in countries with large numbers of resource-poor farmers.

Policy

The policy component of an agricultural technology system can enable or limit extension in ways beyond the reach of extension managers. The principal areas of influence are price signals to farmers and decisions by government that affect public agricultural development organizations.

Agricultural Development

Policy-making bodies of government set development goals and objectives such as achieving food security or surplus agricultural production to stimulate economic growth, or providing health care and education for rural development. It is the task of government extension organizations to help meet these goals by formulating subordinate objectives and strategies to achieve them.

The amounts invested by government and the development community in agricultural development influence the pace and scale of effort. In many instances, spending ceilings for public institutions have been imposed by a structural adjustment programme agreed to by government. Extension planning needs to take these limits into account, especially when dealing with the question of staffing, which has major budget implications.

The endowment levels for public agricultural extension are set by government policy and planning bodies and impose limits on expenditures. This can hamper extension efforts in basic ways. The operating as compared to salary budget amounts is often inadequate, with negative effects. For example, a lack of money to buy vehicles and fuel undermines the mobility of extensionists. In addition, if salaries are too low, extension cannot attract or hold qualified staff, and services will suffer. Careful planning, guided by a strategy and programme of action, can improve the use of resources and the effectiveness of extension work within government funding limits.

Market Intervention

Governments set policies on consumer and producer commodity prices, subsidies for inputs, credit availability, import substitution, export earnings, food self-sufficiency, and natural resource management. These send direct and indirect price signals to farmers and influence their production decisions. High fertilizer prices, for example, cause farmers to use less of this input. Extension organizations should be aware of market signals to ensure that they are recommending technologies that meet farmers' current needs. They also require feedback from farmers to ensure the continued relevance of their activities.

If there is no profit incentive for production of a specific crop, there is little point in developing or transferring improved technology related to that crop. For example, policies that favour the import of cereal grains at concessionary prices on the international market discourage in country production of those crops. Technology generation and transfer outputs that focus on such crops are not likely to interest farmers. The examination of the policy context is important and allows organizations to avoid wasted effort and resources.

Infrastructure

Infrastructure, particularly the condition of transport, communication, and market facilities, affects both farmers and extension. The capacity to move people, inputs, and produce and to send and receive information influences extension activities and capacity. Market infrastructure itself can be lacking or inadequate.

Transportation

There may be many areas that cannot be reached by road, or transport vehicles may be in short supply. In either case, farmers under these conditions are difficult to reach with improved technology, and they will have problems transporting inputs and farm produce. For example, input supply outlets can be far apart, and excessive transport distances can make it difficult or impossible for traditional farmers to adopt improved technology.

Communications

Communication infrastructure can impose additional constraints for extension organizations. Farmer access to mass media such as publications, radios, or television may be limited, thus reducing the options available to extension for communicating its messages. At the same time, extension itself may have little or no access to telephone and radio services for long-range communications. This can severely hamper its ability to organize and carry out field operations.

Two aspects of a country's media organizations, both print and audio-visual, affect the flow of extension messages to farmers (Mathur and Sinha, 1991). One is the attitudes and subject interests of media managers responsible for programming for rural audiences. The other is the organizational climate, especially morale. It is crucial that extension consider the capacity of mass media organizations (newspapers, magazines, radio, and television) as part of its strategy and plans for communicating with farmers.

Institutional Factors

Public agricultural extension organizations in most countries have the task of providing a two-way flow of improved technology and information between research and users, primarily farmers. They operate in an institutional environment that includes other public and private organizations active in agriculture. In particular, those other actors involved in generating and transferring agricultural technology must be examined and understood to improve extension's effectiveness and efficiency.

Research

Agricultural research organizations are extension's closest institutional partners in technology generation and transfer. While these functions are also undertaken by private organizations, public sector organizations have the task of generating technology relevant for all categories of producers and agro-ecological zones. The way research is structured and organized, and the planning and management of research-extension linkages, can limit or enhance extension's effectiveness.

Agricultural research organizations in developing countries confront many problems (Idachaba, 1987; Oram and Bindlish, 1984). These include lack of financial resources, acute shortages of well-trained scientists, lack of farmer feedback to ensure relevance of research results, lack of access to external sources of knowledge, inadequate research facilities and equipment, low staff morale, and inadequate operating budgets, staff incentives, and remuneration. Few of these can be addressed by extension managers, but they can impede the generation of technology, resulting in fewer research outputs for extension to transfer. An understanding of research's problems is an important step in planning extension activities and coordinating them with research.

Poor linkages between research and extension are major constraints in technology flow in many countries (Kaimowitz, Snyder, and Engel, 1990; Eponou, 1993). The linkage problems are of two basic types: those affecting feedback from farmers to research and extension and those relating to coordination and cooperation between research and extension. In both cases, extension managers can improve the situation by developing a linkage strategy and allocating responsibilities and adequate resources for linkage tasks.

Education and Training

Educational organizations that train extensionists are important elements in the institutional context for extension. The work of universities and training institutes in particular has a significant impact on extension organizations. The content of their curricula as well as the numbers and qualifications of their graduates are limiting or enabling factors in any country. In many cases, communication between extension and education organizations is poor. As a result, extension commonly has staffing problems. It is not unusual for extension organizations to have posts that are either vacant or filled by under qualified personnel.

Inadequate numbers and qualifications of staff remain a difficult problem for public sector extension organizations. Salaries and benefits are rarely competitive with those of comparable private and public enterprises, resulting in low morale and high staff turnover. Education levels may be quite low, especially for farmer contact staff. The ability to attract and retain qualified extension staff is limited in most countries by civil service salary scales established by other agencies of government.

The situation can be improved by establishing staffing and training plans. The staffing plan inventories current human resources, identifies staffing gaps, and projects staffing needs over a specified time. The training plan identifies specific types of training (in-service and formal) required to fill skill gaps in human resources and to cover staffing needs for planned operations. The additional step of coordinating training needs with the educational organizations is needed.

Input Supply

Farmers need inputs to increase production, but access to these is often poor in less developed countries. While inadequate transport and marketing infrastructure are often at the root of the problem, there are certain aspects that can be addressed by extension.

Genetic Technology

Among the major outputs of the technology generation subsystem is improved genetic material. New plant and animal varieties with higher yields or resistance to pests or diseases become available. However, farmers need to know when the variety is released, how it performs under farm conditions, and where to obtain seed or breeding material. Extension is responsible for disseminating this information through appropriate mass media and contact methods. But a common constraint on the flow of technology is the availability of genetic material. In some countries, extension is also involved in the multiplication and distribution of seed. Establishing effective linkages with others involved in the process can also help ensure that genetic material is available for farmers.

Agrochemicals and Other Inputs

The performance of new varieties is often improved by, or even dependent on, the availability of agrochemical and other inputs at the farm level. Recommendations for fertilizer types and amounts suited to local soil conditions, for animal feed mixtures and practices, and for the control of plant and animal pests and diseases constitute an indispensable part of extension messages. Lack of access to this information and materials prevents yield maximization, so extension organizations need to ensure that farmers are informed on availability and use. If access is limited because of external factors, extension should plan its campaigns accordingly, with attention given to low-input recommendations.

Private companies and non-profit organizations also provide advice to farmers on agrochemical and other input use. Extension organizations need to develop communication and coordination linkages with these actors because excessive use of agrochemicals can harm human health and the environment. Conservative recommendations and alternative approaches, such as integrated pest management, are in the interests of the public and can be promoted by extension through its farmer contacts and other linkages.

Credit

Access to credit is one way to improve farmer access to new production technology and increase productivity. Farmers' ability to purchase inputs such as improved seed and fertilizer is particularly important. If appropriate technology is available but not being used by farmers, then the way credit is handled by government may be part of the problem.

Understanding the credit context - government and bank policies, availability of credit, and the institutional relationships involved in its delivery - is important for extension. At a minimum, the existing credit situation should be examined so that factors affecting the adoption and use of technology can be identified. These include inequitable access to credit, insufficient amounts of it, and overlap of transfer activities due to credit institution involvement in extension work. The knowledge is useful for extension managers in targeting farmers and in coordinating extension objectives with credit institutions.

Farmer Organizations and Other NGOs

There is growing involvement of the private sector, both non-profit and commercial organizations, in agricultural research and extension. While such organizations may have limited research objectives and restricted regional coverage targets, they are partners in technology generation and transfer. Private corporations such as seed and agrochemical companies play a key role in developing some types of technology, providing inputs, and advising farmers on their use. Ideally, extension should reach agreements with these actors so that duplication of effort is minimized and conflicting messages to farmers are avoided. At a minimum, their activities should be noted and an attempt made by extension to catalogue and uses their successful methods and outputs.

Farmer organizations, particularly grass-roots organizations, are part of the utilization component. They offer an effective channel for extension contact with large numbers of farmers, as well as opportunities for participatory interaction with extension organizations. Feedback on farmer needs, production problems, and the results of adoption from such groups will be increasingly important considerations.

The AKAP Sequence

It is convenient to visualize extension as achieving its ultimate economic impact by providing information and educational or training services to induce the following sequence:

A: Farmer awareness

K: Farmer knowledge, through testing and experimenting

A: Farmer adoption of technology or practices

P: Changes in farmers' productivity

Changes in farmer behaviour will be reflected in quantities of goods produced, the quantities of inputs used, and in their prices. These, in turn, can be measured as "economic surplus," which is the added value of goods produced from a given set of inputs made possible by the extension activities.

Productivity Gaps and Extension

The AKAP sequencing is, as noted above, related to the flow of new technical information and to the existing state of un-adopted technology. We can see this interrelationship more clearly in the context of productivity "gaps." Figure portrays what is meant by gaps and relates these to the technological infrastructure types used in Table.

Figure depicts crop yields, adjusted for fertilizer and other inputs, for five technology infrastructure types. Four yield levels are depicted for each type:

A: Actual yields

BP: Best practice yields

BPBI: Best practice, best infrastructure yields

BPBIRP: Best practice, best infrastructure, research potential yields

These yield levels in turn define three "gaps":

G (P): A practices gap between the best practice (BP) yield and actual (A) farmers' yields

G (I): An infrastructure-institutions gap between the best institutions, best practice (BPBI) yield and best practice (BP) yield

G (R): A research gap between the research potential yield (BPBIRP) and the best practice, best institutions (BPBI) yield

These gaps provide a way to classify the contribution of extension activities and to show how research and extension are linked. A stylized sequence across technology types is depicted. This could also be visualized as a time sequence.

Extension programmes are designed to reduce both the practice gap, G (P), and the institutions gap, G (I). Extension programmes are not the only activities that reduce these gaps. Providing market information to farmers and developing organized farm groups reduce G (I). Information and teaching reduce G (P). Research programmes are generally required to reduce G(R), although extension programmes can facilitate the reduction of G(R) via facilitating the importing and local modification of improved technology developed elsewhere. Research programmes in most developing countries also modify and adopt imported technologies and germplasm.

Two of the gaps are closely linked. When G(R) is closed (that is, when the BPBI yields go up), G (P) is opened. (This may happen with G(I) also, but to a lesser extent.) Further, it should be noted that the size of the gap is an index of the potential impact of research or extension. As extension succeeds in closing G (P), diminishing returns set in. Successful research opens up new potential by increasing G (P).

Needs Assessment Techniques

This section describes different data-collecting techniques available for carrying out needs assessment. The techniques are discussed under four categories: individual, group, secondary source, and rapid rural appraisal.

Individual Techniques

Individual techniques involve collecting data from people one at a time. The people from whom the needs assessment data are collected do not interact with one another in the course of providing data. Individual techniques include face-to-face interviews, key informant interviews, questionnaires, informal personal observations, and formal personal observations.

Face-to-Face Interviews

This technique is appropriate when dealing with less literate audiences or complex issues about which there is little available information. Both structured and unstructured questions are appropriate for face-to-face interviews, depending on the issues involved and the time available for the interviews. Unstructured questions are useful when dealing with complex or sensitive issues which require probing in order to get accurate data. For example, small-scale farmers may not have a direct answer about how they budget their resources. However, if probed about what they do on a typical day or week, they may provide insights into their economic activities and therefore their needs.

Key Informant Interviews

Key informants are people who are considered experts in a given area because of their professional knowledge or their position of influence in the community or organization. Examples include teachers, religious leaders, grass-roots workers, and traditional and political leaders. There is evidence to show that interviewing several of these categories of informant's yields fairly accurate information about the problems and needs of the community at large. Key informants are particularly useful if the needs assessment has to be done fast, using a limited budget.

Questionnaires

This needs assessment technique tends to be more structured than interview schedules and can be administered by phone, mail, or in group settings. Questionnaires are commonly administered in developed countries by conducting telephone interviews; however, in developing countries they are rarely administered in this manner because of the limited availability of telephones. When dealing with literate communities that have access to good mail services (public or private), needs assessment surveys can be conducted by mail. Dillman (1978) provides excellent tips on conducting mail surveys which apply to develop and developing countries alike. Instead of relying on conventional mail services, questionnaires also can be hand-delivered to respondents and collected after they have been completed. Alternatively, one may take advantage of occasions such as annual club or association meetings when potential respondents might come together. In this case, the questionnaires are presented to group members, who are asked to complete and return them before they leave. For best results, the questionnaire should cover pertinent issues and be short enough to be completed in the time the respondents have. When used appropriately, this method may save both time and money necessary for collecting the information. However, one must be cognizant of the high possibility for biases in the information collected. This is because there is always a chance that the assembled group completing the questionnaire is not representative of the audience in mind.

Stevens (1980) indicated that extension workers are often assigned to their positions without proper equipment and adequate preparation. Maalouf and Contado (1983) indicated that there was a need to assess the training needs of extension personnel and establish priorities. When using a questionnaire to collect needs assessment information, let's assume that an agricultural officer has decided to identify the professional skills of extension agents most in need of improvement in their region. One of the first activities of the agricultural officer was to identify the professional skills that extension agents should possess.

Informal Personal Observations

Valuable needs assessment data often are gathered through informal observation. Fieldworkers see or experience a lot as they travel and work with farmers in the field. If noted or remembered, this information can be used in needs assessment. People in the habit of keeping diaries or writing notes to themselves are more likely to provide more accurate observational data for assessing needs.

Formal Personal Observations

This needs assessment technique is based on using rating forms, checklists, or observation schedules for collecting information. Formal observations differ from informal personal observations in that the items to be observed are predetermined. This technique also can be used to collect both quantitative and qualitative data.

Group Techniques

Group techniques allow participants to interact with one another during needs assessment activities. Information can be collected in writing, as in the Delphi technique, or orally in a group setting such as a focus group, m both cases, successful needs assessment depends on competent leadership and on having participants who have both the knowledge and willingness to participate actively in the interactive group process (Caffarella, 1982). This section will discuss the Delphi, focus group interviews, the nominal group, and informal group methods.

Professional Skills Survey of Extension Personnel

This questionnaire lists a number of professional skills that are used by Extension personnel in conducting educational programs. For each of the following skills we would like you to indicate: (1) your competence to perform the skill and (2) the importance of the skill. The following scales are used:

Competence to Perform the Skill	*Importance of the Skill*
1 = Not at all Competent	1 = Unimportant
2 = Little Competence	2 = Little Importance
3 = Moderately Competent	3 = Moderately Important
4 = Fairly Competent	4 = Important
5 = Very Competent	5 = Very Important

Please circle the number that best represents your response for each item. Example:

Competence	*Professional Skills*	*Importance*
1 2 3 4 5	1. Producing educational/teaching materials	1 2 3 4 5
1 2 3 4 5	2. Using audio-visual aids	1 2 3 4 5
1 2 3 4 5	3. Conducting a method demonstration	1 2 3 4 5
1 2 3 4 5	4. Assessing community needs	1 2 3 4 5
1 2 3 4 5	5. Determining information needed for evaluations	1 2 3 4 5
1 2 3 4 5	6. Developing program budgets	1 2 3 4 5

Delph

In this technique, people with exceptional knowledge about a given subject area are involved in repeated questioning and feedback, using written questionnaires, until a consensus is reached on the subject (Miller and Hustedde, 1987). The process begins with specifying the needs to be assessed, who will be involved, and how the information collected will be used. Ten to fifteen people may be used if responses are not expected to vary a great deal. If major differences in opinion are expected, then larger samples are recommended.

Three to four rounds of questionnaires are used to collect the data. The first round contains open-ended questions asking the participants to write their responses. After the first questionnaire is returned, the responses are summarized and developed into a second questionnaire. The second questionnaire asks respondents to rank their concerns, support for or disagreement with the various group responses, and explain their position. It identifies areas of agreement and disagreement on priorities. The responses are analysed by tallying and summarizing the comments made. The third questionnaire is developed from the summary of the second. The respondents are asked to review their prior responses in light of the group responses and make additional comments. Additionally, they are asked to vote on the order of importance of the items listed. The third questionnaire is often the last, but if there are major areas of disagreement, additional questionnaires may need to be administered.

Once the questionnaires have been completed, a final report based on the last questionnaire is prepared and sent to the participants. The report should summarize the goals, process, and conclusions or actions reached as a result of the study. The Delphi has two advantages. First, it avoids the direct confrontation of people with opposing views. Participants do not face the pressure to conform to the majority position. Second, it eliminates the costs of having participants travel to a central place, a major limitation of focus group interviews and the nominal group technique. The technique has limitations too. Several conditions must be met in order to use this technique (Miller and Hustedde, 1987). First, adequate time must be allotted. It is estimated that in the United States it takes a minimum of 45 days from the development of the questionnaire to the final report. In developing countries, the process could take longer because of relatively poor mail services. Second, the participants must have good writing skills because the technique is used in a written context. Lastly, the participants must be highly motivated because no one will be available to encourage them to complete the several rounds of questionnaires.

Focus Group Interviews

Krueger (1994) defines a focus group interview as a technique in which a group of people who possess certain characteristics provide data of a qualitative nature in a focussed discussion. Each interview involves a group of six to eight people who discuss a common topic for one to two hours under the direction of a moderator and assistant moderator. The discussion is recorded on an audio tape and later transcribed and reported as qualitative data. Typically, several group interviews are conducted. In each case, the moderator raises various issues pertinent to the needs assessments, following an outline commonly known as the questioning route. The interviews are recorded and analysed for patterns and trends among participants in one interview, as well as across the entire set of groups involved in the needs assessment.

The following example indicates how the focus group technique could be used in conducting a needs assessment. Suppose a home economics extension agent noticed widespread signs of malnutrition (such as poor health and stunted growth) among children in a rural community of 2,000 households. The community produces a variety of food including poultry, livestock, fruits, and vegetables that are critical in diets for children. The community is served by one public health centre that functions fairly well in providing vaccinations and other basic medical services. Several drilled wells provide water for the residents. The extension agent decided to carry out a needs assessment to explore the problem and to identify practical interventions to address it. The agent set the following objectives of the needs assessment: determining causes of malnutrition among children in the community and identifying measures to improve the situation. Focus group interviews were used because they provided an opportunity for the participants to engage in dialogue about the problem, its causes, and how to go about solving it (Krueger, 1994). Five steps are critical in the use of focus group interviews (Archer and Layman, n.d.; Krueger, 1994), as seen in the example above:

1. *Developing a Questioning Route*

The questioning route was based on the objectives of the needs assessment. Brainstorming with colleagues and prospective users of the information helped to generate the questions. The questioning route, developed in the local language, consisted of four main questions and several questioning probes. The questions were: (a) What problems do parents in this village face in feeding their children? Are the problems similar for all families? (b) What should be done to make sure children get proper nutrition? What can parents themselves do? Can all parents afford these measures? (c) What problems do parents in this village face in keeping their children healthy? (d) What can parents do to improve the health of their children? Does the government have any role to play?

2. *Recruiting the Participants*

The extension agent selected one of her assistant agents to help in conducting the interviews. Five interviews, one in each of five villages in the community, were conducted to help ensure representative data and to minimize travel costs on the part of participants. For each interview, eight participants were recruited to represent women's groups, local politicians, school teachers, religious leaders, health practitioners, and parents. The extension agent contacted the prospective participant personally and explained to them the purpose and timing of the focus group interviews. The participants were selected on the basis of their knowledge of the village situation, ability to discuss freely in a group, and interest in child nutrition and health.

3. *Planning Resources*

The extension agent developed the timetable for the sequence of steps involved, as well as the fiscal plan. She approached the administrators of local schools and secured rooms in which to conduct the interviews. A tape recorder was obtained from a local business person.

4. *Moderating Interviews*

The extension agent moderated the interviews with the help of her assistant extension agent. Moderating is the process of keeping the discussion on track. It involved bringing the conversation back on target when irrelevant topics were introduced. This guidance had to be provided without reducing group enthusiasm and interest in the discussion. There are several personal attributes of a good moderator (Krueger, 1994), including (a) familiarity with group process either from previous experience in working with groups or through training in group dynamics, (b) good listening skills, (c) adequate background knowledge on the topic of discussion, (d) well-developed written and oral communication skills, and (e) a sense of humor.

5. *Data Analysis and Reporting*

Data analysis and reporting followed the interpretative summary format, whereby the data were not only described but also interpreted (Krueger, 1994). The analysis started with a debriefing immediately after the interview ended. The interview tape was played and a summary of the interview was written the next day. Each interview summary included key incidents, strong statements, and frequently occurring responses. Next, the key incidents, strong statements, and frequent responses were classified by question, coded, and grouped. The coding and grouping helped identify the general themes in the responses (Krueger, 1994; Miles and Huberman, 1994). The identified themes were then compared across interviews in order to develop a general picture for the whole community on a question by question basis. Lastly, specific concluding statements were made related to the two objectives of the needs assessment. Recommendations also were made regarding measures that extension and other organizations could take to improve child nutrition and health in the communities and region.

Krueger (1994) provides detailed guidelines and examples of focus group interviews as conducted in developed countries, specifically the United States. Focus group interviews have great potential for use in developing countries.

The Nominal Group

This technique can be used to generate possible items and set priorities in conducting a needs assessment. Although the data are generated in a group setting, verbal communication is minimized. For example, a chairperson of a farmers' association invited their extension agent to facilitate a meeting of selected members assembled to determine the activities for the following year. After learning that the farmers could read and write their native language, the extension agent suggested that the nominal group technique be used to assess needs for the association. The extension agent used the following six steps in conducting the nominal group process:

Stating of Question or Problem

This preceded the group session and involved clearly stating the question or problem to be addressed by the participants. The question addressed was: What activities should the association carry out next year?

Generation of Ideas

The participants sat at a table facing each other and were asked to spend the first several minutes in silence, writing their ideas on a piece of paper. This silence allowed the participants to generate ideas uninterrupted and without being dominated by aggressive members.

Presentation of Ideas

After they had finished writing down their ideas, each member presented one idea from his or her list in a round-robin fashion. The chairperson served as the recorder and wrote the ideas on a flip chart in front of the participants. The chairperson listed the ideas in concise phrases without attempting to change the wording or judge the ideas. The listing continued in order until all the participants had presented their ideas. The ideas suggested included organizing a farm fair, conducting an agricultural tour, acquiring an office for the association, increasing membership, and starting a farmers' market in a neighbouring town.

Clarification of Ideas

The farmers were encouraged to seek clarification of any of the ideas list ed. They were allowed to express their reasons for agreement or disagreement about each item, but argument was discouraged. This step ensured that the ideas listed were clarified without high status or aggressive members dominating the process.

Rating of Priorities

At this stage, the farmers were asked to choose three of the most important ideas from the list and rank them in order of priority. The rating was done on small cards which, after being collected, were shuffled to ensure anonymity. The votes were then tallied and the results disclosed on a flip chart in front of the group. The three highest rated items were: starting a farmers' market in a neighbouring town, conducting an agricultural tour, and increasing member ship in the association.

Discussion and Voting

In the last step, the farmers discussed the vote, made additional clarifications, and voiced their agreements or disagreements. Following this discussion, the farmers decided that they would only start a farmers' market and work to increase their membership. Delbecq (1975) is the original source on the nominal group technique. Miller and Hustedde (1987) have provided some guidelines on the use of the nominal group technique in community needs assessment in the United States. The technique is not widely used in developing countries because, like Delphi, it can only be used with participants who have developed writing skills.

Informal Group Methods

This category includes gathering information at group meetings and social gatherings. It is common for participants at meetings to talk about issues and problems in their family, community, or organization even when they are not part of the agenda. These side conversations may provide insights into the problems facing the organization or the individuals involved, as well as what can be done to address them. Tea-coffee breaks also provide another setting for members of an organization to talk about issues important to themselves, the organization, and the community at large. Important issues or needs of an organization may be identified by simply attending and listening to the conversations going on among its members during these short breaks.

Secondary Sources

Voss, Tordella, and Brown (1987) defined secondary data as "information gathered for the purpose other than the immediate or first application". Secondary data sources include census reports, previous studies, and administrative records and reports. Extension staff rarely uses these data sources in needs assessment, probably because their application is not straightforward (Sofranko and Khan, 1988).

Census Reports

These data provide aggregate national and community statistics on such important sectors as population, housing, agriculture, education, health, and labour. When recent, these data can be-used to identify the general areas of concern in the community and should serve as a good starting point in assessing needs. For example, small numbers of livestock, chicken, and fish would point to high chances of animal protein deficiency in the diets of community members. This might in turn imply poor health, especially among children.

Previous Studies

These reports include studies done by both groups and individuals. Although the issues studied may not be exactly the same as the needs assessment, there is always a chance that another development agency is, or has been, involved with similar kinds of issues. Local libraries and archives should be searched as a first step in needs assessment. In searching for previous studies, do not be restricted to school and alone. Also, do not forget the nonconventional sources of reading materials such as government departments, NGOs, and other local development agencies. In developing countries, local donor agencies (for example, the World Bank, Food and Agriculture Organization of the United Nations, United Nations Development Programme, United States Agency for International Development, and British Council) maintain valuable collections of (consultancy) studies and reports on past and present projects. Such reports are good for providing expert opinion regarding the problems and prospects upon which needs assessments could be based.

Administrative Records and Reports

Both government and nongovernment organizations maintain records on their activities. Most organizations also publish their activities quarterly or annually. When accessible, relevant organizational records and reports should be perused as one of the sources of needs assessment data.

Rapid Rural Appraisal

Rapid rural appraisal refers to the use of several data collection methods to gather practical information on development issues in local communities quickly (Freudenberger, 1994). These might include interviewing key informants, reviewing secondary data sources, mapping exercises, and conducting semi-structured interviews with groups and individuals. Refer to chapter 6 for a detailed description of both rapid rural appraisal (RRA) and participatory rural appraisal (PRA).

Procedures

This section provides some guidelines to help in conducting a needs assessment. Extension personnel should be encouraged to be creative, efficient, and effective in designing their efforts. Consider the following guidelines:

a) Determine the purpose for conducting the needs assessment. Among these purposes are developing responsive programmes, generating awareness of programmes, satisfying official mandates, aiding in improved programme decisions, and promoting citizen participation and action.

b) Define the goals and objectives for the needs assessment. Show what it is you want to find out about and from whom. How are target clientele (including farmers of different resource levels, genders, and ethnicities) involved in the setting of goals and objectives?

c) Select the approach you will take in collecting the information. Summers (1987) suggests four considerations: (a) the reasons for involving the public, (b) the decisions to be made using the information collected, (c) the need to generate representative information, and (d) the cost involved. Decide whether the needed information already exists, if a new data collection effort is needed, or if a combination of approaches is needed.

d) Design the instrumentation and procedures. When you design needs assessment instruments, it is usually best to keep the process simple. Long and complicated instruments discourage responses. Additionally, short instruments are less expensive to produce, distribute, collect, and analyse. Once a draft instrument has been prepared, it should be checked against the original purposes, goals, and objectives to make sure that nonessential information has not been included.

e) Prepare an estimated time line and budget for the needs assessment. The amount of resources available is likely to be one of the major determinants of the technique used for needs assessment. This is particularly critical in developing countries where resources are more limited. Although critical, budget constraints should not be used as an excuse for poor needs assessment.

f) Conduct a pilot test of the instrumentation and procedures. Special consideration should be given to collecting information from farmers of different resource levels, genders, and ethnicities. Many mistakes can be identified and eliminated in a pilot test with small groups of target clientele.

g) Collect the information. Limit the collection time in order to help develop a sense of urgency and keep the needs assessment targeted.

h) Analyse the data and information. If there is a large response, try to have access to a computer to analyse the data. There are also software packages to analyse qualitative data.

i) Prepare a report of the findings. Make it as user-friendly as possible. Do not feel confined to create one long document. It is probably better to divide the report into several brief documents for specific audiences. Consider using "white space" and figures to help communicate important points. Also consider developing audio-visual reports. Video tapes, transparencies, and slides also can be effective in communicating results.

j) Evaluate your efforts. Take time after the needs assessment has been completed to judge its merit and worth. What worked well? What problems were encountered? How could it have been done better? Once the evaluation is complete, share it with others interested in needs assessments. This will provide an opportunity for others to learn from what was done.

k) Use the needs assessment information. In order to have the information used, the following suggestions are offered: (a) issues that users perceive as important must be addressed; (b) the information must be communicated to the appropriate potential users; © groups must feel empowered to design and improve their programmes; and (d) the information must be available in a timely manner and in an understandable form.

3

Human Resource Planning Extension

Human resource planning forecasts the future personnel needs of extension organizations. With the rapid changes in technology, needs of farmers, market situation, and competitive environment, planning for human resources has become an important, challenging task for extension. Human resource planning involves plans for future needs of personnel, their required skills, recruitment of employees, and development of personnel (Miller, Burack, and Albrecht, 1980). Human resource forecasting and human resource audit are the two most important components of this type of planning. Human resource forecasting refers to predicting an organization's future demand for number, type, and quality of various categories of employees. The assessment of future needs has to be based on analysis of present and future policies and growth trends. The techniques of forecasting include the formal expert survey, Delphi technique, statistical analysis, budget and planning analysis, and computer models. The human resource audit gives an account of the skills, abilities, and performance of all the employees of an organization (Werther and Davis, 1982).

Job Analysis

Job analysis traditionally was done for purposes connected with recruitment, pay, administration, and supervision. But the increasing complexity of work has made job analysis an important instrument for developing people in organizations. Job analysis requires a systematic collection, evaluation, and organization of information about the job. This information is collected through interviews, mailed questionnaires, observation, study of records, and similar methods. The collected information becomes a basis for preparing job descriptions and specifications.

The job description, or job profile, is a written statement which includes detailed specifications of duties to be performed, responsibilities, and working conditions and indicates what is expected of a job holder. A job specification is a profile of the human characteristics needed for the job, such as education, training, skills, experience, and physical and mental abilities (Werther and Davis, 1982).

Extension organizations in developing countries do not have clearly defined job descriptions or job specifications for extension personnel. The training and visit system of extension considerably improved the preparation of job charts, work plans, and time-bound work for different categories of extension personnel. However, the actual utility of job descriptions in extension organizations is complicated by factors such as work overload, seasonality of extension, the range of cropping systems, and distribution of extension service over a large area (Hayward, 1990). Studies analysing the role of extension agents reveal that they face work-related problems such as role ambiguity and lack of job authority, expertise, and accountability (Vijayaragavan and Singh, 1989). This shows that job analysis is needed to improve the performance and effectiveness of extension employees. Job analysis can more effectively contribute towards the development of extension personnel by adopting the following procedures which involve identifying key performance areas (KPAs) and critical attributes.

Key Performance Areas for Various Categories of Extension Personnel

A job description consists of many details, but does not specify key areas which need attention. Further, it gives the details of what is expected from the current jobholder. On the other hand, key performance areas are specific and show the critical functions relevant at present and for the future to achieve the objectives (Pareek and Rao, 1992). The identification of key performance areas helps in role clarity as well as in delegation of functions. This in turn aids in performance appraisal and training. Generally, four or five key areas for a job are identified. The core extension personnel of developing countries consist of village extension workers, subject-matter specialists, and supervisory staff or extension officers. Examples of key performance areas of core extension personnel are given below.

Village Extension Workers

People in this category:

- Make regular and systematic visits to villages and farms to develop rapport with the clientele and to understand their problems;
- Undertake educational activities in the form of meetings, campaigns, demonstrations, field days, training sessions, and exhibitions; and
- Provide advisory services to the farmers and solve their production problems.

Subject-Matter Specialists

Their role is to:

- ✰ keep abreast of current recommendations and findings related to farm production by maintaining continuous contact with agricultural research stations;
- ✰ provide feedback to the research system about farmers' problems which need solutions; and
- ✰ train and backstop village extension workers on the latest farm technology and help them in solving field problems.

Supervisory Staff or Extension Officers

People holding these positions:

- ✰ Plan, organize, coordinate, and implement extension programmes and activities;
- ✰ Supervise and monitor the work of field staff, providing guidance, motivation, and evaluation of performance; and
- ✰ Coordinate the programme with inter-and intradepartmental agencies.

Critical Attributes for Extension Personnel

The key performance areas indicate the important roles and contributions of different categories of extension personnel. Once the roles are delineated, they can be analysed to indicate the attributes which can discriminate an effective from an ineffective role occupant. These critical attributes consist of qualities such as educational qualifications, skills, experience, physical characteristics, mental abilities, values, and attitudes needed for extension. The critical attributes needed for field-level and supervisory extension staff is necessary formal training in agriculture, practical skills and experience in farming, and knowledge of modern farm practices. Abilities in group dynamics, human relations, and communication are also important. Basic skills related to management and leadership are needed by extension supervisors. Values and attitudes such as faith in rural people, commitment to agricultural development, and concern for the whole community are important for all extension personnel (Gupta, 1963; Bhasin, 1976). The importance of assessing personal and professional attributes for selecting productive extension personnel has been reported by several researchers (Gupta, 1963; Perumal, 1975). Assessment is essential because an unsatisfactory educational level of extension staff is one of the most serious problems of extension in countries like Bangladesh, Botswana, Kenya, Malaysia, Sudan, and Zambia (Blanckenburg, 1984).

Recruitment and Training of Extension Personnel

Recruitment is important in selecting the right kind of extension personnel. Since the job of extension personnel calls for technical skills as well as commitment and willingness to educate rural people, an appropriate selection system is essential to ensure the right selection. The success of extension depends heavily upon selection of qualified and motivated personnel. Extension organizations in developing countries use two major sources of recruitment: from outside and from within. Entry-level positions such as village extension workers and agricultural extension officers are filled by outside recruitment, using the services of government placement agencies. Other channels of recruitment are advertisements, private placement agencies, professional search firms, and educational institutions. In some countries, farmers are recruited to help extension agents (Adams, 1982). Most of the extension departments in developing countries have the policy of promoting or recruiting within for middle-level and top-level positions. For example, in India, positions like deputy director, joint director, and additional director of extension are filled through promotion (Vijayaragavan, 1994). The advantages of this policy are that it promotes loyalty and provides opportunities for existing extension staff to get high-level positions. However, its greatest disadvantage is that it prevents the lateral entry of talented extension personnel and promotes complacency because seniority ensures promotion.

Methods and Techniques for Selecting Extension Staff

The selection of extension staff starts with making the job opportunities known to all potential applicants through advertisement. The help of extension workers' training centres, agricultural colleges, rural institutions, and local government agencies may be sought to give wide publicity, as well as to inform candidates living in rural areas. This is followed by screening applicants to short-list suitable candidates and by evaluating potential candidates through various tests. A typical selection process consists of the following steps: completed job application, initial screening, testing, in depth selection interview, physical examination, and job offer (French, 1982). In general, extension organizations in developing countries use a simple knowledge test and a brief interview to select extension personnel. By using the above method, it is impossible to discriminate an effective candidate from an ineffective candidate, because selecting extension personnel demands thorough, in depth testing of cognitive and non cognitive abilities. Testing cognitive ability includes a knowledge test, a skill or ability test, and an aptitude test. A non cognitive test is a measure of behavioural dimensions which are important for field-level extension personnel, including concern for and commitment to rural people, empathy, problem-solving orientation, high motivation to influence and educate farmers, ability to work under unsupervised and difficult village conditions, patience and persistence, and team spirit. The candidates to be recruited go through a number of simulation exercises, and an expert assesses their behaviour. The techniques used are a psychological test, role play, in-basket exercise, group discussion, projective test, knowledge test, and interviews.

Training and Development

The training of extension personnel contributes directly to the development of human resources within extension organizations. Training has to start with the identification of training needs through job analysis, performance appraisal, and organizational analysis. Once the training needs of extension personnel have been identified, the next step is to organize training programmes. Methods such as games, role playing, simulation exercises, and case study can be used in extension organizations to create learning situations based on experience (Lynton and Pareek, 1990). Training based on actual field experience should be emphasized. Emerging new farm technologies such as integrated pest management and improved practices in horticulture call for actual field experience. Extension agents need training not only in the technological aspects but also in human relations, problem solving, sensitivity towards disadvantaged groups, and the basic concepts of management (Hayward, 1990).

Management Development Programmes

Management development programmes are meant to improve the managerial skills of senior-level extension officers and to prepare them for future roles. There is a great need for management development programmes in extension organizations because they face complex situations due to changing agricultural scenarios. Further, extension managers have to be exposed to modern management techniques and methods. Management development programmes have to be suited to the needs of top-level extension managers and should be based on needs analysis. Methods such as coaching, job rotation, training sessions, classroom instruction, and educational institute-sponsored development programmes are used to train managers. In India, a separate institute called MANAGE has been established to train senior extension managers in managerial skills and human relations.

Performance appraisal

Performance appraisal is important for effective human resource management. Performance appraisal is a process of evaluating employee performance in order to guide and develop the employee's potential. In many extension organizations which are government departments, the performance appraisal is nothing more than a confidential judgement of work done and a character report used to facilitate disciplinary action or promotion. The employees do not get feedback about their performance. Extension organizations need to have an open appraisal system to provide feedback and opportunities for open discussion with employees on their performance, because they have immense potential to grow and develop. This system can create a healthy working climate and employee motivation. The performance appraisal which aims at facilitating employee development has the following major purposes:

- To provide feedback and guidance,
- To set performance goals,
- To identify training needs, and
- To provide inputs for management of pay administration, rewards, and promotion.

The steps involved in effective performance appraisal are:

- Identification of key performance areas and setting yearly objectives under each KPA,

- Identification of critical attributes for effective performance,
- Periodic review of performance,
- Discussion of performance with employees, and
- Identification of training and developmental needs (Pareek and Rao, 1992).

Potential Appraisal

The potential appraisal is a future-oriented appraisal by which the potential of an employee to occupy higher positions and to assume higher responsibilities is evaluated. The potential appraisal can help the extension staff to know their strengths and weaknesses and can motivate them to further develop their skills. Thus the potential appraisal helps in planning overall career development of employees. Some of the techniques used for the appraisal are self-appraisals, peer rating, the management by objectives (MBO) approach, psychological test and simulated work exercises, case analyses, and leadership exercises.

Performance Review and Counselling

An important purpose of the performance appraisal is to counsel and guide employees towards greater job effectiveness. Thus a system of performance counselling is needed in extension organizations. Performance counselling is provided by the manager to the subordinates to help them in the analysis of job performance, identification of training needs, and finding solutions to the problems which hinder job effectiveness. Counselling is an art of communication involving two people - manager and employee. Counselling differs from training in that the former involves a dyadic relationship and establishes more mutuality and confidentiality. The success of performance counselling depends upon the employee's interest, a climate of openness and mutuality, and the counselling process. Extension managers can use directive, nondirective, and cooperative counselling (Werther and Davis, 1982).

Supervision

Two major functions of supervision are task orientation and concern for employees. Therefore, direction and organization of activities, motivation of employees, and management of work groups are the important functions of extension supervisors.

Direction and Organization

Extension supervisors have to plan the work and maintain a high standard of performance. The whole process of job analysis, identification of key performance areas, and performance appraisal will help in planning and organizing extension work. The training and visit system of extension has introduced mechanisms for defining goals, planning, and scheduling work at the field level with provisions for monitoring and evaluation. Some of the management techniques used by extension organizations in overall planning and management of programmes are the programme evaluation and review technique (PERT/critical path method

(CPM) (Wiest and Levy, 1982), management by objectives (MBO) (McConkey, 1983), programme and performance budgeting system (PPBS), and time management techniques. These techniques have been practised by extension organizations in Asian and African countries with varying success. Personal computers offer good scope for extension managers to increase certain managerial skills. Asia reported "friendship" as the most effective way for supervisors to ensure extension workers' reliable performance (Goodell, 1983). Thus if extension managers are to be effective, they have to give supportive evaluation by way of enhancing employee motivation and improving the functioning of work groups.

Motivating the Extension Personnel

The work motivation and morale of extension staff, as reported earlier, are very poor in many countries. The reasons are many. The bureaucratic structure of extension administration, lack of rewards and incentives, poor facilities, poor promotional avenues, and the low esteem given to extension are the major causes of poor motivation and morale. Extension supervisors should have the ability to motivate and lead the field extension workers so that the field agents perform more than routine jobs, and supervisors should be involved in attaining excellence in extension work. This calls for extension managers having an understanding of various theories of motivation as applicable to frontline extension agents. Therefore, a knowledge of major theories of motivation such as Maslow's hierarchy of needs theory, Herzberg's two factory theory, McClelland's need theory, theory X and theory Y, and expectancy theory of motivation is essential (Stoner and Freeman, 1992). Special training for developing motivation among field-workers has to be undertaken by supervisors.

Work-Group Management

Every organization has formal and informal groups. Formal groups are established by the management, while informal groups are spontaneous and developed to satisfy mutual interest of the members. Because work groups have a considerable influence on the work situation, supervisors should be sensitive to the needs of the group and develop skills to guide and achieve the group's goal, which will benefit the organization and the members. Effective extension supervision can use work groups in problem solving because they can provide many creative solutions. One way to improve supervisory effectiveness in extension work is to develop a leadership style which represents the extension workers' group interest at the higher level of organization. This will increase the confidence and morale of the work group. An understanding of group dynamics and their implications for increasing work-group performance is essential for extension supervisors. For example, in the "Hawthorne Effect," increased performance due to special treatment of the group can be effectively used in extension organizations (Honadle, 1982). Studies have pointed out that well-developed group dynamics result in increased extension performance (Leonard, 1977).

Management of Rewards and Incentives

An important aspect of human resource management which needs special attention in extension organizations is the development of a reward system which will attract, retain, and motivate extension personnel, as well as provide training and promotional opportunities. Extension organizations in Asian and African countries have a poor reward system (Vijayaragavan, 1994; Swanson, Farmer, and Bahal. 1990). The extension agents are not only poorly paid but are paid late and after reminders or visits to head-quarters (Wiggins, 1986). Most of the extension services are run by government agencies and operate under rules and regulations of public administration. These rules do not have provisions for rewarding superior performance or for a wage system based on merit. Promotion criteria are based on seniority and length of service. Thus the bureaucratic structure of extension services is a basic hindrance to designing a better reward system. Among many of the government departments, the agricultural department and extension service have a low public esteem and poor pay structure (Vijayaragavan and Singh, 1992).

The rewards and incentive system can be improved in several ways.

Rewarding Superior Performance

Extension organizations have to develop a reward system which encourages superior performance so that pay and wage administration will be an effective tool to promote performance, motivation, and satisfaction. A clear job description, performance standards, and performance appraisal will help in evaluating extension work and rewarding people for meritorious service. Ways and means have to be found within the existing framework of public administration for basing pay on performance. For example, extension workers on the basis of their performance can be sent for higher education. Nonmonetary rewards such as recognizing the good ideas of field workers or awarding honourable titles will also help in improving performance. Extension personnel may also be encouraged to form professional societies to develop and communicate high standards, as well as to recognize superior performance. A professional monthly journal or newsletter can help extension agents to communicate innovative ideas and reinforce superior performance.

Improved Working Conditions at the Field Level

The reward system must also be internally equitable. The relative importance of field-level extension functionaries has to be realized in terms of pay compensation and other amenities. Lower level extension workers often have to work under unpleasant and isolated conditions. A carefully planned system of field allowance will compensate this (Baxter, 1990). The living conditions of field extension workers must be improved by providing adequate facilities for housing, transport and medical and educational allowances for children.

Career Planning and Development for Extension Personnel

A career refers to all of the jobs that people hold during their working lives. Career planning is the process by which employees plan their career goals and paths. Career development refers to all of the technical and managerial skills employees acquire to achieve their career plans. Career advancement, which gives a picture of future opportunities in terms of promotion, is a motivating factor for performance and development of skills. Unfortunately, no career structure exists for extension personnel in many organizations. In developing countries like India, there are many cases where one joins as a village extension worker and retires in the same position after serving thirty to thirty-five years. Extension personnel should have a salary structure as well as promotion opportunities comparable to other professions like health or engineering. In Kenya, the pay and career opportunities of extension workers are comparable to other government employees (Onyango, 1987). Recommendations have already been made to equate the status of agricultural extension with that of agricultural research by offering an equal salary structure, professional advancement, and incentives and rewards (FAO, 1985).

Improvement of the Quality of Work Life

The earlier approach to human resource development emphasized individual development through training and proper supervision. However, with the increasing complexity of organizations and society, it was soon realized that training individuals plays only a limited role in the development of organizations. The need for improving the quality of work life through making the job more satisfying and productive has been greatly felt. Factors such as the nature of the job or the role and involvement of employees in work decisions are important for improving the quality of work life. The methods used to do so are job enrichment, job design, and role interventions (Pareek, 1993). An understanding of these methods and their application in extension organizations are essential for extension managers to improve the performance of extension agents. Studies have shown that the work environment of extension organizations is poor and needs improvement (Jhamtani and Singh, 1989, 1992).

Job Enrichment and Job Design

Job enrichment refers to detailed analysis of the work to know the factors which make it a satisfying experience. Job enrichment uses the job as the medium of developing employees and changing organizational practices. Some of the factors which increase job satisfaction are a sense of achievement in the job, recognition for the job, the nature of the work itself, and opportunities to learn new things and grow. The principles of job enrichment, according to Herzberg (1966), are removing controls while retaining accountability, introducing new tasks, giving a complete unit of work, granting job freedom, and helping employees to become expert in their tasks. These principles can be practised by extension managers to increase the quality of work and job satisfaction among extension personnel.

Job enrichment programmes were successful in improving the quality of work and job satisfaction. However, it was found that job enrichment had a limited view of the job, and the need for greater emphasis on human values was realized. This led to the concept of job design, which refers to structuring a job to satisfy the technical, organizational, social, and human requirements of the person performing the work (Davis and Taylor, 1979). Based on the humanization of work, job design aims at increasing the quality of work life through treating the employees as human beings and emphasizing their development and involvement in work decisions.

Role Interventions

The study of roles, which are the positions employees hold in an organization, as defined by the expectations of significant persons and the individuals occupying the positions, is a comparatively neglected aspect of organizations. Roles are an important dimension in increasing organizational effectiveness. Through their roles, people are linked with the organization. This linkage increases organizational effectiveness by integrating the individuals with the organization. Such integration increases mental well-being and personal effectiveness (Pareek, 1993). The purpose of role-based intervention is to increase the mutuality of roles in organizations. Role-based interventions are done through learning situations such as process laboratory, group discussion, and use of questionnaires and schedules. Role-based interventions in extension organizations will result in increased work commitment, motivation, creativity, and team spirit.

Organizational Development

An efficient extension organization needs to develop the capability of responding to changes in relation to its environment. Extension organizations have to cope with changes within and outside the organization, such as changes in farm technology, communication methods, needs of farmers, rural situations, export and import of farm produce, and market economy. Organizational development allows for planned changes in the organization's tasks, techniques, structure, and people. Attitudes, values, and practices of the organization are changed so that it can cope with changing situations. The employees also gain greater skills to deal with new problems.

4

Training and Evaluation in Extension

Learning theories are the basic materials which are usually applied in all educational and training activities. The more one understands learning theories, the better he or she will be able to make decisions and apply them to achieving the objectives. The behaviourists, the cognitivists, and the humanists emphasize different aspects of the teaching-learning process in their approaches. While the behaviourists stress external conditions (environment) resulting in observations and measurable changes in behaviour, the cognitivists are more concerned with how the mind works (mental processes such as coding, categorizing, and representing information in memory). The humanists, on the other hand, emphasize the affective aspects (*e.g.*, emotions, attitudes) of human behaviour that influence learning (IRRI, 1990). In extension systems, effective training must be able to take care of all the theories of learning in order to change the action, belief, and knowledge components of a trainee simultaneously. Andragogy (a theory of adult learning) is usually used rather than pedagogy (a theory of child learning) in extension training.

Training Approach

There are three approaches to training:

1. The traditional approach,
2. The experiential approach, and
3. The performance-based approach (Rama, Etling, and Bowen, 1993).

In the traditional approach, the training staff designs the objectives, contents, teaching techniques, assignments, lesson plans, motivation, tests, and evaluation. The focus in this model is intervention by the training staff. In the experiential approach, the trainer incorporates experiences where in the learner becomes active and influences the training process. Unlike the academic approach inherent in the traditional model, experiential training emphasizes real or simulated situations in which the trainees will eventually operate. In this model, the objectives and other elements of training are jointly determined by the trainers and trainees. Trainers primarily serve as facilitators, catalysts, or resource persons. In the performance-based approach to training, goals are measured through attainment of a given level of proficiency instead of passing grades of the trainees. Emphasis is given to acquiring specific observable skills for a task. This performance-based teacher education (PBTE) model, developed by Elam (1971), is mostly task or skill centred and is also applicable to nonformal educational organizations such as extension.

Extension Personnel Around the World in Need of Training

Worldwide, there are currently more than 600,000 extension workers comprised of administrative staff, subject-matter specialists (SMS), fieldworkers, and some multipurpose unidentified people; the Asian and Pacific countries have absorbed more than 70 per cent of them (Bahal, Swanson, and Earner, 1992). The percentage of extension personnel by position, as reported by Swanson, Earner, and Bahal (1990), was 7 per cent administrative, 14 per cent SMS, and 79 per cent field staff, with regional differences. Almost 13 per cent of extension workers are women, with significant regional differences (Bahal *et al.*, 1992). The ratio of SMS to field staff is also low in Asia, Africa, the Near East, and Latin American countries, varying from about 1:11 to 1:14. The ratio for countries of Europe and North America varies from 1:1.5 to 1:1.6. The worldwide ratio of SMS to field staff is 1:11.5 (Swanson *et al.*, 1990).

Deficiencies in knowledge, skills, and ability among extension personnel, particularly those of Asia, Africa, and Latin America, are remarkable. About 39 per cent of the extension personnel worldwide have a secondary-level and 33 per cent an intermediate-level education (Bahal *et al.*, 1992). Moreover, within each region, there is a lot of variation in basic academic qualifications of the frontline extension workers, SMS, and administrators. Differences in training received are also wide. In Africa, most frontline extension workers still have only a secondary school diploma (Bahal *et al.*, 1992). The poor educational background of extension personnel necessitates regular training.

Types of Training

Training may broadly be categorized into two types: pre-service training and in-service training. Pre-service training is more academic in nature and is offered by formal institutions following definite curricula and syllabuses for a certain duration to offer a formal degree or diploma. In-service training, on the other hand, is offered by the organization from time to time for the development of skills and knowledge of the incumbents.

Pre-service Training

Pre-service training is a process through which individuals are made ready to enter a certain kind of professional job such as agriculture, medicine, or engineering.

They have to attend regular classes in a formal institution and need to complete a definite curriculum and courses successfully to receive a formal degree or diploma. They are not entitled to get a professional job unless they can earn a certificate, diploma, or degree from the appropriate institution. Pre-service training contents emphasize mostly technical subject matter such as crops, animal husbandry, and fisheries as well as pedagogical skills to prepare the students to work in agriculture. In general two types of pre-service training are available for agricultural staff. These are (1) degree level (at least a bachelor's degree in agriculture or related field), which is usually offered for four years by a university or agricultural college; and (2) diploma level, which is mostly offered by the schools of agriculture for a period of two to three years. The entry point for the former is normally twelve years of schooling and for the latter ten years of schooling.

In-service Training and Staff Development

In-service training is a process of staff development for the purpose of improving the performance of an incumbent holding a position with assigned job responsibilities. It promotes the professional growth of individuals. In-service training is a problem-centred, learner-oriented, and time-bound series of activities which provide the opportunity to develop a sense of purpose, broaden perception of the clientele, and increase capacity to gain knowledge and mastery of techniques. In-service training may broadly be categorized into five different types: (1) induction or orientation training, (2) foundation training, (3) on-the-job training, (4) refresher or maintenance training, and (5) career development training. All of these types of training are needed for the proper development of extension staff throughout their service life.

Induction or Orientation Training

Induction training is given immediately after employment to introduce the new extension staff members to their positions. It begins on the first day the new employee is on the job (Rogers and Olmsted, 1957). This type of training is aimed at acquainting the new employee with the organization and its personnel. Induction training for all new personnel should develop an attitude of personal dedication to the service of people and the organization. This kind of training supplements whatever pre-service training the new personnel might have had (Halim and Ali, 1988). Van Dersal (1962) said that when people start to work in an organization for the first time, they are eager to know what sort of outfit they are getting into, what they are supposed to do, and whom they will work with. They are likely to be more attentive and open-minded than experienced employees. In fact, the most favourable time for gaining employees' attention and for moulding good habits among them is when they are new to the job.

Foundation Training

Foundation training is in-service training which is also appropriate for newly recruited personnel. Besides technical competence and routine instruction about the organization, every staff member needs some professional knowledge about various rules and regulations of the government, financial transactions, administrative capability, communication skills, leadership ability, coordination and cooperation among institutions and their linkage mechanism, report writing, and so on. Foundation training is made available to employees to strengthen the foundation of their service career. This training is usually provided at an early stage of service life.

Maintenance or Refresher Training

This training is offered to update and maintain the specialized subject-matter knowledge of the incumbents. Refresher training keeps the specialists, administrators, subject-matter officers, extension supervisors, and frontline workers updated and enables them to add to the knowledge and skills they have already. Maintenance or refresher training usually deals with new information and new methods, as well as review of older materials. This type of training is needed both to keep employees at the peak of their possible production and to prevent them from getting into a rut (Van Dersal, 1962).

On-the-Job Training

This is ad hoc or regularly scheduled training, such as fortnightly training under the training and visit (T&V) system of extension, and is provided by the superior officer or the subject-matter specialists to the subordinate field staff. The superior officer, administrator, or subject-matter specialist of each extension department must play a role in providing on-the-job training to the staff while conducting day-to-day normal activities.

Career or Development Training

This type of in-service training is designed to upgrade the knowledge, skills, and ability of employees to help them assume greater responsibility in higher positions. The training is arranged departmentally for successful extension workers, at all levels, for their own continuing education and professional development. Malone (1984) opined that extension services that provide the opportunity for all staff to prepare a plan for career training will receive the benefits of having longer tenured and more satisfied employees, which increases both the effectiveness and efficiency of an extension service. Malone stated that "career development is the act of acquiring information and resources that enables one to plan a program of lifelong learning related to his or her work life". Although extension workers are responsible for designing their own career development education, the extension organization sometimes sets some criteria and provides opportunities for the staff by offering options.

Phases of Training

Training is a circular process that begins with needs identification and after a number of steps ends with evaluation of the training activity. A change or deficiency in any step of the training process affects the whole system, and therefore it is important for a trainer to have a clear understanding about all phases and steps of the training process. In the broadest view, there are three phases of a training process: planning, implementation, and evaluation.

Planning Phase

The planning phase encompasses several activities, two of which - training needs identification and curriculum development - are very important.

Training Needs Identification

Training need is a condition where there is a gap between "what is" and "what should be" in terms of incumbents' knowledge, skills, attitudes, and behaviour for a particular situation at one point in time. This gap is called "a problem," which usually occurs when a difference exists between "desired performance" and "actual performance." The needs identification process assists trainers in making sure that they have matched a training programme to a training problem. For example, Agricultural Development Officers (ADOs) have been giving training to village extension workers (VEWs), but performance of the VEWs is not improving. The reasons may be:

- ✰ The ADOs lack subject-matter knowledge.
- ✰ The ADOs do not conduct training well.
- ✰ The training centre lacks training facilities.
- ✰ The VEWs are organized not to work properly until their demands are satisfied by the government.

The first two problems are related to knowledge and skills and can be solved effectively by a training programme, but the third and fourth problems need government attention to solve.

Training needs identification is possible through different analytical procedures. The major procedures used in determining training needs are the following:

Organizational analysis determines where training emphasis should be placed within the organization and is based on the objectives of an organization. Concerning what one should do in analysing an organization, McGhee and Thayer (1961) suggest four steps:

- ✰ Stating the goals and objectives of an organization
- ✰ Analysing the human resources
- ✰ Analysing efficiency indices
- ✰ Analysing the organizational climate

The results of these analyses are then compared with the objectives of the organization. These comparisons point to specific areas in which training is needed.

Individual Analysis

Individual analysis aims at identifying specific training needs for an individual or group of employees so that training can be tailored to their needs. This analysis centres on individuals and their specific needs concerning the skills, knowledge, or attitudes they must develop to perform their assigned tasks. The possible methods or techniques for individual analysis include performance appraisal, interviews, questionnaires, tests, analysis of behaviour, informal talks, checklist, counselling, critical incidents, recording, surveys, and observations.

Group Analysis

Group analysis includes a number of techniques in which a group of well-informed employees discuss different aspects of the organization, the employees, and the tasks to identify the major discrepancies in achieving predetermined targets for each of them with a view to assessing training needs as distinguished from other necessary changes for removing these discrepancies. The major techniques which are used in this approach are brainstorming, buzzing, card sorts, advisory committee, conferences, problem clinic, role playing, simulation, task forces, workshops, and so forth.

Curriculum Development

This is the most important part in a training programme after a need for training has been identified. The curriculum specifies what will be taught and how it will be taught. It provides the framework and foundation of training. The first phase of curriculum development determines what will be taught, that is, the training content.

Once training needs have been identified and training activities have been decided as part of the solution, a *needs analysis* should be done to determine knowledge, skills, and attitude requirements and performance deficiencies. The needs analysis procedure involves breaking down the "training problem" into its basic parts in different successive phases to identify and understand the important components in each phase. Ultimately it leads to identifying and understanding the training content. The training needs analysis process can be divided into three distinct analytical phases: job analysis, task analysis, and knowledge and skill-gap analysis.

A. Job Analysis

Job analysis is a method of determining major areas of tasks where training may be needed (see JA Worksheet). It involves the dissecting of a job into its component events or parts. This analysis allows a trainer to better understand what an employee does in an organization. Job analysis involves the "task identification" of a particular job (Wentling, 1992). The techniques used in task identification include job questionnaire, interview, participant observation, work sampling, job audit, and small-group discussion. The following steps may provide a guide for completion of job analysis:

a) Identify the job that is to be the subject of the analysis. This involves defining the focal point for the job analysis. It may include the entire job of a group of employees or only a specific segment of their job.

b) Prepare a list of tasks which can be done following different approaches and methods. Four approaches can be used to identify job tasks: (1) expertsidentify and list critical tasks, (2) observations and interviews are conducted with employees, (3) meetings are held with group representatives, and (4) a tentative list of task is reviewed by employees and their supervisors.

c) Verify the tasks. The draft list of tasks should be verified by experts, workers, and supervisors in the analysis process. This can be done through expert review, small-group discussions, and inter views. When the tasks are verified, a final list of job tasks is prepared.

d) Determine the frequency. The workers and super visors can fill in a form indicating how frequently each task in a job is performed. Different scales such as "seldom," "occasionally," "weekly to monthly," "daily to weekly," and "daily" can be used to quantify the intensity of a task accomplished.

e) Determine the importance. Not all tasks are equally important to a job. An occasionally performed task may be very important. Therefore, a relative importance rating is useful along with frequency rating. A scale such as "marginally important," "moderately important," and "extremely important" may be used to determine the relative importance of the job tasks.

f) Estimate the learning difficulty. An estimate of learning difficulty is another dimension of the job-task analysis. It shows the trainer the employees' perception of difficulty, which may be different from the trainer's own perception. A scale such as "easy," "moderately difficult," "very difficult," and "extremely difficult" may be used to determine the difficulty indices of job tasks.

g) Calculate the total score. This can be done by simply adding the scores for frequency, importance, and learning difficulty for each task. The column for total score in a worksheet indicates the priority tasks for training if these are training problems.

h) Review the findings. The results of the job-task analysis should be discussed with significant people in the training system, including government leaders, programme directors, and others interested in related training.

B. Task Analysis

The output of the job analysis is a list of broad job tasks, based on importance, learning difficulty, and frequency of doing the task. Each task is a complex set of procedures in itself, and therefore it needs further analysis to find out which specific segment of the task is critical in designing a training programme. To do this, it is necessary to follow a method called task analysis, which is similar to job analysis. Task analysis procedures include preparing a blank task analysis worksheet, writing down the name of the job at the top of each sheet, and then making copies. Each of these forms will be used for breaking down and analysing each of the most important job tasks. Therefore, it is necessary to write one important task identified for training on each of the task analysis worksheets and to list all component parts of each task on its respective task analysis worksheet. This is followed by the steps used for job analysis to find out the frequency, importance, and learning difficulty for each step of the tasks. Then the score for each component part is put in the "total score" column, and the results are discussed with concerned personnel in the organization. The job analysis and task analysis processes are similar to each other, so the model for both worksheets is the same.

C. Knowledge and Skill-Gap Analysis

The knowledge or skill-gap analysis is a process of determining the training needs of individual employees in relation to the important tasks-steps or components of tasks identified for training. The skill-gap analysis determines how skilled or proficient individual employees are on these tasks-steps or components, how much individuals differ from desired performance, and whether or not they need training. It would be a waste of resources and frustrating to the trainer and trainees to design and deliver training on topics and skills where the trainees are already able and proficient. Identifying the steps-components that appear to have low proficiency is required because there is a gap between what is desired and the current situation. After this, a review is done to ponder whether the gap can be decreased or removed through training or whether training is the most appropriate method. There may be some steps-components for which measures other than training are more appropriate. At this stage, key personnel such as subject-matter specialists, supervisors, and extension-training experts should discuss the findings before finalizing the curriculum. This helps to identify different perspectives and to avoid unnoticed mistakes or biases in curriculum development.

Selecting a Training Method

A training programme has a better chance of success when its training methods are carefully selected. A training method is a strategy or tactic that a trainer uses to deliver the content so that the trainees achieve the objective (Wentling, 1992). Selecting an appropriate training method is perhaps the most important step in training activity once the training contents are identified. There are many training methods, but not all of these are equally suitable for all topics and in all situations. To achieve the training objective, a trainer should select the most appropriate training method for the content to involve the trainees in the learning process. Four major factors are considered when selecting a training method: the learning objective, the content, the trainees, and the practical requirements (Wentling, 1992). According to Bass and Vaughan (1966), training methods should be selected on the basis of the degree to which they do the following:

a) Allow active participation of the learners.

b) Help the learners transfer learning experiences from training to the job situation.

c) Provide the learners with knowledge of results about their attempts to improve.

d) Provide some means for the learners to be reinforced for the appropriate behaviour.

e) Provide the learners with an opportunity to practise and to repeat when needed.

f) Motivate the learners to improve their own performance.

g) Help learners increase their willingness to change.

These criteria indicate that a single training method will not satisfy the objectives of a training programme. A variety of training methods are available to a trainer. The most commonly used methods include:

1. Instructor Presentation

The trainer orally presents new information to the trainees, usually through lecture. Instructor presentation may include classroom lecture, seminar, workshop, and the like.

2. Group Discussion

The trainer leads the group of trainees in discussing a topic.

3. Demonstration

The trainer shows the correct steps for completing a task, or shows an example of a correctly completed task.

4. Assigned Reading

The trainer gives the trainees reading assignments that provide new information.

5. Exercise

The trainer assigns problems to be solved either on paper or in real situations related to the topic of the training activity.

6. Case Study

The trainer gives the trainees information about a situation and directs them to come to a decision or solve a problem concerning the situation.

7. Role Play

Trainees act out a real-life situation in an instructional setting.

8. Field Visit and Study Tour

Trainees are given the opportunity to observe and interact with the problem being solved or skill being learned.

Implementation Phase

Once the planning phase of a training programme is complete, then it is time to implement the course. Implementation is the point where a trainer activates the training plan, or it is the process of putting a training programme into operation. The first step towards implementing a training programme is publicity. Most of the well-established training centres develop training brochures which contain course descriptions, prepare an annual calendar of training opportunities, and inform concerned organizations, agencies, or departments well ahead of time about their training plans. Once the training centre and concerned organizations agree to implement training, the next step is to arrange available resources such as sufficient funds for the course and facilities for food, lodging, transportation, and recreation. All these resources need to be well managed and coordinated to run the programme smoothly.

Evaluation Phase

Evaluation is a process to determine the relevance, effectiveness, and impact of activities in light of their objectives. In evaluating an extension training programme, one needs to consider that most training activities exist in a larger context of projects, programmes, and plans. Thus Raab *et al.* (1987) define training evaluation as "a systematic process of collecting information for and about a training activity which can then be used for guiding decision making and for assessing the relevance and effectiveness of various training components." Kirkpatrick (1976) suggested four criteria to evaluate training programmes: (1) reaction, (2) learning, (3) behaviour, and (4) results. Each criterion is used to measure the different aspects of a training programme. *Reaction* measures how the trainees liked the programme in terms of content, methods, duration, trainers, facilities, and management. *Learning* measures the trainees' skills and knowledge which they were able to absorb at the time of training. *Behaviour* is concerned with the extent to which the trainees were able to apply their knowledge to real field situations. *Results* are concerned with the tangible impact of the training programme on individuals, their job environment, or the organization as a whole.

Types of Evaluation

On the basis of the time dimension, evaluation may be classified as (1) formative evaluation and (2) summative evaluation. *Formative evaluation* involves the collection of relevant and useful data while the training programme is being conducted. This information can identify the drawbacks and unintended outcomes and is helpful in revising the plan and structure of training programmes to suit the needs of the situation. *Summative evaluation is* done at the end of the programme and makes an overall assessment of its effectiveness in relation to achieving the objectives and goals. Raab *et al.* (1987), however, classified evaluation into four major types: (1) evaluation for planning, (2) process evaluation, (3) terminal evaluation, and (4) impact evaluation.

Evaluation for Planning

Evaluation for planning provides information with which planning decisions are made. Training contents and procedures (methods and materials) are usually planned at this stage in order to choose or guide the development of instructional aids and strategies. *Process evaluation* is conducted to detect or predict defects in the procedural design of a training activity during the implementation phase (Raab *et al.*, 1987). Through this process the key elements of the training activities are systematically monitored, problems are identified, and attempts are made to rectify the mistakes before they become serious. Process evaluation is periodically conducted throughout the entire period of the programme.

Terminal Evaluation

Terminal evaluation is conducted to find out the effectiveness of a training programme after it is completed. The objectives of terminal evaluation are to determine the degree to which desired benefits and goals have been achieved, along with the causes of failure, if any.

Impact Evaluation

Impact evaluation assesses changes in on-the-job behaviour as a result of training efforts. It provides feedback from the trainees and supervisors about the outcomes of training. It measures how appropriate the training was in changing the behaviour of participants in real-life situations.

Monitoring extension programmes and resources

The Global Consultation on Agricultural Extension observed that monitoring and evaluation are important yet frequently neglected functions in most organizations (FAO, 1990). In the worldwide survey of national extension systems, it was found that only about one half of all national extension systems have some type of monitoring and evaluation (M and E) capacity. The consultation noted that in many cases the M and E units are weak and are limited to ad hoc studies. Frequently, these M and E units are abandoned when project funding terminates. In many organizations, monitoring and evaluation have a negative image because these units may concentrate on problems, exposing weaknesses and failures. Instead, monitoring and evaluation should be used in a positive manner to improve extension's performance and increase its efficiency. The consultation recommended that national extension systems should be strongly encouraged to establish and use monitoring procedures and evaluation studies both to improve extension performance and to communicate the results of extension programmes to policy makers and clientele being served.

A Conceptual Framework for Monitoring

A conceptual framework for monitoring consists of four principal components an extension organization, a monitoring and evaluation (M and E) unit, an information needs matrix, and a monitoring and evaluation cycle. Top management receives information from the monitoring unit and from other formal and informal sources. This influences programme implementation, leads to better programme planning, and ensures sustain-ability of extension programmes. Ultimately this leads to institutional development, which has been defined as "the process of improving the ability of institutions to make effective use of available human and financial resources" (Israel, 1987).

Approaches to Extension Monitoring

Several approaches to extension monitoring are available. They arose, notably during the 1980s, as a response to the need to improve extension performance.

Basically similar, these approaches, however, differ significantly in their emphasis. All of them advocate simplicity and timeliness, essential requirements of good monitoring. Half a dozen approaches are described briefly below.

Traditional (Administrative) Approach

Based on routine administrative reporting, this approach is concerned with physical and financial achievements in a programme. Its primary weaknesses include multiplicity of reports by programme personnel and absence or neglect of beneficiary contact. It is being increasingly replaced by other approaches.

Conventional Methodologies for Learning

Agricultural institutions of all types have long relied on questionnaire surveys and quick rural visits to gather information on rural people and resources. Samples of people selected from a larger population are asked the same set of questions, and so it is assumed that the interviewers do not influence the process. Many informants are selected to account for all variation, and the resulting data are statistically analysed. Such surveys are used at practically all levels, from the large-scale census to small-scale, village-level research; by governments and NGOs; and for planning, research, and extension. But there are problems with questionnaire surveys. The questionnaire designer has to determine the questions well in advance. Yet those who design these instruments cannot know which issues are important for local people. So they tend to increase the number of questions to ensure that all relevant issues are covered. This leads, in some cases, to forms of absurd length, with several hundred questions taking hours to administer. Rarely is attention paid to the nature of the interviewing process. In the structured survey, many of the contextual grounds for understanding are systematically removed or ignored, and all too often, the ill-trained enumerator further influences the process by prompting with answers. Despite many criticisms of this methodology (Chambers, 1983, 1992c; Fowler and Mangione, 1990; Rhoades, 1990; Gill, 1993), official surveys, such as sample censuses of agriculture or household expenditure surveys, remain remarkably popular.

Alternative Systems of Learning and Action

Partly because of these flaws in conventional approaches, there has been a recent rapid expansion in participatory methods and approaches. These began with the development of data-gathering methods which came to be known as rapid rural appraisal. During the late 1980s, this growing experience was supplemented by drawing upon long-established traditions that had put participation, action research, and adult education at the forefront of attempts to emancipate people. To the wider body of development programmes, these approaches represent a significant departure from standard practice. Methods are being used not just for local people to inform outsiders, but also for people's own analysis of their own conditions (Chambers, 1992b, 1992c; Pretty and Chambers, 1993; Pretty, 1995).

The interactive involvement of many people in differing institutional contexts has promoted innovation and ownership, with many variations in the way that systems of learning have been put together. There are many different terms, but they have the following important common principles (Pretty, 1994):

☆ A Defined Methodology and Systemic Learning Process

The focus is on cumulative learning by all the participants and, given the nature of these approaches as systems of learning and interaction, their use has to be participative.

☆ Multiple Perspectives

A central objective is to seek diversity, rather than to characterize complexity in terms of average values. The assumption is that different individuals and groups make different evaluations of situations, which lead to different actions. All views of activity or purpose are heavy with interpretation, bias, and prejudice, and this implies that there are multiple possible descriptions of any real-world activity.

☆ Group Learning Processes

All involve the recognition that the complexity of the world will only be revealed through group inquiry and interaction. This implies three possible mixes of investigators, namely, those from different disciplines, from different sectors, and from outsiders (professionals) and insiders (local people).

☆ Context Specific

The approaches are flexible enough to be adapted to suit each new set of conditions and actors, and so there are multiple variants.

☆ Facilitating Experts and Stakeholders

The methodology is concerned with the transformation of existing activities to try to bring about changes which people in the situation regard as improvements. The role of the "expert" is best thought of as helping people in their situation carry out their own study and so achieve something. These facilitating experts may be stakeholders themselves.

☆ Leading to Sustained Action

The learning process leads to debate about change, and debate changes the perceptions of the actors and their readiness to contemplate action. Action is agreed upon, and implementable changes will therefore represent an accommodation among the different conflicting views. The debate or analysis both defines changes which would bring about improvement and seeks to motivate people to take action to implement the defined changes. This action includes local institution building or strengthening, thus increasing the capacity of people to initiate action on their own.

The Different Interpretations of Participation

In recent years, an increasing number of analyses of projects have shown that participation by local people is one of the critical components of success in irrigation, livestock, water, and agriculture sectors (USAID, 1987; Reij, 1988; Finsterbusch and van Wicklen, 1989; Bagadion and Korten, 1991; Cernea, 1991; Guijt, 1991; Pretty and Sandbrook, 1991; Uphoff, 1992; Narayan, 1993; World Bank, 1994; Pretty, 1995).The terms "people's participation" and "popular participation" have now become part of the normal language of many development agencies (Adnan, Nurul Alam, and

Brustinow, 1992; Bhatnagar and Williams, 1992). This has brought new dangers. The term "participation" has been used to justify the extension of state control and to build local capacity and self-reliance; it has been used for data collection and for interactive analysis. Participation has often centred on encouraging local people to sell their labour in return for food, cash, or materials. Yet these material incentives distort perceptions, create dependencies, and give the misleading impression that local people are supportive of externally driven initiatives (Reij, 1988; Bunch, 1991; Kerr, 1994). This means that "more often than not, people are asked or dragged into participating in operations of no interest to them, in the very name of participation"(Rahnema, 1992).

A selection of terms and names for alternative systems of participatory learning and action

AEA	Agro Ecosystems Analysis
BA	Beneficiary Assessment
DELTA	Development Education Leadership Team
DPR	Diagnóst co Rurale Participative
FPR	Farmer Participatory Research
GRAAP	Groupe de Recherche et d'Appui pour l'Auto-Promotion Paysanne
MARP	Methode Accélérée de Recherche Participative
PALM	Participatory Analysis and Learning Methods
PAR	Participatory Action Research
PRM	Participatory Research Methodology
PRAP	Participatory Rural Appraisal and Planning
PTD	Participatory Technology Development
PUA	Participatory Urban Appraisal
PfR	Planning for Real
PD	Process Documentation
RA	Rapid Appraisal
RAAKS	Rapid Assessment of Agricultural Knowledge Systems
RAP	Rapid Assessment Procedures
RAT	Rapid Assessment Techniques
RCA	Rapid Catchment Analysis
REA	Rapid Ethnographic Assessment
RFSA	Rapid Food Security Assessment
RMA	Rapid Multi-perspective Appraisal
ROA	Rapid Organizational Assessment
RRA	Rapid Rural Appraisal
SB	Samuhik Brahman (Joint trek)
SSM	Soft Systems Methodology
TfD	Theatre for Development
TfT	Training for Transformation
VIPP	Visualization in Participatory Programmes

A typology of participation: how people participate in development programmes and projects

Typology	*Characteristics of Each Type*
Passive Participation	People participate by being told what is going to happen or has already happened. It is a unilateral announcement by an administration or project management without any listening to people's responses. The information being shared belongs only to external professionals.
Participation in Information Giving	People participate by answering questions posed by extractive researches using questionnaire surveys or similar approaches. People do not have the opportunity to influence proceedings, as the findings of the research are neither shared nor checked for accuracy.
Participation by Consultation	People participate by being consulted, and external agents listen to views. These external agents define both problems and solutions and may modify these in the light of people's responses. Such a consultative process does not concede any share in decision making, and professionals are under no obligation to take on board people's views.
Participation for Material Incentive	People participate by providing resources, for example labour, in return for food, cash, or other material incentives. Much on-farm research falls in this category, as faermers provide the fields but are not involved in the experimentation or the process of learning. It is very common to see this called participation, yet people have no stake in prolonging activities when the incentives end.
Functional Participation	People participate by forming groups to meet predetermined objectives related to the project, which can involve the development or promotion of externally initiated social organization. Such involvement does not tend to be at early stages of project cycles or planning, but rather after major decisions have been made. These instructions tend to be dependent on external initiators and facilitators, but may become self-dependent.
Interactive Participation	People participate in joint analysis, which leads to action plans and the formation of new local institutions or the strengthening of existing ones. It tends to involve interdisciplinary methodologies that seek multiple perspectives and make use of systemic and structured learning processes. These groups take control over local decisions, and so people have a stake in maintaining structures or practices.
Self-Mobilization	People participate by taking initiative independent of external institution to change systems. They develop contacts with external institutions for resources and technical advice they need, but retain control over how resources are used. Such self-initiated mobilization and collective action may or may not challenge existing inequitable distribution of wealth and power.

Source: Pretty (1994), adapted from Adnan *et al.* (1992).

Participatory Methods

The creative ingenuity of practitioners worldwide has greatly increased the range of participatory methods in use (see *RRA Notes*, 1988-1994; IDS/IIED, 1994; Pretty *et al.*, 1995; Chambers, 1992a, 19920; Mascarenhas *et al.*, 1991; KKU, 1987; Conway, 1987). Many have been drawn from a wide range of non-agricultural contexts and were adapted to new needs. The methods can be put into four classes: for group and team dynamics, for sampling, for interviewing and dialogue, and for visualization and diagramming. These methods collected into unique approaches, or assemblages of methods, constitute systems of learning and action. In this section, some key methods are briefly described. Participation calls for collective analysis and good rapport. By working as a group, the investigators can approach a situation from different perspectives, carefully monitor one another's work, and carry out a variety of tasks simultaneously. Groups can be powerful when they function well, because performance and output are likely to be greater than the sum of the individual members. Various workshop and field methods are used to facilitate this process of group formation:

- ☆ **Team contracts**. Team contracts developed by all the team members help to ensure good group dynamics and may include agreements to hold evening discussions and morning brainstorming sessions. One person may be elected to monitor team interactions to provide feedback. The monitor can make a note of each member's location and record interactions by drawing a circle around individuals' names when they speak or an arrow from talker to person being talked to, with duration of speech recorded in seconds.
- ☆ **The night halt**. Rapport between outsiders and villagers is facilitated by staying in the village. Many have made this an essential part of participatory analysis and planning. It provokes change in outsiders' attitudes: they sleep and eat as villagers do; it allows for early morning and evening discussions when people are less busy; and it is an explicit commitment by outsiders to village life.
- ☆ **Work sharing**. When outsiders are taught some-thing by villagers, roles are reversed. Professionals soon learn how much skill is required, say, to plough a furrow, transplant rice, weed, lop tree fodder, cut firewood, dig compost, and wash clothes. Such activities prompt changes in attitude and help to build rapport.
- ☆ **Rapid report writing, with self-correcting notes**. It is essential to record, as a team, the key findings before members disperse to their own organizations. Report writing is made easier by writing a brief summary of how diagrams were constructed and of the key findings. Individuals can be encouraged to keep a private diary or series of notes to focus on things they would like to improve the next time.
- ☆ **Shared presentations**. The key findings should always be presented to villagers and outsiders. This is an important opportunity for cross-checking and feedback. Professionals present and invite comment and criticism. This is a fundamental reversal of roles and is crucial to establishing the trustworthiness of the findings.
- ☆ **Transect walks and direct observation**. These are systematic walks with key informants through the area of interest, observing, asking, listening, looking, and seeking problems and solutions. The findings can be mapped on a transect diagram. Most transect walks result in the outsiders discovering surprising local practices such as indigenous conservation practices, multiple uses of plants, and a great variety of crops.
- ☆ **Wealth rankings and social maps**. Wealth ranking is used to classify households according to relative wealth or well-being. Informants sort

cards, each with one household name on it, into piles. The wealthiest are put at one end, the poorest at the other, and as many piles as desired are made. Wealth rankings are useful for leading into other discussions on livelihoods and vulnerability; producing a baseline against which future intervention impact can be measured; providing a sample frame to cross-check the relative wealth of informants who have been or will be interviewed; and producing local indicators of welfare. This should appear more like a structured conversation than an interview:

- ☆ **Semi-structured interviews (SSI)**. This is guided interviewing and listening in which only some of the questions and topics are predetermined; other questions arise during the interview. Using a guide or checklist, the multidisciplinary team poses open-ended questions and probes topics as they arise. New avenues of questioning are pursued as the interview develops.
- ☆ **Types, sequencing, and chains of interviews**. Many types of interviews may be combined in sequences and chains. These include key informant interviews, by asking who the experts are and then putting together a series of interviews (*e.g.*, men on ploughing, women on transplanting and weeding, shopkeepers for credit and inputs); and group interviews, which may be groups convened to discuss a particular topic (focused or specialist groups), groups comprising a mix of people whose different perceptions illuminate an issue (structured groups), casual groups, and community groups.
- ☆ **Participatory mapping and modelling**. This involves constructing, on the ground or on paper, maps or models, using materials such as sticks, stones, grasses, wood, cigarette packets, tree leaves, coloured sands and soils, rangoli powders, coloured chalk, pens, and paper. Great play is made of the issue of who holds the stick or pen. The person who holds the stick talks about what is most important to him or her. As maps take shape, more people become involved, and so want to contribute and make sequential changes. There are many types of maps: resource maps of catchments, villages, forests, fields, farms, home gardens; social maps of residential areas of a village; wealth rankings and household assets surveys on social maps; health maps, where the health status of each family member is shown on each house, using coloured stickers or other markers (categories might include cases of malnutrition, ear infection, jaundice, and the like); topical maps such as aquifer maps drawn by the water diviner or soils maps by soils experts; impact monitoring maps, where villagers record or map pest incidence, input usage, weed distribution, soil quality, and so forth. Some of the most illuminating maps combine historical views with those of the present.
- ☆ **Seasonal calendars and activity profiles**. Seasonal constraints and opportunities can be diagrammed month by month throughout the year. Ceremonies can be used as a cross-check so that names of months are agreed upon. People use pieces of stick, draw histograms in the dust

or with chalk, or make piles of stones, seeds, or powders to represent relative quantities and patterns of rainfall, soil moisture, crops, labour, food consumption, illnesses, prices, animal fodder, fuel, migration, pests, income, expenditure, debt, children's games, and so on. Seasonal calendars can be drawn in linear fashion with twelve months to show a typical year or eighteen months to illustrate changes between years, or they can be drawn in a circle. Daily pat terns of activity can be similarly explored by charting typical activities for each hour of the day, amount of effort, time taken, and location of work. These can be compared for men, women, the old, the young, and others.

- **Time lines and local histories**. Historical analyses have been found to be a good icebreaker for field exercises and include detailed accounts of the past, of how things have changed, particularly focussing on relationships and trends. These include technology histories and review, crop histories and biographies, livestock breed histories, labour availability, trees and forest histories, education change, and population change. Folklore and songs are valuable resources for exploring history.
- **Venn and network diagrams**. Venn diagrams involve the use of circles of paper or card to represent people, groups, and institutions. These are arranged to represent real linkages and distance between individuals and institutions. Overlap indicates flows of information, and distance on the diagram represents lack of contact.
- **Matrix scoring and pair wise ranking**. These methods are for learning about local people's categories, criteria, choices, and priorities. For pairwise ranking, items of interest are compared pair by pair; informants are asked which of the two they prefer, and why. Matrix scoring takes criteria for the rows in a matrix and items for columns, and people complete the boxes row by row. The items may be ordered for each of the criteria (*e.g.*, for six trees, indicate from best to worst for fuelwood, fodder, erosion control, and fruit supply); or participants may put stones, seeds, or berries into piles for relative scoring.

A Framework for Judging Trustworthiness

- **Prolonged and/or Intense Engagement between the Various Actors**. For building trust and rapport, learning the particulars of the context, and to keep the investigator(s) open to multiple influences.
- **Persistent and Parallel Observation**. For understanding both a phenomenon and its context.
- **Trangulation by Multiple Sources, Methods, and Investigators**: *la triangulation triangulée.* For cross-checking information and increasing the range of different people's realities encountered, including multiple copies of sources of information, comparing the results from a range of methods, and having teams with a diversity of personal, professional, and disciplinary backgrounds.

- ☆ **Analysis and Expression of Difference**. For ensuring that a wide range of different actors are involved in the analysis and that their perspectives and realities are accurately represented.
- ☆ **Peer or Colleague Checking**. Periodic review meetings with peers not directly involved in the original information was constructed and analyzed. Without participant checks, investigators can make no claims that they are representing participants' views.
- ☆ **Reports with Working Hypotheses, Contextual Descriptions, and Visualizations**. These are "thick" descriptions of complex reality, with working hypotheses, visualization, and quotations capturing people's personal perspectives and experiences.
- ☆ **Parallel Investigations and Team Communications**. If subgroups of the same team proceed with investigations in parallel using the same system of inquiry and come up with the same or similar findings, then we can depend on these findings.
- ☆ **Reflexive Journals**. These are diaries individuals keep on a daily basis to record a variety of information about them.
- ☆ **Inquiry Audit**. The inquiry team should be able to provide sufficient information for a disinterested person to examine the processes product in such a way as to confirm that the findings are not a figment of their imaginations.
- ☆ **Impact on Stakeholders' Capacity to Know and Act**. For demonstrating that the investigation or study has had an impact, including participants having a heightened sense of their own realities, as well as an increased awareness and appreciation of those of other people; the report itself could also prompt action on the part of readers who have been directly involved.

***Source: Pretty* (1994).**

Towards a New Professionalism in Extension

Extension has come to mean extending knowledge from a centre of learning to those presumed to be in need of that knowledge. These participatory methods and approaches represent an opportunity to build better linkages between the various actors and to increase the learning from each other. When PRA and other participatory approaches are used, extensionists and researchers have the opportunity to work together on the same team. They exchange knowledge and experiences and reach some consensus with farmers on what is most needed. As a result, all parties come closer together. Farmers become more confident that professionals can help them, without imposing solutions on them.

Developing Appropriate Content

What is appropriate for one farm family may not be appropriate for another, even though both families operate within the same agro ecological zone; or what is appropriate for one country may not be appropriate for another. Even at the national level, it has been shown that agricultural development can be facilitated by the development of appropriate technologies. Hayami and Ruttan (1985) pointed out that in Japan, where land is scarce, technology development was oriented towards biological technologies, for example to improve varieties. In the case of developing countries, their commercial relationship with the developed world predisposes them to accept or develop technologies that are not always appropriate for their farmers. The appropriateness should be defined within the scope of what is technical feasible; economically feasible, socially acceptable and environmentally safe and sustainable.

Technical Feasibility

Technical feasibility should be examined from two perspectives, the first being the ability of farmers to produce a commodity within their environment, and the second being what the farm family can achieve, not what can be achieved on research stations. In the first case, certain commodities do not fit into the production system of small farmers, while others do, for example scale neutral technologies for crops like maize, rice, and bananas. In the second case, a distinction must be made between technological potential and technical feasibility.

Economic Feasibility

The ability of farmers to incorporate a technology or technological package into a farming system has economic implications. The farmers' resource base, both human and financial, must be considered. Do farmers have the financial resources to purchase the inputs to derive the benefits from the technology? Will this technology require the hiring of additional labour, and if so, is it available and affordable? Some Green Revolution technology failed to be adopted because some anticipated users of the technology could not purchase the fertilizer and other inputs which formed part of the package. Economic feasibility can also be examined from the point of view of function within the farming system. Some small banana farmers in the Windward Islands of the Eastern Caribbean continue to grow bananas, although they have been advised against it. The rationale for this advice is that, among other things, soil erosion on the very steep slopes where most small farmers cultivate bananas and the resulting low yields make the cultivation of bananas economically unprofitable to them. However, they contend that banana cultivation allows them to have access to inputs which they can use to increase the profitability of other crops within their farming system, thus making the total farming system economically profitable. As such, banana cultivation within the system is seen by these marginal small farmers as a "facilitator" crop and not primarily for economic profitability. To these farmers, banana production is economically feasible within their farming system.

Social Acceptability

Innovations can be technically sound, but may conflict with the social norms of the end users or even cause societal disequilibrium. For instance, in Zaire, efforts to increase cassava production were constrained by the inability of the women to handle the increased yield. At the time the new varieties were introduced, the women in the targeted communities were already fully occupied. With regards to social disequilibrium, in some cases the adoption of technologies by small groups of individuals has resulted in a shift in power relationship within the society. For example, Campbell (1982) found that in the upland plateau of Ardeche, France, individuals from among the group of farmers who adopted tractors and mechanical harvesters were those who challenged the existing family-linked mayoral structure. Within the social context, there is also what can be termed situational appropriateness, that is, what is most likely to be accepted or rejected by a group at any given time. For instance, it would be unwise to introduce programmes with a crop or livestock component which had recently experienced market failures. Farm families remember such failures and tend to be hostile to such programmes.

Environmental Safety

An extension programme must be cognizant of the effects it will have on the environment. The commercialization of agriculture results in farmers using increasing quantities of fertilizers and agrochemicals. These chemicals for the most part have negative effects on the environment because they contaminate groundwater and may disrupt food chains. For example, birds which feed on insects exposed to chemicals may become contaminated and may die.

Extension's Clientele

Extension's clientele has to be defined within the context of an agricultural policy framework. This framework should provide the boundaries for selecting, from among the broader user categories, the specific groups which are to be targeted. Given today's scope of agricultural development, that framework should not speak of generic small farmers, but should specifically state the groups - women, youth, landless farmers, or other categories that are specific to the situation under consideration. Swanson, Roling, and Jiggins (1984) have cited four major factors which seek to provide a framework for the development of appropriate technology among targeted groups of extension clientele. These factors are agro-ecological zones, access to resources, gender, and age. Ethnic groups can be added to this list.

Agro-ecological Zones

The first step in identifying appropriate technology for extension's clientele is to map the area into agro-ecological zones. Farming systems research (FSR) provides useful methodologies for this exercise (see Hilderbrand, 1986; Shaner, Philipp, and Schmehl, 1982). Mapping allows for the identification of agronomic variables such as soil type, rainfall, slope, and altitude, which will influence the development of location-specific technologies.

Access to Resources

Within each zone, there will be a wide variety of farmers because of socioeconomic factors. These factors explain the differences which exist in terms of access to the factors which facilitate production - land, labour, capital, markets, inputs, tenure, and information. These factors predispose farmers to adopt certain types of innovations. Swanson *et al.* (1984) provide a useful summary of these facilitating and impeding factors which should be considered in developing appropriate technology for extension's clientele:

- ☆ Land. Size of holding: small, medium, large; type of tenure: owner operated, family land, renter-share-cropper
- ☆ Water. Irrigated, no irrigated
- ☆ Labour. Family, hired (cost and availability), communal, or customary
- ☆ Inputs. Availability of improved seeds, agricultural chemicals, fertilizers
- ☆ Markets. Location, availability of storage and transport
- ☆ Capital. Sources and cost of credit, type of collateral needed, and ease of obtaining credit
- ☆ Information. Availability of extension service (worker to farmer ratio), appropriateness of technology
- ☆ Influence. Ability to affect technology development, transfer to be appropriate to user needs such as user control, claim-making capacity

Gender

The role of women in agricultural production is now a topical issue. Several studies have documented women's contribution to both the economic and noneconomic sectors. However, there is need to move from this realization to its practical application in terms of the development and inclusion of appropriate technologies for women in extension programmes. In so doing, it must be taken into consideration that their access to resources and effective technologies is often constrained by gender barriers or blindness (Feldstein and Poats, 1989). Recognition must also be given to the particular functions women perform within the farming system and their work schedule in the farm-household environment. Generally, women play a much greater role in the production of food crops than of export-oriented crops, and within cropping systems they perform certain tasks, for instance, weeding, fertilizing, and harvesting. These factors must be considered in developing appropriate technologies to address this group of players in the agricultural environment.

Age

In most developing countries, agriculture is an important sector, if not the most important sector, and a significant proportion of the agricultural activities takes place in rural areas where 20 to 80 per cent of the country's population may live. This resource is usually characterized by a low level of education, few skills, limited employment opportunities, and a strong desire of young people to leave the area. Extension should consider age as an important characteristic for targeting not only from the point of view of youth but of other age categories. Agriculture usually has a low status because of the low level of technology it employs and low income-earning capacity. Extension programmes should aim at increasing both the level of technology used and income-earning capacity.

Ethnic Groups

In multiracial societies, ethnic groups should be separately targeted because of their different socio-cultural characteristics. Language, food preferences, and religion are some of these characteristics. In cases where ethnic groups are found within the same extension district or zone, extension's programme should reflect these differences.

Incorporating Needs into Programmes

Rapid rural appraisal can be used by research and extension to provide agro-ecological, biological, and socioeconomic data for identifying problems and opportunities for extension's clientele. For effective utilization, these needs should then be categorized on the basis of type of needs; that is, whether they are material needs, knowledge-based needs, or infrastructural support needs. Then the importance of the needs should be ascertained to prioritize them before incorporating them into extension programmes. For instance, the use of nematicide can increase the yield of a commodity by destroying the harmful nematodes, thus increasing nutrient uptake and yield. However, because nematodes are microscopic organisms, their presence in the soil may not be so obvious to the farmers who have not had the education to understand microorganisms. Consequently, an education component will need to be added to the technology to increase its chances of being adopted.

Program Delivery and Implementation

All extension approaches can be classified as either production technology approaches or problem-solving approaches. Production technology approaches tend to emphasize the production targets more than the clientele; technologies used in these approaches are more concerned with addressing production issues than clientele-related problems. Timmer (1982) said that these approaches are geared to produce techno-economic information and favour the larger, more commercial farmers. The small farmers because of their disadvantaged situation do not receive the full benefits from these approaches. The commodity-focussed approach and the technology-centred approach are examples of production technology approaches. The training and visit, community development-cum-extension, and animation rurale approaches fall into this category. As stated earlier, extension's clientele is a heterogeneous group of people differentiated by resources, gender, age, and ethnicity. Communication theories indicate that the trickle-down effect of agricultural innovations rarely takes place among heterogeneous groups, and when it occurs, it does so at an extremely slow pace. The strategy is therefore to develop homogeneous groups of clientele to target extension programmes. The targeting of these groups should be done within the broader groups already defined, namely, agro-ecological zone, access to resources, women, youth, ethnicity, access to information, and age. Extension programmes should then be developed for each of the targeted categories (Swanson *et al.*, 1984). For instance, special programmes should be developed for women which recognize their importance in the agricultural sector, take account of their informal communication network, and analyse their farm-home work schedule. In the case of programmes directed at ethnic groups, efforts should be targeted to the socio-cultural factors which distinguish the group. Language, food preferences, and religious beliefs are among the important points that should be taken into consideration. In addition, extension officers in developing extension programmes should be guided by principles of learning and knowledge of the diffusion process.

The Learning Process

Learning is facilitated by the use of the senses. The more senses that are used, the more accelerated will be the learning. Use of this principle will assist extensionists in selecting methods which will provide an educational experience for their clientele. Learning is also conditioned by the motivation and the abilities of the learner; thus it is important to create an environment conducive to learning, one in which the learners become more responsible for the outcome of their experience. This can be achieved through the selection of teaching methods and supporting material. For example, if farmers are not performing a task effectively, then it will be much more effective to organize a method demonstration than a lecture or group meeting.

The Adoption Process

The other important consideration in selecting methods for the delivery of appropriate content is an understanding of the adoption process. Lionberger (1968) contended that the adoption process consisted of five distinct stages: awareness, interest, evaluation, trial, and adoption. According to Lionberger, the individual in the process goes through each stage within a definable time period. Later works by Lionberger and others have shown that these stages are not as distinct as he first proposed and that some of the stages may become condensed within the individual cognitive processes, thus making them unrecognizable as a behaviour which can be measured over time.

In the **awareness stage**, knowledge of the innovation is critical to the individual. Mass media and popular theatre are the preferred methods because they can reach many people at the same time. In using the mass media, extensionists must pay attention to the characteristics of the audience targeted. For instance, in multiracial societies, an ethnic group with a special language may require programming in that language. The other method, popular theatre, although reaching only smaller audiences, is a very effective means of building awareness because it uses the popular language and rhythms of the people in presenting content to its audiences.

In the **interest stage** of the adoption process, knowledge continues to be important, but building a positive attitude towards the innovation becomes the critical issue. For this reason, the desired methods should include information strengthening and attitude building as their goal. These methods should use the senses of hearing and sight, either individually or collectively. Group meetings, group discussions, and radio forums are recommended for strengthening knowledge, while field days and farm visits will allow individuals to see what they have been hearing, thus providing the opportunity for building the desired attitude towards the innovation.

Evaluation is the most critical stage in the adoption process, because the outcome usually determines whether or not individuals proceed to the trial and adoption stages. At this stage, people need to match knowledge against facts.

Farmers need to be assured that what they heard and saw are indeed workable. Result demonstration, farmer exchange, and field days are recommended because they allow individuals to reinforce their interest by viewing tangible evidence. Within this group of methods, farmer exchange is an important method. Farmers selected for the exchange should be further advanced in the adoption process and within the same reference group as the visiting farmers.

At the **trial stage**, the farmer's technical and management skill should be the main area to be targeted. The individual visit becomes the most preferred method at this stage, and the needs of individual farmers must be taken into consideration. The extensionist has to remember that, although similar farmers are adopting similar techniques, the problems experienced are not always the same. At this stage, methods for reinforcing the farmers' interest by the use of farmer exchange and skill training can be useful in helping individuals to continue adopting.

Once the farmers start **adopting**, extension should continue to support their efforts. Recognition programmes and farmers' competition can be used to encourage farmers to continue adopting. The goals and criteria for these methods should be carefully developed so as not to bring out any negative effects because of poor planning and implementation.

The recommended methods for the different stages of adoption are summarized in the below Figure. To achieve the maximum results from the methods selected at the different stages, extension officers must bring to bear their understanding of the learning process. For instance, in group meetings the extension officer can use a lecture format, supported with a video, and followed with group discussions.

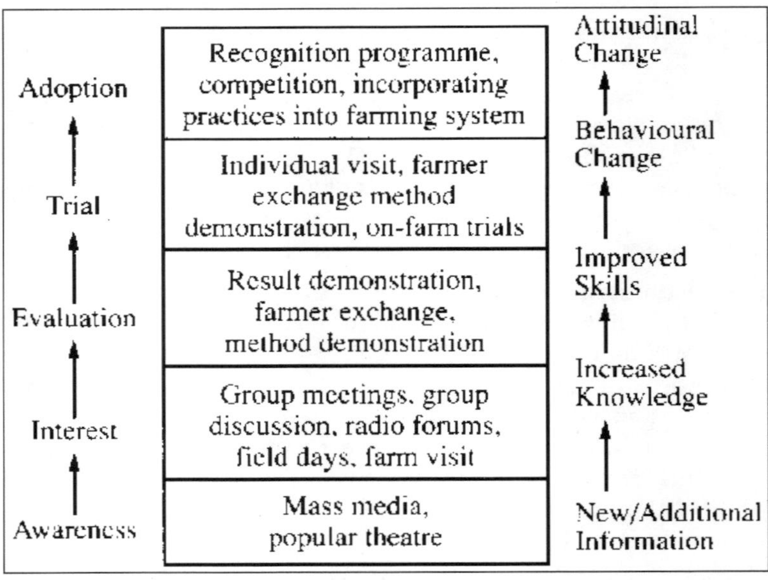

Recommended Extension Methods for Use at different Stages of Adoption.

In the final analysis, the method chosen will depend on the goal, resources, clientele relationship, and skills of the extension officers on the one hand, and on the size and educational level of the target group on the other hand. For instance, if extension officers lack the skills to organize and facilitate group meetings, then they will shy away from their use. Or if extension officers do not have vehicles, then they may not be able to conduct farm visits as frequently as might be desired or needed. Also, if the extension officer's constituency is very large, then it may become impractical to depend too much on individual visits.

As a guide in the selection of extension methods, Van Den Ban and Hawkins (1985) provide us with a useful set of criteria to judge whether the method is well chosen:

- ☆ Is the chosen method adapted to whether we wish to change knowledge, skills, attitude, or behaviour?
- ☆ Are the educational activities clearly specified so that we know what the farmer will see, hear, discuss, and carry out?
- ☆ Are the different methods integrated in such a way that they reinforce each other?
- ☆ Does the planned time scale make it possible to carry out these entire activities well?
- ☆ When choosing learning activities, has the extensionist adequately considered the needs, skills, and means of the target group?

The Need for Gender Analysis

The term gender describes the socially determined attributes of men and women, including male and female roles. (In comparison, sex denotes the physical and biological differences between males and females.) Gender has proven to be an essential variable for analysing the roles, responsibilities, constraints, opportunities, incentives, costs, and benefits in agriculture. Innumerable development projects, government programmes, research studies, and theoretical models have demonstrated that the improvement of women's access to agricultural research and extension services must begin with an analysis of men's and women's participation in the agricultural production process along two related dimensions: their role in agriculture and their role in the household.

5

Gender Analysis in Agricultural Extension

It is now widely demonstrated that rural women, as well as men, throughout the world are engaged in a range of productive activities essential to household welfare, agricultural productivity, and economic growth. Yet women's substantial contribution continues to be systematically marginalized and undervalued in conventional agricultural and economic analyses and policies, while men's contribution remains the central, often the sole, focus of attention.

Women are typically, and wrongly, still characterized as "economically inactive" in statistical surveys of agriculture, a result that tells us more about survey methodology than about reality (Janelid, 1975). Agricultural extension services still do not attach much importance to reaching women farmers or women on the farm. Policy makers and administrators typically still assume (in the face of the empirical data) that men are the farmers and women play only a "supportive role" as farmers' wife (Samanta, 1994).

The official definition of a farmer in Nigeria in 1965, for example, was given as "an adult male. who has the right to the produce of a farm. women are not classified as farmers" (FOS, 1966). Yet among many studies of rural women in Nigeria (WORDOC, 1988), Akor (1990) found that 92 per cent of the surveyed northern rural women gave farming as their primary or secondary occupation. Of these, 74 per cent owned or worked their own separate plots. While the official definition of a farmer in Nigeria has been corrected to be gender neutral, as in most other countries, gender bias is prevalent in official agricultural circles and among field professionals. Similar investigations conducted in selected states in India show that more than 60 per cent of agricultural operations are performed by women farmers, yet the *fact* that "most farmers in India are women" (Shiva, 1991) is simply not reflected in extension provision or training.

Gender Analysis at the Household Level

The rural household typically is conceptualized in extension programmes and agricultural policies as a unit made up of individuals working in similar ways to meet common goals under the direction of a male head. In reality, the household is a more complex and dynamic social entity which may change its composition and goals over time as family members and dependents of varying age groups and sexes engage in various activities to meet the specific responsibilities assigned to each. However, while it is useful to draw attention to the fact that the division of labour along gender lines is a social constant and has profound implications for the organization of agriculture, men's and women's responsibilities and privileges vary along sociocultural and socioeconomic lines specific to a particular time and place. Even within one country, sweeping generalizations are not advisable. Studies in Nigeria have revealed differences in gender relationships even in ethnically similar rural Nigerian communities just kilometres apart (Olawoye, 1985).

It is thus a mistake to view "rural women" as a homogeneous social classification or to derive policies and services for "women in agriculture" that are not based on empirical research which captures the diversity. As we will see later in the chapter, the consequence is that extension services need to be adapted to circumstance rather than designed on the basis of a single universal model. There is no one packaged extension model which can work for all women in all places (Olawoye, 1989; Berger, DeLancey, and Mellencamp, 1984).

The Changing Social Structure in Agriculture

As the composition and structure of rural households change (Snyder, 1990), gender responsibilities are under-going rapid change, typically with rural women becoming more responsible for household food security and children's welfare. One powerful indicator of these changes is the incidence of female-headed rural households, which is on the increase in most developing countries. In sub-Saharan Africa, women head an estimated 45 per cent of rural households in Kenya, 35 per cent in Malawi, 30 to 40 per cent in Zambia, and 15 per cent in Nigeria (ECA, 1973; Keller, 1986; World Bank, 1992a; FAO, 1993).

Typically, female-headed households are among the poorest, with the lowest level of food security (Heyzer, 1992), but in areas where female headship is the norm, as in the Caribbean, female headship can be a poor predictor of agricultural output, household welfare, or income status (Jiggins, 1994). In other cases, where women have had access to agricultural resources and services in their own right, as in parts of the Kenyan Highlands, women farming alone or with only sporadic assistance from migrant husbands have proved themselves more than capable of increasing farm productivity, efficiency, and profit (Jiggins, 1994; Saito and Weidemann, 1990).

Data summarized by the FAO (1993) serve as gross indicators of the potential overall size of extension's female clientele (covering *female farm operators and female-operated farms*) and their distribution among large, medium, and small farms. Case study data from twelve countries (Burkina Paso, Cameroon, Rwanda, Senegal, Bangladesh, China, Indonesia, Philippines, Colombia, Mexico, Egypt, Yemen) indicate the following in the study areas:

- ☆ 25 per cent of large farms were operated by women.
- ☆ 19 per cent of medium-sized farms were operated by women.
- ☆ China and Indonesia both had an exceptionally high proportion of women operators of large and medium-sized farms.
- ☆ The numbers and percentage of women operators were greatest on small farms.
- ☆ Except for Indonesia, Philippines, Senegal, and Yemen, the percentage of female-headed farms was above 15 per cent, and in Mexico (61.46 per cent), Egypt (28.10 per cent), and Kenya (27.09 per cent), they accounted for a quarter to two thirds of all farms studied.

Constraints and Opportunities: Rural Women and Extension Strategies

The constraints affecting rural women's ability to improve yield, profit, and efficiency in agriculture include:

1. Women's legal and cultural status, which affects the degree of control women have over productive resources, inputs such as credit, and the benefits which flow from them (Olawoye, 1989);
2. Property rights and inheritance laws, which govern access to and use of land and other natural resources (Jiggins, 1989a);
3. The relationship among ecological factors such as the seasonality of rainfall and availability of fuelwood, economic factors such as product market failures, and gender-determined responsibilities such as feeding the family, which trade off basic household self-provisioning goals and care of the family against production for the market (Jiggins, 1989b; Horenstein, 1989); and
4. The way that agricultural services are staffed, managed, and designed (FAO, 1993; Saito and Weidemann, 1990; Gittinger *et al.*, 1990).

Improving Women's Access to Extension

Agricultural extension strategies traditionally have focussed on increasing production of cash crops by providing men with training, information, and access to inputs and services. This male bias is illustrated in farmer training centres, which have been established to provide residential training on technical subjects. Most do not provide separate washing and sleeping accommodations for men and women and do not provide facilities for the care of babies or young children, factors which may prevent women from attending the centres. Second, women's daily workloads do not usually allow them to be absent from home for residential training; even attending short courses may cause insuperable problems in arranging substitute care for children or the home. And third, even where attendance of women is quite high as a proportion of the total, women are given instruction mainly in home economics and craft subjects, not technical agriculture (Staudt, 1973; Perraton, Jamison, and Orival 1983).

Further, in the overwhelming majority of countries, extension services have been staffed predominantly by men. Only in countries such as the Philippines have women field staff been deployed in sufficient numbers and with sufficient resources to become effective agents of change among women farmers. On the other hand, it is typical of ministries to assume that home economics services can substitute for agricultural training and information for women. Home economics and agriculture are both important, but they are not substitutes. Where home economics services have been provided, as Aidoo (1988) notes, female home economists worked almost exclusively with rural women, thus reinforcing the institutionalization of gender bias. The introduction of the training and visit system emphasized the selection of *contact farmers* as a mechanism for passing on information to other ("follower") farmers in their area. The recommended selection criteria, such as title to land, literacy, or cooperative membership, as well as male extension staff's assumptions about women's roles in farming, have largely excluded women's involvement (see Aammink and Kingma, 1991 for a Tanzanian case study). In some countries, individual contact has been complemented by *group contact,* especially, but not only, where it may be difficult for male change agents to have any type of contact with individual women other than their own relatives. In many cultural settings, group extension significantly increases women's access (Berger, DeLancey, and Mellencamp, 1984; Ashby, 1981), because the group context calms the fears of male extension agents, husbands, and women about transgressing norms of approved social contact.

This may be particularly true in Islamic areas where women are in partial or total seclusion. Furthermore, in Islamic societies, there are probably not enough qualified adult females who are able to take up the post of change agent at the field level. However, in countries such as Bangladesh, the pioneering efforts of large-scale, non-government, rural development agencies such as BRAC and the Grameen Bank have demonstrated that religion and custom are not necessarily barriers to the hiring and field deployment of female staff, to the mobilization of women's groups and training of women leaders at the group and village levels by male staff, or to the development of efficient savings and credit services for rural women (Jiggins, 1994). Women-dominated and managed agro-industries, notably in poultry and silk, based on individual effort and rewards but mobilized through group mechanisms are becoming sizeable integrated enterprises backed by women's own savings and investments. The group concept is being deepened and expanded in a number of ways, as agro-industry development opens up opportunities for developing functional linkages between women's groups and centralized processing facilities. The production and processing of herbs, medicinal plants, or perfume plants, for example, organized by TATA Industries in the private sector and BAIF in the nongovernment sector, provide examples from India of information and training linkages between women's groups and the agro-industry. The development of women-only cooperatives in the Dairy Development Movement in south India is a notable case (Chen *et al.*, 1986; Jamal, 1994), but numerous other studies document the potential of this approach (Berger, DeLancey, and Mellencamp, 1984).

Changing the Selection Criteria for Contact Farmers in Kenya

The criteria laid down by the World Bank for selecting contact farmers specify that they should (1) represent the local range of farm size, cropping pattern, socioeconomic condition; (2) be regarded by other farmers as worthy of imitation; (3) be active, practising farmers; (4) be willing to adopt extension recommendations on at least part of their land, allow other farmers to observe the new practices, and be willing to explain these to other farmers; (5) to the extent possible, come from different families; and (6) be from geographically dispersed farms.

Three adjustments to selection criteria and the selection process have proven to be useful in Kenya in increasing the percentage of women selected:

- ☆ Encouraging chiefs, ministers, and other leaders to promote women's selection at local meetings and in the media
- ☆ Stressing the importance of selecting women farmers in extension training courses. Emphasizing selection on merit from among those who are actually doing the work

Changing Attitudes and Institutions

Experience has also shown that complementary strategies to bring about changes in attitude and behaviour within institutions are required (Poats, 1991). Gender sensitization training has been developed to initiate the task of attitude change within male-dominated extension and research bureaucracies and donor agencies (Rao *et al.*, 1994). Training materials and methods for gender analysis in agriculture (Paris and Frio, 1994; Feldstein, 1994) have also been developed and are now in widespread use. Specialist material for training of trainers (Rao *et al.*, 1994) has also been developed and is beginning to spread through agriculture training institutes, colleges, and universities.

Farmer-to-Farmer Extension: Women's Participation in Networking

The Farmers' Groups Network in Tanzania (MVIWATA) was formed by group representatives during a workshop hosted by Sokoine Agricultural University in 1993. The workshop was one of a series organized by farming systems researchers at Sokoine Agricultural University who have been working with farmer groups in technology development.

They distilled the following five principles from their experience:

1. Multidisciplinary, in recognition that farmers' problems are multifaceted
2. The use of group approaches, in recognition that decision making is almost always built on group consensus
3. On-farm development of technical innovations to ensure relevance

4. Assistance with removal of critical production bottlenecks
5. Empowerment of farmers through training, facilitation, and networking.

For example, male extension agents were encouraged to ask their male farmer contacts to include their wives during visits, demonstrations, or farmers' meetings. Village leaders (typically male) asked to men identify women needing extension services. Field agents were required to devote a greater percentage of their time to working with women's groups. Women farmers' seminars were organized for women to share with researchers and field staff their solutions to the technical problems specific to women farmers' production systems, and women's field days were organized to celebrate and legitimate women farmers' successes and to promote farmer-to-farmer exchange among women.

Participatory action learning (PAL) also has proved to be a powerful approach to institutional change. The Centre for International Agricultural Development (CIAD) at the University of Beijing and the All-China Women's Federation (ACWF) adopted a participatory learning approach to their work with farm women in Ningjin County (Huang Xiushen, 1995). The Women's Federation (WF) formed a "socialized service network" to provide the backup training, drawing on the training expertise identified in earlier phases. A scientific and technical team was then formed of the WF director and twelve group leaders. It is their responsibility to form a link with each participating household and with technical training resources for the purpose of training women to teach other women.

The Tamil Nadu Women in Agriculture (TANWA) project, India, has been implemented by the state Department of Agriculture with DANIDA assistance since 1986 (Box 3). The project aims to provide small and marginal farm women with improved technical skills and technology through a training-cum-extension program. Over a seven-year period, a core team of thirty qualified women agricultural officers (AOs) has been recruited, and more than 14,000 women have received training, with significant improvements in food security, income, and productivity. The training is village based, given at times suited to women's workload to groups of twenty-five women, and focuses on practical skills and technologies selected and prioritized by the women in consultation with the AOs. The basic training stretches over a two-year schedule and is backed up by two-day training for the group leaders and short courses on special topics. In addition, farm women's conferences allow the women to interact with (male) research and extension officers and with groups from neighbouring villages. The appointment of the women AOs is a breakthrough in Tamil Nadu and follows the earlier success of appointing women dairy development officers in the dairy development movement. The success of women-to-women extension has built confidence among male AOs that farm women are worth contacting, while the group mechanism provides a culturally appropriate means for follow-up contact by male extension staff.

Integrated Pest Management: The Farm Women's Field School, Kolumanivakkam, Tamil Nadu

Since 1994, women farmers have been included in field training on integrated pest management (IPM) in Tamil Nadu. Of the thirty farmers' field schools (FFS) conducted in the 1994-95 rice season in Chengalpattu MGR District, for instance, four have been for women. The FFS runs one half-day a week for thirteen weeks for thirty farmers. They are taught by female agricultural officers to identify the pest and predator insects in rice, to monitor the number of each to ensure that predators are keeping the pests in check, and to observe the life cycles of the key insects and the fluctuations in populations through the production season. Weekly recording and analysis of the agro-ecosystem form an important part of the training. The principles and practices of organic fanning, integrated nutrient management, and the use of bio-fertilizers are included. Cultural, mechanical, and biological pest controls are practised, and participants are trained in the use of bio-pesticides and chemicals as a last resort if predator populations are insufficient to keep pest numbers within economic limits. The women of Kolumanivakkam have composed a *mantra* to express their commitment to becoming experts in their own fields: "I can't lose. Why? I will tell you why. because I have Knowledge, Courage, and Enthusiasm." Prior to the introduction of IPM, farmers in the village regularly used pesticides, accounting for up to 20 per cent of the costs of cultivation. The women would work in the fields after the spraying and reported a range of symptoms such as nausea, headaches, eye troubles, skin rashes, and difficulties in breathing attributable to the chemicals. Now they have stopped spraying their homestead vegetables as well and have begun to monitor pests and predators in their home gardens.

Scaling Up

The Academy for Educational Development (based in Washington, D.C.), the government extension service, and relevant private sector agencies collaborated together in Honduras in adapting social marketing techniques to extension communication (UNIFEM and IWTC, 1990). They developed a model for communication for technology transfer in agriculture (CTTA), which is based on three principles: market research in order to identify user categories and needs, user participation in technology development and dissemination, and strong feedback mechanisms. Women farmers and farm women in various socioeconomic and spatial categories were identified explicitly as "users." The CTTA design focussed on the "four Ps" of social marketing - price, product, place, and promotion - as well as on two contextual factors - policy and politics - in devising more effective ways to involve users in the development, dissemination, and feedback of extension communications. During the pilot phase, the CTTA increased the outreach of extension services in Honduras from 3,000 to 16,000 farmers.

Effective Extension Systems for Rural Women

Building on Present Potentials

It is a mistake to believe that rural women in developing countries do not possess skills and techniques which are an asset to the development process. Where groups already exist, capacity building of existing groups can be more successful than forming a new group to which members are less likely to be committed. Similarly, rural people are less likely to resist adoption of an innovation when the new technique is based upon a concept or procedure they are already familiar with or are currently using. It is important for change agents to acquaint themselves with the organizations and knowledge systems available at the local level to determine how they can be improved, rather than assuming that nothing of significance is currently available. For example, instead of forming entirely new groups for women, local informal work exchange or savings groups could be strengthened through short training exercises on farm bookkeeping and record-keeping, leadership, and democratic procedures. While groups have proven to be a highly successful mechanism, they are not a universal panacea or appropriate for meeting all women's needs in agriculture. The poorest women in particular can find the costs of participation (of time, for example) too high at times of seasonal stress or greatest need.

Institution Strengthening

Several aspects of institution, strengthening, including project integration and gender-sensitization of officials, have been presented. These efforts need to be pursued vigorously and extended. The World Bank (1992b) summarizes four organizational principles which might serve as broader guidelines to institutional development:

- ☆ Situation specificity,
- ☆ Project flexibility,
- ☆ Farmer participation, and
- ☆ Mainstreaming women's programmes.

By incorporating these principles with the other factors considered earlier, agencies concerned with women's projects will be more likely to have a meaningful and sustainable impact.

Interagency Cooperation

Many agencies currently focus on the integration of women in mainstream development efforts. Unfortunately, there seems to be little cooperation between projects administered by the same agency or between agencies, even where these agencies are under the umbrella of the same national government or development policy. In addition, international agencies and NGOs are attempting to achieve similar goals in the same localities, yet typically without any form of coordination or cooperation. The use of human and material resources could be managed more efficiently and with greater impact if development agencies would cooperate with each other (Olawoye, 1991).

Training Programmes

The following suggestions are made for improving and redesigning training programmes for farm women:

- ☆ Adapt programmes to women's needs and skills.
- ☆ Allow sufficient time to enable women to acquire new skills and adjust schedules to fit women's existing workloads.

- ☆ Provide training in agricultural and other productive activities, not just home and family welfare topics.
- ☆ Emphasize activities for which there is an actual income-generation potential.
- ☆ Ensure the involvement and full participation of women from poorer and less educated backgrounds.
- ☆ Use trainers who are not only technically competent and up-to-date, but who empathize with the needs and aspirations of rural women.
- ☆ Provide practical field experience in the use of innovations.
- ☆ Shift more resources to village-based training rather than residential training.

6

Role of Extension in Sustainable Agriculture and Rural Development

SARD and its Policy Implications

Farmers correctly view extension as a form of assistance to help improve their know-how, efficiency, productivity, profitability, and contribution to the good of their family, community, and society. At the same time, politicians, planners, and policy makers in many developing countries view extension as a policy instrument to increase agricultural production, to achieve national food security, and, at the same time, help alleviate rural poverty. In addition, some economists view extension as a policy instrument that will contribute to human capital development and economic growth; therefore, resources allocated to extension are viewed as an economic investment which must produce competitive economic returns. To the practitioner, agricultural extension enhances and accelerates the spread of useful know-how and technologies to rural people. These activities are expected to lead to increased and sustained productivity, increased income and well-being of farm people, and to the promotion of national food security and economic growth. These objectives are to be achieved through non formal education and training programmes and two-way technology transfer and feedback systems where extension has an important contribution to make to agricultural and rural development.

At the same time, agricultural and rural development is no longer a matter of just increasing food and agriculture production; other factors must be addressed by policy makers and support service agencies in formulating and implementing agricultural and rural development policy. These issues include population and environmental concerns, and they have very strong implications for how key support services such as research and extension should be organized and financed.

The Population Factor

An important concern is the rapid population growth in many developing countries. This factor has a direct impact on the demand for food and other agricultural products, and it results in increased pressure on the land and other natural resources. In many countries, the population has more than doubled during the last three decades. During the last decade of the twentieth century, it is projected that more than 850 million people will be added to the world's population. Furthermore, world population growth is expected to increase 57 per cent from 5.3 billion in 1990 to 8.3 billion by the year 2025 (UN-World Population Prospects: The 1994 Revision). Because the bulk of these major population increases are occurring in developing countries, particularly in Africa and Asia, this implies not only an increase in the demand for food, but also more pressure on fragile and marginal lands, increased land fragmentation, and larger numbers of landless people in rural areas.

Natural Resources and Environmental Factors

Population pressure and the demand for increasing food output are now commonly associated with the degradation, depletion, and pollution of soil, water, and other natural resources (Alexandratos, 1995). Numerous actions are required for a society to conserve, protect, rehabilitate, and manage its land, water, and other natural resources; therefore, extension has a central role to play in disseminating sustainable agricultural technology.

The Economics of Extension

Since the 1980s, funding of research and extension has become an increasingly important policy issue, one that has given rise to a progressive decline in financial support for extension. This decline is occurring in a situation where funding of extension has been chronically inadequate (Swanson, 1990). Many economists and development planners believe that public funding of extension should be higher (Wilson in Rivera and Gustafson, 1991, p. 13), and an FAO study on this issue, involving 114 member countries, showed that in most developing countries government support to extension is generally low when compared to agriculture gross domestic product (about 0.5 per cent of AGDP). The Global Consultation on Agricultural Extension recommended that "in countries where more than 60 per cent of the economically active population are engaged in agricultural production, approximately 1 to 2 per cent of the AGDP (depending on the size of the country and factor costs) should be considered the minimum level of financial investment to achieve both human resource development and technology transfer goals of a public sector agricultural extension system" (Swanson, 1990, p. 26-27).

Privatization of extension in developing countries, where the circumstances of farming and farmers is quite different than in the developed countries (*i.e.*, similar to conditions in developed countries some 50 to 100 years ago), is being promoted at the national and international level. This emerging trend is toward a more pluralistic conception of extension, as found in many developing countries today. This approach implies the need for public and private sector cooperation to address the twin problems of poor extension coverage and resource limitations (Mualouf *et al.*, in Rivera and Gustafson, 1991). What is not fully apparent to some planners and policy makers is the relationship between the level of public funding for extension and the economic and social returns to a nation in the form of food security, affordable food for consumers, and the alleviation of widespread hunger and poverty.

Sustainable Agricultural and Rural Development

The integration of the food, population, and environmental nexus has led to a new platform of development, referred to by FAO as sustainable agricultural and rural development (SARD). This new concept of SARD has been defined by FAO as "the management and conservation of the natural resource base, and the orientation of technological and institutional change in such a manner as to ensure the attainment and continued satisfaction of human needs for present and future generations. Such sustainable development (in agriculture, forestry and fisheries sectors) conserves land, water, plant and animal genetic resources; is environmentally non-degrading; technically appropriate; economically viable; and socially acceptable" (FAO, den Bosch Declaration, 1991).

The role of farmers is well recognized in SARD, particularly in the management and conservation of soil, water, and biological resources; maintaining an ecological balance; and applying environment friendly technologies such as integrated pest management (IPM). Farmers are the single largest group of users and managers of land, water, and other biological resources throughout the world. In 1970, about 790 million people were economically active in agriculture, and this number will increase to around 1.13 billion by the year 2000 and to 1.19 billion by 2010 (Alexandratos, 1995). The majority of these men, women, and young farmers will need useful information, appropriate technology, and sound technical advice not only to increase their agricultural productivity and incomes, but also to make farming and rural life richer and more sustainable. Herein lies the important role of agricultural extension.

The Place of Extension Policy in SARD

The first principle to recognize is that extension is one of the most strategically important policy instruments for achieving the twin goals of food security and SARD in developing countries. In countries that already have an effective agricultural extension system, the cost of pursuing SARD with the farming population can be reduced by integrating environmental and sustainable themes into ongoing extension programmes that are relevant and useful to farm people (Stocking, 1994).

Secondly, the new goals of pursuing SARD should be embodied into existing or new extension policies. The FAO-sponsored Expert Consultation on Integrating Environmental and Sustainable Development Themes into Agricultural Education and Extension Programmes noted that "without a persuasive, effective and unambiguous set of directives for good working practice, articulated in the form of high-level policy and agency mandate, the current ad hoc approach to environmental matters will likely continue" (Stocking, 1994). In short, extension policy makers, with the participation of their various stakeholders, need to formulate extension policies that will address both farmers' needs and environmental concerns and that will guide extension managers during the first quarter of the twenty-first century.

Scope of Extension Policy

Agricultural extension policy is a part of national development policy in general and of agricultural and rural development policy in particular. Hence, agricultural extension is one of the policy instruments which governments can use to stimulate agricultural development (Van Den Van in Jones, 1986, p. 91). Extension is very much a part of what Röling refers to as the agricultural development mix. He notes that extension is a weak instrument when it stands alone, but that it becomes powerful when combined with price incentives, input supply, credit, seed multiplication, and so forth (Röling in Jones, 1986, p. 110). The Global Consultation on Agricultural Extension concluded that agricultural extension policy should be consistent with and supportive of national agricultural development policy and goals (Swanson, 1990). Each country should have a comprehensive agricultural extension policy which provides for coordination with research, education, input supply, and credit and marketing systems, as well as some flexibility to reflect the dynamic nature of the agricultural sector. The policy should include the mission and goals for agricultural extension, the responsible agencies and personnel, the clientele to be served, the broad programmatic areas to be addressed, and other relevant guidelines. In developing national agricultural extension policies, representatives of all major groups of farmers should be directly involved and other relevant agricultural organizations should be consulted. "By pursuing a comprehensive policy," the Global Consultation noted, "countries can expect the extension system to contribute to increasing agricultural productivity and farm income, and to improving the quality of life of most rural farm households in pursuit of the general goal of growth with equity.

Extension Policy

This is the most common form of extension policy in most developing countries. In the absence of more formalized extension policies, or at the time when the formally enacted policy has been suspended, a provisional or ad hoc policy comes into play. For example, Mozambique in the early 1980s did not have a national policy for agricultural extension. When the agricultural development policy shifted from a reliance on state farms to the involvement of small family farms, a provisional extension policy was formulated to provide farmers and the cooperative sector with improved training and technology. To develop and test this provisional policy, a UNDP/FAO-supported project assisted the government in defining a national agricultural extension policy and developing a programme of implementation.

Decrees and Proclamations

Decrees and proclamations are policies issued by the head of state or by the executive officer of government. Generally, this approach does not go through the process of consultation and debate involving various stakeholders and beneficiaries. An example of this form of policy was the Brazilian government decree abolishing the national agricultural extension authority and transferring some of its functions and staff to the national agricultural research authority. Also, it empowered the state-level rural extension authorities to continue their respective programmes. In the Philippines, by a presidential proclamation, the name of the Bureau of Agricultural Extension was changed to Agricultural Extension Commission, and the title of "agricultural extension workers" was changed to "farm management technicians." The Bureau of Agricultural Extension, which was created by legislation, was subsequently restored until another proclamation abolished the Bureau of Agricultural Extension and created, in its place, the National Agricultural Training Institute. The agricultural extension function of the Department of Agriculture was then decentralized to the regions and provinces and put under the control of local governments. The decentralization aspect of the proclamation was later superseded by an enactment, in 1991, known as the Local Government Code (Serrano in APO, 1994).

Legislated Extension Policies

Extension policies embodied by the country's highest law-making authority (*e.g.*, congress or parliament) are common in many developing countries. Countries that have enacted extension policy through legislative action tend to have well-organized, financially stable extension systems that have sustained effectiveness and a cumulative impact. Examples of legislated extension policies which have worked well include the following:

1. The legislation that established the Cooperative Extension Service in the United States is known as the Smith-Lever Act of May 8, 1914. Important to policy makers from developing countries is that this policy stimulated the growth and efficiency of American agriculture from the 1920s to the present. Rogers (1995) summarized the worth of the Cooperative Extension System: "The U.S. agricultural extension model is one of the most widely recognized systems for the diffusion of innovations in the world today. Probably no other government or private agency can claim to be more successful in transferring technology."
2. The Japanese Agricultural Promotion Law of 1948 created and provided funding for Japan's Cooperative Agricultural Extension Service. The same extension policy has guided the Japanese extension system from 1948 to the present (Shinji Imai in APO 1994, p. 122). Under this law, the national government is responsible for two-thirds of the salary of extension workers, two-thirds of all operational expenditures for extension programmes and one-half of all expenditures for training extension workers and for rural youth work. The rest of the extension budget is the responsibility of the prefectural government (Agricultural Extension Service in Japan, 1978, p. 83).
3. Agricultural Extension policy in South Korea today is embodied in the 1957 Agricultural Extension Law and in the Rural Development Law of 1962.

It is important to note that, because of "unhelpful interference from the administrative system," the 1962 Rural Development Law put together the Research Bureau, the Extension Bureau, and the Community Development Bureau under the new Rural Development Administration, freeing these two functions from the Ministry of Agriculture's administrative bureaucracy (Yong-Bok Chung and Youl-Mo Dong, 1984 p. 4, 5).

4. Thailand's agricultural extension policy was codified in the 1956 law that created the Department of Agricultural Extension as one of nine departments of the Ministry of Agriculture and Cooperatives. It outlines the functions, scope, organization, and mode of financial support for extension in Thailand.
5. Zimbabwe's Department of Agricultural Technical and Extension Services was established by law in 1981 and, although a relatively young institution, it is gradually building up its extension staff and its government funding. Observations such as these, which demonstrate the relationship between formal extension policy and successful extension systems, led the Global Consultation on Agricultural Extension to recommend that "where possible, agricultural extension policy should be formally enacted through legislative action to provide a stable policy foundation, an explicit mandate, and clear direction for developing and executing programmes" (Swanson, 1990).

Issues of Extension Policy

Although extension has a generic and universal meaning, its mission and goals may need to be adjusted according to national objectives and the context and stage of agricultural and rural development in a given country. Should the mission of extension be to promote agricultural development through technology transfer? Should it give higher priority to human resource development in rural areas, or should it promote sustainable agricultural and rural development? The extension mission should be reflected in the name of the organization, and the preamble for extension policy should be included in the law governing the country's extension system. This mission then should be reflected in a statement of goals and objectives that are agreed upon and assigned to extension in a supporting policy document. This document should be periodically reviewed by policy makers and representatives from stakeholder groups.

Extension Approach and Functions

National extension systems can pursue one of several different extension approaches in implementing extension policy. Most extension systems in developing countries give primary attention to technology transfer, given national agricultural policies that emphasize increasing food production and achieving national food security. An example of a technology transfer approach would be the Training and Visit (T&V) Extension System that has been promoted by the World Bank through its lending programme. Although the U.S. extension system has been particularly effective in technology transfer, its main focus has been on increasing the skills and knowledge of rural farm families, who have become very effective consumers of agricultural technology. Therefore, the extension approach pursued by a country should reflect the mission of extension, and it will define the functions, programmes, and tasks that will be carried out by the extension staff.

Subject-Matter Coverage of Extension

Broadly speaking, the subject matter of extension is implied in the mission statement and even in the title of the extension service. What differentiates between agricultural and rural extension is the subject matter that the extension service will include in its programmes and the target groups to be served among the rural population. Very narrow subject-matter coverage such as promotion of food and cash crops and animal production may invite a costly proliferation of several specialized and uncoordinated extension initiatives. Broader subject-matter coverage such as promoting the entire farming system, sustain-able agricultural and rural development leads to a more unified agricultural extension system. Another issue is whether the extension system should include socioeconomic and sustainable development messages.

Geographical Coverage

Geographical coverage can be an important policy issue because of both political and cost implications. Most political leaders want their jurisdiction to be covered by an effective extension service; therefore, they must find a way to provide funds for extension programmes. If extension funding is to be provided by different levels of government (cost sharing), then the structure of extension must reflect these different sources of funding. Extension personnel will tend to be more responsible to those levels of government that provide extension funding. For example, if local governmental units provide some extension funding, then extension personnel will tend to be more responsive to the needs of farmers and political leaders within these local government units than they are if all funding comes from the national government. In short, having multiple sources of funding, especially from different levels of government, will increase the number of shareholders and result in an extension system that has a broader base of support and that is more responsive to stakeholders at the local level.

On purely economic grounds, some economists believe that agricultural extension should be concentrated in those agricultural areas that are well endowed in terms of both human and natural resources and where the rural infrastructure is already developed. However, to concentrate extension resources on larger, better educated, commercial farmers who frequently control the best land resources in a country will not lead to broad-based agricultural development. Furthermore, the use of only economic criteria in allocating extension resources may result in further degradation of soil and water resources as resource poor farmers continue to exploit marginal land without using appropriate farming practices. Finally, investing in resource poor farm families may increase their technical, management, and leadership skills, thereby enabling them or their children to move into higher paying, nonfarm jobs.

Clientele or Target Beneficiaries

A common criticism of extension services in developing countries is their neglect of the vast number of small-scale farmers in favour of fewer numbers of large farmers, or the very limited attention given to women farmers. This is a policy issue because of its implications for the mission and goals of extension, the priorities for technology generation by research, the cost-effectiveness of extension, and the socio-political goals of growth with equity and poverty alleviation.

The inclusion of women and rural youth in agricultural extension programmes is generally recognized in terms of their numbers and contribution to farming. Worldwide, an estimated 51 per cent of the active population in agriculture are women. In Africa, women's participation in food production is as high as 76 per cent (FAO, 1990) in some areas. In 1995, there were an estimated 1.5 billion rural young people between 15 and 29 years of age, 1.3 billion of them in less developed countries. Given that rural youth may account for up to 60 per cent of the population in developing countries, should they be specially recognized for their crucial role in achieving sustainable agricultural and rural development across the coming generation of farmers?

Organizational Issues

The extension organization embodies different aspects of an extension system, and it provides the management framework for the extension service. This is a policy issue because it affects the scope, magnitude, and structure of the extension system, including factors such as control, cost-effectiveness, and the impact of the extension service. For purposes of illustration, four different forms of extension organization are discussed:

1. *Centralized organization.* Examples include the Department of Agricultural Extension in Thailand and Bangladesh, the Agricultural Extension Bureau of South Korea, and AGRITEX in Zimbabwe. In this form of organization, the national extension office manages and controls extension programme activities and resources at the regional, district, sub district, and village level. Clientele participation and feedback in programme planning are generally limited.
2. *Decentralized organization.* Examples of this form of extension organization are the agricultural extension systems in Brazil, Canada, India, Nigeria, and the Philippines. These systems have almost an invisible national or federal extension office, in that extension programming, management, and the control of activities and resources are vested with state or provincial governments.
3. *Cooperative type of extension organization and funding.* The distinguishing feature of this form of extension organization is the cooperation or partnership between the national, state or provincial, and local governments in funding, programming, and managing the activities and resources of extension. In the United States, extension is a joint undertaking of the U.S. Department of Agriculture (Federal Extension

Service), the state land-grant universities, and the county governments. In Japan, extension is a joint undertaking of the national government and the prefectural government. In China, agricultural extension is a cooperative undertaking of the central, provincial, prefecture, and county governments. Cooperative programming, management, and support are demonstrated at the County Agro-Technical Extension Centre (CATEC), where normally 20 per cent of funding comes from the central government, 30 per cent from the provincial government, and 50 per cent from the county government.

4. *Pluralistic forms of a national extension system.* This is an emerging form of extension organization in many countries, but it is not yet reflected in national extension policy. This structure appears to occur in those countries where the need for extension services is widespread and/or where the public agricultural extension organization can no longer satisfy its clientele because of resource and management problems. As a consequence, many publicly and/or privately funded organizations, including nongovernmental organizations (NGOs), are beginning to conduct agricultural extension programmes. Publicly funded extension organizations may include the crop, livestock, and horticulture departments of the ministry of agriculture, state-funded agricultural colleges and universities, and commodity boards. Privately funded organizations may include rural development-oriented NGOs, agro-business firms (contract extension), and farmer organizations, including cooperatives and commodity associations. Generally, the geographical, subject-matter, and clientele coverage and the standard of work for each of these different organizations are not known. Also, these separate efforts are generally not well coordinated.

Extension Staffing Issues

By the nature of the mission and work that an extension system carries out, its worth to society is largely reflected by the quality and number of the technical and professional staff in the organization. For a national programme of extension, the human resource question that policy makers and extension managers are confronted with is: Given the mission, scope of the work, and available resources, what type of qualifications and how many extension staff should be employed by the extension system? Part of this staff in gmatrix includes other questions: What should be the proportion of subject-matter specialists to field extension workers? What should be the proportion of field extension personnel to the number of farmers, farm households, or other target groups? How should extension staff be deployed, how often should they be transferred, and what incentives should be provided in order to ensure that they work closely with all groups of farmers?

Extension Funding

The most difficult and challenging policy issue facing extension today is to secure a stable source of funding. With the widespread trend to cut government budgets, including structural adjustment programmes, many policy makers have the impression that public extension is both expensive and a drain on the government's limited resources. At the same time, studies carried out in both developed and developing countries indicate that the returns to extension expenditures are high. Therefore, policy makers should examine this issue carefully in deciding what level of public funding is necessary to support extension in relation to the needs of farmers in the country.

While extension managers should look for ways to improve the efficiency and to reduce the cost of public extension (Wilson in Rivera and Gustafson, 1991, and Gustafson in Rivera and Gustafson, 1991), at the same time policy makers should look for ways to increase extension funding to adequate levels of support. In most developing countries, particularly the low-income, food-deficit countries, absolute levels of extension funding are very low. The Global Consultation on Agricultural Extension noted that "the current level of funding for extension in most developing countries is insufficient to provide adequate coverage for all groups of farmers, especially those who are resource poor and at the subsistence level. In countries where funding support to extension is low, the funding for extension should be increased to levels that reflect the anticipated economic rates of return and the social benefits when public funds are properly invested and managed (Alexandratos, 1995; Evenson in Jones, 1987; Pavelia in FES, 1974).

Stability

A good extension policy promotes extension system stability, yet allows sufficient flexibility to reflect the dynamic nature of the agricultural sector. Extension should not be rigid; rather, "It should be responsive to all major groups of farm people and sufficiently inclusive to allow public, private, and non-governmental organizations to contribute fully to the agricultural development goals of the country" (Swanson, 1990). Frequent organizational changes within extension, such as being transferred from one government agency to another, directly impact the organization's effectiveness. Such instability is costly in that trained staffs are poorly utilized and opportunities for improved productivity are forgone. Extension policies in some countries have been successful in preventing disruptive and destabilizing change. For more than 80 years, the U.S. has followed, with some flexibility, the 1914 Cooperative Extension Service law. For almost 50 years, Japan has followed its extension policy, and Thailand has successfully followed its extension policy for the past 40 years. In these countries, agricultural extension is recognized as having contributed significantly to increased agricultural productivity and development.

Extension Policy Formulation

There is no standard formula to be used in formulating agricultural extension policies. It should be noted, however, that most existing laws and policies on extension have been formulated by planners and policy makers in the ministry of agriculture and agriculture committees in the legislative branch of government. Normally, agricultural extension professionals from agricultural universities or from abroad are called on to provide advice and to assist in drafting extension legislation.

A congressional hearing is normally conducted before extension legislation is finally enacted into law.

Improving the Organization and Management of Extension

Managing people effectively in extension programmes is a skill that requires constant planning and development. An extension programme manager can be defined as the person who is vested with formal authority over an organization or one of its sub units. He or she has status that leads to various interpersonal relations, and from this comes access to information. Information, in turn, enables the manager to devise strategies, make decisions, and implement action (Mintzberg, 1988). Management is concerned with the optimum attainment of organizational goals and objectives with and through other people. Extension management organizations are characterized by many strategies, wide spans of control, democracy, and autonomy. Their management practices cannot be reduced to one standard set of operating guidelines that will work for all organizations continually. However, all managers of professional organizations face the same challenge: to manage one's time, objectives, and resources in order to accomplish tasks and implement ideas (Waldron, 1994).

Managers of extension programmes are painfully aware of the need for revision and development of the new skill sets held by today's high performers. They know that a new focus on connecting the heads, hearts, and hands of people in their organization is necessary. Astute managers know what needs to be done but struggle with how to do it. Quite often they prefer to consider themselves as teachers or communicators rather than managers. This results in under-utilization of the increasing amount of literature on management theory and practice. The root of the problem is implementation. They must learn how to motivate others and build an efficient team. More formally defined, management is the process by which people, technology, job tasks, and other resources are combined and coordinated so as to effectively achieve organizational objectives. A process or function is a group of related activities contributing to a larger action. Management functions are based on a common philosophy and approach. They centre around the following:

a) Developing and clarifying mission, policies, and objectives of the agency or organization
b) Establishing formal and informal organizational structures as a means of delegating authority and sharing responsibilities
c) Setting priorities and reviewing and revising objectives in terms of changing demands
d) Maintaining effective communications within the working group, with other groups, and with the larger community
e) Selecting, motivating, training, and appraising staff
f) Securing funds and managing budgets; evaluating accomplishments and
g) Being accountable to staff, the larger enterprise, and to the community at large (Waldron, 1994b).

The management functions listed above can be categorized by using the acronym POSDCORB (Bonoma and Slevin, 1978, from Gulick and Urwick, 1959):

- **Planning:** outlining philosophy, policy, objectives, and resultant things to be accomplished, and the techniques for accomplishment
- **Organizing:** establishing structures and systems through which activities are arranged, defined, and coordinated in terms of some specific objectives
- **Staffing:** fulfilling the personnel function, which includes selecting and training staff and maintaining favourable work conditions
- **Directing:** making decisions, embodying decisions in instructions, and serving as the leader of the enterprise
- **Coordinating:** interrelating the various parts of the work
- **Reporting:** keeping those to whom you are responsible, including both staff and public, informed
- **Budgeting:** making financial plans, maintaining accounting and management control of revenue, and keeping costs in line with objectives

Planning

Planning is the key management function of any extension worker. It is the process of determining in advance what should be accomplished, when, by whom, how, and at what cost. Regardless of whether it is planning long-term program priorities or planning a two-hour meeting, the planning aspect of management is the major contributor to success and productivity. Stated simply, "If you don't know where you are going, then you won't know when you have arrived!" Planning is the process of determining the organization's goals and objectives and making the provisions for their achievement. It involves choosing a course of action from available alternatives. Planning is the process of determining organizational aims, developing premises about the current environment, selecting the course of action, initiating activities required to transform plans into action, and evaluating the outcome. The types of planning that managers engage in will depend on their level in the organization and on the size and type of the organization.

- Generally there are four major types of planning exercises: strategic, tactical, contingency, and managerial.
- Strategic planning involves determining organizational goals and how to achieve them. This usually occurs at the top management level.
- Tactical planning is concerned with implementing the strategic plans and involves middle and lower management.
- Contingency planning anticipates possible problems or changes that may occur in the future and prepares to deal with them effectively as they arise (Marshall, 1992).
- Managerial planning is usually considered as microlevel planning. It helps in combining resources to fulfil the overall objectives of the extension organization.

Long-Range Planning

Long-range planning is vitally important in that it focuses attention on crucial future issues which are vitally important to the organization. It involves studying societal trends and issues, surveying current and anticipated learners' needs, and being aware of long-term research directions and changes in technology. Many extension workers may think that such management is beyond their level of authority, control, or involvement. They may feel that such management is the prerogative of the director, the deputy minister, or the president. However, while senior levels of management must be involved, those who implement the objectives resulting from long-range planning should also be involved.

Strategic Planning

Strategic planning has been defined as that which has to do with determining the basic objectives of an organization and allocating resources to their accomplishment. A strategy determines the direction in which an organization needs to move to fulfil its mission. A strategic plan acts as a road map for carrying out the strategy and achieving long-term results. Occasionally a large gap exists between the strategic plan and real results. To boost organizational performance, people must be a key part of the strategy. A stronger, more capable and efficient organization can arise by defining how its members can support the overall strategy. Strategic planning is different from long-term planning. Long-range planning builds on current goals and practices and proposes modifications for the future. Strategic planning, however, considers changes or anticipated changes in the environment that suggest more radical moves away from current practices. When doing strategic planning, the organization should emphasize team planning. By involving those affected by the plan, the manger builds an organization wide understanding and commitment to the strategic plan (Flemming, 1989). The elements of strategic plans include:

- ☆ Organization mission statement – What
- ☆ Strategic analysis – Why
- ☆ Strategic formulation - Where
- ☆ Long-term objectives implementation - When and How
- ☆ Operational plans - When and How

Changes are essential to better position the extension organization and focus on client needs and moving forward in rural development and sustainability programmes. The strength and resilience of the traditional rural and farm population and the trend towards a decentralized society with more and more urbanites moving to the country suggest that successful rural communities will depend on people's ability to change, to adapt, and to work toward a better future. In the 1990s, facilitating farmer participation is a major extension activity (Chambers, 1993). Updated plans are then implemented by work teams at all levels of management. Work-team objectives include:

- ☆ Involving all levels of staff in consultation

- Designing and implementing a process to develop-goals and objectives for the organization and unit; a strategic process for the next five to ten years
- Defining and clarifying organizational structures and identifying functions, customers, and service delivery models
- Identifying changes and staged approaches needed to move from the current situation to what will be required over the next three to five years
- Identifying and recommending priorities for policy and programme development
- Incorporating goals for expenditure reduction, service quality improvement, workforce management, accountability, technology, and business process improvement
- Stating the start date and first report date

Managerial Planning

If long-range planning can be linked to "macro," then managerial planning can be linked to "micro." Managerial planning is the implementing of the strategic plan; it is the combining of resources to fulfil the overall objectives and missions of the organization. Managerial planning focuses on the activity of a specific unit and involves what needs to be done, by whom, when, and at what cost. The strategic planning process serves as an umbrella over the management planning process which deals with the following:

- Establishing individual goals and objectives
- Forecasting results and potential problems
- Developing alternatives, selecting alternatives, and setting priorities
- Developing associated budgets
- Establishing personnel inputs
- Establishing specific policies related to the unit
- Allocating physical resources
- Appraising how the management unit has succeeded in meeting its goals and objectives

Decision Making

Closely related to both strategic and managerial planning is the process of decision making. Decisions need to be made wisely under varying circumstances with different amounts of knowledge about alternatives and consequences. Decisions are concerned with the future and may be made under conditions of certainty, conditions of risk, or conditions of uncertainty. Under conditions of certainty, managers have sufficient or complete information and know exactly what the outcome of their decision will be. Managers are usually faced with a less certain environment. They may, however, know the probabilities and possible outcomes of their decisions, even though they cannot guarantee which particular outcome will actually occur. In such cases, there is a risk associated with the decision and there is a possibility of an adverse outcome. Most managerial decisions involve varying degrees of uncertainty. This is a key part of a manager's activities. They must decide what goals or opportunities will be pursued, what resources are available, and who will perform designated tasks. Decision making, in this context, is more than making up your mind. It consists of several steps:

Step 1: Identifying and defining the problem

Step 2: Developing various alternatives

Step 3: Evaluating alternatives

Step 4: Selecting an alternative

Step 5: Implementing the alternative

Step 6: Evaluating both the actual decision and the decision-making process

Managers have to vary their approach to decision making, depending on the particular situation and person or people involved. The above steps are not a fixed procedure, however; they are more a process, a system, or an approach. They force one to realize that there are usually alternatives and that one should not be pressured into making a quick decision without looking at the implications. This is especially true in the case of non programmed decisions (complex and novel decisions) as contrasted to programmed decisions (those that are repetitive and routine).

One of the most difficult steps in the decision-making process is to develop the various alternatives. For example, if one is involved in planning a workshop, one of the most crucial decisions is the time, format, and location of the workshop. In this case, one's experience as well as one's understanding of the clientele group greatly influences the selecting of alternatives. Often decision trees can help a manager make a series of decisions involving uncertain events. A decision tree is a device that displays graphically the various actions that a manager can take and shows how those actions will relate to the attainment of future events. Each branch represents an alternative course of action. To make a decision tree it is necessary to:

1. Identify the points of decision and alternatives available at each point,
2. Identify the points of uncertainty and the type or range of alternative outcomes at each point,
3. Estimate the probabilities of different events or results of action and the costs and gains associated with these actions, and
4. Analyse the alternative values to choose the next course of action.

In extension, the decision-making process is often a group process. Consequently, the manager must apply principles of democratic decision making since those involved in the decision-making process will feel an interest in the results of the process. In such a case, the manager becomes more of a coach, knowing the mission, objectives, and the process, but involving those players who must help in actually achieving the goal. The effective manager thus perceives himself or herself as the controller of the decision-making process rather than as the maker of the organization's or agency's decision. As Drucker (1966) has pointed out, "The most common source of mistakes in management decision-making is the emphasis on finding the right answer rather than the right question. It is not enough to find the right answer; more important and more difficult is to make effective the course of action decided upon. Management is not concerned with knowledge for its own sake; it is concerned with performance."

Organizing

Once strategic planning and management planning are implemented, organizing to get the job done is next. Organizing is the process of establishing formal relationships among people and resources in order to reach specific goals and objectives. The process, according to Marshall (1992), is based on five organizing principles: unity of command, span of control, delegation of authority, homogeneous assignment, and flexibility. The organizing process involves five steps:

- ✰ Determining the tasks to be accomplished,
- ✰ Subdividing major tasks into individual activities,
- ✰ Assigning specific activities to individuals,
- ✰ Providing necessary resources, and
- ✰ Designing the organizational relationships needed.

In any organizing effort, managers must choose an appropriate structure. Organizational structure is represented primarily by an organizational chart. It specifies who is to do what and how it will be accomplished. The organizing stage provides directions for achieving the planning results. There are several aspects to organizing - time, structures, chain of command, degree of centralization, and role specification.

Time Management

Managers must decide what to do, when, where, how, and by or with whom. Time management is the process of monitoring, analysing, and revising your plan until it works. Effective planning is a skill that takes time to acquire. It is difficult to implement because you have no one but yourself to monitor how effectively you are using your time. Everyone has the same amount of time - 168 hours per week. How that time is managed is up to the discretion of each person. One extension agent joked that he was so busy taking time management courses, he had little time left to manage. Effective time management involves philosophy and common sense. Time is not a renewable resource - once it is gone, it is gone forever. To function effectively, managers have to be able to prioritize and replace less important tasks with more important ones. Most of us work for pay for only 1,800 hours per year. Effective and efficient time management encourages us to achieve and be productive while developing good employee relations.

Goals should be Specific, Measurable, Attainable, Realistic, and Timely (SMART). Once the goals are known, it is important to think about how they can be achieved. Effective time managers facilitate planning by listing tasks that require their attention, estimating the amount of time each task will take to complete, and prioritizing them - deciding what tasks are most important to do first and numbering them in rank order. It is essential to know what is crucial and what is not. Some activities have relatively low levels of importance in completing a given task. By planning ahead, managers can decide what to do and take the time to come up with ideas on how to do it. They can make their own list of steps to eliminate or reduce time wasters. Maintaining a daily "To Do" list with priorities attached and maintaining a daily, weekly, monthly, and yearly diary is helpful. Managers should analyse their daily activities to see which are directed toward results and which are simply activities. They could learn how to manage meetings more effectively since considerable management time seems to be wasted in non-directional formal meetings.

One of the methods that helps allocate time according to priorities is Pareto's Law or the 80/20 rule: if all items are arranged in order of value, 80 per cent of the value will come from 20 per cent of the items. For example, 80 per cent of the complaining in your department is likely to be done by 20 per cent of your staff. Four suggestions for better time management are (1) never handle the same piece of paper twice; (2) learn how to say "no" without feeling guilty about requests that do not contribute to the achievement of your goals; (3) when a visitor drops in to your office, stand up while you have your discussion to ensure that only a brief period of time will be consumed by the visitor's interruption; and (4) avoid being a slave to the telephone. By managing time well, managers are better able to solve problems quickly, make decisions, avoid frustration, keep from getting bogged down in day-to-day tasks, handle crises, work on their goals and priorities, and manage stress. Guidelines for scheduling time include:

- Always put your schedule in writing.
- Focus on the objectives you are trying to accomplish.
- Continually review objectives, priorities, and scheduled actions to keep on track.
- Schedule around key events and actions.
- Get a productive start by scheduling early-day actions.
- Group related items and actions whenever possible.
- Do not hesitate to take large time blocks for important tasks.
- Be sure to allow enough time for each task, but not too much time.
- Build in flexibility for unexpected events.
- Include some thinking time for yourself.
- Consider how to make waiting and travel time useful or otherwise productive.

- ☆ Try to match your work cycles to your body cycles.
- ☆ Learn to control your unscheduled action impulses.
- ☆ Prepare tomorrow's schedule before you get to the office in the morning.

Structures: Centralized versus Decentralized, Line versus Staff

Working productively and developing feelings of cooperation and effectiveness are related to having the right people doing the right jobs. Structure, then, can be defined as a system of interrelated jobs, groups of jobs, and authority. There is no standard organizational structure, but most organizations and agencies follow the "Christmas Tree" system with the star (*e.g.*, president, minister) at the top, smaller branches at management levels, and bigger branches at the production levels. Some would claim that the lower branches support the upper branches, but as in the tree, the branches are supported by a single trunk, which can be thought of as the organizational mission and objectives. Each part of the tree has its specific function. When all parts work together, the system survives, functions productively, has balance, and is a pleasure to see! There are four primary elements in designing an organizational structure:

- ☆ Job specifications - what each division/office/unit is responsible for
- ☆ Departmentalization - the grouping of jobs and responsibilities in common sectors with the objective of achieving coordination
- ☆ Span of control - a definition of how many job roles should be in each unit and which roles require coordination by a unit manager
- ☆ Delegation of authority - assigning the right to make decisions without having to obtain approval from a supervisor

Chain of Command

Once an organization starts delegating authority, then there is automatically a chain of command, "the formal channel which specifies the authority, responsibility and communication relationships from top to bottom in an organization" (Ivancevich, Donnelly, and Gibson, 1980). Thus authority flows from presidents to vice-presidents to divisional managers, from ministers to deputies to directors, from principals to vice-principals to deans, *etc.* In complex organizations, there may be bridges from one level to another and there will be complex procedures for maintaining the chain of command. Adult and extension educators, if working for an organization or agency, will be part of a structure and part of the chain of command. One cannot often make major changes in these two elements; it is wise, however, to be very aware of the organizational structure and chain of command if you wish to accomplish things efficiently.

Centralized organizations are those in which the key authority and decision-making role is focused on one or a very few individuals. Where authority is distributed among many managers, then one can see a decentralized structure. As the organization's various roles become more diverse in terms of programme, product, or geographical location, one can see a more decentralized organizational structure with authority being delegated to those who are closest to the action. Centralization refers to authority, whereas centrality refers to the proximity to the organization's stated mandate and objectives. One could have a much decentralized organization with each unit being responsible for programmes, staffing, and budget, and yet be very close to the main mission and objectives of the organization.

Another important point in terms of structure is the concept of line and staff functions. Line functions are those involved in creating, developing, and delivering a programme. Staff functions are those that are of an advisory and consultative order. Line functions contribute directly to the attainment of the organization's objectives, and staff functions contribute indirectly.

Staffing

A key aspect of managing an adult and extension enterprise is to find the right people for the right jobs. Much of one's success as a manager is related to appropriate human resource planning, regardless of whether it is the hiring of a secretary or an instructor for a particular work-shop. The staffing function consists of several elements:

- Human resource planning - how many staff resources, with what backgrounds, and at what cost can be considered for objectives implementation?
- Recruitment - how does one proceed to find the person with the appropriate mix of education, experience, human relations skills, communications skills, and motivation? An important component of the recruitment process is writing the job description. The nature of the job, scope, authority, and responsibilities form the core of the job description. Indications of preferred educational background as well as salary range must also be included. In times of high unemployment, one can always expect several dozen applications for any one opportunity for employment. This leads to the next task of staff selection.

Staff Selection

The process of staff selection involves evaluating candidates through application forms, curriculum vitae, and interviews and choosing the best candidate for the specific job responsibility. One can even have a list of criteria and a score sheet for each individual. Even then, successful hiring is often a very intuitive act and involves some degree of risk.

As a means of giving some structure and design to the staffing process, the following guidelines are useful (dark, 1973). Each job interview should be characterized by:

a) A clear definition of the purpose of the interview
b) The presence of a structure or general plan
c) The use of the interaction as a learning experience in a pleasant and stimulating atmosphere

d) The creation and maintenance of rapport between the interviewer and interviewee
e) The establishment of mutual confidence
f) Respect for the interviewee's interest and individuality by the interviewer
g) An effort to put the interviewee at ease
h) The establishment and maintenance of good communication
i) The willingness to treat what is being said in proper perspective
j) The just treatment of each interviewee

Staff Orientation

This is the process of formally introducing the selected individual to the particular unit, to colleagues, and to the organization. The selected person should be aware of the mission and objectives of the unit, the nature of responsibilities and level of authority, the degree of accountability, and the systems and procedures followed to accomplish the tasks associated with the job. A motivated individual will simply ask for such things as personnel manuals, administrative procedures handbooks, and aims and objectives statements. Such orientation tools should be available.

Directing

At one time there was a management emphasis on "directing" in the directorial (autocratic) sense, but in recent times, the concept of directing has become more congruent with leading than with pushing. Thus today, directing is more related to leading and leadership styles. Leadership in this context means the process whereby a work environment is created in which people can do their best work and feel a proprietary interest in producing a quality product or service.

McGregor (1960) proposed that managers might assume that employees are motivated in one of two ways. His dichotomy was labelled theory X and theory Y. Theory X relates to traditional management whereby managers assume that they must control, coerce, and threaten in order to motivate employees. Theory Y, the opposite of theory X, suggests that employees want to do challenging work, that they are interested in accepting responsibility, and that they are basically creative and want to be involved in policy development and objective setting. Today, theories X and Y don't really sound very revolutionary; the problem is that management styles and employee motivation do not fit easily into two theoretical labels. As a result, additional theoretical labels (*e.g.*, theory Z) are being developed.

Coordinating

This important stage consists of interrelating the various parts of the work. It involves coordinating the various job roles and responsibilities of yourself and other staff, of your unit and other units within the same organization, and of your unit with the broader community.

There are two forms of coordination:

1. Vertical reporting to your supervisor(s) and to your staff, and
2. Horizontal reporting to your colleagues and your management team.

Adult and extension educators are usually involved in very complex organizations such as governments, colleges and universities, and boards of education. Because of the size of the organization, the increasing demands for public accountability, the many government regulations and policies, the increasing competition among providers of adult education opportunities, and the changes in technology, it is essential that the coordinating role be given top priority.

How, then, can effective coordination be accomplished?

- ☆ Coordination needs professional, competent leadership, a democratic style that leads to trust, open communication, and ease of information flow.
- ☆ Coordination needs a constant definition and communication of mission and objectives that are understood by all managers.
- ☆ Coordination, to be effective, must have open, two-way channels of communication.
- ☆ Coordination involves a sharing atmosphere as well as commonly agreed on direction. Because effective coordination requires cooperation and communication, the meeting technique is still the most effective format for assuring the interrelationships among the various job responsibilities.

Reporting

This function, closely related to the coordinating function, consists of keeping those to whom you are responsible informed as to what is going on. It is essential that competent managers keep the information flowing, especially in this age when there is so much information being transmitted in so many forms. The reporting function is more than preparing an annual report, quoting statistics, and informing your staff of current developments. The reporting function is almost an evaluation function since it compares how you are doing with what you set out to do. It reviews your objectives and determines to what extent you are meeting your objectives. It consists of more than course numbers or annual statistics, but relates programme direction, policy changes, refinement in objectives, and changes in structures and priorities. It also uses the vertical and horizontal flows of information as presented previously. One of the key elements of the reporting function is the annual report. Such a report gives you the opportunity to summarize programmes, projects, and activities and to provide statistics as well. Such a report can be used as a public information document by having it distributed to other adult education agencies in the community, to your senior levels of management, to your own managers, to your colleagues, and to the press.

Budgeting

This management function includes fiscal planning, accounting and revenue, and expense controls. Budgeting requires specific planning, a thorough understanding of objectives and future programmes, a sixth sense of economic conditions and realities, and a hunch for predicting the unpredictable.

In many cases, an organization specifies the budget system being used. It could be based on:

a) Historical data (what you had last year with variations for the coming year);
b) O-based data where the budget is created and justified on a line-item basis according to programmes and priorities;
c) An MBO system - management by objectives whereby specific objectives are funded; and
d) A PERT system - programme review and evaluation technique - where each programme is reviewed and assessed according to its contribution to specific goals.

These are only a few of the budgeting systems in use. However, the key elements of any budget system consist of:

a) Determining what line items are necessary in terms of objectives;
b) In line with policies, determining the financial amounts for each line;
c) Determining overhead, surplus, and/or profit margins;
d) Determining anticipated revenue from fees, grants, gifts, contracts, *etc.*;
e) Drafting a budget with specific amounts and justifications; and
f) Discussing and making adjustments to produce a working budget.

The budget then becomes a guide which, however, may always be in a state of change. The budget process is not in a vertical something that one does only once a year; it is a continual process of regular review and possible revision. One should always be checking to see how one is doing compared with how one anticipated doing.

Budget management, then, consists of three parts:

a) Budget determination - allocating revenue according to priorities and by line items;
b) Budget accountability - how well the anticipated budget matches reality; and
c) Using a +, 0 - notation in answering the questions and by placing the notations in the boxes on the chart.

In this way, one can get a picture of the predominant types of management modes currently being used. While this may be useful in describing what is, it could be even more useful in describing what could be. It is also useful in providing some clues as to possible areas of role conflict - the scholarly research model would likely collide with the competent practitioner model (Waldron, 1994a).

Systems Theory

Modern management is characterized by two approaches, the systems and the contingency approach. The systems approach views the organization as a total system comprised of interacting subsystems, all of which are in complex interaction with the relevant external environment (Lerman and Turner, 1992). Organizations are pictured as "input-transformation-output systems" that compete for resources. The survival and prosperity of an organization depend on effective adaptation to the environment, which means identifying a good strategy for marketing its outputs (products and services), obtaining necessary resources, and dealing with external threats.

Survival and prosperity also depend on the efficiency of the transformation process used by the organization to produce its goods and services, on worker motivation, and on cooperation. Efficiency of the transformation process is increased by finding more rational ways to organize and perform the work and by deciding how to make the best use of available technology, resources, and personnel. Top management has primary responsibility for designing an appropriate organizational structure, determining authority relationships, and coordinating operations across specialized subunits of the organization (Yuki, 1994). A system can survive only when it delivers an output that can be exchanged for new inputs as well as for maintaining the system. Similarly, an extension service is expected to produce some beneficial output.

Criteria for Model Classification (Waldron, 1994a)

Perceived Status

1. Is the unit headed by a dean?
2. Does the chief manager have tenure?
3. Is the boss a professor?
4. Does the unit have representation on senate, governing council, or the key academic decision-making body
5. Are the employees expected to do research and publish in journals? 6 Is the dean or director expected to have a doctoral degree?

Mission, Mandate, Objectives

a) What model would best apply to your mission, mandate, and objectives?
b) Is your unit perceived as a money maker?
c) Do your objectives give priority to research, innovation, product development, or product delivery?

Reward System

Ways of rewarding you for your effort:

1. Increments negotiated by a union or association

2. Salary negotiated with a senior administrator
3. Bonuses based on meeting targets
4. Ratings by colleagues

Decision-making Authority

1. Does your unit decide on what courses and programmes need to be delivered?
2. Does your unit determine course content?
3. Can your unit allocate and reallocate budgets?
4. Does your unit have a unit-controlled development fund?
5. How independent do you think the unit is in terms of decision making?

Market Orientation

1. Is your unit market driven?
2. Do you publish a market-oriented course calendar?
3. Do you use charge cards for tuition fee payment?
4. Do you publish regular press releases and media information?
5. Do you do market research and needs assessments?

Accountability

1. Do you report to a president or vice president-academic?
2. Do you present proposed policy to a senate or governing council?
3. Do you make use of a community advisory council?
4. To whom would you send an annual report?

Contingency Theory

Theories that explain management effectiveness in terms of situational moderator variables are called contingency theories. The contingency or situational approach recognizes that neither the democratic nor the autocratic extreme is effective in all extension management situations. Different traits are required in different situations. Table describes the major features of five contingency theories and the Vroom and Yetton (1973) normative decision model. The table makes it easier to compare the theories with respect to content and validation. A synopsis of each is presented below:

- ✫ The path-goal theory examines how four aspects of behaviour influence subordinate satisfaction and motivation.
- ✫ Leadership substitute theory identifies aspects of the situation that make leadership behaviour redundant or irrelevant.
- ✫ The multiple linkage model uses a model of group performance with six intervening variables to explain leadership effectiveness.

- Fiddler's LPC model deals with the moderating influence of three situational variables on the leadership between a leader trait (LPC) and subordinate performance.
- Cognitive resources theory examines the conditions under which cognitive resources such as intelligence, experience, and technical expertise are related to group performance.

Comparison of Six Contingency Theories (*Sources*: Yuki, 1994)

Contingency Theory	*Leader Traits*	*Leader Behavior*	*Situational Variables*	*Intervening Variables*	*Validation Results*
Path-Goal Theory	None	Instrumental, Supportive, Participative, Achievement	Many aspects	Expectancies, Valences, Role Ambiguity	Many Studies, some support
Leadership substitutes Theory	None	Instrumental, Supportive	Many aspects	None	Few Studies, inconclusive
Multiple linkage model	None	Many aspects	Many aspects	Effort, ability organization, teamwork, resources, external coordination	Few Studies, inconclusive
LPC Contingency Model	LPC	None	Task Structure, L-M Relations	None	Many studies, some support
Cognitive Resource Theory	Intelligence, Experience	Directive	Stress Group Support	None	Few studies, some support
Normative Decision Theory	None	Decision Procedures	Many aspects	Decision Quality and acceptance	Many studies, moderate support

7

Monitoring and Evaluation for Extension Programmes

Zones of Concentration Approach

The Cernea and Tepping approach, introduced in 1977, concentrates on three zones:

a) Visits, as the final outcome of extension efforts;
b) Recommendations, as the content of the visit and means towards the end yields; and
c) Yields, as the eventual consequence of the development effort (Cernea and Tepping, 1977).

Methodological Approach

The Slade and Feder approach, introduced in 1981, builds upon the zones-of-concentration approach and suggests a monitoring survey early in each cropping season, a monitoring-cum-evaluation survey in each cropping season at the time of harvest, specific indicators, and reporting. Working manuals are a characteristic feature of this approach (Slade and Feder, 1985).

Expanded Monitoring Approach

Casley and Kumar suggested in 1987 an expansion of the monitoring function to cover not only physical and financial information, but also beneficiary contact information and project diagnostic studies. Under this approach, there is greater emphasis on monitoring and less on evaluation. Project diagnostic studies are a novel feature of this approach.

Adoption Rates Approach

Murphy and Marchant suggested in 1988 an approach which concentrates on adoption rates as key indicators. This approach moves away from trying to monitor agricultural results and concentrates on directly monitoring the provision of and response to project services. Under this approach, extension service is treated as the "medium" and the recommendations as the "message."

Marketing Approach

Lee (1990) has suggested an approach which is based on market segmentation, a standard technique in marketing. Under this approach, the need and likely demand for new technology are first assessed, and then target market segments are predicted. For example, 10 per cent of the targeted farmers are likely to readily adopt a new technology; a further 15 per cent are likely to adopt the technology when returns for wheat reach x dollars per ton; 40 per cent of the targeted farmers are likely to adopt the technology when they perceive that their peers have adopted it; and 35 per cent of the target market is not likely to adopt the technology at all in the next five years.

Operationalizing the Definition of Monitoring

Monitoring is a specialized, dynamic, semiautonomous, and institutionalized management resource. Preferably it is computerized. Monitoring helps to ensure the implementation of extension programmes in accordance with their design and takes into account the interests of various stakeholders. The definition of monitoring can be operationalized by establishing principles to follow; setting up a data collection system; establishing relationships among the monitoring unit, management, the extension staff, and extension's clients (farmers); and making appropriate use of non-governmental organizations (NGOs).

Principles of Monitoring

In light of past experience in monitoring and its designated role, it is possible to lay down the following ten principles of monitoring:

a) Monitoring must be Simple

A complex or complicated monitoring system is self-defeating. The basic task of monitoring is to simplify the field-level complexity, sifting the more important concerns from the less important.

b) Monitoring must be Timely

Timeliness is of the essence in monitoring. Management requires input from the monitoring system so that timely action may be taken. Also, timeliness is closely related to the credibility of monitoring itself.

c) Monitoring must be Relevant

It must concern itself only with parameters which are relevant to programme objectives. This also ensures that monitoring does not generate information that is not used or is not usable by management.

d) Information Provided through Monitoring should be Dependable

Management will rely on monitoring findings only if the information is believed to be reasonably accurate.

e) Monitoring Efforts should be Participatory

Effort should be made to ensure participation by all concerned with extension, be they field-level personnel, subject-matter specialists, or extension's clients (the farmers).

f) Monitoring must be Flexible

It is iterative in nature. It also gets routinized with the passage of time. These two features should not, however, lead to rigidity.

g) Monitoring should be Action Oriented

Monitoring often leads to action. Consequently, it should follow pragmatic approaches, keeping the requirements of extension's clients uppermost in view. Generating information for which there is no intended use should be assiduously avoided.

h) Monitoring must be cost-effective.

Monitoring efforts cost money and time. It is therefore essential to make it cost-effective. While principles of simplicity, time-lines, relevance, and accuracy will lead to cost-effectiveness, computerization also can help to make monitoring more cost-effective by reducing staff hours in data processing.

i) Monitoring Efforts should be Top Management Oriented

Monitoring units should keep in mind the requirements of top management when designing and operating a monitoring system. Yet at the same time, monitoring must take into account the fact that those who provide information to the system also must benefit or the quality of the information provided will decline.

j) Monitoring Units represent Specialised Undertakings

Monitoring is not merely concerned with the collection and analysis of data, but with diagnosing problems and suggesting alternative practical solutions.

Organized Relationship in Monitoring and Evaluation Due to Management Intervention

Monitoring and its Main Stakeholders

There are many stakeholders in monitoring efforts. The funding agency, usually the ministry of finance or planning, wants to know that money has been spent properly. The ministry of agriculture wants to be able to say it has been efficient and effective. The target beneficiaries, that is, small and marginal farmers, want help from the organization and want to improve the quality of their lives. And nongovernmental organizations (NGOs) need information generated by monitoring efforts and also want to contribute information. Thus there are four main stakeholders in monitoring - funding agencies, the implementing agency, beneficiaries, and NGOs - and their interests need to be taken into consideration when designing a system of monitoring.

Frequency of Monitoring

Monitoring is an ongoing, continual exercise. Data collection involving crop production should be undertaken twice in a cropping season: at the time of sowing to obtain benchmark information and again at the time of harvest. The period of recall for collection of data should not be more than a month.

Monitoring Unit

The monitoring unit should be staffed by technical personnel having specialized skills. The staffs often consist of extension specialists, economists, sociologist or anthropologists, statisticians, computer programmers, and supporting staff. The head of the monitoring unit may come from any of these disciplines. The staff should be no larger than necessary to accomplish the work of the unit. The leader of the monitoring unit should report to one of the top managers in the organization.

Monitoring Indicators

Indicators, as the term suggests, are variables that help to measure changes in a given situation (ACC, 1984). They are tools for monitoring and evaluating the effects of an activity. Indeed, indicators are the principal means by which a monitoring unit keeps track of extension's capability, effectiveness, and efficiency. Any monitoring system will therefore incorporate the use of appropriate indicators in these four aspects.

There are two approaches to indicator development: the inductive and the deductive. In the inductive approach, a system of social, economic, and demographic statistics is created and a wide range of indicators is developed on the basis of the statistics available. This is the approach followed in the case of United Nations Social Indicators (FAO, 1988). In the deductive approach, the areas of interest are first identified, and then requisite indicators are developed. Socioeconomic indicators for the World Conference on Agrarian Reform and Social Development (WCARD) follow this approach (FAO, 1988). In extension monitoring, both the inductive and the deductive approaches are followed; that is, indicators are developed on the basis of available statistics, and some (for example, performance indicators) are developed as a result of specially collected data.

There are different types of indicators, for example, development indicators, socioeconomic indicators, agricultural development indicators, and extension indicators. They range from general to specific concerns. Indicators can be categorized into direct and indirect or proxy indicators (Clayton, 1983; ACC, 1984); single and unitary or composite indicators; quantitative and qualitative indicators; primary, core, and supplementary indicators (FAO, 1988); input and output indicators; and monitoring and evaluation indicators.

The criteria for selecting indicators depend upon the purpose, resources, and time available. The following criteria are usually suggested:

- *Simplicity*: The indicator should be simple enough to be understood by nonspecialists (FAO, 1988).
- *Unambiguous definition*: It should be clearly defined (Casley and Lury, 1982; Casley and Kumar, 1987).
- *Ready determination*: The data can be obtained without undue difficulty (WHO, 1989). This is also referred to as "timely" (ACC, 1984) and "feasible" (FAO, 1988; Gha, Hopkins, and McGranahan, 1988).
- *Accurate measurement:* The indicator should be measured accurately, which is often difficult when dealing with farming communities (Casley and Lury, 1982).
- *Validity:* The indicator should actually measure what it is supposed to measure (ACC, 1984; FAO, 1988; see also Gha *et al.*, 1988).
- *Relevance:* It should be geared to the specific needs of decision makers (Petry, 1983) and be relevant to project objectives (ACC, 1984).
- *Specifity:* It should reflect changes only in the situation concerned (WHO, 1989) and should measure specific conditions that the project aims to change (Casley and Kumar, 1987).
- *Consistency:* The value of indicators should stay constant so long as they are collected in identical conditions, no matter who does the collecting (Casley and Kumar, 1987). Indicators should be objective and verifiable (FAO, 1988).
- *Sensitivity:* Indicators should be sensitive to changes in the situation being observed (ACC, 1984). They should be sensitive enough to reflect changes in the phenomenon (FAO, 1988).
- *Prioritization:* Indicators should be prioritized and a minimum feasible list prepared (Gha *et al.*, 1988).

Capability, Effectiveness, Efficiency, and Impact

These four concepts are basic to monitoring and evaluation. They correspond, respectively, to operational investment (*e.g.*, investment in agricultural extension per farm family), operational efficiency (*e.g.*, the number of visits, meetings, demonstrations, and trials, per extension worker), technical efficiency (*e.g.*, the number of adopters, hectorage, output, and value added), and extension-induced changes (*e.g.*, production, productivity, income, and income distribution) (Ruthenburg, 1985).Capability, effectiveness, and efficiency fall in the monitoring domain. Impact falls in the evaluation domain.

Capability is the command that extension has over physical, financial, and human resources, enabling it to serve its clients (the farmers). It is reflected by extension's outreach, intensity, technical competence, and physical and financial resources. Extension performance depends directly upon its capability.

Effectiveness is defined by a handbook on productivity management as "the degree to which goals are attained" (Prokopenko, 1987). Agricultural extension has many goals such as social goals (*e.g.*, farmer welfare) and economic goals (*e.g.*, increased income). Among these, operational goals (*e.g.*, physical and financial targets) are of special significance because their attainment makes realization of other goals possible.

Efficiency in extension is usually measured by the rates at which farmers adopt recommended practices. Adoption rates of varying degrees of complexity can be conceived (Casley and Lury, 1982).

Impact in extension can be measured by a simple indicator, like yield of a crop per hectare, or by constructing simple productivity indices. Such indicators provide ultimate tests for the success of extension programmes.

Extension Monitoring Indicators

Extension monitoring indicators can be grouped into two categories: (1) extension capability indicators, and (2) extension performance indicators. Both should be generated by the monitoring unit.

Extension Capability Indicators

Extension capability indicators must be monitored regularly not only to know the status of extension's capability at a certain point in time, but also to determine changes in it over time. These indicators should be calculated annually. They involve only desk work because they are based on in-house data.

Extension Performance Indicators

Extension performance indicators reflect extension's operational and technical efficiency. They can be grouped into two categories.

Extension Effectiveness Indicator

These can again be grouped into two subcategories:

1. Single indicators and
2. Unitary or composite indicators.

By definition, a single indicator will reflect an aspect of extension performance, while a unitary or composite indicator will reflect two or more aspects of extension performance. It may be useful to construct a unitary or composite indicator to provide a consolidated view of extension effectiveness to management, because management is often interested in having an overall view of extension effectiveness. These indicators reflect extension's cooperational efficiency.

Extension Efficiency Indicators

These indicators are based on adoption rates of recommended practices and reflect extension's technical efficiency.

Extension performance indicators, as suggested here, should be calculated separately for contact farmers, other (noncontact) farmers, and all (contact and noncontact) farmers.

Extension Evaluation Indicators

Yield and productivity indices occupy the central position in these groups of indicators. They are calculated on the basis of crop-cutting experiments which are conducted by provincial revenue or economics and statistics departments.

Action Plan and Chain of Events

Monitoring requires collection of data and their analysis. For this, an action plan in a monthly, quarterly, and annual timeframe is essential. Data collections will require choosing appropriate methods, which will depend upon the monitoring unit's time and physical and financial resources such as computerization and trained technical personnel for collecting and analysing data.

Monitoring methodologies can be perceived as a chain of distinguishable and sequential events comprising (1) planning and design of the study, (2) desk research, (3) selection of methods, (4) data collection and analysis, (5) report writing, (6) report presentation, and (7) follow-up action. A detailed study programme should be prepared within the prescribed time limit.

An array of methodologies is now available, ranging from casual, informal interactions to highly structured sample surveys, including emerging methodologies of rapid appraisals. These methodologies can be grouped into two categories: informal and formal.

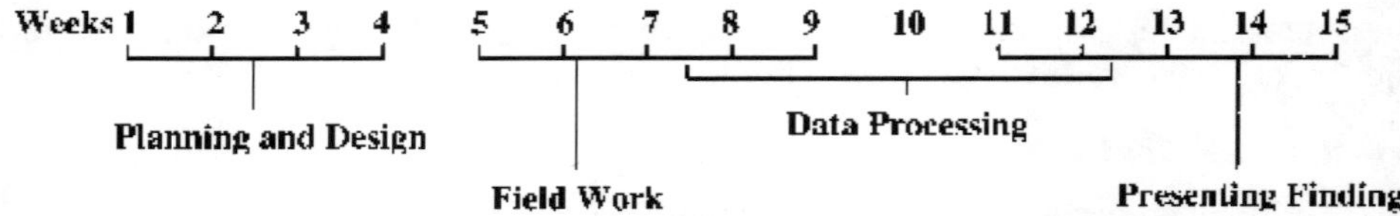

A Typical Summary Programme for the Study (Nichols (1991)

Extension Performance Indicators

(I)	Extension Effectiveness Indicators	
(a)	Single Indicators	
1.	Awareness	Number of farmers aware of Village Extension Worker (percentage)
2.	Visit	Number of visits by Village Extension Worker to farmers a) twice a month, b) once a month, and c) no visit (average)
3.	Field Meetings	Number of meetings of Village Extension Worker with farmers in their fields (percentage)
4.	Regularity	Number of meetings of Village Extension Worker with farmers on the fixed day (percentage)
5.	Field Day	Number of field days organized by Village Extension Worker a) in preceding month, b) quarterly, and c) annually (average)

6.	Demonstration	Number of a) method demonstrations, b) result demonstrations, and c) method-cum-result demonstrations organized by Village Extension Worker(i) in preceding month,(ii) quarterly, and(iii) annually
7.	Supervision	Number of supervisory visits from Agricultural Extension Officers to Village Extension Worker in the field per month (average)
8.	Research-Extension Linkage	Number of research-extension linkage workshops organized per month (average)
9.	Farmer Training	Number of farmers trained in farmers' training centres (institutionalized training courses) per year (average)
(b)	**Unitary or Composite Indicators**	
1.	Extension Effectiveness	Arithmetic average of selected extension effectiveness indicators, say, Awareness Indicator (know the Village Extension Worker), Visit Indicator (number of visits twice a month). Field Indicator (meeting place at farmers' fields), and Regularity Indicator (visit on the same day) (Misra, 1994)
(ii)	**Extension Efficiency Indicators**	
1.	Performance Index	Actual number of farmers reached out of the target number which should be reached (Casley and Lury, 1982, p.7) (percentage)
2.	Penetration Index	Number of farmers adopting the recommended practice out of the actual number reached (ibid. p.37) (percentage)
3.	Achievement Index	Number adopting the recommended practice out of the target number of farmers (ibid. p.37) (percentage). Note that (13) =(12) x (11)
	Extension Productivity Indicators	
1.	Yield	Yield per hectare for main crop(s) (average)
2.	Productivity Index	Increase in yield over base year compared with base year (percentage)

Extension Capability Indicators

1.	Coverage	Area under Cultivation per Extension Worker
2.	Intensity	Number of Farm Families per Extension Worker
3.	Competence	Number of Graduate Extension Workers out of total number of Extension Workers (percentage)
4.	Subject-Matter Specialist	Number of Subject-Matter Specialists per hundred Extension Workers
5.	Research-Extension Ratio	Number of Agricultural Scientists per hundred Extension Workers
6.	Monitoring	Number of Monitoring Unit Personnel per thousand Extension Workers
7.	Gender Ratio	Number of Female Extension Personnel out of total number of Extension Personnel (percentage)
8.	Equity	Number of Small and Marginal Farmers out of total number of Contact Farmers (percentage)
9.	Mass Contact	Number of group meetings held per month per Extension Worker in a year(average)
10.	Computerization	Number of personal computers in Extension Organization per thousand Extension Personnel

11.	Print Media	Number of leaflets/pamphlets distributed per month per Extension Worker in a year (average)
12.	Audio-Visual Media	Number of audio-visual (cinema/television) shows organized per month per Extension Worker in a year (average)
13.	Training	Number of Extension Personnel out of total number of Extension Personnel trained in specialized training courses in a year (percentage)
14.	Finance	Budgetary expenditure on Agricultural Extension out of total budgetary expenditure on agriculture per year (percentage)
15.	Investment	Expenditure on Agricultural Extension as percentage of Agricultural Gross Domestic Product per year
16.	Transport	Number of(i) bicycles, (h) motorcycles, and(iii) 4-wheel vehicles, per thousand Extension Workers

Notes: 1. Extension Personnel means total personnel engaged in extension, including Subject-Matter Specialists, whereas Extension Worker means Village Extension Worker; 2. The data source for all the indicators will be the offices of District Agriculture Officers and Provincial Directors of Agriculture. 3. At the national level, the indicators are required to be aggregated by the Central Monitoring Unit in the Ministry of Agriculture.

Informal Methods

Informal methods include participant observation, case studies, key informants, individual interviews or discussions, group interviews or discussions, oral testimonial and life histories, longitudinal studies, cross-sectional studies, interdisciplinary terms, reconnaissance survey, diagnostic studies, rapid rural appraisal, and participatory rural appraisal (Casley and Lury, 1987; Nichols, 1991; Pratt and Loizos, 1992; Hildebrand, 1981; FAO, 1992; Beebe, 1987; Kumar, 1993).

Formal Methods

Formal methods include using the population and agricultural census; sample surveys such as random sampling, including simple, systematic, stratified, cluster, and multistage; non random sampling, including purposive, quota, and accidental (Shaner, Phillipp, and Schmehl, 1982); and special studies. Among these, the sample survey occupies a unique place. A sample, as a representative part of the population, enables us to draw reasonable inferences about the characteristics of the population. As such, the sample survey will constitute the "handover" of monitoring methodologies. Other methods, notably the emerging rapid rural appraisal, can be used to advantage for exploratory, diagnostic, complementary, and supplementary purposes for decision making.

Establishing a Management Information System

Information is a *critical resource* in the operation and management of organizations. Timely availability of relevant information is vital for effective performance of managerial functions such as planning, organizing, leading, and control. An information system in an organization is like the nervous system in the human body: it is the link that connects all the organization's components together and provides for better operation and survival in a competitive environment. Indeed, today's organizations run on information.

The term *information system* usually refers to a computer-based system, one that is designed to support the operations, management, and decision functions of an organization. Information systems in organizations thus provide information support for decision makers. Information systems encompass transaction processing systems, management information systems, decision support systems, and strategic information systems.

Information consists of data that have been processed and are meaningful to a user. A system is a set of components that operate together to achieve a common purpose. Thus a management information system collects, transmits, processes, and stores data on an organization's resources, programmes, and accomplishments. The system makes possible the conversion of these data into management information for use by decision makers within the organization. A management information system, therefore, produces information that supports the management functions of an organization (Davis and Olson, 1985; Lucas, 1990; McLeod, 1995).

Basic Concepts

Data versus Information

Data refers to raw, unevaluated facts, figures, symbols, objects, events, *etc.* Data may be a collection of facts lying in storage, like a telephone directory or census records. Information is data that have been put into a meaningful and useful context and communicated to a recipient who uses it to make decisions. Information involves the communication and reception of intelligence or knowledge. It appraises and notifies surprises and stimulates, reduces uncertainty, reveals additional alternatives or helps eliminate irrelevant or poor ones, and influences individuals and stimulates them to action. An element of data may constitute information in a specific context; for example, when you want to contact your friend, his or her telephone number is a piece of information; otherwise, it is just one element of data in the telephone directory. Computers have made the processing function much easier. Large quantities of data can be processed quickly through computers aiding in the conversion of data to information. Raw data enter the system and are transformed into the system's output, that is, information to support managers in their decision making.

Characteristics of Information

The characteristics of good information are relevance, timeliness, accuracy, cost-effectiveness, reliability, usability, exhaustiveness, and aggregation level. Information is relevant if it leads to improved decision making. It might also be relevant if it reaffirms a previous decision. If it does not have anything to do with your problem, it is irrelevant. For example, information about the weather conditions in Paris in January is relevant if you are considering a visit to Paris in January. Otherwise, the information is not relevant. Timeliness refers to the currency of the information presented to the users. Currency of data or information is the time gap between the occurrences of an event in the field until its presentation to the user (decision maker). When this amount of time is very short, we describe the information system as a *real-time* system.

Value of Information

Information has a great impact on decision making, and hence its *value is* closely tied to the decisions that result from its use. Information does not have an absolute universal value. Its value is related to those who use it, when it is used, and in what situation it is used. In this sense, information is similar to other commodities. For example, the value of a glass of water is different for someone who has lost his way in Arctic glaciers than it is to a wanderer in the Sahara Desert. Economists distinguish value from *cost* or *price* of a commodity incurred to produce or procure the commodity. Obviously, the value of a product must be higher than its cost or price for it to be *cost-effective.*

The concept of *normative value* of information has been developed by economists and statisticians and is derived from decision theory. The basic premise of the theory is that we always have some preliminary knowledge about the occurrence of events that are relevant to our decisions. Additional information might modify our view of the occurrence probabilities and consequently change our decision and the expected payoff from the decision. The value of additional information is, hence, the difference in expected payoff obtained by reduced uncertainty about the future event.

Information supports decisions, decisions trigger actions, and actions affect the achievements or performance of the organization. If we can measure the differences in performance, we can trace the impact of information, provided that the measurements are carefully performed, the relationships among variables are well defined, and possible effects of irrelevant factors are isolated. The measured difference in performance due to informational factors is called the *realistic value* or *revealed value* of information.

For most information systems, particularly those supporting middle and top management, the resulting decisions often relate to events that are not strictly defined and involve probabilities that cannot be quantified. The decision-making process often is obscure and the outcomes are scaled by multiple and incomparable dimensions. In such cases, we may either attempt to perform a multi attribute analysis or derive an overall *subjective value.* The *subjective value* reflects people's comprehensive impression of information and the amount they are willing to pay for specific information (Ahituv, Neumann, and Riley, 1994).

Information as an Aid to Decision Making

Simon (1977) describes the process of decision making as comprising four steps: intelligence, design, choice, and review. The *intelligence* stage encompasses collection, classification, processing, and presentation of data relating to the organization and its environment. This is necessary to identify situations calling for decision. During the *decision* stage, the decision maker outlines alternative solutions, each of which involves a set of actions to be taken. The data gathered during the intelligence stage are now used by statistical and other models to forecast possible outcomes for each alternative. Each alternative can also be examined for technological, behavioural, and economic feasibility. In the *choice* stage, the decision maker must select one of the alternatives that will best contribute to the goals of the organization. Past choices can be subjected to *review* during implementation and monitoring to enable the manager to learn from mistakes. Information plays an important role in all four stages of the decision process. Figure 1 indicates the information requirement at each stage, along with the functions performed at each stage and the feedback loops between stages.

Classification of Management Information Systems

There are various types of management information systems. Mason and Swanson (1981) describe four categories of management information systems:

1. Databank information system,
2. Predictive information system,
3. Decision-making information system, and
4. Decision-taking information system.

The classification is based on the level of support that the information system provides in the process of decision making. Sachdeva (1990) comprehensively presents these four types of systems:

Databank Information System

The responsibility of this information system is to observe, classify, and store any item of data which might be potentially useful to the decision maker. Examples of the kind of data that might be recorded in such a database for a given village, region, or area are as follows:

- ☆ Number of farms
- ☆ Number of units of arable land (hectares, fedans, acres)
- ☆ Average farm size
- ☆ Amounts of selected farm inputs applied annually
- ☆ Production per year on a unit of land for selected crops

A second example of data that might be recorded in a database (this time involving data internal to the organization) is as follows:

- ☆ Number of extension staff by category and assigned to a particular village, region, or area
- ☆ Number of work hours devoted by staff to selected concerns for a particular village, region, or area
- ☆ Total extension salary costs and other expenses by village, region, or area
- ☆ Number of demonstrations conducted for selected farm technologies by village, region, or area

- ✰ Number of on-farm trials conducted for selected farm technologies by region or area
- ✰ Number of radio, TV, and print media releases regarding selected farm technologies by time period and region or area

Role of MIS in the Management of Agricultural Extension Programmes

National agricultural extension systems, especially in developing countries, tend to be very large. For example, in India, the national agricultural extension system employs about 125,000 people. Extension managers at various levels need relevant information in order to make effective decisions. In the absence of such information, they act only on the basis of their intuition and past experience. Data that have been processed, stored, and presented properly will aid them in analysing situations and to make effective decisions. Management information is the information required by managers as they make their decisions, such as the number of extension personnel employed by category, their training requirements, career development plans, job descriptions, budgets, forecasts, benchmark surveys, reports on socioeconomic conditions of people served, and existing facilities (Ramesh Babu and Singh, 1987). The main purpose of management information systems is to provide management information to decision makers at various levels in the organization. Specifically, in an agricultural extension organization, MIS is needed:

a) To plan the most effective allocation of resources, for example, the allocation of extension personnel under a T&V extension system, the need for communications and training equipment and facilities, mobility, the amounts of required operational resources

b) To choose between alternative courses of action, whether to conduct a study on the impact of the T&V system with the resources on hand or hire an expert to investigate

c) To control day-to-day operations, for example, comparing the actual results achieved and those planned under the T&V system.

Design of a MIS in an Agricultural Extension Organization

The following are suggested steps to follow when designing a MIS for a national agricultural extension system.

1. Assessing Information Needs for Planning, Monitoring, and Evaluation

An investigation needs to be conducted into the types of decisions that extension managers have to make. For example, village extension workers (VEWs) seek solutions to their problems from their supervisors. In turn, supervisors need to be in a position to resolve these problems and to document how problems were solved for future reference. State-level managers also need information to resolve problems. They are concerned with implementing extension programmes district by district. They need information on staffing, transport, research-extension linkages, staff training activities, and successes (or lack of them) in solving technical problems. Feedback is needed from field staff and farmers on farmer problems and on which recommended practices are helpful. State-level managers need to know something about the amounts, kinds, and combinations of media support (*i.e.*, print, radio, television) that have been used for various efforts. They need to know if external factors have limited the success of particular efforts such as supply of credit or farm inputs and they need some assessment of farmers' responses to extension programmes (Raheja and Jai Krishna, 1991).

2. Deciding the Levels of Information Groups, Information Frequency, and Content

The number of information groups within an agricultural extension organization has to be decided because each group potentially will require a different type of information. Data processing consists of identifying each item of data and systematically placing it within a scheme that categorizes data items on the basis of some common characteristic or feature. Data not organized into a meaningful pattern can serve almost no useful purpose to those who must use them to make decisions. A computer can help in processing the data effectively. Rao (1985) suggested the use of computers in agricultural extension in India. He proposed that computer programmes be focused on district and sub divisional levels. In that way, information collected can be viewed in terms of the crops that are likely to be grown, agroclimatic conditions, soil types, irrigation facilities, resources of the farmers, and availability of various farm inputs. Documentation (storage and retrieval) involves storing items of information in an orderly manner. Storing information means recording it on storage media from which it can be made available when needed. Storage media are materials such as ordinary office paper, magnetic tapes, magnetic disks, microfilms, film strips, and a few other devices. Once the information is recorded on these storage media, the system can generate, on demand, information required for making decisions, solving problems, or performing analyses and computations. Presentation of information should be in a form and format suitable to the needs of extension managers. Generally, information is presented in reports, statistical summaries, analyses, and so forth in the form of text, figures, charts, tables, and graphs. The presentation of information should be precise, clear, and appealing.

3. Ensuring System Flexibility and Adaptability

Flexibility means the ability to retrieve information from a system in whatever form it may be needed by decision makers. Therefore, data need to be collected in some detail so that they can be rearranged or summarized according to the needs of managers. But system design should not be too complex because it must first serve the needs of the lowest levels of management (*i.e.*, sub district) that are likely to be instrumental in collecting important components of the original data. In addition, the system also must serve the needs of the district, regional, state or provincial, and national levels. Therefore, considerable care must be taken in assessing what types of information are required by management at the different levels. At the same time, effort must be made to ensure that the information collected meets acceptable standards of accuracy, timeliness, and coverage for each level.

Need for Automation

An automated MIS system contains data just as a manual system does. It receives input, processes input, and delivers the processed input as output. Some input devices allow direct human-machine communication, while others require data to be recorded on an input medium such as a magnetisable material (specially coated plastic flexible or *floppy* disks and magnetic tapes). The keyboard of a workstation connected directly to a computer is an example of a direct input device. Use of automation makes it possible to store immense quantities of information, to avoid many of the errors that find their way into manual records, and to make calculations and comparisons that would be practically impossible in a manual system.

Organization of a Database

Data are usually generated at the field level through transaction-processing systems, but once the data are captured, any echelon along the organizational hierarchy may use them, provided that information requirements have been well defined, appropriate programmes have been implemented, and a means has been arranged for the sharing of the data. This would imply that the same data can be used by different sets of programmes; hence we distinguish between the database (a set of data) and the applications (a set of programmes). In a decision support system (DSS), this set of programmes is the *model base* (Keen and Morton, 1978). The term *database* may refer to any collection of data that might serve an organizational unit. A database on a given subject is a collection of data on that subject that observes three criteria: comprehensiveness (completeness), non redundancy, and appropriate structure. Comprehensiveness means that all the data about the subject are actually present in the database. Non redundancy means that each individual piece of data exists only once in the database. Appropriate structure means that the data are stored in such a way as to minimize the cost of expected processing and storage (Awad and Gotterer, 1992). The idea of a large corporate database that can be flexibly shared by several applications or model bases has been realized by means of software packages specially devised to perform such tasks. These packages, called *database management systems* (DBMSs), are available in the market under different trade names such as ORACLE, SYBASE, INGRES, FOXBASE, and dBASE.

Networking and Interactive Processing

The two principal blocks that facilitate development and use of MIS are DBMS and *telecommunications.* The former makes data integration possible, while the latter brings information closer to the end users, who constitute nodes in a telecommunication network. The notion of telecommunications implies that some geographical distance exists between the computer site and the users' locations and that data are electronically transmitted between them. Remote applications may be executed between two floors in the same building, two offices in the same city, two offices on the same continent, or two places on opposite sides of the globe (Martin, 1990).

System Alternatives and Evaluation: Centralization versus Decentralization

A completely centralized information system handles all processing at a single computer site, maintains a single central database, has centralized development of applications, provides central technical services, sets development priorities centrally, and allocates computer resources centrally. The system's remote users are served by transporting input and output data physically or electronically. A completely decentralized system may have no central control of system development, no communication links among autonomous computing units, and stand-alone processors and databases at various sites. Each unit funds its own information-processing activities and is totally responsible for all development and operation. An advantage of centralized information systems is that they provide for standardization in the collection of data and the release of information. There also are some *economies of scale. A* centralized system reduces the need for multiple hardware, software, space, personnel, and databases. It may be possible to recruit more qualified personnel in a central facility. Observations indicate that user motivation and satisfaction are increased under a decentralized environment. This is attained because users feel more involved and more responsible, systems are better customized to their specific needs, and they usually get better response time in routine operations as well as in requests for changes. It is likely that for national agricultural extension systems, neither a completely centralized nor a completely decentralized system is desirable.

End-user Computing

The widespread use of personal computers and computer-based workstations has brought with it the age of end-user computing. End-user computing is a generic term for any information-processing activity performed by direct end users who actually use terminals or microcomputers to access data and programmes. The manager as end user may be provided with powerful software (like DBMS) for accessing data, developing models, and performing information processing directly. This has brought computing directly under the control of the end users and eliminates their dependence on the information systems specialist and the rigidities of predesigned procedures. They may now make ad hoc queries of information and analyse it in various ways. They may write programmes, or may often use ready-made programmes stored in the computer, using the computing power of a local PC or the mainframe to which it is connected.

Illustrative Computer-based MIS

A national agricultural extension system is a nationwide system managed by the national government. In India, agriculture is a state subject under the division of powers between the national and the state levels. Nevertheless, the national government supplements the financial resources of the states and provides coordination at the national level. The state's administrative machinery is divided into districts, districts into subdivisions, subdivisions into blocks. A block is a group of villages and the basic unit for the administration of an agricultural extension

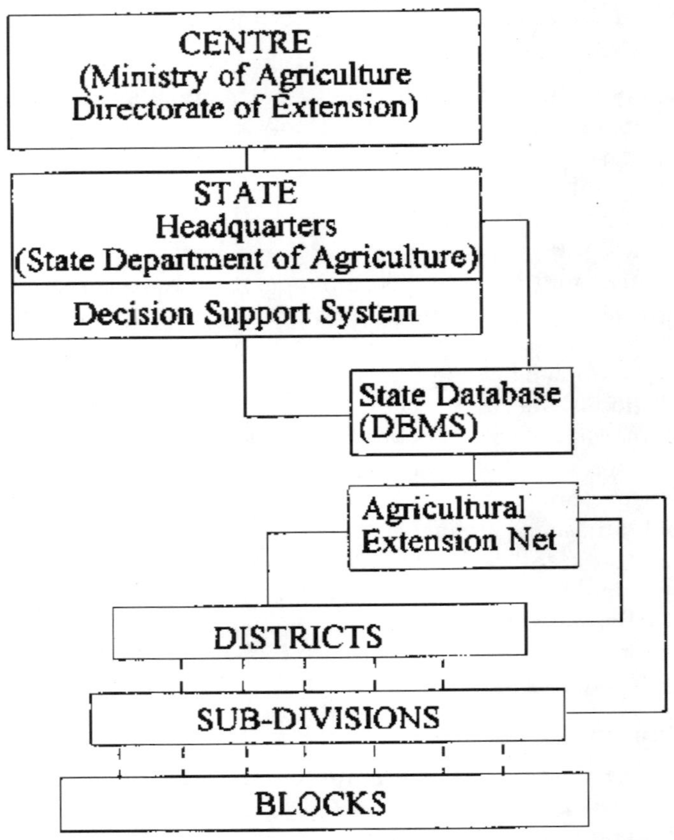

A Typical MIS for a National Extension System.

programme. Data collected at the block level need to be integrated at higher administrative levels to provide an integrated view at the district and state levels to support planning, monitoring, and decision making. An integrated database for the entire state may be supported by a mainframe/minicomputer at the state headquarters. Suitable programmes for the analysis of data may be designed to provide an interactive decision support system at the state level. Each district and subdivision may be provided with a mini/micro computer, depending on the volume of data to be handled. The computers in the districts and subdivisions may be networked with the state computer. The local data may be stored and processed in the district/subdivision, and the shared data with appropriate level of aggregation may be transmitted to the state headquarters to update the integrated database. The districts and subdivisions would have direct access to the integrated database with proper authorizations assigned to them through their passwords. The blocks may have only the input-output terminals connected to the subdivision computer to feed data to the subdivision and make on-line inquiries as and when necessary.

Strengthening Research-Extension-Farmer Linkages

The lack of a close working relationship between national agricultural research and extension organizations, and with different categories of farmers and farm organizations, is one of the most difficult institutional problems confronting ministries of agriculture in many developing nations. Research and extension organizations generally compete over the same scarce government resources and, frequently, leaders of these institutions do not see themselves as part of a broader system: the agricultural technology system (ATS). Instead, they try to increase the flow of resources coming to their respective institutions and to solve day-to-day management problems, rather than ensuring that their respective organizations contribute to the broader goal of getting improved agricultural technology to all major categories of farmers. The objective of this chapter is to outline an approach for identifying research-extension-farmer linkage problems, and then to describe different mechanisms that might be used to solve these problems. Three basic assumptions underlie this chapter. First, agricultural technology is a complex blend of materials, processes, and knowledge. Second, because of the complexity of agricultural technology, different institutional arrangements are needed to transfer different types of technology to technology users. And, third, most small-scale farmers in developing nations operate relatively complex farming systems in each agro ecological zone (AEZ) of the country; consequently, farmers in different AEZs need access to a wide variety of locally validated technologies if they are to increase their productivity.

Agricultural Technology: Some Basic Concepts

Generally, technology is used to improve the human condition, the natural environment, or to carry out other socioeconomic activities. Technology can be classified into two major categories:

1. *Material technology*, where knowledge is *embodied* into a technological product such as tools, equipment, agrochemicals, improved plant varieties or hybrids, improved breeds of animals (*e.g.*, semen from progeny-tested sires used for artificial insemination), and vaccines; and
2. *Knowledge-based technology* such as the technical knowledge, management skills, and other processes that farmers need to successfully grow a crop or produce animal products.

The transfer of *material* technology to farmers generally involves the production, distribution, and sale of seeds, implements, agrochemicals, and other production inputs. Therefore, the transfer process for material technology is generally simpler than training and disseminating technical knowledge and management skills to large numbers of poorly educated farmers who operate in different agro ecological zones (*i.e.*, the extension function). Also, the delivery systems needed for these different types of technologies are generally different. In most cases, the private sector is best suited to produce and distribute material technology. On the other hand, most *knowledge-based* technologies such as improved crop or livestock management practices, integrated pest management (IPM), and soil and water management practices are generally taught through vocational training programs for rural young people or disseminated through a publicly funded extension system for adult farmers. Therefore, material and know-ledge-based technologies are generally closely intertwined. Private sector firms in developing countries have very limited technical capacity to train farmers in these product-related skills and knowledge; therefore, the transfer of most knowledge-based technologies is, by design or by default, left to the national or provincial extension system.

Using Systems Analysis to Identify Linkage Problems

Systems analysis is an effective procedure to use in identifying linkage problems, since it is a *problem-solving methodology.* An outgrowth of operations research during World War II, systems analysis has been successfully applied in numerous fields, especially as a management tool to analyse, design, and implement complex technical processes. In short, systems analysis systematically examines a problem and makes each step of the analysis explicit. For example, in agricultural research, each disciplinary or commodity researcher (*e.g.*, a plant breeder or agronomist) has relatively routine research methods that are commonly used in the process of developing specific technologies for a particular commodity (*e.g.*, determining plant population, fertility, and other technical recommendations for a new cereal variety that has just been released for a particular AEZ). The problem is how to integrate these individual research efforts and relate them to broader system objectives such as increasing the productivity of different groups of farmers throughout the country. To achieve these objectives, the ATS will need to develop and transfer a package of recommendations for all of the economically important commodities being produced within different farming systems in each AEZ. By using a systems approach, it is possible to examine each system component and linkage at different levels within an ATS. In the process, specific system constraints and weaknesses can be easily identified, and then the most appropriate intervention strategy can be determined.

8

Agricultural Extension in Sustainable Agricultural Development

During the past fifty years, agricultural development policies have been remarkably successful at emphasizing external inputs as the means to increase food production. This has led to growth in global consumption of pesticides, inorganic fertilizer, animal feed-stuffs, and tractors and other machinery. These external inputs have, however, substituted for natural processes and resources, rendering them less powerful. Pesticides have replaced biological, cultural, and mechanical methods for controlling pests, weeds, and diseases; inorganic fertilizers have substituted for livestock manures, composts, and nitrogen-fixing crops; information for management decisions comes from input suppliers, researchers, and extensionists rather than from local sources; and fossil fuels have substituted for locally generated energy sources. The basic challenge for sustainable agriculture is to make better use of these internal resources. This can be done by minimizing the external inputs used, by regenerating internal resources more effectively, or by combinations of both.

Evidence is now emerging that regenerative and resource-conserving technologies and practices can bring both environmental and economic benefits for farmers, communities, and nations. The best evidence comes from countries of Africa, Asia, and Latin America, where the concern is to increase food production in the areas where fanning has been largely untouched by the modem packages of externally supplied technologies. In these complex and remote lands, some farmers and communities adopting regenerative technologies have substantially improved agricultural yields, often using only few or no external inputs (Bunch, 1991; GTZ,

1992; UNDP, 1992; Lobo and Kochendörfer-Lucius, 1992; Krishna, 1993; Shah, 1994; SWCB, 1994; Pretty, 1995).

But these are not the only sites for successful sustainable agriculture. In the high-input and generally irrigated lands, farmers adopting regenerative technologies have maintained yields whilst substantially reducing their use of inputs (Kamp, Gregory, and Chowhan, 1993; UNDP, 1992; Kenmore, 1991; van der Werf and de Jager, 1992; Bagadion and Korten, 1991). And in the very high-input lands of the industrialized countries, farmers have been able to maintain profitability even though input use has been cut dramatically, such as in Europe (Vereijken, 1992; Vereijken, Wijnands, Stol, and Visser, 1994; Van Weeperen, Röling, Van Bon, and Mur, 1995; Pretty and Howes, 1993; Jordan, Hutcheon, and Glen, 1993; El Titi and Landes, 1990) and in the United States (Liebhart *et al.*, 1989; NRC, 1989; Hanson, Johnson, Peters, and Janke, 1990; Dobbs, Becker, and Taylor, 1991; Faeth, 1993).

Sustainability and Levels of Action

A necessary condition for sustainable agriculture is that large numbers of farming households must be motivated to use coordinated resource management. This could be for pest and predator management, nutrient management, controlling the contamination of aquifers and surface water courses, coordinated livestock management, conserving soil and water resources, and seed stock management. The problem is that, in most places, platforms for collective decision making have not been established to manage such resources (Röling, 1994a, 1994b). The success of sustainable agriculture therefore depends not just on the motivations, skills, and knowledge of individual farmers, but on action taken by groups or communities as a whole. This makes the task more challenging. Sustainability is commonly seen as a property of an ecosystem. But Sustainability can be seen from other perspectives, which are more relevant for extension. Environmental issues emerge from the human use of natural resources. Sustainability can therefore be defined in terms of human reasons, activities, and agreements. The definition of Sustainability then becomes part of the problem because people need to agree on how they define Sustainability and what priority they will give it (Pretty, 1994b). In this approach, Sustainability is not a scientific, "hard" property which can be measured according to some objective scale, or a set of practices to be fixed in time and space. Rather, Sustainability is a quality that emerges when people individually or collectively apply their intelligence to maintain the long-term productivity of the natural resources on which they depend (Sriskandarajah, Bawden, and Packham, 1989). At the farm level, there is the farm household. At the above-farm level, there are the collective stakeholders, who might or might not be organized for sustainable use of the whole resource unit. In an irrigation scheme, it is common for an irrigators' association collectively to manage water use at the scheme level. But when it comes to watersheds or other vulnerable resource units, it is usually impossible to identify an appropriate "platform" for decision making (Röling, 1994a, 1994b).

Resource Conserving Technology Development and Transfer

Although many resource-conserving technologies and practices have been widely proven on research stations to be both productive and sustainable, the total number of farmers using them is still small. This is because these technologies involve the substitution of management skills, knowledge, and labour for external inputs. The modern approach to agricultural research and extension, however, has been to emphasize comprehensive packages of technologies. Few farmers are able to adopt the whole modem packages of production or conservation technologies without considerable adjustments. Part of the problem is that most agricultural research still occurs on the research station, where scientists experience conditions quite different from those experienced by farmers.

This is true of many sustainability-enhancing innovations. Even though resource-conserving technologies are productive and sustainable, if they are imposed on farmers, then they will not be adopted widely. Alley cropping, an agro forestry system comprising rows of nitrogen-fixing trees or bushes separated by rows of cereals, has long been the focus of research (Kang, Wilson, and Lawson, 1984; Attah-Krah and Francis, 1987; Young, 1989; Lal, 1989). Many productive and sustainable systems, needing few or no external inputs, have been developed. They stop erosion, produce food and wood, and can be cropped over long periods. But the problem is that very few, if any, farmers have adopted these alley cropping systems as designed. Despite millions of dollars of research expenditure over many years, systems that have been produced are suitable only for research stations.

Incorporating Farmer Experimentation

The problem with agricultural science and extension is that it has poorly understood the nature of "indigenous" and rural people's knowledge. For many, what rural people know is assumed to be "primitive," "unscientific," or overtaken by development, and so formal research and extension must "transform" what they know so as to "develop" them. An alternative view is that local knowledge is a valuable and underused resource, which can be studied, collected, and incorporated into development activities. Neither of these views, though, is entirely satisfactory because of the static view of knowledge implied (Chambers, Pacey, and Thrupp, 1989; Röling and Engel, 1989; Warren, 1991; Long and Long, 1992; Scoones and Thompson, 1994). It is more important to recognize that local people are always involved in active learning, in (re)inventing technologies, in adapting their farming systems and livelihood strategies. Understanding and supporting these processes of agricultural innovation and experimentation have become an important focus in facilitating more sustainable agriculture with its strong locality-specific nature.

From Teaching to Learning and a Whole New Professionalism

Teaching implies the transfer of knowledge from someone who knows to someone who does not know. Teaching is the normal mode of educational curricula and is also central to many organizational structures (Ison, 1990; Argyris, Putnam, and Smith, 1985; Russell and Ison, 1991; Bawden, 1992, 1994; Pretty and Chambers, 1993). Universities and other professional institutions reinforce the teaching paradigm by giving the impression that they are custodians of knowledge which can be dispensed or given (usually by lecture) to a recipient (a student). Where these institutions do not include a focus on self-development and on enhancing the ability to learn, they do not allow students to grasp an essential skill in the sustainable management of a complex agro ecosystem. In that case, "teaching threatens sustainable agriculture" (Ison, 1990). The problem for farmers is that they cannot rely on routine, calendar-based activities if they engage in sustainable farming. Technology for sustainable farming must emphasize measurement and observation equipment or services that help individual farmers assess their situations, such as soil analysis, manure analysis, and pest identification (Röling, 1993). Erosion control, water harvesting, biodiversity, access to biomass, recycling waste between town and countryside and between animal and crop production, all require local cooperation and coordination. What becomes important is the social transition, or new learning path, that farmers and communities must take to support sustainable agriculture. Learning for sustainable agriculture involves a transformation in the fundamental objectives, strategies, theories, risk perceptions, skills, labour organization, and professionalism of farming. This learning path has four key elements:

The Information System

Sustainable agriculture must be responsive to changing circumstances, so farmers need to invest in observation, observation equipment, record keeping, and monitoring procedures.

Conceptual Framework

Sustainable agriculture is knowledge intensive, and so farmers must know about life cycles of pests and disease organisms and their recognition, biological controls, ecological principles, soil life processes, nutrient cycles.

Skills

Sustainable farming requires a whole set of new skills, including observation and monitoring, compost making, mechanical weed control, spot application of pesticides, and risk assessment.

Higher System-Level Management

Generally, sustainable management of the farm is not enough, and it is necessary to think at system levels higher than the farm and take part in the collective management of natural resources at those levels.

In educational systems, therefore, the fundamental requirement for sustainable agriculture is for universities to evolve into communities of participatory learners. Such changes are very rare, an exception being Hawkesbury College, which is now part of the University of Western Sydney, Australia (Bawden, 1992, 1994). However, a regional consortium of NGOs in Latin America concerned with agroecology and low-input agriculture recently signed an agreement with eleven colleges of agriculture from Argentina, Bolivia, Chile, Mexico, Peru, and Uruguay to help in the joint reorientation of curriculum and research agendas towards sustainability and poverty concerns (Altieri and Yuryevic, 1992; Yuryevic, 1994). The agreement defines collaboration to develop more systemic and integrated curricula, professional training and internship programmes, collaborative research efforts, and the development of training materials.

The Key Principles of Farmer Field Schools

a) What is relevant and meaningful is decided by the learner and must be discovered by the learner. Learning flourishes in a situation where teaching is seen as a facilitating process that assists people to explore and discover the personal meaning of events for themselves.

b) Learning is a consequence of experience. People become responsible when they have assumed responsibility and experienced success.

c) Cooperative approaches are enabling. As people invest in collaborative group approaches, they develop a better sense of their own worth.

d) Learning is an evolutionary process, and is characterized by free and open communication, confrontation, acceptance, respect, and the right to make mistakes.

e) Each person's experience of reality is unique. As people become more aware of how they learn and solve problems, they can refine and modify their own styles of learning and action.

Sources: Adapted from Kingsley and Musante, 1994; Van de Fliert, 1993; Kenmore, 1991; Stock, 1994.

From Directive to Participatory Extension

Extension has long been grounded in the diffusion model of agricultural development, in which technologies are passed from research scientists via extensionists to farmers (Rogers, 1962, 1983). This approach is exemplified by the training and visit (T&V) system. It was first implemented in Turkey in 1967 and later widely adopted by governments (Benor, 1987; Roberts, 1989). It was designed to be a management system for energizing extension staff, turning desk-bound, poorly motivated field staff into effective extension agents. Extension agents receive regular training to enhance their technical skills, which they then hope will pass to all farmers through regular communication with small numbers of selected contact farmers. But the contact farmers are usually selected on the basis of literacy, wealth, readiness to change, and "progressiveness," and so this sets them apart from the rest of the community. The secondary transfer of the technical messages, from contact farmers to community, has been much less successful than predicted, and adoption rates are commonly very low among noncontact farmers. Without a doubt, T&V is now widely considered as ineffective (Axinn, 1988; Howell, 1988; Moris, 1990; Antholt, 1992, 1994; Hussain, Byerlee, and Heisey, 1994).

Challenges for Supportive Policy Processes

Policy making is commonly considered the prerogative of some central authority that formulates a policy, which is then decreed, imposed, and implemented regardless of conflicting knowledge and concerns. But policy is, in practice, often the net result of the actions of different interest groups pulling in complementary and opposing directions. This is particularly true with environmental problems because they are marked by uncertainty, complexity, and high stakes complexity, and high stakes (Funtowicz and Ravetz, 1993). There is therefore a growing tendency to see policy as a negotiated agreement resulting from interaction among citizens, in which central authorities play a facilitating role (Van der Poel and Van Woerkum, 1994). Policy is only effective if it is based on a widely shared consensus. From this perspective, it is easy to see why so many environmental policies which rely on coercion, control, and transfer have failed (Pretty and Shah, 1994; Pimbert and Pretty, 1994).

The First Steps on the Learning Path in the Netherlands

Predator mites were introduced into Dutch fruit orchards to control the red spider mite, which had become resistant to chemical controls. The use of this biological control meant that growers had to learn how to manage their orchards as biotopes for the predator mite. Soon they were carrying magnifying glasses to study the progress of their little helpers. This made them much more observant and accustomed to investing in regular observation. Furthermore, the health of the predator mites precluded use of broad-spectrum pesticides against other pests. As a result, growers also had to learn alternative controls for those pests.

Establishing and Strengthening Farmer Organizations

Historically, extension has mainly involved technology transfer, with the village extension worker (VEW) transferring knowledge from research stations to farmers by using individual, group, and mass media methods. More recently, extension has been asked to play a "technology development role" by linking research with community group needs and helping to facilitate appropriate technology development. It is in the historical context that many government agencies developed national policies for rural development and designed a policy framework to help rural people become organized so that the delivery of services could be channelled through the various types of farmer organizations or groups. Well-meaning policies also provided blueprint structures for farmer organizations (FO) in the form of cooperatives and commodity organizations in order to provide various input, marketing, and educational services to the farmers. Targets for forming groups and farmer organizations were given to VEWs without training them properly in the theory and principles of community organization. VEWs did not have many skills and not much experience in the process of establishing these organizations. However, because VEWs in many countries lack knowledge, skill and blueprint policies, and high targets, they resorted to shortcut methods to establish farmer organizations and groups. Many VEWs presented government policies in an oversimplified way to rural communities, suggesting that unless they are organized into cooperatives or associations or groups, they will not get government subsidies or access to credit and technical services. As a result, several FOs were established overnight on paper. Many FOs remained active during the period that government subsidies were distributed, but did not actively create cooperatives or partnerships and mobilize local resources to help achieve agricultural development. Community organization and facilitation skills were not part of staff training programmes.

The traditional approaches to organizing farmers and forming cooperatives need to be revised to meet the following development challenges of the twenty-first century:

a) The increasing absolute and relative poverty in many countries
b) The degradation of natural resources such as soil, water, flora, and-fauna
c) The low involvement of women in health, agriculture, and other development programmes
d) The poor health and education facilities in rural areas
e) The increasing socio-political unrest among the communities

New Roles for Extension

Several extension roles can be conceptualized to help rural communities get organized, but we will focus on four important roles.

Empowerment Role

The empowerment role can be a cornerstone of the new approach to extension. Extension personnel need to develop a new philosophy where their role is to help farmers and rural communities organize themselves and take charge (empowerment) of their growth and development. The term *empower* means to enable, to allow, or to permit and can be viewed as both self-initiated and initiated by others. For VEWs, empowering is an act of helping communities to build, develop, and increase their power through cooperation, sharing, and working together. The power in empowerment comes from releasing the latent energy hidden in the community and building collective actions for the common good, rather than from merely redistributing power from the haves to the have-nots. FOs can help harness this synergetic power for its members' survival, growth, and development. Empowered FOs can act as convergent points or platforms for solving local problems and mobilizing human and financial resources for sustainable development.

Community-Organizing Role

Village extension workers must learn the principles of community-organizing and group management skills (Chamala and Mortiss, 1990) in order to help the community, especially the poor or weaker sections, to organize itself for development. Understanding the structures, by-laws, rules, and roles will help leaders to plan, implement, and monitor their programmes and to perform this new role effectively. Skills in conflict resolution, negotiation, and persuasive communication help VEWs to develop FO leaders and members.

Human Resource Development Role

The human resource development approach empowers people and gives new meaning to all other roles. Development of technical capabilities must be combined with management capability. Training modules are now available (Chamala and Mortiss, 1990; Mortiss and Chamala, 1991) to help develop individual and group management skills. The entire philosophy of human capacity building is to encourage rural communities to understand their personal and group styles of managing themselves and to improve their planning, implementation, and monitoring skills.

Problem-Solving and Education Role

Problem solving is an important role, but the role is changing from prescribing technical solutions to empowering FOs to solve their own problems. This is achieved by helping them to identify the problems and seek the right solutions by combining their indigenous knowledge with improved knowledge and by using their resources properly. Similarly, there is a shift in the education role from lectures, seminars, and training to learning by doing and encouraging farmers and FOs to conduct experiments and undertake action-learning projects.

Farmer Organizations

Types of Organizations

Farmer organizations can be grouped into two types: one is the community-based and resource-orientated organization; the other is the commodity-based and market-orientated organization.

Community-Based, Resource-Orientated Farmer Organizations

This type could be a village-level cooperative or association dealing with inputs needed by the members, the resource owners, to enhance the productivity of their businesses based on land, water, or animals. These organizations are generally small, have well-defined geographical areas, and are predominantly concerned about inputs. However, the client group is highly diversified in terms of crops and commodities. There are many primary-level agricultural cooperatives in the developing world, but the majority of them have been financially vulnerable and ineffective. This group of organizations can generate income from the sale of inputs and outputs. The income can then be put back into the organization by spending it on extension, data generation, business planning, and administration. It is essential to have professional and honest management with constant monitoring and periodic rounds of evaluation (Gupta, 1989).

Commodity-Based, Market-Orientated Farmer Organizations

These organizations specialize in a single commodity and opt for value-added products which have expanded markets. They are designated as output-dominated organizations. Not specific to any single community, they can obtain members from among the regional growers of that commodity who are interested in investing some share capital to acquire the most recent processing technology and professional manpower. These FOs are generally not small and have to operate in a competitive environment. Research, input supply, extension, credit, collection of produce, processing, and marketing are all integrated to maximize the returns on the investments of the members who invested in the collective enterprise. The rate of success of these organizations is determined by their capacity to arrange for major investments and a continuous flow of raw materials. This requires the competent and convincing management of both enterprise-related and member-related aspects. The profits generated are used to provide supplementary and supportive services at reduced cost to encourage members to use them. VEWs can consult and work with other governmental agencies and nongovernmental organizations. However, each farmer organization will need to define its own BASE (basic activity sustaining the enterprise). In India, or for that matter in many developing countries, there is tremendous potential for expansion of commodity-based FOs. One rule of thumb suggests that any commodity which accounts for more than 50 per cent of the costs of the raw material can be considered for value-added processing by a farmer organization (V. R. Gaikwad, personal communication, 1994).

Issues Influencing Participation in FOs

The following issues will influence the extent of participation:

a) The degree of the farmer's dependence on the outputs of the organized activity.

b) The degree of certainty of the availability of the outputs.

c) The extent to which the outputs will be available only as a result of collective action.

d) The extent to which the rewards associated with the collective action will be distributed equitably.

e) The extent of availability of rewards within a reasonable time frame.

f) The extent to which the rewards are commensurate with the costs associated with continued participation (Shingi and Bluhm, 1987).

The Role of Extension in Farmer Organizations

The role of extension will vary with the role of the organization, the sectors in which the organization operates, the services offered, and the organizational form used. In community-based organizations, extension is used as a supplementary or supportive activity to realize the objectives of the BASE function of the organization. In commodity-based organizations, extension is integrated with all the other aspects of the organization to maximize the returns on the investment of the collective enterprise. Extension is taken seriously by both the organization and its members because both derive direct and measurable benefits from it. The following issues need to be considered when developing the extension role, especially for farmer organizations which are set up to specialize in the extension function:

a) Is there an identifiable need for extension in specific commodities in the area covered by the FO?

b) Would the FO be able to generate enough revenue from the extension activity alone (with farmers willing to pay for these services) to meet the FOs expenses and to provide satisfactory rewards to its members for their monetary and nonmonetary contributions? It will be important to anticipate the high potential for unresolved conflict over the issues of equity and charges for extension services.

c) How sustainable will the extension activity be over time, and therefore how sustainable will the organization be? It is possible for advice to be converted to freely transmittable knowledge which can be transferred to anyone without payment. The cost of the extension advice limits access to this knowledge. Therefore, the revenue earned from the extension activity could decline, especially with a client group which has a low resource base and is primarily concerned with a subsistence economy.

d) Can the advice given be actually put into practice and produce tangible benefits to the FO members? The FO would need to control or arrange for the supply of necessary inputs to ensure this; otherwise, the extension organization will fail, as has happened in the past in many developing countries. This means that the extension function needs to be integrated backwards with research recommendations and forward with the supply of inputs.

e) The organization will need to provide specific information in addition to the general information available from research centres. To do this and to survive, the organization will need a research linkage with government and university research institutions. It is beneficial if the FO can employ qualified and committed scientists who have active contacts in research organizations or who can act as consultants to groups of members. This would increase the cost of extension advice to members if the FO is supposed to be financially self-supporting.

f) It is necessary to appreciate that "extension markets" are governed by factors such as agro climatic variations, infrastructure development, and the strength of market forces. FOs operating in desert regions, single-crop rainfed areas, and predominantly irrigated areas will have different occupational and extension needs; therefore, variable response patterns to extension have to be anticipated (Gupta, 1981, 1985). Similarly, FOs operating in food-deficit and food-surplus stages will have different roles, expectations, and returns.

Steps in Establishing Farmer Organizations

Some of the principles and steps for social action models, community organizations, and management are used in designing the simplified step-wise approach in establishing FOs (Chamala, 1990). Care needs to be taken in selecting a person or a group of people who will take the lead in establishing community-based or commodity-based farmer organizations. This person - from another FO or government or other organization, such as a village extension worker - needs to be acquainted with and convinced of how the FO can be used to increase the income of small producers. In the following steps, the term VEW will be used to refer to the person or group of people selected to establish the FO.

Step 1. Understanding the Village Community

VEWs should enter the community with an open mind and understand the community structure. They must understand the community power structure, problems, and opportunities for development. Walking around and talking to key people can help ease their entry into the community. VEWs should also develop an under-standing of the entire community, including the poor and marginal farmers and women. Some of the techniques that can help to understand the community are using key informant techniques, participating in the community meeting and tea stalls, and walking around. Understanding the ownership of community resources and people's attitude, knowledge, and skills in the development of agricultural production will help the VEW learn about the community. The VEW can also understand the situation in more depth by collecting both qualitative and quantitative information on the levels of income and productivity, costs of cultivation, post-harvest losses, output utilization, and the likelihood of making striking improvements to each of these factors. Step 1 is crucial and cannot be done in a hurry. VEWs must spend some time (up to a month or so) to get the feel of the village community structure, politics, groups, and past experience in FOs.

Step 2. Identifying Potential Leaders in the Community

The usual tendency is for the VEWs to talk to formal leaders and commercial sections of the community. By using the sociometry method they can identify potential leaders. However, it is also important to be sensitive to the leadership structures operating in that culture and to the knowledge and skills needed to be a successful leader of a farmer organization. In some cultures, it is wise to search for and contact middle-aged leaders of the area who are not too young or too old. Preferably, they should come from better-off families, have social status and respect, and be from farming households. Case studies of leaders who have established farmer organizations show that these leaders initially made considerable sacrifices, experienced financial losses, and had to be supported by their families. These leaders also had to struggle and negotiate with bankers, bureaucrats, politicians, critics, and others including their own farmer members (Seetharaman and Shingi, 1992).

Sociometry

Ask a few people informally the following questions:

1. I am new to the village. Could you tell me three names of leaders of the village? (These are positional leaders who are currently holding leadership positions.) Write down their names or remember them and write them down later.

   Positional Leaders

   .

   .

   .

2. Name three leaders whose opinions or ideas on agriculture have influence on their friends and community leaders. (Make sure to get these names from each section of the society: the larger, medium, and small farmers because opinion leaders exist in each socioeconomic strata of the community.)

Opinion Leaders

Higher.

Medium.

Lower or poorer.

Step 3. Talking to the Identified Leaders and Seeking Cooperation from Other Agencies

VEWs might talk to these leaders on general agricultural development and get ideas and information on FOs in the village. They might also seek cooperation from government agencies and especially from NGOs (if there are any in the area) to help establish FOs and support them in achieving success. Does the community have a farmer organization? If not, is there a need for such an organization? If the community has an FO, what is its structure and history of performance? How could the FO play a role in village or community development? VEWs can canvass for ideas from the community and add their own ideas on the need for and the role of FOs in the entire process of broad-based agricultural development. They can explain how FOs in other communities helped them in development. It is important to provide facts and figures to convince potential leaders of the possibilities and approaches for increasing the income of a sizeable number of farmers and for contributing to the economic development of the region. Good FO leaders might also receive political advantages through satisfied farmers and people living in the region. The leaders' support should be secured and key people encouraged consulting among themselves about the pros and cons of forming FOs in the community.

☆ ***Steps in Developing an Organizational Structure***

The core group leaders' main goal is to understand the appropriate organizational structure, composition, and working rules for the efficient management of their farmer organization.

a) Leaders should secure the relevant guidelines of other FOs and study them carefully. An FO can be initiated by the government or by farmers and the community. If it is government initiated, the leaders should obtain copies of guidelines and rules. If it is community initiated, they should talk to other FOs and get their guidelines and constitution. The core group should study them carefully and discuss how they can fit into their community needs.

b) Leaders should then draw up a tentative organizational structure and working rules for their farmer organization. They should consider various models or types of organizations that serve their special needs for agricultural development and discuss them with other experienced leaders from that village or neighbouring villages. The structure should serve the functions. Are subgroups needed to achieve each task?

Step 4. Helping Local Leaders to Call Community Meetings

VEWs can help enthusiastic local leaders to call for community meetings. Sometimes more than one meeting may be needed to discuss the need for and the role of FOs in agricultural development. FO leaders from neighbouring villages can be invited to speak at these meetings. Farmer-to-farmer information exchange helps them. Sometimes smaller meetings can be held for low-resource (small and marginal) farmers and minority groups. They may prefer having a separate FO to serve their specific needs. There is no harm in having more than one FO in a village. It is important that producers from all sectors in the village participate in these meetings because the success of an enterprise-based FO depends on the volume of raw material procured from member producers.

Step 5. Nominating Core Group Leaders to Develop or Establish the FO

From the community meetings, core group leaders are elected or nominated to design the FO with further community consultation. In some cultures, however, this approach may lead to the appointment of leaders who are unable to cope with the complexities of a farmer organization. If leaders are elected or nominated, VEWs should consider the issues raised in step 2 when assessing each candidate for election or nomination. In situations where it is more appropriate to appoint a leader, the search for a suitable person is critical and requires time and patience. It is also critical that the selected leader be involved in the search for a potential agro-based enterprise to act as a BASE (basic activity sustaining the enterprise) for economic development. This process requires time, patience, and interactive and intellectual inputs. The VEW can help the appointed FO leader in this process.

Step 6. Developing an Organizational Structure for the FO

The VEW can help the core group of leaders in developing an organizational structure for their farmer organization. In the past, the "blueprint" approach was taken without understanding the function that the FOs structure plays in its performance. Group discussions help to highlight the need for careful planning. The structure should serve the organization's functions and goals. Understanding various types of FOs is useful. Should they be commodity-based organizations, cooperatives, partnerships, groups, or syndicates? Should they be multipurpose? Should there be one FO for the entire village or several to cater to the needs of special-interest groups (low-resource farmers, women, craftsmen, small businesses, and the like)? Should they have subgroups and an advisory committee? At this stage, the VEW should, as far as possible, play a passive role because the leaders are the ones who are building the FO. NGOs also may share their experience and help leaders to develop an organizational structure. Several less exacting chores also need to be taken care of at this stage. These include locating premises and negotiating for land, money, technology, personnel, construction, and infrastructure services such as power lines and telephones (if available).

Step 7. Developing the FO's Management through Education and Action Learning

An essential part of community empowerment is to help educate the leaders and members in management principles covering planning, implementing, and monitoring their projects and programmes. The following empowerment methods may be useful:

a) *Educating.* Organize formal and informal learning activities.

b) *Leading.* Help the leaders to lead and to learn from their actions by reflection as a team.

c) *Mentoring and supporting.* Help the members initially by mentoring or supporting them in their planning and implementation stages.

d) *Providing.* Obtain the services of other stake holders, FOs, and VEWs in providing various services to nurture the FO in the early stages of development.

e) *Structuring.* Help the FO to structure its meetings and various participative planning activities and to learn from their experience through reflection.

f) *Actualizing.* Help them to reflect on the process of managing their FO. Learning by doing can help them in self-actualization.(For more details on empowerment, see Vogt and Murrell (1990).

Step 8. Gearing up for Action

In this step, FOs examine their action plans, and task groups are set up to mobilize human and financial resources. Understanding the participative action management (PAM) planning process is useful. A detailed six-stage PAM planning cycle can help FO leaders in designing an inclusive and participative planning process. This is the stage to start considering the timing, scale, and content of the extension and research input of the FO.

Step 9. Implementing Selected Projects

In this step, the village extension worker can help the FO leaders implement the projects they have chosen.

The following process may be useful:

a) Start implementing the selected project(s).

b) Secure resources and allocate tasks.

c) Develop a calendar of activities to achieve the goal.

d) Develop monitoring processes for reflecting on events and activities regularly, either formally or informally.

e) Ask committee members to meet to discuss actions periodically and report to general members regularly to keep them informed and involved.

Step 10. Monitoring and Evaluating the FO's Progress

Usually evaluation is done annually to meet formal requirements. But VEWs can encourage FOs to reflect on their activities more frequently so that they learn and improve their management skills. They need to watch for people who want to take over the FOs for their self-interest. It is important to take action against any negative influences. These monitoring or reflection processes help strengthen FOs and avoid self-defeating problems. Learning organizations are created through collective reflection and openness on financial and the other managerial matters (Senge, 1990).

How to Strengthen Existing FOs

Many farmer organizations that go through a high activity phase become inactive or defunct over a period of time. This is a normal process for many groups or FOs. VEWs can help the community to understand the reasons and causes for decline. By reviewing the literature on groups and management, Chamala (1995b) identified twenty-six factors that factors, (2) service agency factors, (3) community factors, and (4) other external factors.

VEWs need to understand the factors that influence group or organization effectiveness or success. Many FOs have failed because of corruption, mismanagement, conflict, and lack of clear goals. It is important that VEWs understand these forces that influence their functioning. Several steps are suggested below to revive or revitalize existing FOs.

Different Factors Influencing Community Group Effectiveness

Group Internal Factors

Several factors are identified under this category:

a) Group composition
b) Group structure and size
c) Group atmosphere
d) Cohesion
e) Group standards and norms
f) Leadership styles
g) Balance between group maintenance needs, individual needs, and task needs
h) Level "group think" characteristics in the group
i) Development phase of the group
j) Group culture: empowering or controlling or a balance

Service Agency Factors

Government and nongovernment agencies can influence the effectiveness:

a) Technical capabilities of extension staff

b) Staff's "people skills" in managing groups

c) Staff attitudes and commitment to groups

d) Types of planning methods used: directive or participative, top-down or bottom-up, or balance of methods to maximize participation

e) Means or ends distinction: some groups are formed as means for development, while others are formed to harvest government subsidies. A group could get help, but it needs to mobilize its resources too.

f) Support for field extension officers

Community Factors

Groups and organizations are part of the community in which they exist. Hence the community influences a group's success.

Step 1. Understanding the Village Community and Gaining their Trust

This step is similar to step 1 for establishing farmer organizations. The key factor here is for the VEW to gain the community's trust in order to help them develop.

Step 2. Talking to Leaders about Why FOs are Less Active

Getting the multiple perspectives of identified leaders in the community informally is the next step. Putting pieces of the jigsaw puzzle together to understand why FOs are inactive requires diplomatic and detecting skills. The historical perspective of the FOs and various local factors that influenced their inaction must be obtained. The factors influencing community group effectiveness can be used as a guideline by selecting the relevant factors and ignoring the irrelevant ones.

Step 3. Seeking Ideas on Strengthening and Revitalizing Community FOs

Again, VEWs can talk to key leaders and seek their ideas on how to strengthen or revitalize the existing FOs or how to create new ones. It is the community's organization for their development, so the leaders' opinions and support should be obtained.

Step 4. Encouraging Local Leaders to Call Community Meetings

The VEW can help the leaders in setting up a community meeting to strengthen FOs or to form a new one. The VEW can unobtrusively provide the format and make suggestions on how to conduct meetings and how to strengthen FOs.

Step 5. Establishing a Core Group of Leaders to Draft a Proposal

A core group of leaders or a steering committee will further deliberate, using the suggestions made by the members to design and develop a strategy to strengthen or revive the farmer organization.

Step 6. Getting Comments on and Community Endorsement of the Strategy

Leaders should be careful not to let the meetings and the community mood for "head hunting" or "apportioning the blame for the FO's inefficiency" to take control. It is necessary to guide the community in a positive direction to develop working strategies. Several methods are available, such as SWOT (strengths, weaknesses, opportunities, and threats) analysis, prioritizing, and action-planning methods.

Increased Demand for Services

Will strong farmer organizations create more demand for services, and if so, how can this issue be resolved? Some policy makers believe that strong farmer organizations could create more demand on research and extension. This is partly true, but if the overall policy framework emphasizes empowerment, self-help, or user-pay systems, then more local resources could be mobilized and help increase the correct way of using current natural resources on a sustainable basis. Several low-resource farmer groups working with other stakeholders like banks, agribusiness, NGOs, private consultants, and religious and philanthropic agencies will bring more resources to the development process. Retired professional teachers, nurses, and agricultural scientists could be encouraged to work as volunteers with FOs. In many developed countries, volunteers are a major force in providing services to the community.

Competition for Services

Will organized low-resource farmers compete for extension services with the existing commercial farm sector, and if so, how can this issue be resolved? In principle, farmer organizations with full empowerment reduce the pressure on "routine extension" activities, which take up a lot of the VEW's time. For example, "Land care" groups in Australia and some successful FOs in Malaysia, the Philippine, Thailand, India, and Africa are helping other farmer groups or farmer organizations to be effective not for egalitarian or welfare reasons, but because they see that the commercial sector is dependent on low-resource farmers or upland farmers in solving their salting, erosion, water quality, and pest and weed management problems. Ecologically, the commercial sector is dependent on resource-poor or other minority tribes and groups for their survival. Just as in farmer-to-farmer extension methods, VEWs should help link the commercial sector with the low-resource farm sector so that win-win projects can be developed. Similarly, the commercial sector needs peace and prosperity to avoid social unrest and to stop thefts and other problems. VEWs can play a positive role in developing a common or shared vision for sustainable development. Again, several techniques like future research, problem census, and SWOT analysis will help develop a shared vision for the commercial sector as well as the low-resource farming sector or minority groups. Collective action is crucial for survival and sustainable development. Community organization is essential for rural development. One should not take a blueprint approach, but rather a learning-process approach within an action-learning framework (Korten, 198).

Privatizing Agricultural Extension

Diverse agricultural extension funding and delivery arrangements have been undertaken since the mid-1980s by governments worldwide in the name of "privatization." This chapter reviews these actions and their implications. When agricultural extension is discussed, *privatization* is used in the broadest sense - of introducing or increasing private sector participation, which does not necessarily imply a transfer of designated state-owned assets to the private sector.

Forces for Change

The evolution of public agricultural extension arrived at a worldwide turning point in the 1980s, one that represented the end of a major phase in the growth of publicly funded extension in both the developed and developing world. Agricultural extension increasingly has become defined as one or other of (apparently) differentiated activities of technology transfer or rural development. In many situations, the transfer of technology, heretofore considered the purview of public sector systems, has been reconceived. Such changes suggest a refocusing of paradigms for the delivery of public sector extension.

Global Competition

The consequence of the ratification of the General Agreement on Tariffs and Trade (GATT) is that countries will have to more actively develop comparative agricultural advantages in the production and marketing of food and fibre. Coincidental with a shift toward more conservative political ideologies and free-market economics, global developments suggest increased competition in agriculture. While countries will focus more on their comparative advantages, they also, in many cases, still face national food security concerns and abject rural poverty.

Reassessment of Public Extension

While "modern" extension has existed since the nineteenth century, agricultural extension is quite young worldwide as a formal institution, with the majority of countries initiating such services since the 1950s and 1960s. Even in high-income countries where extension began at earlier dates, fiscal commitment took significant upswings following World War II when a backlog of science and technology had accumulated. In an FAO survey of 207 agricultural extension organizations in 115 countries (Swanson, Farner, and Bahal, 1990), 50 per cent of these organizations had been established or were reorganized in the previous two decades. Because extension worldwide has large numbers of staff, the recurrent costs of extension are of significant magnitude. In the FAO *Report of the Global Consultation on Agricultural Extension,* Swanson *et al.* (1990) reported that there were approximately 600,000 agricultural extension personnel worldwide, with 95 per cent of these working in public agricultural extension systems. In the United States, there are about 9,000 extension agents, 4,000 subject-matter specialists, and 1,000 directors and administrative support personnel (USDA, 1993 data). While the unit cost of extension staff in many countries is low, large staff sizes translate into large government outlays. As a result of financial concerns, many countries have examined alternative structural arrangements, including the feasibility of reducing public sector extension expenditures (with associated staff reductions), changes in tax raising, charges for government extension services, and commercialization and privatization (Howell, 1985). A number of countries have moved towards reducing, recovering, or shifting the burden of the costs associated with provision of public sector agricultural extension, particularly transferring "private good" functions to private industry.

Strategies for Change

Public sector extension, facing criticism for its cost and its lack of efficiency and for not pursuing programmes that foster equity, is confronted with a number of possibilities for change.

There has been a trend, perceptible throughout various extension systems undergoing adjustment, of greater flexibility and multiple partners in funding agricultural advisory services (OECD, 1989; Le Gouis, 1991). Le Gouis observed three major policies adopted by government and farm organizations regarding privatization of extension:

a) Public financing by the taxpayer only for the kinds of services that are of direct concern to the general public

b) Direct charging for some individual services with direct return (in the form of improved income)

c) Mixed funding shared between public and private professional association contributions for some services where the benefits are shared (Le Gouis, 1991).

A pervading development in new forms of financial support for extension is the trend to mixed sources of funding, reflecting strategies to gain access to additional sources of funding. In several developing countries, public-private extension coordination is already established. Alternative patterns indicate a fostering of private corporate initiative, encouraging cooperative ventures by farmers, coordinating public-private extension services, and privatizing the public system (Wilson, 1991).

Revitalization

The United States Cooperative Extension Service, when criticized for lack of relevance and vision (Dillman, 1986), regrouped and reviewed the criticisms. Its Extension Committee on Organization and Policy (ECOP) organized a Futures Task Force to review issues and put forward recommendations with a view to revitalizing the system (ECOP, 1987), which has led to various alterations structurally and programmatically.[1] Meanwhile, the advancement of electronic information systems is resulting in increased privatization, with important implications for the future structure of U.S. agriculture (Goe and Kenney, 1988).

Commercialization

New Zealand's Ministry of Agriculture and Fisheries' (MAF) agricultural advisory service now operates under user-pay, commercial criteria (Hercus, 1991). The MAF advisory service, renamed MAF Consulting and, subsequently. Agriculture New Zealand, has remained (temporarily) a public agency, although its employees have given up a number of public employment benefits and now receive commissions for consulting work undertaken. The agency depends for its annual budget on consulting fees received from farmers and contractual arrangements with government for the supply of policy information and rural intelligence to government.[2]

Cost Recovery

Other public extension systems have moved toward cost-recovery approaches. Mexico has developed a fee-based system among large-scale farmers in the northwest region and plans the development of a similar arrangement among small-scale farmers in the south central region (Wilson, 1991). The Agricultural Development and Advisory Service (ADAS) in England and Wales, notionally "commercialized," operates on a partial cost-recovery basis. Clients of ADAS pay a fee for advice which formerly was free of charge. This process of cost recovery, introduced in 1987, was directed towards the agency receiving 50 per cent of its income from commercial fees by 1993-94 (Bunney and Bawcutt, 1991; Harter, 1992).

Voucher Systems

Some countries have replaced public extension delivery systems with vouchers, distributed by government services, for farmers to use in hiring private extension consultants (as in Chile). Coupons attached to agricultural bank loans, committing a certain percentage of the loan for extension services, have been used in Colombia.

Gradual "Privatization"

In 1990 The Netherlands "privatized" approximately one-half of its public extension service by transferring field extension personnel, with initial government financial support, to the farmer associations. The elements of the extension service responsible for linking research and the privatized extension services, policy preparation, implementation, and promotion and regulatory tasks remained under the aegis of the Ministry of Agriculture (Le Gouis, 1991). The "privatized" extension service is governed by a board on which farmers' organizations and the government are equally represented (Proost and Röling, 1991).

Dutch farmers make a partial contribution to the cost of the new organization through membership subscriptions to farmer associations, as well as through direct payment for individual analyses. Farmers will eventually contribute 50 per cent of the cost of the service: special services such as individual analyses will be fully paid for by the farmer clients. The Dutch government has established new government-funded structures for integrating subject-matter specialists into extension teams to facilitate the transfer of information and knowledge and for the provision of information on government policy (Bos, Proost, and Kuiper, 1991; Proost and Röling, 1991).

A gentler form of "privatization" has been proposed for the delivery of government extension services in the Australian state of Victoria. A review of extension services determined that, for government-provided services conferring essentially private benefits to individuals, rather than cost recovery by government fee charging, it is more desirable and more efficient that private advisers deliver such services. However, because of the complexities of extension service delivery and the varying nature and levels of development of different agricultural sectors, a number of constraints were identified which precluded universal application of such a principle (Watson *et al.*, 1992; Cary, 1993a).

In order for rural industry organizations to take a greater responsibility for technology transfer, the Victorian government has proposed "outsourcing" for delivery of future extension programmes. Outsourcing means that the government extension agency will retain a core pool of extension project staff and "buy in" private sector professional services with skills that the agency considers unnecessary to maintain. Agricultural consultants and contract staff will be employed to help deliver services in specific projects funded by rural industry and the federal government. Such projects are likely to be broad and industry wide and not tailored to individual farm circumstances.

Other Arrangements

Some countries have never developed public sector agricultural extension services, leaving the function of agricultural extension to private sector commodity enterprises or industry agencies, albeit often with some government financial subsidy. In France, while chambers of agriculture and private sector companies provide extension services, the former are substantially supported financially by public funds. In New Zealand, extension services to the dairy industry for many years have been delivered by the Dairy Board consulting service, financed by the dairy industry. In other cases, nongovernmental organizations have been used to supplement public sector extension services, especially in the area of rural development (Amanor and Farrington, 1991). This arrangement has certain advantages for increasing extension coverage and encouraging farmer participation in technology systems, but it also has certain inherent limitations.

Alternative Funding and Delivery

Diverse directions have been taken and multiple means of payment (public and private) have emerged as governments have opted for alternative financial and delivery arrangements to pay for and deliver public sector agricultural extension services. Extension provision is often multi-institutional and organized in ways that are not necessarily independent.

Where the public sector provides extension, the alternative funding arrangements include:

a) *General tax-based public funding* for agriculture, including funding of agricultural extension services, that is, the traditional public sector mode of funding extension

b) *Commodity tax-based public funding* (through cess or parafiscal tax), for example on an agricultural commodity such as coffee, as in El Salvador

c) *Fee-based public funding,* in which fees are charged, usually to large farmers for extension service, for instance in Mexico's grainrich northern region

d) *Contract-based commercialisation of public services,* whereby contract-based arrangements are made between farmer and public sector extension services, as in New Zealand

Where the *private sector* provides extension, the alternative funding arrangements include:

a) *Government revenue-based vouchers,* provided to farmers who then contract with private sector agents for extension information provision, as in Chile
b) *Public credit revenue-based coupon schemes* attached to agriculture loans, obligating the farmer-borrower to use a percentage of the loan for extension advising purposes
c) *Membership and fee-based, including commodity tax-based funding,* whereby farmers pay member ship and service fees, and the private organization (*e.g.*, a chamber of agriculture) also receives funds through a public cess or parafiscal tax charged on agricultural commodities, which funds are then transferred to the private sector organization; the private sector then provides the extension services - although public sector officials generally sit on the chamber's governing board
d) *Membership fee plus commercial sponsorship by groups of input suppliers,* where farmer groups are provided non advisory, educational extension services by a consortium of privately employed agricultural consultants with partial financial support from rural sector commercial sponsors - such groups can operate on a large scale, with coordinated extension objectives[3]
e) *Privatization,* whereby provision and, eventually, agent salary payments are shifted to a farmers' association or other private entity

The Context for Extension "Privatization"

The debate on the role of the public sector is not limited to the context of agricultural extension, but encompasses the larger concerns of public policy and institutional and organizational development. Indeed, the degree of government versus private involvement in an economy is an enduring philosophically and politically vexing question. The move toward privatization and efforts to decentralize government functions relate to this theme.

The Privatization Debate

There are two themes in the broader privatization debate: first, a "political economy" consideration of the role and size of government in an economy, which focusses on whether or not there is a failure of private markets; and, secondly, an expressed need to reduce government outlays. While many reassessments of publicly funded extension have reflected the second theme, it is worth considering the rationale for public versus private activity in an economy. In mixed economies, the prevailing economic justification for government involvement in an activity such as agricultural extension is market failure, whereby the market mechanism alone cannot perform all economic functions for appropriate resource allocation. Market failure may arise because some goods or services are public goods (such as publicly funded agricultural research knowledge) which can be consumed in a non rival fashion by all members of society without any individual's consumption reducing the amount available for other individuals. Because the benefit of providing such goods cannot be appropriated by individuals, individuals generally will not provide such goods in a society even though there may be significant gains for producers and consumers. Some extension activities are clearly concerned with public goods subject to market failure. Other activities (such as individually tailored advice) confer appropriable private benefits which could be adequately supplied by private markets.

The Debate with Respect to Extension

While much of the public policy debate related to extension has focussed on either so-called privatization or commercialization as means of reducing government out-lays, other aspects need consideration. The commercialization experience of Agriculture New Zealand (Walker, 1993), while not without its problems, provides examples of some of the arguments for commercialization. Commercialization is perceived to have had a positive effect on moving "beyond the farm gate" into an involvement of the extension staff in the entire production-processing-transporting-marketing chain. There also has been the shift in focus to a client orientation and a concern to identify and produce results rather than simply to engage in activities (Hercus, 1991). In economically developed countries with a predominance of larger-scale commercial farming, increasingly the technologies of modem, industrialized farming are being developed by nongovernment industrial institutions; such technologies are appropriable for private marketing and generally have little need for government extension. In developed economies, it is more difficult to argue for publicly funded extension for rural industries containing fewer producers who are closely linked and integrated with research systems.

Institutional Considerations

The search for appropriate institutional arrangements for different situations echoes the larger debate currently under way on creative use of the private sector for supplanting or supplementing public services. Privatization represents one position in the debate over how public functions should be organized. Wise (1990) has observed that "privatization. is not necessarily a simplifying strategy. the responsibilities of public organizations does not disappear, they merely change." The primary issue may not be whether a certain function should be entrusted to public or private organizations, but, rather, what configuration of organizations, both public and private, is needed and what arrangements between them provide the most effective outcomes. In some instances, central government bureaucracies are seen as unresponsive and inefficient, and the diffusion of responsibility arises out of a concern that the public sector should be reduced in size. In other cases, however, the emphasis is less on reducing the size of the public sector and more on sharing authority among different units (White, 1989). The question of what role the government should play within an increasingly complex institutional arena is itself complex and not one to which, necessarily, there are simple answers.

Summary of Rationales

The rationale for private sector provision of agricultural extension services is generally based on an expectation of increased efficiency with the operation of private markets and with the resulting efficiencies contributing to the growth of a country's GNP. In contrast, the rationale for public provision of agricultural extension services is based on the following points:

a) Much agricultural information is a public good;
b) Only government extension services are likely to promote concern for natural resources management;
c) Public sector extension may enhance the education of farmers who often lack adequate access to educational institutions;
d) The public service often provides information that reduces risk to farmers;
e) Tservice may provide information that reduces transaction costs; and
f) An extension service may be concerned with community health issues related to possible human hazards such as accidents and poisonings linked to agricultural chemicals.

The argument for privatization is based upon:

a) More efficient delivery of services
b) Lowered government expenditures
c) Higher quality of services

Privatization may have some attendant disadvantages because of unequal access to resources and because of a diversity of "agencies" and the associated difficulty of coordinating external groups and other government departments. Private delivery agents will be less responsive to government policy direction, and there may be linkage problems with public applied research organizations. While the process of information transfer amongst farmers traditionally has been characterized by a cooperative, free exchange of information, industrial information traditionally has been a private good characterized by patent rights, process licensing, the use of paid consultants, and differentiated production and marketing processes. In developed economies with commercialized agriculture sectors, many of these features of industrial information transfer are becoming more common in agriculture. The trend to privatization will be stronger the more such circumstances exist. The range of different circumstances prevailing in agricultural extension worldwide suggests that a wide variety of approaches should prevail.

Implications of Extension "Privatization"

In general, a more commercialized approach broadens the focus of extension personnel and makes an extension service more responsive to client needs and changing economic and social conditions. But other immediate implications of privatization appear to include:

1. The tendency toward a reduction of linkages both among organizations and among farmers in the exchange of agricultural and other relevant information;
2. The tendency to enhance large-scale farm enterprise to the detriment of small-scale farming;
3. The diminishing emphasis on public-good information and the advancement of knowledge as a saleable commodity; and
4. The trend toward agricultural development services that cater primarily to large-scale farming.

The Netherlands' experience in moving to a partially privatized system highlights some of the implications for agricultural extension, particularly in developed countries. The Netherlands' approach reduced government outlays as well as the government agency role conflict between concern for farmers' interests and the implementation of increasingly stringent environmental policies (Bos *et al.*, 1991; Proost and Röling, 1991). With farmers paying for an increasing share of the extension services, their representatives have more influence on the direction of the extension service. New organizational structures and linkages have had to be established to link the "privatized" and private extension services with the research institutes, experiment stations, and regional experiment farms. Consequent upon, or in parallel with, the changed Dutch arrangements, other changes have taken place in the Netherlands' extension system. There is some evidence, at least for the vegetable greenhouse sector, that the high level of cooperation among extension information organizations in both the public and private sectors no longer exists (Huang, 1992). The more commercial orientation of the system appears to be creating tensions between extension workers and their clients in a less "open" knowledge and information system, with farmers who used to share information during study-group meetings now being more reluctant to do so. Those extension services that have adopted a commercialization or privatization strategy most vigorously have traditionally employed an advisory approach to extension delivery. The advice given is more likely to be a private good. As well, the extension advisers are more likely to be able to adapt to providing services commercially. However, some staff will not make such a transition easily, new commercial skills will be required by newly commercialized advisers, and the dynamics of any change will have to be planned carefully. Le Gouis (1991) has noted that government "commercial" fees should be set at the market rate so as not to compete unfairly with existing private consultants.

Institutional Implications

The new developments highlight greater institutional pluralism. Extension, interpreted broadly, now is often a mixed system or a "complex" where services are provided by private and public sector entities. The larger context in which a mix of public and private services operates presents a new challenge with new potential roles and responsibilities for the public sector. A major premise of this chapter is that policy makers must consider the entire agricultural extension complex when planning to allocate funds or seeking alternative funding arrangements for the public sector.

By taking into account the complex of extension services, their purposes, and audiences, governments can better reconsider the public sector role. In particular, to what extent should the public sector:

a) Attend targeted audiences un-served by the private sector?

b) Coordinate multiple extension providers?

c) Serve as the final reference or arbitrator of conflicting information?

d) Assure accountability of both public and private extension services to the public?

e) Facilitate the operation of the complex through regulation and information provision?

Considering public sector extension as only one or one set among many providers of extension services facilitates analysis and informed decisions on how best to provide extension within the complex of diffused responsibility.

9

The Role of Non-Governmental Organizations in Agricultural Extension

In recent years, many observers have suggested that agricultural and rural development strategies would benefit from increased collaboration between government research and extension organizations and nongovernmental development organizations, hereafter called GOs and NGOs, respectively (Can-oil, 1992; de Janvry *et al.*, 1989; Jordan, 1989; Korten, 1987). Donors in particular have begun to call for more NGO involvement in programmes that have traditionally been implemented through the public sector, and there has been a recent upsurge of donor interest in direct-funding south-based NGOs (World Bank, 1991a, 1991b; Farnworth, 1991; Bebbington and Riddell, 1994).

These advocates of closer NGO-GO collaboration have tended to underemphasize:

a) The wide range of interaction that currently exists, not all of it collaborative; much involves pressure by one side or the other.

b) The limitations facing efforts to work together.

c) The preconditions for successful collaboration; in particular, the prior informal contacts necessary to build up mutual trust.

d) Limitations as well as successes of NGO action.

e) The extent to which certain functions relating to, for example, "public goods" will remain more cost-effectively performed by the public sector than by NGOs. Analysis of how GOs might work with NGOs must be accompanied by continuing attention to ways of improving public sector management, an area in which structural adjustment reforms have not had the success expected.

NGO Characteristics

NGOs are defined here as non membership development-oriented organizations. Our concern here is with the stronger of the south-based NGOs that provide services either directly to the rural poor or to grass-roots membership organizations, and with the local branches of international NGOs that enjoy varying degrees of autonomy. They are therefore distinct from (but, as discussed below, often linked with) formal and informal membership organizations such as farmers' associations. But even within this definition, there exists wide diversity of origins and philosophy. Some NGOs were set up by left-leaning professionals or academics in opposition to the politics of government or its support for or indifference to the prevailing patterns of corruption, patronage, or authoritarianism. Some are based on religious principles, others on a broadly humanitarian ethos, and yet others were set up as quasi-consultancy concerns in response to recent donor-funding initiatives. Some NGOs reject existing social and political structures and see themselves as engines for radical change; others focus on more gradual change through development of human resources (usually through group formation) to meet their own needs or to make claims on government services; yet others focus more simply on the provision of services (*e.g.*, advice, input supply) largely within existing structures. Their ideological orientations also differ widely in relation to agricultural technology: many are concerned with low external input agriculture, 1 others pursue fundamentally organic approaches, 2 and, especially in the case of Andean societies, some are concerned to strengthen or reinstate traditional agricultural practices which formed the basis of social organization (CAAP, 1991).

NGO Strengths

a) The majority of NGOs are small and horizontally structured with short lines of communication and are therefore capable of responding flexibly and rapidly to clients' needs and to changing circumstances. They are also characterized by a work ethic conducive to generating sustainable processes and impacts.

b) NGOs' concern with the rural poor means that they often maintain a field presence in remote locations, where it is difficult to keep government staff in post.

c) One of NGOs' main concerns has been to identify the needs of the rural poor in sustainable agricultural development. They have therefore pioneered a wide range of participatory methods for diagnosis[3] and, in some contexts, have developed and introduced systems approaches for testing new technology, for example in Chile (Sotomayor, 1991). In some cases, these approaches have extended beyond fanning systems into processing and marketing, as with soya in Bangladesh (Buckland and

Graham, 1990), sesame in the Gambia (Gilbert, 1990), and cocoa in Bolivia (Trujillo, 1991).

d) NGOs' rapport with farmers has allowed them to draw on local knowledge systems in the design of technology options and to strengthen such systems by ensuring that the technologies developed are reintegrated into them (Chaguma and Gumbo, 1993).

e) NGOs have also developed innovative dissemination methods, relying on farmer-to-farmer contact, whether on a group or individual basis (*e.g.*, Sollows, Thongpan, and Leelapatra, 1993).

f) In some cases, NGOs have *developed* new technologies such as soya production in Bangladesh (Buckland and Graham, 1990) or management practices such as the sloping agricultural land technology in the Philippines (Watson and Laquihon, 1993), but more often they have sought to *adapt* existing technologies, such as PRADAN's efforts in India to scale down technologies developed by government for mushroom and raw silk production and so make them accessible to small-scale farmers (Vasimalai, 1993).

Undoubtedly, one of the main strengths of NGOs has been their work in group formation. This has been in response to perceived needs at several levels:

- ✰ To meet the technical requirements of certain types of innovation. Thus, Action for World Solidarity in India worked with grass-roots organizations to achieve simultaneous action in an integrated pest management programme (Satish and Vardhan, 1993). In the Gambia and Ethiopia, NGOs helped farmers to organize local informal seed production in ways to avoid undesirable cross-pollination (Henderson and Singh, 1990).
- ✰ To manage "lumpy" assets. In Bangladesh, NGOs have helped to organize landless labourers to acquire and operate water-pumping technology (Mustafa, Rahman, and Sattar, 1993).
- ✰ To manage common property resources. Many examples exist of formal and informal associations, often supported by NGOs, which manage irrigation water. In other cases, NGOs have supported group efforts in soil and water conservation, whether on private land or on a micro-watershed basis involving both private and common land (Fernandez, 1993a). They have also helped in managing common grazing and forest land in a sustainable fashion in relation both to technology and the creation of a capacity to make demands on government over, for example, access issues (Femandez, 1993b).

NGO Weaknesses

1. NGOs' small size means that their projects rarely address the structural factors that underlie rural poverty. Small size, independence, and differences in philosophy also militate against learning from each other's

experience and against the creation of effective forums, whether at national or provincial levels.

2. Some "fashionable" locations have become so densely populated by a diversity of NGOs that problems have arisen not merely of competition for the same clientele, but of some undermining the activities of others (Ayers, 1992).
3. NGOs have limited capacities for agricultural technology development and dissemination and limited awareness of how to create effective demand-pull on government services.
4. Some NGOs are more accountable to external funding agencies than to the clientele they claim to serve. Donor pressure to achieve short-term impacts, combined with a lack of cross-learning, has led in some cases to the promotion of inappropriate technology, such as protected horticultural systems in the Bolivian Andes (Kohl, 1991).
5. Many NGOs place great emphasis on voluntarism. Whilst such concepts as "volunteer extension workers" have great intuitive appeal and reflect widely commended values, they are sometimes promoted at the expense of financially sustainable alternatives. This was evident in SIDA's farm-level forestry project in North Vietnam, for instance, where the scope for supporting an emerging private nursery sector in the provision of technical advice was ignored, and complex and largely voluntary advisory services at the village level were promoted instead (author's observation, April 1994).

NGO-GO Configurations for Providing Technical Advice and Feedback

This is argues that the extreme institutional position in which all extension services are provided by the public sector is likely to be inefficient. At the other extreme, only in very specific circumstances can government hand over large parts of the extension function entirely to NGOs. This has been done in Chile, where government has contracted private technology companies to cater to the larger commercial farmers, and NGOs for small subsistence-oriented farmers.

However, similar attempts in India have been largely unsuccessful. A proposal in the Eighth Plan to hand over the entire range of technology transfer and training activities to NGOs in parts of Gujarat, Rajasthan, Orrisa, Kerala, and West Bengal, with some technical support from the state agricultural universities and departments of agriculture, has generated only a lukewarm response from NGOs. One reason for this is the NGO concern that many of the technical recommendations from GOs that they would be expected to disseminate are not relevant to small-scale farmers. Another reason is that mechanisms for bottom-up feedback in existing technologies and for the articulation of demands for new technologies remain weak.

An attempt by the secretary of state for agriculture in Rajasthan to hand over responsibility for extension to NGOs in Udaipur District - renowned for its high density of NGOs - provoked a reaction that is likely to be typical of NGOs in many countries: namely, that by doing so the state is abrogating what is properly its responsibility to ensure a regular supply of technologies relevant to small-scale farmers.

Models developed elsewhere to provide a division of tasks more closely corresponding to the respective comparative advantages of the two sides have been more successful. One of the best-known models is in eastern Bolivia, where public sector extension services have long been characterized by chronic weakness (Thiele, Davies, and Farrington, 1988). Under a new strategy devised in 1989, Centre for Tropical Agricultural Research (CIAT) established a coordination unit - the Technology Transfer Department (DTT) - whose role was not to work directly with farmers, but with various intermediate users of technologies who had their own local extension services. NGOs are one of the most important types of intermediate users.

The DTT has subject-matter specialists and zonal specialists whose work is supported by a communications section. The subject-matter specialists (SMSs) are in regular contact with their corresponding CIAT researcher and collaborate on some research work. They package research information for delivery to intermediate users and are mandated to transmit feedback on farmer needs to the researcher.

Frequently, technologies developed in the experimental centres are still not ready for transfer. SMSs therefore carry out on-farm adaptive trials, in addition to ensuring that extensionists pass on the appropriate messages to farmers. Other dudes of the SMSs include preparing technical bulletins for extensionists, enhancing feedback and advice to extensionists (for instance, on the establishment of demonstration plots), and explaining how to give talks to farmers.

Informal collaborative arrangements rely heavily on the initiative of GO staff to feed lessons back into the next round of the research and extension agenda. In Bolivia, feedback was encouraged through a range of instruments, including NGO representation on the research planning committee of the local research station, and consultation with a number of zonal substations, part of whose function was to assemble users' views on the technologies being made available.

A different type of formal arrangement being developed in Udaipur District of India is a quarterly forum hosted by a "hybrid" NGO-GO institution - a government Farm Science Centre located in an NGO - in which interested NGOs and GOs participate. In essence, it is intended to promote familiarization by allowing cross-visits to be set up to each other's programmes, to allow training courses to be designed to meet NGOs' requirements, and to allow NGOs to "feedback" on currently available technologies and to voice further technology needs that are not currently being met. The forum is also intended to develop lessons for future interaction from efforts to monitor the process and impacts of current NGO-GO interaction within the district. The forum is beginning to make progress in some of these areas, but a major difficulty remains in the form of pre-programmed "targets" that the Department of Agricultural Extension has to meet each season in terms of, for example, the number of demonstrations of a given type. The forum has had the added benefit of facilitating substantive interaction among GOs (departments of agricultural extension, of animal production and health, and of horticulture) which otherwise meet on a monthly basis, but only for administrative purposes, and among the large number of NGOs in the district.

What remains to be addressed is the scope for similar formal links between NGOs, the membership organizations they are working with, and the GOs (such as the revenue and forest departments) which control large areas of land suitable for grazing or fodder collection. Although the central government has approved wider access by villagers (often with the assistance of NGOs) to such land under "joint management" arrangements, parts of the land have been encroached upon by wealthier farmers, and local-level officials find it difficult to rectify this so as to make the land accessible to village groups.

NGO-GO Configurations in Training

Some of the farmer training conducted by GOs is linked more strongly with GOs' programmes and targets than with farmers' needs. Much training is given in a classroom environment, without the practical content necessary to engage farmers' interests. NGOs have sought to work with GOs to address these shortcomings in several contexts:

1. In Gujarat, India, the Aga Khan Rural Support Project (AKRSP) identified village training needs through discussions with farmer groups (Shah and Mane, 1993). Initially, AKRSP organized government provision of this training, but the courses were formal in style (lectures in a classroom), and farmers' evaluations showed that they had learned little of practical value from them. In response, AKRSP developed an alternative needs-based training and dissemination methodology which it tested over several areas. Government staffs were then brought in to observe, participate in, and finally adopt the methodology. Successful adoption was reinforced by informal networks and exchange of experience at workshops and consultations. AKRSP, along with Myrada (Femandez, 1993a), has also been instrumental in training GO staff in participatory methods.
2. In a different context, the International Institute for Rural Reconstruction in the Philippines brought together resource people from NGOs and GOs at a one-week workshop, the objective of which was to produce a completed *Agro forestry Resources Training Manual.* The manual is now widely used (Gonsalves and Miclatteves, 1993).
3. Clearly, there are also many instances in which NGO staff benefit from the skills which GO staff are able to impart; training in grafting techniques, for instance, has been found useful by a number of NGOs (Giordano, Satish, and Farrington, 1993).

NGO-GO Configurations in Group Formation

Substantial scope exists for GOs to benefit from NGOs' group-organizing skills. In India, for instance, recent modifications to the training and visit extension system now require village-level extension workers to interact with groups of approximately twenty farmers instead of with individual "contact farmers." However, extension workers are not trained in group formation skills, and groups that they form are unlikely - if they survive at all - to become interested in anything other than the testing of immediately available technology. The examples cited above illustrate how NGOs can effectively organize groups around integrated pest management, soil and water management, and the management of common property resources and capital assets.

What Extension Services can do to Further Collaborate with NGOs?

The foregoing has several implications for extension services which aim to develop closer links with NGOs:

- Explicit recognition of the wide diversity of NGO types will be necessary. Not all Many NGOs seek to support the establishment and growth of membership organizations capable of meeting their technology requirements over the longer term either from their own resources or by creating demands on government services or by a combination of both. Thus in seed supply, Cromwell and Wiggins (1993), for instance, quote numerous examples of ways that NGOs have supported local groups to produce seed, including vegetable and soya bean seed production in Bangladesh, and the multiplication of planting material for potatoes in the Ecuadorian Andes. In other countries (*e.g.*, Nepal, the Gambia) local seed production initiatives have arisen plant breeding focused more fully on the needs of the rural poor, and the facilitation of linkages among the various agencies concerned with seed production and distribution. Some of these efforts see viable commercial arrangements as an essential feature of long term sustainability. Thus PRADAN, in India, in an effort spanning several years to support the introduction of chrome-leather tanning by a local group, encouraged links with commercial lending organizations and private leather traders, not least because the latter could give accurate feedback on product quality. In a more complex example of experimenting over several years with several types of women's groups for poultry production, the Bangladesh Rural Advancement Committee finally devised a multitier structure embracing rearing, local feed production, and health care by women paravets. These women drew on government for the necessary vaccines, earned a living by charging for injections, and provided elements of the extension function by giving advice on management and nutrition to those who paid for vaccinations (Box 2) will be willing to enter into a collaborative "service delivery" relationship with government, and those that do will do so only if GOs have something to offer appropriate to their clients' needs. Extension services therefore need to search for links with NGOs from a position of confidence that the research-extension system already has, or at least has the capacity to generate, something useful to NGOs and their clients.

- ☆ Close interaction will be impossible if extension departments expect NGOs merely to assist in fulfilling preset targets such as the achievement of a given number of demonstrations of a given kind each season. GOs will have to bring an open agenda into the relationship, where possible keeping some resources "unallocated" in order to be able to respond to needs as they are articulated by NGOs.
- ☆ Very specific efforts will have to be made to convey both feedback on existing technologies and NGOs' requirements for new technologies to researchers. In many GOs, reward systems provide no incentive among either researchers or extensionists to respond to feedback.
- ☆ GO and NGO staff can jointly participate in training courses (ideally led by a joint team) in the action-oriented methods such as participatory rural appraisal favoured by NGOs. The relevance of these to individual NGO staff will vary, but their capacity to enhance awareness of farmers' perspectives is important.
- ☆ Depending on their philosophy, NGOs are concerned to develop local capacities for experimentation which build solely on farmers' indigenous knowledge or on this and relevant "outside" ideas. This strategy may contribute to rural advancement in its own right, and the capacity it creates may prove a useful independent source of innovations in the absence of usable technologies from government. Alternatively, where GOs are willing and able to work with the poor, it will be a useful complement to what extension service can offer.

The Role of the NGOs and the Private Sector

NGOs have pioneered many of the initiatives subsequently incorporated in public sector extension services. NGOs continue to play a lead role in ensuring that women farmers and women on the farm receive training, information, and improved technologies. Their services often are increasing in scope and scale, either as complementary support to government efforts or to fill the gaps created as government expenditures and capabilities decline. An important emphasis which recently has been highlighted in NGO programmes is their support for membership-based community and farmer organizations. Women as well as men benefit from the expanding opportunities to develop farmer-to-farmer extension and training networks and to form partnerships with agricultural researchers and development agencies.

Strategic Extension Campaign: What and why

A "strategic extension campaign" (SEC) methodology developed by FAO has been introduced in Africa, the Near East, Asia, and Latin America. This methodology emphasizes the importance of *people's participation* (*i.e.*, intended beneficiaries such as small farmers) in strategic planning, systematic management, and field implementation of agricultural extension and training programmes. Its extension strategies and messages are specifically developed and tailored based on the results of a *participatory* problem identification process on the causes or reasons for farmers' non adoption, or inappropriate practices, of a given recommended agricultural technology or innovation. The SEC technology transfer and application approach is needs based and demand driven and has a problem-solving orientation.

The SEC programme follows a systems approach, which starts with a farmers' Knowledge, Attitude, and Practice (KAP) survey whose results are used as planning inputs and benchmark-baseline for summative evaluation purposes. In addition, a series of practical and participatory approach workshops are conducted to train extension personnel, subject-matter specialists, trainers, and farmer leaders together on the skills of extension programme planning, strategy development, message design and positioning, multimedia materials development, pretesting and production, as well as management planning, implementation, monitoring, and evaluation. One of the strengths of this approach is in orienting and training relevant extension personnel to apply *a systematic, rational, and pragmatic approach to planning, implementing, managing, monitoring, and evaluating regular or routine programmes* of an agricultural extension service.

Empirical evaluation studies (using information recall and impact surveys, focus group interviews, and management monitoring surveys) of strategic extension campaign methods applied to specific FAO-supported extension activities conducted, for instance, in Bangladesh and Malaysia (on rat control), Thailand (on pest surveillance system), Malaysia (on weed management), Zambia (on maize production), Malawi, Jamaica, and Morocco (on population education), reported positive changes in farmers' knowledge, attitudes, and practices vis-a-vis the recommended technologies as well as significant economic benefits.

This SEC method has been replicated with FAO assistance in many countries in Asia, Africa, the Near East, and the Caribbean, with topics such as line-sowing method of rice cultivation, maize production, cocoa cultivation, tick-borne disease control, contour tillage, population education, and ploughing with draught-animal power. In addition to various SEC replications within a country, the multiplier effects of its method are felt beyond national boundaries. For example, extension specialists from Ghana, Malawi, Ethiopia, France, Malaysia, Thailand, and the Philippines who had been trained by FAO on this SEC method and implemented such programmes have now served as consultants and resource persons to train their counterparts, and/or have assisted in similar SEC replications in Sri Lanka, the Philippines, Malaysia, Thailand, China, Liberia, Zambia, Malawi, Kenya, Uganda, Morocco, Tunisia, Rwanda, Burundi, Guinea, Jamaica, and Honduras.

The Usefulness of SEC

The Strategic Extension Campaign (SEC) is *not* an alternative to the conventional extension programme or activity. SEC is, and should be, an integral part of the programmes of an agricultural extension service. The effectiveness and efficiency of such a service could be increased because of SEC's emphasis on its problem-solving orientation, participatory planning approach, intensive extension personnel training, multimedia materials development, and extension management, monitoring, and evaluation procedures. Its activities should be carried out by extension personnel and to support the ministry of agriculture's policies, strategies, and priority programmes.

The strategic extension campaign is useful and important to an agricultural extension service for the following reasons:

- ✰ It advocates a participatory planning approach.
- ✰ It has a needs-based and demand-driven orientation.
- ✰ It uses a strategic planning and integrated systems approach.
- ✰ It considers the human and behavioural dimensions.
- ✰ It has a problem-solving orientation.
- ✰ It employs a cost-effective multimedia approach.
- ✰ It provides specific extension support materials and training.
- ✰ It has built-in process documentation and evaluation procedures.
- ✰ Its method is applicable to other extension programmes.

Participatory Planning Approach

This participatory approach extension method is responsive to intended beneficiaries' agricultural development problems and information needs because its extension objectives, strategies, methods, messages, and multimedia materials are specifically developed on the basis of survey results of their knowledge, attitudes, and practices (KAP) vis-a-vis the recommended agricultural technologies. Such a participatory approach in planning SEC activities increases the degree of relevance, and thus acceptability, of extension messages or recommendations among intended *beneficiaries who are consulted* during the planning process regarding their priority concerns and needs. It does not assume the target beneficiaries (*i.e.*, farmers) to be ignorant or requiring all the information there is to know. Rather, it tries to understand and assess farmers' local indigenous knowledge, values, and belief system on farming practices, which may be good, need to be improved, or perhaps need to be discouraged. In short, it follows the well-known principles of rural reconstruction: "Start with what people already know," and "Build on what they already have."

Needs-Based and Demand-Driven Orientation

In order to make the best use of available extension resources, SEC activities concentrate on meeting the information, education, and training needs of intended target beneficiaries. Rather than providing them with the spectrum of information and skills related to a given recommended technology, SEC activities are geared to *narrowing the gaps* between knowledge, attitudes, and/or appropriate practice levels of the target beneficiaries vis-a-vis the technology recommendations. Furthermore, the focus of SEC activities is to create a demand (through information and motivation approaches) and/or to satisfy the demand (through education and training) among the intended target beneficiaries for the necessary relevant knowledge and skills for adopting the recommended technologies. Such a method needs to apply bottom-up and participatory planning procedures which will give *high priority in meeting the interests and needs of the target beneficiaries.* Tailoring the SEC messages and activities to the specific needs of the intended beneficiaries would not only increase the chances of success, but also would increase the efficiency in resources utilization.

Strategic Planning and Integrated Systems Approach

The SEC method advocates an integrated and holistic approach in extension strategy development, programme planning and management, training, media and materials development, and monitoring and evaluation. To ensure its relevance to audience needs and to utilize its resources efficiently, it relies heavily on both quantitative data and qualitative information obtained from target beneficiaries (*i.e.*, farmers) to assist in problem analysis, objective formulation, strategy development, and management planning. It applies a *strategic planning* approach in programming and managing its activities to achieve *maximum* outputs or results using *minimal* inputs or resources in the *shortest time* possible. SEC activities such as surveys, strategy and management planning, multimedia materials design and development, training, field implementation, monitoring, and evaluation are integrated as a system, which is also an integrated part of a larger extension programme which has linkages with relevant agencies or units dealing with research, inputs-supplies, training, and marketing.

Human and Behavioural Dimensions

In order to minimize the heavy technology bias of many extension activities, the SEC method gives adequate consideration to *human behavioural* aspects, such as socio-psychological, socio-cultural, and socioeconomic factors which may facilitate or impede adoption, or continued practice, of recommended technologies by farmers. Without sufficient understanding of their positive or negative attitudes and behaviour towards a given technology, the technology transfer process would be slow and ineffective, especially if the extension emphasis is on appropriate technology application by farmers. There is considerable evidence to suggest that non-adoption of a recommended agricultural technology or innovation is often related to, or caused by, *non-technological* factors such as social, psychological, cultural, and economic problems.

The SEC method gives due attention to human and environmental factors which may influence the important decision-making process related to agricultural technology adoption and practices. It employs a behavioural science analysis, based on a participatory needs assessment and problem identification of the target audience, in developing appropriate strategies and tactics to overcome or minimize human related constraints affecting the agricultural technology transfer and application process.

Problem-Solving Orientation

The SEC is particularly distinguished in that it normally focuses on *specific* issues related to a given agricultural technology recommendation. Its main aim is to solve or minimize problems which caused non adoption of such a recommendation by intended target beneficiaries (*i.e.*, farmers). Unlike more conventional extension programmes or activities, it does not "extend" the whole gamut of information on the recommended technology package. Instead, it *selects*, prioritizes, and utilizes only the most relevant and necessary information or facts which can maximize the effectiveness of extension efforts to minimize or solve the identified problems of non adoption of a recommended technology. It stresses the need to provide strategic, critical, and "quality" information, which must also include non technological information as the reasons why non adoption of agricultural technologies are often related to socio-psychological, socio-cultural, and socioeconomic factors. Appropriate human behavioural science principles are thus applied to extension problem solving and in information positioning and utilization, which is *responsive* rather than prescriptive in nature.

The *segmentation* or classification of extension problems, objectives, strategies, and information needs according to a target audience's levels of knowledge, attitudes, and practices (KAP) in regards to a given recommended technology is not only conceptually important, but practical and useful as well. Problems related to low knowledge level require different solutions than those related to attitudinal problems. Similarly, strategies for changing negative attitudes on a recommended technology are likely to be different than those for solving incorrect practices in technology application or convincing people to try to practise a recommended technology. The implications of the different KAP levels would greatly influence the development of problem-solving strategies, message design, selection of multimedia mix (including when and how to utilize group and interpersonal communication channels, such as extension workers), and materials development. Application of a behavioural modification approach using information based on the KAP levels of the target audience for message development, media selection, and materials development alone could significantly increase the cost-effectiveness of extension activities.

Cost-Effective Multimedia Approach

One of the most common problems or constraints of a national extension service is the shortage of field extension personnel to reach large numbers of farmers in widely spread geographical areas with inadequate transportation. Moreover, extension workers are usually overburdened with an unnecessarily heavy workload which includes almost everything that has to do with farmers at the village level. Such an overreliance on extension workers is neither technically sound nor operationally efficient. Some extension functions for certain purposes such as awareness creation, information delivery, and motivational campaigns can be more effectively and efficiently performed by other means, channels, or non extension groups under the coordination and supervision of extension workers.

Extension workers' workload could be reduced by mobilizing appropriate rural and community-based resources, including the increasingly accessible and low-cost mass communication channels (*e.g.*, local radio stations, rural press, folk/traditional media, posters, flip charts, silk-screened printed materials, audio-cassettes, slide-tape presentations, leaflets, comics) to disseminate standardized and packaged extension messages, and by using local volunteers (*e.g.*, school teachers, children, local and religious leaders) to serve as intermediaries in reaching farmers.

Such an approach does *not* imply that extension workers can or will be replaced by these community resources. Rather, it is a rational approach of using available resources most efficiently for certain tasks, such as using extension workers for educational or instructional purposes, which require two-way interactions, field demonstrations, group discussion, and so forth that cannot be done as effectively by mass communication channels.

The SEC method employs a multimedia approach whereby a cost-effective *combination of mass, personal, and group communication channels* (including extension workers and trainers) and materials are efficiently utilized to reduce extension cost and efforts and to increase its effectiveness in dealing with a larger number of target audiences more rapidly.

Specific Extension Support Materials and Training

Most extension services in developing countries suffer from the lack of relevant and practical extension and training materials to support the field activities of their extension workers. Many extension workers rely primarily on their interpersonal communication skills, and thus their time during farmers' meetings may not be used as effectively as it should be.

Providing specifically designed and relevant training support materials to extension workers will not only facilitate their tasks and reduce their heavy workload. It will also ensure a certain degree of *quality control* in the delivery of technical information or extension message contents. Experience has also shown that extension workers' motivation, enthusiasm, confidence, and credibility increase when they are given relevant and attractive multi-media support materials which they can use to improve the effectiveness of their extension and training work.

Applicability to Other Extension Programmes

Most, if not all, of the important principles and techniques employed in planning, implementing, and managing SEC activities are applicable for developing and implementing any extension programme. The SEC's process, operational phases, and implementation steps are essentially similar to that of a regular (but well-designed) extension programme. SEC could thus be considered a microcosm of an agricultural extension programme. It may be safe to assume that if SEC activities can be carried out successfully in a campaign context, which has a very short time period, then SEC processes, methods, and techniques, either partially or holistically, can be incorporated effectively into a regular and institutionalized extension programme which has a longer time span.

Given appropriate training in various SEC principles and techniques and through direct involvement in undertaking a planned SEC programme, trained extension staff could help in applying a systematic, rational, and pragmatic approach to planning, implementing, managing, monitoring, and evaluating regular/routine programmes of an agricultural extension service. As indicated in a number of SEC evaluation studies, many extension staff who had been trained and have implemented SEC activities have continued to use their skills in developing and implementing other institutionalized extension programmes for various agricultural technologies. Some have replicated the complete SEC process, while others have applied only certain SEC principles or techniques. These efforts have been appreciated and welcomed by many senior officials of the ministry of agriculture where SEC activities were undertaken because they could see its concrete outputs, results, and impact.

Operational Definition of Strategic Extension Campaign

In the context of agricultural extension, a campaign is one of the methods of extension which can reach a large number of target beneficiaries in a short time period. To complement and improve programmes of a national agricultural extension service, the SEC method gives special emphasis to the following:

- ☆ A campaign is purposive, problem-solving oriented, and focuses on a specific issue or recommended technology.
- ☆ Its goals are consistent with, and guided by, the overall agricultural development policies and extension programme objectives.
- ☆ Campaign objectives are specific and formulated based on intended beneficiaries' felt needs and problems identified through a baseline survey of their knowledge, attitudes, and practices (KAP) vis-a-vis the recommended technology.
- ☆ A specific campaign strategy is developed with the aim of solving problems that caused non adoption, or inappropriate or discontinued practice, of the recommended technology.
- ☆ A strategic planning approach is applied in the process of target audience segmentation, multi media selection, message/information positioning and design, and extension/training materials pack aging, development, and production, with a view to obtaining maximum output-impact with the least effort, time, and resources.
- ☆ Formative evaluation in the form of field pretesting of prototype multimedia campaign materials is conducted before they are mass produced.
- ☆ A comprehensive and detailed campaign management plan is an integral and vital part of the SEC process. It will not only spell out the implementation procedures and requirements, but will also be used to develop a management information system, including monitoring and supervision procedures.
- ☆ Special briefing and training for all personnel who are involved in SEC activities must be undertaken to ensure that they understand their specific tasks and responsibilities and have the necessary skills and support materials to perform such tasks effectively.
- ☆ Process documentation and summative evaluation on SEC implementation and impact are conducted, and those results are used to improve its ongoing performance (through Management Monitoring Survey) and to determine SEC's results and overall effectiveness (through Information

Recall and Impact Survey, Focus Group Interviews, *etc.*), as well as to draw lessons learned from such experiences for future replications.

Integrated Process and Systems Approach

The conceptual framework of the strategic extension campaign (SEC) follows a generic model originally proposed by Adhikarya (1978) and described in detail in several of the author's other publications (Adhikarya, 1994; Adhikarya and Posamentier, 1987). The SEC method advocates carrying out extension activities in a *systematic, sequential, and process-oriented manner,* rather than on an ad hoc basis. It is a *planned* extension programme with interrelated activities to be carried out following a management implementation plan by well-trained personnel within a given time schedule.

Staff Training as an Integral Part of SEC

Applying such a systems approach points out the need to train staff to master the whole extension process, rather than only some elements of the process or part of the activities. It is, therefore, imperative that in the process of introducing or replicating the SEC method in a country, training a national group of extension planners, managers and trainers, subject-matter specialists, communication support staff, and field extension officers on relevant SEC methods and techniques must be an integral part of the SEC programme planning and implementation. Significantly higher cost and a longer time are required for the initial implementation of SEC activities, but such an investment in human resources development is perhaps one of the most cost-effective inputs which could significantly contribute towards institutionalization of an SEC approach in improving and strengthening a national agricultural extension system and service.

KAP Surveys, Evaluation Studies, and Follow up Actions

In addition, operations research such as baseline Knowledge, Attitude, and Practice (KAP) surveys and other means of formative and summative evaluation must be built in to the extension process, programme, and methodology. The most vital part of the process is the actual field action and implementation activities which will be required for the preparation and follow up of strategic extension campaign training and operations research activities.

Participatory Method of Planning and Implementation

The SEC method relies heavily on the participation of those who are involved in campaign activities, including campaign staff as well as target beneficiaries, in providing relevant inputs or suggestions throughout the campaign process. By so doing, SEC activities are more likely to address relevant problems and issues, as well as to provide practical solutions which are based on target beneficiaries' needs. Not only should the intended beneficiaries of a campaign be consulted, but potential campaign implementers such as concerned field staff, trainers, and community leaders should also be involved and given a role to play in different aspects of campaign planning, implementation, and management.

Through a careful and systematic method of soliciting fairly specific ideas and suggestions from a sample of concerned groups of people involved in SEC activities, a useful *campaign map* can be drawn to determine its direction, strategy, types of message contents, training needs, and field implementations requirements. A problem-solving and demand-driven strategic extension campaign also needs relevant "feed-forward" information from its intended beneficiaries and concerned campaign personnel. It is thus imperative that a participatory oriented method of planning be applied to ensure a needs-based extension campaign which can realistically be implemented and provide benefits to its intended clientele.

Suggested Conceptual Framework for Strategic Planning of Extension Campaigns

Planning is defined as a process of identifying or defining problems, formulating goals, thinking of ways to accomplish goals, and measuring progress towards goal achievements (Middleton and Hsu Lin, 1975). Such a plan must outline the management actions to be taken in implementing the strategy. Thus, in this context, campaign planning has to include both strategy planning (*i.e.*, what to do) and management planning (*i.e.*, how to make it happen).

Strategic planning can be defined as the best possible use of available resources (*i.e.*, time, funds, and staff) to achieve the greatest returns (*i.e.*, outcome, results, or impact). A strategic extension campaign plan should provide specific guidelines and directions in developing and making information, education, and communication activities operational. It must be constantly reviewed, especially at the implementation stage. Modification of the plan may be required because of specific local conditions and problems or alteration of the policies or objectives which guided the original plan. The plan should be flexible and ready for necessary modification as suggested by feedback results (*e.g.*, through process and formative evaluation, including pretesting) in order to improve the strategy or management of campaign activities.

Creating Positive Evaluation Metaphors

The term "evaluation" can be threatening. If evaluations are to be positively embraced by extension organizations, it may be necessary for those initiating them to be positive in the language they use to describe the proposed effort. Negative metaphors that are sometimes used for evaluation, such as "under the gun," "a mine field," "an invasion of inspectors," "last rites before hanging," and "defend the fortress" reflect past negative experiences with evaluation. Why not create more positive metaphors for the evaluation task? For instance, we could describe evaluation in medical metaphors such as "complete programme physical" to identify physical soundness and potential programme health risks. The term "autodiagnostico" has been used in the Dominican Republic and "self-strengthening" in Sri Lanka. Or we could use a business metaphor, such as "taking stock" of profits and liabilities. We could use sports metaphors, such as "watching video tapes" of yesterday's game to improve our "game plan." We can use "mirrors" to see ourselves as others see us. We could consider evaluation to be a "learning process" or a "sabbath" day of reflection, renewal, gratitude, and celebration. Being self-conscious and deliberate about the language that is used regarding evaluation may help reduce unnecessary resistance to the effort and may build positive support.

Exploding Myths about Evaluation

Several evaluation myths have often discouraged extension managers from engaging in useful evaluation.

Myth #1: *Evaluate only when mandated.* Many funded programmes require evaluation as a form of accountability. However, it is a myth that evaluation should occur only if it is mandated. On the contrary, evaluations that are self-initiated are more likely to be taken seriously for immediate programme improvement. Programmes become responsible and excellent just as often through self-initiated evaluation.

Myth #2: *Evaluation is an add-on.* It is a myth that evaluation is an add-on activity or at most a pre-test with a post-test. It is most meaningful when it is integrated with decision making at every stage of programme planning and operation (Patton, 1991).

Myth #3: *Evaluation is an activity for experts.* It is a myth that evaluation should be undertaken only by technical experts. Yes, complex methods can be used; however, systematic evaluation can be undertaken by inexperienced managers, and specialists and educators themselves can be helped to critique their own work. This chapter is intended to help managers and specialists to do evaluation themselves.

Myth #4: Outside evaluators are best. It also is a myth that evaluation should be done only by external, outside, objective evaluators. Yes, external evaluators are often useful in challenging insiders to address what they have overlooked because of their nearsightedness. However, internal, self-initiated, and subjectively oriented evaluations also can be rigorous and valuable. In fact, because they often are participatory in generating, analysing, and interpreting data, they may result in greater acceptability of the findings and recommendations.

Myth #5: *There is one best evaluation approach.* Still another myth is that there is one best way to conduct an extension evaluation. Some approaches are probably better than others for addressing particular types of questions or concerns. However, the many types of evaluation approaches have their own strengths and limitations. Some situations require quantification and measurement, while others require qualitative, descriptive, and subjective data. Alternative approaches will be briefly described later in this chapter.

Myth #6: *Quantitative data are best.* A mixed-methods approach combining qualitative and quantitative methods can lead to better understanding and appreciation of phenomena under evaluation and provide triangulation, convergence, and corroboration of results from different methods. Qualitative methods are best for understanding the nature of something, while quantitative methods help in appreciating its extent. If we do not know the nature of something, we should conduct qualitative studies. After measuring something, we may still need to use qualitative methods to learn about variations and unique forms. The use of mixed methods can increase user responsiveness to evaluation information (Greene, Caracelli, and Graham, 1989).

Major Elements in Evaluation

There are at least five major elements in most evaluations:

1. Focus questions,
2. Objects or events to be evaluated,
3. Data or evidence,
4. Analysis and interpretation using judgement perspectives, and
5. Judgements, conclusions, or findings. Purposes and approaches or models may vary, but these elements will be present in one form or another.

Selecting Evaluation Purposes

There are many purposes for undertaking evaluations in any particular situation. It cannot be assumed that all stakeholders (farmers, extension staff, administrators, funders) share common purposes. Although people may never completely agree about the purposes, one of the tasks at the outset is to seek clarity about evaluation purposes through discussions with major stakeholders. Appreciating different purposes at the outset can reduce conflicts and disappointments down the road. Consider the following purposes.

Pseudo Self-Serving Purposes

Since organizations, including extension systems, have a self-serving tendency, it is not unreasonable to expect that some staff members, especially those in the highest places, may want a pseudo evaluation that will postpone, buy time, or avoid threatening change. In these cases, evaluators are not taken seriously, and the evaluation becomes a meaningless political diversion. In other cases, some members of organizations want evaluations as excuses for evading or avoiding administrative responsibility or to provide a scapegoat for criticism. Evaluations that are undertaken only to make the programme look good ("whitewash job") or to make someone or some aspect of a programme look bad ("hatchet job") are pseudo and illegitimate.

Enhance Accountability Purposes

It is quite common for external donors to expect that evaluation will provide accountability through evidence of impact, or to document cost-benefits, or to measure efficiency-effectiveness. In some cases, this evaluative evidence is considered in decisions to continue the programme; or to propose change, expansion, or reduction of a programme; or to change a policy, organizational structure, philosophy, or design. The potential for negative findings and the threat of discontinuing funding has led to "hiding the mistake," a dysfunctional practice. However, evaluations rarely provide a single basis for political decisions. They often are used by funders, administrators, or policy makers to justify their decisions even when the evidence of benefits is weak.

Improve Performance Purposes

This purpose of evaluation is sometimes called "formative" because the results are intended to help improve the programme during its formative stages. This is in contrast to "summative evaluations" when the purpose is to sum up or summarize the accomplishments at a point in time. When evaluations are to improve programmes, lessons learned about strengths and limitations of the programme are mined from the data so that changes can be made immediately. Sometimes the intent is to discover new approaches and alternatives or to adjust the programme to changing situations or client groups. Evaluation also is used to understand multiple reasons for apparent failure or to improve the management or operation of a programme.

Social Learning and Communication Purposes

Sometimes evaluations are intended to stimulate political dialogue or to resolve political conflicts intelligently. For example, an evaluation of extension in a country could provide an opportunity to debate the need to hire more women agents to respond to an increase of women in small-scale agriculture or to extend the extension network to subsistence farmers not being served. Often the most significant contribution of an evaluation is the creation of new expectations, new organizational arrangements, new linkages, and new purposes and goals. Evaluation may give visibility to a good idea and new language that can communicate new ways of viewing extension to others who also may want to share an experiment.

Evaluation purposes tend to vary, depending upon where one stands within a system like extension. External funders often want an accountability purpose, while field staff are more likely to favour a programme improvement purpose. Policy makers and programme administrators can often appreciate an evaluation that contributes to new ways of thinking about extension or new forms of extension. Farmers want an evaluation to improve the benefits they may receive from extension staff.

Recognizing the Politics of Evaluation

Evaluations are never value neutral. Political implications are always present because stakeholders have interests to defend. In this sense, all evaluations are political. Therefore, it is very important to know the values, expectations, and interests of the stakeholders at the outset. Failure to understand these interests may make the findings irrelevant or may cause unnecessary conflict and rejection of the findings. Knowing the interests also helps in the focus and design of the evaluation.

Values, expectations, and interests are reflected in stakeholders' visions regarding what "good" looks like in extension. Some may favour an extension system that empowers the poorest of farmers in their struggle for more land and a larger share of market benefits. Multinational agribusinesses may view the success of extension as increasing productivity for export. What a "good" or "excellent" extension system should look like will vary according to these different values and criteria. What different funders want may differ from what various groups of farmers may want in an extension system. Assumptions of stakeholders also vary regarding the relationship of extension education to larger social and political visions. For some, the extension system is intended to reproduce existing power relationships between government and farmers.

Alternative Approaches and Models

What are several alternative approaches to programme evaluation? Choosing among these approaches is important because for each there are different assumptions about what data to collect, how to collect them, and how to make judgements about success. The following seven major approaches will provide a sufficient choice for most extension evaluation situations:

1. Expert model,
2. Goal-free model,
3. Attainment of objectives model,
4. Management decision model,
5. Naturalistic model,
6. Experimental model, and
7. Participatory evaluation model.

Major Models for Programme Evaluation

Expert Model

This approach relies on expert judgement (Eisner, 1983). Usually, documentation is prepared in advance of experts' visits. The experts then interview, analyse documents, and make judgements using their own judgement perspectives or those set as standards by the outside organizations or stakeholders. Typically this type of evaluation brings in a team of experts from FAO or extension systems from several countries to make judgements and comparisons regarding strengths and limitations.

Goal-Free Model

This approach assumes that outside evaluators do not know, or need to know, what the programme has intended to accomplish, but that it is the task of the evaluators to uncover what is actually happening relative to farmers' interests regardless of stated goals and intentions. The focus point is to identify environmental and farming conditions and then to compare these needs with what people are actually experiencing as a result of the extension programme. The gap is then viewed as a starting point for making changes in the programme. An example is an evaluation that describes conditions of indigenous farming groups cultivating fragile hillside soils and comparing these conditions with access to and appropriate content of knowledge from existing extension services. This approach relies heavily on open-ended interviewing and observation by persons who do not have a vested interest in the programme (Scriven, 1972).

Attainment of Objectives Model

This approach assumes that the success of a programme can be determined by measuring a programme's outcomes against its own goals and objectives. This type of evaluation begins with clarifying measurable objectives and then gathering data that validate the extent to which these objectives have been met. For this model to be credible, an essential feature should be added, namely, the evaluation of the appropriateness of goals and objectives, given the circumstances and needs of farmers. For example, an extension system may have adequately met its objectives of increasing production of maize among large landholders, but at the same time it may have neglected to question its lack of commitment to small landholders or tenant farmers. If an attainment of objectives evaluation is anticipated, programmes are often tempted to set goals quite low so that outcomes will be met easily, thus appearing to be successful while ignoring major challenges. This model also has a "black box" limitation in that it tends to ignore the extension process, thereby failing to provide explanations for outcomes (Provus, 1971).

Management Decision Model

The purpose of this model is to provide relevant information as a management tool to decision makers. It assumes that evaluation should be geared to decisions during programme initiation and operation stages to make results more relevant at each particular stage. Participation of stakeholders is central to the process because evaluation should serve their decisions. Sometimes cost effectiveness and operations monitoring are included (Stufflebeam, 1971; Tripodi, Pellin, and Epstein, 1971; Gold, 1988). One limitation of this model is the tendency for the decisions of major stakeholders to be viewed as more important than those of various types of farmers, especially women in agriculture who may not benefit directly from such an evaluation unless care is taken.

Naturalistic Model

This model assumes that a programme is a natural experiment and that the purpose of evaluation is to understand how the programme is operating in its natural environment. There is an assumption that programmes are negotiated realities among the significant stakeholders and that evaluation serves this value-laden negotiation (Cronbach, 1981; Guba and Lincoln, 1989). Data should be collected and analysed from multiple perspectives. The outcome of the evaluation is dialogue concerning disagreements about objectives, expectations, problems, opportunities, policies, procedures, and suggested changes in methods or activities. Many positive collaborative changes can be made through this model of evaluation if conflict resolution skills are combined with evaluation. Another purpose of this model is to diagnose or to identify the causes for certain behaviour on the part of some farmers, agency staff, or other development actors (Murphy and Marchant, 1988).

Participation Evaluation Model

Why use participatory approaches to evaluation? Involving people who are on the receiving end is likely to assure the most efficient allocation of scarce resources and the early identification of ineffective or wasteful use of resources. People on the receiving end are ultimately the best judges of impact, whether benefits have been produced or not (Uphoff, 1989, 1992). Being included in planning, implementation, and evaluation will show farmers that they are regarded as responsible, capable individuals and not simply passive "beneficiaries" or a "target group." Participatory evaluation is self-educating. It can encourage the development of human capacities among farmers. Without participation in evaluation, sustainable agriculture is unlikely. Participatory evaluation also can decrease the paternalistic, directive, impatient, or insensitive relationships among officials, technicians, and farmers by improving staff attitudes toward working with persons having less status and education. Although an initial cost of time is required, participation in evaluation can speed implementation when participants take greater ownership of efforts.

Part II

Society with Sociology and Psychology

10

Rural Sociology

Introduction of Sociology

What is Sociology?

The word sociology is derived from the Latin ward. Socius literally meaning friend, companion, associates but in broader sence group or society and the Greek Word logus means science or study. Thus sociology means basically as a science or study of Society.

According to Rogers, sociology is defined as a study of ways in which social experiences function in developing, maturing, refreshing human beings through interpersonal stimulation. Some important definitions of sociology have been given by different sociologists are as follows.

Kimball young defines sociology as a systematically and orderly study of man in society that is of group life and of the customs, traditions, institutions, and *ways* of thinking and living which are linked to group life.

R. M. Maclver has defined sociology as the study of the nature of the human relationship. It is the study of human relationship between the individuals. It aims at the discovery of the principles of cohesion and harmony in a social structure, which conditions the activities of social individuals.

J. B.Gitter has defined "Sociology as the study of forms and processes of human togetherness". Sociology stidies basic principle of human associations.

According to F.H. Fiddings "Sociology is the science of associations of minds, sociology tries to explain the origin growth and structure of society."

Sociology helps us in understanding society, its various institutions human relationship and their manifestation man's social interactions, and socialization.

From the above definitions we get three views about what is sociology *i.e.* sociology is the science of society, it is a study of social relationships and it is a study of the form of social relationships. When the field of sociology is limited to a particular area it can be named as:

1. **Urban Sociology :** Dealing with urban peoples
2. **Rural Sociology:** Dealing with rural people.
3. **Industrial Sociology:** Dealing with industrial people
4. **Sociology of family and so on.**

Rural Sociology

- ☆ Rural sociology it is made of two terms. Rural arid sociology, sociology it is limited to the study of various aspects of rural society, Rural area where agriculture is the main occupation of the people. Rural area is the area where population is less than 5000.
- ☆ Rural sociology studies the rural life and the development in rural life. Rural sociology is the science, which studies the structure and functioning of rural society. Different sociologist and social thinkers have defined rural sociology in different ways.
- ☆ As defined by Sanderson "Rural sociology is the sociology of rural life in the rural environment.
- ☆ Bertrand in its broadest definition, defined rural sociology is the study of human relationship in rural environment.
- ☆ According to F Stant Chapin has defined "The sociology of rural life is the study of rural population, rural social organization and the social processes operative in rural society".
- ☆ Supporting this view T. L. Smith says "Some investigators study social phenomenon that are present only in or largely confined to the rural environment to persons engaged in agricultural occupation. Such sociological facts and principles as are derived from the study of the rural social relationship may be referred to as rural sociology".
- ☆ According to A. R. Desai "Rural sociology is the science of rural society. The laws of the structure and development of rural governing society"
- ☆ Rural sociology is the study of the sociology of life in rural environment which systematically studies rural communities to discover heir conditions and tendencies and to formulate to discover principles of progress. Rural sociology is therefore the study of rural environment and social facts and social interactions that are found in rural society. As a scientific study it studies the social organizations, structures and set up. It provides us in that knowledge about the rural social phenomenon which can helps us in making contribution to the development of rural sociology.

Nature of Rural Sociology

There is a controversy whether the sociology can be regards as science with his own subject matter. Science may be defined as body of systematized knowledge or synthesized body of knowledge. In the collection of facts for any knowledge, when we apply a scientific method it is called as a science. Science goes with the method and not with the subject matter. Scientific method consists of systematic observation, classification and interpretation of data. It is believed that rural sociology employs the scientific method. The nature of rural sociology as a science can determine on the basis of following facts:

- **Use of Scientific Method:** It is uniform fact that rural sociology employs the scientific method. Almost all the methods of scientific study *viz.* observations, interview schedule, questionnaire method, case study, statistical methods *etc.* are employed in the study of rural sociology. In the absence of scientific approach the village problems cannot be studied. On such studies we formulate generalized principles and laws on which we forecast future trends.
- **Factual Study:** It studies the social events, social relationship and process in a factual manner. It also studies and analyzes the facts and the underlying general principles and theories.
- **Discovery of Cause and Effect:** Rural sociology formulates its theories and laws on the basis of cause and effect relationship.
- **Universal Laws:** The laws formulated by rural sociology are universal in nature. Because under normal or similar condition they prove to be correct and produce the same result.
- **Predictions:** Since the laws formulated by rural sociology by rural sociology are based on cause and effect relationship. It is possible to predict the result. On the basis of conditions enumerated above, it is said that rural sociology is by nature a science. However, there are some factors which limits the scientific sociology is by nature a scientific nature of the subject.

Limitations of Rural Sociology

Lack of Objectivity

It is not possible to have objectivity in the study of rural sociology as in case of natural science. While in the study of rural sociology and its problems the investigator continues to remain as a part of the society he is studying. He has own ideas and are influenced by the subject matter.

Lack of Laboratory

Rural sociology is not studied in laboratories as a natural science. Because of this it is not possible to verify and test the theory and principles of rural sociology.

Lack of Measurement

There is no definite and standard measurement for measuring units of rural sociology.

Lack of Exactness

It is lack of objectivity and different to follow its laws and principles universally. They are not acceptable at every point.

Lack of Prediction

- ☆ Because of lacks objectivity and exactness the principles formulated by the rural sociology are not always correct. As such predictions are not possible.
- ☆ It is not possible to draw a line between the rural and urban areas. There is no sharp demarcation to tell where rural area ends and urban area begins.
- ☆ The science of rural sociology is not fully developed.

Importance of the Study of Rural Sociology

- ☆ The importance of rural sociology can be evaluated properly when it realize the importance of rural society. Rural society presents a scientific picture of rural life. Villages are important because they are the springs to feed urban areas. The importance of rural sociology can be put under following heads.
- ☆ Man has an urge to know human relationship and this can be satisfied through rural sociology.

Rural Population is in a Majority

In almost all the Countries of the world majority or the world resides in villages in villages. It is true that over 80 per cent population of India resides in villages.

It Gives Complete Knowledge of Village Life

Rural sociology gives us complete knowledge of village life. Village is the first unit of development in country. It is a centre of culture of any country.

Rural Reformation

Rural reformation is the primary aim of rural sociology. In this context it helps in following works.

Organization

The Village units which are un-organized and they can be organized through rural sociology. It improved in the co-ordination of various units and helps in bringing an improvement in economic, social and health conditions.

Economic Betterment

Through detailed study of village problems and observation rural sociology gives stress on the importance of increasing the quantity and quality of production. This results in to raising the standard of living.

Provide Technology and Systematic Knowledge and reforms in Farm Production

Main occupation of 80 per cent population of village is agriculture. In order to improving this main occupation of rural people, the earlier researches in rural sociology was made in agricultural college.

Solutions of Pathological Social Problems

Rural sociology examines the social pathological problems and it suggests ways for the improving these problem.

Education

The improvement, the development of any community depends on its education. Rural sociology lays stress on education in rural problems.

Planning for Development

Rural sociology encourages the development of various plans for any rural development programme. The work must be carried out according to these plans for the progress in rural society.

Rural Sociology Development Relationships of Village with Industry

Rural Sociology is most important in Agricultural Countries

About 90 per cent of world progress is based on agriculture. It is only in agricultural countries that people realize the importance of rural sociology. India is mainly agricultural country. For its all sided development the development of rural sociology is very important.

Importance of Rural Sociology in India

India is predominantly an agricultural and rural Country from ancient times. Village is the basic and important unit in the organization of Indian Society. After independence the process of reconstruction of the country has been started and the importance of rural sociology was recognized. If India is to achieve the progress, the villages are to be improved and education has to be spread in these villages. In order to this every attempt is being made to improve the villages of this country in community development programme. This object can be achieved only when the administrators and planners have the correct knowledge of the rural life.

The importance of the Study of Rural Society in India can be Evaluated Under the Following Heads:

- The village is the basic source of Indian culture. India lives in villages. The birth and development of its culture took place from the villages. India's progress is based on the progress of villages. Thus it is necessary to develop the villages.
- India is an agricultural country and poverty of this country can only be removed when the agriculture is improved.
- Solution of rural problems can bring about change in rural society.
- The country and its society can be reconstructed only through rural reconstruction.
- In the interest of democratic decentralization, village community is to be studied in a totality.
- Rural sociology can helps in organizing the disorganized Indian rural structure.

Scope of Rural Sociology in India

India is an agricultural country and village is a basic and important unit of the society. After independence the process of rural reconstruction was started and importance of the rural sociology was recognized. The need of development of the villages and speed of education in the villages were understood and to achieve this community development programme was started. It can be achieved when the planners and administrator correct knowledge of the rural life.

- It is very essential to develop village because India's development depends upon the progress of the villages.
- India is agriculture country and poverty can be removed through improvement in agriculture.
- Solutions of rural problems can bring the change in the rural society.
- The country and its society can be reconstructed only through rural developments.
- For successful implementation of democratic decentralization the village community is to be studied in detail.
- Rural sociology can help to organize the disorganized Indian in detail.
- The extension worker must know the rural culture, rural institutions, problems, resources *etc.* for successful transfer of technology for improvement of agriculture. It can be achieved through the study of rural sociology.
- Through the technology and communication methods are known to the extension workers. The study of rural sociology helps the extension worker to transfer the technology.
- For successful implementation of the community development programmes the knowledge of rural sociology is very essential.

Characteristics of Rural Life

Rural people are different from those in living in urban areas. The distinctive characteristics of rural life can easily be shown by comparing them with those of urban (city) life. The difference in cultures or subcultures of these two areas differences in attitude and behaviour and perhaps in dress and dialect (language use in a region or area). The cities are large, impersonal and complex in social structure, while rural communities are small, intimate and simple in organization. Though there are differences, there are also common attitudes and behaviours that both rural and urban people share as a member of larger culture. Common elements between urban and rural life is:

- ✰ Common language, literature, philosophy *etc.*
- ✰ Common institutions like religion, education, family life business and political organizations *etc.*

Both rural and urban people have own characteristics. The characteristics of rural life can best and most easily recognized by comparing those or urban life. Following are the some rural – urban differences:

Difference between Rural and Urban Life

Rural Life	*Urban Life*
Environment: Close/direct contact with nature. Preliminaries influenced by natura environmental elements like rain, heat, drought, frost, sow *etc.* over which there is no control.	Greater isolation from nature. Predominance of manmade (artificial environment).
Occupation: Agricultural is the fundamental occupation. Majority of population is engaged in agriculture. Neighbors of Agricu turist are also agriculturist	No fundamental occupation. Most of people engaged in principally in manufacturing, mechanical pursuits, trade commerce, professions and other non-agricultural occupations.
Size of Community: Size of community is very small in size. Agriculturalism and size of community are negatively co-related.	Size of community is large in size. Urbanity and size of community are positively co-related.
Density of Population: Density of population is lower. Density and rurality are negatively co-related.	Size of community is large in size. Urbanity and size of community are positively co-related.
Homogeneity and heterogeneity of population: More homogenous in social, racial and psychological traits. Negative co0-relation with heterogeneity. (Most are agriculturists are directly connected with agriculture).	More heterogeneous than rural. Urbanity and heterogeneity are positively co-related (Different type of population is seen in cities, different places, religions, caste, class race, community, economic and cultural differences, occupations and behavioral pattern also different).
Social Differentiations: Low degree of social differentiation	High degree of social differentiation
Social Stratification: More rigid Fewer economic, occupational, and socio-political classes. Less social stratification than urban.	Less rigid Urban community is much more strategic than the rural with having much more economic, occupational and social political classes.
Social Mobility: Mobility is less intensive. Territorial, occupational and other forms of social mobility of the population are less intensive. They follows same occupation, stay in the same village	Social mobility is more intensive. People change occupation and even leave places in search of new and better occupation

Rural Life	*Urban Life*
Social Interaction: Less numerous contacts. The area of interaction system is narrower. More professional, simple, face to face. Informal, sincere relations.	More numerous contacts. Area of interactions is wider, the relation are superficial and short-lived. The popular are more formal and showy.
Social Solidarity: Social solidarity or cohesiveness and unity are stronger/greater than urban. Common traits, similarity of experiences, common aims and purposes, common customs and traditions are the basis of unity in village. Strong sense of belonging and unity.	Social solidarity is less strong than rural, dissimilarities, division of labour, interdependence, specialization, impersonal, strictly formal relationships results comparatively less sense of belonging and unity.
Social Control: Social pressure by community is strong. Conformity of norms is more by informal social pressure.	Control is more by formal impersonal means of laws, prescribed rules and regulations.
Social Change: Rural social life is relatively static and stable.	Urban social life is under constant and rapid social change
Culture: Sacred (Religious) culture.	Secular (Non-religious) culture.
Leadership Pattern: Choice of leadership more on the basis of known personal qualities of individual, due to greater face to face contacts and more intimate knowledge of individual.	Choices of leadership is comparatively less on the basis of know personal qualities of individual
Group: Rural society is simple Unit-group society	Urban society is complex multi-group society.
Social Institutions: Most of the institutions are natural outgrowth of rural social life. Less enacted institutions.	Numerous enacted institutions.
Standard of Living: Home conveniences, public utilities, educational recreational religious, medical, communication and other facilities for living can be provided if supported by sufficient population base	In urban areas such conveniences and facilities are provided due to greater density of population
Standard of living is low.	Standard of living is high.

Rural Sociology and Educational Psychology

What is Culture?

Man alone of all the animals creates and transmits a social environment. It is his unique characteristic; other animals can acquire behavior pattern by initiating other animals or by resorting to trial and error. They do not transmit their learned activities to future generations through languages and other symbolic systems neither they can make physical objects to use as tools in their efforts to get food, shelter and other goods nor they pass them on to succeeding enervations. Man in adjusting to the natural environment – and to his fellows – fashions vast quantities of physical and biological materials and large bodies of through and action pattern. All these uses to fulfil the desires and needs which make it possible for satisfy present and expanding wants, to satisfy new values, to express creative desire, to store product for future use and to transmit succeeding generations.

The total content of this produced social environment of a society its culture. It is the integrated, interdependent whole of these socially produced and inherited action patterns built around a body of socially created physical and biological material. The culture of society therefore the total environment of the physio-social, bio-social and psycho-social universes, man has produced and socially created mechanisms through these social products operate.

Culture has been defined in a Number of Ways

Tylor defined "Culture as that complex which includes knowledge beliefs, morals, laws, customs and any other capabilities and habits acquired by a man as member of society".

According to Maclver "Culture is the expression of our nature, in our modes or living and thinking, inter course, in our literature, in religion, in recreation and enjoyment".

Hepple says "Culture includes all the material and non material traits in our environment which are manmade or modified by man and transmitted from one generation to another by the process of learning".

Culture can be defined as historically derived pattern of living of an individual. It includes traditions, customs, folkways and mores. All these ideas integrated together for pattern of living called culture.

Cuber defined "Culture is the continually changing patterns of learned behavior and products of learned (including attitudes, values, knowledge and material objects) which are shared by and transmitted among the member of society".

Culture is a socially standardized, way of feeling, thinking and acting that man acquires as a member of society.

Culture includes everything that man learns or acquired as a member of particular society.

To summarize, culture is continually changing it is learned behavior which has been organized into patterns, it includes products of learned behaviors – ideas, knowledge, values and material objects and is shared and transmitted among members of society.

Characteristics of Culture

- Only human beings posses the culture, other animals do not posses it. Man has created the culture during the process of controlling himself, others and nature.
- All the societies in the world have culture but each society has a different culture from one another. Indian culture – African culture – Western culture.
- Culture is learned or acquired after birth and through life consciously or unconsciously through agents influencing directly or indirectly on individuals. It is not innate.
- Culture is a social and not individual heritage of man. One has to learn the tradition, customs as the society in which he is living.

- ☆ Culture is transmitted from one generation to another generation during the process of socialization and is disseminated among members of society.
- ☆ Culture is adoptive – It must adjust itself to external forces of various kinds.
- ☆ Culture is integrative – culture has general tendency towards consistency and integration so as to held society together.
- ☆ Culture is idealistic – it stands for ideal norms of human behavior toward which persons are expected to strive.
- ☆ Culture is shared – Learned behaviour and products of learned behavior is nor exclusive property of a single individual or group but belong to all members of society and are shared by a large proportion of the.
- ☆ Culture is gratifying. It satisfies the man's bio-logical and socio-culture needs- foods, clothing, shelter and for various relationship with other individuals and group.
- ☆ Culture is accumulating. It goes on increasing in size, act to the funds of his knowledge, regarding the science or literature in each generation. This knowledge is preserved in the form of books, films and pictures.
- ☆ Culture is not static but dynamic culture is continuously changing. The commonly way of thinking, feeling, acting are continuously changing. *e.g.* observing untouchability, praying god for cure of diseases.

Importance of Culture for an Individual and the Group

The Culture forms an important element in social life of a man. It is the culture that makes human animal to man. It regulates his conduct and prepares him for group life. It teaches him what type of food he should take and what manners, how he should cover himself and behave with the fellows, how he should speak and influence the people, how he should co-operative and compete with other. Man has acquired these qualities required to live and social behaviour even for complicated situations.

There would not have been group life without culture. Culture regulates the behavior of the people and satisfies their primary drives *i.e.* hunger, shelter and sex, which enable him to maintain group life. People behave the way of society. Culture has provided a number of checks upon irrationally or irresponsibly. It has kept social relationship intact.

Culture has given new vision to the individual by providing him a set of rules of co-operation. Culture teaches him to think of himself as a part of large whole. It provides him with the concept of family, state, nation, class and makes possible the co-ordination and division of labours. Culture also creates new needs and drives for example thirst for knowledge and arranges for their satisfaction. It satisfies the aesthetic, moral and religious interest of members of group.

In brief, the culture gives the individuals or groups the feelings of unity with the group. It enables them to live and work together without too much confusion and mutual interference.

Functions of Culture

a) Culture provides us with design for living. It is always learned and acquired.

b) Culture provides a series of pattern by which biological and socio-cultural demands of group members are met *e.g.* food, shelter, and reproduction and relationship with group and individuals.

c) Culture provides a set of rules to ensure co-operation of the individuals of a group in adjusting environmental situation.

d) Culture provides individual a set of ready-made definition of situation.

e) Culture helps in understanding and predicting the human behavior and also it provides channels of interaction for individuals within the group.

f) Culture provides us a guidepost or kind of map for all our life activities. It defines the pattern of behaviour for individuals so that he acts according to the behavior pattern prescribed and defined by culture.

g) Culture acts as a means of social control through norms, folkway, and moves laws.

h) Thus culture functions to deeply influence control and direct behaviour and life of the individual and of group. All these functions are controlled through norms or rules made in society. These norms may be governed by sanctions or punishment, so the study of norms, mores, folkways, laws *etc.* is necessary for knowing the culture of society.

Social Group

Man's life is enormous extent a group life. Human group is the elemental unit for functioning in societies. By group we mean any collection of human beings who are brought in to social relationship with one another. Social relationship is the relationship in which members involved are aware of each other's presence. When people cannot achieve any objective individually they come together and work towards the common objective with joint efforts. Social groups are vital for human existence.

A social group can be defined as two or more persons who came together to achieve common objectives.

According to Maclver "By group we mean any collection of human beings who are brought into social relationship with one another."

According to Ogburn and Nimkoff "Whenever two or more individuals come together and influence one another, they may be said to constitute a social group."

Group consists of leaders and following "Leader is a person who influence the group in its efforts towards the common objectives."

Characteristics of Group Life

a) Members of the group are related to each other.

b) Group involves sense of unity and feeling of love and sympathy between the members.

c) Members of the group have a sense of we feeling.

d) The interests, ideals, values of the group members are common.

e) Groups exercise a control over the actions of their members and direct them towards common goal or principles.

f) Social groups are dynamic and not static. They change in activity.

g) They may give up one function and take another. Human nature is not same, it differ from man to man. The different natures naturally lead to privation of different groups. Man from various group to satisfy their various needs. Groups have been classified on different basis.

Classification of Social Groups

1. According to the Degree of Organization.
2. According to the Nature of the Interaction.
3. According to the Nature of Membership.
4. According to the Size of the Group.
5. According to the Territorial Limitations.
6. According to the Profession of Occupation.
7. Classification According to Duration.
8. Classification Based on Social Class.
9. Classification Based on Personal Feeling of Belonging.
10. Classification Based on the Type and Quality of Relationship.

1. According to the Degree of Organization

A well-organized group has well defined objectives. The functioning of such groups is governed by number of rules and regulations, which govern the functioning of such groups. According to this criterion a group can be classified into two types.

a) Formal Groups

These groups have more rules and regulations to govern their functioning. The relationship of the members is governed by these rules. These groups are generally large in size and the members have many restrictions *e.g.* College, Government Departments, Army *etc.*

b) Informal Groups

In those types of groups there are no much formalities, rules and regulations. The degree of organization is less in the sense that the members have not be undergo confinements and strict limitations for behavior and actions. The members have many liberties and very less control as exists in Friendship group. Gossip group and family group *etc.*

2. According to the Nature of the Interaction

In this classification the nature of the interaction or the relationship between the members of the group is taken into consideration. According to this criteria the groups can be classed into

a) Primary Groups and

b) Secondary Groups.

Like family and friendship the relationship is direct and face to face. Generally such groups are small in size. Their relations are always personal. In secondary groups the relationship is indirect and such groups are bigger in size and they are formal in nature *e.g.* Government departments, Industrial organizations *etc.*

Difference between Primary Groups and Secondary Groups

Primary Group	*Secondary Group*
Mutual relations of the members are direct and more intimate.	Mutual relations of the members are indirect and there is task of face to face relations.
There is continuity of relationship among members and they know each other from childhood.	There is no continuity or permanence of relationship of the members and the knowledge of the members is limited only to his work.
The position of each person is determined on the basis of family	The position of member is determined by his work or function.
Relationship is natural.	Relationship is formal.
The behavior of members is controlled by the Inner powers.	The behavior of members is controlled by external.
Individual depends upon the group for fulfilment of his needs.	Individual is self reliant from the point of view of fulfilment of his need.
The membership is compulsory and continues for lifelong.	The membership is optional. He can resign from the membership at any time.
The aims of the members are identical or same. There is love and sympathy.	The aims or interests of the members are not identical or same. Lack of co-operation and sympathy.
The size is small.	The size is big and relations are impersonal.
Found mostly in the village.	Found mostly in the cities.
Controls even personal life of its members.	Do not exercise any control over the personal of its members.
The individual becomes its member since his childhood it cannot be acquired.	The individual becomes its member later on and its membership is required

3. According to the Nature of Membership

Here nature or type of membership is taken into consideration *e.g.* whether the membership is optional or voluntary or compulsory. This depends upon the mode of entrance to the members. The members limit themselves to their own interest.

a) Voluntary Groups

In these type of groups, the membership is voluntary and members have no compulsion to participate in the activity of the group. The withdrawal from such groups is also on voluntary basis. *e.g.* Gossip groups friendship group, youth organization, and library.

b) Non-voluntary Groups

In these types of groups membership is compulsory and members have no choice. Social conversions and traditions rather than personal choice determine the relationship as observed in a family. Every one borne in a family has to function as a member of a family by compulsion *e.g.* Caste group, National group, Religion group, Sex group, Age groups *etc.*

4. According to the Size of the Group

Here number of number in the group is taken into consideration. According to these criteria groups can be classified as:

a) Small Groups

Member in this type is considerably less. Each member an identify each other and can establish close or direct relationship. In such groups feeling of co-operation and sympathy can be achieved individually. Size is limited *e.g.* family, Sport Club, Bhajani Mandal *etc.*

b) Big Groups

This type of group is bigger in size. Number of members in a group is considerably large. *i.e.* 61 to 1000 *e.g.* University, Army *etc.* In larger group relationship is not direct and involves in to association.

5. According to the Territorial Limitations

Here the limit territory in which the group is functioning is taken into consideration. These groups work in defined territorial limits and they have fixed boundary.

a) Natural Territorial Group

These are the groups were territorial limits have been fixed by nature. Those groups are formulated by their natural similarities and boundaries. There boundaries and limits are fixed by geographic and climatic situations. *e.g.* Region

b) Artificial Territorial Group

Here the territory is fixed artificially by man on functional basic *e.g.* Village, Taluka, District, State *etc.*

c) Non-Territorial Groups

Here the natural and artificial territorial limit do not play any part. This group has a common functioning in all most all parts of the world and they do not take into consideration any limitation of other types *e.g.* UNO, FAO, Red Cross, International Trade Unions *etc.*

6. According to the Profession of Occupation

These groups are formulated on the basis of the professional or the occupations of the members *e.g.* Religious, Educational Groups, Political Groups, Artisans *etc.*

7. According to Duration

The stabilized and relatively unorganized or short duration group can be of two types:

a) Permanent

The members remain together for great length of time and aware of the objective of the group. There are formalities and defined roles to play. The members are tied together by potential ties and formulates. The groups as a whole has permanent existence *e.g.* Family, Government Department *etc.*

a) Temporary

Temporary groups are the mere collections of physical bodies and congregate in casual way on the street or on the stations. The size of such group is indefinite and they are unorganized. They do not remain together for great length of time. It has no division of labor and no system of statutes. The participants are all on one level because their attention is focused on one thing and interaction is uncontrolled. The individual as a member of the group is more aware of himself and his own interests and less aware of the group interests *e.g.* Crowd, Audience, and Mob.

8. Based on Social Class

People carry on their group relations with others influenced by class structure. Accordingly groups are classified into two types:

a) Horizontal Groups

These groups often organize themselves of a personal from the same level of society. The persons are alike in the status or position in the class system of society. Farmers, blacksmith, carpenter would be the members of their respective occupations belong to a horizontal group.

b) Vertical Groups

Vertical groups are those groups that are composed of members from different social strata. Its membership cuts vertically across the horizontal groupings in society *e.g.* political parties. The persons of different classes (*i.e.* lower and upper) work together in close relationship to promote their parties interest.

9. Based on Personal Feeling of Belonging

According to this criteria the group has been classified into

a) In-group

In group is a group either primary or secondary towards which persons have a solidarity, loyalty, friendliness and co-operation. Such groups are characterized by the expressions "We belong, we believe, we fell, we act or my family my neighborhood, my club, my association. In-group attitude contra constrains some elements of sympathy and sense of attachment or obligation to the other members of our group.

b) Out-group

An out group is defined by the individual with relation to the in-group usually expressed in the contrast between them and us. Every group is conscious that other groups are those to which we do not belong or not with us. We are democrats. They are Communist, we are Hindus, we are Muslims, we are Brahmins, and they are Harijans. A person has no sense of loyalty, sympathy, co-operation while they have sense of indifference even antagonism to the members of out-groups.

10. Based on the Type and Quality of Relationship

Similar to the concept of primary and secondary groups, the group has been classified into:

a) Gemeinschaft Group

Here is the society which most relationship are traditional or personal or after both *e.g.* Zamindar system in India. In which landlords had his tenants who were personally known to him and who had obligation for their welfare and to whom tenants fulfilled certain obligations. In such groups written documents or contracts were not present while traditional pattern existed and was accepted by society.

b) Gesellschaft Group

It is a society in which neither personal attachment nor important or traditional rights obligations and duties. Relationships are based on bargaining and clearly defined agreement. This society flourishes in urban area and business organizations or associations of whole salers.

c) Reference Groups

Reference Group may be defined as a group with which the individuals feel identified, the norms and objectives of which he accepts. A reference group may therefore by any group primary formal, horizontal or otherwise and strongly influences the behavior of an individual. A family neighbourhood reference group may influence a farmer to accept or reject an adoption of improved farm practice. Individual assumes his point of view and attitudes toward various aspect of life. Individual may not be a member of his reference group but he identify with a reference group and aspire to be a member. A member of lower social class may have higher class group as his reference group, identify with it and aspire to be a member of it.

Rural Social Institutions

Social institution is the structure and machinery through which a human society organizes, directs and executes the multifarious activities required to satisfy human needs.

When man relates himself with others he creates what have been described as forms or structures in order to enable him meet his needs and function in other ways of life. These forms constitute the framework, channels or means through which he functions in society. Thus, the man is creator of forms in society, the formulator of rules, regulations, procedures, and forms or behavior, is governed and controlled by them, in order to function effectively in fulfilment of needs.

Social institutions have been created by man form social relationships in society to meet basic needs such as stability, law and order, clearly defined roles or authority and decision making.

There are five major institutions in rural society. These institutions are present in all societies so there are called as a basic institutions.

1. **Family:** As an institution provides for care, protection and nature of children.
2. **Educational:** Institutional imparts knowledge, skills and society acceptable attitudes.
3. **Political:** Government Institution provide for law orders, settlement of disputes, administration affairs.
4. **Religious:** Institution prescribes prayers and worship as a part of relation with God.
5. **Economic (Occupation):** Institution provides basic physiological needs of the body- food, shelter and clothing. For this society provides farming and industry.

Rural Family

Family can be defined as a relatively permanent and socially sanctioned group of parent, and children. It may all so include near relatives of husband or wife as in case of joint family. The members of the family are relate to each other by blood or by marriage. The family is a basic unit of all the societies. Almost all the human beings are born into family. When they matured they marry and establish families of their own. Membership in the family is permanent except for the withdrawal of the adult children or in case of family disorganization by divorce. Sometimes the term family includes all the members of group of person other than mother, father and their children who occupies a common dwelling place.

According to Maclver "family is a group defined by sex relationship sufficiently precise and enduring to provide for procreation and upbringing of children". Ogburn and Nimkoff say, "Family is more or less durable association of husband and wife with children or of a man and women alone."

Classification of Family

The families can be classified according to various criteria:

1. On the Basis of Lineage

a) **Patrilineal Family:** When properly inheritance and reckoning descent along the male line (father).

b) **Matrilineal Family:** When it is along the female (mother) line.

2. On the Basis of Headship

a) **Patriarchal Family:** In this case father is head of the family.

b) **Matriarchal Family:** In this case mother is head of the family.

3. On the Basis of Transfer of Bride Groom

a) **Patrilocal Family:** There are the families where wife transfer to the husbands house after marriage.

b) **Matrilocal Family:** Where husband transfer to the wife's house after marriage. Matrilineal families are matrilocal families.

4. On the Basis of Number of Mates

a) **Monogamous Family:** In this case the husband marries only one wife.

b) **Polygamous Family:** In this case the husband can marry more than one wife.

c) **Polyandrous Family:** In this case the wife can marry more than one husband.

Nuclear or Conjugal or Individual Family

Such a family consists of married couple and their children, and is well separated from other relatives who may pay short visits if at all.

Consanguineal Family

Such a family consists often of grandparents, their sons, their sons 'wives and even their sons' grand children. Consanguincal literally means "of one blood". Eldest male member is the head of family.

Functions of Family

a) It helps in perpetuating the race through procreation.

b) It provides for a means to satisfy sexual needs of man and woman.

c) It helps in socializing the children and make them acceptable to the society.

d) It means the basic needs of its member.

e) It is the starting point of division of labor.

f) Family as a primary economic unit ensures that economic activities are shared by the family members.

g) It provides for education and vocational training of members so that they can share the economic burden.

h) It arranges to regulate the sex activities of unmarried members of a family in line with the restraints imposed by the society.

i) Family provides the children with a status till he becomes economically independent.

j) Children learn about customs, mores, superstitions and religion while they live in family and thus these are passed on from generation to generation.

k) The family carries out certain recreational activities for the benefit of young children and thus educates and broadens their outlook.

l) Family provides a sense of security to its members.

m) Family contributes towards preservation and enrichment of cultural heritage.

n) Family looks after its aged and disabled members.

o) Family disciplines its members so that the family is held in high esteem.

p) Family nourishes its member with mutual love and affection.

Characteristics of Indian Rural Family

1. The Rural Family is Patriarchal in Nature

Father is head of the family. He shoulders the responsibility towards, government, societies *etc.* He possesses some authority and takes all the decisions. He settles the marriages of other sub-ordinate members of the family and distributes the work. If father is not alive, then mother or elder brother act as a head of family.

2. Rural Family is Orthodox in Nature

They generally do not want to change their ways of behavior because of limited education, limited sources, and their occupation depending upon nature. They cannot afford to take risk and hence they follow the old ways of living.

3. Family Works as the Unit of Production and Unit of Consumption

All the members of the family members work as a team. The adult family members carry out heavy and major operations in the field like ploughing, sowing, transplanting *etc.* while children help than in to field operations like watching crops, looking after cattle, picking of fruits, harvesting of vegetables *etc.* Ladies also helps by carrying the meals to fields. Whatever produces family gets from farm is kept for his own consumption.

4. Dominance of Family Age

In the rural area there notices what is known as feminization: The rural people are religious as well as family minded. They always think about the well being of the family. The family pervades in to all acts of individual members of family. They try to protect food name of the family even with the biggest sacrifice. In Urban area people are less attached to the family. Feminization is that act in which the name of the family pervades in to all action.

5. Rural Family is Joint Family

This joint pattern is showing sends of breaking still however it is a common form in rural area.

6. Rural Family is Based on Peasant Households

The family has also peasant neighbourhood. All the neighbours of rural family are farmers. They always come in contact with each other and talk is generally cantered around the agriculture and allied subject.

Rural family is more homogenous, stable, integrated and organically functioning than urban family. The ties, bindings of the members of rural families for instance the husbands and wife parent and children are stronger and last longer than. There is greater discipline and interdependence in the rural families.

Joint Family System

The joint family is prevalent in India since the times immemorial. This institution was introduced in India and some other countries by Aryans. It consists of a group of people of common descent, living together under one roof, who worship a common deity, hold common property. It comprises of male members having a common male ancestors, unmarried female children and females married to male members. The eldest male member is the head of family and its absolute rural. The members of family contribute their labor and receive their share of produce to meet their basic minimum needs.

Characteristics of Joint Family

a) The head of family is its absolute rural.

b) The family owns a common property and the head of family is the trustee of the property.

c) The landed property is prevented from fragmentation. An agricultural family finds it economically profitable to sow larger lands using the lager manpower by joint family.

d) In a joint everyone is assured of meeting his basic minimum needs in turn everyone performs the work allotted to him by the head of family.

e) Commonality of place of living saves lot of expenses which would have otherwise to be made on houses rent of cost of constructing house and cost of the articles of daily use.

f) There is little chance of children going astray as there is always somebody to look after them, even when parents are away for work.

g) In a joint family every member has to do limited work. Everybody gets more time for recreation and leisure.

h) The family has clear-cut division of labor. The members do their work efficiently.

i) In rural India joint family is still considered a matter of pride.

Disadvantages of Joint Family System

a) The joint family under the strict control of head of family tends to be conservative and orthodox. It cannot change with times.

b) Member of family lack initiative because fruits of their initiative will ultimately be shared by other members who may be lazy and idle. Therefore, nobody puts in hard labor, which is so necessary for advancement in standard of living.

c) Due to lack of initiative in its members, the economic condition of the joint family goes on deteriorating.

d) There is disharmony due to generation gap in the family which affects the progress of the family adversely.

e) In a joint family the women look after household chores. Thus their talents are wasted and their lives are rendered miserable

f) The common property is looked after by none and its condition worsens for want of proper attention.

g) Due to diverse temperaments, there is always conflict between males and females and young and old.

h) As a large family has to live in limited space, there is no privacy.

i) Members tend to spend more and more due to the feeling that expenses will be scared by the whole family This makes the economic condition of family worse.

j) Because of presence of numerous other members of family a harmonious relationship based of mutual love and affection cannot develop.

k) As the responsibility for upbringing of children is shared, there is uncontrolled reproduction. As a result of this the family economy goes down.

l) When a joint family disintegrates, the commonness of property leads to family founds and often to litigation in courts of law.

Changes in and Dis-integration of the Joining Family

The joint family structure came into being in India because or its agricultural economy and because of the needs of the agriculture. Now the situation has changed. The country is getting industrialized. The population has gone up. The houses have become shall and the process of urbanization has changed the socio-economic conditions. Because of all these factors, the institution of joint family has started changing. In certain respects it has even dis-integrated. There are reasons for it. The important factors that have brought about the change in the joint family structure are:

a) Industrial development or industrialization.

b) Development of the means of communication and scientific and technological development.

c) Increase in population.
d) The process of urbanization.
e) Impact of western civilization and culture.
f) Education and development of individualism.
g) Dis-integration of village industries.
h) Social movements.
i) Social legislations, particularly those pertaining to the status of women.

Caste System

Every society has its caste system, which is an integral part of its social system. It is more or less impossible to away with the system. Caste system is an important characteristic of Indian society. It determines the relationship between the various groups and individuals. It is therefore necessary to have some knowledge about caste system in order to understand the structure and functioning of the Indian society. Caste system in India is based upon the birth while in other advanced countries. It is based upon the colour or the skin and wealth *etc.*

What is Caste System? According to Risley, "A caste may be defined as a collection of families or group of families bearing a common name which usually denotes or is associated with specific occupation claiming common descent from a mythical ancestor, human or divine, professing to follow the same professional calling and are regarded by those who are competent to give an opinion as forming single and homogeneous community". A. W. Green while defining caste says, "Caste is a system of stratification in which mobility, mobility, movement up and down the status ladder at least ideally may not occur".

Characteristics of Caste System in India

1. Hereditary

Caste status of an individual is determined strictly by his heredity, *i.e.* the caste into which one is born. No amount of personal accomplishments or efforts can alter his caste status.

2. Endogamous

It endogamous character strictly prohibits inter-caste marriages. Accordingly a person born in low caste can never hope to marry someone in higher caste. Each individual is supposed to marry within his caste and sub-caste. Marrying outside caste makes an individual or 'without a caste which is the lowest category even below Shudra'.

3. Hierarchal

Caste system has a system of superiority and subordination. According to Hindu Caste hierarchy, Brahmin occupies the highest followed by kshatriya, Vaishya and Shudra.

4. Fixed Occupations

Members of any caste are obligated to adopt the professions of their caste. Having developed from Varna system the occupation in caste system is definite; son of blacksmith peruses the occupation of his father, son of carpenter becomes carpenter and so on. (With development of industries people belong to many castes have lost their occupation and have taken agriculture or some other occupation).

5. Restricted Food Habits

Higher castes try maintaining their traditional purity by different food habits. Thus Brahmins will only take 'Satwil' or 'Pure' food. Kshtriya and Vaishya will take 'Royal' food. A Shudra takes 'Tamsi' food. Each individual caste has its own laws which govern the food habits. There is no restriction against fruit, milk, butter, dry fruit *etc.* but food can be accepted only from the members of ounces own or higher caste.

6. Untouchability

In Indian caste system Shudra and out castes are considered to be untouchables. In certain times of day even seeing a shudra is considered to be pollution. Even if shadow of a low caste falls on a Brahmin, latter is said to have been polluted.

7. Absence of Vertical Mobility

In a caste system, there is no mobility movement of its members, up or down, the social status ladder. A person's status at birth is his life time status.

8. Reinforcement by Religious Beliefs

Religious beliefs have played a significant role in making caste system unavoidable. Religion has described Brahmin as sacred and also an element of reverence and awe is attached to him. In absence of religious support such rigid caste system was not possible.

Advantage or Merits of Caste System

a) **Preservation of Culture:** Caste system helped in preservation of culture and these were passed on from generation to generation.

b) **Preservation of Purity:** Caste system, because of its endogenous nature, permitted marriage only within the caste and thus preserved purity of each caste.

c) **Division of Labor:** Caste system required each individual to do the work prescribed for his caste, *i.e.* Brahmins job was to teach, Kshatriyas to fight war, Vaishyas to run trade and Shudras to serve other castes. This division of labor ensured smooth functioning of society.

d) **Co-operation within Castes:** Caste system fastened cooperation with each caste. They co-operated with each other to preserve their culture and protect it from degradation by other castes.

e) **Absence of Competition:** As social status was hereditary and no amount or personal accomplishments could change it, there was no competition to improve status. People, therefore, utilized their energies for general benefit of society rather than Personal advancement.

f) **Panchayat System:** Panchayat system is an outcome of caste system. All disputes within a caste were referred to the caste Panchayat. Panchayats though secular in outlook exist even now in rural India.

g) **Increased Professional Proficiency:** Caste system helped in increasing proficiency in each vocation because accumulated experience of and store was handed over by father to son.

h) **Healthy Social Life:** Caste system made people disciplined and co-operative. People carried out the duties assigned due to their caste with responsibility. There was social amity between members of same caste.

i) **Protected the Society:** Rigidity of our caste system was responsible for protecting our society from cultural invasion by alien societies.

j) **Permanency and Continuity:** Each casts had a permanent body of constitutions to guide his social behaviors and action. The customs and beliefs were passed on from generation the generation. Thus caste system had permanency and continuity.

k) **Improved Living Standards:** Each caste struggled hard to maintain and improve living standards of its members to retain its superiority over subordinate castes. This resulted in overall progress and prosperity.

l) **Social Life Not Dependent on Political Conditions:** Caste system was predominantly influenced by religion. It was totally independent of prevailing political conditions and, therefore provided stability in society.

Disadvantage or Demerits of Caste System

a) **Undemocratic:** Caste system denied equal opportunity for advancement to persons belonging to different castes. It is, therefore, undemocratic.

b) **No Vertical Mobility:** Caste system expected every individual to adopt the occupation prescribed for his caste, which purely hereditary. Any amount of struggle could not change cast status. Therefore, these were no upward or downward mobility of labor.

c) **Encouraged Un-touchability:** Caste system prohibited any physical Contact or communication between Brahmin and Shudra. Even if a Shudra was sighted by a Brahmin at certain house it was considered pollution. Shadow of Shudra failing on Brahmin would defile him. This it encourages un-touchability.

d) **Created a Class of Idlers:** Brahmins who were well entrenched at the top of social hierarchy stopped devoting themselves to study and teaching and started living on alms provided by other castes. Thus they become parasites of society.

e) **Oppression of Low Caste People:** Restrictions were imposed on low caste people regarding their place of living, their movements and other activities by high caste people so that former may not defile the latter. This resulted in low caste people being huddled together in a limited area having no access to community and facilities leading a miserable life. Brahmins became virtually tyrants.

f) **Encouraged Conversion:** The caste system was solely responsible for large conversion of people to Christianity and even Islam to extent. Low castes people who no hope of any honourable life had found that by conversion to Christianity they could lead a respectable life.

g) **Against Integrity of Nation:** Caste system made people more faithful and devoted to their castes than to their nation. Thus caste system made national unity and integrity difficult task.

h) **False Sense of Superiority and Inferiority:** In caste system Brahmin was superior being, simply because he was born as a Brahmin. This was not necessarily matched by his intellectual development and wisdom. On the contrary, any one born in a low caste was inferior. Members of high caste considered themselves sole possessors of intellect and wisdom and expected the lower castes to seek their advice before doing anything. This created bad feeling between castes.

i) **Hind Erance to Socio-economic Progress:** As there was no social mobility. People lost all initiative and this resulted stagnancy in socio-economic conditions.

j) **Nation Not Benefited:** Affairs of the nation were run by a few high caste people. Thus the nation was deprived of benefit of wisdom of the masses.

k) **Encouraged Treason:** Continuous oppression of low caste people by high caste ones made some of them rebets in the society. They even went to the extent of committing treason and helping invaders against their nation.

Factors Responsible for Existence of Caste System

1. Widespread illiteracy.
2. Influence of religion.
3. Slow rate of industrialization.
4. Influence of foreign rule.
5. Geographical isolation.
6. Inadequate progress in means of transportation and communication.
7. Exploitation by political Parties.

Factors Unfavourable to Existence of Caste System

1. Provisions of Indian constitution.
2. Industrialization and urbanization.
3. Spread of education.
4. Modern means or transport and communication.

5. Change in social value.
6. Social transformation.
7. Democracy.
8. Class and status consciousness.

Erosion of Caste System in India

Caste system in India especially in villages was very rigid in past traditional castes were functioning. Caste orientated people were orthodoxical. But the greater exponent of the rigid caste system was Mahatma Gandhi who wanted abolish caste. In India the caste system is becoming weaker due to the influence of the factors like industrialization, urbanization, and increase mean of transportation, democratic government and laws abolishing untouchability. The Republic of India 1950 provides for justice, liberty, equality and fraternity, caste, colour and creed has no special consideration due to increase in desire for money caused by the influence of the materialistic tendencies. The sense of superiority or inferiority is now based on wealth and social power rather than on caste. Social welfare departments, backward class wings are trying to remove catechism. There are reservations of admissions or posts in appointment for lower caste. This way the caste system is now vanishing.

Difference between Class and Cast

A principle difference between class and caste is that class in open for all and social mobility is possible. In the caste system the vertical mobility is not possible. Caste in India has religious background and everybody tries to fulfil the caste duties, but in class system of social stratification religion has place. There the physical and mental qualities are more important.

Caste and Class jointly determine the position of an individual in social strain particularly in rural communities where caste system has maintained its rigidity. It forms the basis for economic and special life. In a single village there may be as many as 24 castes and of these are interdependent. Even in the urban society a constant tendency to make caste distinction is observed in the upper and middle classes. Thus the castes have maintained their importance in class system of social stratification.

Difference between Caste and Class

Caste	*Class*
Membership of a cast is hereditary and no amount or struggle and change it.	A person is placed class by virtue of his acquisition of education, wealth or other achievement.
There is no social mobility.	Social mobility is possible, *i.e.* it is possible to improve social status
Members are normally not conscious of their social status.	Members are generally conscious of their social status.
Caste system expects members of follow certain customs, folkways, rituals *etc.*	Social class has no prescribed customs rituals and folkways

Caste	Class
Inter-caste marriage is not possible because it will earn wrath of society	Marriage between two individuals belonging to different classes is possible without earning displeasure of the society
Caste system is based on inferiority or superiority of human beings. Therefore, does not promote democracy.	Social classes are based on superiority or inferiority of social status of an individual. Social classes help in working of democracy.
In caste system the members must follow a particular religion.	Members of social classes may follow any religion
Caste system is a closed class system in which hereditary status is the life time status.	Social classes are open class system in which movement from one class to another is completely unrestricted.
In caste system, there is no occupational mobility, *i.e.* one has to follow occupation of ancestors and it cannot be changed	As a member of social class one can adopt any occupation and change it at will.
Social gap between members of different castes is too wide.	Social gap is not so wide as in caste system.
Caste system is supported on religious grounds as a manifestation of God's will.	Social classes have no such religious a support.

Rural Religion

Religion exists in every society in some form or other. Religion plays very important part in the life of the people on the whole in India. India is basically country of religious. Every aspect of life is governed by the religion. Rural society or the village people are more favourable inclined towards religion. Religion is very old institution. It is not clear as how and when it originated. Origin or religion is essentially out of mans ignorance to find out the causes of many natural phenomenon which occurring in the universe. There are so many things in Universe. Which the present science is unable to explain and when reasons fails faiths continues beliefs continues, and all these beliefs continues, and all these beliefs of supernatural power, rites, ceremonies all these combined together brought religion in to existence. But it is clear that after its origin it greatly influenced the course of our social action and behaviour.

Definitions

"Religion is a set of beliefs regarding the relationship of a man to the supernatural power called God". Green says, "Religion is a system of belief and symbolic practices and objects governed by faith rather than by knowledge, which relates to man to unseen supernatural realm beyond the known and beyond the controllable."

Why to Study Rural Religion?

Rural people have greater affinity for the religion than the urban people.

a) The dependence of agriculture on natural forces like rains *etc.* and absent of scientific culture which provides the correct understanding of the natural and social world among the rural people are two main reasons for greater degree of religiosity among them.

b) The religious outlook of rural people dominates their all aspects of life. Their family life, caste life, general social life, economic and even recreational life are more or less governed by the religious approach and religion norms.

c) In most of the educationally and economically backward countries the leadership of village life was provided by priestly group. In India the Brahmins. The mores which this group lay down for individual behaviour as well as for social control were determined by the traditional religious concepts. Hence the life of village aggregates village people in all spheres was mould in the spirit of religious ideas and was controlled by the religious institution and leaders.

d) Though secular centralised state has taken ever the administration of village from village Panchayat and caste council who had a necessarily a religious outlook. The religious still continue to exercise a powerful hold over the minds or rural people and determine the behaviour in number of secular fields.

Basic Function of Religion

a) It prescribes different prayers and worships as a part of relation with god.

b) It provides idea about the religious life and teaches the people how to live in the life.

c) It teaches the moral to the people.

d) It acts as a social control.

e) It helps in transmitting culture.

Social Function and Social Roles of Religion - as Motivating and Restraining Agent

a) In primitive societies religion maintained social cohesion and controlled individual conduct.

b) In civilized societies while it unites group belonging to a particular religion, it is also a cause for conflict between such groups forming part of a large society. Religion provides set of doctrines and beliefs to unite people.

c) Religion has also influence on economic order.

d) In India, Hinduism through its caste system was responsible for systematic division of labour.

e) Religion teaches people to obey the men in authority with reverence and live peacefully and harmoniously.

f) Religion has encouraged literary activity.

g) Religion provides relief from human suffering and pain by explaining these as the type will of the God it soothes human mind and thus it saves people from frustration.

h) Religion plays an important role in shipping other social institutions. Its presence is felt in all activities, be it social political economic or domestic.

i) It encourages people to becomes benevolent and help their poor and needy fellowmen. It promotes tolerance and brotherhood.

j) It helps in creating fear of God in minds of people and thus restrains them from any anti-social activities.

k) It strengthens and rationalizes the group values and thus checks the disintegrating tendency.

Disadvantages or Demerits of Religion

a) Religion has also supported at times such practices as cannibalism, suicide, incest, murder slavery and is responsible for many aberrations in social behaviour.

b) It created hatred and war between people belonging to different religions.

c) Religion made people believers in their destiny and people did not make efforts to improve their lot.

d) Religions created an idle class of people *viz.* priests whose only aim in life was to preach religion. They lived like parasites in society and lot of power and esteem.

e) Religion has most of the times, resisted the spread of modern scientific though and has regarded various products or scientific research as tools devil.

f) Religion is also responsible to large extent for corruption in the society, particularly in priestly class.

Main Religious of the Word

There are four main religious in the world.

1. **Hinduism,**
2. **Islam,**
3. **Christism,**
4. **Buddhism.**

In India we find all these religious besides these there are small of Jains, Shikhs, Parshis and Jues.

Through there are so many religious there are many things common.

1. All religions induce the concept of 'God'.
2. All religions prescribe the way in which the man relates himself to the God.

3. All the religions include moral principles and influence the social value of God.

4. All religions proscribed the way of salvation.

Characteristics of Rural Religion

a) Rural religion is more orthodox. It has tendency to all the people to believe in old things, traditions and customs.
b) The rural religion is stricter; it means that if a person is not following religious teaching then some action will be taken against him.
c) Rural religion is less kind towards the other people.
d) It is more emotional and sentimental.

What is Education?

Education can be defined as production of desirable changes in the behavior of people or individual. The changes which are not desirable are not considered as a education. What is desirable and what is undesirable is decided by social values. Education is continuous process in the life of an individual. This process begins immediately after birth and continuous till death. He may be conscious of it or he may not be. In everyday life a man learned so many things but he is not conscious (aware) of this type of education.

Education is important for all. If village people are properly educated, they shall be able to know, what is expected of them and what way, they can be useful members of the society.

It is Important for the Rural People because of the Following Factors:

a) Backwardness of the village life and the agriculture.
b) For village progress and reconstruction.
c) For proper training of the 70 per cent of the population of the country which depends on agriculture.
d) To remove their superstitions.
e) To educate them about rural indebtedness which is responsible for various drawbacks of the village society?

Type of Education

Formal Education

Formal education is generally imported in well organized institutions like school, college other research institutions or specially organized school. These institutions have definite procedure of work and are generally governed by various rules and regulations.

Informal Education

In this type of education there is no definite course of procedure no rules and regulations, *e.g.* education given by parent, extension education.

Difference between Formal and Informal Education

Formal	**Informal**
There are definite rules and regulations and generally they are in written form. Disobeying of this result to severe consequences.	There are no rules and regulations and hence the question of disobeying does not arise.
There is definite time and fixed place for teaching. The procedure of teaching is also fixed.	It can be given at any time. No fix procedure for teaching.
There is fix syllabus about the subject matter.	There is no fix syllabus.
The level of educational qualifications, experience and age of learner is almost same	There is considerable difference in the level of age, experience and educational qualification.

In a village generally primary school acts as an educational institution. The basic knowledge like reading, writing, arithmetic *etc.* is imparted to rural children in school.

At present primary education in the villages is looked after by the Zilla Parishad.

Different Problems Faced by the Village Primary School

1. Lack of proper school building.
2. Lack of adequate and trained teaching staff.
3. Lack of adequate teaching and sport material.
4. Lack of interest on the part of parents.
5. Lack of interest on the part of students.

Present System of Education and Its Utility for Indian Rural Society

The present system of education is more or less futile for the Indian villages. It has no relevance to Indian life, much or less to the rural life. If it has to suit the needs of the rural society, it shall have to be-organized and change in many respects. The defects and drawbacks of the present system of education are discussed below:

a) **Impractical and Expensive:** The present system of education is impractical and too is based on bookish knowledge. It has no relationship and with the objectives of the society and the life. It does not teach the young boys and girls, the ways through which they can earn their livelihood.

b) **Lacks of practical utility**, it is very expensive; it is beyond the means of poor people to acquire education. The education facilities are more expensive and it is not possible for ordinary people to afford it. Villagers are very poor and the present day education is much beyond their reach.

c) **No Relevance to the Interest of the Students:** The present system of education is neither interested nor aptitude oriented. Only that education can be effective, useful and helpful which is based on the interest and the aptitude of the children. From this point of view, the Indian education is not at all useful for the needs of the Indian rural society.

d) **Low Standard and Defective Examination System:** The education system today is of very low standard. It means that the education Institution side not has all the equipments that are required for proper education. Therefore, the boys are not able to take full advantage of the education facilities that exists.

e) **The examination system** which is most important aspect of the present day education is also defective. It is neither an effective method of evaluating the competence of the education nor does it provide an opportunity to the child to go through his studies in scientific manner. Rural society needs a system of education which can be indigenous and which should provide for the need of the rural people. All these things are lacking form the Indian system of education.

f) **Lack of Facilities for Education:** In rural society, there is dearth of facilities for education. There are no institutions nor there any arrangement through which cans the people be educated. Whatever education facilities are available is less in preparation to the requirements of the people.

11

Rural Government (Gram Panchayat)

Government is a group of people who possess the power to rule in given area.

Necessity of Government

Government is a political institution, Government administers the regulatory function of law and order and maintains security in society, it provides both the means for regulating the behaviour of individuals within society in accordance with required norms and protection from external accordance with required norms. It is essential to have all over the control over individual, group organization and institution societies. It is possible to handle most of the things when societies are small however as society increases in size of area and population it become difficult to control and there is need of Government is increased.

Forms of Government:

There are two forms of Government:

1. **Democratic:** In democracy the political power is in hands of people. There are various political organizations and they trying to capture the local control.
2. **Automatic:** In autocracy power is in hands of one or few person. The concentration of power is in one or few person. It is not dependent upon the consent of the people but upon the capacity of few persons to achieve the control by various ways.

Rural Local Government (Gram Panchayat)

Under the three tier system of democratic decentralization of Panchayat Raj Gram Panchayat function at village level. It is the primary unit of local self government. Now the Panchayats are the agency for planning and executing the local development programme.

Democratic Decentralization (Panchayat Raj System)

The world "democracy" is derived from the greet roots – 'cracy' meaning 'rule of' and 'demos' meaning 'the people. It is governance of the people, by the people, for the people The emphasis is on the 'people 'as distinct from 'offer'.Rule by majority is, no doubt, an important feature of this system of governance;but the more important ingredient is rule by consultation ...consultation between the people's representatives on one hand, and consultation with hey officers on the other. It is, in essence, a pooling of the intelligence and the experience of all concerned administration. As corollary it also implies and even enjoins on them an implicit acceptance of the decisions taken by the body of members, and a resolve to implement them.

'Decentralization' means devolution of central authority among local units close to the areas served.Where authority devolves by this process on people's institutions, it is democratic decentralization. The team headed by Balwantrai Mehta (1956) recommended 'democratic decentralization'. According to this committee to delegate the powers, responsibility and resources for planning and execution of the development programme to people's institutions. As per the recommendations of the committee it was aimed to secure the maximum participation of the rural people in their own development programmes. The committee was of the opinion that one of the major reasons for the failure of the community development programme which are implemented after India's independence was the apathy and the non-participation of the people in the developmental programme.

The study team recommended a basic pattern of democratic decentralization with the Gram panchayat (Village Council) at the village level, the Panchayat Samit at the block level and the Zilla Parishad (district council) at the district level. The Gram panchayat is at the bottom of the Panchayat Raj System and the Zilla Parishad is at the apex. The Panchayat Samit constitutes the middle tier of this three-tier new set up of rural administration.

The study team's recommendations as approved by the National Development Council were communicated by the Center to the State for implementation. In the middle of 1958 the State of Madras had started as an experimental measure a pilot block of democratic decentralization in her own way as early as 1957. The experiences of this block were there Andhra State started in July 1958, twenty pilot blocks more or less on the same lines as recommended by the study team, one in every district of Andhra. Inspired by these experiences the State of Rajastan become the pioneer to bring the whole of Rajastan under democratic decentralization on October 2, 1959 On Nov, 1, 1959, Andhra Pradesh state introduced this scheme of democratic decentralization. In the entire state, then it was adopted by Mysore, Tamil Nadu,

Orissa, Assam, Punjab, Uttar Pradesh and other states. Now all the state of India has introduced the system with minor variations so as to make it suitable to their own conditions and requirements.

Panchayat Raj

There is something anomalous and contradictory in the term "Democratic Decentralization". It was also not easily understood by our people in India. At the instance of the Prime Minister it was seceded to give this process of the three-tier administration a strictly Indian name "Panchayat Raj' evolved as the natural expression that fitted admirable to the situation. Panchayat Raj means a system of Government horizontally it is a network of village panchayat. Vertically, it is an organic growth of the Panchayat rising up to the national level. Panchayat Raj was accepted as the new concept because it meant administration by mutual consultation, Consent and consensus.

A three tier system of democratic decentralization recommended the system operates at the district, block and village level, as follow:

District level -

Zilla Parishad

Block level -

Panchayat Samiti

Village level - Village panchayat (Gram Panchayat)

What is Gram Panchayat

A Gram Panchayat is a local self government which is responsible mainly for administration of the village and which also looks after the welfare of the people in the village.

It is basic, first formal democratic institution at the village level. The chairperson of this unit is called as Sarpanch. It is primary unit of local self-government. Gram panchayat is a cabinet of the village elders, directly elected by the adult citizens of the village. There are 8 to 10 ward punches, two or three coopted members, who constitute the body of Gram Panchayat; they are consists of 8 to 10 villages. There are three functional sub committees which meet four times every month to decide agricultural production programme, cottage industry programme, finance and budget and social amenities in Gram panchayat area.

The members of the Gram Panchayats have tenure of five years and are directly elected from wards while the Sarpanch is elected by the members. There is provision for reservation of seats for women and for scheduled casts and scheduled tribes. There is Gram Sabha for each panchayat and the Sarpanch is required to conduct Gram Sabha meetings at least once in six months. Also Sarpanch is required to conduct meeting of the members of the Panchayat once in a month.

Powers and Functions of the Sarpanch

1. To convene and preside over the meetings of the Gram Panchayat.

2. To conduct and regulate and be responsible for the proper maintenance of the records of the proceedings of the meetings.
3. To execute documents relating to controls on behalf of the Gram Shasan.
4. To be responsible for the proper custody of all records and documents, all valuable securities and assets belonging to the Gram panchayat or belonging to the Gram shasan.
5. To be responsible for the proper working of the Gram panchayat as required by or under the Act.
6. To exercise supervision and control over the acts and proceedings of all officers and employees of the Gram panchayat.
7. To have authority to enter into correspondence on behalf of Gram panchayat.
8. To order preparation of all statements and reports.
9. To exercise such other powers, discharge such other duties and perform such other functions as may be conferred or imposed on or assigned to him.

Power of Gram Panchayat Members

1. To supervise during office hours, the records of the Gram Panchayat after giving due notice to the Sarpanch.
2. To move resolution or question on any office bearer on matters connected with the administration of the Gram Panchauyat.
3. To exercise inspection overt all works undertaken by the Gram Panchayat.
4. To supervise all institutions working under the direction, management and control of the Gram panchayat.
5. To bring to the notice of the Sarpanch the irregularities if any, noticed during such inspection.

Functions of Gram Panchayat

There are number of functions perform by Gram Panchayats. These functions are divided into two categories, namely, the obligatory functions and the options functions. Following are the obligatory or the compulsory functions.

Following are the Obligatory or the Compulsory Functions

1. Construction, repairs, maintenance, alteration and extension of village roads, provisions of lights on the roads and other places of public resort and removal of encroachments and obstructions on the roads and other public places.
2. Construction, maintenance and cleaning of drainage system and provision of sanitation in the village by the removal of filth and clearance of marshy areas.

3. Supply of drinking water to the villages.
4. Adoption of preventive measures against epidemics and other dangerous diseases, prevention of obnoxious and dangerous trade, registration of births and deaths and the preparation of the necessary records for the purpose.
5. Maintenance of common pasture and other public institutions.
6. Preparation of census records of men and animals, maintenance of relevant records and submission of periodic records and returns.
7. Management of Panchayat properties as assets.
8. Spread of primary education and its management.
9. Social conservation.
10. Control of Markets, ferries, fairs, Ghats and other public places.
11. Adoption and encouragement of improved methods of cultivation.
12. It has to perform such other functions which are given to the Gram panchayat on a compulsory basis.

Optional Functions

In addition to the compulsory functions, each Gram Panchayat is also required to perform certain optional functions for the development of the rural people. These functions are given below:

1. Planning and maintenance of trees
2. Development and maintenance of village forests.
3. Development of the livestock.
4. Construction, management and control of slaughter houses.
5. Reclamation of cultivable wastes and follows land.
6. Organization and management of multi-purpose co-operative societies.
7. Famine relief measures.
8. Establishment and maintenance of village libraries.
9. Marketing of agricultural produce
10. Organization of the Fire services and protection of life and property in case of fire.
11. Maternity and child welfare and establishment of centers of the purpose.
12. Establishment and management of village Akharas and clubs.
13. Establishment and maintenance of works for providing employment in time of scarcity and establishment and granaries.
14. Organization, management and promotion of cottage industries.
15. Organization and maintenance of industrial and agricultural exhibitions.
16. Construction and maintenance of Dharmasalas and Rest houses.

17. Provision of adult education, establishment of primary schools with the prior approval of the panchayat samiti.
18. Prevention of gambling and implementation of prohibition.
19. To keep the records about the unemployed persons.

Importance of Gram Panchayat

1. It helps to train the villagers in the art of governing themselves.
2. It trains and develops local leadership.
3. It helps in securing participation of local people.
4. It trains rural people in democratic procedure.

i) Procedure for Forming Gram-Panchayat

Every village having a population of 500 and more must have a Gram Panchayat. In smaller villages a group gram panchayat for two or three villages is established. The size of population covered by each panchayat varies from state to state.

The number of members of a gram panchayat depends on the size of population. It varies from 7 to 17. Members of gram panchayat are elected by villagers for a period of five years. An adult with an age of 21 years and above is eligible for contesting the election and for voting in the same. One of the members is elected as Sarpanch and another Deputy Sarpanch.

Gram panchayat has a paid secretary (Gramsevak) who is appointed by Zilla Parishad.

a) The Chief executive officer makes the inquiry and collects the information in prescribed form and put in the meeting of standing committee.
b) After approval of standing committee the proposal is send to commissioner.
c) The commissioner passes the order to district collector to form separate Gram panchayat.
d) After declaration of separate Gram panchayat, state eleven Commissioners has to conduct the election of new Gram panchayat within a period of six months.
e) After election the functioning of the Gram panchayat is started.

ii) Election of Gram-Panchayat

a) Elections of the gram panchayat are conducted by stats election Commissioner through District Collector and Tahasildar.
b) The elections are conducted for every five years. The date of 1st meeting of elected members is taken as a date of starting of the functioning of that gram panchayat.
c) Tahasildar announces the date of election.
d) The number of members is fixed according to population of village.

e) The wants are prepared considering number of members.

f) List of voters is published and the election is conducted as general selection.

g) All the villagers having age more than 18 years are the voters.

h) The villagers having age more than 21 years is eligible for contesting the election if he fulfils the other requirements.

Population (Nos)	*Members*
Up to 1500	7
1501 to 3000	9
3000 to 4500	11
4500 to 6000	13
6000 to 7500	15
above 7501	17

iii) Reservation of Sets of Members of Gram-Panchayat

a) 30 per cent seats are reserved for women.

b) 27 per cent seats are reserved for backward class. The election Commissioner fix the number of seats reserved for Other Backward Class (OBC), Nomadic tribes (NT) and Vimukth Jamati (VJ).

iv) Election of Sarpanch and Deputy Sarpanch

a) Election of Sarpanch and Deputy Sarpanch is conducted in 1st meeting of Panchayat under Chairmanship or representative of Collection. The 1st meeting is called as early as possible after election.

b) The notice of election is given to elected members at least before three days.

c) The names in prescribed form are accepted forms are accepted before 2 hours of meeting from elected members.

d) The election is conducted by open election method but if any member desires, the secret ballet method is used.

v) Reservation for the Post of Sarpanch

a) 1/3 of the seats are reserved for women.

b) 27 per cent of the total seats of Sarpanch from the districts are reserved for backward class members.

c) 27 per cent seats are reserved for other backward class (OBC) members.

d) In proportion of population of NT&VJ to total population the seats of the Sarpanch are reserved.

The reservation of seats is fixed by Collector by draw method.

vi) Working of Gram Panchayat

a) The meeting of members is arranged once in a month to take decisions about different activities of panchayat.

b) Gram Sabha is arranged twice in a year. The first meeting is arranged in the month of April or May and the second Gram-Sabha is arranged in the month of November. If required more meeting can be arranged. All the voters are invited for meeting. In the first meeting progress report, account and audit report of last year and in 2nd meeting budget and activities of the next year are discussed.

vii) Source of Income

a) Land revenue of village.

b) Collection of various taxes such as house tax, water tax, marriage tax, professional tax, vehicle tax, entertainment tax *etc.*

c) Income from stray cattle pen (Kondwade) fines and sale of compost.

d) Grants from Zilla Parishad.

e) Annual fair and festival tax.

f) Marketing tax.

viii) Functions

a) Supply of water for domestic use.

b) Construction and repairs to public roads, drains, small irrigation bandharas.

c) Maintaining sanitation and public health.

d) Construction, repairing and maintaining public buildings, grazing lands, forest public wells and tanks in good condition.

e) Lighting on roads and public places.

f) Controlling fairs, bazaars (public market) bullets cart stands.

g) Controlling and marinating village cemetery.

h) Taking part in Agricultural development.

i) Maintaining a library and opening elementary school.

j) Providing and maintaining a playground for children.

k) Construction and maintaining public latrines.

l) Watch and ward work.

m) Planting and preservation of trees on the sides of the public roads in the village.

n) Collection of taxes.

o) Providing recreational facilities through establishment of T.V. unit.

Panchayat Samiti

This is the next tier of administration at the Block level. It consists of:

a) Sarpanchas of all panchayat.

b) Local M.L.A. and M.L.C. with right to vote but not to hold office.

c) One person nominated by District Collector for every panchayat for which no Sarpanch has been elected

Reservation and Cooption

a) Two women.

b) One from scheduled castes.

c) One form scheduled tribes.

d) Two persons with experience in administration and public life.

The president and vice-president of the samiti are elected form among the village panchayat presidents. Block Development Officer appointed by the Government is the chief executive of the samiti and function as the leader of the team of block level officials.

Every Panchayat Samiti Normally ahs Seven Standing Committees as follow

Standing Committee

a) Planning and Production

b) Co-operation and Industries

c) Education

d) Women Welfare

e) Social Welfare

f) Communication

g) Taxation and Finance

Each standing committee consists of 7 members including the Samiti President who is the Chairman of all standing committees. Each standing committee has independent powers of sanction within specified limits and arranges for executing the works and schemes within its jurisdiction,

Powers and Functions of Panchayat Samiti

a) The President and members of the Panchayat Samiti have to endeavour to instill among the people within their jurisdiction a spirit of self-help and initiative and harness their enthusiasm for raising the standard of living.

b) They have to enlist the whole-hearted support of the people for the implementation of the Development Programmes, not only of those which relate to the community for which Government assistance is forthcoming but much more so of those which relate to individuals and which as mainly based on self-help.

c) They have to exercise all the powers conferred on and perform all the functions entrusted to the Panchayat Samiti by the Government.

d) In particular all the activities of Community Development Programme are taken over by Panchayat Samiti. The activities concerning the rural welfare and development in the field of agriculture, Animal Husbandry, Health and Sanitation, Elementary Education, cottage industries, social welfare *etc.* which were being carried on by the normal Development Departments are now entrusted to the Panchayat Samiti. These schemes have been transferred with all the institutions, staff and funds to the Panchayat Samiti.

e) The powers and functions of the District Boards, with institutions, staff and assets and liabilities have been transferred to the Panchayat Samiti with effect from 1, Dec, 1959. The Presidents and members will have to see to the proper implementation of tall the programmes and working of the institutions entrusted to them.

f) The Panchayat Samiti can borrow funds for carrying out the purposes of the Act subject to the conditions laid down in the Andhra Pradesh Panchayat Samiti and Zilla Parishad Loans Rules 1959.

g) The powers of the various authorities to accord administrative and financial sanction in respect of the works and schemes of Panchayat Samiti are embodied in the rules issued by Government.

h) Government of Andhra Pradesh have given funds to the Panchayat Samiti provided under the head "Loans for irrigation and rural housing" in the Community Development Programme. These funds will be recovered from the Panchayat Samiti in instalments. The Panchayat Samiti will have to sanction and disburse loans to individuals and will have to recover them from the loanees.

i) The loan funds available with some of normal development departments such as Agriculture. Animal Husbandry and Industries *etc.* are also made over to Panchayat Samiti to be similarly spent and recovered.

j) The funds available under the Village Housing Project are also entrusted to Panchayat Samiti. All these loans will have to be recovered from the loanees by the Panchayat Samiti Themselves.

Zilla Parishad and its Function

This is the third tier of Panchayat Raj operating at the district level. It consists of:

a) All Presidents of Panchayat Samiti in the district.

b) The district collector.

c) M.L.A.s of the District
d) M.L.Cs. with right to vote but not to hold office.
e) M.P.s of the District
f) Two women representatives.
g) One representative of Scheduled Casts.
h) One representative of scheduled Tribes.
i) Two persons interested in rural development.
j) The members of the Parishad elect a chairman and a vice-chairman.

The District heads of development departments take part in the proceedings of the Parishad and its standing committees. There will be a secretary appointed by the Government, who attends all meetings of the Parishad and its standing committees. Official members are not entitled to vote. Every Zilla Parishad has normally seven standing committees. As follow:

Standing Committee

a) Planning and Production
b) Co-operation and Industries
c) Education
d) Women Welfare
e) Social Welfare
f) Communications
g) Taxation and Finance

Powers and Functions of the Zilla Parishad

1. Zilla Parishad should function as advisory body over the Panchayat Samiti with powers to a) approve their budgets b) co-ordinate their plans and c) distribute funds given by the Government among the blocks.
2. It has to prepare plans for all items of developmental activities in the district including Municipal areas.
3. It has to secure execution of plans *etc.* which are common to two or more blocks.
4. Secondary education is the responsibility of Zilla Parishad.
5. The parishad should perform such of the powers and functions of the District Board as are transferred to it by the Government.
6. It should also perform the functions of Panchayat Samiti in respect of non-samiti blocks.
7. It should advise the Government in all matters relating to rural development in the district.

8. It should discuss and review at it's ordinary meetings the progress made or the results achieved under-various items. Similarly the District Officer of every Development Department furnishes to the Parishad a brief note on the achievements in the schemes of his department. Such notes will be periodically reviewed by the Parishad.

12

Society with Sociology

A **society** is a group of individuals involved in persistent social interaction, or a large social group sharing the same geographical or social territory, typically subject to the same political authority and dominant cultural expectations. Societies are characterized by patterns of relationships (social relations) between individuals who share a distinctive culture and institutions; a given society may be described as the sum total of such relationships among its constituent of members. In the social sciences, a larger society often evinces stratification or dominance patterns in subgroups.

Insofar as it is collaborative, a society can enable its members to benefit in ways that would not otherwise be possible on an individual basis; both individual and social (common) benefits can thus be distinguished, or in many cases found to overlap. A society can also consist of like-minded people governed by their own norms and values within a dominant, larger society. This is sometimes referred to as a subculture, a term used extensively within criminology.

More broadly, and especially within structuralist thought, a society may be illustrated as an economic, social, industrial or cultural infrastructure, made up of, yet distinct from, a varied collection of individuals. In this regard society can mean the objective relationships people have with the material world and with other people, rather than "other people" beyond the individual and their familiar social environment.

The term "society" came from the Latin word societas, which in turn was derived from the noun socius ("comrade, friend, ally"; adjectival form socialis) used to describe a bond or interaction between parties that are friendly, or at least civil. Without an article, the term can refer to the entirety of humanity (also: "society in general", "society at large", *etc.*), although those who are unfriendly or uncivil to the remainder of society in this sense may be deemed to be "antisocial".

However, the Scottish economist, Adam Smith taught instead that a society "may subsist among different men, as among different merchants, from a sense of its utility without any mutual love or affection, if only they refrain from doing injury to each other."

Used in the sense of an association, a society is a body of individuals outlined by the bounds of functional interdependence, possibly comprising characteristics such as national or cultural identity, social solidarity, language, or hierarchical structure.

Society, in general, addresses the fact that an individual has rather limited means as an autonomous unit. The great apes have always been more (Bonobo, Homo, Pan) or less (Gorilla, Pongo) social animals, so Robinson Crusoe-like situations are either fictions or unusual corner cases to the ubiquity of social context for humans, who fall between presocial and eusocial in the spectrum of animal ethology.

Cultural relativism as a widespread approach or ethic has largely replaced notions of "primitive", better/worse, or "progress" in relation to cultures (including their material culture/technology and social organization).

According to anthropologist Maurice Godelier, one critical novelty in society, in contrast to humanity's closest biological relatives (chimpanzees and bonobos), is the parental role assumed by the males, which supposedly would be absent in our nearest relatives for whom paternity is not generally determinable.

In Political Science

Societies may also be structured politically. In order of increasing size and complexity, there are bands, tribes, chiefdoms, and state societies. These structures may have varying degrees of political power, depending on the cultural, geographical, and historical environments that these societies must contend with. Thus, a more isolated society with the same level of technology and culture as other societies is more likely to survive than one in closer proximity to others that may encroach on their resources. A society that is unable to offer an effective response to other societies it competes with will usually be subsumed into the culture of the competing society.

In Sociology

The social group enables its members to benefit in ways that would not otherwise be possible on an individual basis. Both individual and social (common) goals can thus be distinguished and considered. Ant (formicidae) social ethology.

Sociologist Peter L. Berger defines society as "a human product, and nothing but a human product, that yet continuously acts upon its producers." According to him, society was created by humans but this creation turns back and creates or molds humans every day. Sociologist Gerhard Lenski differentiates societies based on their level of technology, communication, and economy: (1) hunters and gatherers, (2) simple agricultural, (3) advanced agricultural, (4) industrial, and (5) special (*e.g.* fishing societies or maritime societies). This is similar to the system earlier developed by anthropologists Morton H. Fried, a conflict theorist, and Elman Service, an integration theorist, who have produced a system of classification for societies in all human cultures based on the evolution of social inequality and the role of the state. This system of classification contains four categories:

- Hunter-gatherer bands (categorization of duties and responsibilities).
- Tribal societies in which there are some limited instances of social rank and prestige.
- Stratified structures led by chieftains.
- Civilizations, with complex social hierarchies and organized, institutional governments.

In addition to this there are:

- Humanity, mankind, upon which rest all the elements of society, including society's beliefs.
- Virtual society, a society based on online identity, which is evolving in the information age.

Over time, some cultures have progressed toward more complex forms of organization and control. This cultural evolution has a profound effect on patterns of community. Hunter-gatherer tribes settled around seasonal food stocks to become agrarian villages. Villages grew to become towns and cities. Cities turned into city-states and nation-states. Many societies distribute largess at the behest of some individual or some larger group of people. This type of generosity can be seen in all known cultures; typically, prestige accrues to the generous individual or group. Conversely, members of a society may also shun or scapegoat members of the society who violate its norms. Mechanisms such as gift-giving, joking relationships and scapegoating, which may be seen in various types of human groupings, tend to be institutionalized within a society. Social evolution as a phenomenon carries with it certain elements that could be detrimental to the population it serves. Some societies bestow status on an individual or group of people when that individual or group performs an admired or desired action. This type of recognition is bestowed in the form of a name, title, manner of dress, or monetary reward. In many societies, adult male or female status is subject to a ritual or process of this type. Altruistic action in the interests of the larger group is seen in virtually all societies. The phenomena of community action, shunning, scapegoating, generosity, shared risk, and reward are common to many forms of society.

Types

Societies are social groups that differ according to subsistence strategies, the ways that humans use technology to provide needs for themselves. Although humans have established many types of societies throughout history, anthropologists tend to classify different societies according to the degree to which different groups within a society have unequal access to advantages such as resources, prestige, or power. Virtually all societies have developed some degree of inequality among their people through the process of social stratification, the division of members of a society into levels with unequal wealth, prestige, or power. Sociologists place societies in three broad categories: pre-industrial, industrial, and postindustrial.

Pre-industrial

In a pre-industrial society, food production, which is carried out through the use of human and animal labor, is the main economic activity. These societies can be subdivided according to their level of technology and their method of producing food. These subdivisions are hunting and gathering, pastoral, horticultural, agricultural, and feudal.

Hunting and Gathering

The main form of food production in such societies is the daily collection of wild plants and the hunting of wild animals. Hunter-gatherers move around constantly in search of food. As a result, they do not build permanent villages or create a wide variety of artifacts, and usually only form small groups such as bands and tribes. However, some hunting and gathering societies in areas with abundant resources (such as people of Tlingit) lived in larger groups and formed complex hierarchical social structures such as chiefdom. The need for mobility also limits the size of these societies. They generally consist of fewer than 60 people and rarely exceed 100. Statuses within the tribe are relatively equal, and decisions are reached through general agreement. The ties that bind the tribe are more complex than those of the bands. Leadership is personal charismatic and used for special purposes only in tribal society. There are no political offices containing real power, and a chief is merely a person of influence, a sort of adviser; therefore, tribal consolidations for collective action are not governmental. The family forms the main social unit, with most members being related by birth or marriage. This type of organization requires the family to carry out most social functions, including production and education.

Pastoral

Pastoralist is a slightly more efficient form of subsistence. Rather than searching for food on a daily basis, members of a pastoral society rely on domesticated herd animals to meet their food needs. Pastoralists live a nomadic life, moving their herds from one pasture to another. Because their food supply is far more reliable, pastoral societies can support larger populations. Since there are food surpluses, fewer people are needed to produce food. As a result, the division of labor (the specialization by individuals or groups in the performance of specific economic activities) becomes more complex. For example, some people become craft workers, producing tools, weapons, and jewellery. The production of goods encourages trade. This trade helps to create inequality, as some families acquire more goods than others do. These families often gain power through their increased wealth. The passing on of property from one generation to another helps to centralize wealth and power. Over time emerge hereditary chieftainships, the typical form of government in pastoral societies.

Horticultural

Fruits and vegetables grown in garden plots that have been cleared from the jungle or forest provide the main source of food in a horticultural society. These societies have a level of technology and complexity similar to pastoral societies. Some horticultural groups use the slash-and-burn method to raise crops. The wild vegetation is cut and burned, and ashes are used as fertilizers. Horticulturists use human labor and simple tools to cultivate the land for one or more seasons. When the land becomes barren, horticulturists clear a new plot and leave the old plot to revert to its natural state. They may return to the original land several years later and begin the process again. By rotating their garden plots, horticulturists can stay in one area for a fairly long period of time. This allows them to build semi permanent or permanent villages. The size of a village's population depends on the amount of land available for farming; thus villages can range from as few as 30 people to as many as 2000. As with pastoral societies, surplus food leads to a more complex division of labor. Specialized roles in horticultural societies include craftspeople, shamans (religious leaders), and traders. This role specialization allows people to create a wide variety of artifacts. As in pastoral societies, surplus food can lead to inequalities in wealth and power within horticultural political systems, developed because of the settled nature of horticultural life.

Agrarian

Agrarian societies use agricultural technological advances to cultivate crops over a large area. Sociologists use the phrase agricultural revolution to refer to the technological changes that occurred as long as 8,500 years ago that led to cultivating crops and raising farm animals. Increases in food supplies then led to larger populations than in earlier communities. This meant a greater surplus, which resulted in towns that became centers of trade supporting various rulers, educators, craftspeople, merchants, and religious leaders who did not have to worry about locating nourishment. Greater degrees of social stratification appeared in agrarian societies. For example, women previously had higher social status because they shared labor more equally with men. In hunting and gathering societies, women even gathered more food than men. However, as food stores improved and women took on lesser roles in providing food for the family, they increasingly became subordinate to men. As villages and towns expanded into neighbouring areas, conflicts with other communities inevitably occurred. Farmers provided warriors with food in exchange for protection against invasion by enemies. A system of rulers with high social status also appeared. This nobility organized warriors to protect the society from invasion. In this way, the nobility managed to extract goods from "lesser" members of society.

Cleric, knight and peasant; an example of feudal societies

Feudal

Feudalism was a form of society based on ownership of land. Unlike today's farmers, vassals under feudalism were bound to cultivating their lord's land. In exchange for military protection, the lords exploited the peasants into providing food, crops, crafts, homage, and other services to the landowner. The estate of the realm system of feudalism was often multigenerational; the families of peasants may have cultivated their lord's land for generations.

Industrial

Between the 15th and 16th centuries, a new economic system emerged that began to replace feudalism. Capitalism is marked by open competition in a free market, in which the means of production are privately owned. Europe's exploration of the Americas served as one impetus for the development of capitalism. The introduction of foreign metals, silks, and spices stimulated great commercial activity in European societies. Industrial societies rely heavily on machines powered by fuels for the production of goods. This produced further dramatic increases in efficiency. The increased efficiency of production of the industrial revolution produced an even greater surplus than before. Now the surplus was not just agricultural goods, but also manufactured goods. This larger surplus caused all of the changes discussed earlier in the domestication revolution to become even more pronounced.

Once again, the population boomed. Increased productivity made more goods available to everyone. However, inequality became even greater than before. The breakup of agricultural-based feudal societies caused many people to leave the land and seek employment in cities. This created a great surplus of labor and gave capitalists plenty of laborers who could be hired for extremely low wages.

Post-industrial

Post-industrial societies are societies dominated by information, services, and high technology more than the production of goods. Advanced industrial societies are now seeing a shift toward an increase in service sectors over manufacturing and production. The United States is the first country to have over half of its work force employed in service industries. Service industries include government, research, education, health, sales, law, and banking.

The term "society" is currently used to cover both a number of political and scientific connotations as well as a variety of associations.

Western

The development of the Western world has brought with it the emerging concepts of Western culture, politics, and ideas, often referred to simply as "Western society". Geographically, it covers at the very least the countries of Western Europe, North America, Australia, and New Zealand. It sometimes also includes Eastern Europe, South America, and Israel. The cultures and lifestyles of all of these stem from Western Europe. They all enjoy relatively strong economies and stable governments, allow freedom of religion, have chosen democracy as a form of governance, favor capitalism and international trade, are heavily influenced by Judeo-Christian values, and have some form of political and military alliance or cooperation.

Information

Although the concept of information society has been under discussion since the 1930s, in the modern world it is almost always applied to the manner in which information technologies have impacted society and culture. It therefore covers the effects of computers and telecommunications on the home, the workplace, schools, government, and various communities and organizations, as well as the emergence of new social forms in cyberspace.

One of the European Union's areas of interest is the information society. Here policies are directed towards promoting an open and competitive digital economy, research into information and communication technologies, as well as their application to improve social inclusion, public services, and quality of life. The International Telecommunications Union's World Summit on the Information Society in Geneva and Tunis (2003 and 2005) has led to a number of policy and application areas where action is envisaged.

Knowledge

As access to electronic information resources increased at the beginning of the 21st century, special attention was extended from the information society to the knowledge society. An analysis by the Irish government stated, "The capacity to manipulate, store and transmit large quantities of information cheaply has increased at a staggering rate over recent years. The digitisation of information and the associated pervasiveness of the Internet are facilitating a new intensity in the application of knowledge to economic activity, to the extent that it has become the predominant factor in the creation of wealth. As much as 70 to 80 per cent of economic growth is now said to be due to new and better knowledge."

The Second World Summit on the Knowledge Society, held in Chania, Crete, in September 2009, gave special attention to the following topics:

a) Business and enterprise computing;
b) Technology-enhanced learning;
c) Social and humanistic computing;
d) Culture, tourism and technology;
e) e-government and e-democracy;
f) Innovation, sustainable development, and strategic management;
g) Service science, management and engineering;
h) Intellectual and human capital development;
i) ICTs for ecology and the green economy;
j) Future prospects for the knowledge society; and
k) Technologies and business models for the creative industries.

Society: Sociologist Views, Characteristics and Definitions

Wallerstein, in his World Systems Analysis (1974) writes: "No concept is more pervasive in modem social science than society, and no concept is used more automatically and unreflectively than society, despite the countless pages devoted to its definition." In popular speech the word 'society' has several meanings. Scores of definitions of the word 'society' exist and the word has a range of meanings extending far beyond sociology, including history, economics and political science. In everyday life this term is used for various kinds of social units or social aggregates as if it exists 'out there' and beyond the individual subject such as Indian Society, French Society, American Society, Capitalist Society, *etc.* At many times, we associate this term for secondary associations Indian Sociological Society, The Theosophical Society, Society for the Prevention of Cruelty to Animals or to Children.

Likewise, in everyday speech, there is probably little distinction being made between society and nation, whereas in sociology such distinction would be significant. Not only this, the word 'society" is interchangeably used for the community also. Such a usage has its problems. Because of these problems Wallerstein argued that the concept of 'society' should be dropped from social analysis. Some symbolic interactionist says that there is no such thing as society. It is a useful covering term for things we do not know about or understand properly. Others, such as Emile Durkheim, treat society as a reality in its own right.

How Sociologists View Society?

As against its commonsense usage, sociologists use this term in a specific sense and in a precise way. In social sciences since nineteenth century there is a long debate about the use of the concept 'society'. It was taken to mean as tissues of manners and customs that hold a group of people together. In some sense, 'society represented something more enduring and deeper than the 'state', less manipulative and certainly more elusive.

Sociologists have defined society with two angles

a) In abstract terms, as a network of relationships between people or between groups.

b) In concrete terms, as a collection of people or an organisation of persons.

An earlier social scientist, L.T. Hobhouse (1908) defined society as "tissues of relationships". R.M. Maclver (1937) also defined it in more or less the same terms as "web of social relations which is always changing". Refining this definition, MacIver, along with his co-writer Charles Page, later on defined it in his new book Society: An Introductory Analysis (1949) thus: "It (society) is a system of usages and procedures, of authority and mutual aid, of many groupings and divisions, of controls of human behaviour and of liberties. This ever changing, complex system we call society." For Maclver and Page, society is an abstract entity as they write, "We may see the people but cannot see society or social structure but only its external aspects ... society is distinct from physical reality". Talcott Parsons (Encyclopedia of the Social Sciences, 1934) wrote: "Society may be regarded as the most general term referring to the whole complex of relations of man to his fellows."

Characteristics

1. Society is Abstract

If society is viewed as web of social relationships, it is distinct from physical entity which we can see and perceive through senses. As written earlier, Maclver argued, "we may see the people but cannot see society or social structure, but only its only external aspects". Social relationships are invisible and abstract. We can just realize them but cannot see or touch them. Therefore, society is abstract. Reuter wrote: "Just as life is not a thing but a process of living, so society is not a thing but a process of associating".

2. Likeness and difference in Society

Society involves both likeness and difference. If people are all exactly alike, merely alike, their relationships would be limited. There would be little give-and-take and little reciprocity. If all men thought alike, felt alike, and acted alike, if they had the same standards and same interests, if they all accepted the same customs and echoed the same opinions without questioning and without variation, civilisation could never have advanced and culture would have remained rudimentary. Thus, society needs difference also for its existence and continuance. We can illustrate this point through the most familiar example of family. The family rests upon the biological differences between the sexes. There are natural differences of aptitude, of capacity, of interest. For they all involve relationships in which differences complement one another, in which exchange take place. Likeness and difference are logical opposites but for understanding likeness, comprehension of its relation to the other is necessary. Society exists among those who have some degree of likeness in mind and in body. F.H. Giddings called this quality of society as "consciousness of kind" (a sense of likeness). Though likeness and difference both are necessary for the society to exist, but difference is always subordinated to likeness in society. Likeness has a predominant share in the constitution of society.

3. Cooperation and Conflict in Society

Cooperation and conflict are universal elements in human life. Society is based on cooperation but because of internal differences, there is conflict also among its members. This is why; MacIver and Page observed that "society is cooperation crossed by conflict". We know from our own experience that a person would be handicapped, showed down, and feels frustrated if he is expected to do everything alone, without the aid of others. "Cooperation is most elementary process of social life without which society is impossible" (Gisbert, 1957). Though cooperation is essential for the constitution of society but modem conflict theorists (such as Marx) have highlighted the role of conflict in society. If there is no conflict, even in small measure, society may become stagnant and people may become inert and inactive. However, the expression of disagreement in the form of conflict must always be held within tolerable bounds.

4. Society is a Process and not a Product

"Society exists only as a time sequence. It is becoming, not a being; a process and not a product" (MacIver and Page, 1956). In other words, as soon as the process ceases, the product disappears. The product of a machine endures after the machine has been scrapped. To some extent the same is true not only of material relics of man's past culture but even of his immaterial cultural achievements.

5. Society as a System of Stratification

Society provides a system of stratification of statuses and classes that each individual has a relatively stable and recognisable position in the social structure.

6. Society in Concrete Terms: "A Society"

When society is viewed from the point of view of persons who constitute it, it takes the shape of 'a society' instead of 'society" in general terms. A society is the largest number of human beings who interact to satisfy their social needs and who share a common culture. "A society may be defined as a network of interconnected major groups viewed as a unit and sharing a common culture" (J.H. Ficther, Sociology, 1957).

A Similar definition of society is offered by Ian Robertson (Sociology, 1977)

"A society is a group of interacting individuals sharing the same territory and participating in a culture." This definition of 'a society is quite rearer to the definition of 'community explained in the following pages. Thus, a society is different from society in general; 'a society is any organisation that enables people to carry on a common life. It is concrete, having physical reality and is an aggregate of persons while society is abstract and is something more than an aggregate of individuals and the sum of individuals. It refers to the whole constellation of associations that characterise a people. When we talk about Indian society, French society, or American society, we generally have the idea of 'a society' in our mind.

Defining Society Mike O'Donnell (1997) Writes

"A society consists of individuals belonging to groups which may vary in size." Anthony Giddens (2000) states; "A society is a group of people who live in a particular territory, are subject to a common system of political authority, and are aware of having a distinct identity from other groups around them." This definition is a mixture of characteristics of a community and a nation-state. From this view, some societies, like those of hunters and gatherers are very small; others are very Iarge involving millions of people such as modern Indian society.

Difference between Single-Purpose and Multi- Purpose Society

Single-Purpose Society	*Multi-purpose Society*
Single-Purpose Co-operative Society is organized by certain persons for fulfilling their one common need.	Multi-purpose Co-operative Society is organized for fulfillment of more than one common need.
Generally Single-Purpose Co-operative Society is organized for fulfilling the needs of credit, irrigation, milk *etc.*	Multi-purpose Co-operative Society is organized with the object of fulfilling or various needs such as credit, supply of improved seeds, agricultural implements, healthy sanitation *etc.*
The liability of the members of Single-Purpose Co-operative Society may be limited as well as unlimited	The liability of members of Multi-Purpose Co-operative Society is normally limited
The field of activity of a single purpose co-operative is very wide.	The scope of multi-purpose co-operative society is limited.

Functions of Multi-Purpose Co-operative Society

The multipurpose co-operative society has large number of functions to discharge. As the name indicates its responsibility for different purpose for which it has been set up. It can work for arranging credit, improved seeds, agricultural implements, fertilizers, sanitation, health *etc.* for its members. Normally it discharges the following functions.

1. Making Arrangement for Credit

The main task of these co-operative societies is to arrange for short term credit for its members in order to help them to improve their agriculture and other trade and business.

2. Encouraging the Improved Method of Agriculture

These societies arrange supply of to the members seeds, fertilizers, agriculture implements *etc.* They also arrange for irrigation facilities so that the members may be able to have higher agricultural yields.

3. Marketing and Business

These societies not only help the farmers to improve their agriculture but also help them for marketing of the goods. Through such facilities they get proper price for the agricultural products and other goods produced by its members.

4. Set up of Subsidiary Cottage and Small Scale Industries

These societies help their members to set up other trades and industries. Though these cottage industries and small-scale industries, the economic condition of the members is improved and they are able to improve their standard of living.

5. Helping Members to Increase their Standard of Living

These co-operative societies by providing various types of facilities for their members, improve their standard of living. They also carry out the programme of health, education and social education. They make arrangements for health and economic facilities, education and recreation for their members.

6. Encouraging the Member for Saving

These societies act as saving agencies and encourage their members to save a part of their income for their future needs. Through these savings, the members of such co-operative societies are able to get loans and also solve their problems.

7. Other Functions

The multi-purpose Co-operative societies apart from these functions discharge the following functions:

a) Purchase various things for their members at a reasonable price and supply them.

b) In order to convert the un-economic holding into economic these societies take a consolidation of holdings and co-operative farming.

c) Try to solve the disputes between members through arbitration and adjudication.

d) Thus multi-operative society discharge large number of functions for their members. The functions depend upon the interest and the object of the members. If the members are anxious to do more things, the multi-purpose co-operative societies can take to other function also.

13
Culture and Society

Culture consists of the beliefs, behaviors, objects, and other characteristics common to the members of a particular group or society. Through culture, people and groups define themselves, conform to society's shared values, and contribute to society. Thus, culture includes many societal aspects: language, customs, values, norms, mores, rules, tools, technologies, products, organizations, and institutions. This latter term institution refers to clusters of rules and cultural meanings associated with specific social activities. Common institutions are the family, education, religion, work, and health care.

Popularly speaking, being cultured means being well educated, knowledgeable of the arts, stylish, and well mannered. High culture generally pursued by the upper class–refers to classical music, theatre, fine arts, and other sophisticated pursuits. Members of the upper class can pursue high art because they have cultural capital, which means the professional credentials, education, knowledge, and verbal and social skills necessary to attain the "property, power, and prestige" to "get ahead" socially. Low culture, or popular culture generally pursued by the working and middle classes refers to sports, movies, television sitcoms and soaps, and rock music. Remember that sociologists define *culture* differently than they do *cultured, high culture, low culture*, and *popular culture.*

Sociologists define society as the people who interact in such a way as to share a common culture. The cultural bond may be ethnic or racial, based on gender, or due to shared beliefs, values, and activities. The term *society* can also have a *geographic* meaning and refer to people who share a common culture in a particular location. For example, people living in arctic climates developed different cultures from those living in desert cultures. In time, a large variety of human cultures arose around the world.

Culture and society are intricately related. A culture consists of the "objects" of a society, whereas a society consists of the people who share a common culture. When the terms *culture* and *society* first acquired their current meanings, most people in the world worked and lived in small groups in the same locale. In today's world of 6 billion people, these terms have lost some of their usefulness because increasing numbers of people interact and share resources globally. Still, people tend to use *culture* and *society* in a more traditional sense: for example, being a part of a "racial culture" within the larger "U.S. society."

Defining Crime

Any discussion of deviance remains incomplete without a discussion of crime, which is any act that violates written criminal law. Society sees most crimes, such as robbery, assault, battery, rape, murder, burglary, and embezzlement, as deviant. But some crimes, such as those committed in violation of laws against selling merchandise on Sundays, are not deviant at all. Moreover, not all deviant acts are criminal. For example, a person who hears voices that are not there is deviant but not criminal.A society's criminal justice system punishes crimes. Punishment becomes necessary when criminal acts are so disruptive as to interfere with society's normal functioning.

Limitations of Criminal Statistics

The FBI's annual *Uniform Crime Report* contains the official crime statistics drawn throughout the United States. Significant biases exist in the reporting and collecting of crime data, and problems occur when people interpret these criminal statistics. Some of these biases include the following:

- Many crimes in the United States go unreported, which makes the validity of crime statistics limited at best.
- Victims are often unwilling to cooperate with authorities.
- Complaints do not always translate into reported crimes. That is, while some victims of crime may complain to police, this does not mean that their complaint ends up reported in the *Uniform Crime Report*.
- Many people do not know how properly to interpret social science statistical data, including criminal statistics. For example, one very common error is attributing cause and effect to correlational data.
- White collar crime, committed by high status individuals during the course of business, tends not to appear in the *Uniform Crime Report*. Typical white collar crimes include embezzlement, bribery, criminal price fixing, insurance fraud, Medicare theft, and so forth.
- Some police and government officials exaggerate or downplay criminal statistics for political purposes. An incumbent politician may report "less crime" statistics in a re election campaign, while a social service agency may report "more crime" statistics in a proposal for funding.

Types of Crime

The types of crimes committed are as varied as the types of criminals who commit them. Most crimes fall into one of two categories crimes against people or crimes against property.

Organizational Models

Like groups in general, formal organizations are everywhere. Thus, sociologists have been quite interested in studying how organizations work. To learn more about how organizations operate effectively, sociologists develop organizational models. Some models describe the actual characteristics of organizations, while others describe the ideal characteristics for achieving their goals. To date, no single model has adequately described the fully complex nature of organizations in general.

Bureaucratic Organizations

Max Weber noted that modern Western society has necessitated a certain type of formal organization: bureaucracy. According to Weber, who believed that bureaucracy is the most efficient form of organization possible, the essential characteristics of a bureaucracy include

- ☆ Written regulations and rules, which maximize bureaucratic operations and efficiency.

  A highly defined hierarchy of authority, in which those higher in the hierarchy give orders to those lower in the hierarchy. Those who work in bureaucratic settings are called bureaucrats.
- ☆ Bureaucratic authority resting in various offices or positions, not in individuals.
- ☆ Employees being hired based on technical know how and performance on entry examinations.
- ☆ Formal and impersonal record keeping and communications within the organization.
- ☆ A paid administrative staff.

Although a bureaucracy itself may be specialized and impersonal, bureaucrats still retain their humanity. Within any bureaucracy, informal relationships invariably form, which can increase worker satisfaction, but only to a point. Informal groups can become disruptive to the efficiency of the bureaucracy.

Critics of Weber note that bureaucracy can also promote inefficiency. A bureaucracy can only formulate rules based on what it knows or expects. Sometimes novel situations or extenuating circumstances arise that the rules do not cover. When the unusual happens, rules may not be of much help.

Pros and Cons of Bureaucracy

Even though many Americans dislike bureaucracy, this organizational model prevails today. Whether or not they wish to admit it, most Americans either work in bureaucratic settings, or at least deal with them daily in schools, hospitals, government, and so forth. Hence, taking a closer look at the pros and cons of bureaucracy is important.

Pros of Bureaucracy

Although the vices of bureaucracy are evident (and are discussed in the next section), this form of organization is not totally bad. In other words, benefits to the proverbial "red tape" associated with bureaucracy do exist. For example, bureaucratic regulations and rules help ensure that the Food and Drug Administration (FDA) takes appropriate precautions to safeguard the health of Americans when it is in the process of approving a new medication. And the red tape documents the process so that, if problems arise, data exists for analysis and correction. Likewise, the impersonality of bureaucracies can have benefits. For example, an applicant must submit a great deal of paperwork to obtain a government student loan. However, this lengthy and often frustrating process promotes equal treatment of all applicants, meaning that everyone has a fair chance to gain access to funding. Bureaucracy also discourages favoritism, meaning that in a well run organization, friendships and political clout should have no effect on access to funding. Bureaucracies may have positive effects on employees. Whereas the stereotype of bureaucracies is one of suppressed creativity and extinguished imagination, this is not the case. Social research shows that many employees intellectually thrive in bureaucratic environments. According to this research, bureaucrats have higher levels of education, intellectual activity, personal responsibility, self direction, and open mindedness, when compared to non bureaucrats.

Parkinson's Law, named after historian C. Northcote Parkinson, states that work creates more work, usually to the point of filling the time available for its completion. That is, Parkinson believed that bureaucracies always grow–typically 6 per cent annually. Managers wish to appear busy, so they increase their workload by creating paper and rules, filling out evaluations and forms, and filing. Then they hire more assistants, who in turn require more managerial time for supervision. Moreover, many bureaucratic budgets rely on the "use it or lose it" principle, meaning the current year's expenditures determines the following year's budget. This provides a deep incentive to spend (even waste) as much money as possible to guarantee an ever increasing budget. Parkinson's views remain consistent with those of conflict theorists, who hold that bureaucratic growth serves only the managers, who in turn use their increasing power to control the workers.

Social Groups

Social groups and organizations comprise a basic part of virtually every arena of modern life. Thus, in the last 50 years or so, sociologists have taken a special interest in studying these scientific phenomena from a scientific point of view. A social group is a collection of people who interact with each other and share similar characteristics and a sense of unity. A social category is a collection of people who do not interact but who share similar characteristics. For example, women, men, the elderly, and high school students all constitute social categories. A social category can become a social group when the members in the category interact with each other and identify themselves as members of the group. In contrast, a social aggregate is a collection of people who are in the same place, but who do not interact or share characteristics.

Psychologists Muzafer and Carolyn Sherif, in a classic experiment in the 1950s, divided a group of 12 year old white, middle class boys at a summer camp into the "Eagles" and the "Rattlers." At first, when the boys did not know one another, they formed a common social category as summer campers. But as time passed and they began to consider themselves to be either Eagles or Rattlers, these 12 year old boys formed two distinct social groups.

In-Groups, Out-Groups, and Reference Groups

In the Sherifs' experiment, the youngsters also erected artificial boundaries between themselves. They formed in groups (to which loyalty is expressed) and out groups (to which antagonism is expressed).

To some extent every social group creates boundaries between itself and other groups, but a cohesive in group typically has three characteristics:

- Members use titles, external symbols, and dress to distinguish themselves from the out group.
- Members tend to clash or compete with members of the out group. This competition with the other group can also strengthen the unity within each group.
- Members apply positive stereotypes to their in group and negative stereotypes to the out group.

In the beginning, the Eagles and Rattlers were friendly, but soon their games evolved into intense competitions. The two groups began to call each other names, and they raided each other's cabins, hazed one another, and started fights. In other words, loyalty to the in group led to antagonism and aggression toward the out group, including fierce competitions for the same resources. People may form opinions or judge their own behaviors against those of a reference group (a group used as a standard for self appraisals). Parishioners at a particular church, for instance, may evaluate themselves by the standards of a denomination, and then feel good about adhering to those standards. Such positive self evaluation reflects the normative effect that a reference group has on its own members, as well as those who compare themselves to the group. Still, reference groups can have a comparison effect on self evaluations. If most parishioners shine in their spiritual accomplishments, then the others will probably compare themselves to them. Consequently, the "not so spiritual" parishioners may form a negative self appraisal for not feeling "up to par." Thus, reference groups can exert a powerful influence on behavior and attitudes.

Primary and Secondary Groups

Groups play a basic role in the development of the social nature and ideals of people. Primary groups are those in which individuals intimately interact and cooperate over a long period of time. Examples of primary groups are families, friends, peers, neighbors, classmates, sororities, fraternities, and church members. These groups are marked by primary relationships in which communication is informal. Members of primary groups have strong emotional ties. They also relate to one another as whole and unique individuals.

In contrast, secondary groups are those in which individuals do not interact much. Members of secondary groups are less personal or emotional than those of primary groups. These groups are marked by secondary relationships in which communication is formal. Members of secondary groups may not know each other or have much face to face interaction. They tend to relate to others only in particular roles and for practical reasons. An example of a secondary relationship is that of a stockbroker and her clients. The stockbroker likely relates to her clients in terms of business only. She probably will not socialize with her clients or hug them.

Small Groups

A group's size can also determine how its members behave and relate. A small group is small enough to allow all of its members to directly interact. Examples of small groups include families, friends, discussion groups, seminar classes, dinner parties, and athletic teams. People are more likely to experience primary relationships in small group settings than in large settings.

The smallest of small groups is a dyad consisting of two people. A dyad is perhaps the most cohesive of all groups because of its potential for very close and intense interactions. It also runs the risk, though, of splitting up. A triad is a group consisting of three persons. A triad does not tend to be as cohesive and personal as a dyad.

Leadership and Conformity

The pressure to conform within small groups can be quite powerful. Many people go along with the majority regardless of the consequences or their personal opinions. Nothing makes this phenomenon more apparent than Solomon Asch's classic experiments from the 1950s and 1960s.

The pressure to conform is even stronger among people who are not strangers. During group think, members of a cohesive group endorse a single explanation or answer, usually at the expense of ignoring reality. The group does not tolerate dissenting opinions, seeing them as signs of disloyalty to the group. So members with doubts and alternate ideas do not speak out or contradict the leader of the group, especially when the leader is strong willed. Group think decisions often prove disastrous, as when President Kennedy and his top advisors endorsed the CIA's decision to invade Cuba. In short, collective decisions tend to be more effective when members disagree while considering additional possibilities.

Two types of leaders normally emerge from small groups. Expressive leaders are *affiliation motivated*. That is, they maintain warm, friendly relationships. They show concern for members' feelings and group cohesion and harmony, and they work to ensure that everyone stays satisfied and happy. Expressive leaders tend to prefer a cooperative style of management. Instrumental leaders, on the other hand, are *achievement motivated*. That is, they are interested in achieving goals. These leaders tend to prefer a directive style of management. Hence, they often make good managers because they "get the job done." However, they can annoy and irritate those under their supervision.

Social Organizations

Secondary groups are diverse. Some are large and permanent; others are small and temporary. Some are simple; others are complex. Some have written rules; others do not.

Colleges, businesses, political parties, the military, universities, and hospitals are all examples of formal organizations, which are secondary groups that have goal directed agendas and activities. In contrast to official organizations, the informal relations among workers comprise informal organizations. Studies have clearly shown that quality informal relations improve satisfaction on the job and increase workers' productivity. However, professionals seem to place more importance on their relations with their co workers than blue collar workers do, perhaps because professionals' jobs require more interaction with co workers.

Organizations use rational planning to achieve their goals. They identify needs, generate alternatives, decide on goals, figure the most effective means to achieve the goals, decide who is best capable of achieving the goals, and then implement a specific plan of action. All of this usually requires strict adherence to policies, which can make large organizations seem businesslike and removed.

Relationships in Middle Adulthood

By middle age, more than 90 per cent of adults have married at least once. Married people often describe their marital satisfaction in terms of a "U curve." People generally affirm that their marriages are happiest during the early years, but not as happy during the middle years. Marital satisfaction then increases in the later years after finances have stabilized and parenting responsibilities have ended. Couples who stay together until after the last child leaves home will probably remain married for at least another 20 years as long as their intent was not to wait until the last child leaves the home to divorce.

Divorce

Middle adults do not exhibit an immunity to problems in relationships. About 50 per cent of all marriages in the United States end in divorce, with the median duration of these marriages being about 7 years. And of those that do last, marital bliss is not always a prominent feature. Why do so many marriages dissolve, and can spouses do anything to ensure that things work out?

Relationships dissolve for as many reasons as there are numbers of relationships. In some cases, the couple cannot handle an extended crisis. In other cases, the spouses change and grow in different directions. In still others, the spouses are completely incompatible from the very start. Long term relationships rarely end because of difficulties with just one of the partners. Conflicts, problems, growing out of love, and "empty nest" (feeling a lack of purpose in life or emotional stress in response to all the children leaving home) issues inevitably involve both parties.

The course of love changes over time, and these changes may become evident by middle adulthood. The ideal form of love in adulthood involves the three components of passion, intimacy, and commitment called consummate love, or complete love. This type of love is unselfish, devoted, and most often associated with romantic relationships. Unfortunately, achieving consummate love, as Sternberg noted, is similar to losing weight. Getting started is easy; sticking to it is much harder.

For many middle age couples, passion fades as intimacy and commitment build. In other words, many middle adults find themselves in a marriage typified by companionate love, which is both committed and intimate but not passionate. Yet love need not be this way, nor do such changes necessitate the end of a long term relationship. In contrast, many middle adult couples find effective ways of improving their ability to communicate, increasing emotional intimacy, rekindling the fires of passion, and growing together. The understanding that evolves between two people over time can be wonderful.

For others, the end of passion signals the end of the relationship. Passion enamors some people to such a degree that they do not approach their loving relationships realistically. This observation especially holds true for those who base their relationships on infatuation or the assumption that "true love" takes care of all conflicts and problems. When the flames of passion die out (which is inevitable in many cases) or the going gets rough, these spouses decide to move on to a new relationship. Divorce and extramarital relationships are but two consequences of marital unhappiness and dissatisfaction.

Interpersonal disagreements may increase as the couple becomes better acquainted and intimate. People who never learned how to communicate their concerns and needs effectively with their spouse or how to work through conflicts are more likely to become separated or divorced. Most couples quarrel and argue, but few know how to work at resolving conflicts equitably. What characteristics predict if a loving relationship will thrive or die? Long term relationships share several factors, including both partners regarding the relationship as a long term commitment; both verbally and physically expressing appreciation, admiration, and love; both offering emotional support; and both considering each other as a best friend. Essential to preserving a quality relationship is the couple's deciding to practice effective communication. Communication establishes and nurtures intimacy within a relationship, helping partners to better relate to and understand each other. Intimacy helps them feel close, connected, and loved, and creates an atmosphere of mutual cooperation for active decision making and problem solving. Communicating realistically leads to a satisfying and healthy relationship, regardless of the relationship's level of development.

Friends

In all age groups, friends provide a healthy alternative to family and acquaintances. They offer support, direction, guidance, and a change of pace from usual routines. Although many young adults manage to maintain at least some friendships, family, school, and work can become greater concerns for middle adults. Life responsibilities reach an all time high, so time for socializing is often at an exceptional premium. For this reason, middle adults generally maintain fewer close friendships than their newlywed and retired counterparts, although this is not always the case. Yet where quantity is lacking, quality predominates. People often nourish some of the closest ties between friends during middle adulthood.

Children

As adults wait later to marry and start families, more and more middle adults find themselves raising small children. This is not the typical pattern, however. By the time most parents reach middle age, their children are at least of adolescent age. Ironically, middle adults and their adolescent children often both experience emotional crises. For adolescents the crisis involves the search for their own identities as separate from their family members; for middle adults, the search is for generativity, or fulfilment through such activities as raising children, working, or creating. These two crises are not always compatible, as parents try to deal with their own issues as well as those of their adolescents (for example, discovering identity). Some middle adults begin to "live out" their own youthful fantasies through their children. They may try to make their teenage children into improved versions of themselves. Witnessing their children on the verge of becoming adults can trigger a midlife crisis. The adolescent journey into young adulthood reminds middle age parents of their own aging processes and the inescapable settling into middle and later adulthood. As a result, parents may experience depression or seek to recapture their youth through age inappropriate behavior and sexual adventures. Some teenagers ignite so much tension at home that their departure to college or into a career acts as a relief to parents. Other parents experience the empty nest syndrome after all of their children leave home. Without the children as a focal point for their lives, they have trouble reconnecting to each other and rediscovering their own individuality separate from parenthood.

In recent decades, Americans have witnessed the phenomenon of grown children staying or returning home to live with their parents. Whether they choose to stay at home for financial or emotional reasons, adult children who live with their parents can cause difficulty for all parties. Parents may delay their own "getting reacquainted" stage while managing a "not so empty nest," and their adult children may have to adjust to social isolation and problems establishing intimacy with significant others of their own age. Adult children living at home may also shirk necessary adult responsibilities.

This "adult child living with the parents" arrangement tends to work best when both parties agree upon it as a temporary situation, and when the child is less than 25. Middle age parents typically maintain close relationships with their grown children who have left home. However, many parents report feeling as if they continue to give more than they receive from their relationships with their children. This can be all the more the case for "sandwich" generation middle agers who must also tend to the needs of their own aging parents.

Parents

Most middle adults characterize the relationship with their parents as affectionate. Indeed a strong bond often exists between related middle and older adults. Although the majority of middle adults do not live with their parents, they usually maintain frequent and positive contact. And, perhaps for the first time, middle adults see their parents as fallible human beings. One issue facing middle adults is that of caring for their aging parents. In some cases, adults, who expected to spend their middle age years traveling and enjoying their own children and grandchildren, instead find themselves taking care of their ailing parents. Relationships with older adult parents vary a great deal. Some parents remain completely independent of their adult children's support; others partially depend upon their children; and still others completely depend upon them. Daughters and daughters in law most commonly take care of aging parents and in laws. Even though the death of a parent is never welcome, some long term adult caretakers express certain ambivalent feelings about the event.

Relationships in Older Adulthood

Given increases in longevity, today's older adults face the possibility of acquiring and maintaining relationships far longer than during any other time in modern history. For instance, nearly 1 in 10 adults over the age of 65 has a child who is at least age 65. Nurturing long term family relationships can bring both rewards and challenges. Over the decades, sibling rivalry may disappear and give way to peaceful relationships, while younger adults may feel the strain of trying to care for their aging and ailing parents, grandparents, and other relatives. Still, most young people report satisfying relationships with their older family members.

Marriage and Family

People sometimes refer to older adult marriages and families as "retirement marriages" or "retirement families." In such families the following demographics typically hold true:

- ☆ The average age of the wife is 68, and the husband, 71.
- ☆ Their previous marriages had lasted for more than 40 years, and they had high levels of marital satisfaction.
- ☆ They have three grown children, the oldest being about 40.
- ☆ Even though they consider themselves retired, 20 per cent of the husbands and 4 per cent of the wives continue to work.

Widows

Widowhood, or the disruption of marriage due to death of the spouse, presents by far the most devastating event in older adult marriages. Nearly 3 per cent of men ("widowers") and 12 per cent of women ("widows") in the United States are widowed. In the 75 and older age group, approximately 25 per cent of men and 66 per cent of women are widowed. Widows and widowers commonly complain of the difficulty that they experience finding a new spouse or partner. This especially holds true for widows, who must contend with the social stigmas of being "old" and "asexual." Widows tend to outnumber widowers in retirement communities, assisted living facilities, and nursing homes.

Relationships with Adult Children

The majority of older Americans 80 to 90 per cent have grown children, and enjoy frequent contact with them. Contrary to popular misconception, although the elderly enjoy these contacts they do not want to live with their grown children. Instead, they want to live in their own homes and remain independent for as long as possible. Typically, they would rather move into a private room in an assisted living facility or group home than move in with their children. At any one time, only about 5 per cent of adults over age 65 live in an institution. More than 75 per cent of institutionalized older adults, however, live within an hour's drive of one of their children. As for the quality of the relationship between older adults and their grown children, most research suggests that the elderly rate their experiences as positive. This particularly holds true when they have good health, enjoy common interests (such as church, holidays, hobbies), and share similar views (on politics, religion, child rearing, and so forth) with their children. The elderly do not necessarily rate as positive frequent contacts with their children when these contacts come from long term illness or family problems (a daughter's divorce, for example).

The potential for elderly abuse, or the neglect and/or physical and emotional abuse of dependent elderly persons, creates one very disturbing aspect of older adulthood. Neglect may take the form of withholding food or medications, not changing bed linens, or failing to provide proper hygienic conditions. Physical abuse may occur as striking, shoving, shaking, punching, or kicking the elderly. Emotional abuse may take the form of verbal threats, swearing, and insults. Estimates are that approximately 5 per cent of American older adults receive abuse each year.

Elderly abuse can occur in institutions, but more commonly takes place in the homes of the older person's spouse, children, and grandchildren. The typical victim is an older adult who is in poor health and lives with someone else. In fact, the person who lives alone has a low risk of becoming a victim of this form of abuse.

Both victims and abusers require treatment, whether individual, family, or group therapy. The main goal, however, is ensuring the safety of the elderly victim. The law requires many licensed professionals, such as clinical psychologists, to report known cases of elderly abuse to the authorities.

Grand Parenting

On average, men become grandfathers at age 52, and women become grandmothers at age 50. Therefore, grand parenting hardly restricts itself to older adults.

Although idealizing grand parenting is easy to do, the quality of grandparent grandchild relationships varies across and within families. Generally, the majority of grandparents report having warm and loving relationships with their grandchildren.

In addition to helping their grandchildren develop an appreciation for the past, positive grand parenting helps older adults avoid isolation and dependence while finding additional meaning and purpose in life. Grand parenting also facilitates personality development in later life by allowing older adults opportunities to re-examine and rework the tasks of earlier psychosocial stages.

Friendships

Having close friends in later life, as in any other period of life, consistently corresponds with happiness and satisfaction. Friends provide support, companionship, and acceptance, which are crucial to older adults' sense of self esteem. They provide opportunities to trust, confide, and share mutually enjoyed activities. They also seem to protect against stress, physical and mental problems, and premature death. Because older men more likely rely on their wives for companionship, older women typically enjoy a wider circle of close friends. Older men, however, develop more other gender friendships, probably because women generally live longer that men, meaning more women than men are available for such friendships. When older women can find available men with whom to be friends, they may hesitate to become too close. This is especially true if either is married. These women may also worry about what others are thinking, as they do not want to appear improper or forward.

Homosexual and Bisexual Relationships

Some couples do not fit into traditional models of marriage and family. A person who is attracted to members of the opposite sex is heterosexual, or "straight." A person who is attracted to members of the same sex is homosexual, or "gay" (male) or "lesbian" (female). And a person who is attracted to members of both sexes is bisexual, or "bi."

Although the United States does not legally recognize same sex marriages or unions, homosexuals and bisexuals (like heterosexuals) enjoy a variety of relationships. They live together, alone, with family or friends, or have housemates. They also come from all walks of life, socioeconomic levels, and backgrounds. Making generalizations about homosexual or bisexual relationships romantic, social, working, or otherwise is difficult, as is making generalizations about heterosexual relationships.

Relationship Categories

Nonetheless, in the 1970s, researchers A. P. Bell and M. S. Weinberg studied nearly 1,000 gays and lesbians and found that about 75 per cent fell into one of the following relationship categories:

- **Asexual homosexuals** are single, have few or no sex partners, and would probably be considered sexually withdrawn.
- **Dysfunctional homosexuals** are single, have many sex partners, but also have sexual problems and/or regrets about being homosexual.

- ☆ **Functional homosexuals** are single, have many sex partners, and have few or no sexual problems or regrets about being homosexual.
- ☆ **Open coupled relationships** involve one on one relationships, but both partners have a number of sex partners.
- ☆ **Close coupled relationships** resemble monogamous heterosexual marriages. As the above categories show, homosexual relationships are diverse.

Sociologists estimate that between 40 and 60 per cent of homosexuals maintain committed relationships. Gay and lesbian couples may stay together for as many years as heterosexual couples. Also like heterosexuals, they may or may not find satisfaction in such relationships.

Children

Many homosexual couples decide to raise children. Researchers suggest that children brought up in homosexual homes do not differ from children brought up in heterosexual homes in terms of intelligence, gender role, gender identity, or general life adjustment. Researchers also suggest that children from homosexual homes are no more likely to become homosexuals themselves than are children from heterosexual homes.

Adult Development

Adulthood is primarily a time of determining lifestyles and developing relationships. Among other things, most adults eventually leave their parents' home, develop a long term romantic relationship, and start a family, creating a new home. Research by sociologist Daniel Levinson determined the stages of adult development, as presented in

Lavinson's Stages of Adult Development

Age (Years)	*Levinson's Stage*
17 - 33	Novice stage of early adulthood
17 – 22	Early adult transition
22 – 28	Entering the adult world
28 – 33	Age – 30 transition
33 – 45	Culmination of early adulthood
33 – 40	Setting down
40 – 45	Midlife transition
45 – 50	Entering middle adulthood
50 – 55	Age – 50 transition
55 – 60	Culmination of middle adulthood
60 – 65	Late adult adulation
65+	Late adulthood

These stages are generally accepted by researchers today in seeking to explain and evaluate adult development.

Relationships in Early Adulthood

Love, intimacy, and adult relationships go hand in hand. Psychologist Robert Sternberg proposed that love consists of three components: *passion, decision/commitment,* and *intimacy*. Passion concerns the intense feelings of physiological arousal and excitement (including sexual arousal) present in a relationship, while decision/commitment concerns the decision to love the partner and maintain the relationship. Intimacy concerns the sense of warmth and closeness in a loving relationship, including the desires to help the partner, to self disclose, and to keep the partner in one's life. People express intimacy in three ways:

- ✰ **Physical intimacy** involves mutual affection and sexual activity.
- ✰ **Psychological intimacy** involves sharing feelings and thoughts.
- ✰ **Social intimacy** involves enjoying the same friends and types of recreation.

The many varieties of love described by Sternberg consist of varying degrees of passion, commitment, and intimacy. For example, infatuation, or "puppy love" so characteristic of adolescence involves passion, but not intimacy or commitment.

In addition to love and intimacy, sexuality is realized during young adulthood within the context of one or more relationships, whether long or short term. Although adolescent sexuality is of a growing and maturing nature, adult sexuality is fully expressive. The following sections discuss some of the more familiar types of adult relationships.

Singlehood

Today, many people are choosing singlehood, or the "single lifestyle." Regardless of their reasons for not marrying, many singles clearly lead satisfying and rewarding lives. Many claim that singlehood gives them personal control over their living space and freedom from interpersonal obligations. Today the number of singles in the United States remains at about 26 per cent of men and 19 per cent of women in the 1990s staying single for at least a portion of adulthood. Eventually, approximately 95 per cent of Americans will marry. Most singles date; many are sexually active, with the preferred sexual activities for singles remaining the same as those for other adults. Some singles choose celibacy abstaining from sexual relationships.

Cohabitation and Marriage

Cohabitation and **marriage** comprise the two most common long term relationships of adulthood. Cohabitors are unmarried people who live together and have sex together, more than 3 million Americans (most between the ages of 25 and 45) cohabitate. Many individuals claim that they cohabitate as a test for marital compatibility, even though no solid evidence supports the idea that cohabitation increases later marital satisfaction. In contrast, some research suggests a relationship between premarital cohabitation and increased divorce rates. Other individuals claim that they cohabitate as an alternative to marriage, not as a trial marriage. The long term relationship most preferred by Americans is marriage. More than 90 per cent of Americans will marry at least once, the average age for first time marriage being 24 for females and 26 for males.

Marriage can be advantageous. Married people tend toward healthier and happier lives than their never married, divorced, and widowed counterparts. On average, married males live longer than single males. Marriages seem happiest in the early years, although marital satisfaction increases again in the later years after parental responsibilities end and finances stabilize.

Marriage can also be disadvantageous. Numerous problems and conflicts arise in long term relationships. Unrealistic expectations about marriage, as well as differences over sex, finances, household responsibilities, and parenting, create only a few of the potential problem areas.

As dual career marriages become more common, so do potential complications. If one spouse refuses to assist, the other spouse may become stressed over managing a career, taking care of household chores, and raising the children. As much as Americans may hate to admit this fact, women in our culture still bear the primary responsibilities of child rearing. Conflicting demands may partly explain why married women with children leave their jobs more often than childless and single women. Multiple roles, however, can be positive and rewarding. If of sufficient quality, these roles may lead to increased self esteem, feelings of independence, and a greater sense of fulfillment.

Extramarital Relationships

Severe problems in a marriage may lead one or both spouses to engage in extramarital affairs. *Nonconsensual* extramarital sexual activity (not agreed upon in advance by both married partners) constitutes a violation of commitment and trust between spouses. Whatever the reasons, nonconsensual affairs can irreparably damage a marriage. Marriages in which one or both partners "cheat" typically end in divorce. Some couples may choose to stay together for monetary reasons or until the children move out. On the other hand, *consensual* extramarital sexual activity ("swinging") involves both partners consenting to relationships outside of the marriage. Some couples find this to be an acceptable solution to their marital difficulties, while others find it to be detrimental to the long term viability of their marriage.

Divorce

When significant problems in a relationship arise, some couples decide on **divorce**, or the legal termination of a marriage. About 50 per cent of all marriages in the United States end in divorce, the average duration of these marriages is about 7 years. Both the process and aftermath of divorce place great stress on both partners. Divorce can lead to increased risk of experiencing financial hardship, developing medical conditions (for example, ulcers) and mental problems (anxiety, depression), having a serious accident, attempting suicide, or dying prematurely. The couple's children and the extended families also suffer during a divorce, especially when disagreements over custody of the children ensue. Most divorcees, their children, and their families eventually cope. About 75 per cent of divorcees remarry, and most of these second marriages remain intact until the death of one of the spouses.

Friends

Friends play an important role in the lives of young adults. Most human relationships, including casual acquaintances, are *nonloving* in that they do not involve true passion, commitment, or intimacy. According to Sternberg, intimacy, but not passion or commitment, characterizes friendships. In other words, closeness and warmth exist without feelings of passionate arousal and permanence. Friends normally come from similar backgrounds, share the same interests, and enjoy each other's company. Although many young adults feel the time pressures of going to school, working, and starting a family, they usually manage to maintain at least some friendships, though perhaps with difficulty. As life responsibilities increase, time for socializing with others may be at a premium. Adult friendships tend to be same sex, non romantic relationships. Adults often characterize their friendships as involving respect, trust, understanding, and acceptance typically the same features as romantic relationships, but without the passion and intense commitment. Friendships also differ according to gender. Females tend to be more relational in their interactions, confiding their problems and feelings to other females. Males, on the other hand, tend to minimize confiding about their problems and feelings; instead, they seek out common interest activities with other males. Friends provide a healthy alternative to family members and acquaintances. They can offer emotional and social support, a different perspective, and a change of pace from daily routines.

Starting a Family in Early Adulthood

As young adults enter the culminating phase of early adulthood (33–45), they enter the settling down (33–40) stage. By this time, they have established a career (at least the first one!) and found a spouse. If the couple have not already done so, they will probably decide to have one or more children and start a family. People generally think that parenthood strengthens marriages, even though research indicates that marital satisfaction often declines after the birth of the first child. This need not be the case, however. If the marriage is already positive and the spouses share equally in parenting duties, they can minimize the hassles of parenthood and keep it from significantly interfering with marital happiness. Regardless of the many joys of parenthood, new parents are not always prepared for the responsibility and time commitment that raising a child requires, especially when the pregnancy is accidental rather than planned, or when the child is "difficult" and prone to irritability and excessive crying.

The postponement of marriage and childbearing until the 30s makes for an interesting trend in today's world. Two advantages of waiting are the emotional maturity of both partners and the stability of their relationship. A more mature and stable couple possesses the necessary tools for weathering the storms of parenthood. Another advantage is financial stability due to more years on the job, promotions, and long term savings.

Another interesting trend is an increase in non-traditional family units. Examples of these include blended families (or "stepfamilies," in which children from previous marriages are "blended" into a new family), single parent families, and same sex families. Some couples choose to remain childless. Couples who have children do not necessarily regard themselves as more "fulfilled" than couples who do not. The critical factor seems to be the couples' ability to choose their lifestyle. Sources of information about family planning, conception, birth control, and other pregnancy options include the Planned Parenthood Federation of America, the National Right to Life Committee, and National Abortions Rights Action League.

Population and Demographic Variables

Humans throughout history have generally favored large families-for the most part to assure survival of a particular family line or racial group. High death rates from plagues, predators, and wars led people to produce as many offspring as possible. However, the situation has changed dramatically in the 20th century as technological advances of one sort or another have caused a global "population explosion," with the world currently gaining 90 million people each year (most of this increase in poorer countries). Given this trend, the global population will exceed 6 billion in the early 2000s, and 8 billion by the 2020s. Understandably, sociologists around the world exhibit urgent concern about increases in the global population.

Demography (from the Greek word meaning "description of people") is the study of human populations. The discipline examines the size and composition of populations, as well as the movement of people from locale to locale. Demographers also analyze the effects of population growth and its control. Several demographic variables play central roles in the study of human populations, especially *fertility* and *fecundity*, *mortality* and *life expectancy*, and *migration*.

Fertility and Fecundity

A population's size is first affected by fertility, which refers to the number of children that an average woman bears during her reproductive years from puberty to menopause. People sometimes confuse the term fertility with fecundity, which refers to the number of children an average woman is capable of bearing. Such factors as health, finances, and personal decision sharply affect fecundity.

To determine a country's fertility rate, demographers use governmental records to figure the crude birth rate (the number of live births for every thousand people in a population). They calculate this rate by dividing the number of live births in a year by the total population, and then multiplying the result by 1,000. As one might expect, the governmental records used in this type of research may not be completely accurate, especially in third world countries where such records may not even exist.

While the world's average fertility rate is about 3 children per woman, its fecundity rate is about 20 per woman. The highest fertility rate (nearly 6 children per woman) in the world occurs in Africa, whereas the lowest occurs in Europe (about 1.5). The fertility rate for women in the United States is about 2.

Mortality and Expectancy

Mortality, or the number of deaths in a society's population, also influences population size. Similar to the crude birth rate, demographers calculate the crude death rate, or the number of deaths annually per 1,000 people in the population. Demographers calculate this figure by dividing the number of deaths in a year by the total population, and then multiplying the result by 1,000. The crude death rate in the United States normally stays around 8 or 9.

Infant mortality rate, which is the number of deaths among infants under age one for each 1,000 live births in a year, provides demographers with another measure. Compared to other countries, North American infant mortality rates tend to be low. Still, the figures can vary considerably *within* a society. For example, African Americans have an infant mortality rate of about 19 compared to those of whites who have a rate of about 8.

A low infant mortality correlates with a higher life expectancy, which is the average lifespan of a society's population. U.S. males and females born today can look forward to living into their 70s, which exceeds the life expectancy of those in low income countries by 20 years.

Migration

Finally, **migration** (the movement of people from one place to another) affects population size. While some migration is involuntary, such as when slaves where brought to America, other migration is voluntary, such as when families move from cities into suburbs.

Migration into an area, called immigration, is measured as the immigration rate, which is the number of people entering a region per each 1,000 people in the population. Migration out of an area, or emigration, is measured as the emigration rate, which is the number leaving per each 1,000 people in the population. Internal migration is the movement from one area to another within a country's borders.

Population Growth

Fertility, mortality, and migration all influence the size of a society's population. Poorer countries tend to grow almost completely from internal causes (for example, high birth rates due to the absence of reliable contraception), while richer countries tend to grow from both internal causes and migration. Demographers determine a population's natural growth rate by subtracting the crude death rate from the crude birth rate. The world's low growth nations tend to be more industrialized, such as the United States and Europe. The high growth countries tend to be less industrialized, such as Africa and Latin America.

Population Composition

Demographers also take an interest in the composition of a society's population. For example, they study the gender ratio (or sex ratio), which is the number of males per 100 females in a population. The sex ratio in the United States is about 93 males for every 100 females. In most areas of the world, the gender ratio is less than 100 because females normally outlive men. Yet in some cultures that practice female infanticide, such as among the Yanomamo, the ratio can reach well above 100.

Malthusian Theory

The field of demography arose two centuries ago in response to the population growth of that day. Thomas Robert Malthus (1766–1834), English economist and clergyman, argued that increases in population, if left unchecked, would eventually result in social chaos. Malthus predicted that the human population would continue to increase *exponentially* (1, 2, 4, 16, 256 ...) until the situation is out of control. He also warned that food production would only increase *arithmetically* (1, 2, 3, 4, 5 ...) because of limitations in available farmland. To say the least, Malthus provided a disturbing vision of the future that included massive, global starvation as a consequence of unrestrained population growth.

As it turned out, Malthus' predictions were mistaken because he failed to account for technological advancements and ingenuity that would increase agricultural and farm production, not to mention the increasing development and acceptance of birth control methods. Yet Malthus' forebodings do not lack merit. As noted by the New Malthusians, a group of demographers, assets such as habitable and fertile land, clean air, and fresh water are finite resources. And with medical advances increasing fertility and lowering death rates, the global population continues to grow exponentially with no end in sight.

Demographic Transition Theory

Replacing Malthus' ideas today, demographic transition theorydefines population growth in an alternating pattern of stability, rapid growth, and then stability again. This theory proposes a three stage model of growth.

- **Stage 1: Stable population growth.** In this stage, birth and death rates roughly balance each other. Most societies throughout history have stayed at this stage.
- **Stage 2: Rapid population growth.** Death rates fall sharply while birth rates remain high in Stage 2. Most poor countries today fit into this stage. Malthus formed his ideas during one such high growth period.
- **Stage 3: Stable population growth.** In this stage, fertility falls because high living standards make raising children expensive. Women working outside the home also favor smaller families, brought about by widespread use of birth control. Death rates drop because of technological advances in medicine. With low birth rates and death rates, the population only grows slowly, if at all. It may, in fact, witness population shrinkage, in which deaths outnumber births in a society.

Stage 3 suggests that technology holds the key to population control. Instead of the out of control population explosions that Malthus predicted, demographic transition theory claims that technology will ultimately control population growth and ensure enough food for all.

Population Control: The Importance of Family Planning

Historically, many groups and societies have discouraged contraception (the prevention of conception, or birth control) to assure survival of its members and humanity as a whole. Certain religious groups strongly disapprove of sexual activity that does not culminate in coitus and the possibility of conception. Other groups place little importance on the matter of contraception. The Yanomamo of South America, for instance, harbor little or no concept of contraception. Instead, they parent as many children as possible, and then kill off those they view as the undesirable, such as some females and deformed infants.

Modern medicine has spread throughout different parts of the world, and people of all ages now live longer, causing the world's population to explode in growth. In fact, at five billion today, the world's population doubles, on average, every 35 years, with most of this growth occurring in developing countries. Given this population crisis, certain governments, like that of China, regulate the number of births allowed per household. Besides the issue of controlling overpopulation, other benefits to practicing contraception exist. For example, a young couple may want to postpone having children until their finances improve. Or an unmarried, sexually active teenager may wish to finish her education or get married before starting a family, thereby reducing her chances of eventually relying on the government for financial support. Family planning also plays an important role in protecting the physical health of both mother and child. The older or younger a woman is, and the closer together she bears children (that is, more frequently than every two years), the greater the risk of pregnancy and birth complications, early infant mortality, and maternal death. For example, women over age 40 or under age 19 have an increased risk of bearing a child of low birth weight, and thus a variety of birth defects and even outright death. Estimates say that approximately one million teenage women in the United States become pregnant each year.

Urbanization and its Historical Stages

By the early 1900s both Great Britain and the United States had become predominantly urbanized nations; since that time, urbanization has been occurring around the globe at a rapid rate. Today, as many as 50 per cent of the world's population lives in urban areas, compared to only a few per cent just 200 years ago. Sociologists studying urbanization trends note three distinct historical stages in the development of cities: *preindustrial, industrial,* and *metropolitan megalopolitan* stages.

Preindustrial Cities

For the vast majority of human history, as far as anyone knows, people roamed about in search of sustenance. While they gathered edible plants, fished, and hunted, our ancestors could never find enough food in one area to sustain themselves for an extended period of time. Consequently, they had to keep moving until they could find another place in which to settle temporarily. Eventual technological improvements such as simple tools and information on how to farm and raise animals allowed people to settle in one place. They built villages, with perhaps only a few hundred people living in each, and, for the following 5,000 years, produced just enough food for themselves with nothing more in reserve. About 5,000 years ago, however, humans developed such innovations as irrigation, metallurgy, and animal drawn plough. These developments allowed farmers to produce an excess of food beyond their immediate needs. The resulting surplus of food led some people to make their living in other ways: for instance, by making pottery, weaving, and engaging in other non-agricultural activities that they could sell or exchange with others for the surplus food. As a result, people moved off the farms, commerce developed, and cities began to form.

Preindustrial cities, which first arose on fertile lands along rivers in the Middle East, Egypt, and China, were quite small compared to today's cities. Most preindustrial cities housed fewer than 10,000 inhabitants. Others, like Rome, may have contained as many as several hundred thousand people. Preindustrial cities differed significantly from today's cities. The residential and commercial districts were not as sharply separated as they are today. Most traders and artisans worked at home, although people with the same trades tended to live in the same areas of town. People in cities also segregated themselves from one another according to class, ethnicity, and religion–with little or no chance for social mobility or interaction with other groups.

Industrial Cities

Between 1700 and 1900, increasing numbers of people moved into cities, resulting in an urban revolution. For example, in 1700 less than 2 per cent of British people lived in cities, but by 1900 the majority of them did so. The United States and other European countries soon attained similar levels of urbanization, driven by the *Industrial Revolution*. Industrialization produced the mechanization of agriculture, which, in turn, limited the amount of work available on farms. This lack of employment forced farm laborers to move to cities to find work. This migration of workers from rural to urban areas then gave rise to the industrial city. The industrial city was larger, more densely populated, and more diverse than its preindustrial counterpart. It contained many people of varying backgrounds, interests, and skills who lived and worked together in a defined amount of space. The industrial city also served as a commercial center, supporting many businesses and factories. The latter attracted large numbers of immigrants from other countries hoping to better themselves by securing stable work and finding a "fresh start."

Metropolis and Megalopolis Cities

As larger and larger industrial cities spread outward in the early 1900s, they formed metropolises (large cities that include surrounding *suburbs,* which are lands outside the city limits, usually with separate governance). While some suburbs become distinct cities in and of themselves, they retain strong geographic, economic, and cultural ties to their "parent" city. Many metropolitan areas house a million or more residents.

The upper and middle classes ultimately brought about the so called flight to the suburbs. As economic woes increasingly plagued cities in the latter half of the 1900s, many families decided to move out of their inner city neighbourhoods and into the suburbs. The ability to afford an automobile also influenced this migration. Beginning in the 1970s, most suburbs were largely "bedroom communities," which means that suburban residents commuted into the city to work and shop, and then returned to the suburb at night. Commuting presented a downside, but most people felt that escaping "urban ghettoization," or the tendency for the quality of life in inner cities to decline, was well worth any hassles, given the fact that suburbs tended to offer nicer and larger homes, better schools, less crime, and less pollution than cities provided.

Today, suburbs continue to grow and develop. Many have become economic centers in their own right. Offices, hospitals, and factories coexist with shopping malls, sports complexes, and housing subdivisions. In this way, many suburbs have essentially become small (and, in some cases, not so small) cities. Demographically, suburbs tend to attract "whiter" and more affluent residents than do cities. Yet not all suburbs and *suburbanites* are alike. Even within a suburb, families of varying ethnic and religious backgrounds exist.

Because of all this growth, many suburbs have developed "urban" problems, such as air and water pollution, traffic congestion, and gangs. To escape these problems, some people have chosen to move to rural areas. Others have chosen to return to and revive their cities by renovating and remodelling buildings and neighbourhoods. Such an interest in urban renewal (also called *gentrification*) has turned some slums into decent areas in which to live, work, and raise a family.

The vast urban complex known as a megalopolis was created as suburbs continued to grow and merge with other suburbs and metropolitan areas. That is, some suburbs and cities have grown so large that they end up merging with other suburbs and cities, forming a virtually continuous region. One example of a megalopolis is the hundreds of miles of almost uninterrupted urbanization from Boston to Washington, D.C. The typical megalopolis consists of literally millions of people.

14

Educational Psychology

Educational Psychology in concerned primarily with the study of human behavior as it is changed or directed under the social process of education, and secondarily with those studies of processes that contribute to an increased understanding of now behavior is changed and directed through education.

Aim of Educational Psychology

Understanding, predication, and control of behavior are the goal of psychology and educational psychology.

Educational Psychology is Designed To

a) Enhance and enrich the lives of the learners.

b) To furnish students with the knowledge and understanding that will help them for improvements in the quality of instruction. The subject matter includes the topic of growth and development, mental processes, learning, motivation, individual differences, adjustments, guidance and other aspects of behavior that appear in learning situations.

The Role of Educational Psychology in Education may be Summarized

a) As to evaluate education theories.

b) To study and evaluate contemporary practices in education.

c) To evaluate different methods of teaching in the light of established principles of learning and motivation.

d) To help teacher to solve their personal problems.

e) To furnish methods for scientific study of problems in the general are of educational psychology.

General Nature of Learning

- ☆ Learning is a modification of behavior through experience and training. So it is a growth and development process.
- ☆ The need arises, goal is set-up, adjustment begins, and behavioral charges take place and stabilize. This whole process is called as learning.
- ☆ Learning involves problems solving *i.e.* understanding and discovering relations between different contents in a situation. Learning and activity aroused by the personal and social demands. Why should one learn? Because it is the demand of individual. Individual leans many things *i.e.* desirable and undesirable also. But aim of education is to bring about desirable changes in the behavior of an individual and to eliminate the undesirable changes. Thus learning is the process of improvement.
- ☆ Learning is a change in behavior. Learning is a reconstruction, combined thinking, skill, information and appropriation in single unity process and it is characterized by flexibility.
- ☆ Learning consists of the ways in which we modify our responses to that stimulus that upset us but there is not a response to every organic state of the organism. The process by which we come to make the responses, which will overcome the threatening stimuli, is called as learning
- ☆ Learning is concept of describing changes in the behavior which result from practice.

There are Three Phases of Learning

a) Practice or learning itself.

b) A time delay.

c) Performance test of what was learned.

Learning can never be observed directly. Changes in the performance of some bit of behavior are the indication of learning. Sometimes changes in the behavior are also observed due to motivation, adaptation and fatigue.

Learning is certainly a universal experience. Every one learns at every stage in his life. In facts you must learn to talk, to dress and feed themselves. Adolescents must learn acceptable social habits of the community. Adults must learn how to perform their job and how to meet the responsibility of their family life. Our daily life is the succession of major and minor problems that have to be solved by learning. To learn means to discover or invent, to commit to memory and become efficient.

Definition of Learning and Principals of Learning

Definition of Learning

Learning is that process, which is the result of training and experience leads to new a changed responses.

Learning is the process by which a person becomes changed in his behavior through self activities.

Learning is that process by which the physical and mental activity of the individual leads to new or changed response.

Principals of Learning

The principals of learning process are well established and have useful application in extension education. The principals provide real foundation for any educational efforts.

1) Learning is Growth Like and Continuous

Teaching should begin from where the learner is *i.e.* level of knowledge, understanding and interest. The new idea must be related to old hence it is growth like and continuous.

2) Learning should be Purposeful

Learning varies directly with meaningfulness of the subject. Learner must fell that it is useful to him. The objective of learning must be clear and meaningful to learner. The objective must specify the change to be brought out and the subject matter.

3) Learning Should Involve the Maximum Number of Senses

Messages reach the human mind through the senses *i.e.* seeing, hearing, feeling testing, smelling *etc.* more than one sense is to be included in the process of learning so that learning becomes effective.

4) Learning Must Be Challenging and Satisfying

Learning must be challenging. It should give satisfaction to the learner. It should be based on needs and wants of the learner. Teacher's motivation, timely recognition and favorable attitude accelerate the learning while negative attitude retard learning.

5) Learning Must Result in Functional Understanding

In useful learning the learner should not only acquire new facts and ideas but also understand how to apply them in real condition knowledge is prerequisite for understanding.

6) Learning is Affected by Physical and Social Environment

The general physical and social environment should be suitable for learning; physical factors include light, ventilation, furniture, sitting temperature *etc.* If the social environment is encouraging it helps for the learning process.

7) Learning Ability Varies Among Individuals

The level of understanding varies according to intelligence of the individual. The level of communicator and the level of the learner must be in the line. So the subject can reach to the learner easily.

8) Learning is General and Gradual Process Which Requires Several Exposures

Learning is a slow process and it increases gradually. For complete understanding of the subject several exposures are required complete understanding of the subject bring the change in the learner.

9) Learning Capacity is significantly high in Adult Life

Adult normally posses a more capacity to learn the speed of learning decline after middle age. Learning capacity increases rapidly till the age of 20 years then it remain constant up to around 50 years.

10) Learning is Active Process On the Part of Learner

Learning is individual or personal matter. It is active process and depends upon the interest of the learner and stimulus of the teacher. The teacher can set up the learning situation and stimulate the person to learn.

11) Learning Requires Effective Communication

Communication is the process by which two or more person exchange ideas for common understanding. Communication is a essence of learning and teachings.

12) Theory and Practical must be Related

Theory and practical have always relationship. Theory relates to why and practical relates to how. Theory explains the structure and practical to its application.

13) Learning and Teaching is an Intentional Process

The aim of teaching is to promote learning which will bring the desired change. The objective of teaching and objective of learning are not clear in learning situation, than it will lead to confusion.

14) Learned Behavior Results in to Changes Farms and Home Practices

Types of Learning

1) Imitation

Much is learn in the life by way of imitation. Children learn many more new things by way of imitation. Imitation may be conscious or unconscious. Through unconscious imitations the customary behavior and moral codes of society the learnt and made permanent part of the child's character. The conduct of the parents the personality of the teacher, literature, school environment, movies, all are very effective models for imitation. Fashion and customs are largely through limitation. The imitation is an important, moral, intellectual and special agent, in formal education and for throughout life. But for learning newer and more complex things imitation played a little role. Fashion is imitation of contemporaries while custom and tradition is the imitation ancestors.

2) Trial and Error (Selective Learning)

This type is also known as learning by selection of successful variant. When no ready-made solution of problem is available to the learner, he adopts the method of trial and error. He first tries one solution. If it does not help him, he rejects it. Then he tries for another. In this way he eliminates errors and finally discovers the correct solution. Thus in the trial and error method the learner makes random activities and finally reaches to the goal accidentally. Trial and error method is not very much useful for teacher. There is much wastage of time and energy in learning by this method. It is useful for subjects like math's, grammar, memorizing poetry *etc.*

3) Learning by Insight

Learning is the organization and reorganization of the behavior which arises from the interaction or a maturing organism and its environment. This is brought about through new forms of perception, imagination, Co-ordination and other organic behavior. Sudden appearance of the solution is a peculiar characteristic of insightful learning. Insight is the perception of a relationship between at least three factors, an agent, a goal and intervening conditions or obstacles. In-sight learning has three principles of insight, the law of whole and sudden appearance of the solution. This type of learning is very important in education field and it does not consume much time. It emphasizes upon meaningfulness, trains imagination, reasoning and thinking abilities. In insight can be brought about through maturation experience and good arrangement of environment forces. Learning by insight gives emphasis upon encouraging, aiding and directing behavior.

4) Learning by Conditioning

This is the learning from the behaviouristic view point. According to this learning is formation of conditioned reflexes or acquisition of involuntary anticipatory adjustment or a habit formation so that behavior may become automatic. Many things of the school subjects are learnt more adequately through condition learning. Like reading, writing, discipline, good habits *etc.* Direct method or teaching English is a process of conditioning. We learn many things through this process.

Curves of Learning

Learning involves acquisition (improvement) or material, Acquisition is not sudden. It progresses steadily and sometimes acquisition and learning both go side by side. All these activities acquisition, retention and forgetting can be measured graphically. Graphic representation of the progress and rate learning, rate of learning, rate of forgetting and retention are termed as learning curves.

Important of Learning Curves in Education

a) They give a rough generalization of the rate of progress.

b) Helps to teacher to know where his students as regards the progress in learning.

c) It helps teacher to know which difficulty is faced by the student and why the progress is blocked. So that it can be removed by the teacher and teaching can be adjusted accordingly.

Limits of Learning

The limit of learning means the attitude of one's power to do a certain work. Each person reaches a limit in practice of learning below his actual or physiological or the critical limit. This is known as practical limit, very few in physiological limits. Mostly people reach the practical limit. Between the critical or physiological limit and practical limit lies the optimum limit. This is reached by efficient work habit and by the most efficient management of time and efforts. Many are below it. Teacher should stimulate the students to achieve this optimum limit.

Learning Situation

An effective learning experience can only be had in well structured and skill fully executed learning situation. Extension teaching consists basically or arranging situations that promote desired learning. The essential role of extension worker is to create learning situations that stimulates and guide the learning activity. The effective extension worker is one who can create and manage learning situations in which learners have effective learning experience. A learning situation is the condition or environment in which all the elements necessary for promoting learning are present.

An Effective Learning Situation Consists of Following Five Essential Elements

a) An effective Instructor or Leader.

b) Learners who want and need to learn.

c) Content or subject matter that is useful to learners.

d) Appropriate teaching equipments and material.

e) An appropriate physical environment

The Elements of Learning Situation

The above diagram shows that the learner is the central element in the learning situation. This is significant since the entire purpose is to cause them learn. Therefore, the learning on the part of the learner becomes the objective to be attended while other four elements become the means for achieving this end product. To have an effective learning situation, these five elements should satisfy the following conditions.

I) Instructor – He should

a) Have clear objective.

b) Know the subject matter and have it well organized.

c) Be able to communicate with learner.

d) Allow learners participation, ask for it.

e) Use a definite leaching plan.

f) Speak loudly and clearly so that all can hear and understand.

g) Be skilful in use of teaching materials and equipments.

II) Learner: Should

a) Have need for information.

b) Be interested.

c) Be capable of learning.

d) Use the information again.

III) Subject Matter should be

a) Related to learners need.

b) Applicable to real situation.

c) Thought intellectual levels of the learners.

d) Well organized and logically and clearly presented.

e) Challenging and satisfying.

f) Fulfilling overall objectives.

IV) Physical Facilities

a) Free from outside destruction.

b) The room should be well lighted.

c) Adequate space for the group.

V) Teaching Equipments

a) Meet the needs effectively.

b) Readily available.

c) In working order.

The nature of each of these elements their relationship to each other, their role in education process must be thoroughly understood by the instructor and skill developed in handling them. Effective learning situation are created through the skilful use of appropriate teaching methods and equipments.

Factors Influencing Learning

1. Kind of Material

It is experienced that some material is harder to members *e.g.* poetry is easier to remember that prose.

2. Sense Modality

General option seems to favour vision over learning.

3. Age of the Learner

Brain and nervous system loss their plasticity and sensitiveness to new impressions in middle and old age.

4. Sex of the Learner

Girls show superiority over boys of the same age in test of memorizing and in school learning.

5. Motivation

Greater the motivation more is impressed upon the brain to be remembered.

Extension Teaching and Learning

The objectives of Extension Education are educational in nature. The effort of the change agent is to change the knowledge, skill and attitudes of the farmers so that their personality undergoes change. This changed 'man' is expected to make use of modern advances in Science for solving the problems surrounding him. It is therefore, necessary to understand the teaching and learning processes in extension education.

1. **Extension Teaching**
2. **Extension Learning**

Extension Teaching

It is the process of arranging situations in which the important things learned are called to the attention of the learners, their interest developed, desire aroused, and action promoted.

Principles of Extension Teaching

The principles which need to be followed for making the extension teaching effective are discussed here:

1. Extension Teaching Requires Specific and Clearly Defined Objectives

While deciding the objectives of teaching, following aspects are to be considered:

a) People to be taught.
b) Behavioural changes to be developed in people.
c) Content or subject matter to bring the desired change in behaviour.
d) The life situation in which the action is going to take place.

2. Extension Teaching Requires a Suitable Learning Situation

The learning situation consists of the following five inter-related elements:

a) Instructor (Extension Worker)
b) Learners (Farmers, Women, Youth)
c) Subject Matter (*i.e.* Agriculture, Horticulture, Social Forestry)
d) Teaching Materials (Seeds, Plants, Audio-visual aids).

3. Extension Teaching Requires Effective Communication

Extension teaching, to the effective, must take into consideration proper functioning of the elements of communication process.

a) Communicator
b) Message
c) Channel
d) Treatment
e) Audience

Effective communication can motivate people to act.

4. Extension Teaching Requires Both Content End Method

Content means the subject matter. Method means the delivery system. Content should be relevant and method should be appropriate.

5. Extension Teaching must be looked upon as an Intentional Process

It should be properly planned on the basis of existing situation and available research findings. There is no scope for haphazard thinking and action.

6. Extension Teaching must Result in Effective Learning

The topic taught should be understood by the learners. As per the situation, combinations and teaching methods should be done. It should lead to maximum behavioural change among the farmers.

7. Extension Teaching Requires Careful Evaluation of Results

The changes brought about in the behaviour of people need to be assessed. Later, steps need to be taken to meet to the deficiencies identified in the process.

Extension Learning

It is the process by which an individual, through his own efforts and abilities, changes his behaviors.

Principles of Learning

With a new to provide effective learning experience to rural people, the extension worker should know the characteristics of learning. These are called as Principles of learning.

1. Learning is Growth Like and Continuous

The extension worker should try to understand the level of the farmer and should try to develop his understanding. The things already known by and also new ideas should be clarified. The new ideas may be presented intermittently to emphasis their importance.

2. Learning should be Meaningful

The extension worker should place the objectives of learning very clearly and meaningfully before the farmers. While doing this, it should be seen that the subject does not go over the heads of the farmers. The extension education programme should be within the physical and economic limits of the people.

3. Learning should be able to Make Use of Maximum Number of Sense Organs while Learning

Seeing hearing, touching, smelling and testing are the five senses through which an individual learns new idea. Extension believes in **'Learning by doing'** and **'seeing is believing'**. So, demonstrations, visual aids, and other symbols should used by the extension worker, to make the learning experience more effective.

4. Learning should be Challenging and Satisfying

Abilities acquiring through learning should help the farmer to solve the problems, to overcome the difficulties and gradually lead to a more satisfying life.

5. Learning should Develop Functional Understanding of Learners

Only acquiring knowledge is not enough, it must be understood and applied in real life situation. To develop the functional understanding of the farmers, it is necessary that he should understand the whole subject, as well as different topics and their inter-relationships.

6. Learning is Affected by Physical and Social Environment

Good physical and social environment creates a favorable back ground for effective learning. The student should be required to spend minimum energy in adjusting with the environment. The teacher should create a suitable environment.

7. Learning Ability Varies Widely Among Individuals

The teaching should be tailored according to the level of understanding of the farmers. So also, the subject should be within the capacity of the farmers' ability to learn.

8. Learning is a Gradual Process, Requiring Several Exposures Before Change is noticed

The new ideas have to be vividly presented through different media in order to make the learning meaningful. Research shows that for convincing the necessity of innovation, it has to be presented at least six to eight times through different media.

9. The Adults have Learning Capacity

The rate of learning declines about one per cent a year, after the age of 35 years. The main reason for this is declining physical ability, especially functioning of sense organs. Reluctance to learning, fear of failure, old habits and impact of particular ideology are also the reasons for it. The extension workers should show the things clearly, speak clearly by choosing words carefully, proceed step by step, repeat important ideas, encourage good ideas, and avoid ridicule.

10. Learning is an Active Process on the Part of Learner

The kind of learning which takes place is a result of the kind of experience one has. The experiences must be compelling and directed on the part of learners. The farmer himself will have to practice the skills to learn them, change the present attitude to create new attitude, and appreciate new things. Learning is group facilitates this.

11. Learning Requires Effective Communication

Learning becomes effective if the extension worker puts the message in an interesting way with the help of appropriate audio-visual aids and teaching methods.

12. Theory and Practice should be Related in Learning

As far as possible, the farmers should be taught both, theory as well as practice. A proper balance in these two helps in acquiring high professional competency.

Steps in Extension Teaching

Extension teaching is a planned and deliberate act on the part of the extension worker. The extension worker has to move step by step in a scientific and logical way to impact training to the clients who are farmers, farm women and rural youth. The role of the extension worker is that of a facilitator and motivator, though the details of the procedure may vary from situation to situation.

There are some steps which are basic to extension teaching. These are presented following by Wilson and Gallup (1955).

SATISFACTION

ACTION

CONVICTION

DESIRE

INTEREST

ATTENTION

Steps in Extension Teaching

1. Attention

The first step in extension teaching is to make the people aware of the new ideas and practices. The people must first know that a new idea, practice or object exists. This is the starting point for change. Until the individual's attention have been focused on the change that is considered desired, there is no recognition of a problem to be solved or a want to be satisfied.

Mass methods like Radio, Television, Exhibition *etc.* and personal contact by the Extension workers, contact through Local Leaders are important at this stage.

2. Interest

Once the people have come to know of the new idea, the new step is to stimulate their interest. This may be done by furnishing them more information about the topic in a way they will be able to understand and use. It is necessary to present one idea at a time relevant to their needs. Personal contact by the extension worker, contact through local leaders, farm publications, radio, television *etc.* are important at this stage.

3. Desire

It means un-freezing the existing behaviour and motivating the people for change. At this stage it is necessary to emphasize on the advantages of the new idea or practice. Visit to demonstrations, farm publications, personal contact by the extension worker, group discussion *etc.* are important at this stage.

4. Conviction

It is a stage of strong persuasion so as to convince the people about the applicability of the new idea or practice in their own situation and that it would be beneficial for them. The people are furnished with adequate information about the idea and how it works. Field day or Field Visit, Slide Show, personal Contact by the Extension worker and Trainings are important at this stage.

5. Action

This is the stage of the idea or practice into operation. Small scale demonstration with supply of critical inputs may be set up in real life situation of the individuals who come forward. This provides the opportunity of direct experience on the part of the learners. At this stage it is necessary to collect evidence of change such as change in yield, income, employment *etc.* Demonstration, personal contact by the extension worker, supply of critical inputs and ensuring essential services are important at this stage.

6. Satisfaction

To produce lasting change, the extension efforts should produce satisfying results. Satisfaction may come from high yield, more income, better health *etc.* Satisfaction reinforces learning and develops confidence, which generates motivation for further change. To sustain the changed behaviour, it is necessary to furnish new and relevant information about the practice on the continued basis till change in the practice itself is felt necessary.

Use of mass media, local leaders and personal contact by the extension workers are important at this stage. Availability of critical inputs and essential services are also to be ensured. The steps in Extension Teaching are to be synchronized with the innovation-decision process of the learners. The teaching methods and aids mentioned here are by way of example.

Extension Teaching Methods

An Extension Worker is a Teacher. He must anticipate the varied learning situations with which he will be faced. He must plan his teaching activities. His teaching tools must be most appropriate to the teaching situations. Success in bring in about desired changes in the behaviour of learners frequently depends on the Teacher's skill in arranging the best learning situations and in using the most effective methods of teaching in that situation.

Definition of Teaching Methods

"Teaching methods may be defined as the devices or tools used to create learning situation in which effective communication can take place between the teacher and the learner".

Functions of Extension Teaching Methods

a) To provide communication so that the learner may see, hear and do the things to be learnt.

b) To provide stimulation that causes the desired mental or physical action on the part of learners.

c) To take the learner through one or more steps of teaching, learning process *e.g.* Awareness, Interest, Evaluation, Trial, Adoption.

Objectives of Extension Teaching Methods

The Extension Teaching Methods are useful for:

a) Approaching village people.

b) Working with village people.

c) Encouraging village people.

d) Influencing village people.

e) Attracting their attention towards improved practices.

f) Arousing and developing their interest.

g) Convincing the people to adopt improved practices.

h) To provide communication so that the learner may see, hear and do the things to be learnt.

i) To provide stimulation that causes the desired mental and physical action.

j) To take the learner through one or more steps of teaching, learning and adoption process *i.e.* to arrange situation in which the farmer, farm, woman, boy or girl may see or hear about and to the things to be learnt and adopted.

k) To involve village people to build their organization.

l) To guide village people to build their organization.

m) Spread of information about new practices.

n) To improve extension education and increase opportunities for information to people.

Classification of Extension Teaching Methods

Extension education is mainly concerned with the dissemination of useful and practical information relating to rural life and helping the rural people in the practical application of such knowledge to rural situations. The extension worker can achieve this objective. It has been shown by research that extension workers.

a) Using only one extension teaching method was able to influence a little over 1/3rd of the village families to improve practice in various fields.

b) But where the workers used three methods, including demonstration, visual aids and the written words, almost 2/3rd of the families were influenced to use better practices.

c) Where five methods of teaching were used almost "9 out of 10" families changed and

d) Where nine methods were used 98 per cent of the families were led to change their practices.

Before planning the strategy of extension teaching methods the extension worker should know the details of the methods available for use in the rural situation. From this point of view the extension teaching methods are classified and are presented in the following table:

Classification of Extension Teaching Methods

1			**According to use**	**Indirect Influence**
	a.		**Individual contacts**	
		i)	Farm and home visits;	
		ii)	Office calls	
		iii)	Telephone calls;	
		iv)	Personal letters	
		v)	Result demonstration	
	b.		**Group contacts**	
		i)	Methods demonstration meetings	
		ii)	Leader training meetings	
		iii)	Lecture meetings	
		iv)	Conferences and discussion meetings	
		v)	Meetings at result demonstrations	
		vi)	Tours	
		vii)	Schools	
		viii)	Miscellaneous meetings	
	c.		**Mass contacts**	
		i)	Bulletins	
		ii)	Leaflets	
		iii)	News letters	
		iv)	Circular letters	

		v)	Radio	
		vi)	Television	
		vii)	Exhibits	
		viii)	Posters	
2			**According to form**	**Indirect Influence**
	a.		**Written**	
		i)	Bulletins	
		ii)	Leaflets	
		iii)	News articles	
		iv)	Personal letters	
		v)	Circular letters	
	b.		**Spoken**	
		i)	General and special meetings of all kinds	
		ii)	Farm and home visits	
		iii)	Office calls	
		iv)	Telephone calls	
		v)	Radio	
	c.		**Visual or Objective**	
		i)	Result demonstrations	
		ii)	Exhibits	
		iii)	Posters	
		iv)	Motion pictures, charts, slides, and other visual aids	
	d.		**Spoken and visual**	
		i)	Method demonstration motion	
		ii)	Meetings at result demonstrations	
		iii)	Meetings involving motion pictures, charts and other visual aids	
		iv)	Television	

While using the foregoing methods the extension worker uses these methods independently or takes the help of certain audio-visual aids. The word audio-visual comprises three words namely:

Audio – refers to sense of hearing, Visual – refers to sense of seeing and Aid – instructional device.

The audio visual aids are classified as under:

Audio	*Visual*	*Audio-visual*
Tape recorder	Flash cards	Cinema projector
Radio	Black board	Television
Recording	Pictures	Drama

Another method of classification is to divide the audio visual aids in to projected and non-projected aids.

Projected Aids	*Non Projected Aids*
Cinema Projector	Flash cards
Slide Projector	Flannel graph
Overhead Projector	Charts
Opaque Projector	Pictures
	Black board
	Bulletin board
	Models
	Posters
	Graphs
	Specimens
	Exhibits
	Photographs

Teaching, Principals of Teaching and Steps in Extension Teaching

Teaching

"Teaching is a process of providing the situation in which learning takes place". Teaching is arranging the situation in which the things to be learnt are the Attention of the learner, their interest is developed, action promoted, desire aroused, action promoted and satisfaction ensured.

Teaching is a process of guided interaction between the teacher, learner and material of instruction. It is a process of creating situation that facilitates learning process.

The situation created should help in bringing desired change in the behavior of the learner.

Principals of Teaching

a) The student should understand the purpose of learning.

b) The learner should want to learn.

c) The teacher should keep informal and friendly climate in learning situation.

d) Physical facilities should be favorable and appropriate.

e) The teacher should involve the learner so that they can participate and accept the responsibility.

f) The teacher should make use of learner's experience.

g) The teacher should prepare well for the class keep his teaching aids ready.

h) The method of instruction should be appropriate.

i) The teacher should follow learner.
j) The teacher should have flexible plan for the course.
k) The teacher should able to learn at their own place.
l) The learner should aware about his own progress.

Steps in Extension Teaching

a) Getting the attention of the learner.
b) Stimulating the learner's interest.
c) Increase the learner's desire for information.
d) Convincing the learner about the act.
e) Getting action by the learner.

Intelligence

Children as well as adults recognize differences among people in ability to meet situation in which habitual behavior of routine responses are inadequate. According their success or failure in everyday activity individuals are described as bright, quick, smart, dull, slow *etc.* The person who applies these terms to another is comparing his behavior with others.

Meaning

Intelligence is a concept. It is difficult to define. Psychologists have defined the term intelligence in different way. The term is usually applied to those mental activities that constitute higher mantel processes.

A person's intelligence is evaluated in terms of overt behavior that reflects the Functioning of complex elements comprising mantel capacity. The concept of intelligence is readiness to learn, ability to do abstract thinking and ability to adopt to novel stations with speed and accuracy. In the mental process it is said that it is a degree of effectiveness in perception, memory, reasoning and imagination. It is general capacity of an individual conscious to adjust his thinking to new requirements. Stoddard says, "Intelligence is the ability to undertake activities that are characterized by difficulty, complexity, abstractness, speed, adaptivenss to goal, social value innovativeness and to maintain such activities under conditions that demand a concentration of energy and resistance to emotional forces. Other says that, it is ability to adapt adequately to the environment, general tendency toward achievement. Some says that it is the mixture of particular abilities called as amplitudes.

Intelligent Behavior is Classified into Three Kinds

a) Social Intelligence: Understanding of people and the ability to act wisely in human relationship.
b) Mechanical Intelligence: It means the skill in manipulation tools managing the working of machines.

c) Abstract Intelligence: It is the ability to handle symbols and ideas such as words, numbers, formulas and scientific principles.

Theories of Intelligence

Various theories have been formulated in connection with the structure and organization of the intelligence. Out of these three theories are important.

a) Connection Theory (By Thorndile) or Multi-Factor Theory.

b) Two-Factor Theory (By Spearman).

c) Group factor or Primary mantel abilities Theory. (By Thurstone).

1) Connection Theory or Multi-Factor Theory (By Thorndile)

According to him the basis of intelligence are neural connections between stimulus and response that under lie behavior. According to this theory a person might be able to earn success in abstract learning but not in social relationships or in mechanical activities. Successful achievements in two or three areas would result from overlapping of the intelligence rather than functioning of general intelligence. Intelligence is a multitude of atoms of mental ability. This theory is also called as atomistic theory. The main assumption of this theory is that mental activity involves a number of atoms of mental energy which function together.

2) Two Factor Theory (By Spearman)

He says that 'intelligent' behavior is affected by two mental ability factors:

a) General intelligence (G. Factor)

b) Specific abilities (S. Factors)

The G and S factors are working together in a unit. Generally G factor of intelligence is not found in the same proportion among all individuals. Variations in G factors are due to individual differences. Concept of a G factor is the physical energy of the individuals. G factor is a general energy that functions in all mental activity, but such as mathematics, language, science or any special ability. G factor is gifted ability of the individual.

3) Group Factor or Primary Mental Ability Theory (By Thrustone)

This theory is most popular because it has led to construction of famous chieqge tests and primary mantel abilities. According to him there is a primary factor operation in different mental activities of a common nature. Thus it is assumed that there are number of group of mental abilities having their own primary factors is to provide unity and cohesion to group of mental activities. Thurstone evolved a multiple factor theory including thirteen factors.

They are:

a) Ability to visualize space,

b) Number facility,

c) Memory,

d) Ability in verbal relations,

e) Ability to reduce from presented data,

f) Speed of perception,

g) Problem solving.

Measurement of Intelligence

One of the oldest and most useful of the tests of behavior is that which is intended to measure and predict the intellectual capacity of man. The test of intelligence holds a very practical origin in selecting grade school children who would profit from different training. Intelligence tests are useful for assessing differences among adults. The intelligence test has been analyzed for its contribution to the understanding of various performance differences among the men.

From the beginning of intelligence testing there have been two parallel trends, one towards performance tests and other towards verbal test-tests in which the instructions are given in words and in which a good deal of the performance is in the farm of words. Performance tests can be given to people who do not know the language or who have not been to school. They are often blind, people with poor vision and hearing or to children whose handicap seem more severe in verbal than in other areas.

Binet test of General Intelligence

He noted that the intellectual capacity increase with age, the intelligent person would be less susceptible to distraction, more like to adopt the situation to achieve a goal and likely to criticize his own work.

Mental Maturity

If you give a child same test twice with enough time between two tests, you will find that his scores would improve as he grew older. He would certainly do better at the age of 18th than at the age of 4 years. However, a time would come when his ability to answer the test questions would no longer improve. A person does not stop learning even he reaches to mental maturity. Generally mental maturity reached somewhere between 14 to 18th age. The generally accepted figure is 15th year.

Mental Age

It is a degree of intelligence exhibited by an individual in relation to other of his age group. Binet and Simson have development a scale called as mental age scale for measuring brightness and dullness among the individuals.

Binet testing of school children classified individuals into levels of mental development, which is called as mental age. Binet and Simson scale was applied as follows- A child who passes all the eight year old test is regarded as 8 year old mentally, however his actual age (chronological age) may be 6 year. So he is mentally 2 years in advance. Another child with mental age of 8 may be 11 years old means in this case he is retarded in three years, so brightness or dullness in terms of number of years, advance of retardation.

Intelligence Quotient (I.Q.)

William Sterm says that you can get a number that would show how intelligent a person. This is by dividing his mental age by his chronological age. The I.Q. is obtained by dividing mental age by chronological age and multiplying by 100. Formula of I.Q. is

$$\text{Intelligent Quotient} = \frac{\text{Mental Age}}{\text{Chronological Age}} \times 100$$

$$= \frac{\text{M. A.}}{\text{C. A.}} \times 100$$

The concept of mental maturity raises an interesting problem. Example, think of 15 years old with a mental age of 15, clearly his I.Q. is 100 or his mental age is still presumable 15 year. Therefore his I.Q. sunk from 100 to 75 I.Q. we must take the denominator age as 15. For avoiding these difficulties Wechsler-Bellevue adult Intelligence scale is used. Here final score is not calculated from mental age values of the tests passed by the person being examined, but from the total number of points made by him on all tests.

Normal Distribution of Intelligence Levels

I.Q.	*Description*	*Present of Population*
0-19	Idiot	1
20-49	Imnecile	1
50-69	Moron	2
70-79	Inferior	6
80-89	Dull	15
90-109	Average	46
110-119	Bright	18
120-129	Superior	8

Normal desires and emotions but do have intelligence to control behavior. The superior, very superior and genius have capacity to make outstanding success.

Factors Affecting Intelligence

1) Heredity and Environment

Heredity provided the physical body to be developed with certain inherent capabilities while environment provides for the maturation and training of the organism. In 1940 Newman concluded that variations in I.Q. were determined about 68 per cent by heredity and 32 per cent by environment. During an individual's life time variation in I.Q. is due to environment, since the heredity cannot change. Improved nutrition health, an stimulus situations would account for this change. Environment of children ages 2 to 4 years appears to be critical since the child normally learns language at this time. Better homes, school, medical facilities, less economic discrimination, it provides a mean for achieving a mentally able population.

2) Age

A person who is bright or dull in childhood tends to remain bright or dull throughout his life. Growth I intelligence can continue through the early twenties, person achieves his maximum 10 at about 20 years are and remain stable till 10 years and decreases after that. Some abilities remain constant while others decline rapidly due to decline in physical efficiency. It is observed that the more intelligent person has more rapid of mental growth and continues to develop longer than does the mentally less able.

3) Race and Nationality

There is no prominent evidence to show that race in a factor for determining intellectual level. Difference exist between families are due to environment. The differences in races are also due to opportunity for training in early years.

All evidence indicates that there is little or no difference in inherited intellectual capacity due to race. There are differences, mostly due to opportunities for training and learning. The idea of inferior races due to heredity is not true.

4) Culture

Answer in intelligent test question is highly cultural. Cultural will determine the degree of a person's attitude and abilities.

5) Health and Physical Development

Physical and mental health is related to one's ability to gain desired achievement. A mental activity with delicate health, one may not possess enough energy to engage in mental activity to the extent that he achieves success. Physical defects such as in complete maturation of brain cells sensory and physical handicaps many interfere with observable intelligent behavior. Diseases also affect intelligence. Emotional bodies may be interfering one's ability; unfavourable health affects mental status of the individual.

6) Sex

It is a popular belief that boys are suppose to be more intelligent than girls. The research studies have shown that there is nc significant differences between the sexes, on the average, girls seems to show slight superiority in language, memory and appreciation. Differences in intelligence are caused partly by environmental conditions

7) Social and Economic Conditions

Home plays a significance role in the early developmental years. Home conditions influence on behaviours attitude. The financial status or the parents, neighbourhood and environmental conditions also affect on the intelligence but due to their socio-economic status. Mentally defective, People (Feeble – minded): Individuals who's I.Q.'s are below 70 they are called as mentally defective or feeble – minded people.

Part III

Rural Development

15

Rural Development

What is Rural Development

There is no universally accepted definition of rural development. The term is used in different ways in vastly divergent contexts. As a concept, it connotes overall development of rural areas with a view to improve the quality of life rural people. As a phenomenon, it is the result of various physical, technological, economic, socio-cultural and institutional factors. As a discipline, it is multi-disciplinary in nature representing an intersection of agricultural, social, behavioural and management of sciences. In short, rural development is a process that aims at improving the standard of living of the people living in the rural areas.

Rural development may be defined as overall development of rural areas to improve the quality of life of rural people. It is an integrated process, which includes social, economical, political and spiritual development of the poorer sections of the society.

Rural development can be defined as, helping rural people set the priorities in their own communities through effective and democratic bodies, by providing the local capacity; investment in basic infrastructure and social services, justice, equity and security, dealing with the injustices of the past and ensuring safety and security of the rural population, especially that of women.

According to Robert chambers, rural development is a strategy to enable a specific group of people, poor rural women and men, to gain for themselves, and their children more of what they want and need. It involves helping the poorest among those who seek a livelihood in the rural areas to demand and control more of the benefits of rural development. The group includes small scale farmers, tenants and the landless.

Thus, the term rural development may be used to imply any one of the above-mentioned connotations. The avoid ineffective floundering among the myriad definitions, we shall define rural development as A Process leading to sustainable improvement in the quality of life of rural people, specially the poor.

The Concept of Rural Development

The meaning of rural development has been the subject of much debate and little agreement. The definition of rural development varies from one point of view to the other. The definition or rural development may be centered around income criterion in which the concept is made to address the problem of rural poverty. Or it may be defined in sociological concept in which the rural poor represents a reservoir of untapped talent a target group that should be given the opportunity to enjoy the benefits of development through improved education, health and nutrition. This is one of the most important definitions of rural development as the provision of social infrastructures could provide the catalyst that would transform the rural areas.

Rural development may also be seen as an ideology and a practice. It may mean planned change by public agencies based outside the rural areas such as the national Government and International organization; It may also be the bringing of the countryside into an active state, as well as the transformation of the inferior nature of the country side into something more superior in terms of activities.

Rural development is the improvement in the standard of living of the rural dwellers, by engaging them in productive activities, such as the establishment of rural industries which will increase their income. It is seen by these scholars as a means of raising the sustainable living of the rural poor by giving them the opportunity to develop their full potentials.

Rural development can be distinguished from agricultural development which it entails and transcends. In essence Rural Development may imply a broad based re-organization and mobilization of rural masses in order to enhance their capacity to cope effectively with the daily task of their lives and with changes consequent upon this. According to the World Bank Rural Development must be clearly designed to increase production. It recognizes that improved food supplies and nutrition, together with basic services, such as health and education, not only directly improve the physical well-being and quality of life of the rural poor, but can also indirectly enhance their productivity and their ability to contribute to the national economy.

Rural Development ensures the modernization of the rural society and the transition from its traditional isolation to integration with the national economy. It is concerned with increased agricultural production for urban and international markets. This is essential so as to generate foreign exchange, and to attract revenue to finance public and private consumption and investment. In order to encourage increased production rural development may offer a package of inputs and welfare services for the rural masses. Such inputs and welfare services include physical inputs (such as the provision of feeder roads, water and electrification), social inputs– (namely health and educational facilities) and institutional inputs such as credit facilities, agricultural research facilities, rural expansion services among others.

Objective of Rural Development

The main objective of the Rural Development is improving the living standards of rural people by utilizing the easily available natural and human resources. The other objectives of rural development programmers are as follow:

1. Development of agriculture and allied activities.
2. Development of village and cottage industries and handicrafts.
3. Development of socio-economic infrastructure which includes setting up of rural banks, co-operatives, schools *etc.*
4. Development of community services and facilities *i.e.* drinking water, electricity, rural roads, health services *etc.*
5. Development of Human resource mobilization.

Importance of Rural Development

Improvement in the quality of life of rural people is the important agenda of rural development programme. In India – a country where the number of people living in rural areas, rural development programme is necessary aspect.

Rural development implies both the economic betterment of people as well as greater social transformation. The basic objective of all rural development endeavors/programmes has been the welfare of the millions. In order to achieve this, planned attempts have been made to eliminate poverty, ignorance and inequality of opportunities. A wide spectrum of programmes has been undertaken so far, to alleviate rural poverty and ensure improved quality of life for the rural population especially those below the poverty line. In the initial phase of planned rural development, the concentration was on sectors of agriculture industry, communication, education and health. The Ministry of Rural Development places importance now on health, education, drinking water, housing and road so that the quality of life in rural areas improves and the fruit of economic reform are shared by all sections of the society.

With time and experience, it is realized that accelerated and meaningful development can be achieved only if people of the grass root are involved, "people's participation" has become the keyword in rural development programmes. The participation of the people is necessary to provide the rural people with better prospects for economic development.

Problems in Rural Development

As we know the 60-70 per cent of rural population in India lives in primitive conditions. This sorry state exists even after 60 years of independence. So that Rural Development programmes have urgency in the present condition also. There are many obstacles in the rural development programmes which are as under:

1. In 21st Century, there is no electricity supply in many villages.
2. Now also many rural peoples using primitive methods of cooking, living and farming and they have trust on these methods.

3. By using primitive cook stoves, around 300,000 death/year takes plan due to pollution.
4. 54 per cent of India's population is below 25 years and most of them live in rural areas with very little employment opportunities.
5. Literacy is the major problem in rural development programme.
6. The poor extension linkage causes slow growth of rural development.
7. Untrained, unskilled, inexperienced staff in extension linkage cannot provide satisfactory help to rural peoples.
8. Every one want to go to the cities, so that rural people's remains as ignores part by the policy makers also.
9. Privatization concept is useful for rural development but, government not praying much attention to this aspect.
10. Policy makes prepared policies, programmes for betterment of rural people but, if these programmes are not implemented very well then have no used.

Historical and Emerging Perspective of Agricultural Extension in India

The philanthropist and the enlightened people had started extension work as voluntary service. Cooperative extension service of USA is the first organized extension work in USA. The approach of extension in India changed from time to time from general, overall development to target group approach and training and visit system. It has been reformed and strengthened with the help and support from World Bank. It has now become the part and parcel of rural development strategy in most nations of the world. The extension activities started in India from preindependence era.

Attempts in the Pre-British period

Because the villagers were self-contained, self-sufficient and self-governed units, there was little need for social welfare. The social system provided these in times of need. This was disturbed by the invasions of foreigners and Mughal rule when people felt the need for rural reconstruction work. The Panchayat system provided some of these facilities but the development of centralised seats of the government made these unimportant.

Efforts by British Government

Till the last quarter of the 19th century, nothing was done by these foreign rulers. But the series of famines, from 1875 to 1901, numbering 18 out of a total of 33 during the whole 19th century, forced the Government to appoint some Commissions. They recommended rural development work. There followed certain acts like: Land Improvement Loans Act, and Agricultural Loans Act of 1888, Cooperative Act of 1904 and then the amendment in the Cooperative Act in 1912, establishment of

Development Departments like Agriculture, Animal Husbandry and Veterinary, Irrigation, *etc.* Some irrigation projects were also taken up to control famines.

But these Development Departments could not take up any comprehensive programme. Their activities were not based on detailed studies of programme planning, or the needs and resources of the people. It never emphasised the people's participation, cooperation and involvement.

Development Programmes of Pre-independence and Post Independence Era

The present day human institution has not developed in a day or two nor in the creations of one man. They represent the cumulative fruit of the endeavor, experience, thoughtful planning and patient labours of large number of people through generations. To comprehend, understand and appreciate the present day human institutions adequately, it is necessary to have a back ground knowledge of the course of its growth and development since the roots of these institutions lie deeply buried in the past, Organized extension work in India was started during post independence the alien rule under East India Company was largely concerned with maintenance of law and order rather than development. However, concern for villagers and woes of toiling masses moved saints, sheers and social leaders to launch voluntary efforts at rural reconstruction. One can find resemblance of extension work in these endeavors of pre-independence era.

16

Developmental Programmes of Pre-Independence Era

1) Work at Shantiniketan

In 1908, Shri Rabindra Nath Tagore, under his scheme of rural development work, started youth organisations in the villages in the Kaligram Pargana of his Zamindari. He tried to create a class of functionary workers who could learn to identify themselves with the people. In 1921, he established a Rural Reconstruction Institute at Shantiniketan. A group of eight villages was the centre of the programme. The activities of the institute were development of agriculture, cooperatives, industries and education through village organisations.

Under its agricultural programme, the institute conducted demonstrations on farmers; holdings on improved practices; established a dairy to supply pure milk and better animals to the farmers for breeding; and established a poultry farm for the same purpose. The students and workers from the institute trained weavers, organised their cooperatives and provided facilities for training in tanning, pottery, embroidery, tailoring, *etc.*

Early effort at rural development was initiated by Shri Rabandranath Tagore in 1908 by establishing youth organization in the Kaligram Progana of his Zamindari, He tried to create a class of functionary workers who could learn to identify themselves with the people. In 1921 he established a Rural Reconstruction Institute at Shantiniketan in West Bengal. A group of eight villages was the centre of the programme. This project, co-incidentally, had many elements of extension education in both spirit and action. Activities like demonstration on scientific methods of agriculture, training of youths, adult education and health co-operatives were important aspects of the work aimed to make a group of villages self-reliant.

This was a very comprehensive programme combining culture, health, education and economic aspects of village life together. Concept of village level workers and regeneration of village organization were put to work. This project was closely guided by Mr. Leonard Elmhirst, an Englishman trained in economics from USA.

Objectives of the Programme

1. To create a real interest in people fcr rural welfare work.
2. To study rural problems and to translate conclusions into action.
3. To help villagers develop their resources
4. To improve village sanitation.

These objectives were desired to be achieved by creating a spirit of self-help, developing village leadership, organizing village scouts called Brati Balika, establishing training centers for handicrafts and establishing a demonstration centre at Shantiniketan.

These demonstration centers conducted demonstration or farmer's holding for improved practices. Under this programmes establishment of dairy to supply pure milk and better animals to the farmers poultry farm for development of farmers. The students and worker of the institute were provided facilities for training in tanning, pottery, embroidery tailoring *etc.* This institute also had a mobile library and runs night schools film shows in the rural areas.

Though the institute could not get much help from the government it could not conduct research work on the lines initially planned by R. N. Tagore and so its work remained limited to the eight villages only. But in the course of history, the Government of Independence India did recognize it as an important pioneering centre of extension research in India.

2) Marthandam

The work was commenced by Dr. Spencer Hatch an American Agricultural expert in Travancore under the auspicious of young Men's Christian Association (YMCA) in 1921. The aim of the project was to bring more abundant life for rural people. It was intended to symbolize the three-fold development, not only spiritual, mental and physical but also economic and social. The essential technique of the centre was 'Self-help with intimate expert counsel'. From the demonstration centre at Marthandam, about hundred villages were covered through Y.M.C.A. centers in villages. The extension secretary was appointed supervise the activities of the group.

Marthandam was in a strategic position to serve the villages. It kept prize bulls and goats, model bee-lives, demonstration plots for improving grain and vegetable seeds, poultry runs with prize laying-hens, a weaving shed, *etc.* Inside the centre, there was equipment like honey extractors, health charts and the items needed for other cottage vocations. At the centre, cottage vocations were taught and agricultural implements tested. The emphasis throughout was on self-help and co-operation. The successful output of this project was the Egg-selling Club. In 1939 which became a self governing body. Another co-operative society was honey club, where the villagers were taught the use of modern bee-hives and extracted honey scientifically. The honey was cured and marketed co-operatively. There were Bull clubs, weaver's blub also. The activities conducted at centre could meet the mental, physical and spiritual needs of the villagers. The main shortcomings of the project were inadequate funds and governmental help. The activities were mainly organized the Marthandam and the village workers did not stay in villages. The religious bias of the institution was also a major hindrance in its activities.

3) Gurgaon Experiment

Rural upliftment movement on amass scale was first started by Mr. F. L. Brayne, Deputy Commissioner in the Gurgaon district of Punjab state. He was prompted by the backwardness, poverty and misery of the people. A village guide had been posted to act as a channel through which the advice of the experts in various departments could be passed on to the villagers. The programme of introducing improved seeds, implements, the methods of cultivation *etc.* was started throughout the district. As the village guides were not technical men, very little permanent value was achieved in fact. The project could not develop leadership in the villages that would continue work when the village guides had left the villages.

The work again gathered momentum, after 1933, where Mr. Brayne was appointed Commission of Rural Reconstruction in the Punjab. 1935-36. Government of India granted Rs.1 crore for various rural works which acted as a stimulus. Nevertheless the project could not make much headway ass the local talent was not utilized for development process. Most of the work done by exercising authority over the people rather than by voluntary participation of local people.

4) Gandhian Constructive Programme/Sewagram

Self contained and self sufficient village life was the dream of Gandhiji. He was aware about the grassroots' problems of India, rural set up and he wanted to solve these problems without intervention of any outside agency. He wanted to solve these problems by local people and through local resources. People know Gandhiji not only as a Mahatma or political agitator, but also as a social and economic reformer. He made people to understand that India lives in villages and that the common man's upliftment is the upliftment of the country.

Regarding development work in the country, he emphasized that the "salvation of India lies in cottage industries." They key-words of his economy are:

Decentralized production and equal distribution of wealth.

Self-sufficiency of Indian villages. For equal distribution of wealth, cruel process of extermination was not followed but throughout the heart of the owners by persuasion and appeal to the better sense of man.

According to him self-sufficiency of Indian villages can be achieved by eradicating middlemen, so that the farmer could get the full price for his produce. He wanted that the tiller should be able to consume his own products like fruits, milk, vegetables *etc.*

For better of people he formulated an 18 point programme, which includes the promotion of village industries, basic and adult education rural sanitation, uplift of backward tribes, uplift of women, education in public health and hygiene, propagation of natural language, love for the mother tongue, economic equality, organization of kisans, labour and students and so on.

He wants to make villagers self-sufficient and also want to develop stamina which is useful against oppression and injustice. The important institutions, which were organized to foster his ideas were; all India Spinner Association, All India Village Industries Association, Gandhi Ashram at Tiruchungodi, Gandhi Niketan at Kallupatti, Gandhi Gram at Dindigal, Gandhi Sewa Sadan at Porur (Malawar), Kasturba Ashram in Trichr, Kerala. Truly speaking, the Gandhian constrictive Programme was became big institutions and simple ideas became philosophies. His emphasis on Khadi became the Charka movement and then, the All India Khadi a Village Industries Board. His thought, against untouchability and caste system, resulted in the organization of Harijan Sewak Sangh and many like this. He created leaders like Vinoba Bhave, Nehru, Jayaprakash Narayan, Mira Ben *etc.* who came form common stock, but got inspiration from Gandhi.

All the people engaged in reconstructive programme felt that their work was needed in a great programme for their country reconstruction. They were soldier of the Grand Army of the Father of Nation. They were builders of a new society and torchbearers of new civilization in this country which, due to their efforts, has again been recognized as an important country.

5) Servants of India Society, Poona

At Mayapur village, in Madras, this organisation started a centre to impart training in agriculture and cottage industries to vilage boys and girls. It also started centres in U.P. and M.P. and published booklets on subjects like basic education, labour problems, indebtedness, *etc.*

6) Sir Daniel Hamilton's Scheme of Rural Reconstruction

In 1903, Sir Daniel Hamilton formed a scheme to create model villages, in an area in Sunderban (Bengal), based on cooperative principles. He organised one village of this type and set up one Cooperative Credit Society which functioned upto 1916. In 1924, he organised a Central Cooperative Bank and Cooperative Marketing Society and established a Rural Reconstruction Institute in 1934. The Institute provided training facilities in cottage and subsidiary industries.

7) Rural Reconstruction Work by Christian Missions

The activities of these Missions can be accounted under two heads: (1) Education and Rural Reconstruction.

Education

The aim of their education system has been to impart education which one in harmony with the rural environment and site. Their system has in view character-building, the raising of the standard of general intelligence, spread of literacy, recreation, cultural activities and instruction in subsidiary and constructive employment. The educational institutions run by these Missions, number about 2000 elementary schools, 325 high schools, and 38 colleges. In order to impart both theoretical and practical training to the students and to equip them for life, Central Institutes, attached to Rural Reconstruction Centres, have also been established.

Rural Reconstruction

Several agricultural demonstration centres at Marthandam, Ramnathpuram, Patanchery, and the other Y.M.C.A. and Y.W.C.A. institutions have been doing useful work in rural areas.

17

Developmental Programmes of Post-Independence Era

1. Firka Development Project

This project was government sponsored and aided at the attainment of the Gandhian ideal of Gram Swaraj by bringing about not only educational, economic, sanitary and other improvements in villages, but also by making the people self-confident. The scheme was launched in 1946 in 34 Firkas (group of 5 villages) throughout the state, and on April 1, 1950, it was extended to another 50 additional Firkas at the rate of two Firkas for each state. The selection of the Firkas was based on general backwardness of area and the possibilities for initiating the production of handloom cloth and other cottage industries to give a encouragement to Rural Reconstruction.

This scheme was aimed at attacking the problem of rural people as well as short term plans for the development of rural communication, water supply, formation of panchayats, organization of co-operatives and sanitation programme. In the long term plans to make the area more self-sufficient through agricultural, irrigational and livestock improvements and developing the Khadi and other Cottage Industries.

The Collector was primarily was primarily responsible to see the working of the scheme in the district. Then he was assisted by rural welfare officer of the rank of Naib Tahsildar. He was put incharge of 2-3 selected Firkas. Each Firkas was divided into 5 to 10 group of villages which were put in the charge of Gram Sevaks who were of the rank of Revenue Inspecrtors. Each Firka or group of Firka was provided with special staff like agricultural field men, administrative officers, Mistries, P. W. D. supervisors and minor irrigation overseers. To associate the people with the implementation of the programme, Development committees, consisting of officials and non-officials, were constituted in each Firka. At the state level, there was a State Rural Welfare Board comprising the heads of the Departments and influential and constructive social workers.

For stimulating healthy competition between the official and non-official agencies, the Government of Madras decided to entrust the development schemes to non-official agencies were selected and paid grants for doing Firka development of: a) Rural Reconstruction, b) Drinking water facilities, c) Sanitation d) Agriculture and Khadi and other village industries.

At the end it was realized that these efforts were restricted in scope and lacked co-ordination. There was lack of direction, support and encouragement from the central authority. However, it was provide that no extension programme which is implemented without the help and co-operation of local people could continue for long and have a desired impact on the Rural Reconstruction.

2. Etawah-Pilot Project

The ideal of this project was conceived and born in 1947. Actually this projected was put into action in September, 1948 with headquarter a Mahewa village about 17 miles form Etawab (U. P.) First 64 villages which were then increased to 97 were covered under it. Lt. Col. Albert Maya was the originator of this project. He started this project with the aim of introducing work on the rural reconstruction front. The Government of U.P. helped him in setting up machinery at district level and with extra staff for the project. The point 4-programme of America also provided finances. This project had a widespread effect on educating the villagers and broadening their mental horizons. The experiment proved not only that the material was moldable, but that the saying that the villager is ignorant, conservation and incapable of improvement was an outmoded one. The project handled the rural problem by Efforts to broaden the mental horizon of the villagers so that he might accept new and tested ideas which might then become self-generating and self-perpetuating.

The method of approach used under this project was educative and persuasive rather than coercive. For getting faith and confidence of village people, it was essential to extension worker to live in the village and prove themselves as friend of rural people.

The project started as a pilot project for introduction of improved agricultural technology. It also included general awakening of all round village upliftment activity, so that the panchayats get on a sounder-footing.

The most effective achievement of this project was that the entire area was under improved wheat crops. The area under vegetables was extended and diseases like Rinderpest and Hemorrhagic Septicemia controlled. The other programmes taken on were the construction of roads, soak pit, adoption of improved agricultural practices *etc.* In this project, all round development in the village life, in terms of social, economic, health and hygiene *etc.* were reported form the area.

3. Adarsh Sewa Sangh, Pohri (Gwaliro)

This plan of rural reconstruction was put into operation in 232 villages, falling in the Jagirdari of Cor. Shitole, It aimed at increasing the per capita income of villagers.

In each village, a Village Reconstruction Society was formed and the important items of work were compost making, deep ploughing, improved breeding and management of cattle, *etc.* The Sangh published a monthly journal *"Rural India"* Which was devoted to Planning and Community Projects.

4. I.V.S. (Indian Village Service)

Its founder was Mr. Arther T. Mosher of New York, and Shri B.N. Gupta who established it in 1945. The objectives of this organization were to assist village people to realise the best in their own villages by developing individuals, volunteer leaders and local agencies, and enabling them to be effective in helping themselves and others. The object was to assist the Government in developing villages.

For the realisation of the above objectives, the organisation adopted techniques like, personal contacts, informal group discussions, use of volunteers, demonstrations, use and production of visual aids, exhibitions, tours and trips, dramas, books and periodicals, *etc.* It was financially supported by contributions and donations.

5. Sarbodaya Programme

It was a Gandhian concept and evoked great enthusiasm in Bombay State. The main features were simplicity, non-violence, sanctity of labour and reconstruction of human values. It aimed at raising the standard of living, scientific development of agriculture, promotion of cottage industries, spread of literacy, medical and health facilities and the development of Village Panchayat.

6. Firka Development Scheme of Madras State

It was Government-sponsored and aimed at the attainment of the Gandhian ideal of Gram Swaraj by bringing about not only educational, economic, sanitary and other improvements in villages, but also by making the people self-confident. The scheme was launched in the last quarter of 1946 in 34 Firkas throughout the State, and on April 1, 1950, it was extended to another 50 additional Firkas, at the rate of two Firkas for each district. The selection of the Firkas was based on considerations of the general backwardness of the area and the possibiliti9es for initiating the production of handloom cloth and other cottage industries.

The scheme, which aimed at attacking the rural problems as a whole, as well as in parts, consisted of short-term plans for the development of rural communications, water supply, formation of Panchayats, organization of cooperatives and programmes for sanitation, as also long-term plans to make the area self-sufficient through agricultural, irrigational and livestock improvements, and the development of Khadi (hand-made cloth) and other Cottage Industries.

In order to effectively stimulate healthy competition between the official and non official agencies, engaged in doing constructive work. Five non-official agencies were actually selected and paid grants for doing Firka development of:

a) Rural Reconstruction,

b) Drinking water facilities,
c) Sanitation,
d) Agriculture, and
e) Khadi and other village industries.

It was later realized that these efforts were restricted in scope and lacked coordination. They proved ineffective owing largely to lack of direction, support and encouragement from the central authority.

7. Nelokheri Experiment

Nelokheri was the part of State of Punjab and witness of displacement and destitution due to partition. It was started to rehabilitate 7000 displaced persons from Pakistan and later integrated with the 100 surroundings villages into what came to be a rural-cum-urban township. It was built round the vocational training centre that was transferred from Kurukshetra, in July 1948 to the 1100 acres of swampy land on the Delhi-Ambala highway. The central figure of this project was Shri. S. K. Dey, later Union Minister for Communal Development and Co-operatives up to 1965. This scheme was called urban township in all the essential requirements of life, The colony has school, an agricultural farm, polytechnic training centre, dairy, poultry farm, piggery farm, horticulture, garden, printing press, garment factory, engineering workshop, soap factory and so on. This was probably as ideal experiment which provides to be comparatively more successful addressing the immediate problems of providing help to the displace people of Punjab.

The Programme was not New

The Community Development Programme was a novel idea. The beginning was made in 55 selected projects. In 1952, each such project covered an area of about 500 sq miles, with 300 villages and a population of about 0.2 million. In 1953, the C.D. programme was extended to additional areas with a plan to cover the entire rural area of the country. Under this new scheme, new blocks of about 100-150 villages, covering an area of about 150-200 miles and a population of 60 to 70 thousand people were created. Each block was divided into Village Level Workers' circles.

Organization up to 1958

Under this new approach, in 1953, a Central Office under the administrator for C.D., which later on became the Ministry for C.D. and Panchayati Raj and Cooperatives, was created. Shri S.K.Dey, who formerly worked as Project in-charge at Nilokheri, was appointed to it. In each State, a post of Development Commissioner was created to coordinate the activities of the various Development Departments like Agriculture and Animal Husbandry, Health, Education, P.W.D., AETC. At the district level, the Collector was given the job of coordinating the activities of various District Development Officers. At the Block level, a team of various subject-matter specialists in Agriculture, Animal Husbandry, Cooperatives, Cottage Industries, Health, Social Education, and panchayats was created. For using their technical advice was the village level worker who was a first-aid man trained for the job.

Formerly, his training was for six months, then it was raised to one year, then to 1½ years and after 1958 to two years. The specialists were employees of various subject-matter Departments of the State.A post of Block Development Officer for each Block was created under the Development Department. The VLW was also an employee of the Development Department.

Organization Between 1958-65

Stages of the Programme: From April, 1958, the programme had three phases: (1) Pre-Extension Phase which was for 1 year, confined to agricultural development with a budget of Rs. 18,000. (2) Stage I: This was for intensive development for five years with a provision of Rs.1.2 million. (3) Stage II. For another five years with an allocation of Rs. 0.5 millions. After the completion of Stage II, the Block was supposed to become the permanent unit of planning and development and an established channel for developmental expenditure. Where this was not achievedc to any substantial extent, the State Government provides Post Stage II Blocks with a minimum outlay of Rs. 0.1 million a year.

The second major change was democratic decentralization through Panchayati Raj as recommended by the Committee on Plan Projects.

Extension Need of Intensive Efforts at National Level

In 1949, the Constitution of India pledged to the people: social, economic and political justice, liberty of thought, belief, faith and worship, equality of status and opportunity and to promote among them all fraternity, assuring the dignity of the individual and the unity of the nation. The main aim was to create a welfare state. This created the need for community development. In 1951, India prepared its First Five Year Plan. As a part of total planning came the Scheme of Community Development Projects on October 2, 1952, the birthday of Mahatma Gandhi, the great leader who helped India achieve political independence.

18

Community Development Programme

The community development programme was started in India just after independence (1952). It was a multi project programme with the aim of an overall development of rural people. This programme consisted of agriculture, animal husbandry, irrigation, cooperation, public health, education, social education, communication, village industries *etc.* In fact all these aspects of life relate to the 80 per cent of India's farming population. There are officials for each activity at district level to plan, execute and evaluate the programme up to the village level.

Community development is an exclusive term. It is frequently used to encompass any and every effort towards the advance of community interests. A variety of interpretations are therefore easily available. Community development is a compound term. It is useful, therefore, to consider its components.

The Community

A community is a group of people, who live in a geographical area and have interests in each other for the purpose of making a living. It is a farm of social organization existing between the family and state.

A community, while in itself consisting of several parts, is also a part of a larger social system. It is a dynamic social unit which is subject to change of internal or external origin. Some of the important characteristics of the community are:

1. Communities are close-knit
2. Their customs are interrelated

3. These communities are complexes of sub-group relationship and
4. There is a discernible leadership within the community.

Development

The term development connotes growth or maturation. It implies gradual and sequential phases of change. By understanding the above terms, we can say that community development programmes means a programme for gradual change in a group of people living in a geographical area and have interest in each other for the purpose of making a living.

Concepts of Community Development

1. Community development is a movement designed to promote better living for the whole community with the active participation and on the initiative of the community.
2. Community development is a balanced programme for stimulating the local potential for growth in every direction. Its promise is of reciprocal advance in both wealth and wealth and welfare, not on the basis of outside charity but by building on the latent vitality of the beneficiaries themselves with the minimum of outside aid.
3. Community development is technically aided and locally organized self help.
4. The term community development has come into international usage to denote the process by which the efforts of the people themselves are united with those the governmental authorities to improve the economic social and cultural conditions of the communities, to integrate these communities into the life of the nation and enable them to contribute fully to national progress.
5. Community development is the term used to describe the technique which many governments have adopted to reach their village people and to make more effective use of local initiative and energy for increased production and better living standards.
6. Community development is a process of social action in which the people of a community organize themselves for planning and action, define their needs and problems.

Community development has now set the pattern for the development of the rural people and the rural areas. The objectives of development and the new approach it makes to the solution of the problem of rural reconstruction, the comprehensive nature of the programme that it is promoting. The approach to the programme is twofold, educational and organizational. The rural people are to be educated in the art of better living, for bringing about a change in their attitude, for breaking away from primitive methods of production, unhygienic says of living based on tradition and for the adopting of progressive ways based on science and technology.

Size of Unit

For each community project, as at present planned, there will be approximately 300 villages with a total area of about 450 to 500 square miles, a cultivated area of about, 1,50,000 acres and a population about 2,00,000. The project area is conceived as being divided into 3 development blocks, each consisting 50,000 to 70,000. The development block, is, in turn, divided into groups of 5 villages each, each group being the field of operation for a village level worker.

Location of Units

The initial programme has been started with approximately 55 projects of rural development located in select areas in the several states of India, A certain degree of flexibility is allowed in the actual allotment of projects. Thus, while many are complete projects of about 300 villages each, some are also independent development bocks of about 100 villages each, depending upon the needs and conditions of the particular areas chosen for development

Feature Community Development Programme

The community development programme having a principle feature which gives the ideas about programme.

1. It has undertaken a countrywide programme of rural development. The limited pilot approach with which the programme started was abandoned soon after in preference to a countrywide programme. Therefore, a limited pilot approach in the matter of educating the rural people becomes inadequate.
2. An extension service well manned and trained is being placed right in the village and at the block level to assist and guide the rural people.
3. A big scheme for orientation and training of personnel has been undertaken to provide trained workers of or the community development programme.
4. The promotion of local leadership through education and training on a scale commensurate with the programme of community development is now urgently called for.
5. The promotion of youth programmes as an integral part of the community development programme is being taken up and will have much contribution to make to the education of rural youth.

Main Lines of Activity in Community Development Programme

The main lines of activity which will be undertaken in a community project can be briefly divided into the following:

a) Agriculture and Related Matters

The programme includes reclamation of available virgin and wasteland; provision of commercial fertilizers and improved seeds; the promotion of fruit and vegetable cultivation, of improved agriculture technique and land utilization; supply of technical information, improved agricultural implements, improved marketing and credit facilities; provision of sol surveys and prevention of soil erosion, encouragement of the use of natural and compost manures and improvement of livestock, the principal emphasis here being on the establishment of key villagers for breeding pedigree stock and the provision of veterinary aid, as well as artificial insemination centers. For attaining this objective, agricultural extension service will be provided at the rate of one agricultural extension worker for every 5 villages.

b) Irrigation

The programme visualizes provision of water for agriculture through minor irrigation works, *e.g.* tanks, canals, surface wells, tube wells *etc.* the intention being that at least half of the agricultural land, if possible, be served with irrigation facilities.

c) Communication

The road system on the country side is to be so developed as to link every village within the project area up to a maximum distance of half a mile from the village, the latter distance being connected by feeder roads through voluntary labour of the villagers themselves, only the main roads being provided for and maintained by the state or other public agencies.

d) Education

It has been realized that the full development of a community cannot be achieved without a strong educational base, alike for men and women. The community projects have been planned to provide for social education, expansion to basic type, provision of educational facilities for working children and promotion of youth welfare.

Vocational and technical training will be emphasized in all the stages or the educational programme. Training facilities will be provided for imparting improved techniques to existing artisans and technicians, both in urban and rural areas. Training centers which already exist in any area, will be strengthened and developed, and new ones established to meet the requirements of the project area.

e) Health

The Health Organization of the project area will consist of 3 primary health unit equipped with a hospital and a mobile dispensary at the headquarters of the project area and serving the area as a whole. It would aim at the improvement of environmental hygiene, including provision and protection of water supply; proper disposal of human and animal wastes; control of epidemic diseases such as Malaria, Cholera, small-box, Tuberculosis *etc.* provision of medical aid along with appropriate preventive measure, and education of the population in hygienic living and in improved nutrition.

f) Supplementary Employment

The unemployed and the under-employed persons in the village community will be provided with gainful employment to such extent as in possibly by the development of cottage and small-scale industries, construction of brick-kilns and saw mills and encouragement of employment through participation in the tertiary sector of the economy.

g) Housing

Apart from the provision of housing for community projects personnel, steps will be taken, wherever possible, to provide demonstration and training in improved techniques and designs for rural housing. In congested villages, action in the direction of development of new sites, opening of village parks and playgrounds and assistance in the supply of building materials, may also be necessary.

h) Training

The training of village level workers, project supervisors and other personnel for the community development programme will be carried out in 30 training centres which have been set up with the assistance of the Ford Foundation of America. Each training centre will have facilities for about 70 trainees. Each centre will have double training staff so that the trainees can be divided into two groups. One group will be getting practical and supervisory work experience, while the other group will be utilizing the centers' facilities for lectures, demonstrations and discussions. The training period will be limited to six months. Other than this, step will be taken for the training of the agriculturists, panches and village leaders.

i) Social Welfare

There will be provision for audio-visual aid for instruction and recreation, for organizations of community entertainment sports activities and meals.

Organizational Set-up for Community Development Extension Service

The organizational set-up for Community development Programme runs from the national level through state, district and block levels to the village level and there are three main constituents of this new set-up.

1. The direct-line staff such as State Development Commissioner, B.D.O and Village Level Worker.
2. The auxiliary or specialist staff, such as different heads of technical departments at the state and district levels and extension officers at the block level.
3. Panchayati Raj System- The Zila Parishads, Block Samitis and Village Panchayats.

A) National Level

At the National level programme, the policies are formulated by the National Development Council presided over by the Prime Minister of India. Membership of the Council consists of the Central ministers of the concerned ministers, chief ministers of all states, and members of the Planning Commission. The Planning

Commission provides guidance for Plan formulation and gives it approval to annual and Five-year Plans of the states as well as of the Centre. The Ministry of Agriculture and Irrigation is responsible for giving national guidance, policy formulation and technical assistance in regard to Agriculture Extension and Community Development (now Rural Development programmes). In the Agriculture department, the Agricultural Commissioner, Government of India, assisted by a number of assistant commissioners and directors, with the supporting staff, is in charge of all agricultural development programmes at the national level Within this Department, special mention may be made of the Directorate of Extension Training responsible for the training of Extension officers, VLWs, instructors of Village-Level Workers Training Centers and others and the Directorate of Farm Information which is concerned with the dissemination of new agricultural technology and innovations through various media.

B) State Level

At state level also, there is usually a State Development Committee presided over by the Chief Minister of the state with the other concerned ministers as its members. This Committee is responsible for the state's plan and programmes and for fixing the targets for regions and districts. Besides this committee, there are usually a number of other advisory or technical committees.

As regards the actual administrative functioning the State Developments Commissioner is the top-level executive responsible for directing, coordinating and providing overall guidance for development programmes and maintaining a two-way channel of communication between the state governments and the Central government. He co-ordinates the activities of different de3velopment departments, such as agriculture, animal husbandry, co-operation, panchayati raj, health, education, irrigation, power and electricity. The heads of these technical departments are responsible for planning and implementing the technical programmes and for providing the necessary technical guidance, manpower and support.

C) District Level

At the district level also, there is usually a District Development or District Planning Committee presided over by the District Collector or Deputy Commissioner. The other members of this committee are the heads of the departments in the district, chairman and vice-chairman of the district boards, representatives of voluntary organizations, local bodies and members of parliament and state legislatures. In the states, where the Panchayati Raj is operating, the Zila Parishads are responsible for planning, co-ordinationg and consolidating the development programme in the district. The District collector is the key official who co-ordinates the activities of all development departments at the districts level. The district-level technical heads of agriculture, animal husbandry, co-operation, panchayats, public health, irrigation, education and rural industries are responsible for planning and implementing the development programmes relating to their departments, Administratively, they are responsible to the district collector on one hand and to their state heads of development departments on the other.

D) Block Level

A district is subdivided into a number of community development programmes. The block development officer is the head of the block team, and co-ordinates all the activities of the development departments at the block level. He is assisted by eight extension officers form different fields, namely agriculture, animal husbandry, health, co-operation, panchayats, engineering, social education and rural industry. At the non-official level in the states, where the Panchayati Rah has been implemented, the Panchayati samiti (also called the Block), this Samiti) has the statutory powers for formulating and executing development programmes. The Samiti is assisted by the B.D.O. and the extension officers. Wherever the panchayati Raj is not working, there are block development advisory committees.

E) Village Level

At the village level, the multi-purpose village-level worker is the main extension staff. He is the last extension functionary in the administrative hierarchy and is the main contact person. He is responsible for all developmental work at the village level, and forms a connecting link between the various technical departments and the rural people. Usually, in a normal community development block, there are 10 village-level workers. Their number has been double in the intensive Agricultural Development Programme (IADP) blocks. On the non-official side, usually there is Panchayat in every village or for a cluster of villages, and is responsible for planning and implementing the community development programmes and ensuring people's participation in them.

Philosophy of Community Development Programme

The philosophies on which the community development programme should be based are as follows:

1. Work based on "felt needs": The programme should help the community to solve some of the problems which it feels are existent.
2. Work based on assumption that people want to be free form poverty and pain:

   It is assumed that the members of the community want a standard of living that allows them to be free form pain caused by lack of sufficient social side it is assumed that the people have four basis wishes a) security b) recognition c) response, d) new experience.
3. It is assumed that people wish to have freedom in controlling their own lines and deciding the forms of economic, religious, education and political institutions, under which they will live.
4. People's values given due consideration: It is presumed that co-operation, group decision-making, self initiative, social responsibility, leadership, trustworthiness and ability to work are included in the programme.
5. Self-Help: The people actually plan and work on the solution of their problems themselves. If the problems of the community are entirely ameliorated through the efforts of some outside agency, then the

development of such things as group decision-making, self-initiative, self-reliance, leadership *etc.* will not be forthcoming and it cannot be said that the community is developing.

6. People are the greatest resource: It is by getting the participation of the people in improvement activities that they become developed.
7. The programme involves a change in attitude, habits, ways of thinking relationship among people in the level of knowledge and intellectual advancement of people, changes in their skills, *i.e.* practices of agriculture health *etc.*

Principles of Community Development

Canadians Michael and Julie Bopp described the principles of community development as follow:

1. Harnessing Community Tensions

For changes to occur there must be a tension in people's mind between the way things are and the way people want them to be. There must be a desire for change and a desire to work together to create the change.

2. Facilitating Consultation about Community Realities and Needs

As noted in detail below, for community development to work successfully, there has to be awareness about community needs as decided upon by the community themselves, not as decided upon by outsides.

3. Maintaining Unity and Healthy Human Relations

It is important to build respectful healthy relationship between members of the group and to work on managing conflicts as they occur so as to maintain a collective vision.

4. Developing a Common Vision of Sustainable Future

If community development is to work, it must be sustainable and the process must include everyone. Stories abound of development processes failing because the resources or the process were not sustainable or well devised or were done without consultation. We have all seen pictures of computers lying unused because the hardware or software broke or just because people in the community didn't actually see need for computers.

5. Supporting Core Group Development

It is essential to encourage and support your group. Get to know the people in your group. What are their interests and skills? How much time can they dedicate to the group's activities? Everyone needs to be clear about their capacities and the limits to their time and energy that may exist. Encourage and build up a strong support base so that the group is not dependent upon only one or two people. Make sure the group is as representative of the community as it can be-that means including and listening to the needs of women and children as well.

6. Personal Revitalization and Healing

Good community development processes provide an opportunity for people to become empowered, energized and to undo some of the harms of the past.

7. Facilitating Learning

The entire development process is a learning experience. People can become aware of their environment, their social and political reality and can learn how to make the steps necessary to change and improve their situation.

8. Building Effective Organizations

Formalizing the initial volunteer group is crucial part of the development process. Creating and implementing formal organization structures creates legitimacy and respect for the organizations and lets the community know that you are serious about your endeavours. It helps keep people unified and accountable and is part of the process of developing and implementing plans and activities.

9. Networking with Resources and Allies

Creating the networks between likeminded individuals who then form a group who then network with other groups in order to learn and share skills and resources and knowledge is a crucial part of the community development process. The most effective groups are those that are well networked and work on co-operation with others in their area. Find out whether your area has a friendship school or rotary club or other group that is already working.

10. Programme Development

It is important to have a clear programme. An as hoc approach to community development processes will not work. A programme must be developed in conjunction with the community and be about community needs and priorities. They must adhere to realistic timelines and have inbuilt monitoring and evaluation processes so that changes can be tracked. The prorgramme acts as a map to signal where it is the group going.

11. Reflection on the Process: Monitoring and Evaluation

Monitoring and Evaluation is very important as it allows the group to track progress, identify problems, and improve upon community development processes.

12. Protecting the Process

Community development processes are dependent upon a rant of other processes: the good will of those involved, the continuation of funds, a secure political environment, a lack of violence. Once any of these dynamics is disrupted, then negative change and conflict may occur. It is important to have risk management strategies built into your process Less well developed processes might have fallen victim to the conflict and insecurity.

Stages of Community Development

There are three stages that community development processes go through:

Stage One: Relationship Building

Building relationships of trust and developing friendships are crucial to successful community development. Getting to know the people in your own group and the groups with which you are working is the first stage in the process of creating a sustainable working relationship. Take time to get to know the people in your Friendship group.

Stage Two: Process Development

Once good relationships have been developed, groups can begin to plan activities. This will involve sharing ideas, dreams, aspirations and visions, and learning about the community in which you are working. What are their strengths? What are their needs? What are their aspirations?

Stage Three: The Consolidation of Structures and Mechanisms

At this stage commissions, committees and task forces or interest groups are formed to plan and implement the activities and programs that have been collectively decided upon. These structures are also responsible for keeping the information flowing between the groups and out into the wider community, and often carry out the monitoring and evaluation processes that keep the programs and activities vibrant and sustainable.

Objectives of Community Development Programme

The community development programme has for its objectives economic development, social change and democratic growth. These three objectives are to be promoted jointly and in such a manner that they support one another. In India, the objective behind the community development programme is to develop the resources of the people and to assist each village in planning and carrying out the integrated agricultural production. Like this, the major objectives of this project are:

1. To change the outlook of all village people.
2. To improve existing village crafts and industries and organizing new ones, providing minimum essential health services and improving health practices.
3. Providing required educational facilities for children and adults as well as recreational facilities.
4. Improving housing and family living conditions of villagers.
5. To develop a responsive village leadership, village organization and institutions.
6. To develop village people so that they become self reliant and responsible citizens.
7. To help people that they can increase their income and quality of life.

8. Organizing or arranging trainings for voluntary local leaders like members of panchayats, village and block advisory committees *etc.* and professional community development workers like village level workers, extension officers, block development officers.

Difference between Community Development Programme and Extension Education

Community development programme and extension education both are for welfare of the rural peoples. These programmes are having more similarities than dissimilarities between these two concepts. The difference between extension and community development, as revealed by the various statements of objectives, processes, forms and principles.

Extension Education

Objectives

1. Emphasis on the individual.
2. Education aimed at individual development to obtain economic and social improvement.
3. Have as its main theme the individual's needs.
4. Emphasizes decision-making for change by individuals and families.

Form

1. Extension is an educational arm of Government, usually through educational institutions or other Government departments.
2. Emphasizes an organization that either carries out educational services directly or transmits knowledge from other resources to people.
3. Usually represents a transfer of responsibility form administering Government organization to another "Educational Group."
4. Permits cooperation between departments and agencies.
5. Essentially a "Branch" of the Department of Agriculture or directly involved in promotion of local units of the Government.

Community Development

Objectives

1. Emphasis on cooperation.
2. Education aimed at groups of individuals to work collectively to obtain economic and social improvement.
3. Has as its main theme the community's needs.
4. Emphasizes decision – making by groups and representation of groups.

Form

1. Community development is usually a direct Government approach to straight line organization.
2. Emphasizes the coordination of services by a working team made up of representatives of different services.
3. Usually a tight control held by a Government administering agency to cut across participating Governmental departments:
4. To recognize and include the various departments that must provide service.
5. To eliminate departmental reluctance to participate.
6. Forces departments and agencies to participate.
7. Essentially a branch of Government serving several departments of the Government.
8. Tied into promotion of local units of the Government.

Limitations

1. Most of the people are still illiterate and it is a difficult task to train about million people living in villages. For such a population the extension methods like demonstration, individual approaches, exhibitions, group meeting and training classes require thousands of extension workers.
2. Lack of communication channels, lack of roads, lack of vehicles are the major limitation of community development programmes.
3. Limitations of funds and staff for training farmers. In some backward state, where the villages are small and far away from each other, the extension worker have to act as technical advisor as well as act as a he has to do some administrative and clerical job also, and this reduces their working day and also efficiency.
4. A traditional society, with old ways and practices does not want to take risk unless it sees the result. Because change creates insecurity and uncertainty so there was resistance in the early years.
5. In an illiterate, traditional society the real leadership could not come forward. People with vested interests tried to get advantage by fooling the real client system, and the extension agency.
6. Preaching to rural people and educating them in new practices requires very careful handling and needs highly skilled workers who have knowledge of "how change take place" and the skill to induce them. It is very difficult to produce a large number of such workers.
7. In the beginning, such limitations may not have been realized and objectives were very vaguely formulated. Thus, when goals are not fulfilled, to the extent they were formulated, there is public criticism which leads to further weakening the support to community development.

8. The change-over form an attitude of heavy dependence of the Government to one of self-dependence takes place very slowly, and since there is lack of properly trained personnel and weaknesses in the supply line in the initial stage there is disappointment among the people.

National Extension Service

Rural development activities under Government sponsorship was started with the introduction of community development programme on 2nd October 1952. Within a few months of the launching of these pilot projects, it was prominently experience that the people were ready even keen, for the programme. The people in all the project areas responded enthusiastically and indeed much beyond the expectations o0f the Government and the sponsors of the programme. This fact emphasized the need for a rapid extension of the programme to other parts of the country. But the country's resources were not sufficient to sustain a comprehensive plan of the same magnitude as contemplated in the first 355 projects. The Government therefore decided to launch alongside the community development programme another programme which was somewhat less intensive in character, called the National Extension Service programme. The National Extension Service programme was formulated in April 1953 and it was inaugurated one year after the 55 community projects that is, on October 2, 1953.

It was a major development in the sphere of rural reconstruction in India. Since the basic idea underlying both the community development and National Extension Service programme was the same, the two were integrated under one agency at the Centre as well as in the states. Bothe the programmes were complementary and interwoven and ran concurrently. The idea behind the National Extension Service Programme was to cover the entire country within a period of about 10 years, that is to say, by 1960-61.

The inter-relation between the community development programme and Nation Extension Service described as follows:

It is necessary to explain the inter-relation between the community development programme and the National Extension service. The movements have identical aims. The National Extension Service is a permanent organization and will cover the whole country. It provides the basic organization, official, non-official and a minimum financial provision for development. Further funds will be found from the central Government and the State's own allotments under different heads. National Extension Service blocks in which successful results have been achieved with the maximum popular co-operation are selected for intensive development for a period of three years. This intensive development will depend on the available financial resources and local support and local support and enthusiasm.

The National Extension Service and the community development programmes have uniform unit of operation which is called a development block. It represents on an average 100 villages, with a population of 60,000 to 70,000 persons spread over an area of 150 to170 square miles. But the N.E.S. blocks are not developed with the same intensity as areas under the community development blocks. Out of the areas developed as National Extension Service Blocks, selection is made periodically for intensive development work under the community development programme and the block which are selected are C.D. blocks. Only those blocks are selected which in their working showed good results and where people's participation had been in abundance.

19

Organization Set Up in Agricultural Development Programme

1. Intensive Agricultural District Programme (IADP)
2. Intensive Agricultural Area Programme (IAAP) (1964)
3. High Yielding Varieties Programme (HYVP) (1964-65)
4. Institution Village Linkage Programme (IVLP) (1995-96)
5. Small Farmers Development Agency (SFDA) (1969)
6. Marginal Farmers and Agricultural Labours (MFAL) (1975)
7. Drought Prone Area Programme (DPAP) (1975)
8. National Rural Employment Programme (NREP) (1977)
9. Integrated Rural Development Programme (IRDP) (1978-79)
10. Training for Rural Youth for Self Employment (TRYSEM) (1979)
11. Watershed Development Programme (W D P)
    a) State Watershed Programme Implementation and Review Committee
    b) Sanctioning of Watershed Project, Funding Mechanism and Flow of Funds
12. National Agriculture Technology Project (N A T P)
13. Agricultural Technology Management Agency (A T M A)
14. Agriculture Technology Information Center (A T I C)

15. National Horticulture Mission (N H M)
16. National Agricultural Innovation Project (NAIP)

1. Intensive Agricultural District Programme (IADP)

It was felt that the increase in agriculture production under the community development programme was for less that necessary to feed the rapidly increasing population of this country. To tackle this urgent problem the government in collaboration with Ford Foundation launched the intensive agricultural district programme (1960-61) which is popularly known as the package programme. The significant feature of this programme is that the cooperative institutions have become the agency for distribution of credit and supply of agricultural inputs which were essential for implementing the programme.

The district selected throughout the country under this programme are pali, thanjavur, West-Godavari, Shahabad, Raipur, Aligarh, Ludhiana, Aleppey, palght, Mandga, Surat, Sambalpur, Bardwan, Bhandeva and Cochar.

Objectives

a) To increase the income of the cultivator and his family.
b) To increase the economic resources and potential of the village.
c) To create employment facilities.
d) To demonstrate the most effective ways of expansion of the national food production technology by co-operative efforts between officials and not-officials, villagers and individual cultivators.

The following points were kept as the minimum criteria for selection of the district for IADP:

a) Districts have adequate supply of water.
b) Should have minimum natural hazards.
c) They have well developed village industry.
d) They have maximum potential to increase agricultural and animal production.

Where Started

The programme was started in July 1960 in 7 selected districts in various States. These were: (1) West Godavari (A.P.), Shahabad (Bihar), Tanjore (Madras), Raipur (M.P.), Ludhiana (Punjab), Pali (Rajasthan) and Aligarh (U.P.). The selection of the places was made on the basis of their high potentiality for increasing the yield in a shorter time. They were assured of irrigation facilities, were exposed to a minimum of natural hazards and did not have a drainage problem or soil conservation problem. They had well developed Panchayats and Cooperatives.

Extended to other Places

Upto June 1963, the programme covered 120 blocks of gthose districts 45 per cent of the villages and 30 per cent of the total cultivated area in the seven districts. The newly covered districts were Surat in Gujarat, Alleppey and Palghat in Kerala, Mandya,in Mysore, Sambelpur in Orissa, Burdwan in West Bengal, Bhandra in Maharashtra, six blocks in Jammu and Kashmir, Cachar in Assam and two districts in M.P. (Blaghat and Narsinghpur).

Financing

About Rs. 9 crores for the farmers of the first seven districts and Rs. 8.5 crores for other districts were estimated to have been spent. This was shared by the Central Government, State Government and Ford Foundation. This was supposed to provide facilities for additional staff, implement workshops, soil testing laboratories, information units, training of workers *etc.*

Objectives

The short-term objective was, by providing the wherewithal for production, to aim at augmenting production to the extent of 50-60 per cent in the districts covered by the programme, simultaneously developing good livestock and poultry. The long-term objective was to evolve a dynamic pattern of productivity which could be extended to other places and help the rural economy grow on its own strength and resources.

I.A.D.P. versus General Community Development

(1) I.A.D.P. made intensive efforts to make use of improved seeds and fertilizers in the entire cultivated area and to encourage adoption of improved farming techniques and agricultural implements. (2) The cultivators were provide with all supplies *and* services at the appropriate time and in adequate quantities. (3) Under the General Community Development Programme, credit was offered only to credit-worthy farmers, while in I.A.D.P. districts all farmers, who had production plans and participated in the programme, were offered this facility. (4) Marketing and storage facilities were developed within bullock-cart distances. (5) Technical assistance was provided in preparing production plans.

Farm Production Plans

The process of change from traditional farming to scientific farming, from subsistence to commercial farming, from extensive to intensive and from low income to high income enterprises warranted a phased development of farm planning. The I.A.D.P. envisaged three stages in this process.

a) **First phase:** This consists of implementing simple but scientifically prepared farm production plans which involve the use of a package of improved and tested agricultural practices on important crops.

b) **Second Phase:** It comprises further elaboration of the process and its application to all crops raised on a farm.

c) **Third phase:** It envisages long-term planning and the most economic utilization of the farm resources so as to earn the maximum net income. This involves further research in farm management.

Demonstrations

Demonstrations on improved practices of important crops were conducted on farmers' holdings to show how new techniques were applied to crops, and to indicate levels to which the production could be stepped up, if all the recommended practices were adopted simultaneously, the economics of improved techniques was worked out in order to show the net profits and to stimulate others. The measures under this programme included (1) Advance arrangement for timely supply of seeds, fertilizers, pesticides, implements, spraying and dusting equipment and other requirements. (2) Wide publicity so that a greater number of farmers could see the methods and results at various stages. (3) In addition demonstrations were arranged in every crop season to display the use of improved implements, plant protection equipment, *etc.*, and to show the visible effects of new things and practices. (4) The schedule of practices for each crop and each tract was prepared for the I.A.D.P. district jointly by the research and extension staff, based on the results of trials conducted previously and also on the experiences of progressive farmers. One VLW was in charge of 4-5 composite demonstrations in each crop season.

Supply Credit

The increase in the area under improved practices called for additional credit to the farmers. The package programme strived to meet the requirements of credit for productive purposes. For this, cooperative societies were organized in these districts. To enable the primary cooperative societies to play their role effectively, the Cooperative Banks were strengthened and the rules regarding the sanction and disbursement of loans were simplified.

Cooperative Marketing and Storage Facilities

In addition to credit facilities, efforts were also made to organize cooperative marketing and storage facilities.

For the certification of quality seeds, seed testing laboratories were set up in each such district. Agricultural implement workshops costing about five lakhs of rupees were established in every district. These had tested and developed prototype implements and also trained village artisans for their repair and upkeep. These workshops developed implements, manufactured prototypes, serviced implements and vehicles and trained artisans and farmers' sons. Such workshops had been set up in each such district. Other services included plant protection, soil testing, water use and management.

2. Intensive Agricultural Area Programme (IAAP) (1964)

The Agriculture Production Board (1964) agreed with the midterm appraisal of the third five year plan that much greater emphasis should be given to the development of scientific and progressive agriculture in an intensive manner and 20 to 25 per cent of the cultivated area of country should be selected for intensive agriculture production.

The IAAP started in March 1964 also followed the package approach – The use of interrelated factors-physical, social and institutional – in strategic competitions that were likely to produce an impact on agricultural production.

3. High Yielding Varieties Programme (HYVP) (1964-65)

It was introduced in 1964-65 with a new dimension of agricultural production created in the community development project. HYV of Wheat, Paddy and introduced in selected 100 districts but later on it spread in other area also. The objective of this programme is to adopt HYV for maximum production.

For optimizing the yields of the available high yielding varieties of rice, it has been found necessary to advance their sowing time. Efforts were directed to educate the farmers to raise rice nurseries in advance of the main kharif season. A special programme for timely supply of seedlings, by raising community nurseries at tube-well points and on government farms, was undertaken in three command areas in Bihar. This programme has given encouraging results and is being extended to Assam, Uttar Pradesh, Madhya Pradesh, Bihar, Orissa and West Bengal during 1975 kharif.

4. Institution Village Linkage Programme (IVLP) (1995-96)

It is an innovative programme initiated by the Indian council of Agricultural Research (ICRA) on a pilot basis form 1995-96 which was later brought under World Bank funded National Agricultural Technology project (NATP) since 1999. It is different form the earlier first line extension efforts of ICAR, in sense that it lays emphasis on the research aspect through the participation of farmers to be carried out by the multidisciplinary team of scientists, Moreover, IVLP is a production system oriented project with agro-ecosystem analysis of the adopted villages as the basis of identify problems, priorities them and final out technological intervention point which are further developed into action plants to overcome the problems through assessment and refinement of technologies.

Objectives

The significance of client oriented projects received higher attention among had policy makers, which led to the concept, Technology Assessment and Refinement through IVLP. The specific objectives of Technology Assessment and Refinement programme are as under:

a) To introduce technological interventions with emphasis on stability and sustainability along with productivity of small farm production system.

b) To introduce and integrate the appropriate technologies to sustain technological interventions and their integration to maintain productivity and profitability taking environmental issues into consideration in a comparatively well defined farm production systems.

c) To introduce and integrate the appropriate technologies to increase the agricultural productivity with marketable surplus in commercial on and off farm production system.

d) To facilitate adoption of appropriate post-harvest technologies for conservation and on-farm value addition of agricultural products, by products and wastes for greater economic dividend and national priorities.

e) To facilitate adoption of appropriate technologies for removal of drudgery increased efficiency and higher income of farm women.

f) To monitor socio-economic impact of the technology intervention for different farm production system.

g) To identify extrapolation domains for new technology modules based on environmental characterization at meso and mega level.

5. Small Farmers Development Agency (SFDA, 1969)

The scheme is primarily one of supervised credit and supply of inputs. Resources are made to farmer's participants in the shape of 25 per cent subsidy from the agency and the balance as loans. No extra government body, agency, money from state government/central government. This programme was mainly for small farmers.

6. Marginal Farmers and Agricultural Labours (MFAL, 1975)

Marginal farmers are agricultural laborers are assisted in each selected area with subsidized dairy, poultry, fishery, beggary, sheep rearing and horticultural operations. Provision has also been made in this scheme for giving direct wage employment in the off-season to agricultural laborers in rural work of a productive manner. (The scheme is primarily market oriented and will also provide employment to landless agricultural laborers and rural artisans). Resources are made available to the farmer participants in the shape of 33.3 per cent subsidy from the agency and balance as loans.

In the sphere of administrative action the agency for the above 2 schemes in conceived as a separate corporative body. However, the state govt. is to be fully involved. The agency by itself will not execute only of the programme but will work through the existing organizations and rely on their support. The collector or authority in-charge of development in the district heads it. This is also because this programme is a part of total development effort in the area. The agency does not have its own separate hierarchy. It has a minimum compliment of staff who contact the appropriate development department of the govt. at the district level and ensure that the programme are carried to according to schedule.

To ensure continuity of finance and flexibility in its operations, funds are flowing directly from the center to the agency by passing the state govt. budget.

7. Draught Prone Area Programme (DPAP, 1975)

The overall productivity of draught prone area is low (The main thrust of DPAP is to restore a proper ecological balance in the Draught Prone areas). The main strategy is (i) development of irrigation sources (b) soil and moisture conservation, a forestation (c) restricting of cropping pattern and pasture development (d) live stock development (e) change in agriculture practices (f) development of small and marginal farmers and agriculture labours again under P.O., D.M. president.

8. National Rural Employment Programme (NREP)

It was initiated in the year 1977 with the aim of providing work for the rural poor, particularly during the period of slack employment during the year with the aim of creating durable community assets.

9. Integrated Rural Development Programme (IRDP)

It was started in 1978-79. District Rural Development Agency (DRDA) was setup at district level for implementation of the programme. The main aim behind it is to assist selected families of target group in the rural areas to cross the poverty line by taking up self-employment ventures to the people identified below poverty line. It aimed at achieving stated objectives by providing the aspects of generate income. At least 30 per cent of the families to be benefited should belong from SC and ST.

10. Training of Rural Youth for Self-Employment (TRYSEM)

It is initiated by Central Government in the department of Rural Development on August 15, 1979. It has been the facilitating component of the IRDP. It aims at providing technical skill to the rural youths from the families belonging below the poorly line to make them able for self-employment in agriculture and related fields as well as industries, and business activities midterm review of IRDP indicated that women members of target group households was very marginal which resulted a programme of **Development of Women and Children in Rural Areas (DWCRA)** was started as a sub-scheme of IRDP in 1982-83. The basic purpose of DWCRA is providing the income generating activities to women.

11. Watershed Development Programme (WDP)

Watershed development refers to the conservation regeneration and the judicious use of all the resources – natural (like land, Water plants, animals) and human – within the watershed area. Watershed Management tries to bring about the best possible balance in the environment between natural resources on the one side and man and animals on the other. Since in is the man which is primarily responsible for degradation of environment, regeneration and conservation can only be possible by promoting awakening and participation among the people who inhabit the watersheds.

Man and his environment are interdependent. The changes in the environment directly affect the lives of the people depending on it. A degraded environment means degraded quality of life of the people. Environmental degradation can be tackled effectively through the holistic development of the watershed. A watershed provides a natural geo-hydrological unit for planning any developmental initiative.

Geographical Details of India

Area	(m.ha)
☆ Total Geographical Area	329 mha.
☆ Records available	304 mha.
☆ Area fit for vegetation	264 mha.
☆ Are under Crops	142 mha.
☆ Area under forest	67 mha.
☆ Degraded Area in villages	35 mha.
☆ Degraded Area with farmers	20 mha.

The New Guideline for Watershed Development

a) The IWDP scheme is being implemented on the basis of new Guidelines for Watershed Development from 1.4.1995. The new common guidelines envisage the bottling up approach whereby the Users' Group themselves decide their work programme.

b) The strength of the Guidelines lies in the decentralization of decision making process by involving local Panchayati Raj Institutions, NGOs, Government Departments and the watershed community at the grass root level It is an effort on the part of the Govt. to remove the stumbling blocks that have delayed the process of development. In fact, the initiatives taken by the DOWD aim at establishing a system under which village people can actually involve themselves in planning, implementation and monitoring of watershed development programmes. In preparation of the Watershed Development Plan, Users and Self Help Groups and other people directly depending on the watershed are actually involved.

c) Another strength of these guidelines lies on the flexible approach followed in the method of release of funds, the area to be covered in each watershed as well as choice of components.

d) The new guidelines attempt to make the projects sustainable by establishing Watershed Development Fund and involving people in deciding equity issues and usufruct sharing mechanism.

Institutional Arrangements

To make the programme successful, proper Institutional arrangement has been provided in the Guidelines from state level to village level. These institutions help in making the programme broad based, sustainable and equitable. These institutions are given below:

State Watershed Programme and Review Committee

☆ Zila Parishadas/DRAS

☆ Watershed Development Advisory Committees

- Project Implementation Agencies (PIA)
- Watershed Development Team (WDT)
- Watershed Committee (WC)
- User Groups
- Self Help Groups

State Watershed Programme Implementation and Review Committee

This is an apex organization at a state level under the chairmanship of Chief Secretary/Addl. Chief Secretary/APC. Representatives of prominent NGOs, SIRDS, and Heads of Department of related Departments are also member of the Committee. The Committee undertakes monitoring, review and evaluation of Watershed Development projects. It is an important link between DRDAS and Department of Wastelands Development. Success of programme depends on functioning of State Watershed Programme Implementation and Review Committee.

I. District Rural Development Agency/ZP

DRDA/ZP is a key institution in the programme execution. The project is sanctioned in favour of the DRDA and funds are released to it directly from Government of India. The DRDA is responsible for successful implementation of the project as per guidelines and submission of various reports and returns to DOWD as well as State Government.

II. Project Implementing Agency

The PIA is an organization having sufficient exposure and experience in the field of community organization as well as watershed development activities. These organizations can be reputed NGOs having proven credibility or technical officers like DFO Soil Conservation Officer, Horticulture Officer, *etc.* The PIA is an important link between the villagers and the DRDA. It imparts technical know ho to the villagers with the help of Watershed Development Team and ensure that programme is executed as per Guidelines and funds are spec judiciously. It compiles information from Watershed Committees and set to DRDA.

III. Watershed Development Team

The Watershed Development Team is multi-disciplinary team responsible for technical and financial supervision of the project activities. The team consists of field level officials draw from various disciplines like forestry, soil conservation, horticulture, soc science *etc.* These officials are key functionaries in sensation of Self help Groups/User Groups and villagers at large.

IV. Watershed Development Association

Watershed Development Association (WA) consists of all members of the village whose land situated in the watershed area called user group (UG) and those entire member who drive sustenance form the watershed area called self help group (SHG).

V. Watershed Committee

Watershed Committee (WC) is the key institution at Watershed level consisting of about 2-3 representatives, each of USHG, panchayat and women *etc.* Committee also appoints a Watershed Secretary preferably a local man graduate from the same area.

Sanctioning of Watershed Project, Funding Mechanism and Flow of Funds

The IWDP Scheme is 100 per cent Grant-in-Aid form Government of India. The project are sanctioned on getting basic information from DRD as about the watershed be treated and capability of the PIA and over all situations in the area. A Well criteria for selection of watersheds has been provided in Para 27 of the Guideline. The watersheds selected of the projects should be as far as possible contiguous and there should be preponderance of wasteland. Acute shortage of drinking water, low wage rates and non over lapping of watersheds with any other project is other requirements.

Funding Mechanism and Flow of Funds

Funds from DRDA flow to Watershed Committees who open an account in the nearest bank to be operated jointly by WDT member and local man *i.e.* Secretary of WC. The Watershed committee is the primary unit which is directly involved in implementation of programme righty form preparation of action plan and check measurements of works and payment of wages. The WC also decides usufruct sharing mechanism and post project sustenance arrangements. The WC also maintains an account called Watershed Development Fund Account, where contribution realised from the members of Watershed Associations are deposited for utilisation in post project maintenance of assets. Contribution in shape of labour, cash and kind are valued and kept in interest bearing account.

Flow of Funds

- From DOWD to DRDA
- From DRDA to PIA
- From PIA to Watershed Committee
- Watershed Committee to Execute the Projects

Funding Pattern (Distribution)

- ASSISTANCE - Rs.4,000 per ha.
- ADMN COST - 10 per cent
- COMMUNITY ORGANISATION - 5 per cent
- TRAINING - 5 per cent
- WORKS - 80 per cent

12. National Agriculture Technology Project (NATP)

This project was launched by the ICAR on June 30, 1998, with the support of the World Bank to strengthen and complement the existing resources and to argument the output National Agricultural Research System (NARS).

Objectives

The major objective of this component is:

1. To accelerate the flow of technology form research, and extension to farmer.
2. Improve the dissemination of location specific and sustainability enhancing technologies.
3. Decentralize technical and decision making authority to the district level.
4. Create a more effective and financially sustainable public extension system.
5. Step up the privatization of certain technology transfer activities.

Salient Features

- ✰ Pilot testing new institutional arrangements for technology dissemination at the district level and below through establishment of district Agricultural Technology Management Agency (ATMA).
- ✰ Moving towards integrated extension delivery.
- ✰ Bottom up planning procedures for setting the Research Extension agendas.
- ✰ Addressing gender concern in agriculture and
- ✰ Increasing use of information technology for effective dissemination.

13. Agricultural Technology Management Agency (ATMA)

The ATMA at district level would be increasingly responsible for all the technology dissemination activities at the district level. It would have linkage with all the line departments, research organizations, non-governmental organizations and agencies associated with agricultural development in the district. Research and Extension units within the project districts such as ZRS or substations. KVKs and the key line Departments of Agriculture, Animal Husbandry, Horticulture and Fisheries *etc.* would become constituent members of ATMA. Each Research Extension(R-E) unit would retain its institutional identity and affiliation but programmes and procedures concerning district-wise R-E activities would be determined by ATMA Governing Board to be implemented by its Management Committee (MC).

Location of the Office

The registered office of the ATMA shall be located at district Collectorate premises.

Aims and Objectives

The aims and objectives for which the ATMA is formed are:

a) To identify location specific needs of farming community for farming system based agricultural development;

b) To set up priorities for sustainable agricultural development with a Farming Systems Approach;

c) To draw plans for production based system activities to be undertaken by farmers/ultimate users;

d) To execute plans through line departments, training institutions, NGOs, farmers organizations and allied institutions;

e) To coordinate efforts being made by various line departments, NGOs, farmers organizations and allied institutions to strengthen research extension-farmers linkages in the district and to promote collaboration and coordination between various State funded technical departments;

f) To facilitate the empowerment of farmers/products through assistance for mobilization, organization into associations, cooperatives *etc.* for their increased participation in planning, marketing, technology dissemination and agro-processing *etc.*

g) To facilitate market interventions for value addition to farm produce;

To Attain this Objective, the Society May

1. Take steps to ensure that problems, constraints and needs to the farming system based agriculture development are identified and diagnosed periodically.
2. Draw up plans for an integrated research-extension linkage approach for farming systems based agriculture development.
3. Ensure that lime departments/institutions draw up integrated development plan based upon resources available with them and incremental/supplementary resources mobilized by the ATMA.
4. Forge or develop systematic linkages between national/state/district institutions of excellence in the field of agriculture and marketing.
5. Ensure capacity building of manpower engaged in overall agricultural development and strengthen infrastructural support for the benefit of the farmer/producer.
6. Create suitable mechanism to ensure location specific adaptive, Indigenous knowledge based research.
7. Ensure adequate linkages and frequent interaction between scientists, extension functionaries and technicians and farmers, in order to prepare an integrated plan to effectuate their linkage, support each other, better understanding and appreciation of their problems, means adopted to sort out problems and plans *etc.* and to develop a mechanism of feed back;

8. Ensure capacity building of the ultimate users- the farmers in terms of physical, financial and skill resources base by way of adequate financial support channelized through credit institutions, private investments and training for skill up gradation.
9. Facilitate farmers' organization to take lead role on mobilizing support services and resources.
10. Facilitate private investments for infrastructure development, private institutions have to take lead in the delivery of goods and services to ultimate users (farmers).
11. Facilitate the processing and marketing activities of the agricultural, livestock, dairy, poultry, silk and allied produce of the farmers with the help of private sector institutions.
12. Receive and expend project funds, maintain revolving accounts, enter into contracts and agreements, receive donations and provide services and deliver goods to beneficiaries.
13. Accept grants of money, securities or property of any kind and undertaken and accept the management of any endowment, trust funds or donations riot inconsistent with the objectives of the ATMA, on such terms and conditions as may be fitted with the objectives of the ATMA and be prescribed by the Government of India form time to time.
14. Generate resource in order to bring financial sustainability through charging for selected services rendered to beneficiaries by ATMA.
15. Create administrative, technical, ministerial and other posts in the ATMA and make appointments thereto in accordance with the rules and regulations of the State Government.
16. Make rules and bye-laws for the conduct of the affairs of the ATMA and add to amend vary or rescind them from time to time.
17. Do all such other lawful acts and things either alone or in conjunction with other organizations or persons as the ATMA may consider necessary, incidental or conductive to the attainment of the above objectives.
18. To do all such lawful acts and things whether incidental to objectives tin force or not as may be requisite in order to furtherance of the objectives of the ATMA.
19. Sell, lease, exchange and otherwise transfer of any portion and the properties of the society (ATMA).
20. Do all other such things as may be considered by the society (ATMA) and may be incidental or conducive to the attainment of its objectives.

14. Agriculture Technology Information Center (ATIC)

The importance of an appropriate information package and its dissemination as an input has assumed added emphasis in this "information age". The kind of information and the way it is to be used are critical factor to the growth of agriculture.

It is also worth noting that it is no longer enough for research to generate information alone. The required information is also to be delivered to the end user at one place. This information must be direct, clear and easily understandable and without any room for distortion.

There is a greater need for coordination between researchers and technology users. A higher degree of integration needs to be achieved by having a formal management mechanism linking scientists or department in charges of different disciplines (though engaged in interdependent tasks) on the one hand to the technology users on the other. The linkage mechanism should be with formal, permanent, mandated, facilitated and designated function. The establishment of an agricultural Technology Information Centre will provide such a mechanism beyond the individual unit of a research institution to contribute to the dissemination of the information. This will serve as a single window delivery system for services and products of research for the areas in which the concerned institute is involved.

The cornerstone of India's agricultural revolution has been the availability of improved varieties of cereals, oilseeds, pulses, *etc.* breed of livestock including poultry and fisheries; horticultural plant materials, and improve management practice for increase productivity, sustainability and stability of various crops and livestock enterprises. This has raised the search by farmers for future availability of seed, planting materials and other materials, easy accessibility to diagnostic services for soil fertility and plant protection, availability of appropriate information through leaflets and pamphlets and increased scope in sale of consultancy services. Often the farmers are not aware as to whom and where to approach for field problems. It is felt that the facility of a **'single window'** approach at the entrance of the ICAR Institute/State Agricultural Universities will enable the farmers to have the required information for the solution to their problems related to the areas in which the concerned institute is involved. Because of the dominance of small and resource poor farmers and concentration of poor people in several sectors, public institutions like ICAR institutes and SAU's will continue to play a vital role in supply of information for increasing the overall productivity in agriculture.

The rationale for establishment of ATIC are

a) To provide diagnostic services for soil and water testing, plant and livestock health.

b) To supply research products such as seeds and other planning materials, poultry strains, livestock breeds, fish seed, processed products, *etc.*, emerging from the institution for testing and adaptation by various clientele.

c) Providing information through published literature and communication materials as well as audio visual aids.

d) Providing an opportunity to the institutes/SAU/s to generate some resource through the sale of their technologies.

The important criteria of Agricultural Technology Information are

a) Availability (or accessibility) of new technologies,

b) Relevance of new technologies;
c) Responsiveness of new technologies to the needs of different categories of farmers; and
d) Sustainability of such unit within the overall institutional system

The Agricultural Technology Information Centre (ATIC) is a "single window" support system lining the various units of a research institution with intermediary users and end users (farmers) in decision making and problem solving exercise.

By building on the past investment in infrastructures in these institutions considerable farm worthy techniques/technologies/knowledge material have been developed in the institutions which can provide the techniques, technologies, seeds and planting materials to the farmers and other organizations for taking up the frontier technologies, to the field. This will facilitate in dealing effectively with the complexity and diversity of information system and channels. Such information will be useful for:

1. Farmers;
2. Farmer-entrepreneurs;
3. Extension workers and development agencies;
4. NGOs; and
5. Private sector organization.

These centers will provide a 'balanced scorecard' in terms of

a) Financial- Resource generation and financial sustainability;
b) Customers-measures on performance of the technology from the customer's point of view;
c) Process- the performance of key internal processes in terms of providing quality services, seed and plant materials, *etc.* linkage with district extension system and, spread of improved technology and productivity in the area: and
d) Learning- the ability of the research organization to improve continuously and innovate in its products, services and processes.

Objectives

The objectives for establishment of such centers as a single window system are as follows:

a) To provide a 'single window' delivery system for the products and species available form an institution to the farmers and other interested groups as process of innovativeness in Technology Dissemination at the institute level.

b) To facilitate direct the farmers access to the institutional resources available in terms of technology, advice, technology products, *etc.* for reducing technology dissemination losses; and

c) To provide mechanism for feedback from the users to the institute

15. National Horticulture Mission (NHM)

Recognizing the importance of horticulture sector in the growth of Indian agriculture, Honourable Finance Minister announced launching of National Horticulture Mission in his budget speech on July 8, 2004. The horticulture sector includes fruit, vegetables, spices, medicinal and aromatic plants, flowers, mushroom and a variety of plantation crops like coconut, areca nut, cashew nut and cocoa which has been contributing significantly to the GDP in agriculture (28.5 per cent form 8.5 per cent area). National Horticulture Mission is a centrally sponsored scheme in which Government of India provide 100 per cent assistance to the state mission during the year 2005-06 (During XI plan, the assistance form Government of India will be 85 per cent with 15 per cent contribution by the State Government.

Objectives

To develop horticulture to the maximum potential available in the State and to augment production of all horticultural production of all horticultural products (Fruits, Vegetable, flowers, Plantation crops, Spices, Medicinal Aromatic plants) in the state is to be considered.

a) To provide holistic growth of the horticulture sector through an area based regionally differentiated strategies

b) To enhance horticulture production, improve nutritional security and income support to farm households;

c) To establish convergence and synergy among multiple on-going and planned programmes for horticulture development;

d) To promote, develop and disseminate technologies, through a seamless blend of traditional wisdom and modern scientific knowledge.

e) To create opportunities for employment generation for skilled and unskilled persons, especially unemployed youth;

In pursuance of the above goals, the National Horticulture Mission will focus in the areas of horticultural research, development, post harvest management,, processing and marketing. The programmes under horticulture research will concentrate on technology generation appropriate to each region/state keep in in view their specific agro-climatic and socio-economic conditions. Emphasis will be given for effective transfer of technologies in production, which are already available in India and abroad. The Council of Agricultural Research (ICAR) in association with State Agriculture Universities (SAU's) and other research institutes/organizations in the public and private sector having capabilities will be involved in the research programme. The programme under the horticultural development aims at increasing the production and productivity of all horticultural crops through adoption of improved technologies in crop production. Under this programme, special emphasis will be given for regionally differentiated crops, which are most suitable for the state/region. This programme will be implemented by the Horticultural Departments of the State Government, which may also include cooperative organizations, self-help group, NGOs and economically organizations for achieving the targeted production and productivity of identified crops. Risk management in the form of crop insurance is also proposed

Post harvest management would include creating suitable infrastructure for efficient post harvest management and marketing of horticulture produce such as handling, transfer, storage and markets *etc.* besides taking up market promotional activities such as dissemination of market information to the farmers, processors, traders, and consumers. Special trust would be provided to promote the export of horticultural produce through establishment of AEZs, for which there is potential global market. National Horticulture Board, Directorate of Marketing and Inspection, National Cooperative Development Corporation, NERAMAC, TRIFED and Agriculture and Processed Food Products export Development authority will be involved for the purpose. It is proposed to make use of the existing schemes of the National Horticulture Board (NHB), Directorate of Marketing and Inspection (DME) and National Cooperative Development Corporation (NCDC) to maximum possible extent.

The Mission will also focus on promoting processing of horticultural produce and value addition buy providing incentives for setting up of horticulture processing industries and food parks in potential areas and to encourage linkages between the markets for the horticulture produce and processing industry. This activity will be supported by the Ministry of Food Processing Industry (MFPI) and implemented through agencies under the administrative control of MFPI and other organizations and the concerned departments of the State Governments. These programmes would be credit-linked through NABARD/IDBI/State Financial Corporations.

16. National Agricultural Innovation Project (NAIP)

The Government of India has launched the National Agricultural Innovation project with a credit support of the World Bank. The project will run up to June 2012. The ICAR is operating the Project. The overall objective of the project is to facilitate accelerated and sustainable transformation of Indian agriculture for rural poverty alleviation and income generation by the application of agricultural innovations through collaboration among public research organizations, farmers' groups, NGOs, the private sector and the civil societies and other stakeholders.

The India National Agricultural Innovation Project contributes to the sustainable transformation of Indian agricultural sector to more of a market orientation to relieve poverty and improve income. The specific aim is to accelerate collaboration among public research organizations, farmers, the private sector and stakeholders in using agricultural innovations. The project has four objectives:

a) Strengthens the Indian Council of Agricultural Research (ICAR) as the catalyzing agent for managing change in the Indian National Agricultural Research System (NARS) by focusing on: 1.1 Information, communication and dissemination system; 1.2 Business planning and development; 1.3 Learning and capacity building; 1.4 Policy and gender analysis and visioning; 1.5 Remodelling financial management and procurement systems; and 1.6 Project implementation.

b) Funds research on production-to-consumption systems.

c) Funds research on sustainable rural livelihood security.

d) Supports basic and strategic research in the frontier areas of agricultural science features

The project will have a strong and transparent governance strategy for efficient working. Institutional and implementation arrangement will be fully streamlined to follow modern financial management, procurement system, knowledge management, and a results framework and monitoring which will ensure continuous progress and achieving the expected output. Systematic economic and financial analysis will be pursued along with close monitoring of environmental and social safe guards. Another major component of the project is a strong institutional learning and capacity building plan for self-renewal of National Agricultural Innovation System. The plan includes comprehensive training need assessment, harnessing modern ICT in knowledge and education dissemination management for agriculture, capacity building to deal with globalize agricultural market and economy, capacity building for visioning and foresight *etc.*

20

Basic Elements of Rural Development

There are at least three basic elements which are considered to constitute the 'true' meaning of rural development. They are as follows: (1977).

1. Basic Necessities of Life

People have certain basic needs, without which it would be impossible (or very difficult) for them to survive. The basic necessitates include food, clothes, shelter, basic literacy, primary health care and security of life and property. When any one or all of them are absent or in critically short supply, we may state that a condition of 'absolute underdevelopment' exists. Provision of the basic necessities of life to everybody is the primary responsibility of all economies, whether they are capitalist, socialist, or mixed. In this sense, we may claim that economic growth is a necessary condition for improvement of the 'quality of life' or rural people, which is rural development.

2. Self Respect

Every person and every nation seeks some sort of self-respect, dignity or honour. Absence or denial of self-respect indicates lack of development.

3. Freedom

In this context, freedom refers to political or ideological freedom, economic freedom and freedom from social servitude. As long as society is bound by the servitude of men to nature, ignorance, other men, institutions and dogmatic beliefs, it can not claim to have achieved the goal of 'development'. Servitude in any from reflects a state of underdevelopment.

Social Justice and Poverty Alleviation Programmes

- ☆ Integrated Tribal Development Project (ITDP)
- ☆ Integrated Rural Development Programme (IRDP)
- ☆ Swarnajayanti Gram Swarojgar Yojana (SGSY)
- ☆ Prime Minister's Employment Yojana (PMEY)

a) Integrated Tribal Development Project (ITDP)

The Programme on Integrated Tribal Development Projects (ITDP) under Tribal Sub-Plan (TSP) is being implemented since the Fifth Five Year Plan wit specific objectives of reducing poverty, improving educational status an eliminating exploitation of the tribal families. The programme was evaluated by the programme Evaluation Organisation (PEO) in 1996 to assess the effectiveness and performance of planning and monitoring arrangements, and impact of the programmes on the well being of the Scheduled Tribes. The report was released in July, 1997.

The findings of the study indicate that a large majority of the tribal fees that they are better off now in terms of possession of productive and utility assets and access to food, clothing, transport facilities, electricity, schools *etc.* than they were 10 years ago. Though, the administrative arrangements to execute this programme existed in all the states as per Central guidelines, the system of decentralised planning, implementation and monitoring as envisaged in the guidelines for the scheme were not operational in most of the states. The delivery system of the programme was not effective and a very large number of tribal farmers were found to use irrigation water, High Yielding Variety (HYV) seeds, fertilizers and other inputs form private sources, even though a significant proportion of funds in being spent on free delivery of such inputs. Inadequate facilities ofr primary education and health were also observed.

In order to improve the situation it has been suggested that ITDP Project Officer should be made more effective in planning and implementation of Tribal Development Programmes. It is also necessary to assess the deprivation of tribal in all the areas of social concern for adoption of an appropriate principle for determining the size of Tribal Sub-plan and allocation of funds across various sectors. This must be backed by a strong monitoring mechanism.

b) Integrated Rural Development Programme (IRDP)

IRDP launched on October 2nd. 1980 all over the Country and accordingly all the 15 Blocks of Boudh- Kandhamal district have been covered under the Scheme. Since then, prior the above period, IRDP was in operation in 8 blocks of the district since 1978-79. The I.R.D.P. continues to be a major poverty alleviation programme in the field of Rural Development. The objective of I.R.D.P. is to enable identified rural poor families to cross the poverty line by providing productive assets and inputs to the target groups. The assets which could be in primary, secondary or tertiary sector are provided through financial assistance in the form of subsidy by the Govt. and term credit advanced by financial institutions. The programme is implemented in all the blocks in the country as a centrally sponsored scheme funded on 50:50 basis by the Centre and State. The Scheme is merged with another scheme named S.G.S.Y. since 01.04.1999.

Limitations

a) In the integrated rural development, the village has been treated as a homogeneous concept and as a unit of development which is not there. Thus there is a serious problem of uneconomic and non-viable villages or rural settlements that can form a more economically efficient base for integrated rural development.

b) No attempt was made to consider the policy of distribution of land or for more equitable distribution pattern and revitalizing the possessing of other productive assets in the rural areas by limiting size of individuals units.

c) Inability of science and technology to solve, by itself, the problem of rural poverty.

c) Swarnajayanti Gram Swarojgar Yojana (SGSY)

This programme was launched in April, 1999. This is holistic programme covering all aspects of self employment such as organization of the poor into self help groups, training, credit, technology, infrastructure and marketing.

Objective

The objective of SGSY is to provide sustainable income to the rural poor. The programme aims at establishing a large number of micro-enterprises in the rural areas, based upon the potential of the rural poor. It is envisaged that every family assisted under SGSY will be brought above the poverty-line with in a period of three years.

Scope

This programme covers families below poverty line in rural areas of the country. Within this target group, special safeguards have been provided by reserving 50 per cent of benefits for SCs/STs, 40 per cent for women and 3 per cent for physically handicapped persons. Subject to the availability of the funds, it is proposed to cover 30 per cent of the rural poor in each block in next 5 years.

Funding

SGSY is a Centrally Sponsored Scheme and funding is shared by the Central and State Governments in the ratio of 75:25 respectively.

Strategy

SGSY is a Credit-cum-Subsidy programme. It covers all aspects of self-employment, such as organization of the poor into self-help groups, training, credit technology, infrastructure and marketing. Efforts would be made to involve women members in each self-help group. SGSY lays emphasis on activity clusters.

Four-five activities will be identified for each block with the approval of panchayat Samities. The Gram sabha will authenticate the list of families below the poverty line identified in BPL census. Identification of individual families suitable for each key activity will be made through a participatory process. Closer attention will be paid on skill development of beneficiaries, known as swarozgaris, and their technology and marketing needs.

How to Seek Assistance

For assistance under the programme, District Rural Development Agencies and Block Development Officers may be contacted. The objective of Swarnajayanti Grama Swarojagar Yojana (S.G.S.Y.) is to provide sustainable income to the rural poor. The programme aims at establishing a large number of Micro-enterprises in the rural areas building upon the potential of the rural poor. It is envisaged that every family assisted under SGSY will be brought above the poverty line in a period of three years. This scheme is launched on 1 st April, 1999, the programme replaces the earliar self Employment and allied programmes IRDP, TRYSEM, DWCRA, SITRA, GKY and MWS, which are no longer in operation. The programme covers families under below poverty line in rural areas of the country within this target group, special sage guard have been provided by reserving 50 per cent of benefits of SC/STs, 40 per cent for women and 3 per cent for physically handicapped persons subject to availability of funds. It is proposed to cover 30 per cent of the rural poor in each block in the next five year. S.G.S.Y. is a credit cum subsidy programme. It covers all aspects of aspects of self employment such as organization of the poor into self-help groups training, credit technology, infrastructure and marketing. SGSY is a centrally sponsored scheme and funding shared by the Central and State Government in the ratio of 75:25.

d) Prime Minister's Employment Yojana (PMEY)

Prime Minister Employment Yojana for providing self-Employment to Educated Unemployed Youth was announced by the Prime Minister on 15th August, 1993 to provide self-employed opportunities to one million educated unemployed youth in country. The Scheme has been formally launched on 2nd October, 1993.

Objectives

The PMEY has been designed to provide employment to more than a million Person by setting up of 7 lakhs micro enterprises by the educated unemployed youth. It relates to the setting up of the self-employment ventures through industry, service and business routes. The scheme also seeks to associate reputed non-governmental organizations in implementation PMEY scheme especially in the selection, training of entrepreneurs and preparation of project profiles.

Educated constitute nearly 40 per cent of the unemployed. Incidence of unemployment among the educated lab our force is 11.8 per cent against 3.8 per cent for educated and uneducated taken together. Further, the incidence is much higher *viz.* 26.7 per cent among women than 9.8 per cent among educated men. Women constitute 27 per cent of the educated unemployed level of education. Projection made using alternative techniques put the estimate of educated unemployed at 6-7 million in 1992. This is out of an estimated educated labour force of 52 million. Of the mew employment opportunities that are being generated in the economy in recent years, about 45 per cent are estimate to be going to the educated. But, though relatively high, the employment growth of the educated still falls short of the growth of labour force by about 7 lakhs in a year. The trend of a higher growth of their employment is likely to continue with the introduction of modern technologies There may, in fact be a shift towards employment of the educated in the activities which hitherto were the preserve of the uneducated. Hence, the problem of the educated needs a special focus within the overall strategy for tackling unemployment. Prime Minister's Employment Yojana (PMEY) seeks to address itself to this problem.

New Trends in Extension Privatization

Extension has been, and still is, under attack from a wide spectrum of politicians and economists over its cost and financing. As a result, Extension Systems have had to make changes, by restating the system's mission, developing a new vision for the future, and formulating plans for the necessary transition to achieve the desired change.

At least three scenarios have been suggested by government and farm organizations with regard to privatization of extension:

- ☆ Public financing by the taxpayer only for the kinds of services of direct concern to the general public.
- ☆ Direct charging for some individual services that produce direct return in the form of improved income, with the possibility of differential rates for specific situations or target groups.
- ☆ Mixed funding shared between public and private professional association contributions for services, with delayed return or collective services, such as applied research, training of farmers and agents, and improvement in Extension methods and tools.

Extension service has been traditionally organized and delivered by the public sector all over the world, which led to a situation wherein, whenever one refers to extension, it denoted public extension service. Similarly, whenever private sector is referred to, there is a tendency to consider only the corporate sector in the category. However, private extension has a broader canvas including all relevant private groups than the narrow canvas of corporate sector.

Privatization of extension services does not aim at substituting private sector for public extension service. In fact, privatization has adopted a variety of forms involving different stakeholders. The paper portrays the major stakeholders, viz, private corporate firms, credit institutions, farmer's association non-governmental organizations and media organizations and analyses their participant configurations.

The success of an extension service depends on the effectiveness of planning at four levels policy, programmes, projects and strategy. Policy and programmes must be decided by the public extension system, while projects and strategy can be formulated by the private extension organizations. When the private extension organizations get involved in providing extension support to farmers, it is likely there will be competition among the various extension providers, which will result in more efficient and demand-driven service. Both technical and allocative efficiency which are basically economic in nature are well take care of by the private extension agencies, resulting in cost minimization, profit maximisation and optimal use of resources, which are warranted in a competitive environment.

Public extension service often views sustainability of programmes only in terms of continuity. Sustainability is different form continuity, which has both ecological and equity dimensions. The private extension agencies, especially NGOs and media organizations provide valuable service in ensuring sustainability of programmes in terms of the above two dimensions.

The private extension system in India offers the following services for farmers – terms of sharing, augmenting and supplementing the public extension efforts besides offering unique and innovative initiatives, which the public extension service can also emulate. Some of the Services are:

- ✰ Cost sharing by farmers' groups
- ✰ Cost recovery on selected services offered to farmers
- ✰ Contracting services to small groups
- ✰ Paid extension services for affordable farmers
- ✰ Value addition by agro-processing firms
- ✰ Consultancy services (both technical and managerial)
- ✰ Privatised service centres for farmers
- ✰ Self Help Groups of farmers
- ✰ Information support through media organizations

Private extension system can offer a variety of services for farmers in a competitive environment which the public sector may not be able to. Hence, it is suggested that public sector extension may limit its activities only to regulatory and enabling functions, and should mainly focus on educational activities, which are unattractive to private sector delivery.

Women Development Programme

a) Development of Women and Children in Rural Areas (DWCRA)
b) Integrated Child Development Scheme (ICDS)
c) Mahila Samridhi Yojana (MSY)
d) Mahila Arthik Vikas Mahamankal (MAVIM)

a) Development of Women and Children in Rural Areas (DWCRA)

The Development of Women and children in rural areas (DWCRA) programme was launched as a sub-component of IRDP and a centrally sponsored scheme of the Department of Rural Development with UNICEF cooperation to strengthen the women's component of poverty alleviation programmes. It is directed at raising the income levels of women of poor households so as to enable their organized participation in social development towards economic self reliance. The DWCRA's primary thrust is on the formation of groups of 15 to 20 women form poor household at the village level for delivery of services like credit and skill training, cash and infrastructural support for self employment. Through the strategy of group formation, the programme aims to improve women's access to basic services of health, education, child care, nutrition and sanitation. It is merged with S.G.S.Y. SWARNAJAYANTI GRAMA SWAROJAGAR YOJANA since 01.04.1999.

a) The special scheme of Development of Women and Children in Rural Areas (DWCRA) aims at strengthening the gender component of IRDP.

b) It was started in the year 1982-83, on a pilot basis, in 50 districts and has now been extended to all the districts of the country.

c) DWCRA is directed at improving the living conditions of women and, thereby, of children through the provision of opportunities for self-employment and access to basic social service.

Strategy

a) The main strategy adopted under this programme is to facilitate access for poor women to employment, skill up gradation, training, credit and other support services so that the DWCRA women as a group can take up income generating activities for supplementing their incomes.

b) It seeks to encourage collective action in the form of group activities that are known to work better and are more sustainable than the individual effort. It encourages the habit of thrift and credit among poor rural women to make them self-reliant.

c) The programme also envisages that this target group would be the focus for convergence of other services like family welfare, health care nutrition, education, childcare, safe drinking water, sanitation and shelter to improve the welfare and quality of life of the family and the community.

b) Integrated Child Development Scheme (ICDS)

Launched on 2nd October 1975 in 33 Community Development Blocks, ICDS today represents one of the world's largest programmes for early childhood development. ICDS is the foremost symbol of India's commitment to her children. India's response to the challenge of providing pre-school education on one hand and breaking the vicious cycle of malnutrition, morbidity, reduced learning capacity and mortality, ion the other.

It is an inter sectoral programme which seeks to directly reach out to children, below six years, especially from vulnerable and remote areas and give them a head-start by providing an integrated programme of early childhood education, health and nutrition. No programme on Early Childhood Care and Education can succeed unless mothers are also brought within it ambit as it is in the lap of the mother that human beings learn the first lessons in life.

Objectives

a) Lay the foundation for proper psychological department of the child
b) Improve nutritional and health status of children 0-6 years
c) Reduce incidence of mortality, malnutrition and school drop-outs
d) Enhance the capability of the mother and family to look after the health, nutritional and development needs of the child
e) Achieve effective coordination of policy and implementation among various departments to promote child development.

Services

The Scheme provides an integrated approach for converging basic services through community-based workers and helpers. The services are provided at a centre called the 'Anganwadi'. The Anganwadi, literally a courtyard play centre, is a childcare, located within the village itself. A package of following six services is provided under the ICDS Scheme:

a) Supplementary nutrition
b) Non-formal pre-school education
c) Immunization
d) Health Check-up
e) Referral services
f) Nutrition and Health Education

The three services namely immunization, health check-up and referral are delivered through public health infrastructure *viz.* Health Sub Centers, Primary an Community health Centers under the Ministry of Health and Family Welfare.

c) Mahila Samridhi Yojana (MSY)

The Mahila Samridhi Yojana (MSY) was launched on 2nd October, 1993 with the objective of empowering the rural women through building thrift habit, self-reliance and confidence. During the first two years of its operation the performance of scheme, as measured by achievement against the all India target and the amount of the money deposited in MSY accounts was not encouraging, and the cost of the scheme was also on the higher side, Programme. Evaluation Organisation was asked to evaluate the performance, imple-mentation and impact of the scheme, to identify the gaps, if any, and suggest measures that would improve the performance of the scheme. The report was submitted in April, 1996.

It was observed in the PEO study that the factors that had adversely affected the performance of MSY related primarily to the method of implementation and complexity/ambiguity in the rules for operating MSY accounts. The operational cost of MSY was on the higher side and far outweighed the benefits received by the beneficiaries. The high cost of operation might not have affected performance, but it certainly raised the question of sustainability of the scheme. There was wide variation in performance across various states and districts due to interaction of a large number of factors such as the levels of socio-economic development, the effectiveness of the publicity media and implementing agencies. The operational rules of MSY were complex and the Branch Post Masters, the key persons for operation of the schemes, lacked their understanding due to poor quality of training and educational background. There was poor monitoring and supervision of the scheme also. The scheme, in general, had not been successful in developing thrift habit among the rural women.

It was observed that the fundamental problem in the implementation of the scheme was adoption of the target oriented and top-down approach without actual participation of the rural women in the programme. Therefore, it was suggested that the task of popularising the scheme be entrusted to the grass-toot level women. Self Help Groups, Mahila Mandals, DWCRA, NGOs *etc.* There was an urgent need for drastic simplification of the operational rules for MSY accounts and reducing the operational costs to ensure sustainability of the scheme. There was also need to strengthen the monitoring mechanism for the scheme. It is learnt that Department of Women and child Development have taken follow-up actions on the findings of the evaluation study of PEO.

With the objective of providing economic security to the rural women and to encourage, the saving habit among them, the Mahila samridhi Yojana was started on 2 October 1993. Under this plan, the rural women of 18 years of above age can open their saving account in the rural post office of their own area with a minimum Rs. 4 or its multiplier. On the amount not withdrawn for 1 year, 25 per cent of the deposited amount is given to the depositor by the government in the form of encouragement amount. Such accounts opened under the scheme account opened under the scheme account opened under the scheme are provided 25 per cent bonus with a maximum of Rs. 300 every year. Up to 31 March 1997 2.45crore accounts were opened under this scheme with a total collection of Rs 265.09crore. The Department of Women and Child Development, the nodal agency for MSY, decided in April 1997 that now new MSY accounts should be opened form 1 April 1997 onwards but the existing account could be maintained.

d) Mahila Arthik Vikas Mahamandal (MAVIM)

On behalf of the Government of Maharashtra, the Mahila Arthik Vikas Mahamandal (MAVIM) has taken up the cause of overall development of women – half the part of our society. MAVIM has taken tireless efforts for over a decade to initiate, expand and establish the self-help group movement in Maharashtra. The progressive Government of Maharashtra has always supported the cause of women's empowerment and has form time to time devised supportive policy mechanisms and assures them that even in the future the Government will continue to support them.

International Women's Day 2017 was celebrated by MAVIM along with the Dept. of Women and Child development and Dept. of Rural Development with Anganwadi Sevikas. Hon. Mr. Devendra Fadanvis the Chief Minister of Maharashtra, Hon. Ms. Pankaja Munde the Minister of Rural Development and Women and

Child Development, Hon. Ms. Vidya Thakur, the State Minister Women and Child Development, Hon. Mr. Dadaji Bhuse, State Minister of Rural Development, Ms. Vinita Singhal The Secretary, Women and child Dev. Department and Ms. Kusum Balsaraf General Manager, MAVIM were present. During this event the Hon. Chief Minister of Maharashtra Mr. Devendra Fadnavis inaugurated the Portal of Rent a Room for Women; MAVIM's new Initiative for women. The district team of MAVIM Thane presented cultural performance which was highly appreciated.

Celebration of International Women's Day was done by MAVIM in all districts of Maharashtra. Various programs like mahila melawa, cultural programmes, Rallies, Voters registration in coordination with election department *etc.* has been done in districts. In Chandrapur district Hon. Minister of State Home Affairs Mr. Hansraj Ahire, and in Nagpur district Hon. Minister of Energy New and Renewable Energy and Guardian Minister Mr. Chandrashekhar Bavankule presided over the function. In all other districts The district Collectors, Deputy Collectors, NABARD, District Rural Development Agencies (DRDA), Social Welfare Officers, District Election Officer, District Rural Development Department, Mayor, LDM, Bank Officials, College lecturer, Director Rural employment Training Institute, Sarpanch, Police Officials were present in various parts of Maharashtra.

Participation at Regional Consultation Organized by IFAD

IFAD organized a Regional Consultation in the preparation of the third global meeting of the Indigenous People's forum. The theme of the workshop is Economic Empowerment of Indigenous peoples with focus on women and Youth. A case story on **Ecologically Sensitive Approach to Livelihoods and sustainable approach**, which was submitted by MAVIM and the story talk about Lac cultivation of **Yeshaswi** micro livelihood groups of Chandanmeta village of Gondia District under the Tejaswini Programme supported by IFAD. The above case study has been nominated as one of the best story at Asian country level. Smt. Vandana Uikey has been participated in third Global meet as a representation of her group on 22nd -25thNov 2016 at Cambodia. She shared about the economic upliftment gained from this activity.

E-Catering initiative

E-Catering Initiative is a first pilot project by Self Help Group in India initiated by Mahila Arthik Vikas Mahamandal established in Maher Loksanchalit Sadhan Kendra at Sawantwadi block of Sindhudurg district where SHGs federation will provide local delicious cuisine in train through e-Catering service of IRCTC. National Bank for Agriculture and Rural Development (NABARD) under LEDP project has provided fund to e-Catering. Lupin Human Welfare and Research Foundation, Non-Government Organization (NGO) has provided technical support to this project. Hon. Railway Minister of India Mr. Suresh Prabhu inaugurated this project on 3rd December 2016. On this occasion Shri. Rajendra Kulkarni – Hon. Chief General Manager of NABARD, Shri Dr. A. K. Manocha- Hon. Chairman- IRCTC and Smt R. Vimala- Hon. Vice Chairman and Managing Director of MAVIM were present.

Different Important Component of Mahila Arthik Vikas Mahamandal

Capacity Building Training: These trainings provide information, knowledge to women which help them to understand their own situation and how to face the situation. This training also helps to change their attitude. MAVIM organize these trainings on different level in different stages of the life cycles of self help groups.

SHG Concept Training – At the beginning of group formation MAVIM Sahayoginies conduct SHG concept training for all the SHG members. These trainings help women to understand importance of SHG, function of SHG, their role in SHG.

Book Keeping/Accounts Training – Saving is one of the important activity of the Self Help Groups which bring women together initially. The proper record of savings and internal lending help to develop trust among each other. Account training helps women to know how to keep account records properly.

Leadership Development training – Leaders plays an important role in group development. Each group has three leaders, their responsibility to lead the group. Leaders need to understand group dynamic, development stages and role of leader. Leadership training provide all the inputs required to a good leader.

Gender Sensitization Training – In the process of women empowerment, gender sensitization training plays an important role by making women aware about their own status and importance.

Functional Literacy – This training help women to understand the functions of different institutions and organisations which play important role in their day to day life such as Grampanchayat, Zilla parishad, bank, post office, police patil *etc.*

Panchayat Raj Institutions (PRI) – This training gives information to SHG women in functioning of Panchayat Raj and the role of women in panchayat raj which leads to the increased participation of women in panchayatraj.

Monitoring and Evaluation – Self Help Groups develop through various stages. Each stage is very important for development of SHG. These trainings make SHG women understand how to monitor and evaluate their own groups at different stages.

Village Level Committee – VLC is an umbrella body of all SHG s formed in a village. This training gives focus on roles and responsibilities of VLC's, functions of VLC and how to develop village development plan in convergence with Grampanchayat.

CMRC – This is an umbrella body of all formed SHGs in a compact of geographical cluster. This training gives conceptual training on CMRC concept, roles and responsibilities and their functions.

Apart from these stipulated trainings, MAVIM organizes different need base training for the SHG members such as health training, *etc.*

Resource Pool

MAVIM is engaged in SHGs movement since last fifteen years. MAVIM has several experience staff working at various levels. These trainers are experts in grassroots Institution building, Livelihoods, Women empowerment and Micro Finance services.

21

Reorganized Extension System (T&V System)

This system was introduced in 1974 with the World Bank assistance. It was presumed that transfer of technology trough the 'contact farmers' shall benefit all farmers. This system was introduced in Rajastan Canal area in Rajasthan and Chambal Command area in Madhya Pradesh. The system has since been extended to 13 states and 4 more are in pipe-line. This system will be further extended in the remaining states of the country by 1985. In two and a half decades, T&V became the dominant method of restructuring the extension services in over sixty countries in Asia, Africa and Latin America. The system tries to achieve changes in production technologies used by the majority of farmers through assistance form well trained extension agents who have close links with agricultural research and supported by supply, service and marketing facilities.

Objectives

The working objectives of the system as follows:

1. Coordinating research, training and extension activities effectively.
2. To make research more effective by catering to the local needs and situation.
3. To evolve an intensive training program on a systematic basic for extension workers and farmers and to ensure effective supervision and technical support.

T&V was regarded as an improved management system of agricultural extension and had the following key features-

1. Professionalism

Each extension agent is fully and continuously trained to handle one's particular responsibilities

2. Single Line of Command

The extension service must be under a single line of technical and administrative command within the Ministry/Department of agriculture.

3. Concentration of Effort

All extension staff works only on agricultural extension. They are not responsible for any other activity not directly related to extension. In training session, attention is concentrated on important major points.

4. Time-bound Work

Messages and skills are taught to farmers in a regular and timely fashion. The village Extension worker (VEW) must visit the farmers regularly on a fixed day, usually once each fortnight. All other extension staff must make timely and regular visit to the field.

Recommendations for a specific area and for particular farming conditions for each two – fortnight periods are discussed and learned by Subject Matter Specialists (SMSs) at regular monthly workshops, the recommendations are then presented to VEWs and Agricultural Extension Officers (AEOs) at the next two fortnightly training sessions.

5. Field and Farmer Orientation

The contact with the farmers must be on a regular's basis, on a schedule known to farmers, and with a large number of farmers representing all major farming and socio-economic types.

6. Regular and Continuous Training

Regular and continuous training of extension staff is require both the teach, and discuss with them, and the prepare specific production recommendations required by farmer for the coming fortnights and the upgrade and update their professional skills

7. Linkages with Research

Problems faced by farmers that cannot be resolved by extension agents are passed on to researchers of either an immediate solution of investigation. Seasonal and monthly workshops, joint field visit, training of extension are some of the means by which linkages with research and maintained.

Merits and Limitations of Reorganized Extension System (T&V System)

Merits

1. Increasing the cultivated area under High Yielding Varieties Programme.

2. Dissemination of new knowledge of farming for increasing the cropping intensity.
3. Increasing employment of family labour.
4. Raising marginal value of productivity of all inputs.
5. Accelerating the extent of adoption of recommended practices

According to Ray et.al (1979), there was more impact of this system on the small holdings as compared to large ones. Thus T&V system has a considerable positive impact on the farming economy.

Limitations

Croot (1997) analysed certain limitations of T&V system.

a) The T&V model has served to raise the productivity of crops such as rice and wheat form about 1.5 to 4.5 tones per hectare. It has worked especially in areas with relatively uniform, high potential agro-ecosystems with a conducive institutional support such as effective input supply, a good market and an attractive urban demand. Nevertheless, the conditions for small holder agriculture in rainfed areas are often not so conducive foe agricultural development. Small holder's agriculture is featured by complex, high risk, variable and diverse agro-cosystems, not only at the farm level but also at the watershed or landscape levels. Moreover, T&V approach is not the most effective for dealing with natural resources management issues.

b) Researchers develop now technology which extension then delivers to farmers. Farmers are expected to adopt these technologies to improve their farming practices. Researchers are extension, to close the innovation development cycle. However, practice teaches us that in hierarchically organized government institutions this hardly occurs. Moreover, experiences show that innovations are for example, which pass form one individual to another or accumulate on the shelf.

c) Research does not appear to be a constant source of new technologies as assumed in the T&V model. Most new ideas do not originate form research, but form practice itself. Farmers themselves are active information seekers, keen experimenters for promoting collective change. The development of innovations appears to be a much more complex process than the "technology transfer" model leads us to believe.

d) The concept of contact farmers does not work very well. In some countries, a modified T&V type has changed his concept into the use of contact groups which has also not worked well.

e) Experiences shows that the ideas of the same message as being appropriate for each farmer located in the same recommendation domain does not make sense. Even in on village, the conditions and opportunities for farming may vary so much with regard to types of soil, labour availability, income or land size, that it can be more relevant to increase the problem solving capacity of farmers than to provide one message supposed applicable to all farmers.

f) Last but not least limitation was that after the World Bank loan for the T&V system expired; many governments were not able or willing to continue to finance this system.

Single Window System of Extension in Maharashtra

Agriculture, Horticulture and Soil conservation and Watershed Development Department were re-organized and a single agency responsible for agricultural development at village level-'Single Window System' was implemented on 1st July, 1998, the birth anniversary of Late. Vasantrao Nike (Legendary figure in green revolution).

In the earlier system only 64 per cent staff was available at grass root level and 36 per cent at apex level. Now, due to single window system, 81 per cent staff will be made available at grass root level and 19 per cent at apex level.

Salient Features of Single Window System

1. Formation of separate offices at Circle and Taluka level.
2. Integrated planning of agriculture, soil conservation and watershed management and watershed management and horticulture departments by single extension worker.
3. Maximum staff available at village level for transfer/dissemination of Agricultural technology.
4. Increased participation of Agricultural Universities in dissemination and training in agricultural technology.
5. To enhance commercial, business minded farming, separate counter/ office at district level.
6. More stress on technology transfer for horticulture department.
7. Additional area and officers/workers for fruit/horticulture nurseries.
8. Soil testing laboratories in all districts.
9. Project investigation department in all division for implementing future plans/projects.
10. Establishment of precaution team at State level for quality control.
11. Computerization facility upto sub-divisional level in the first phase.
12. Right/authority of input quality control upto taluka level.
13. To solve the problems of officer/worker staff regarding services, immediately staff welfare department constituted.
14. All the above aspects in reorganization achieved through earlier existing/ sanctioned designation only.
15. Realistic, long term agricultural development plan for all districts.

Broad Based Extension (BBE)

Economic growth is an essential component of development; it is not the only as development is not a purely economic phenomenon. The common man in India, as also in other developing countries, expects a higher standard of living for himself, his family, his community and his nation. The village peoples also expect such standard of living. For that purpose farmers are not concerned with crop production alone. They undertake a number of land based activities. Therefore, the focus of agricultural extension needs to be shifted form commodity production or crop production oriented to farmers' income oriented. That means, for development of farmer it is necessary to develop their economic status by improving other production with crop production. To improve their production capacity they have to improve resources, technical knowledge and skills. On this basis this broad based extension concept was emerge out.

The process of broadening scope of agricultural extension on bringing a number of farm activities in addition to the crop production under the umbrella of agricultural extension is known as broad based agricultural extension.

For improving the overall development of farmer's regular flow technology through unified extension services together with specialized support to grass root level farmer is necessary.

Emerging Trend of Agricultural Extension

In the early years of this century, extension services were in their formative stage; they were relatively small in scale and limited in the scope of their work and contact with farmers, and their organization was often somewhat haphazard even though based on legislation. They were organized predominantly either by central or local governments, or by agricultural colleges, usually in close association with experiment stations, or by farmers' organizations (agricultural societies, cooperatives, farmers' unions, or chambers of agriculture), combinations of these parent bodies. As the century has progressed, the organizations have matured. Changes have often occurred to their parent affiliations, government funding has become relatively more important, their objectives have become broader, especially in "the North," and the extension workers have become better trained and more professional. In addition, several other kinds of organizations have developed comparable work : agriculture-related commercial companies; agricultural commodity marketing boards, concerned to assure the supply and quality of their specific product; agricultural development projects, many of considerable territorial scale; and a variety of nongovernmental organizations (especially religious and charitable) involved in agricultural and rural development.

Agricultural extension ahs now become recognised as an essential mechanism for delivering information and advice as an "input" into modern farming. Since commercial farmers can derive direct financial benefits from these inputs, there is a trend towards the privatization of the extension organizations, often as parasttatal or quasigovernmental agencies, with farmers being required to pay for services which they had previously received free of charge. This trend is strong in the North, and there are examples of it beginning in the South.

The pace of change in the organization, functions, strategies, and approaches of agricultural extension is clearly accelerating.

The need for agricultural and rural information and advisory services is likely to intensify in the foreseeable future. In much of the world, agriculture faces the challenge of keeping pace with rapidly increasing population with few reserves of potentially cultivable land. Farmers will have to become more efficient and specialized. From government perspectives, whatever priority is given to production, extension will remain a key policy tool for promoting ecologically and socially sustainable farming practices.

With the collapse during the past decade of socialist forms of economic organization, the (dominant) role of the public sector in national economies has become questionable, with a strengthening trend to reduce levels of public spending. Thus government extension services and those which are largely publicly financed are, and will continue to be, under pressure to become more efficient, to reduce their expenditure and staff, and to pass on (some of) the costs of provision to their clients who directly benefit financially. This is particularly the case in countries where the farm population forms a small minority and agricultural production is in surplus. The case is weaker, but not absent, in less developed countries where farming households form a high proportion of the total population and where increasing food production is still important.

In the future, rural populations will undoubtedly be progressively better educated, while their exposure to the mass media will continue to reduce their isolation and detachment from information, ideas, and an awareness of their situation within a national and international context. However, this exposure will not reduce the need for extension. Rather, given the changing demands on agricultural producers from population growth, increasing unbanization, legislative changes, and market requirements, the more knowledgeable farming population will require different kinds of extension services. Social and economic trends within rural areas will therefore necessitate more highly trained, specialized, and technically competent workers.

A recognition of the locale-specific nature of farming systems and the agricultural information systems which support them is an important source of the pressure towards the debureaucratization and develution of extension services. This recognition also implies that extension workers and farmers be jointly involved in the verification and adaptation of new technology, and thus that the extension workers respect farmers as experimenters, developers, and adapters of technology and devote more energy on communication within their local areas.

The continuing rapid development of telecommunications and computer-based information technology (IT) is probably the biggest factor for change in extension, one which will facilitate and reinforce other changes. There are many possibilities for the potential applications of the technology in agricultural extension (FAO, 1993; Zijp, 1994). IT will bring new information services to rural areas over which farmers, as users, will have much greater control than over current information channels. Even if every farmer does not have a computer terminal, these could become readily available at local information resource centres, with computers carrying expert systems to help farmers to make decisions. However, it will not make extension workers redundant. Rather, they will be able to concentrate on tasks and services where human interaction is essential – in helping farmers individually and in small groups to diagnose problems, to interpret data, and to apply their meaning (Leeuwis, 1993).

The future will call for more able, more independent, more client-oriented extension workers. The emphasis will be on the quality of interaction between agent and client rather than on the movement of "messages" through a hierarchical system.

Flexibility and adaptability will be seen as virtues rather than aberrations. Paradoxically, these trends will bring us full circle to the early manifestations of modern extension in Europe. The itinerant agricultural teachers, unencumbered by large bureaucracies and tall hierarchies, will find their modern counterparts in the computer-carrying extension workers who are at ease helping farmers to identify the information they need in order to realize the potential of their farming operations. Looking back, we can regard the period from 1970 to 1995 as a necessary but expensive stage in the evolution of extension systems, after which extension agents were able to settle down to their main task – bringing together the expertise of farmers and the best available scientific knowledge to develop farms and local agricultural economies.

Part IV

Participatory Development

22

Participatory Technology Development

Participatory technology development (PTD) is an approach to learning and innovation that is used in international development as part of projects and programmes relating to sustainable agriculture. The approach involves collaboration between researchers and farmers in the analysis of agricultural problems and testing of alternative farming practices.

Origins

The origins of Participatory Technology Development can be found in a number of approaches, concepts and techniques that became popular in the 1970s and 1980s:

- Farming systems research and extension, an approach that introduced social-economic issues into organisations previously dominated by biological and chemical scientists. FSR/E also promoted the idea on 'on-farm' trials;
- Appropriate technology, a concept that recognises the importance of matching the design of technology to the resources and culture of the user.
- Indigenous technical knowledge, a concept that recognised the importance of local expertise and traditional wisdom;
- Participatory rural appraisal, a set of techniques that promoted the involvement of rural people in decision-making.

Process

One of the leading authorities on this process is the Centre for learning on sustainable agriculture - ILEIA based in the Netherlands. ILEIA has described PTD as "a process between local communities and outside facilitators which involves:

- ☆ Gaining a joint understanding of the main characteristics and changes of that particular agro-ecological system;
- ☆ Defining priority problems;
- ☆ Experimenting locally with a variety of options derived both from indigenous knowledge ... and from formal science, and
- ☆ Enhancing farmer's experimental capacities and farmer-to-farmer communication".

Participatory rural appraisal (PRA) is an approach used by non-governmental organizations (NGOs) and other agencies involved in international development. The approach aims to incorporate the knowledge and opinions of rural people in the planning and management of development projects and programmes.

Origins

The philosophical roots of participatory rural appraisal techniques can be traced to activist adult education methods, such as, those of Paulo Freire and the study clubs of the Antagonish Movement. In this view, an actively involved and empowered local population is essential to successful rural community development. Robert Chambers, a key exponent of PRA, argued that the approach owes much to "the Freirian theme that poor and exploited people can and should be enabled to analyze their own reality."

By the early 1980s, there was growing dissatisfaction among development experts with both the reductionism of formal surveys, and the biases of typical field visits. In 1983, Robert Chambers, a Fellow at the Institute of Development Studies (UK), used the term rapid rural appraisal (RRA) to describe techniques that could bring about a "reversal of learning", to learn from rural people directly. Two years later, the first international conference to share experiences relating to RRA was held in Thailand. This was followed by a rapid acceptance of usage of methods that involved rural people in examining their own problems, setting their own goals, and monitoring their own achievements. By the mid 1990s, the term RRA had been replaced by a number of other terms including participatory rural appraisal (PRA) and participatory learning and action (PLA).

Robert Chambers acknowledged that the significant breakthroughs and innovations that informed the methodology came from community development practitioners in Africa, India and elsewhere. Chambers helped PRA gain acceptance among practitioners. Chambers explained the function of participatory research in PRA as follows:

The central thrusts of the [new] paradigm ... are decentralization and empowerment. Decentralization means that resources and discretion are devolved;

turning back the inward and upward flows of resources and people. Empowerment means that people, especially poorer people, are enabled to take more control over their lives, and secure a better livelihood with ownership and control of productive assets as one key element. Decentralization and empowerment enable local people to exploit the diverse complexities of their own conditions, and to adapt to rapid change.

Overview of Techniques

Over the years techniques and tools have been described in a variety of books and newsletters, or taught at training courses. However, the field has been criticized for lacking a systematic evidence-based methodology.

The Basic Techniques Used Include

Understanding group dynamics, *e.g.* through learning contracts, role reversals, feedback sessions

- ☆ Surveying and sampling, *e.g.* transect walks, wealth ranking, social mapping
- ☆ Interviewing, *e.g.* focus group discussions, semi-structured interviews, triangulation
- ☆ Community mapping, *e.g.* Venn diagrams, matrix scoring, eco-grams, timelines

To ensure that people are not excluded from participation, these techniques avoid writing wherever possible, relying instead on the tools of oral communication and visual communication such as pictures, symbols, physical objects and group memory. Efforts are made in many projects, however, to build a bridge to formal literacy; for example by teaching people how to sign their names or recognize their signatures. Often developing communities are reluctant to permit invasive audio-visual recording.

Developmental Changes in PRA

Since the early 21st century, some practitioners have replaced PRA with the standardized model of community-based participatory research (CBPR) or with participatory action research (PAR). Social survey techniques have also changed during this period, including greater use of information technology such as fuzzy cognitive maps, e-participation, tele presence, social network analysis, topic models, geographic information systems (GIS), and interactive multimedia.

Participatory development (PD) seeks to engage local populations in development projects. Participatory development has taken a variety of forms since it emerged in the 1970s, when it was introduced as an important part of the "basic needs approach" to development. Most manifestations of PD seek "to give the poor a part in initiatives designed for their benefit" in the hopes that development projects will be more sustainable and successful if local populations are engaged in the development process. PD has become an increasingly accepted method of development practice and is employed by a variety of organizations. It is often presented as an alternative to mainstream "top-down" development. There is some question about the proper definition of PD as it varies depending on the perspective applied. Two perspectives that can define PD are the "Social Movement Perspective" and the "Institutional Perspective":

The "Social Movement Perspective" defines participation as the mobilization of people to eliminate unjust hierarchies of knowledge, power, and economic distribution. This perspective identifies the goal of participation as an empowering process for people to handle challenges and influence the direction of their own lives. Empowerment participation is when primary stakeholders are capable and willing to initiate the process and take part in the analysis. This leads to joint decision making about what should be achieved and how. While outsiders are equal partners in the development effort, the primary stakeholders are primus inter pares, *i.e.*, they are equal partners with a significant say in decisions concerning their lives. Dialogue identifies and analyzes critical issues and an exchange of knowledge and experiences leads to solutions. Ownership and control of the process rest in the hands of the primary stakeholders.

The "Institutional Perspective" defines participation as the reach and inclusion of inputs by relevant groups in the design and implementation of a development project. The "Institutional Perspective" uses the inputs and opinions of relevant groups, or stakeholders in a community, as a tool to achieve a pre-established goal defined by someone external to the community involved. The development project, initiated by an activist external to the community involved, is a process by which problem issues in a community can be divided into stages, and this division facilitates assessment of when and to what degree a participatory approach is relevant. From an institutional perspective, there are four key stages of a development project: Research Stage, Design Stage, Implementation Stage, Evaluation Stage that are defined in later sections of this article. The institutional perspective can also be referred to as a "Project-Based Perspective".

Advocates of PD emphasize a difference between participation as "an end in itself", and participatory development as a "process of empowerment" for marginalized populations. This has also been described as the contrast between valuing participation for intrinsic rather than purely instrumental reasons. In the former manifestation, participants may be asked to give opinions without any assurance that these opinions will have an effect or may be informed of decisions after they have been made. In the latter form, proponents assert that PD tries to "foster and enhance people's capability to have a role in their society's development".

Participatory development employed in particular initiatives often involves the process of content creation. For example, UNESCO's Finding a Voice Project employs ICT for development initiatives. Local content creation and distribution contributes to the formation of local information networks. This is a bottom-up approach that involves extensive discussions, conversations, and decision making with the target community. Community group members create content according to their capacities and interests. This process facilitates engagement with information and communication technology (ICT) with the goal of strengthening individual and social development. This participatory content creation is an important tool for poverty reduction strategies and creating a digitally inclusive knowledge society.

Each project issue in participatory development can be divided into stages, and this division facilitates assessment of when and to what degree a participatory approach is relevant. From an institutional perspective, there are four key stages of a development project:

1. **Research Stage** is where the development problem is accurately defined. All relevant stakeholders can be involved in this process. The research around the development problem can include studying previous experiences, individual and community knowledge and attitudes, existing policies and other relevant contextual information related to socio-economic conditions, culture, spirituality, gender, *etc.*
2. **Design Stage** defines the actual activities. A participatory approach helps to secure the ownership and commitment of the communities involved. Active participation by local citizens and other stakeholders aims to enhance both the quality and relevance of the suggested interventions.
3. **Implementation Stage** is when the planned intervention is implemented. Participation at this stage increases commitment, relevance and sustainability.
4. **Evaluation Stage** participation ensures that the most significant changes are voiced, brought to common attention and assessed. For a meaningful evaluation, indicators and measurements should be defined in a participatory process at the very beginning of the initiative involving all relevant stakeholders.

Forms of Participatory Development

- ☆ **Passive participation** is the least participatory of the four approaches. Primary stakeholders of a project participate by being informed about what is going to happen or has already happened. People's feedback is minimal or non- existent, and their participation is assessed through methods like head counting and contribution to the discussion (sometimes referred to as participation by information).
- ☆ **Participation by consultation** is an extractive process, whereby stakeholders provide answers to questions posed by outside researchers or experts. Input is not limited to meetings but can be provided at different points in time. In the final analysis, however, this consultative process keeps all the decision- making power in the hands of external professionals who are under no obligation to incorporate stakeholders' input.
- ☆ **Participation by collaboration** forms groups of primary stakeholders to participate in the discussion and analysis of predetermined objectives set by the project. This level of participation does not usually result in dramatic changes in what should be accomplished, which is often already determined. It does, however, require an active involvement in

the decision-making process about how to achieve it. This incorporates a component of horizontal communication and capacity building among all stakeholders a joint collaborative effort. Even if initially dependent on outside facilitators and experts, with time collaborative participation has the potential to evolve into an independent form of participation.

- **Empowerment participation** is where primary stakeholders are capable and willing to initiate the process and take part in the analysis. This leads to joint decision making about what should be achieved and how. While outsiders are equal partners in the development effort, the primary stakeholders are primus inter pares, *i.e.*, they are equal partners with a significant say in decisions concerning their lives. Dialogue identifies and analyzes critical issues and an exchange of knowledge and experiences leads to solutions. Ownership and control of the process rest in the hands of the primary stakeholders.

Variations of Participatory Development

Manifestations

There are many different manifestations of Participatory Development. PD has been promoted as a way to improve the "efficiency and effectiveness" of "formal" development programs. This method usually involves external and local actors working together on a particular project. GTZ (Deutsche Gesellschaft für Technische Zusammenarbeit), a German development agency, describes participation as "co-determination and power sharing throughout the program cycle". By involving those who will benefit from the programs in their development and having local and international groups work together, it is hoped that development projects will be made more sustainable and successful.

Enabling "mutual learning" is another way that PD is conceptualized. The goal is to enhance "communication, respect, listening and learning between development workers and those they serve" in order to achieve more applicable, "useful outcomes". Participatory Rural Appraisal (PRA) is one example of mutual learning, a form of research which acknowledges that "illiterate, poor, marginalized people [can] represent their own lives and livelihoods. do their own analysis and come up with their own solutions".

Some hope that PD will be able to cause a shift in power relations by "valorizing. voices" that usually go unheard by political and development groups. This speaks to the idea that PD has the potential to increase a population's ability to be self-determining. Those who promote this view of PD would like to see local communities making, rather than only contributing to, important decisions. These activists hope that PD will lead to better civil engagement, whereby people are able determine the ways their own communities function. In these cases, international organizations can support and draw attention to the efforts of groups working for self-determination.

Implementation

Some theorists have highlighted a difference between "invited" and "claimed" spaces for PD. Invited spaces are usually formal events where local communities are asked by development agencies to share their thoughts. There is often a goal of coming to an agreement. Conversely, claimed spaces are created when marginalized individuals step in and "[take] control of political processes". The Zapatista Army of National Liberation movement can be viewed as an example of local people "claiming" space to advocate for political change.

Benefits

Research conducted by several development agencies (World Bank, CIDA, USAID, IRDP) suggests that there are many benefits to be gained through the use of PD. These studies suggest that while PD projects may have high start up costs, they will be less expensive and more sustainable in the long run. These studies also found that PD projects are better at addressing local needs and are generally more relevant to local populations than traditional development projects.

Community participation is also thought to increase the efficiency of development projects. Participation can also contribute towards more equitable outcomes so long as elite capture of participatory mechanisms is avoided.

Criticisms

When compared with traditional forms of development, PD is sometimes criticized for being costly and slow. A project may take longer if one has to engage, work and come to a consensus with local communities, than if one did not have to do these things. PD may also have higher start up costs than traditional development. In addition, PD is criticized for reaching a smaller population than traditional development. Community dialogue and augmentation may initially involve only a few individuals, whereas dropped food aid reaches hundreds of people.

More radical development thinkers have put several criticisms forward. PD projects have been accused of treating communities as if everyone in them is the same. This issue has been raised most specifically with regard to gender. Critics suggest that while many organizations acknowledge the importance of including women in PD projects, the history of success has been limited. This may be because PD projects seek to address women's immediate needs "without addressing underlying aspects of gender subordination such as the unequal division of reproductive labour, restrictions on female mobility, domestic violence, women's lack of autonomy and so on". Critics have also said that PD protects fail to adequately address other inequalities such as class and caste. In trying to give voice to communities, development agencies may connect only with elite members of a group, thereby re-enforcing local inequalities. PD projects have also been accused of enabling tokenism, where a few "handpicked" local voices are allowed to speak as a "rubber stamp to prove.participatory credentials". This view suggests that organizations only include local voices to improve their image, without really seeking to engage the population with which they are working.

Participatory Action Research (PAR) is an approach to research in communities that emphasizes participation and action. It seeks to understand the world by trying to change it, collaboratively and following reflection. PAR emphasizes collective inquiry and experimentation grounded in experience and social history. Within a PAR process, "communities of inquiry and action evolve and address questions and issues that are significant for those who participate as co-researchers". PAR contrasts with many research methods, which emphasize disinterested researchers and reproducibility of findings. PAR practitioners make a concerted effort to integrate three basic aspects of their work: participation (life in society and democracy), action (engagement with experience and history), and research (soundness in thought and the growth of knowledge). "Action unites, organically, with research" and collective processes of self-investigation. The way each component is actually understood and the relative emphasis it receives varies nonetheless from one PAR theory and practice to another. This means that PAR is not a monolithic body of ideas and methods but rather a pluralistic orientation to knowledge making and social change.

Overview

PAR has multiple progenitors and resists definition. It is a broad tradition of collective self-experimentation backed up by evidential reasoning, fact-finding and learning. All formulations of PAR have in common the idea that research and action must be done 'with' people and not 'on' or 'for' people. It counters scientism by promoting the grounding of knowledge in human agency and social history (as in much of political economy). Inquiry based on PAR principles makes sense of the world through collective efforts to transform it, as opposed to simply observing and studying human behaviour and people's views about reality, in the hope that meaningful change will eventually emerge.

PAR draws on a wide range of influences, both among those with professional training and those who draw on their life experience and those of their ancestors. Many draw on the work of Paulo Freire, new thinking on adult education research, the Civil Rights Movement, South Asian social movements such as the Bhoomi Sena, and key initiatives such as the Participatory Research Network created in 1978 and based in New Delhi. "It has benefited from an interdisciplinary development drawing its theoretical strength from adult education, sociology, political economy, community psychology, community development, feminist studies, critical psychology, organizational development and more". The Colombian sociologist Orlando Fals Borda and others organized the first explicitly PAR conference in Cartagena, Colombia in 1977. Based on his research with peasant groups in rural Boyaca and with other underserved groups, Fals Borda called for the 'community action' component to be incorporated into the research plans of traditionally trained researchers. His recommendations to researchers committed to the struggle for justice and greater democracy in all spheres, including the business of science, are useful for all researchers and echo the teaching from many schools of research:

"Do not monopolise your knowledge nor impose arrogantly your techniques, but respect and combine your skills with the knowledge of the researched or grassroots communities, taking them as full partners and co-researchers. Do not trust elitist versions of history and science which respond to dominant interests, but be receptive to counter-narratives and try to recapture them. Do not depend solely on your culture to interpret facts, but recover local values, traits, beliefs, and arts for action by and with the research organisations. Do not impose your own ponderous scientific style for communicating results, but diffuse and share what you have learned together with the people, in a manner that is wholly understandable and even literary and pleasant, for science should not be necessarily a mystery nor a monopoly of experts and intellectuals."

In the UK and North America the work of Kurt Lewin and the Tavistock Institute in the 1940s has been influential. However alternative traditions of PAR, begin with processes that include more bottom-up organising and popular education than were envisaged by Lewin.

PAR strategies to democratize knowledge making and ground it in real community needs and learning represent genuine efforts to overcome the ineffectiveness and elitism of conventional schooling and science, and the negative effects of market forces and industry on the workplace, community life and sustainable livelihoods. These principles and the ongoing evolution of PAR have had a lasting legacy in fields ranging from problem solving in the workplace to community development and sustainable livelihoods, education, public health, feminist research and civic engagement. It is important to note that these contributions are subject to many tensions and debates on key issues such as the role of clinical psychology, critical social thinking and the pragmatic concerns of organizational learning in PAR theory and practice. Labels used to define each approach (PAR, critical PAR, action research, psycho-sociology, socio-technical analysis, *etc.*) reflect these tensions and point to major differences that may outweigh the similarities. While a common denominator, the combination of participation, action and research reflects the fragile unity of traditions whose diverse ideological and organizational contexts kept them separate and largely ignorant of one another for several decades.

The following review focuses on traditions that incorporate the three pillars of PAR. Closely related approaches that overlap but do not bring the three components together are left out. Applied research, for instance, is not necessarily committed to participatory principles and may be initiated and controlled mostly by experts, with the implication that 'human subjects' are not invited to play a key role in science building and the framing of the research questions. As in mainstream science, this process "regards people as sources of information, as having bits of isolated knowledge, but they are neither expected nor apparently assumed able to analyze a given social reality". PAR also differs from participatory inquiry or collaborative research, contributions to knowledge that may not involve direct engagement with transformative action and social history. PAR, in contrast, has evolved from the work of activists more concerned with empowering marginalized peoples than with generating academic knowledge for its own sake. Lastly, given its commitment to the research process, PAR overlaps but is not synonymous with Action learning, Action Reflection Learning (ARL), participatory development and community development recognized forms of problem solving and capacity building that may be carried out with no immediate concern for research and the advancement of knowledge.

Organizational Life

Action research in the workplace took its initial inspiration from Lewin's work on organizational development (and Dewey's emphasis on learning from experience). Lewin's seminal contribution involves a flexible, scientific approach to planned change that proceeds through a spiral of steps, each of which is composed of 'a circle of planning, action, and fact-finding about the result of the action', towards an organizational 'climate' of democratic leadership and responsible participation that promotes critical self-inquiry and collaborative work. These steps inform Lewin's work with basic skill training groups, T-groups where community leaders and group facilitators use feedback, problem solving, role play and cognitive aids (lectures, handouts, film) to gain insights into themselves, others and groups with a view to 'unfreezing' and changing their mindsets, attitudes and behaviours. Lewin's understanding of action-research coincides with key ideas and practices developed at the influential Tavistock Institute (created in 1947)) in the UK and National Training Laboratories (NTL) in the US. An important offshoot of Tavistock thinking and practise is the socio-technical systems perspective on workplace dynamics, guided by the idea that greater productivity or efficiency does not hinge on improved technology alone. Improvements in organizational life call instead for the interaction and 'joint optimization' of the social and technical components of workplace activity. In this perspective, the best match between the social and technical factors of organized work lies in principles of 'responsible group autonomy' and industrial democracy, as opposed to deskilling and top-down bureaucracy guided by Taylor's scientific management and linear chain of command.

NTL played a central role in the evolution of experiential learning and the application of behavioral science to improving organizations. Process consultation, team building, conflict management, and workplace group democracy and autonomy have become recurrent themes in the prolific body of literature and practice known as organizational development (OD). As with 'action science', OD is a response to calls for planned change and 'rational social management' involving a normative human relations movement and approach to work life in capital-dominated economies. Its principal goal is to enhance an organization's performance and the work life experience, with the assistance of a consultant, a change agent or catalyst that helps the sponsoring organization define and solve its own problems, introduce new forms of leadership and change organizational culture and learning. Diagnostic and capacity-building activities are informed, to varying degrees, by psychology, the behavioural sciences, organizational studies, or theories of leadership and social innovation. Appreciative Inquiry (AI), for instance, is an offshoot of PAR based on positive psychology. Rigorous data gathering or fact-finding methods may be used to support the inquiry process and group thinking and planning. On the whole, however, science tends to be a means, not an end. Workplace and organizational learning interventions are first and foremost problem-based, action-oriented and client-centred.

Psycho-sociology

Tavistock broke new ground in other ways as well, by meshing general medicine and psychiatry with Freudian and Jungian psychology and the social sciences to help the British army face various human resource problems. This gave rise to a field of scholarly research and professional intervention loosely known as psycho-sociology, particularly influential in France (CIRFIP). Several schools of thought and 'social clinical' practise belong to this tradition, all of which are critical of the experimental and expert mindset of social psychology. In addition to this humanistic and democratic agenda, psycho-sociology uses concepts of psychoanalytic inspiration to address interpersonal relations and the interplay between self and group. It acknowledges the role of the unconscious in social behaviour and collective representations and the inevitable expression of transference and counter transference language and behaviour that redirect unspoken feelings and anxieties to other people or physical objects taking part in the action inquiry.

The works of Balint, Jaques,) and Bion are turning points in the formative years of psychosociology. Commonly cited authors in France include Amado, Barus-Michel, Dubost, Enriquez, Lévy, Gaujelac, and Giust-Desprairies. Different schools of thought and practice include Mendel's action research framed in a 'socio-psychoanalytic' perspective and Dejours's psychodynamics of work, with its emphasis on work-induced suffering and defence mechanisms. Lapassade and Lourau's 'socianalytic' interventions focus rather on institutions viewed as systems that dismantle and recompose norms and rules of social interaction over time, a perspective that builds on the principles of institutional analysis and psychotherapy. Anzieu and Martin's work on group psychoanalysis and theory of the collective 'skin-ego' is generally considered as the most faithful to the Freudian tradition. Key differences between these schools and the methods they use stem from the weight they assign to the analyst's expertise in making sense of group behaviour and views and also the social aspects of group behaviour and affect. Another issue is the extent to which the intervention is critical of broader institutional and social systems. The use of psychoanalytic concepts and the relative weight of effort dedicated to research, training and action also vary.

Community Development and Sustainable Livelihoods

PAR emerged in the post war years as an important contribution to intervention and self-transformation within groups, organizations and communities. It has left a singular mark on the field of rural and community development, especially in the Global South. Tools and concepts for doing research with people, including "barefoot scientists" and grassroots "organic intellectuals", are now promoted and implemented by many international development agencies, researchers, consultants, civil society and local community organizations around the world. This has resulted in countless experiments in diagnostic assessment, scenario planning and project evaluation in areas ranging from fisheries and mining to forestry, plant breeding, agriculture, farming systems research and extension, watershed management, resource mapping, environmental conflict and natural resource management, land rights, appropriate technology, local economic development, communication, tourism, leadership for sustainability, biodiversity and climate change. This prolific literature includes the many insights and methodological creativity of participatory monitoring, participatory rural appraisal (PRA) and participatory learning and action (PLA) and all action-oriented studies of local, indigenous or traditional knowledge. On the whole, PAR applications in these fields are committed to problem solving and adaptation to nature at the household or community level, using friendly methods of scientific thinking and experimentation adapted to support rural participation and sustainable livelihoods.

Literacy, Education and Youth

In education, PAR practitioners inspired by the ideas of critical pedagogy and adult education are firmly committed to the politics of emancipatory action formulated by Freire, with a focus on dialogical reflection and action as means to overcome relations of domination and subordination between oppressors and the oppressed, colonizers and the colonized. The approach implies that "the silenced are not just incidental to the curiosity of the researcher but are the masters of inquiry into the underlying causes of the events in their world". Although a researcher and a sociologist, Fals Borda also has a profound distrust of conventional academia and great confidence in popular knowledge, sentiments that have had a lasting impact on the history of PAR, particularly in the fields of development, literacy, counter hegemonic education as well as youth engagement on issues ranging from violence to criminality, racial or sexual discrimination, educational justice, healthcare and the environment.

Community-based participatory research and service-learning are a more recent attempts to reconnect academic interests with education and community development. The Global Alliance on Community-Engaged Research is a promising effort to "use knowledge and community-university partnership strategies for democratic social and environmental change and justice, particularly among the most vulnerable people and places of the world." It calls for the active involvement of community members and researchers in all phases of the action inquiry process, from defining relevant research questions and topics to designing and implementing the investigation, sharing the available resources, acknowledging community-based expertise, and making the results accessible and understandable to community members and the broader public. Service learning or education is a closely related endeavour designed to encourage students to actively apply knowledge and skills to local situations, in response to local needs and with the active involvement of community members. Many online or printed guides now show how students and faculty can engage in community-based participatory research and meet academic standards at the same time.

Collaborative research in education is community-based research where pre-university teachers are the community and scientific knowledge is built on top of teachers' own interpretation of their experience and reality, with or without immediate engagement in transformative action.

Public Health

PAR has made important inroads in the field of public health, in areas such as disaster relief, community-based rehabilitation, accident prevention, hospital care and drug prevention. Because of its link to radical democratic struggles of the Civil Rights Movement and other social movements in South Asia and Latin America (see above), PAR is seen as a threat to their authority by some established elites. International alliance university-based participatory researchers, ICPHR, omit the word "Action", preferring the less controversial term "participatory research". Photo voice is one of the strategies used in PAR and is especially useful in the public health domain. Keeping in mind the purpose of PAR, which is to benefit communities, Photo voice allows the same to happen through the media of photography. Photo voice considers helping community issues and problems reach policy makers as its primary goal.

Occupational Health and Safety

Participatory programs within the workplace involve employees within all levels of a workplace organization, from management to front-line staff, in the design and implementation of health and safety interventions. Some research has shown that interventions are most successful when front-line employees have a fundamental role in designing workplace interventions. Success through participatory programs may be due to a number of factors. Such factors include a better identification of potential barriers and facilitators, a greater willingness to accept interventions than those imposed strictly from upper management, and enhanced buy-in to intervention design, resulting in greater sustainability though promotion and acceptance. When designing an intervention, employees are able to consider lifestyle and other behavioral influences into solution activities that go beyond the immediate workplace.

Feminism and Gender

Feminist research and women's development theory also contributed to rethinking the role of scholarship in challenging existing regimes of power, using qualitative and interpretive methods that emphasize subjectivity and self-inquiry rather than the quantitative approach of mainstream science.

Civic Engagement and ICT

Novel approaches to PAR in the public sphere help scale up the engaged inquiry process beyond small group dynamics. Touraine and others thus propose a 'sociology of intervention' involving the creation of artificial spaces for movement activists and non-activists to debate issues of public concern. Citizen science is another recent move to expand the scope of PAR, to include broader 'communities of interest' and citizens committed to enhancing knowledge in particular fields. In this approach to collaborative inquiry, research is actively assisted by volunteers who form an active public or network of contributing individuals. Efforts to promote public participation in the works of science owe a lot to the revolution in information and communications technology (ICT). Web 2.0 applications support virtual community interactivity and the development of user-driven content and social media, without restricted access or controlled implementation. They extend principles of open-source governance to democratic institutions, allowing citizens to actively engage in wiki-based processes of virtual journalism, public debate and policy development. Although few and far between, experiments in open politics can thus make use of ICT and the mechanics of e-democracy to facilitate communications on a large scale, towards achieving decisions that best serve the public interest.

In the same spirit, discursive or deliberative democracy calls for public discussion, transparency and pluralism in political decision-making, lawmaking and institutional life. Fact-finding and the outputs of science are made accessible to participants and may be subject to extensive media coverage, scientific peer review, deliberative opinion polling and adversarial presentations of competing arguments and predictive claims. The methodology of Citizens' jury is interesting in this regard. It involves people selected at random from a local or national population who are provided opportunities to question 'witnesses' and collectively form a 'judgment' on the issue at hand.

ICTs, open politics and deliberative democracy usher in new strategies to engage governments, scientists, civil society organizations and interested citizens in policy-related discussions of science and technology. These trends represent an invitation to explore novel ways of doing PAR on a broader scale.

Ethics

Calls for norms of ethical conduct to guide the relationship between researchers and participants are many. Norms in research ethics involving humans include respect for the autonomy of individuals and groups to deliberate about a decision and act on it. This principle is usually expressed through the free, informed and ongoing consent of those participating in research (or those representing them in the case of persons lacking the capacity to decide). Another mainstream principle is the welfare of participants who should not be exposed to any unfavourable balance of benefits and risks with participation in research aimed at the advancement of knowledge, especially those that are serious and probable. Since privacy is a factor that contributes to people's welfare, confidentiality obtained through the collection and use of data that are anonymous (*e.g.* survey data) or anonymized tends to be the norm. Finally, the principle of justice equal treatment and concern for fairness and equity calls for measures of appropriate inclusion and mechanisms to address conflicts of interests.

While the choice of appropriate norms of ethical conduct is rarely an either/or question, PAR implies a different understanding of what consent, welfare and justice entail. For one thing the people involved are not mere 'subjects' or 'participants'. They act instead as key partners in an inquiry process that may take place outside the walls of academic or corporate science. As Canada's Tri-Council Policy Statement: Ethical Conduct for Research Involving Humans suggests, PAR requires that the terms and conditions of the collaborative process be set out in a research agreement or protocol based on mutual understanding of the project goals and objectives between the parties, subject to preliminary discussions and negotiations. Unlike individual consent forms, these terms of reference (ToR) may acknowledge collective rights, interests and mutual obligations. While they are legalistic in their genesis, they are usually based on interpersonal relationships and a history of trust rather than the language of legal forms and contracts.

Another implication of PAR ethics is that partners must protect themselves and each other against potential risks, by mitigating the negative consequences of their collaborative work and pursuing the welfare of all parties concerned. This does not preclude battles against dominant interests. Given their commitment to social justice and transformative action, some PAR projects may be critical of existing social structures and struggle against the policies and interests of individuals, groups and institutions accountable for their actions, creating circumstances of danger.

On the matter of welfare, empowerment through recognition and 'being heard' may be more important to the research than are privacy and confidentiality. It is important to strike a balance between allowing privacy and confidentiality, and respect for individuals and groups who wish to be heard and identified for their contribution to research. The former may be hard to reconcile with PAR. The latter can be shown through proper quoting, acknowledgements, co-authorship, or the granting of intellectual property rights.

By definition, PAR is always a step into the unknown, raising new questions and creating new risks over time. Given its emergent properties and responsiveness to social context and needs, PAR cannot limit discussions and decisions about ethics to the design and proposal phase. Norms of ethical conduct and their implications may have to be revisited as the project unfolds. This has implications, both in resources and practice, for the ability to subject the research to true ethical oversight in the way that traditional research has come to be regulated.

Challenges

PAR offers a long history of experimentation with evidence-based and people-based inquiry, a groundbreaking alternative to mainstream positive science. As with positivism, the approach creates many challenges as well as debates on what counts as participation, action and research. Differences in theoretical commitments (Lewinian, Habermasian, Freirean, psychoanalytic, feminist, *etc.*) and methodological inclinations (quantitative, qualitative, mixed) are numerous and profound. This is not necessarily a problem, given the pluralistic value system built into PAR. Ways to better answer questions pertaining to PAR's relationship with science and social history are nonetheless key to its future.

One critical question concerns the problem-solving orientation of engaged inquiry–the rational means-ends focus of most PAR experiments as they affect organizational performance or material livelihoods, for instance. In the clinical perspective of French psychosociology, a pragmatic orientation to inquiry neglects forms of understanding and consciousness that are not strictly instrumental and rational. PAR must pay equal attention the interconnections of self-awareness, the unconscious and life in society.

Another issue, more widely debated, is scale how to address broad-based systems of power and issues of complexity, especially those of another development on a global scale. How can PAR develop a macro-orientation to democratic dialogue and meet challenges of the 21st Century, by joining movements to support justice and solidarity on both local and global scales? By keeping things closely tied to local group dynamics, PAR runs the risk of substituting small-scale participation for genuine democracy and fails to develop strategies for social transformation on all levels. Given its political implications, community-based action research and its consensus ethos have been known to fall prey to powerful stakeholders and serve as Trojan horses to bring global and environmental restructuring processes directly to local settings, bypassing legitimate institutional buffers and obscuring diverging interests and the exercise of power during the process. Cooptation can lead to highly manipulated outcomes. Against this criticism, others argue that, given the right circumstances, it is possible to build institutional arrangements for joint learning and action across regional and national borders that can have impacts on citizen action, national policies and global discourses.

The role of science and scholarship in PAR is another source of difference. In the Lewinian tradition, "there is nothing so practical as a good theory". Accordingly, the scientific logic of developing theory, forming and testing hypotheses, gathering measurable data and interpreting the results plays a central role. While more clinically oriented, psycho-sociology in France also emphasizes the distinctive role of formal research and academic work, beyond problem solving in specific contexts. Many PAR practitioners critical of mainstream science and its overemphasis on quantitative data also point out that research based on qualitative methods may be theoretically-informed and rigorous in its own way. In other traditions, however, PAR keeps great distance from both academic and corporate science. Given their emphasis on pluralism and living knowledge, many practitioners of grassroots inquiry are critical of grand theory and advanced methods for collaborative inquiry, to the point of abandoning the word "research" altogether, as in participatory action learning. Others equate research with any involvement in reflexive practice aimed at assessing problems and evaluating project or program results against group expectations. As a result, inquiry methods tend to be soft and theory remains absent or underdeveloped. Practical and theoretical efforts to overcome this ambivalence towards scholarly activity are nonetheless emerging.

Participatory technology development (PTD) is an approach to learning and innovation that is used in international development as part of projects and programmes relating to sustainable agriculture. The approach involves collaboration between researchers and farmers in the analysis of agricultural problems and testing of alternative farming practices.

Objectives of Participation

Participation activities may be motivated from an administrative perspective or a citizen perspective on a governmental, corporate or social level. From the administrative viewpoint, participation can build public support for activities. It can educate the public about an agency's activities. It can also facilitate useful information exchange regarding local conditions. Furthermore, participation is often legally mandated. From the citizen viewpoint, participation enables individuals and groups to influence agency decisions in a representational manner. The different types of political participation depend on the motivation. When a group is determined to work to solve a community problem, there can be led marches to work for candidates. Most immigrant racial groups have higher motivation since there is an increase in geographical dispersion and are faster growing racial groups. How well participation can influence the relation between citizen and their local government, how it increases trust and boosts peoples willingness to participate Giovanni Allegretti explains in an interview using the example of participatory budgeting.

Classifying Participation

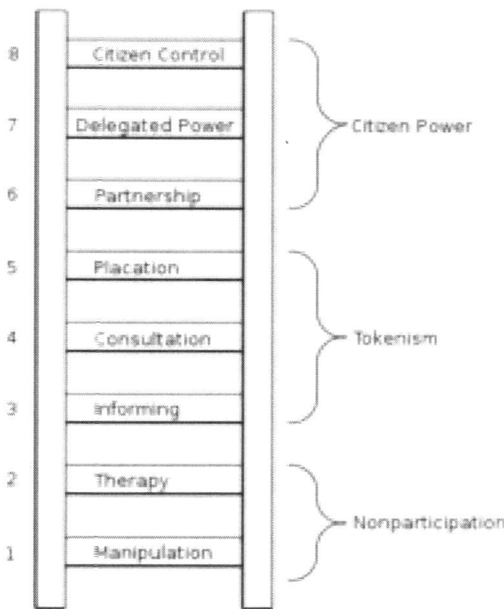

Ladder of Citizen Participation by Sherry Arnstein

Sherry Arnstein discusses eight types of participation in A Ladder of Citizen Participation (1969). Often termed as "Arnstein's ladder", these are broadly categorized as:

- Citizen Power: Citizen Control, Delegated Power, Partnership.
- Tokenism: Placation, Consultation, Informing.
- Non-participation: Therapy, Manipulation.

She defines citizen participation as the redistribution of power that enables the have-not citizens, presently excluded from the political and economic processes, to be deliberately included in the future.

Robert Silverman expanded on Arnstein's ladder of citizen participation with the introduction of his "citizen participation continuum." In this extension to Arstein's work he takes the groups that drive participation into consideration and the forms of participation they pursue. Consequently, Silverman's continuum distinguishes between grassroots participation and instrumental participation.

Archon Fung presents another classification of participation based on three key questions: Who is allowed to participate, and are they representative of the population? What is the method of communication or decision-making? And how much influence or authority is granted to the participation? Other "ladders" of participation have been presented by D.M. Connor, Wiedemann and Femers, A. Dorcey *et al.*, Jules N. Pretty and E.M. Rocha.

Community-led total sanitation (**CLTS**) is an approach to achieve sustained behavior change of people who participate in a guided process of "triggering"; the triggering is intended to lead to spontaneous and long-term change of social behaviours, in particular the abandonment of open defecation. The concept was first developed by Kamal Kar for rural areas in Bangladesh in around the year 2000. The concept originally involved mainly provoking shame and disgust about poor sanitation in order to bring about change. It has been further developed since then by applying the lessons learnt from large scale applications in different rural and urban settings, focussing more on aspects of pride. CLTS has spread throughout Bangladesh and to many other Asian and African countries (however not significantly in Latin America) with support from the Water and Sanitation Program of the World Bank, DFID and other bilateral donors, Plan International, Water Aid, CARE, UNICEF, SNV and other large INGOs and many national NGOs. CLTS was successfully included in holistic programs which also included sanitation marketing, governance and sanitation for public institutions with a strong focus on schools using children as messengers of change.

When CLTS was first introduced in a country, NGOs were often in the lead, though India was an exception. Since about 2011, CLTS is an established approach. Governments in many countries have applied it or have even included it in their national policies on sanitation. This change has created a new situation with additional challenges.

The original concept of CLTS - which made it quite unique at the time - did not include the use of subsidies for hardware (such as toilets) and building toilets for the villagers. CLTS proponents believed that by provoking behavior change in the people this will ensure that they take ownership of their own sanitation situation, construct their own toilets (often pit latrines) and pay for necessary improvements themselves. Challenges of CLTS may include possible human rights infringements within the communities, low levels of toilet standards and issues with long-term usage rates. This has been counteracted in some programs by successfully applying subsidies for a limited number of users. Rewards may also be given for successful declaration of "open defecation free" (ODF) villages.

Definitions

Open defecation is the practice of defecating outside and not into a designated toilet.

"Open defecation free" (ODF) is a central term for community-led total sanitation (CLTS) programs and primarily means the eradication of open defecation in the entire community. However, it can also include the following additional criteria:

- ✰ Household latrines are hygienic, provide the safe containment of feces, offer privacy, have a lid on the defecation hole or a water seal and a roof to protect the user.
- ✰ All household members and all members of the community use these toilets.
- ✰ A hand washing facility is nearby with water, soap or ash, and is used regularly.

Even more stringent criteria which may be added to achieve "ODF status" for a community might include:

- ✰ Safe drinking water and storage
- ✰ Food hygiene
- ✰ Grey water disposal
- ✰ Solid waste management
- ✰ Provision of toilets at schools, markets and for passers-by

Overview

A facilitator and the community during a triggering in Malda District, West Bengal, India

School-led total sanitation "triggering" event

These school children in West Bengal, India are looking at a glass of water and fresh feces. Flies will pass from the water to the feces and back, which demonstrates how water can get polluted with pathogens.

This is what CLTS tries to stop: Open defecation in rural Bihar, India

CLTS focuses on community-wide behavioural change, rather than merely toilet construction. The process raises the awareness that as long as even a minority continues to defecate in the open; everyone is at risk of disease. CLTS uses community-led methods, such as participatory mapping and analyzing pathways between faces and the mouth (fecal-oral transmission of disease), as a means of teaching the risks associated with OD.

A tool called triggering is used to propel people into taking action. The official handbook used by the CLTS Foundation explains that this takes place over a day with a team of facilitators. The team visits a community which is identified as practicing open defecation and encourages villagers to become aware of their own sanitation situation. This aims to cause disgust in participants, and the facilitators help participants to plan appropriate sanitation facilities.

Using the term "shit" (or other locally used crude words) during triggering events or presentations - rather than feces or excreta - is a deliberate aspect of the CLTS approach, as it is meant to be a practical, straight forward approach rather than a theoretical, academic conversation.

Phases

Pre-triggering

Pre-triggering is the process by which communities are assessed to be suitable for CLTS intervention. This involves visits and a number of different criteria, which are used to identify communities likely to respond well to triggering. During pre-triggering, facilitators introduce themselves to community members and begin to build a relationship.

Triggering

The "CLTS Handbook" from 2008 states that there is no "one way" of doing triggering in CLTS. A rough sequence of steps is given in this handbook which could be followed. Facilitators are encouraged to modify and change activities depending on the local situation.

The UNICEF manual approved for use of CLTS in Sierra Leone suggests the following steps for the triggering process:

- ✰ Visit the community, emphasising that it is for learning about their sanitation situation
- ✰ Facilitate 'Kaka Mapping' - which involves drawing the main sites in the village then the main sites for defecation
- ✰ Pretend to leave
- ✰ Facilitate the 'Walk of Shame' - walking with community to the sites of Open Defecation
- ✰ Take a piece of faeces in a bag
- ✰ Put faeces on the floor in front of the community and discuss the way flies move between food and faeces
- ✰ Wait for the shocked understanding that the community is 'eating' the faeces
- ✰ Put some faeces into a water bottle and ask community if they would drink it
- ✰ Calculate how much faeces is produced each day and asks where it goes

- ☆ Ignition
- ☆ Wait for the emergence of "Natural Leaders" to work with in order to develop a plan of action.

The idea of the triggering process is to stimulate disgust in the villagers by physical demonstration of the sanitation problems. At the 'ignition' phase, the villagers are expected to realise that there is a real sanitation problem and that they need to do something about it. Natural Leaders are people from the community who are engaged by the process and who are seen to be people who can drive change.

Post-triggering

After a positive response to the ignition phase, NGO facilitators work with communities to deliver sanitation services by providing information and guidance relevant to the local situation.

Comparison of different CLTS Programmes

One small study compared different CLTS programmes. Participants from NGOs involved in delivering CLTS reported that although they included some of the activities described in the guidance materials, they often omitted some and included others depending on the local situation. Some reported that subsidies were included, and some offered specific design and construction options.

Applications to Urban Situations, Schools and other Settings

More recently, CLTS has been adapted to the urban context (for example in Kenya by the NGO Plan and Practical Action), and has even been used schools and the surrounding communities, which is referred to as "school-led total sanitation". Increasingly, there is also discussion about how CLTS could be adapted to post-emergency settings. There has been some experience with this in Haiti, Afghanistan, Pakistan, Philippines and Indonesia. UNICEF has reported in 2014 positive outcomes with CLTS in "fragile and insecure contexts", namely in Somalia and South Sudan.

Outcomes and Health Aspects

Millions of people worldwide have benefitted from CLTS. A positive outcome is typically measured by the declarations of "ODF villages", where ODF stands for "open defecation free".

Community-based participatory research (CBPR) is a partnership approach to research that equitably involves, for example, community members, organizational representatives, and researchers in all aspects of the research process and in which all partners contribute expertise and share decision making and ownership. The aim of CBPR is to increase knowledge and understanding of a given phenomenon and integrate the knowledge gained with interventions and policy and social change to improve the health and quality of life of community members.

History

The historical roots of CBPR generally trace back to the development of participatory action research by Kurt Lewin and Orlando Fals Borda, and the popular education movement in Latin America associated with Paulo Freire.

In CBPR projects, the community participates fully in all aspects of the research process. CBPR projects start with the community. Community is often self-defined, but general categories of community include geographic community, community of individuals with a common problem or issue, or a community of individuals with a common interest or goal. CBPR encourages collaboration of "formally trained research" partners from any area of expertise, provided that the researcher provide expertise that is seen as useful to the investigation by the community, and be fully committed to a partnership of equals and producing outcomes usable to the community.

Equitable partnerships require sharing power, resources, credit, results, and knowledge, as well as a reciprocal appreciation of each partner's knowledge and skills at each stage of the project, including problem definition/issue selection, research design, conducting research, interpreting the results, and determining how the results should be used for action. CBPR differs from traditional research in many ways. One of the principal ways in which it is different is that instead of creating knowledge for the advancement of a field or for knowledge's sake, CBPR is an iterative process, incorporating research, reflection, and action in a cyclical process.

Examples

Methods in Community-Based Participatory Research for Health This thoroughly revised and updated second edition of Methods for Community-Based Participatory Research for Health provides a step-by-step approach to the application of participatory approaches to quantitative and qualitative data collection and data analysis. With contributions from a distinguished panel of experts, this important volume shows how researchers, practitioners, and community partners can work together to establish and maintain equitable partnerships using a Community-Based Participatory Research (CBPR) approach to increase knowledge and improve the health and well-being of the communities involved.

- ☆ The Detroit Community-Academic Urban Research Center (Detroit URC) is a long-standing Community-Based Participatory Research partnership working to provide economic and public health benefits focused on eliminating health inequities in Detroit. The Detroit URC focuses on enhancing understanding of the relationship between the social and physical environmental determinants of health, and translating that knowledge into public health interventions, programs, and policies that build upon community resources and strengths.
- ☆ Cofounds aims at establishing a prediction market for funding of participatory research.
- ☆ Community-Campus Partnerships for Health

- Progress in Community Health Partnerships: Research, Education, and Action- The mission of the Journal is to facilitate dissemination of programs that use community partnerships to improve public health, to promote progress in the methods of research and education involving community health partnerships, and to stimulate action that will improve the health of people in communities. Communities, as defined by the Journal, may be based on geography, shared interests, or social networks. The Journal is dedicated to supporting the work of community health partnerships that involve ongoing collaboration between community representatives and academic or governmental partners. This area of research and evaluation may be referred to as community-based participatory research (CBPR). The W. K. Kellogg Foundation defines CBPR as a collaborative approach to research that equitably involves all partners in the research process and recognizes the unique strengths that each brings. CBPR begins with a research topic of importance to the community and has the aim of combining knowledge with action and achieving social change to improve health outcomes and eliminate health disparities.
- Journal of Health Care for the Poor and Underserved The Journal of Health Care for the Poor and Underserved (JHCPU) is an academic journal founded in 1990 by David Satcher, MD, PhD then President of Meharry Medical College (later, U.S. Surgeon General). JHCPU is published by Johns Hopkins University Press for Meharry, and is affiliated with the Association of Clinicians for the Underserved. It is the premier journal covering the health and health care of medically underserved populations in North and Central America and the Caribbean.
- Participatory Research, Community Development and Drug Misuse (1) (2)
- Community-Based Participatory Research (CBPR): A Partnership Approach for Public Health is a free training resource available from the Michigan Public Health Training Center and the Detroit Community-Academic Urban Research Center. This training resource is intended as an introduction to CBPR for people who are in the early stages of using or considering using CBPR. It is designed for academic researchers from multiple fields, including public health, education, nursing, medicine, social work, urban planning, and for health and human service practitioners, and members of community-based organizations.

Asset-Based Community Development (ABCD)

Asset-based community development (ABCD) is a methodology for the sustainable development of communities based on their strengths and potentials. It involves assessing the resources, skills, and experience available in a community; organizing the community around issues that move its members into action; and then determining and taking appropriate action. This method uses the community's own assets and resources as the bases for development; it empowers the people of the community by encouraging them to utilize what they already possess. The ABCD approach was developed by John L. McKnight and John P. Kretzmann at the Institute for Policy Research at Northwestern University in Evanston, Illinois. They co-authored a book in 1993, Building Communities from the Inside Out: A Path Toward Finding and Mobilizing A Community's Assets, which outlined their asset-based approach to community development.[4] The Community Development Program at Northwestern University's Institute for Policy Research established the Asset-Based Community Development Institute based on three decades of research and community work by John P. Kretzmann and John L. McKnight.

Principles

Whereas needs-based community development emphasizes local deficits and looks to outside agencies for resources, asset-based community development focuses on honing and leveraging existing strengths within the community. Related to tenets of empowerment, it postulates that solutions to community problems already exist within a community's assets. Principles that guide ABCD include:

a) Everyone has gifts: Each person in a community has something to contribute.

b) Relationships build a community: People must be connected in order for sustainable community development to take place.

c) Citizens at the center: Citizens should be viewed as actors–not recipients–in development.

d) Leaders involve others: Community development is strongest when it involves a broad base of community action.

e) People care: Challenge notions of "apathy" by listening to people's interests.

f) Listen: Decisions should come from conversations where people are heard.

g) Ask: Asking for ideas is more sustainable than giving solutions.

h) Inside-out organization: Local community members are in control.

i) Institutions serve the community: Institutional leaders should create opportunities for community-member involvement, then "step back."

Tools

The ABCD approach utilizes several tools to assess and mobilize communities.

Capacity Inventory

Skills Information: lists the many skills that a person has gained at home, work, in the community, or elsewhere. Examples of these skills can include internet knowledge, hair-cutting, listening, wallpapering, carpentry, sewing, babysitting, *etc.*

- ✰ Community Skills: lists the community work in which a person has participated to determine future work they may be interested in.

- ✰ Enterprising Interests and Experience: lists past experience in business and determine interest in starting a business.
- ✰ Personal Information: lists minimum information for follow-up.

Asset Mapping

There are five key assets in any given community: individuals, associations, institutions, physical assets, and connections. These assets are broken down into three categories: Gifts of individuals, Citizens' Associations, and Local Institutions. Asset maps are used in lieu of needs maps which focus solely on negative aspects of communities. Asset maps, on the other hand, focus on community assets, abilities, skills, and strengths in order to build its future.

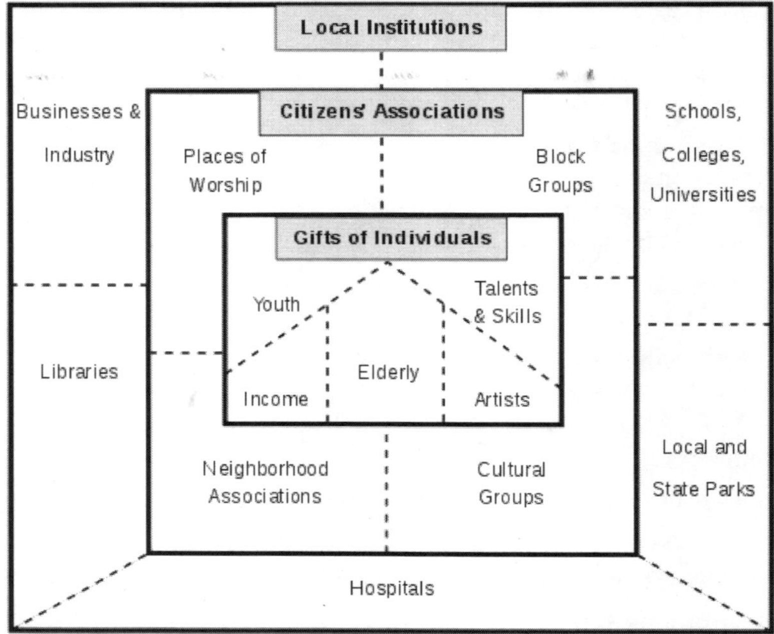

Time Banks

Time banks are an example of using community assets to connect individuals' assets to one another. Neighbors and local organizations share skills with one another and earn and spend 'Time Bank Hours' or 'credits' in the process, allowing an hour of child care to equal an hour of home repair or tax preparation.

Ethics

Because ABCD relies on existing community assets to create change, it has been criticized for implying that disadvantaged communities have all the resources they need to solve community problems. According to the ABCD Institute, however, ABCD methodology recognizes that systemic injustice may require disadvantaged communities to seek assistance from outside the community. ABCD maintains that interventions from exterior sources will be most effective when a community's assets are leveraged at full capacity. ABCD is described as a more sustainable model of community development than needs-based community development, because needs-based approaches may perpetuate community problems by emphasizing deficiencies and the necessity for reliance on outside assistance. By contrast, ABCD aims to build capacity within communities by expanding their social capital. By working with outside resources and simultaneously building trust within the community, more members can make use of a wider array of strengths.

Participatory Poverty Assessment (PPA)

Participatory poverty assessment (PPA) is the approach to analyzing and reducing poverty by incorporating the views of the poor. PPAs attempt to better understand the poor, to give the poor more influence over decisions that affect their lives, and to increase effectiveness of poverty reduction policies. PPAs are seen as complements to traditional household surveys by helping to interpret survey results, and aim to capture the experiences of the poor by being more open-ended.

History

Before the 1990s, household-survey approaches were the main way to measure poverty. This shifted to participatory activities in the 1990s after critiques of these traditional methods.

The term "participatory poverty assessment" was coined within the World Bank in 1992, and was seen as a participatory form of poverty assessments (PAs, *i.e.* survey-based approaches).

Early experiences of PPAs were in sub-Saharan Africa in the early 1990s.

By 1998, 43 PPAs had been conducted: 28 in Africa, 6 in Latin America, 5 in Eastern Europe, and 4 in Asia.

Early PPAs focused most on making higher quality information available to policy makers, rather than seeing the poor as itself being involved in the process of reducing poverty.

Later in the 1990s, "second generation" PPAs appeared, which tried to allow the poor to influence policy decisions.

As of 2012, the World Bank's Voices of the Poor is the largest PPA ever to have been conducted.

Rationale and Comparison to Previous Methods

Both PPAs and traditional surveys contain both quantitative and qualitative information, as well as "subjective" and "objective" parts, so this is not necessarily a distinguishing feature of PPAs. Rather:

Traditional survey data can be used to count, compare, and predict. The strength of the PPA is not in counting but rather in understanding hidden dimensions of poverty and analyzing causality and processes by which people fall into and get out of poverty.

PPAs are also distinguished from anthropological studies because the latter analyze few communities in depth, whereas PPAs analyze various communities of a country. PPAs also focus on policy decisions (which anthropological studies don't need to).

Types of Methodologies and Activities

Within PPAs, various methodologies exist. Some PPAs use more than one methodology.

a) Field work and conducting interviews
b) Rapid rural appraisal
c) Participatory rural appraisal (includes use of visuals, such as matrices and diagramming)
d) SARAR (self-esteem, associative strength, resourcefulness, action planning, and responsibility)
e) Beneficiary assessment
f) Semi-structured interviews and focus groups

PPAs can last from several weeks to several months.

Organizations Using this Approach

Various organizations have used PPAs as part of their work.

a) World Bank
b) United Nations Development Programme (UNDP)
c) Department for International Development
d) Action Aid
e) Asian Development Bank
f) Danish International Development Agency
g) Oxfam UK
h) Thailand Development Research Institute

In addition, "secondary stakeholders" (*i.e.* organizations that do not conduct PPAs directly, but participate in PPAs) such as UNICEF, various NGOs (*e.g.* Save the Children and Oxfam), academic institutions, *etc.*, have also participated in PPAs.

Results

Participatory poverty assessments confirmed the multidimensional nature of poverty (*i.e.* that the poor deal not just with lack of money, but with various problems like lack of resources, poor health, physical violence, social isolation, *etc.*) that was seen from traditional household questionnaire-based methods. PPAs have also been able to access information that was not obtained in household surveys by building trust. For instance PPAs have obtained information on sensitive topics such as child prostitution, drug use, and domestic violence in relation to poverty.

In Zambia, PPAs discovered that the timing of school fees caused economic stress for households, so new legislation to change this timing was proposed.

In Niger, a PPA found that the poor see education only as a seventh priority for poverty reduction due to "the concern that what children learn had little relevance to getting people out of poverty". As a result of this PPA, food security was prioritized.

PPAs have also informed World Bank's World Development Reports.

Participatory video is a form of participatory media in which a group or community creates their own film. The idea behind this is that making a video is easy and accessible, and is a great way of bringing people together to explore issues, voice concerns or simply to be creative and tell stories. It is therefore primarily about process, though high quality and accessible films (products) can be created using these methods if that is a desired outcome. This process can be very empowering, enabling a group or community to take their own action to solve their own problems, and also to communicate their needs and ideas to decision-makers and/or other groups and communities. As such, PV can be a highly effective tool to engage and mobilise marginalised people, and to help them to implement their own forms of sustainable development based on local needs.

Process

Participants (men, women and youth) rapidly learn how to use video equipment through games and exercises.

a) Facilitators help groups to identify and analyse important issues in their community by adapting a range of Participatory Rural Apraisal (PRA)-type tools with participatory video techniques (for example, social mapping, action search, prioritising, *etc.*)

b) Short videos and messages are directed and filmed by the participants.

c) Footage is shown to the wider community at daily screenings.

d) A dynamic process of community-led learning, sharing and exchange is set in motion.

e) Completed films can be used to promote awareness and exchange between various different target groups. Insight Share, for example, has worked with pastoralists, farmers, marginalised communities and youth in rural and urban settings, street children, refugees and asylum seekers, people with mental health problems, learning difficulties and physical disabilities.

f) Participatory video films or video messages can be used to strengthen both horizontal communication (*e.g.* communicating with other communities) and vertical communication (*e.g.* communicating with decision-makers)

Comparison to Documentary Filmmaking

Whilst there are forms of documentary filmmaking that are able to sensitively represent the realities of their subjects' lives and even to voice their concerns, documentary films very much remain the authored products of a documentary filmmaker. As such, the subjects of documentaries rarely have any say (or sometimes have some limited say) in how they will ultimately be represented. By contrast, in PV the subjects make their own film in which they can shape issues according to their own sense of what is important, and they can also control how they will be represented. Additionally, documentary films are often expected to meet stringent aesthetic standards and are usually made with a large audience in mind. The PV process, on the other hand, is less concerned with appearance than with content, and the films are usually made with particular audiences and objectives in mind.

Origins

The first experiments in PV were the work of Don Snowden, a Canadian who pioneered the idea of using media to enable a people-centered community development approach. Then Director of the Extension Department at Memorial University of Newfoundland, Snowden worked with filmmaker Colin Low and the National Film Board of Canada's Challenge for Change program to apply his ideas in Fogo Island, Newfoundland, a small fishing community. By watching each other's films, the different villagers on the island came to realise that they shared many of the same problems and that by working together they could solve some of them. The films were also shown to politicians who lived too far away and were too busy to actually visit the island. As a result of this dialogue, government policies and actions were changed. The techniques developed by Snowden became known as the Fogo process. Snowden went on to apply the Fogo process all over the world until his death in India in 1984.

The first community-made video in Canada was the 1969 Challenge for Change video VTR St-Jacques, filmed in a poor Montreal neighbourhood. In order to make VTR St-Jacques, directors Dorothy Henault and Bonnie Sherr Klein trained community members in video to represent their struggle for affordable and accessible medical care. VTR St-Jacques was shown across Canada and the U.S., inspiring other projects.

There has been no uniform movement to promote and practise PV but different individuals and groups have set up pockets of PV work, usually molding it to their particular needs and situations. PV has also grown with the increasing accessibility of home video equipment.

An early and significant book on participatory video was published in the UK in 1997 by Clive Robertson and Jackie Shaw, Directors of Real Time Video, and has informed many subsequent books and articles, including the book this article has drawn from. Real Time are an educational charity that pioneered many of the techniques and methodologies still used today, and have been working in the participatory video field since 1984.

Applications

In combination with other methodologies such as Participatory Learning in Action (PLA) techniques, Participatory Rural Appraisal and others PV has been successfully applied to projects focussing on; community development; promoting local innovation and endogenous development; therapeutic work; a voice for marginalised groups; a catalyst for community-led action; a tool for communicating with policy makers; a means of involving users in their own research for example action research, participatory research, user-led research; also for programme monitoring and evaluation or Social impact assessment.new possible applications are being continually developed.

23

Participatory Extension and Participatory Research: Concept, Implication and Modalities

In course of quest for bringing about development in the rural areas, numerous official and non-official attempts have been made in the country to mobilize people to participate in the activities initiated under the programmes of directed change through varied categories of interventions for bringing about structural, institutional and technological changes in the rural milieu. A review of all these attempt is made till date, provides an interesting account of shift in strategies followed for mobilizing people' participation and their varied impact on the socio-politico-economic fabric of the society. Depending on the similarity in approaches, these attempts can be categorized under several perspectives:

1. Sporadic attempts;
2. Extensive extension;
3. Intensive extension;
4. Target group oriented extension;
5. Mono-purpose extension; and
6. Privatized extension.

Attempts are being made in the following paragraphs to analyze the programmes of action initiated under each of these perspectives.

Perspective I: Sporadic Attempts of Extension

Micro Level Attempts

Need for carrying out development efforts in the rural areas were felt even during the pre-independence era with the colonial regime attempting to provide an administrative infrastructure, sporadically in selected pockets across the country. Attempts for mobilizing rural people through development programmes of government agencies, date back to pre-independence days. While Sriniketan project was the forerunner, Gurgaon project, Marthandam project, Nilokheri project, Etawah Pilot Project were a few of such sporadic attempts of rural development. While all of these were essentially micro level attempts under individual leadership, most of them lacked sustainability and were discontinued with the departure of the leader.

Grow More Food Campaign

Towards the end of the colonial regime, acute food crisis, crescendoed with the infamous Bengal famine, led the policy makers to initiate a programme, known as Grow More Food campaign across the country. Keeping in view the need for increasing food production, the campaign was basically an attempt of increased productivity. Through the expansion of area was under crops, thereby making it an exclusively agriculture oriented programme. Thus, the programme was characterized by restricted scope of operation with short term objective and lacked any integration of efforts towards addressing the problems of rural areas.

Perspective II: Extensive Extension

During the post-independence era, the popular government attempted to seek people's participation in rural development in a more concerted and systematic manner. Immediately after independence, the very first step was to ensure social justice through the abolition of Zamindari system. The government enacted legislation towards land reform and redistributed surplus land among the land less to confer land ownership with them. Though primarily a structural intervention, loaded with a series of administrative and legislative measures, it created an initial feeling of enthusiasm and acted as a motivating force for securing people's participation. However, such popular enthusiasm could not be sustained in view of lack of administrative and political will.

In order to develop a mechanism of delivery system, a nationwide network of Blocks was created in 1952 under the programme, better known as Community Development (CD), with the organizational set up of National Extension Service (NES). The Blocks were hence termed as NES blocks. All the revenue villages in the country were covered under this network and a comprehensive programme of action, involving the major facets of rural life was offered for the benefit of the millions of families, inhabiting the countryside. These programmes were offered to all the families in the villages, by following a whole village approach. In order to carry out the programmes of action, the multipurpose village level worker (VLW) was entrusted with the cutting edge responsibility of liasoning with members of the rural community. The issue of people's participation was the central concern in community development approach, which was addressed by way of involving community organizations through the application of extension education process.

In an effort towards facilitating grass roots participation in the development process, the concept of democratic decentralization was introduced in 1953 and Panchayats became the institution for involving people's representatives in the programme. A three tier Panchayati Raj system was launched for managing the affairs of rural life. A Panchayat – Block interface was introduced through the instrument of block and village production plans, wherein the officialdom coordinated their efforts with the people's representatives of the community organizations.

Perspective III: Intensive Extension – Multipurpose

In order to combat the problem of food shortage, which attained very serious proportion during the mid sixties, a number of programme of actions were dovetailed within the frame work of extensive extension (through community development), focusing upon the national priority of increased agricultural production, particularly in the food front. The emphasis on agriculture in the block programmes of actions was intensified through introduction of a number of area (IADP/IAAP) as well as individual intensive (HYVP/ND/MCP/FTC) schemes, all aiming at increased food production.

A perusal of the components of these schemes lead one to identify a few common elements of approaches, which can be spelt out as follows :

- Most of these schemes were highly input intensive in terms of requirements of seeds, fertilizers, pesticides, irrigation, credit *etc.*
- As a result, mostly the resource – endowed farms/farmers could capitalize on the gains of such technologies.
- In view of the compulsion of increased food production, cereal (wheat and rice) oriented production policy was the mainstay of most of these scheme.
- As a result, in an effort to motivate the resource – endowed farmers to adopt input intensive cereal production technologies, the price policies had also the cereal bias.
- In view of almost near total emphasis on production of cereals, the traditional crops/technologies started getting ignored/withered.

Perspective IV: Target Group/Area Specific Extension-State Interventions

The allegation leveled against green revolution era of accelerating inter personal/regional disparities led he policy makers to give a new twist to the whole village approach of community development programme. In order to help the resource poor pepole/areas, a series of target group (SFDP/MFAL/NREP/RLEGP/IRDP/JRY/SJRY *etc.*) and area (DPAP/HAD/CADP *etc.*) specific programmes were introduced within the block organizational structure. The basic philosophy behind these programmes of action was to extend the benefits of state interventions to bridge the resource gaps of the resource poor individuals/areas for increasing their absorptive capacities. All these programmes were brought under the umbrella of poverty alleviation.

These programmes brought in sweeping changes in respect of rural development strategy from those designed with the initiation of community development programme, immediately after independence. The changes can be summarized as per the following details:

- ☆ Contrary to the whole village approach of community development, the emphasis was placed on resource poor areas/persons in the poverty alleviation programmes.
- ☆ While community development primarily aimed at developing a spirit of self reliance among the rural communities, the poverty alleviation programmes were designed on the basic assumption that given to themselves, these resource poor areas/persons wee not in a position to improve their lot. Hence, the state interventions.
- ☆ In the poverty alleviation programmes, a near total shift was evidenced from the community to individual approach
- ☆ Schemes were drawn by the development agencies and people were asked to participate, thereby relegating the practices of developing block/individual production plants through participatory decision making into the background.
- ☆ With the introduction of poverty alleviation programmes, the people were reduced to mere beneficiaries, looking ahead only for the state largesse.
- ☆ While the introduction of target group/area specific programmes of action, institutional finance was made a complementary input of development administration in the rural areas.

Perspective V: Mono Purpose Extension

Along with the shift in community development approach for accommodation the resource poor areas and individuals in the mainstream development process, another significant change took place in the latter half of seventies in respect of integrated character of the community development programme. Agriculture, an integral component of the Block administration, was separated out and formed an independent unit with single line of command from the Block to the State headquarter. This was done at the behest of the recommendation of the World Bank, the funding agency, which undertook such reorganization of agricultural extension delivery system in fifty odd developing countries across the globe. The main arguments put forth in such recommendation in favour of cessation of agriculture department from the multipurpose block organization were the dilution of agricultural extension efforts due to too many desperate activities, poor research extension linkages, outdated knowledge base of extension personnel, excessively large size of area of command for the Village Level Workers and predominance of supply and service related workload for them, leading towards neglect of real extension job like information dissemination and capacity building among the farmers.

In order to combat this array of problems, the Agriculture departments across the country were reorganized under the aegis of training and Visit (T&V) systems and the salient features of the changes were as follows:

- ✰ Unlike the multipurpose nature of extension offered in community development approach, in T&V system, it was single purpose approach, centering around agriculture only.
- ✰ Under T&V system, the VLWs no longer remained multipurpose functionary and was required to perform agricultural extension only.
- ✰ The most important departure of T&V system from the earlier approach of Community Development was in respect of very strong research – extension linkage, inbuilt and institutionalized into the reorganized extension mechanism.
- ✰ Regular updating of knowledge of entire extension machinery, through training, was another inbuilt component of the T&V system.
- ✰ Instead of geographical area of jurisdiction, prescribed for the VLWs under the CD, T&V system made it very specific in terms of certain number of farm families (800-1250 under irrigated and 500-600 in dry land areas).
- ✰ The extension task of he field functionaries, specially the VLWs, was redefined as information dissemination only, leaving aside he supply, services and regulatory jobs, otherwise forming the components of job chart for them under CD.

The reorganization of the agriculture department under T&V opened a floodgate of disintegration and many of the development departments, started placing their respective field functionaries with specific tasks related to their departments. The Community Development blocks, on the other hand, is mainly concentrated on poverty alleviation programme components. Mono purpose extension, on the other hand, though continuing on paper, had been modified in several states in several ways and the funding agency has been contemplating to revise the system by way of broad basing the organizational structure through incorporation of other land based departments in addition to agriculture. Towards such broad basing, among various organizational alternatives, the Agricultural Technology Management Association (ATMA) is currently being put to trial run in pilot scale.

Perspective VI: Privatized Extension

In view of globalization of Indian agriculture and the emerging cost consciousness among the policy makers, the desirability of public extension system is increasingly being questioned on the grounds of economic as well as administrative efficiency. While the public agricultural extension services are funded by the state and is widely viewed as supply driven, rather than demand driven, privatization of agricultural extension is argued to be the answer to achieve the goal of exportable surplus in agriculture. The Indian experience of privatized extension is in a very rudimentary stage and need to be documented, analyzed and disseminated for wider understanding and implementation.

Private extension service is provided either through individual initiative or through organizational efforts, other than that of the government. In the Indian context, a wide range of such service providers can be identified. They are unemployed agricultural graduates, agricultural consultants/firms, farmers' organizations/cooperatives, agri-business companies, input dealers, newspapers and agricultural magazines, private television channels *etc.*

However, the applicability of privatized extension in India has to be decided, keeping in view the composition of the farming community. India has seventy to eighty per cent of small and marginal farmers characterized by illiteracy and subsistence farming. This is in contrast to the western countries, wherein the privatization is well entrenched. With only five to ten per cent of the population professing farming, they can afford to ignore the social welfare dimension in their development planning. But with the majority of Indian farmers growing crops for sheer reasons of ensuring food security, blanket application of privatized extension by replacing public sector one is bound to neglect their interest. Privatization, generally favour progressive and wealthy farmers, who can afford to practice high tech agriculture, with a view to get back the investments made on consultancy charges.

Nevertheless, privatized extension can be practiced by way of making some adjustments with the nationwide network of public agricultural extension. In the fields of agro-processing, diagnostic services and commodity based production systems, privatized extension organized in collaboration with public sector extension, are playing important role in different parts of the country. Success stories like *NDDB, Pepsi, UPASI, Sugar Cooperatives, "Farmer's One Stop Shop",* introduced by *Mahindra and Mahindra,* the tractor giant, Tata Kisan Kendras *etc.* are instances of some of the effective privatized extension efforts.

Thus, with the passage of time, extension in agriculture and rural development had undergone series of changes. While the focus on self reliance and community participation had undergone total dilution through high dose of state intervention, introducing widespread dependence syndrome, not only among the resource poor, but also the resource rich alike, the composite character of the extension delivery system, developed during the CD era was disintegrated. Populism, political expediency and bureaucratic red tapism met the extension system vulnerable to nepotism and inefficiency. Seeing the ill effects of this entire phenomenon, the policy makers started advocating adoption of participatory approaches, which however, till now remained elusive.

The knowledge and experiences of people should be the starting point to solve community's problems. Farmers are being viewed as active partners in development of innovations and their dissemination in the community. Farming Systems Research and extension, Participatory Technology Development and farmer First Extension are new waves of extension calling for active roles of farmers and building structures for farmer-to-farmer extension. Extension in the changed scenario is more farmers - propelled and interactive.

The extension must certainly be more than simple propagation of technology and distribution of inputs although this will usually be included. The participatory aspects of extension come into play when villagers are given a chance to use the extensionists as a resource with special skills for facilitating villagers in order to implement their own ideas. Participatory extension consists of contents or subject matter, planning, information, adult education, assist in solving problems, encouragement, inputs and services, storage and marketing, on-farm research, infrastructure, evaluation, control and law enforcement. The concept, principles and sequence of actions of people's participation may be derived as:

Concept of People's Participation

The term people's participation is used to connote different things by different people in different contexts. By people, it may mean the target population, the beneficiaries, the men and women, the old and young, the formal and informal leaders, people of different segments and strata of the community, depending upon the mandate of the specific development programmes and activities.

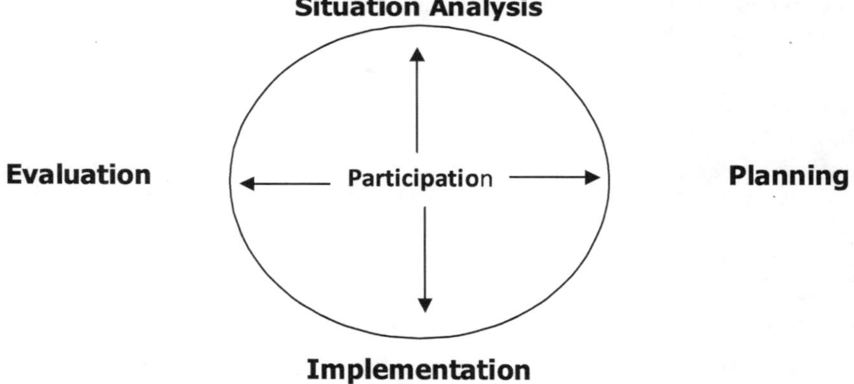

People's participation in extension, in a broader sense, means the involvement of people in the programmes of directed social change, initiated by the development agencies by way of analyzing the situations, deciding on the problems to be tackled, fixing up of priorities, drawing plans of action, taking initiatives in implementing activities of the project as partners through contributing their ideas, materials, resources, labour and time *etc.* and finally evaluating the results thus accrued to themselves.

In the ultimate sense, the whole purpose and process of people's participation in the context of directed social change is human resource development – the development of human and inner material resources, with stimulus and support external to the community. The process of people's participation in rural development involves transfer of administrative and financial powers from 'haves' to 'have not's and sharing of technical and legal information with the local people, whose participation is sought. Drawing upon the experiences, some of the rationale for participatory extension vis-à-vis rural development can be identified as follows:

- Reduction in development cost to the government and other development agencies.
- Building awareness regarding the problem associated with the people.
- Enhancing the capacity of the rural communities to deal with their problems.
- Correction of mistakes made by project authority in designing and implementing the programmes of action.
- Increase in the level of political awareness of the people.
- Reaching the programme benefits to all the legitimate claimants.
- Decrease in perpetual dependence of people on government and thereby making the programme self sustaining and local people self-reliant.
- Gaining access to and control of resources
- Easier mobilization of local resources
- Gradual empowerment of socially and economically disadvantaged people in the community.

Principles of Participatory Development

After the above questions have been asked (and answered) the next step is to establish the basis or the set of principles, which will provide the overall framework of approach, which the project will adopt. This is an equally critical step since, without such a framework, a participatory project will not be able to construct an appropriate strategy nor determine how the project is to be implemented. Participatory development is not 'blue-print' development but it is a strategy, which constructs its approach in relation t the demands of the project context. In this respect, some basic principles will help to determine what this approach should be. In the practice of participatory development to date, the following are the kinds of key principles which have been seen to be more important:

- **The Primacy of People** whatever the purpose or ultimate goal of the project, people's interests, their needs and their wishes must be allowed to underpin the key decisions and actions relating to the project. It is not a question of including people as and when it is felt by project management to be convenient; people must sit centre-stage and their interests taken into consideration during the whole course of the project.
- **People's Knowledge and Skills** must be seen as a potentially positive contribution to the project: a project which does not seek to make use of local knowledge and skills may not only be less effective but will also be squandering a useful resource. A participatory project should seek every possibility to base its activities upon local resources, both to avoid situations of dependence on external ones and also to help develop local capabilities, which will be important if the development is to be sustained.

Participation is to do with developing people's capacities and this can best be achieved by building on and strengthening their existing knowledge and expertise.

- ☆ **People's Participation must empower Women** participatory development should seek to improve gender inequalities through providing a means by which women can take part in decision making. Women's participation must be transformative, not be merely tokenistic; while there are often enormous social and cultural barriers which hinder women's participation, participatory development should seek to bring about change and to create the circumstances where women's voices can also be heard. This is a sensitive and critical issue, but efforts to involve women in an appropriate manner must be central.
- ☆ **Autonomy as opposed to Control** as far as it is realistic to do so, seek to invest as much responsibility as possible for the project with the local people, and thus avoid having absolute control in the hands of project staff. Such a principle is not always realistic and it should not be pushed too far in such circumstances; but there must be a discernible move to minimize the control of project staff and to maximize the potential responsibility of local people.
- ☆ **Local Actions as opposed to local Responses** encouraging local people to make decisions and to take action within the broad parameters of the project, as opposed to merely responding passively to initiatives proposed by others. The more than these kinds of local initiatives can be encouraged, the more a sense of ownership will develop among local people; but the major obstacle may see project staffs that are unwilling or unable to promote local ownership of project activities and outcomes.
- ☆ **Allow for some Spontaneity in project Direction** promoting people's participation will mean that, as far as it is reasonably possible, the project should be allowed to develop in accordance with the abilities of local people to play an increasing role and to begin to assume some responsibility. Unless the project is able to plan for an to accommodate people's participation, it will remain a straight jacket and may push on to predetermined objectives regardless of whether local people are on board or not. Participatory projects do often take longer but, if they are truly participatory, the outcome is more positive and sustainable.

Participation as a Sequence of Actions

It is widely understood that participation is not a one-off input into a project but it is a process which should be an intrinsic part and characteristic of a project throughout its duration. This process evolves through a series of stages, but these stages will vary according to the nature and purpose of the project. There is no universal or common set of stages in the process of participation and development agencies at different levels will structure a series of stages which are most appropriate to the project. Already there are a considerable number of examples of how development agencies have gone about promoting participation as a series of actions at different stages during the lifetime of a project. There should be a logical sequence of actions and events, which is an important feature of participatory development. Participation has to be developed and this development has to be projected over a period of time and the appropriate action taken at its different stages. This perspective is fundamental to participatory development and it will be essential for country office staff to adopt this approach in order to map out the likely evolution of people's participation. The danger is that staff will see participation merely as an input to be brought into activities as appropriate and will not recognize the critical importance of allowing it to develop over a period overtime. Given that there is no universal model of the stages of a process of participation, the following is an example which is intended to illustrate the kinds of stages which have been identified:

- ☆ Creating an awareness of the issues
- ☆ Developing community leadership
- ☆ Sharing in decision-making
- ☆ Learning together
- ☆ Analyzing options
- ☆ Planning together
- ☆ Evaluation (Christian Aid, Kenya)

The above example shows the range of actions, which are undertaken in projects to promote participation. The purpose is not to suggest that such and such a way is how participation is promoted, but to underline the importance of seeing participation as a process with several dimensions and of the need to develop a plan for its promotion. The examples illustrate the range from a more technocratic approach, which sees participation more as a means, to a less structured looser approach, which sees participation more as part of an overall goal of the project. All these examples, however, reinforce the argument that participation has to be planned and developed and that it should be seen as important as the other activities associated with project development. Indeed the central issue is that participation should be the basic operating mechanism within the project and should characterize all of the project's activities ranging from initial analysis, needs determination, planning, monitoring, evaluation and so on.

Need for Participatory Extension

The form of extension, which had brought about spectacular production increase in food production front, was primarily following a scientist led top down approach in transfer of technology among the farming population. A demand for resorting to participatory extension for more sustainable results in terms of generation and transfer of need based technological options relevant under the varied resource profiles of different categories of farmers had been created. For such process of participatory extension to succeed, it is necessary that the process of technology generation should start with identification of technology requirements at farmer's levels in direct consultation with them. The extension system should help the scientists by applying suitable participatory extension techniques to understand the farming systems at the village level in the context of the agro-ecological features of the area as a whole. The overall approach of participatory extension is based upon the philosophy that the farmers know their systems best and any intervention should be in consultation with them, so that they become more relevant and sustainable in the long run.

Participatory Extension Tools

While in previously stated array of extension and communication media also will be effective in communicating relevant messages, certain tools of participatory learning and planning will be helpful in knowing the farming systems and their characteristics better. Some of such participatory learning techniques can be generally stated to be the following:

1. **Participatory rural appraisal (PRA) -** The way of enabling communities to define, influence and evaluate their economic, environmental health and educational status. It is a methodology for interacting with the villagers, understanding them and learning from them. It is an intensive and systematic learning experience carried out in a community by a multidisciplinary team, which includes the community members.

   Principles of PRA –

   - ✰ Optimal ignorance- Knowing what is worth knowing
   - ✰ Seeking diversity- Analysis of difference
   - ✰ Offsetting biases and triangulation- Probing
   - ✰ Listening and learning –Progressively through participation.

2) **Agro-ecosystem analysis (AESA):** A technique of understanding the situations of certain agro-socio-ecological complex in a comprehensive manner. It may be done by resorting to participatory learning exercises like PRA and PLA.

3) **Gender analysis:** To assess how do the women involve with a project and anticipate the effect of the project on women. The framework includes the components like a) activity profile, b0 access and control profile, c) analysis of factors influencing activity, access, control and d) project cycle analysis.

4) **Goal Oriented Project Planning (GOPP):** An effective technique in assisting through a participatory communication process, all parties involved to identify and analyse those problems, they see as inhibiting progress and prepare a concrete and realistic plan.

Once the ground level realities are analysed by a way of application of o these techniques, extension strategy to be followed can be more pragmatically decided and implemented. The results of such development will be more environment and people friendly.

Participation in the Context of Rural Development

A huge majority of the population constituted of ruralites, programmes for facilitating development in the rural areas, became major areas of concern in their development administration. Each one of them evolved comprehensive framework of rural development administration, which attempted to organize various categories of development interventions for the benefit of the rural community. On the basis of such details of understanding about the concept and typologies of participation, attempts are now being made to delineate a common frame of guidelines for organizing people for rural development programmes. These are:

- ☆ The whole planning process in rural development should be based upon bottom up approach, *viz.* start from people.
- ☆ In order to enable people to effectively participate in policy planning, people need to be provided with sufficient and objective information.
- ☆ To bring about people's participation, they need to be conscientised, so that they become aware of their right and privileges.
- ☆ In any community naturally available group structures exist on the basis of certain characteristics like religious, political, peer group *etc.*
- ☆ For maximizing participation of people, keeping in view the composite nature of the production systems in most of the developing economies integrated nature of development is warranted.
- ☆ Developing a sense of ownership and responsibility for a programme in the community is needed.

Participatory Research

Previously in extension education conventional research was followed but now day's scientists are using participatory research to get more interactive and iterative information. So, now we can discuss regarding need of participatory research. Participatory research methods have been used in natural resource management like, soil and water conservation, forestry, fishery, wildlife, village planning *etc.*; agriculture, poverty and gender analysis; health and food security.

Need of Participatory Research

- ☆ The researcher does not know all about the people and their situation. A more meaningful research therefore requires the involvement of local people.
- ☆ More to gain an understanding of complexities of the topic, rather than to gather highly accurate statistics on a list of variables.
- ☆ To obtain a differentiated understanding of the population's attitude, beliefs, and behaviours.
- ☆ To apply most effectively in small and homogenous rural communities.
- ☆ Short duration and low cost.
- ☆ Possible to apply a series of research methods.

- ✰ A way of learning from and with the community members to investigate, analyze and evaluate constraints and opportunities, and to make informed and timely decisions regarding development projects.

Features of Participatory Research

a) Triangulation: A form of cross-checking.

b) Multidisciplinary team: The members of the participatory research should have different skill and back ground.

c) Mix of techniques: Using wide range of tools.

d) Flexibility and informality: It is semi-structured, revised, adapted and modified.

e) Optimal ignorance and appropriate imprecision: Avoids unnecessary, detail, accuracy and over collection of data.

f) On the spot analysis: Learning takes place in the field.

Methods for Participatory Research

Among the several methods some are expressed here:

- ✰ **Review of secondary sources**: Secondary data are sources of information which are relevant to the area or subject of the planned participatory research and are available in published or unpublished form.
- ✰ **Direct observation**: Systematically observing objects, events, relationship or people, and recording these observations.
- ✰ **Semi-structured schedule**: a form of guided interviewing where only some of the questions are predetermined.
- ✰ **Individual interview**: Interviews are conducted with an opportunitysample of purposively selected individual respondents.
- ✰ **Key informant interview**: Key informant is anyone who has special knowledge on a particular topic.
- ✰ **Group interview**: To obtain community level information.
- ✰ **Focus group discussion**: Suitable for discussing specific topic in detail with the target group.

Tools of Participatory Research

Amongst the tools of participatory research we can classify it as:

- ✰ **Space analysis**: A type of explanatory analysis of a particular area to observe minute details of the area. These are social map, hydrological map, land use map, micro farming situation, transect walk, diversity study *etc.*
- ✰ **Time analysis**: A chronological study of any area or social system regarding its historical attributes, qualitative and quantitative development over time. These are time line, time trend, seasonality diagram, crop calendar *etc.*

- ☆ **Flow analysis**: Effluence of village activities, opportunities over time. These are mobility map, crop profile, Venn diagramming *etc.*
- ☆ **Decision analysis**: Opinion judgment. These are Matrix ranking, wealth ranking, pair wise matrix ranking, Log frame analysis *etc.*

Part V

Training and Development

24
Training

"Man always learns, he must always be a student". This particular advice opens deep spheres of thinking in the minds of people who are really keen to learn. Perhaps what is hidden in the advice is the incompleteness in each one of us. If man wants to be complete one, the gap has to be filled in between 'what he is' and 'what he wants to be', this is achieved through Training.

Training

Training refers to a planned effort by a company to facilitate employees' learning of job-related competencies. These competencies include knowledge, skills, or behavior that are critical for successful performance. The goal of training is for employees to master the knowledge, skills, and behaviors emphasized in training programs and to apply them to their day-to- day activities.

A training need is the gap between the knowledge, skill and attitude that the job demands, and the knowledge, skills and attitudes already possessed by the trainee. It exists at all levels of the organizations. It is only an employee on one or the other aspect which changes whether one is growing a new chairman or as instructing an operator, given knowledge skill and attitudes or the chosen criteria around which all jobs based in the assessment of training needs. It is possible to over stress the need for overall view if the company s aims, objectives and man power needs, given the exercise that determines the priorities for implementing training which can off course be critical.

There are three major area in which the staff displays such gaps as knowledge skills and attitude that the job demands and the knowledge, skills and attitudes already possessed.

1. To match the employee specifications with the job requirements and organizational needs: when their performance in their present position

does not match up with the required standards this could not be a fault of their own *E.g.*:-new job.

2. Organizational viability and the transformation process when the requirement of job changes due to changing circumstances.
3. When the present job Indus job holder to the change the job created which can be filled up only through training.
4. Technological advancement: Even organization in order to survive and to be effective should adopt the latest technology like mechanization, computerization, and automation.
5. Organizational complexity: With the emergency of increased mechanization, and automation, manufacturing of multiple product or dealing in services diversified lines, extension of operation to various reason of the country or in overseas countries organization of the most companies have become complex. This leads to growth in number and kinds of employees and layer in organization hierarchy. This creates the complex problem of coordination and integration of activities adaptable for and adaptable to the expanding and diversifying situations.
6. Human relations: -The trends in approach towards personnel management has changed from the commodity approach to partnership approach crossing the relation approach. So, today's management of most of the organization has maintained human relations and sound industrial relation all though. So training in human relation is necessary to maintain the industrial peace and deal with human problems.
7. Change in job assignment: - Training is also necessary when the existing employee is promoted to the higher level in the organization and when there is some new job or Occupations due to transfer. Training is also necessary to equip the old employee with the advance disciplines, techniques or technology.

Need of the Training

Training is the most important activity or plays an important role in the development of human resources. To put the right man at the right place with the trained personnel has now become essential in todays globalize market. No organization has a choice on whether or not to develop employees. Therefore training has nowadays become an important and required factor for maintaining and improving interpersonal and intergroup collaboration.

Human resources are the life blood of any organization. Only through well-trained personnel, can an organization achieve its goals.

At a glance, we find that training gives the following results:

1. Growth, Expansion, Modernization cannot take place without trained person.
2. Its increase productivity and profitability, reduces cost and finally enhances skill and knowledge of the employee.

3. Prevents obsolescence.
4. Helps in developing problem solving attitude.
5. Give people Awareness of rules and procedures.
6. Training makes people more people competent.
7. Personnel become committed to their job resulting in proactiveness.
8. People trust each other more.

Finally, our approach to training must be to increase productivity and profitability and secondly to initiate personal growth and development.

The training should be proved more effective if it is need base and it will give the positive impact on both industrial development and human resource development:

A Needs Assessment is a systematic exploration of the way things are and the way they should be. These "things" are usually associated with organizational and/or individual performance.

- Why design and conduct a Needs Assessment? We need to consider the benefits of any Human Resource Development (HRD) intervention before we just go and do it.
- What learning will be accomplished?
- What changes in behavior and performance are expected?
- Will we get them?
- What are the expected economic costs and benefits of?

Concept of Training

According to J.P. Champbell

"Training refuges only to instruction in the technical and mechanical questions specifically stated training to organization procedure by which people learns knowledge and skill for definite purpose".

According to L.L. Steinmetz

"Training is short term process utilizing a systematic and organized procedure by which non managing personnel learn technical knowledge and skill for definite purpose".

The efficiency of an organization depends directly on the capability and talent of its personnel, and how motivated they are. Capability of a person depends on his ability to work and the type of training he receives. While his personal capability is evaluated through proper selection procedure, his training is taken care of by the organization after he has been employed by the organization. Since training inputs may vary from on-the-job experience to off-the-job training, most of the organizations undertake some kind of training for their employees. In Indian organizations, training and development activities have assumed high importance in recent years because of their contributions to the achievement of organizational objectives.

As face with the cutthroat competition, it becomes highly necessary for an organization to be dynamic in plan and action it has to respond effectively and timely to the change in the business climate. This is possible only when the employees in the business climate. This is possible only when the employees in the organization are capable enough to cope with the pressure of changing environment. There is no other alternative them to subject them to various training programs.

Training is a tool to improve effectiveness; it is a plan deliberate learning process in broader sense. Training is thus an avenue for the growth of the organization.

With the recognition of training as an important avenue for development, the scope and application of training has been considerably widened. While largely it still remains a key tool for enhancing job related performance and organizational effectiveness its value is being increasingly acknowledged in engendering behavioral changes and in developing life skill leading to personal growth. Thus, training has influenced every facet of an individual's life. As a result of these developments, human relation and personal development training now occupy significant places in the realm of training.

Training is a process of attitudinal change that integrates with life work and self development. It could be a useful aid in improving the transformation process that takes place in an organization in terms of the processing of inputs to outputs. Training needs have to be related in terms of the organizations demands and that of the individuals. Thus, arises the need of identifying training needs.

Identification of training needs is the first step in designing a training program. This exercise takes into consideration not only the existing situation in the organization, but to some extent the post training scenario as well. The process o f identifying training needs is, therefore complex and intensive as well as time consuming.

Training needs are identified through a systematic process which helps meet the needs of organization and its culture. This process should enable the organization to determine and keep under review its priority for training should enable quick reaction to problems and opportunities includes both formal and informal activities and encourage and assist managers and others to play their own roles in providing a learning environment. Training is an act of increasing the knowledge and skill of an employee for doing a particular job. Training is the systematic acquisition of skills, rules, knowledge, habit and attitudes that have specific applicability to a certain set of situation. Non-managerial persons learn technical knowledge and skills for definite purpose. The verb 'to train' is derived from the old French work,' trainer' meaning, to dray' along. To instruction, practice, exercise *etc.* according to Chinese HRD scholar, Dr. Lee Yeng- "If you wish to plan for a year sow seeds. If you wish for a ten years; plant a trees and if you wish to plan for a life time, develop a man." These sentences itself shows the significance of the training. In the words of 'Decenzo and Robins,' "training is learning experience, in that it seeks a relatively permanent change in an individual that will improve his ability to perform on the job." Basically, training can involve changing of skills, knowledge and attitude or social behaviors. About the concept of training, Harris Daniyal of Oxford University has observed, "training of any kind should have as its objectives the redirection or improvement of behavior so that the performance of the trainee worker becomes more useful and productive for himself, and for the organization of which he is a part. Training practices normally concentrates on improvement of either operative skills (decision making skills, how to arrive at the most satisfactorily to other) or a combination of these."

Thus, the objective of training is to develop specific and useful knowledge, skills and technique. It is intended to prepare workers to carry out pre-determined tasks in well defined job contexts. Training is basically a task oriented activity, aimed at improving performance in current or future jobs. The term 'workers-training' connotes equipping managers with such knowledge, skills and techniques as are relevant to managerial tasks and functions. Development encompasses the whole complex process by which workers as individuals learn, grow, improve the abilities to perform a wide variety of roles within and outside organizations and acquire socially desirable attitudes and values. Workers training and development is aimed at improving one's abilities to perform professional tasks. It involves at learning on the job through experience. Participation in formal training or educational programmes is an integral part of overall workers development.

The need for training of workers on a continuing basis in organized sectors of human activity is no longer a matter of debate. The need has been recognized as an essential activity not only of worker but also of trade union leaders. The primary concern of an organization is its viability, and hence its efficiency. There is continuous environment pressure for efficiency, and if the organization does not respond to this pressure; it may find itself rapidly losing whatever share of market it has. Workers training therefore, imparts specific behavioral skill, technical skills, and knowledge to workers in order that they contribute to the organizations efficiency and be able to cope up with the pressures of changing environment. From the above discussion we come to know the significant place of training programmes in HRD activities. It is relevant here to discuss in detail the significance of training programmes.

Historical Perspective of Training

The history of training can be traced back to the Stone Age, as human beings invented weapons, clothing, shelter, and language and need for training become an essential ingredient in the mach of civilization. What is important is that human beings had the ability to pass on to the others the knowledge and skill gained in mastering these circumstances. Skill in fishing, hunting and self-protections, were passed on to others in that era. With time knowledge and skill passed by artisans, soldiers, and priests were taught to youngsters and a kind of Apprenticeship system was formed whereby an experienced person along knowledge passed to the trainee.

Training grew with the great industrial expansion. As early as 1809, the Masonic Grand Lodge of New York, under the leadership of De Witt Clinton, established vocational training facilities. Manual training began in the United States about 1825.

However, most of the manual training schools that sprang up after 1825 were more disciplinary than vocational schools.

One of the Factory school was established in 1872 at Hoe and Company a manufacturer of printing presses in New York City. Similar Factory schools were established at Westinghouse in 1988, at General electrical company and Baldwin locomotive works in 1901, and at international Harvester Company in 1907. Other companies saw the advantages of having factory schools and they soon became a common practice.

Since the 20th century, training programme has been increasing constantly both in terms of their number and the areas in which training can be offered. Training began being more specialized covering a wide range of areas that may be of significance to the organization.

Our society as whole is also vitally interested in training and educational programme in order to promote employment and utilize the talents of its citizens. One of the earlier Legislative attempts in his regard was the manpower development and Retaining Act of 1962, which was designed to assist in the conversion to new skills of those through out of work by changing job requirements. In 1964, The Economic opportunity Act was directed towards providing training assistance for young entrants in the job market. Neighbourhood youth corps and collage work-study programmes were formed to help younger persons whose unemployment rates are typically double the average. In 1973, because of the confusion generated by literally thousands of publicly financed training programmes, the comprehensive Employment and Training Act was passed. After an experimental programme in the late 1970s, the job Training Partnership Act of 1982 allocated $ 3 billion to private industry councils to implement training for the hard-core unemployment.

Training and Development Learning Theories

There are numerous theories that have implications for training and development; however several are most frequently included in discussions regarding training and development (Noe, 2008). As emphasized by Lewin (1951), it is important to examine these theories not only from a scholarly perspective, but from a practical perspective as well. Training development and learning theories should have clear implications for practice. Additionally, Argyris and Schon (1974) have demonstrated that practitioners who do not have interest or background in academic literature are still likely to be guided by theory-in-use that is ideas about the war training works. The theories or frameworks examined below include action learning, andragogy, critical action research, facilitation theory, individual learning styles, and social constructivism.

Training and development of human resources in industry has been recognized as an important tool for the development of desirable motives effective work habits and methods of work and thereby improves job performance, reduces waste and accidents, prepare individuals for modified jobs, avoids unnecessary turnover, improved the quality of the product and so on. It means no only merely improved production but also equips the, with capabilities for promotion and health relations. New programmes may encourage employers and managers to accept organizational change to gain a better understanding of organizational goals and philosophy and evaluate economic and social consideration. "Training,' 'Education' and 'Development' are three terms frequently used. On the face of it, three might not appear some differences between them, but when a deep thought is given, there appear some differences between them. In all 'training' there is some 'education' there is some training. And the two processes cannot be separated from 'development'. Precise definitions are not possible and can be misleading; but different persons have used these activities in different Training is essentially the instructing of others in information new to them and its application. It may, and often does, involve the teaching of new skills, methods and procedures. Very few people are bom trainers, and most of those who wish to be trainers require training. Even those few who are bom trainers benefit from training, and their effectiveness is enhanced as a result. The most important element in a training situation is the trainer. The trainer who is enthusiastic, energetic and genuinely interested in both the subject and getting his or her message across will evoke the greatest response from the trainees. The trainer who lacks interest in training, who has little or no enthusiasm for the subject of the training and who merely goes through the motions of training is a failure. Such a trainer wastes not only his or her own time but also that of the trainees. The inept trainer is quickly identified by the trainees, who react with inattention, lassitude, undisciplined behaviors and absence from training sessions. Successful training - that which produces the desired result - lies almost entirely in the hands of the trainer. In the trainer's hands lies the heavy responsibility for ensuring that the trainees achieve the maximum possible from the training.

A measure of the success of training is the relationship that develops between trainer and trainees. In a sound, productive training situation there is mutual respect and trust between them, with the trainer taking care to ensure that even the weakest trainee performs to the highest possible level, and the trainees feeling a desire within them to achieve. In this situation the trainer is the motivator and the trainees are the motivated.

Significance of Training

Training has become more of a perk and less of a 'competency- building' instrument. In recent years with increased emphasis on cost cutting, customer service and efficient management of all system, there is a need for more training to achieve these goals, while on the one hand the deterioration of training function has been recognized. On the other hand, the need for good training for all workers is being recognized in the environment, particularly the need to train the workforce and all the workers in addition to the managers has also been recognized. Organizations are likely to get higher return on their investments in training if they move in the direction of skill development programmes.

In Indian industry the training inputs have concentrated largely on managerial personnel. The vast and vital backbone of the organization namely workforce has been ignored many times. Training for them only means technical or trade skill. Attitudinal part of training has been ignored. Actually, it goes hand in hand with; to ensure total employee effectiveness. In fact, attitudinal part of training gives job-satisfaction and it will create a sense of belonging and work-culture. Building of new work culture for today's globalization era will be the most urgent, important and long drawn affair. The challenge of HRD would be to create an environment of resilience; which can accommodate and assimilate successfully, changes in systems structures, technology methods *etc.* people's team and collectivises would have to describe the right meaning to the change process. An organization's success is determined as much by the skill and motivation of its workers as by almost any other factor. While this has always been true, the pace and volume of modern change is focusing attention on ways of training through HRD activities, can be used to ensures organizational members have what it takes to successfully meet their challenges.

Today, every organization is for Total Quality Management (TQM). There is a tremendous race to achieve ISO certification. Terminologies like TQM, productivity improvement, and kaizen have become watchwords of the business and industrial organizations. To face the new challenges of globalization, managers and workers understand the value of training. Training practices would obviously envisage and all round training and development of personality of the workers. In the absence of adequate training and development efforts, there may by frustration, dissatisfaction towards job and apathy among the workforce, thus, training practices at shopfloor level is very important for the purpose of workers involvement. Training activities at shopfloor level should gear up to bring about changes in attitudes of workers and not merely try to improve their skills. Training should be given to all workers in the company. Management must be willing to provide "space" for workers to participate and contribute channels of communication. Free communication removes the hurdles in the grievance handling procedure. It will be effective and quick. To achieve workers' involvement, it is necessary that the management style puts a strong focus on the "human side". An open environment provides opportunity to the workers to contribute and develop a sense of pride. There is a need for a "Nurturing" style to faster team work and respect for every member.

Harris Delmark has stated that, training is a tool for attitudinal change. He says, "key values are critical in helping to manage change but only if people live up to them. To accept any change, the shift in the mindset is absolutely necessary with a view to unfreeze and realize one's own potential and creativity and to bring about a change in their perceptions and attitudes.

The Role of Training Activities through HRD Programmes in Today's Industrial World

No organization has a choice of whether to train its employees or not, the only choice is that of methods. The primary concern of an organization is its viability, and hence its efficiency. There is continuous environmental pressure for efficiency, and if the organization does not respond to this pressure, it may find it self rapidly losing whatever share of market it has. Training imparts skills and Knowledge to employees in order that they contribute to the organization's efficiency and be able to cape up with the pressures of changing environment. The Viability of an Organization depends to a considerable extent on the skills of different employees, especially that of management cadre, to relate the organization with its environment.

Bass has identified three factors which necessitate continuous training in an organization. These factors are technological advances, organizational complexity, and human relation. All these factors are related to each other. The training can play the following roles in an organization.

1. **Increase in efficiency**: Training plays active role in increasing efficiency of employees in an organization. Training increases skills for doing a job in better way. Through an employee can learn many better if he learns how to do the job. This becomes more important especially in the context of changing technology because the old method of working may not be relevant. In such a case, training is required even to maintain minimum level of output.
2. **Increase in Morale of Employees:** Training increases morale of employees. Morale is a mental condition of an individual group which determines the willingness to co-operate. High morale is evidenced by employee enthusiasm, voluntary conformation with regulations, and willingness to co-operate with others to achieve organizational objectives. Training increases employee morale by relating their skills with their job requirements. Possession of skills necessary to perform a job well often tends to meet such human needs as security and ego satisfaction. Trained employees can see the jobs in more meaningful way because they are to relate their skills with jobs.
3. **Reduced Supervision:** Trained employees require has supervision. They require more autonomy and freedom. Such autonomy and freedom can be given if the employees are trained properly to handle their jobs without the help of supervision. With reduced supervision, a manager can increase his span of management. This may result in to lesser number of intermediate levels in the organization which can save much cost to the organization.
4. **Increased Organizational Viability and Flexibility:** Viability relates to survival of the organization during bad days, and flexibility relates to sustain its effectiveness despite the loss of its key personnel and making short-term adjustment with the existing personnel. Such adjustment is possible if the organization has trainee people who can occupy the positions vacated by key personnel. The organization which does not prepare a second line of personnel who can ultimately take the charge of key personnel may not be quite successful in the absence of such key personnel for whatever the reason. In fact, there is no greater organizational asset than trained and motivated personnel, because these people can turn the other assets in to productive whole.

Training is a continuous process and comprehensive system by itself. As stated earlier, of all the factors of production, man has the highest priority and is the most significant factor of production and plays a pivotal role in areas of productivity and quality. If we think deeply over the meaning of term, "manpower" or "human resources," it is obvious that man is an individual human being distinguishable physiologically. He possesses a highly developed mind, a keen and perceptive intellect and exalted soul. The word 'manpower' combines the words 'man' and 'power', which refers to the power-physical, mental, intellectual and spiritual inherent in him.

Progressive organizations know that change due to globalization is inevitable and must be managed effectively and creatively. Thus, many organizations are making efforts to remain flexible in order to accommodate change. New principles of personnel management/development are therefore, turning towards shared vision, building teams, internal customer orientation *etc.* Industrial sector is changing with the dramatic changes in technology, production process and quality of products. This is turn calls for simultaneous changes in the skills, responsibilities, and attitudes of the workers. This necessitates training to workers through HRD programmes.

Survival and success of any industrial organization depends on its people. HRD through training deals with human aspect. Human resource is important in improving the productivity and productivity depends upon the quality the workforce and quality of workforce depends upon the achievement, orientation, motivation, participatory abilities professional skills, organizational commitment, responsiveness to change *etc.* following factors play an important role in upgrading the quality of human resource and bringing about desired productive behavior in the workforce *i.e.*

1. Knowledge
2. Skills
3. Work attitude and behavior
4. Opportunities.

Here training activities through HRD programmes assumes importance.

Improvement in man power productivity depends upon the management and management should look to these above stated factors. Thus, empowering people is the key to success of business, through systematic and planned training activities at shop floor level. Many industrial organizations are spending substantial amount of money and time and efforts on training activities to upgrade the quality of workforce of shop floor. Training activities are playing pivotal role and functions in creating healthy receptive and motivated climate for work. For this purpose on integrated approach is needed at organizational level to improve quality of people and product. This necessitates planning through involvement of people at all levels and strengthening every link chain from board level to shop floor level of organization for its survival and success. HRD is introducing different concepts, tools to achieve its objectives as training and development, counselling, quality circles, Kaizen, Total Quality Management *etc.*

As we know, in changing scenario of liberalization of Indian economy training activities through HRD programmes has to play very significant role. On one hand interdependence among countries is increasing and on the other hand many countries all over the world have to face the competition because of globalization. Rapidly growing global industrialization demanded newer skills, trained employees and it is the knowledge, and training that made it possible for men to acquire the skills in various trades appropriate to the industry. HRD is therefore a field of knowledge that deals with all those aspects of human beings as are concerned with his creative ability. In simple terms the fundamental concern of any HRD effort is to get the best out of the people in any given situation, in any given organization, be it the state, the defence, the public sector organizations, the private sector organizations or any other form of organization. In the past, people and personnel were seen as expenses that must be minimized and controlled. In the context of emerging human resource models workers have to be viewed as assets that require investment.

Here, it is necessary to know about the new environment created by HRD; which calls for

1. Workers participation in progress
2. Job boundaries to go
3. Holistic approach in work
4. Skill boundaries to go
5. Exposure to potentials
6. Flexibility in management operations
7. Better knowledge levels
8 Team work and better adoptability

The role of 'training' through HRD programmes in meeting the above stated factors are examined below:

1) Workers Participation in Progress

This concept calls for not only attitudinal changes of workers, but also of management and unions. Since last four-five decades much has been talked about workers participation in management but very little has happened. In fact, this will become the emphasis of business particularly because there could be no difference causing strife, if industries were to survive strict competition. In fact, it will be to our advantage if this concept is encourage, so that, in case the foreign partners chose to leave at any time, the industrial organizations could survive. In fact, the emphasis on workers contribution to organizations without restriction of roles and departments is encouraged trough the training of Total Quality Management (TQM)

2) Job Boundaries to Go

Need of flexible manufacturing system is realized now by Indian industries. Because, then the process could be modified to match the environmental needs and market pressure. Hence, many Indian industrial organizations have started to redefine the job boundaries, newer organizations talk of skilled workers, and skilled position with reference to machine. There is need to train the workers about it. Such method could be to identify new areas of operation, Secondly to develop the vision of the workers for genuine growth of the organization and of him.

3) Holistic Approach in Work

Here, the realization of each individual worker about his contribution to the organizations success is important. It has been proved that, this is one of the best motivational tools. Here, through training, efforts will be to make workers aware of total organization, its goals the prospects and problems periodically and to ensure a great degree of openness and transparency of operations.

4) Skill Boundaries to Go

It is here that, maximum resistance will be faced. The skill boundary practiced for ages by us cannot go over night, while one can integrate related trades and provide special training to the workers to accept the job in related areas; at least in case of new workers proper job name could be given. If the fight between specializations versus integration of skills, it is now being realized that a person of greater advantages, same organizations have prescribed for promotions to higher grades a need of the worker to have expose of more than one or two skill areas.

5) Exposure of Potentials

This is the most fundamental principle of the 'training'. Here, the efforts of the HR manager will be to – unshackle the minds of the workers and make them to know the total processes and problems and also make them realize and come out with their full potential as over a long period of limited working, they could be thinking of their ability to be limited. Many organizations today are utilizing these tools to coping with globalization.

6) Flexibility in Management Operations

This is essential in order to survive in competition. To achieve this, one rout is to encourage workers through training to come out with suitable suggestions for changes. Again by bringing a lot more transparency and clarifying the intention of the changes, organizations could be flexible.

7) Better Knowledge Level

Most organizations today understand that better knowledge level of workers is an assets and not liability contrary to the belief of the past. A number of Organization base embarked on specifically designed in-house programmes for training of workers not only about the organization, but also to make them more knowledgeable.

8) Team Work and Adaptability

An effective tool to achieve this is a well planned job rotation and training activities in the organization, as compared to older organizations where people continue on some job machines for life time.

Principles of Training

A successful training programme presumes that sufficient care has been taken to discover areas in which it is needed most and to create the necessary environment

Training Process

for its conduct. The selected trainer should be one who clearly understands his job and has professional expertise, has an aptitude and ability for teaching, possesses a pleasing personality and a capacity for leadership, as well-versed in the principles and methods of training, and is able to appreciate the value of training in relation to an enterprise.

No one can dispute the role and importance of training in imparting knowledge and information. It is often said the experience is best method of learning. Training can never substitute experience but certainly shorten the period required for experiencing. Training causes learning, therefore, to make it more effective, the fooling essentials or principles of a good training programme must be noted.

1. **Motivation:** Trainee must be motivated to improve his skill or knowledge by increased remuneration, promotions, recognition, *etc.* His desire to improve his job performance will enable him to grasp more quickly new skill and Knowledge. Regardless of the quality of instruction, an in different student will learn little, because human behavior is goal- oriented, a student must be motivated in order to learn. Relating results to pay promotability and responsibility can motivate the employees to learn.
2. **Progress Report:** Research has demonstrated that people learn faster when they are informed of their progress by means of examinations or through

the instruction's comments providing the trainee with progress reports facilitates the learning process.

3. **Active Involvement:** Learning is most efficient when the learner is involved in the process rather than just listening is more complete if he actually performs the task to be learned in the subject matter presented to him in form of theories and concepts through case studies, laboratory, experiments, classroom discussion and role playing.
4. **Instructions in part:** Rather than presenting the whole training program at one time, it is better to break instructions down in to parts, thus crating a series of sub goals for the learner.
5. **Reinforcement:** Training of employee's results must be supported by means of rewards and punishments. Successful employees must automatically obtain pay increases, Promotions or recognition, *etc.*, on completion of training employee should be increase in pay or status otherwise, he will lose faith in training programs.
6. **Individual Differences:** Training of employees is usually given in groups, but it must provide for differences in individual intelligence, abilities and aptitude. It should be noted that there are individual differences, some take more time to understand concepts and principles, where as others leam faster. Again, some are equipped with more knowledge and information as compared to others. Therefore, groups or batches should be made taking in to consideration their level of understanding and skills. Highly qualified, Experienced and matured persons must provide training with a balanced personality. Expert trainers command lot of respect and attention from the learners.
7. **Ideal Place**: There must be an ideal place of training however; the place of training depends up on a number of factors. If it is on the job training then it is to within the factory premises. Off the job training should be preferably be provided outside the company premises.
8. **Training Period:** It should be neither too long nor too short period; trainee does not get a chance to learn much.
9. **Training system:** Training need and objective must be clearly defined.
   - ✰ There should be a proper balance between theory and practical.
   - ✰ Training Material should be simple and meaningful.
   - ✰ Use of films, audio-visuals, makes training more interesting.

25

Methods of Training

The most widely used methods of training used by organizations are classified into two categories: On-the-Job Training and Off-the-Job Training.

1. **On-the-Job Training** is given at the work place by superior in relatively short period of time. This type of training is cheaper and less time-consuming. This training can be impacted by basically four methods: -
2. **Coaching** is learning by doing. In this, the superior guides his sub-ordinates and gives him/her job instructions. The superior points out the mistakes and gives suggestions for improvement.
3. **Job Rotation:** In this method, the trainees move from one job to another, so that he/she should be able to perform all types of jobs. *e.g.* In banking industry, employees are trained for both back-end and front-end jobs. In case of emergency, (absenteeism or resignation), any employee would be able to perform any type of job.
4. **Off the Job Training:** is given outside the actual work place.
   a) **Lectures/Conferences:** This approach is well adapted to convey specific information, rules, procedures or methods. This method is useful, where the information is to be shared among a large number of trainees. The cost per trainee is low in this method.
   b) **Films:** can provide information and explicitly demonstrate skills that are not easily presented by other techniques. Motion pictures are often used in conjunction with Conference, discussions to clarify and amplify those points that the film emphasized.
   c) **Simulation Exercise:** Any training activity that explicitly places the trainee in an artificial environment that closely mirrors actual working conditions can be considered a Simulation. Simulation activities

include case experiences, experiential exercises, vestibule training, management games and role-play.

d) **Cases:** present an in depth description of a particular problem an employee might encounter on the job. The employee attempts to find and analyze the problem, evaluate alternative courses of action and decide what course of action would be most satisfactory.

e) **Experiential Exercises:** are usually short, structured learning experiences where individuals learn by doing. For instance, rather than talking about inter-personal conflicts and how to deal with them, an experiential exercise could be used to create a conflict situation where employees have to experience a conflict personally and work out its solutions.

f) **Vestibule Training**: Employees learn their jobs on the equipment they will be using, but the training is conducted away from the actual work floor. While expensive, Vestibule training allows employees to get a full feel for doing task without real world pressures. Additionally, it minimizes the problem of transferring learning to the job.

g) **Role Play:** It's just like acting out a given role as in a stage play. In this method of training, the trainees are required to enact defined roles on the basis of oral or written description of a particular situation.

h) **Management Games:** The game is devised on a model of a business situation. The trainees are divided into groups who represent the management of competing companies. They make decisions just like these are made in real-life situations. Decisions made by the groups are evaluated and the likely implications of the decisions are fed back to the groups. The game goes on in several rounds to take the time dimension into account.

i) **In-Basket Exercise**: Also known as In-tray method of training. The trainee is presented with a pack of papers and files in a tray containing administrative problems and is asked to take decisions on these problems and is asked to take decisions on these within a stipulated time. The decisions taken by the trainees are compared with one another. The trainees are provided feedback on their performance.

Factors for Successful Training Methods

No doubt Training is a very powerful tool for the smooth functioning of the organization, but it needs to be used with care in order to derive all the benefits. Here are **seven** recommendations for getting the best out of this tool:

1. Learn about the needs and proficiency of each and every employee before an organization invests its effort, time and money on training. Its better to identify the needs and shortcomings in an employee before actually imparting training to him/her.

2. Experienced and skilled trainer, who possesses good amount of knowledge and understanding about the organization's objectives, individual abilities and the present environment, should give training.
3. Active participation from the trainees should be encouraged. There should be a two-way communication between the trainer and trainee.
4. Feedback should be taken from the trainees after the training is over, so that the organization comes to know about the deficiencies in the training program and also suggestions to improve upon the same.
5. Focus of training should be on priority development needs and to produce strong motivation to bring change in employees.
6. The cost incurred on the training program should not exceed its benefits.
7. The method or type of training should be very cautiously selected by the organization depending upon the organizations' resources and an employee's individual need for training.

Thus, training is a vital tool to cope up with the changing needs and technologies, and ever-changing environment. It benefits both the organization as well as the employees.

To inform participants of the methods of training available to them, with particular attention given to the lecture, the lecture/discussion, the skill lesson and the on-the-job session

Suggested Methods of Instruction

- ☆ Lecture
- ☆ Discussion
- ☆ Demonstration
- ☆ Exercise

Aids

- ☆ Overhead transparencies
- ☆ Demonstrations
- ☆ Handouts

Time Frame

- ☆ One hour lecture/discussion
- ☆ One hour of five-minute mini-lectures

Content

- ☆ The different methods of training
- ☆ Selecting the right method
- ☆ The lecture

- The lecture/discussion
- The skill lesson
- On-the-job training (the four-step method of instruction)

Approach

This module lends itself to a lively presentation by the trainer. The trainer must be capable of demonstrating personally the methods of training selected for special attention. These methods are believed to be the most appropriate for use in training in food control practices including GMPs and HACCP. It is acknowledged that case studies also have their use, but considerable time is required in their preparation.

The trainer should spare no effort to make this module effective. The methods are the tools the trainees will use when they became trainers. It is essential that the presentation of the module provide them with a base for effective training, on which the trainees can build by practising to improve performance.

Exercise

Ask participants to give a five- to seven-minute mini-lecture on a subject of their own choice that is related to food quality control. Instruct the participants to prepare a point outline on the subject of their lecture for use during their presentation.

The Different Methods of Training

You have a choice of the following methods to prepare for effective training:

- Lecture
- Lecture/discussion
- Skill lesson
- On-the-job training (the four-step method)

There are other methods of training, but their effective use is specific to special training situations and will not be discussed in this lecture. Some of those methods include:

- Role play
- Assignment
- Case study
- Training games
- Group exercises
- Programmed learning

Selecting the Right Method

All the resources at your command must be used to make your instruction real and vital for your trainees. The number and types of training methods you use during any presentation depend on many factors, and you must therefore have answers to the following questions before you decide how you will present your material.

- ✰ What is the ability and level of knowledge of the group?
- ✰ How many trainees are in the group and why are they there?
- ✰ How much time do you have to prepare your material?
- ✰ Can you cover your topic fully in the time available?
- ✰ What aids do you require?
- ✰ Do you have the experience to use these aids with confidence?
- ✰ Are you aware of the limitations of aids?

Your method of presentation will depend on the answers to these questions.

The Lecture

Use

- ✰ When the group is large - say 30 or more
- ✰ When knowledge or understanding is to be imparted by an expert
- ✰ When a body of factual information has to be communicated in a short time
- ✰ When information is not readily available to group members

Delivery

Essentials of good delivery:

- ✰ Words must all be clear
- ✰ Words must be spoken at a suitable pace
- ✰ Pauses should occur at logical places
- ✰ Variety should be used: emphasizing important points in a deliberate manner, connecting parts and using illustrations in a conversational way

Preparation and Lecture Notes

Preparation is important. The lecturer's notes need to be designed to facilitate efficient delivery. Distinction is needed between lecture outlines (showing matter only) and lecture notes (showing method and matter).

Notes may be too brief. The lecturer may then improvise, and he or she may be vague or may forget important elements. On the other hand, notes may be too extensive. The lecturer will then read them, and this is undesirable.

Given an outline of the material, prepare the notes by asking these questions:

- ✰ What is it safe to assume that the listeners know?
- ✰ What are they likely to find difficult?
- ✰ Hence, what will require special care or illustration?

- What will the illustrations (in detail) be? Can they be misunderstood or misinterpreted?
- What demonstrations will be appropriate? Will everyone see clearly? (Demonstrations are used to illustrate really important points. The more important the point, the more spectacular the demonstration should be.)
- What new terms will be introduced? What unusual names? Mark these in the notes. They will need to be written on a blackboard, whiteboard, chart or overhead transparency.
- What precisely should everyone know at the end of the lecture? (This is really a re-examination of the outline and a restatement of the important points.)

Structure

Introduction

- Statement of aims
- Relation of this lecture to those that came before and are to follow
- Establishment of goal (which gives purpose and direction) by linking aims with participant needs
- Outline of thoughts that are to be developed

Body of Lecture

- Step-by-step building up of subject matter
- Logical development
- A few well-developed steps, strongly made (more effective than many steps)
- Appropriate use of aids and questions to stimulate student interest and activity
- Appropriately spaced summaries of material covered

Conclusion

- Summary of lecture material
- Restatement of the relationship of this lecture to others in the series
- Reference to additional material that should be read or seen
- Setting of any assignments

Disadvantages

- Lecturer bombards students with considerable information (saturation may occur)
- Participants sit passively without interaction

The Lecture/Discussion

Use

- ☆ When the group is small - say 20 or less
- ☆ When the members know one another well enough to risk making errors
- ☆ When the material is of a kind that can be assimilated readily, at least in part, or when there is some prior knowledge of it

Lecture

Refer to preceding section.

Discussion

The most useful starting point for the discussion is the question. Some uses of questions:

- ☆ At beginning of lecture: to find out what trainees already know and to discover opinions
- ☆ During lecture: to find out whether the participants understand and are following the lecture
- ☆ End of lecture: to recapitulate and test the participants' knowledge and understanding

Desirable Features of Questions

- ☆ They should be clear
- ☆ They should be brief
- ☆ They should lead to some constructive statement rather than to a nod or a grunt
- ☆ They should stimulate thinking, rather than suggest the answer

Pitfalls

- ☆ Repeating the answer (Do not repeat. Move on.)
- ☆ Holding a dialogue with a single answerer (Bring in the group, *e.g.* "Would anyone like to add to that?")
- ☆ Trampling the incorrect answerer
- ☆ Asking too many questions (Adults do not like to be cross-examined.)
- ☆ Letting the discussion take too long (Guide it carefully. Remember the objective of your discussion.)

Structure

- ☆ Introduction
- ☆ Body of lecture

- ☆ Discussion
- ☆ Conclusion

The Skill Lesson

Aims

- ☆ To teach correct and safe job methods
- ☆ To develop confidence in job performance
- ☆ To achieve accuracy and speed
- ☆ To encourage conscientious effort

Structure

Introduction

- ☆ Development (body of skill lesson)
- ☆ Demonstration by trainer (complete)
- ☆ Demonstration and trainee practice of each stage, in sequence
- ☆ Practice of demonstrated job skill Conclusion

On-The-Job Training (The Four-Step Method of Instruction)

Step 1

- ☆ Prepare the worker
- ☆ Put the worker at ease
- ☆ State the job and find out what the worker already knows about it
- ☆ Stimulate the worker's interest in learning the job
- ☆ Place the worker in the correct position

Step 2

- ☆ Present the operations
- ☆ Tell, show and illustrate one important point at a time
- ☆ Stress each key point
- ☆ Instruct clearly, completely and patiently, but teach no more than the worker can master

Step 3

- ☆ Try out the worker's performance
- ☆ Have the worker do the job, and correct errors
- ☆ Have the worker explain each key point to you as he or she does the job again

- ☆ Make sure the worker understands, and continue until you are certain of this

Step 4

- ☆ Follow up
- ☆ Put the worker on his or her own
- ☆ Designate to whom he or she should go for help
- ☆ Check frequently
- ☆ Encourage questions
- ☆ Taper off extra coaching and reduce follow-up

Example of an on-the-job training session: training workers in the correct method of hand washing

Workers in fish processing units must maintain a high degree of personal cleanliness. In order to educate the workers in better hygienic practices, the correct hand washing method is one of the topics demonstrated in fish processing units.

The main objective of washing hands is to avoid contaminating the material with organisms from the hands. Unwashed hands transmit microorganisms. It is therefore essential that hands be washed thoroughly. The following procedure for washing hands is recommended:

- ☆ Wet palms and arms, from the elbow down, with fresh water
- ☆ Apply soap
- ☆ Work lather on and around fingers, nails and arms from the elbow down
- ☆ Rinse palms and hands with fresh water
- ☆ Wipe palms and hands dry using a clean towel

How to make Questions

To provide guidance to the trainee-trainers on how to ask questions, and to make them aware of the do and don'ts of questioning

Suggested Methods of Instruction

- ☆ Lecture/discussion
- ☆ Discussion
- ☆ Handouts

Aids

- ☆ Overhead transparencies
- ☆ Handouts

Time Frame

- ☆ One hour presentation

Content

- ☆ Importance of questioning
- ☆ Types of questions
- ☆ Purpose of questions
- ☆ How to ask questions
- ☆ Preparation of questions
- ☆ Dos and don'ts of questioning
- ☆ Questions asked by trainees

Approach

This module is of great importance, as skilful questioning is essential to a trainer's effectiveness.

Learning Outcome

The participants should have the knowledge and ability to utilize questioning to support effective training.

Importance of Questioning

To be effective, trainers must be skilled questioners. Carefully devised questions, skilfully asked, are the basis of the lecture/discussion method of training, and questions should also feature prominently in other methods of training. Few people question well, and to do so requires careful preparation and practice.

- ☆ Questioning is one of the essential skills for any good trainer.
- ☆ Unless you question properly you cannot hope to know how much (if any) of your message is getting across.

Types of Questions

Rhetorical

A rhetorical question is a question to which no answer is expected. Examples:

- ☆ Now that is simple enough, isn't it?
- ☆ What could be clearer?
- ☆ Anybody could understand that, don't you agree? Do not overuse this type of question.

Direct

A direct question to a named person can be a useful management device in a class situation.

Example

- ☆ Prakash, what detergent would you use for washing fish crates?

Do not overuse direct questions.

Overhead

An overhead question is asked to the whole group, and then a person named to answer.

Example

- ☆ What detergent is used for washing fish crates? Prakash, do you know?

Leading

A leading question suggests the answer.

Example

- ☆ If chlorine kills microorganisms in water, what is it likely to do to them elsewhere?

Leading questions are of limited use.

Purpose of Questions

Questions are used for all sorts of purposes in training. Some of the more common purposes are:

- ☆ Getting trainees to participate
- ☆ Checking on a trainee's understanding
- ☆ Attracting a trainee's attention
- ☆ Testing a trainee's knowledge of the subject
- ☆ Breaking the ice and initiating a discussion
- ☆ Stimulating confidence in shy trainees
- ☆ Reviewing earlier work
- ☆ Changing the topic

How to Ask Questions

- ☆ Ask the question in a friendly and natural way to the group. Pause, then name one individual to answer.
- ☆ Vary tempo with pauses.
- ☆ Spread questions throughout the group at random.

Preparation of Questions

- ☆ Prepare questions before the lesson, but use them flexibly.
- ☆ Introduce questions with such words as: what, when, explain, compare, how, why, outline, contrast, define, trace, describe, illustrate.

An Effective Question

- ☆ Is simple and direct
- ☆ Is clear and well expressed in a complete sentence
- ☆ Contains one main thought
- ☆ Has only one correct answer;
- ☆ Requires more than a "yes" or "no" answer

Dos and Don'ts of Questioning

Do

- ☆ State questions clearly, concisely and audibly
- ☆ Ask in a friendly and natural way
- ☆ Use questions carefully and time them appropriately - to create interest, lift attention and evaluate
- ☆ Involve the whole group
- ☆ Include one main thought in each question
- ☆ Know the answer

Don't

- ☆ Interrogate people
- ☆ Embarrass people
- ☆ Trick people
- ☆ Get sidetracked by answers
- ☆ Ask questions with more than one correct answer
- ☆ Answer your own questions
- ☆ Ask more than one question at a time
- ☆ Ask questions with a "yes" or "no" answer

Questions Asked by Trainees

Genuine

Answer if you can. Do not bluff. If you cannot answer, say so, but indicate that you will try to find out the answer.

Ulterior

A trainee may be trying to embarrass the trainer or someone in the group. The options are:

- ☆ Ignore the question
- ☆ Reply negatively: "Let's leave that. I don't think it's entirely relevant."
- ☆ Relay the question "Would somebody in the group like to answer that?"
- ☆ Reverse to questioner: "What do you think the answer might be?"

26

The Trainer's Role and Responsibility

To introduce the participants to the basic principles of training in the simplest possible way and to establish fully the responsibility of the trainer.

Suggested Method of Instruction

- ☆ Lecture/discussion
- ☆ An exercise in identifying the role of each of the senses in learning
- ☆ An exercise in planning a skill training session

Aids

- ☆ Overhead transparencies
- ☆ Handouts
- ☆ Time frame
- ☆ One hour lecture/discussion

Content

- ☆ The process of learning
- ☆ Factors that hinder learning
- ☆ Obtaining and holding the learners' attention
- ☆ Facilitating understanding
- ☆ Steps in skill training

Approach

This module is important because it is essentially the introduction to the training modules that follow.

Discussion should play a major part in the presentation. Because of their life-experiences the trainees will be familiar with learning, even though they may never have analysed the process. Therefore the major task of the trainer is to plan a sequence of questions that will lead the trainees to an identification of the elements and steps in the learning process and the factors that hamper learning. Trainees should be encouraged to recall the good trainers and teachers they have known and to identify the skills that made their training and teaching memorable.

The material in the lecture and overhead transparencies is in point form and requires explanation by the trainer.

Exercises

- ☆ Ask the participants to give a demonstration of a skill related to an industry or food control operation with which they are associated. Have the participants set down in logical order the steps in which they would present the demonstration.
- ☆ List ways in which each sense may be used by the trainer to make a sufficient impression to get the message across. For example:

Sense	*Use Of The Sense*	*Applications*
Sight	Written word *etc.*	*Textbooks* *Notes* *Handouts* *Blackboard summaries etc.*
Hearing		
Touch		
Smell		
Taste		

Learning Outcome

The participants should be aware of and understand the trainer's role and responsibilities.

The Process of Learning

The successful trainer possesses insight into the process of learning. The learning process conforms to the following pattern: external sensations stimulate the sense organs - ears, eyes, body (touch), nose and tongue - and the nervous system conveys impressions to the relevant sections of the brain. The brain then transmits impulses to the muscles and organs of movement and speech, and the end result is a reaction.

Creating an Impression

Receiving an impression is the first step in learning. Therefore, the trainer must ensure that the trainee receives strong impressions. The strength of the impression will depend on:

- ☆ The number of senses involved
- ☆ The vividness of the impression
- ☆ Whether the impression registers

Observing the Learners

The only way the trainer can know if people have learned the material is by observing their behaviour:

- ☆ Their actions
- ☆ Their written impressions
- ☆ Their speech

Factors that Hinder Learning

- ☆ The learning plateau: at intervals the rate of learning flattens out as the brain rests
- ☆ Saturation: if the message is overloaded the receiver rejects the excess and learning stops
- ☆ Fatigue: a tired receiver is not as receptive as an alert one
- ☆ Inability to concentrate: the longer the message, the more concentration decreases from beginning to end

Obtaining and Holding the Learners' Attention

Before people can learn any material they must focus their voluntary attention on it. The desire to learn comes from within; it is spontaneous.

The good trainer tries to gain and maintain voluntary attention in every session he or she presents.

- ☆ Relate what you aim to teach to those subjects in which you know the trainees are interested.
- ☆ Introduce the session in such a way that the trainees will not only see and become interested in this relationship, but will want to learn more about it.
- ☆ Begin with a good story to which the trainees can relate. An effective trainer makes it his or her business to know the background of the trainees.
- ☆ Having done these things, maintain the trainees' attention by doing all that is possible to facilitate their understanding and absorption of the material.

- Ensure that the trainee's learning is an active process in which the trainer and trainees are equal partners in terms of participation.

Facilitating Understanding

To facilitate understanding, the trainer proceeds from:

- Known to unknown
- Simple to complex
- Whole to part and back to whole
- Concrete to abstract
- Particular to general
- Observations to reasoning
- Point to point in logical order

To facilitate absorption, remember that trainees learn only by impressions received through their senses.

Steps in Skill Training

Having learned a skill, trainees must reinforce its acquisition by using it. Learning by doing is the basic principle underlying the acquisition of any skill.

When teaching skills, the trainer most often achieves the best results by keeping the talk short and by working through a set sequence of discrete steps, as follows:

- Show the trainees the actual skill they are to acquire.
- Demonstrate and explain, step by step, the operations involved (this requires an analysis of the total procedure by the trainer).
- Have trainees imitate the necessary actions.
- Have trainees practise performing the operations.
- Devote at least 50 per cent of the session to trainee practice time.

Prioritising Training

Given the vast range of skills and other competencies which can be developed in people it is useful for some sort of prioritising to take place so that training focuses on the areas which will yield best benefit, in other words, return on investment (typically in terms of organizational performance, although the needs of teams and individuals can also be very significant in prioritising training and development, depending on the situation.)

In addition to the skill-sets and training needs analysis tools on this website, here are three other examples of methods for prioritising training:

Essential/Desirable - simply and quickly define each activity (skill, competency, whatever) according to whether it is essential or desirable for the job purpose and organizational performance. Training priority is obviously given to developing essential competencies.

Importance/Competency matrix - the highest training priorities are obviously the activities (skills, competencies, whatever) which are high importance (of task to organizational performance) and low competence (of trainee skill level).

Training Aids

Suggested Methods of Instruction

- ☆ Lecture
- ☆ Demonstrations
- ☆ Exercises
- ☆ Handouts

Time Frame

- ☆ One hour presentation
- ☆ One hour making and showing aids

Content

- ☆ Why use training aids?
- ☆ Classification of instructional aids
- ☆ Selection of aids
- ☆ Principles to follow in adopting a visual approach
- ☆ Charts and diagrams
- ☆ Handouts
- ☆ Overhead transparencies
- ☆ The computer pallet
- ☆ Colour slides
- ☆ Videos

Exercises

You are to give a training session to workers in a processing plant about one of the following:

- ☆ Plant hygiene
- ☆ Vermin control
- ☆ A processing technique
- ☆ Moisture control
- ☆ The life cycle of an insect
- ☆ Factors adversely affecting quality
- ☆ The importance of food control
- ☆ The application of HACCP

Prepare a chart that may be used in the session you elect to give, or alternatively, in your presentation during the conclusion of this course.

- Prepare a single-page handout on a topic of your own selection (perhaps for use as part of your presentation during the conclusion of the course) and for an audience you designate, *e.g.* quality control officers, factory workers, management, export inspectors.
- Prepare an overhead transparency on a topic of your choice (perhaps for use as part of your presentation during the conclusion of the course), keeping in mind the rules of overhead transparency preparation.

Why Use Training Aids?

All learning is through the senses. The more senses are brought into use, the more effective is the learning; 97 per cent of learning is achieved through simultaneous appeal to the eye and ear. It is because of this that we should make use of audiovisual aids in training.

Effective use of audiovisual aids can be included in any sort of presentation. Charts, slides, videos, overhead transparencies and films can be used to add interest as well as supplement verbal explanations.

Proper use of instructional aids saves time, adds interest, helps trainees learn and makes your job easier. But remember that aids to training are aids only. They are not substitutes for training.

Trainers should use training aids to supplement their training rather than to replace all or part of it.

Classification of Instructional Aids

Projective

- Motion pictures
- Videos
- Colour slides
- Overhead projector transparencies
- Computer pallet

Non-projective

- Chalkboard
- Whiteboard
- Charts and diagrams
- Models
- Exhibits
- Handouts
- Tape recorder

Selection of Aids

In selecting aids, take into account the following:

- ☆ Practicability
- ☆ Attractiveness and interest; vividness
- ☆ Suitability
- ☆ Complexity
- ☆ Clarity
- ☆ Portability
- ☆ Serviceability
- ☆ Availability
- ☆ Location
- ☆ Preparation and presentation
- ☆ Time factor

Principles to Follow in Adopting a Visual Approach

- ☆ Anything that can be quantified or is factual can be presented visually
- ☆ Obtain and select the necessary data; confusing data and confusing information will result in confusing visuals
- ☆ Know clearly what you want to say in your visuals; write it down
- ☆ Plan your visuals; know what you want to include (Sketch an outline of ideas you think will work.)
- ☆ Try the visuals out on others before you use them

Tips to ensure the trainees do not go to sleep during presentation of your visuals

- ☆ Make your visuals visible
- ☆ Ensure that the trainees can see them
- ☆ Use colour for headings
- ☆ Take care with drawings; they can be misinterpreted
- ☆ Make them simple; eliminate details
- ☆ Ensure the key feature occupies a prominent part of the screen or display
- ☆ Minimize reflection
- ☆ Show all the key points (oral presentation includes everything necessary to sell the key points through the ears; visual presentation includes everything necessary to sell the key points through the eyes)

Preparation

- ☆ Plan carefully the use of instructional aids

- ☆ Make sure that the aids can be seen clearly from all areas of the room
- ☆ If you write, write clearly
- ☆ Use colour for emphasis

Charts and Diagrams

These fall in two main categories:

Bold and simple

These are for use during a training session. They should:

- ☆ Be large enough to be seen by all
- ☆ Not necessarily be self-explanatory
- ☆ Be functionally coloured
- ☆ Include only the essentials

Detailed

These are for close study at leisure. They should:

- ☆ Be more or less self-explanatory
- ☆ Be of medium or small size
- ☆ Be suitable for semi-permanent display
- ☆ Be artistically produced

Handouts

Handouts are specially prepared sheets and notes. They are used:

- ☆ For reference purposes during the session or course
- ☆ To substitute for note taking
- ☆ To retain as a permanent record for reference after the course

A Handout can

- ☆ Introduce a topic
- ☆ Provide revision
- ☆ Provoke discussion

Handouts Should

- ☆ Be brief and sharp/containing only essential details
- ☆ Be accurate and complete
- ☆ Be designed clearly and attractively, with good use of white space
- ☆ Include diagrams if appropriate
- ☆ Always have a title

- ✰ Be planned
- ✰ Be of a standard size
- ✰ Be presented in a logical sequence
- ✰ Be pitched at a level appropriate to the audience

Why use Handouts?

- ✰ They carry the stamp of authority
- ✰ They provide a record of important information
- ✰ They supply data to reflect the presentation
- ✰ They can provide background documentation (longer and more comprehensive)
- ✰ They can be studied at the reader's own pace
- ✰ They convey with certainty the same data to a number of people
- ✰ They appeal to the sense of sight

When should Handouts be Distributed?

- ✰ Before the presentation
- ✰ During the presentation
- ✰ At the end

Overhead Transparencies

The overhead projector is one of the most useful training aids. It can replace the need for chalkboards, whiteboards and charts. The overhead projector can be used for presentation to a group of any size.

All material for use on an overhead projector needs to be reproduced on to transparencies using either special pens or printers with either non-permanent or permanent ink (the latter if the trainer wants to keep and reuse the transparencies). It is also possible to make either black and white or colour transparencies using a specially designed photocopier. Computer-generated transparencies can be excellent.

Design of Overhead Transparencies

- ✰ Keep them simple
- ✰ Include only essentials
- ✰ Make sure lettering is of sufficient height (>5 mm)
- ✰ Use colour on colourless film or contrasting colours on coloured film
- ✰ Do not clutter (no more than seven principle points to a transparency)
- ✰ Illustrations can be useful

Using the Overhead Projector

- ☆ Make sure the projector is placed so that all can see
- ☆ Focus correctly
- ☆ Use masking technique: cover part of the transparency so only the material you are discussing is shown

The overhead projector is probably the most flexible of the aids available to the trainer. Used correctly, it will enhance trainee learning by making presentations more interesting and explanations clearer.

The Computer Pallet

The computer pallet is a device that replaces the computer screen. It is placed on top of an overhead projector, allowing the instructor to project material that has been prepared and stored on a computer disk.

The same basic principles that apply to the design of overhead transparencies also apply to the preparation of material on a computer for use on a computer pallet. The benefits of using a computer pallet include flexibility and the ability to amend material easily. Particular computer programs, if available to the instructor, can provide a large selection of graphic materials and presentation packages.

At present this technology is not widely available. An instructor who wishes to utilize a computer pallet should be trained and familiar with its use.

Colour Slides

Main Features

- ☆ Slides are relatively inexpensive to procure
- ☆ They are easily used
- ☆ They facilitate study of a topic one step at a time
- ☆ All trainees get the same clear view
- ☆ Each frame can be studied and discussed at leisure during the screening
- ☆ They can be used in conjunction with a tape-recorder (tape/slide sequence)

How to Use Slides Effectively

- ☆ Do not treat as entertainment
- ☆ Select slides that are relevant
- ☆ Plan your presentation
- ☆ Include an introduction and conclusion
- ☆ Do not prolong the presentation
- ☆ Ensure the equipment is sound and well set up before the presentation

Videos

- ☆ Make sure videos are directly related to the subject; do not use them merely for entertainment or to give yourself a rest
- ☆ Make sure all trainees can see the monitor
- ☆ The video should be introduced; trainees should be told what it is about and what they should look for
- ☆ Review the video in a discussion after screening

Recommendations for an Effective Presentation

Always

- ☆ Allow ample time for preparation: sufficient time to plan and construct and sufficient time to rehearse
- ☆ Make a file copy of your visuals
- ☆ Check on your worst seats, those on the extreme right and left
- ☆ Mount screen high enough for all to see
- ☆ Remove competing attractions; competition will reduce impact of your visuals
- ☆ Check all arrangements before you go on, even if it means going without your breakfast, lunch or dinner; make sure you have done everything possible for a smooth presentation
- ☆ Maintain constant contact with your audience; know your visuals well enough that you do not have to break your commentary to check points
- ☆ Time your visuals to coincide with your comments; mistiming is distracting
- ☆ Make your presentation straightforward; be sincere and win the confidence of your audience
- ☆ Keep your visuals moving; parallel the flow of your words with the flow of visuals
- ☆ Use only the required number of words; avoid excessive wordage
- ☆ Use only well-trained assistants who know the visuals as well as you do
- ☆ Keep your visuals; they may be needed again

Designing and Delivering a Presentation

To guide and advise participants in the skill of planning a presentation and delivering it.

Suggested Method of Instruction

- ☆ Lecture/discussion

Aids

- ☆ Overhead transparencies
- ☆ Handouts

Time Frame

- ☆ One hour presentation

Content

- ☆ Importance of planning
- ☆ Steps in planning a presentation
- ☆ Delivering a presentation

Learning Outcome

Participants should be aware of the importance and methods of effective presentation planning and delivery.

Importance of Planning

Every presentation in a training programme should be planned. The trainers who do not plan their presentations are not doing their job properly Unless the trainer is particularly gifted, it is most unlikely that the presentation will be successful and effective if it is unplanned. All effective trainers plan their presentations. They know precisely how the presentation will flow before they begin. The trainer who does not plan presentations is attracting trouble. Trainees are quick to sense a lack of planning, and their response will reflect their disdain for the trainer.

- ☆ The most important part of the presentation is preparation
- ☆ The first step in preparation is to have a plan
- ☆ Take a sheet of paper and commence planning

Steps in Planning a Presentation

Determine what the trainees need to learn

Examples

- ☆ How to achieve a high level of personnel and plant hygiene
- ☆ How to peel a prawn correctly
- ☆ How to use a thermometer
- ☆ How to disinfest cashews correctly
- ☆ How to apply HACCP
- ☆ How to establish export/import food control

State Objective

Select the objective to be reached in satisfying the trainees' needs. Select it on the basis of what you expect the trainees will know or be able to do at the end of the training session. Consider:

- ☆ Time available in which to conduct the session
- ☆ Facilities available
- ☆ Prior knowledge and abilities of the trainees
- ☆ Relationship to the needs of the trainees
- ☆ Are you concerned with knowledge, skills or attitudes?

Choose an Appropriate Training Method

- ☆ Appropriate to objective: is the objective to solve problems? to change attitudes? To teach skills?
- ☆ Appropriate to the trainee
- ☆ Appropriate to you, the trainer
- ☆ Appropriate administratively

Organize your Material

- ☆ Decide on what the trainees must know
- ☆ Decide on what the trainees should know
- ☆ Decide on what the trainees could know
- ☆ Select key ideas
- ☆ Arrange (sequence) key ideas in logical order, *e.g.* known to unknown, simple to complex, concrete to abstract

Write the Session Plan

- ☆ Needs
- ☆ Objectives
- ☆ Method of training
- ☆ Organization of material: introduction, body, conclusion (keeping in mind time, questions, subheadings, details)
- ☆ Aids to be used
- ☆ Questions to be asked

Prepare and Check Aids

- ☆ Overhead transparencies, chalkboard plan, prepared charts, film, slides, tapes
- ☆ Check readability sequence, volume levels, suitability
- ☆ Handouts: notes, supplements

Presentation Checks

- ☆ Check session plan, particularly timing and trainee activities
- ☆ Check equipment, aids, ventilation, seating and lighting

Evaluation

- ☆ Obtain feedback on students' learning during the session; use test questions
- ☆ Self-evaluation (How can I improve the session?)
- ☆ Write down on the session plan any comments for future reference before you forget them

Presentation Planning Checklist

Use the above outline as a checklist to be sure you follow all the necessary steps.

Use a plan - you will be more confident and training will be more effective.

Delivering a Presentation

Remember that the trainer's job is to help the trainee learn and remember. This is hard work. The trainer must be fully committed to the job. Trainees quickly sense whether a trainer is positive about what he or she is doing or is simply going through the motions. The trainer must motivate the trainees and spark their desire to learn new ideas and skills.

How is this done through delivery? The personal variations in manner and style employed by trainers are important in that they make instruction animated and interesting. Good delivery requires practice!

Influence Attitudes through your Manner

- ☆ Project enthusiasm regarding the training - through use of voice (modulation), body language (gesticulation) and eye contact
- ☆ Be confident and positive (this comes from sound session planning and preparation and from sound knowledge of what you are doing)
- ☆ Be firm
- ☆ Be energetic and lively - do not stay glued to one spot, move around the trainees (but don't overdo it)
- ☆ Introduce humour, if you can do so effectively (forced humour may be counterproductive, and too much humour may be damaging)
- ☆ Be pleasant, relaxed, and sympathetic to trainees; do not talk down to them
- ☆ Involve trainees in the training process, taking them into your confidence

Desirable Qualities of Speech

- ☆ Clear

- Pleasant
- Fluent
- Varied
- Coordinated with gesture

Set an Example

- Convey a sense of the importance of the subject
- Take the subject seriously

Avoid

- Distracting mannerisms
- Not giving your best at every presentation

Training Evaluation

Suggested Method of Instruction

- Lecture/discussion

Aids

- Overhead transparencies
- Handouts
- Model course evaluation questionnaires

Time Frame

- One hour presentation

Content

- The need for evaluation
- Guidelines for course evaluation
- Course evaluation questionnaires
- Trainer self-assessment questionnaire for use before the session
- Trainer self-assessment questionnaire for use after the session

Learning Outcome

The participants should be aware of the importance of evaluation in training and of methods that can be used to evaluate the effectiveness of training.

The Need for Evaluation

It is not good enough for a trainer to feel self-satisfied with his or her training performance without evaluating it. All effective trainers not only evaluate or measure the degree of success of their course; they also evaluate their personal performance at the conclusion of each session or at least at the end of each training day.

Neglecting to make any attempt at evaluation reflects disinterest and lack of professionalism and is symptomatic of a non-caring attitude. Evaluation is a must; it is an integral part of effective training.

Purpose

- ☆ To improve training by discovering which training processes are successful in achieving their objectives (to "sort out the good from the bad")

Evaluation Affects Learning

- ☆ If we set examinations at the end of a course we affect the nature of learning
- ☆ If we study trainees' job behaviour after a course we have generally changed their job behaviour
- ☆ Since testing affects learning we can use it as a training aid

Two Aspects of Evaluation

- ☆ Course evaluation
- ☆ Trainer evaluation (self-evaluation)

Guidelines for Course Evaluation

Break evaluation into clear, achievable steps:

Evaluating Reaction

How well did the trainees enjoy the session(s)/course?

- ☆ Find out how well the trainees liked a particular training session or sessions or the course as a whole
- ☆ Does not include measurement of learning

Evaluating Learning

What principles, facts and techniques were learned?

- ☆ Written test questions, oral test questions, skill tests
- ☆ Avoid questions like "Have you learned anything?"

Evaluating Behaviour

What changes in job behaviour resulted from the training?

- ☆ Best evaluated through appraisal by on-the-job supervisors
- ☆ Remember: good trainers have on-the-job experience; they know the best way of doing things

Evaluating Results

What were the tangible results of the training in terms of improved job performance?

- ☆ Some types of training results are easily measured (*e.g.* typing)

- ☆ Others are not easily measured (where management and attitudes are involved)

Course Evaluation Questionnaires

- ☆ Determine what you want to find out
- ☆ Use a written comment sheet covering the steps above
- ☆ Obtain honest reactions by making the form anonymous
- ☆ Allow trainees to write additional comments not covered by questions

Two model course evaluation questionnaires are included at the end of this module. Model 1 is intended for evaluation of a complete training course. Model 2 can be used to evaluate either a specific training session or module or the overall training course.

Trainer Self-Assessment Questionnaire for Use before the Session

Preparation

- ☆ Do the notes show clearly the limited, definite scope of this training session?
- ☆ Is my session planned to enable my specific purpose to be fully accomplished?
- ☆ Have I allowed for an adequate introduction; a presentation with participant activity; and a recapitulation which will clinch the chief points?
- ☆ Have I arranged for all necessary equipment/materials and teaching aids?

Introduction

- ☆ Will this step excite the interest of the trainees from the start - is it original or linked strongly with an emotion-stirring activity, or some matter of topical or personal interest?
- ☆ Will it pave the way for what is to follow so that the presentation will not discourage or bore by excessive difficulty?
- ☆ Will it provoke curiosity and interest for what is to come, generating a need which will be satisfied?
- ☆ Does it provide adequate revision of what has gone before?

Body

- ☆ Is the instruction broken up into steps of reasonable length?
- ☆ Will each step offer maximum trainee participation and activity?
- ☆ Will each step win trainee interest and attention?
- ☆ Will each step offer some way of evaluating the trainees' comprehension before the next step is undertaken?

- ☆ If there is a written exercise to be done, have I something useful ready to occupy the quicker trainees so that slower ones may finish comfortably?
- ☆ Is there adequate provision for holding the interest of the strongest trainees and giving them worthwhile activity?
- ☆ Have I allowed for a period of relief for trainees and myself after a period of intense concentrated work?

Conclusion

- ☆ Will this step adequately recall and test the vital points of the session?
- ☆ Have I timed my session so that there is time for this important step?

Chalkboard Summary

- ☆ Will my chalkboard or whiteboard summary show what I expect to appear on the chalkboard at the end of the session?
- ☆ Is the arrangement (use of colour, diagrams, *etc.*) attractive?
- ☆ Have I thought out ways of obtaining the maximum help from the chalkboard with a minimum loss of contact with my group during the session?
- ☆ Are there any parts of the chalkboard that I should not use because they are not clearly visible because of poor lighting, shining sun, *etc.*?
- ☆ How will arrangement of any other visual aids fit in with my use of the chalkboard?

General

- ☆ Are there any other aids that will assist me?
- ☆ What rabbits have I ready to pull out of a hat if interest flags?
- ☆ Have I taken into consideration the intellectual level of the group, the time of day the session will take place and interruptions?
- ☆ Have I thought out how this session will fit into the general syllabus for the group?
- ☆ Am I sure of the correct pronunciation of unusual words that I will be using during the lesson?
- ☆ Am I sure of my subject-matter and of the correctness of the questions I intend to use?
- ☆ Am I sufficiently familiar with my questions and steps to be able to carry on the session at maximum effective speed without allowing the thin edge of the wedge of inattention to be inserted?

Trainer Self-Assessment Questionnaire for Use after the Session

Voice

- ☆ Was my voice clearly audible in all parts of room?

- ☆ Was it restrained enough not to irritate trainees or disturb other session leaders?
- ☆ Did I vary the speed, pitch, volume and tone so as to give maximum interest to whatever I said?

Manner

- ☆ Was my manner reasonable, brisk and alert?
- ☆ Did I sincerely convey a sense of earnestness and enthusiasm for what I was instructing?
- ☆ Was my manner reasonably pleasant and general without being affectedly so?

Group Management

- ☆ Did I get off to a clean brisk start, stimulating the group from the beginning?
- ☆ Did I stand in such a position that I could be seen and heard by all trainees?
- ☆ Did I keep all trainees under my eye and control whenever necessary?
- ☆ Did I take steps to see that no trainee disturbed the work of the group or failed to take adequate part in the session?
- ☆ Did I see that at the beginning of the lesson the floor and chalkboards were clean, the desks in order, the windows open and the class settled and ready?
- ☆ Did I have the trainees pulling with or against me?
- ☆ Did I refuse to be sidetracked?

Questioning

- ☆ Were my questions audible to all trainees?
- ☆ Were most questions easy enough for all trainees to be able to attempt an answer?
- ☆ Were there some particularly stimulating questions?
- ☆ Where the response to a question was unsatisfactory, did I take measures to improve the response (*e.g.* reframing the question) rather than waste a good question by immediately giving an answer?
- ☆ Did my questions follow rapidly without hesitation and uncertainty?
- ☆ Did I insist on answers being given loudly and clearly?
- ☆ Did I refrain from unnecessarily repeating answers?
- ☆ Did I distribute questions widely, encouraging weak trainees?

General

- ☆ Did I cover the steps of my session adequately?
- ☆ Was my recapitulation or other final step unhurried?

- ☆ Did I maintain my aim throughout the session?
- ☆ Did I keep as far as possible to the plan of my lesson?
- ☆ Did my trainees and I enjoy the session?
- ☆ What did the trainees gain from this session?
- ☆ What have I learned by leading this session?

Course Evaluation/Reaction Questionnaire

To assist in the planning of future courses it would be of great value if you would complete the sections that follow. Please be frank with your responses. Remember, only your honest reactions will enable adjustments and improvements to be made. The questions asked may not cover all of the aspects about which you wish to comment. For that reason a space headed "General comments" has been provided, and it is hoped that you will use it if appropriate.

Conditions

- ☆ Were you comfortable?
- ☆ What improvements, if any, do you suggest for the accommodation of future courses?
- ☆ Were the seating arrangements satisfactory?
- ☆ Could you see and hear satisfactorily?
- ☆ Were the morning and afternoon sessions well balanced?

Course Content

- ☆ Were the subjects covered the ones you expected would be?
- ☆ Were there any surprises? Why?
- ☆ Was the coverage sufficiently wide? If not, what subjects would you have liked included?
- ☆ Was each subject covered in sufficient depth? Name any that you think were not.

Was the course sufficiently practical in the sense that you will be able to apply knowledge and skills taught?

- ☆ Did the subjects sustain your interest?
- ☆ What additional subjects would you suggest for future courses?
- ☆ What subjects would you omit from future courses?

Presentation

- ☆ Were all sessions presented in a clear and interesting way?
- ☆ Were there any sessions that left you confused or uncertain? Please specify.

- Do you think trainers could have done more to improve their presentations? If so, what?
- Were the lengths of sessions satisfactory?
- Did the aids used help sustain your interest and understanding? Name any particular aid that impressed you.

27

Training Module or Course Evaluation

Instructions

You have just completed the training. Now we would like you to tell us about your feelings on what has just been presented. This information is valuable in helping us make following training sessions more interesting and useful to you. Below you will find a number of questions dealing with the just completed training session. Most questions can be answered by circling a number on the scale to the right of the question. Where a written response is required, please write your reply clearly in the space provided. Please consider your responses carefully and answer truthfully. Everything you say will be held in strictest confidence. The information will be used only to help us make this training activity more responsive to your needs.

Topic Discussed:____________________

I. Content

1. Relevance of the topic to your job	Not relevant				Relevant
	1	2	3	4	5
2. Clarity of the module's objectives	Not clear				Very clear
	1	2	3	4	5
3. Level of instruction	Too basic				Too advanced
	1	2	3	4	5
4. Lecture coverage	Inadequate				Very comprehensive
	1	2	3	4	5
5. Time allotment	Too short				Too long

	1	2	3	4	5
6. Emphasis on details	Too brief				Too detailed
	1	2	3	4	5
7. Organization and direction	Disorganized				Well organized
	1	2	3	4	5
8. Treatment of the topic	Abstract				Practical
	1	2	3	4	5
9. Additional comments you may have on these or other aspects of the content of this training module/ session					

II. Training Aids and Handouts

1. Effectiveness of teaching aids	Not effective				Very effective
	1	2	3	4	5
2. Readability of	Not readable				Very readable
______________________*	1	2	3	4	5
3. Clarity of message of	Not clear				Very clear
______________________*	1	2	3	4	5
4. Appeal of	Not appealing				Very appealing
______________________*	1	2	3	4	5
5. Usefulness of	Not useful				Useful
______________________*	1	2	3	4	5
* Here you would insert the names of instructional aids used: handouts, slides, videos, overhead transparencies, *etc.*					
6. Additional remarks you may have on these or other aspects of the teaching methods, aids, and handouts used in the training session					

III. Instructor Effectiveness

1. Mastery of the subject	Not knowledgeable				Knowledgeable
	1	2	3	4	5
2. Ability to transfer/communicate information and knowledge effectively	Very poor				Excellent
	1	2	3	4	5
3. Ability to arouse and sustain interest	Very poor				Excellent
	1	2	3	4	5
4. Openness to ideas of trainees	Not receptive				Receptive
	1	2	3	4	5

5. Encouragement of trainee participation	Did not encourage				Encouraged
	1	2	3	4	5
6. Time management	Very poor				Excellent
	1	2	3	4	5
7. Speed in talking	Too slow				Too fast
	1	2	3	4	5
8. Clarity of speech	Not clear				Clear
	1	2	3	4	5
9. Additional remarks on these or other aspects of the instructor's effectiveness					

IV. General

1. Please state the three most important ideas or concepts that you have learned from this session
2. Suggestion(s) to improve the session

V. Training Logistics/Administration

1. Quality of the meals	Very poor				Very good
	1	2	3	4	5
2. Quality of accommodation	Very poor				Very good
	1	2	3	4	5
3. Quality of transportation	Very poor				Very good
	1	2	3	4	5
4. Contact with staff members	Very poor				Very good
	1	2	3	4	5
5. Quality of training facilities	Very poor				Very good
	1	2	3	4	5
6. Please use the space below to indicate any suggestions you might have that will help us to improve the facilities and administration					

Testing Trainee Trainers - Individual Presentations

To evaluate the training capability of trainees following their training.

Suggested Methods of Instruction

- ☆ Clear instructions to trainees of what is expected of them - both orally and in writing
- ☆ Personal assistance with preparation of presentations

Aids

- ☆ Handouts

Time Frame

- ☆ Presentations of 30 minutes each

Content

- ☆ Individual presentations as a means of testing trainees
- ☆ Notification of presentation requirements
- ☆ Discussion to aid in preparation of presentations

Learning Outcome

Participants should demonstrate knowledge of the subject material and skills developed based on the foregoing training.

Individual Presentations as a Means of Testing Trainees

There are a number of ways of testing trainees: written and oral examinations, seeing how well they have mastered the skill taught, demonstrating the new on-the-job procedure. A proven method of testing trainees who are training to become trainers is to see how they perform in a training situation.

The material that follows is self-explanatory and is offered as a guide only. It is important that the trainee discuss his or her proposed presentation with the trainer to ensure that pitfalls that often confront novice trainers (*e.g.* inappropriate choice of subject, topic too broad for coverage in the time available, insufficient time in which to prepare suitable aids) are avoided. The trainer should be sure to be available at all times to assist and advise trainees on the preparation of their presentations.

Notification of Presentation Requirements

Trainees should first be advised of the presentations at the beginning of the course, and the following notification should be distributed halfway through the course.

Guidelines for Individual Presentations

During the concluding days of the course each trainee will present a 20-minute training session on a selected topic related to the HACCP system or another nominated area of food control. Topics will be chosen by the trainer (or trainees). As part of the presentation the trainee will:

- ☆ Produce a plan for the session
- ☆ Produce a handout for distribution to other course members following the presentation
- ☆ Prepare visual aids (charts, overhead transparencies, slides, *etc.*) for use in the presentation

The presentation may be pitched at any level, that is, plant employees, inspection staff, quality control personnel, plant management, interested organizations, *etc.*

In the presentation trainees will be expected to make use of training skills (including communication techniques, visual aids, questions, *etc.*) and technical knowledge learned during the course.

Presentations should be developed progressively by trainees during the course, and maximum guidance in their presentation should be available from the trainer.

Each presentation will be followed by a ten-minute evaluation by the group.

Discussion to Aid in Preparation of Presentations

Trainees should be requested to complete a form comprising the following questions for discussion with the trainer before preparing the presentation.

- ☆ What is your subject?
- ☆ Have you given thought to a plan for your presentation? (brainstorming)
- ☆ Have you jotted down thoughts, especially main points you wish to emphasize?
- ☆ Do you need assistance with information?
- ☆ Have you determined at what level you wish to pitch your presentation?
- ☆ How do you propose making the greatest impact?
- ☆ What aids do you propose to use?
- ☆ Do you propose giving a demonstration?
- ☆ Do you propose using handouts?
- ☆ Have you in mind to ask questions?
- ☆ Have you discussed your subject or its presentation with others? Do you intend to do so?

Organizing and Managing a Training Course

Suggested Methods of Instruction

- ☆ Lecture/discussion using the course of which this module is a part as a model

Aids

☆ Handout

Time Frame

☆ One hour lecture/discussion

Content

☆ Organization and management checklists

Learning Outcome

Participants should be aware of the considerations for the planning and management of a successful training course.

Organization and Management Checklists

The need for the trainer to be informed about course organization and management is self-evident. The details that follow amount to little more than a checklist. However, it is valuable for the trainer to go through each list item by item/stressing the relative importance of each and relating the whole to the course of which the module is a part.

Organization

☆ Establish whether the course is part or full time

☆ Establish duration

☆ Establish content

☆ Plan syllabus and timetable

☆ Identify and engage appropriate instructors

☆ Secure suitable training accommodation (well-lit, well-ventilated room)

☆ Select and notify trainees, through appropriate channels, of the dates, time and place

☆ Select and brief session leaders

☆ Select and review preliminary reading materials

☆ Prepare course documentation

☆ Arrange training equipment: lectern, microphone, chalkboard and chalk, writing materials, visual aids (slide projector, video equipment, screen, spare bulbs, *etc.*), other aids

☆ Arrange training room: seating arrangements, namecards, position of chalkboard, screen, *etc.*

Don't forget to arrange coffee and lunch breaks.

Management

☆ Remind session leaders

☆ Arrange transport for outside speakers/trainers

- ☆ Introduce and thank session leaders
- ☆ Meet emergencies (give session yourself or rearrange sessions)
- ☆ Check facilities and equipment (projectors, boards, charts, *etc.*)
- ☆ Ensure trainees receive documentation
- ☆ Introduce visitors
- ☆ Coordinate all aspects of the course
- ☆ Evaluate training (based on trainees', session leaders' and own observations)
- ☆ Leave room tidy; return equipment and aids to proper place
- ☆ Prepare thank-you letters
- ☆ Prepare reports on course
- ☆ Prepare any statistics
- ☆ Each day appoint a "Monitor for the day" from among the trainees to assist with the conduct of the course

The main elements of a mentoring programme that carry quantifiable cost would be:

- ☆ **Training of mentor(s)** - comfortably achievable for £1,000/head - it's not rocket science, but selection of suitable mentor is absolutely critical - good natural mentors need little training; other people who are not ready or able to help others can be beyond any amount of training.
- ☆ **Mentor time away from normal activities** - needs to be a minimum of an hour a month one-to-one or nothing can usefully be achieved, up to at most a couple of hours a week one-to-one, which would be intensive almost to the point of overloading the mentoree. That said, there may be occasions when the one-to-one would necessarily involve a whole day out for the mentor, for instance client or supplier visits. Say on average a day a month including the associated administration work, particularly where the mentoring is required to be formalised and recorded.
- ☆ **Overseeing the program, evaluating and monitoring activity, progress and outputs** - depends on the size of the program, ie., number of mentors an number of 'mentorees' - if the mentoring is limited to just a single one-to-one relationship then it's largely self-managing - if it's a programme involving several mentors an mentorees then estimate an hour per quarter (3 mths) per one-to-one mentoring relationship - probably the responsibility of an HR or training manager. If this person with the overview/monitoring responsibility needs external advice you'd need to add on two or three days external training or consultancy costs.
- ☆ (**Mentoree time away from normal activities** - effective mentoring should ideally integrate with the mentoree's normal activities, and enhance productivity, effectiveness, *etc.*, so this is arguably a credit not a debit.)

Mentoring Principles and Techniques

Rather than simply give the answers, the mentor's role should be to **help the 'mentoree' find the answers for him/herself**. While giving the answers is usually better than giving no help at all, helping the mentoree to find the answers for him/herself provides far more effective mentoring, because the process enables so much more for the mentoree in terms of **experience of learning**. Give someone the answers and they learn only the answers; instead mentors need to facilitate the experience of discovery and learning. The mentor should therefore focus mentoring effort and expectations (of the person being mentored especially, and the organisation) on helping and guiding the mentoree to find the answers and develop solutions of his/her own.

Accordingly, many of the principles of mentoring are common to those of proper coaching, which are particularly prominent within life coaching. You should also refer to aspects of NLP (Neuro-Linguistic Programming), and Sharon Drew Morgen's Facilitative Questioning methodology.

Mentors need to be facilitators and coaches, not tutors or trainers. Mentorees need simply to open their minds to the guidance and facilitative methods of the mentor. The mentor should not normally (unless in the case of emergency) provide the answers for the mentoree; instead a mentor should ask the right questions (facilitative, guiding, interpretive, non-judgemental) that guide the mentoree towards finding the answers for him/herself.

If a mentor tells a mentoree what to do, then the mentoree becomes like the mentor, which is not right nor sustainable, and does not help the mentoree to find his/her own true self.

The mentor's role is to help the mentoree to find his/her own true self; to experience their own attempts, failures and successes, and by so doing, to develop his/her own natural strengths and potential.

We can see parallels in the relationship between a parents and a child. If a parent imposes his or her ways, methods and thinking upon a child, the child becomes a clone of the parent, and in some cases then falsifies his or her own true self to please and replicate the model projected by the parent. The true self might never appear, or when it begins to, a crisis of confidence and purpose occurs as the person tries to find and liberate his or her true self.

When we mentor people, or when we raise children, we should try to help them develop as individuals according to their natural selves, and their own wishes, not ours.

Tips on Establishing a Mentoring Service or Programme

There are very many ways to design a mentoring programme, whether within an organization, or as a service or help that you provide personally to others.

Here are some questions that you should ask yourself. The answers will move you closer to what you seek to achieve:

- ☆ What parameters and aims have you set for the mentoring activity?

- What will your mentoring programme or service look and feel like?
- What must it achieve and for whom?
- What are your timescales?
- How will the mentoring programme or activity be resourced and managed and measured?
- What type of design and planning approach works best for you? (It makes sense to use a design and planning approach that works for you.)
- What are your main skills and style and how might these influence the programme design?
- What methods (phone, face-to-face, email, *etc.*) of communication and feedback are available to you, and what communications methods do your 'customers' need and prefer?
- What outputs and effects do you want the programme to produce for you, and for the people being mentored?
- How might you build these core aims, and the implied values and principles, into your programme design?
- How can you best measure and agree that these outputs - especially the agreed expectations of the people being mentored - are being met.
- How can you best help people in matters for which you need to refer them elsewhere?
- What skills, processes, tools, experience, knowledge, style do you think you will need that you do not currently have?
- What do your 'customers' indicate that they want in terms of content, method and style or mentoring - in other words what does your 'target market' need?, and what parts of those requirements are you naturally best able to meet?
- Mentoring is potentially an infinite demand upon the mentor so you need to have a clear idea of the extent of your mentoring 'offering'.
- Establishing clear visible parameters enables proper agreement of mutual expectations.

General Training Tips

These tips apply essentially to traditional work-related training - for the transfer of necessary job- or work-related skills or knowledge.

These tips do not apply automatically to other forms of enabling personal development and facilitating learning, which by their nature involve much wider and various development methods and experiences.

When planning training think about:

- Your objectives - keep them in mind all the time
- How many people you are training

- ☆ The methods and format you will use
- ☆ When and how long the training lasts
- ☆ Where it happens
- ☆ How you will measure its effectiveness
- ☆ How you will measure the trainees' reaction to it

When you give Skills Training to Someone Use this Simple Five-Step Approach

1. Prepare the trainee - take care to relax them as lots of people find learning new things stressful
2. Explain the job/task, skill, project, *etc.* - discuss the method and why; explain standards and why; explain necessary tools, equipment or systems
3. Provide a demonstration - step-by-step - the more complex, the more steps - people cannot absorb a whole complicated task all in one go - break it down - always show the correct way - accentuate the positive - seek feedback and check understanding
4. Have the trainee practice the job - we all learn best by actually doing it - ('I hear and I forget, I see and I remember, I do and I understand' - Confucius)
5. Monitor progress - give positive feedback - encourage, coach and adapt according to the pace of development

Creating and using progress charts are helpful, and are essential for anything complex - if you can't measure it you can't manage it. It's essential to use other training tools too for planning, measuring, assessing, recording and following up on the person's training.

Breaking skills down into easily digestible elements enables you to plan and manage the training activities much more effectively. Training people in stages, when you can build up each skill, and then an entire role, from a series of elements, keeps things controlled, relaxed and always achievable in the mind of the trainee.

Establishing a relevant 'skill set' is essential for assessing and prioritising training for any role. It is not sufficient simply to assess against a job description, as this does not reflect skills, only responsibilities, which are different. Establishing a 'behaviour set' is also very useful, but is a more difficult area to assess and develop.

More information and guidance about working with 'Skill-Sets' and 'Behaviour Sets', and assessment and training planning see training evaluation, and performance appraisals, and other related linked articles on this site. Using Skill-Sets to measure individual's skills and competencies is the first stage in producing a training needs analysis for individuals, a group, and a whole organisation. You can see and download a free Skill-Set tool and Training Needs Analysis tool the free resources page.

This will not however go beyond the basic work-related job skills and attributes development areas. These tools deal merely with basic work training, and not with more important whole person development, for which more sophisticated questioning, mentoring and learning facilitation methods need to be used.

Psychometric tests (and even graphology - handwriting analysis) are also extremely useful for training and developing people, as well as recruitment, which is the more common use. Psychometric testing produces reliable assessments which are by their nature objective, rather than subjective, as tends to be with your own personal judgement. Your organisation may already use systems of one sort or another, so seek advice. See the section on psychometrics. Some of these systems and tools are extremely useful in facilitating whole-person learning and development.

Some tips to make training (and learning, coaching, mentoring) more enjoyable and effective:

- ☆ Keep instructions positive ('do this' rather than 'don't do this')
- ☆ Avoid jargon - or if you can't then explain them and better still provide a written glossary
- ☆ You must tailor training to the individual, so you need to be prepared to adapt the pace according to the performance once training has begun
- ☆ Encourage, and be kind and thoughtful - be accepting of mistakes, and treat them as an opportunity for you both to learn from them
- ☆ Focus on accomplishment and progress - recognition is the fuel of development
- ☆ Offer praise generously
- ☆ Be enthusiastic - if you show you care you can expect your trainee to care too
- ☆ Check progress regularly and give feedback
- ☆ Invite questions and discussion
- ☆ Be patient and keep a sense of humour

Induction Training Tips

- ☆ Assess skill and knowledge level before you start
- ☆ Teach the really easy stuff first
- ☆ Break it down into small steps and pieces of information
- ☆ Encourage pride
- ☆ Cover health and safety issues fully and carefully
- ☆ Try to identify a mentor or helper for the trainee

As a manager, supervisor, or an organisation, helping your people to develop is the greatest contribution you can make to their well-being. Do it to your utmost and you will be rewarded many times over through greater productivity, efficiency, environment and all-round job-satisfaction?

Remember also to strive for your own personal self-development at all times - these days we have more opportunity and resource available than ever to increase our skills, knowledge and self-awareness. Make use of it all.

Leadership and Management Training and Development - Processes Overview

Here's an overview of some simple processes for training and developing management and leadership skills, and any other skills and abilities besides. Use your own tools and processes where they exist and are effective. Various tools are available on the free resources section to help with this process, or from the links below.

Refer also to the Coaching and Development Process Diagram

1. Obtain commitment from trainees for development process. Commitment is essential for the development. If possible link this with appraisals and career development systems.
2. Involve trainees in identifying leadership qualities and create 'skill/ behaviour-set' that you seek to develop. Training and development workshops are ideal for this activity.
3. Assess, prioritise and agree trainee capabilities, gaps, needs against the skill/behaviour-set; individually and as a group, so as to be able to plan group training and individual training according to needs and efficiency of provision. Use the skill/behaviour-set tool for this activity. Use the training needs analysis tool for assessing training needs priorities for a group or whole organization.
4. Design and/or source and agree with trainees the activities, exercises, learning, experiences to achieve required training and development in digestible achievable elements - *i.e.* break it down. Use the training planner to plan the development and training activities and programmes. Record training objectives and link to appraisals.
5. Establish and agree measures, outputs, tasks, standards, milestones, *etc.* Use the SMART task model and tool.

Training and development can be achieved through very many different methods - use as many as you need to and which suit the individuals and the group. Refer to the Kolb learning styles ideas - different people are suited to different forms of training and learning.

Exercises that involve managing project teams towards agreed specific outcomes are ideal for developing management and leadership ability. Start with small projects, then increase project size, complexity and timescales as the trainee's abilities grow. Here are examples of other types of training and development. Training need not be expensive, although some obviously is; much of this training and development is free; the only requirements are imagination, commitment and a solid process to manage and acknowledge the development. The list is not exhaustive; the trainer and trainees will have lots more ideas:

- On the job coaching
- Mentoring
- Delegated tasks and projects
- Reading assignments
- Presentation assignments
- Job deputisation or secondment
- External training courses and seminars
- Distance learning
- Evening classes
- Hobbies - *e.g.* voluntary club/committee positions, sports, outdoor activities, and virtually anything outside work that provides a useful personal development challenge
- Internal training courses
- Attending internal briefings and presentations, *e.g.* 'lunch and learn' format
- Special responsibilities which require obtaining new skills or knowledge or exposure
- Video
- Internet and e-learning
- Customer and supplier visits
- Attachment to project or other teams
- Job-swap
- Accredited outside courses based on new qualifications, *e.g.* NVQ's, MBA's, *etc.*

Management Training with no Guarantee of a Management Job

Training people, especially graduates, young rising stars and new recruits, is commonly linked to the veiled promise of or allusion to management opportunity. But what happens when the organisation is unable to offer a management promotion at the end of the training programme? This is a familiar pattern and challenge in many organisations. How can you encourage people into a management development programmes, with no assurance of a promotion into management at the end of it?

The problem lies in the mismatched expectations at the outset: the trainee hopes (which develops into an expectation) for promotion. The organisation cannot (quite rightly) guarantee that a management job will be offered. No wonder that it often ends in tears, and what should have been (and actually still is) a positive experience, namely the learning and experience achieved, turns into a crisis for HR to diffuse, because the trainee feels let down and disappointed.

Here's a different Way to Approach Management Development

First, come back a few stages and consider the values, beliefs and real nature of the emotional, spiritual and personal development that these people (the management trainees) might need and respond to most. Then you'll find it easier to define an honest set of expectations on each side (the graduates and the employer).

If the 'training' is positioned as a possible step towards a management promotion, people will become focused on the wrong expectations and aims, and when, as most of them will do, people fail to achieve a promotion they will feel they have failed, and the experience turns sour.

Better to design the 'learning' as a 'significant personal development experience' in its own right, with absolutely no promise of a job or a promotion at the end of it. That way everyone's (employer and employees) expectations match openly and honestly, and people are all focused on enjoying and benefiting from the learning as the central aim, rather than continually hoping that the management job happens, or in the case of the employer and program manager, preparing to defend and appease folks at the end when there's no job.

Added to which, by defining and designing the programme as personal development, enrichment, experience, life-learning, *etc.* (there are many highly appealing and worthy ways to specify and describe a programme like this) - and not being afraid of doing so - you will attract the right sort of people into it; ie., the more emotionally mature and positive ones, who want to do it for the learning and experience, rather than purely for the chance of a promotion into management.

The irony of course is that students, who respond to learning and personal enrichment opportunity per se, with no guarantees or allusions to management promotion, will be the best management candidates of all.

Tips for Assessing Organizational Training Effectiveness

Look at and understand the broad organizational context and business environment: the type, size, scale, spread, geography, logistics, *etc.*, of the business or organization. This includes where and when people work (which influences how and when training can be delivered). Look also at the skills requirements for the people in the business in general terms as would influence training significance and dependence - factors which suggest high dependence on training are things like: fast-changing business (IT, business services, healthcare, *etc.*), significant customer service activities, new and growing businesses, strong health and safety implications (chemicals, hazardous areas, transport, utilities). Note that all businesses have a high dependence on training, but in certain businesses training need is higher than others - change (in the business or the market) is the key factor which drives training need.

Assess and analyse how training and development is organized and the way that training is prioritised. Think about improvements to training organization and planning that would benefit the organisation.

Review the business strategy/positioning/mission/plans (and HR strategy if any exists) as these statements will help you to establish the central business aims.

Training should all be traceable back to these business aims, however often it isn't - instead it's often arbitrary and isolated.

Assess how the training relates to the business aims, and how the effectiveness of the training in moving the business towards these aims is measured. Often training isn't measured at all - it needs to be.

Look at the details and overview of what training is planned for the people in the business. The training department or HR department should have this information. There should be a clear written training plan, including training aims, methods, relevance and outputs connected to the wider aims of the business.

Look also at how training relates to and is influenced by appraisals and career development; also recruitment, and general ongoing skills/behavioural assessment. There should be process links between these activities, particularly recruitment and appraisals, and training planning. Detailed training needs should be driven substantially by staff appraisals. (It goes without saying that there should be consistent processes and application of staff appraisals, and that these should use suitable job performance measures that are current and relevant to the operations and aims of the business.)

Look particularly at management training and development. The big business generally the bigger the dependence on the management of training and development.

Look at new starter induction training - it's critical and typically a common failing in situations where anything higher than a low percentage of new starters leave soon after joining.

Look for the relationships between training, qualifications, job grades and pay/ reward levels - these activities and structures must be linked, and the connections should be visible to and understood by all staff.

Look especially at staff turnover (per cent per annum of total staff is the key indicator), exit interviews, customer satisfaction surveys, staff satisfaction surveys (if they exist) for other indicators as to staff development and motivational needs and thereby, training deficiencies.

Look for any market research or competitor analysis data which will indicate business shortcomings and weaknesses, which will imply staff training needs, obviously in areas of the most important areas of competitive weakness in relation to the business positioning and strategy.

Look to see if there is director training and development - many directors have never been trained for their roles, and often hide from and resist any effort to remedy these weaknesses.

Base Training Recommendations and Changes on Improving Training Effectiveness in Terms of

- ☆ Relevance to organizational aims
- ☆ Methods of staff assessment

- Training design/sourcing
- Training type, mix and suitability, given staff and business circumstances (consider all training options available - there are very many and some are relatively inexpensive, and provide other organizational benefits; in-house, external training courses and seminars, workshops, coaching, mentoring, job-swap, secondment, distance-learning, day-release, accredited/qualification-linked, *etc.*)
- Remedies for identified organizational and business performance problem areas, *e.g.*, high staff turnover, general attrition or dissatisfaction levels, customer complaints, morale, supplier retention and relationships, wastage and shrinkage, legal and environmental compliance, recruitment difficulties, management and director succession, and other key performance indicators of the business (which should be stated in business planning documents)
- Comparative costs of different types of training per head, per staff type/level
- Measurement of training effectiveness, and especially feedback from staff being trained: interview departmental heads and staff to see what they think of training - how it's planned, delivered, measured, and how effective it is.

Measuring and Increasing Training Days or Hours per Person

Measuring training hours per person as an average across the organization, typically per year, is often a useful training and development KPI (Key Performance Indicator) of the training function - more training acronyms here. If you can't measure it you can't manage it, the saying goes.

The degree of difficulty in measuring training time per person depends on what you define as training: training time per person on training courses is relatively easy to measure, but on-the-job coaching, informal mentoring, personal reading and learning - these are less quantifiable - you'd normally need to get this data from the employees via a survey or other special report.

It is possible to manage 'training time per person' aims and data via annual appraisals, when training past and future could be quantified - this could be a relatively simple add-on to whatever appraisal system you are using currently, and could relatively easily be cascaded via managers.

Your previous year's total training course time - *i.e.*, 'person-days' spent on training courses - divided by number of employees in the organization is an easy start point. This will give you the average training course time per employee, and if you have no other benchmarks is as good a start point as any. Then perhaps agree a sensible target uplift on this, assuming the training requirement is linked to organisational aims and personal development, rather than training for the sake of it just to increase the hours per person. You can make this calculation for a team, a job grade, a department or a whole organization.

You could also survey the managers as to their estimate of how much on-the-job-coaching they provided per person as an average during a week. This gives another benchmark, albeit it an estimate, for which you can target an uplift and then monitor via managers reporting back every month or quarter. Remind managers to include, and if possible to categorise all the different sorts of training and coaching that takes place, as they will tend to forget or ignore certain types, for example; job cover, training at meetings, taking on new tasks and responsibilities, delegated tasks, shadowing, *etc.* Training comes in various forms - if you are measuring it make sure you don't underestimate the level of activity.

Training Planning Factors

These guidelines essentially deal with conventional work skills training and development. Remember that beyond this, issues of personal development and learning, for life, not just work, are the most significant areas of personal development to focus on.

To plan traditional training of work skills and capabilities that links to organizational performance improvement you must first identify the organizational performance needs, gaps, and priorities. These are examples of typical training drivers which give rise to training needs. It is rare to use all of these aspects in determining training needs - select the ones which are most appropriate to your own situation, the drivers which will produce the most productive and cost-effective results, in terms of business performance and people-development:

Examples of Training Drivers

- ☆ Customer satisfaction surveys
- ☆ Business performance statistics and reports.
- ☆ Financial reports and ratios.
- ☆ Competitor analysis and comparison, *e.g.* SWOT analysis.
- ☆ Management feedback on employee needs, including from appraisals.
- ☆ Training audits, staff assessment centres.
- ☆ Staff feedback on training needs.
- ☆ Director-driven policy and strategic priorities.
- ☆ Legislative pressures.
- ☆ Relevant qualification and certification programmes.

Use the results and indicators from the chosen driver(s) to produce prioritised training needs per staff type, which will logically enable staff and management to achieve improvements required by the organization.

There are several free training needs analysis and planning tools on the free resources section which might help you assess and analyse staff training needs, and then construct training plans.

Potential Conflict between HR/Training Function and Business Management

Conflict can arise between HR/Training and other parts of the organization, commonly due to differing priorities among performance management functions within a business, and notably relating to training, development and welfare of staff. If so, you need to identify conflict and manage it. Conflict is often caused by the different aims of the departments, and you need to facilitate understanding and cooperation on both sides. This is especially important in order to achieve successful training needs assessment, training design, planning, delivery and optimal take-up and implementation. Aside this there is very much deeper implications for organizations seeking to be truly cohesive, 'joined-up', and aligned towards common set of corporate aims and values. If you see any of the following symptoms of conflict, consider the root cause and facilitate strategic discussion and agreement, rather than limit your activity to simply resolving or responding only to the symptom.

- ☆ management resisting release of staff for training due to day-to-day work demands
- ☆ short-term needs of performance management vs. long-term outlook of HR
- ☆ HR have no line authority over trainees therefore cannot control training take-up
- ☆ Training is rarely well followed-through once delegates are back in jobs, despite HR efforts to achieve this via managers
- ☆ HR budgets are often cut if profits come under pressure

Generally conflict would stem from the values and priorities of directors, managers and staff involved, and the aims and processes of the different HR functions. Here are some subject headings that serve as a checklist to see that the aims and priorities of HR/Training align optimally with those of other departments (the list is not exhaustive but should enable the main points of potential misalignment to be addressed):

- ☆ Profit, costs, budgets
- ☆ Well-being of staff
- ☆ Ethics and morality in treatment of staff
- ☆ Legal adherence
- ☆ Business strategy
- ☆ Training and development needs (skills, knowledge, EQ, *etc.*)
- ☆ Succession planning
- ☆ Assessment and appraisals
- ☆ Promotion
- ☆ Recruitment
- ☆ Age, gender, disability

- Policies
- Harassment
- Counselling
- Workforce planning
- Management structure
- Decision-making and approval processes
- Outsourcing
- Contracts of employment
- Corporate mission and values
- Acquisitions and divestments
- Premises
- Pay and remuneration plans and market positioning
- Use of agencies
- Advertising and image

Positioning Statement and Introduction to Training Courses and Materials for Groups of Mixed Abilities

In many training and teaching situations it is not possible to identify and assemble groups of delegates whose needs, experience and ability levels closely match each other.

Groups will therefore often comprise of trainees and learners who have different levels of experience, and/or abilities, styles, expectations, needs, aims, *etc.*

This places additional demands on the training provider/facilitators to ensure that the needs of all delegates are met, while not causing any frustration or boredom for delegates who already know or possess certain parts of the information and abilities (or think they do) that the teaching seeks to transfer.

As such it is often helpful for trainers and delegates to acknowledge and accept this situation at the beginning of the course or training session, with the purpose of reducing potential frustrations and negative reactions and effects as far as possible.

Here is a suggested introductory statement, which aims to achieve a commitment to understand the needs of others. You will notice that the statement is designed to appeal to the mature and responsible nature that exists in virtually all people. The challenge is to tap into this at the outset, in order to set a positive constructive atmosphere and standard of behaviour for the training. Adapt it to suit your own situation.

This special training introduction is additional to any other introduction that you'll be using to outline the training aims, domestic arrangements, fire-drill, *etc.*

The statement or an adapted version can also be included within the introduction section of training course notes and manuals.

Adapt this training course introduction to suit the situation. It is more relevant to mixed groups of delegates from different experience and skills backgrounds than to groups which have been selected according to closely matching needs and ability levels.

This sort of statement can be included at the beginning of course notes, or given as a separate handout (as a sort of philosophical scene-setter), and/or explained and discussed verbally with the group.

In any event it's good also to seek agreement from the group that the concept of making the most constructive use of time and everyone's ability to contribute is the right and proper approach.

The message to training course delegates is effectively: that learning new things is an enjoyable rewarding part of life and personal development, and so too is helping others to do the same.

Resources for Training and Development - Building your Own Resources - and Helping others do so

- We all need to maintain and develop our value in the marketplace.
- Then we will always be in demand.
- Two generations ago, jobs were for life - now some careers last just five or ten years.
- The world is changing faster.
- Organisations and everyone individually, must be able to assess their capabilities, and re-skill when necessary.
- Trainers, teachers, coaches, managers and leaders are central to these assessing and re-skilling processes.
- Whether you are a trainer, specialist, manager, leader, entrepreneur, whatever, building your own **resources** will enable you to maintain and grow your capabilities and value, and to help others do the same.
- Here are some questions and answers about building training and development resources.

Building Training and Development Resources

(I am grateful to Dawn Barclay of Potential Developments for raising the subject of building personal resources, prompting this additional section and the Q and A format.)

Q. What do we mean by resources in the context of learning and development?

A. Resources are

Materials and tools of various types, which:

- Describe
- Define

- Explain
- Summarise
- Teach
- And/or enable the acquisition, improvement, or delivery of -
- Skills
- Knowledge
- Methods
- Techniques
- Attitude
- And/or behaviour
- And thereby, performance, results, fulfilment, well-being, and other good outcomes.

Resources can therefore be all sorts of things. For example, a single tiny inspirational quotation is a resource. And a big organisational learning and development manual is a resource.

More examples of **resources** are:

- Team building games or exercises
- Testing instruments for individuals and teams (psychometrics and other assessments)
- Guides to a concepts or theories or models
- Spreadsheets or other analytical tools
- Case studies and best practice examples (good case studies are always in demand)
- Samples and examples - of anything relevant to your field or specialism
- Templates and forms
- Surveys and especially survey results
- Statistics and reports
- Contracts and legal documents
- Manuals and guides
- Specifications and project briefs
- Plans of all sorts
- Diagrams, pictures, cartoons
- Books, magazines, journals, newsletters and newspapers (especially newspaper cuttings)
- Films, videos and clips
- Pieces of music
- Puzzles, tricks, and games

- Quizzes and questions and answers
- Websites or a webpages (favourites or links)
- CDs and DVDs
- Physical props - real samples, or props as metaphors like a hammer or a lemon
- Items of curiosity and collectibles - diversity and history are powerful perspectives for teaching and learning
- Personal contacts, or a network of contacts - yes people are resources too.

Tips for Starting your Own New Training Business

- Here are some simple tips for starting your own new training business.
- Much more detailed business start-up help is available on other pages, listed below.
- These are just a few important tips especially for starting a new training business.
- When choosing the type of training to offer think carefully about it and avoid making assumptions or being drawn into too many areas.
- Starting your own training business is in some ways a simple transition from being employed as a trainer, coach, team leader, manager, *etc.*, however a big difference is now that you have to find the work before you can do it.
- On this point, your previous employer can easily be a prime prospect for you. Even if you leave on less than perfect terms, a previous employer is a good opportunity for securing freelance training work, not least because when people leave an organization, there is usually a gap and a period of uncertainty regarding the leaver's previous responsibilities. Lots of employers fail to ask leavers if they can fill in for a while on a contract basis. Ask. In any event, especially if you were well-regarded know their systems, you'll be a safe choice for them if they need some help, so keep in touch and (assuming you are not immediately stacked out with work from other customers) let your previous employer know you are happy to fill gaps in provision after you've gone.
- Expect to negotiate a (sometimes significantly) higher freelance day-rate compared with your previous employed wage. Organizations account for ad-hoc freelance training quite differently to employed staff costs. Many newly self-employed trainers offer themselves too cheaply. See the negotiation page. If in doubt, see what they offer before you suggest a rate - you could be very pleasantly surprised, particularly if they are in a bit of a panic and need a 'safe pair of hands' quickly.
- Aside from your previous employer(s), finding new training contracts or selling training courses entails marketing and advertising - in competition with others operating in the same market place. This could be a new

and significant consideration for you. Marketing and selling training is different to designing and delivering training, and involves different issues.

- ☆ You must now consider what you can market and sell successfully, as well as it being something that you can design and deliver successfully. This requires you to consider the market place, not just the quality of your training.

So when you choose what training to offer and especially how to package, describe and deliver it, ask yourself questions based on the following points, so that you develop training types, services, offerings and delivery which:

1. You can offer with very appealing uniqueness and passion
2. Ideally have good and increasing demand
3. Are not strongly served by competitors
4. Are relevant to industries you are comfortable with, and
5. Can be marketed in a very specific focused way, to decision-makers that you can reach cost-effectively.

Whether a website and online marketing will feature strongly in your business approach or not, 'Google Trends', and Google's Adwords keywords tracker (to access it open an Adwords account), are two excellent tools for evaluating online search trends and relative volumes in training (and for anything that people look for), which greatly assists answering some of the questions above, especially understanding demands, trends and what people are looking for and how they describe it (all of which can be quite different to what you imagine).

From a vital personal perspective, also look at the 'passion-to-profit' process/ template on this website because this helps consider how best to combine your greatest personal potential with a business proposition. You may choose not to use the process in detail, but consideration of its underlying meaning is fundamentally important towards building a sustainable thriving business in any area of product/ service provision.

Choose a business name carefully. Many people successfully use their own name along with a generic word or a few words related to training, because:

- ☆ This usually avoids any future problems with copyright (especially the potentially disastrous and easily made mistake of breaching someone else's rights or trademark)
- ☆ And it says that you are the boss and have the confidence and integrity to have your name as the business name.

If you choose a 'clever' or obscure business name, think very carefully about it because it will have risks (like this website name, which might have failed without the luxury of many years to become established), either or both in terms of copyright protection/breach, and/or misinterpretation or confusion.

Although copyright and trademark law is complex, broadly descriptive business names are less easy to protect, and also less likely to breach someone else's trademark. Non-descriptive business names need to be checked against existing use, especially registered names, which means that when secured they tend to be easier to protect. The UK government intellectual property website is a useful information and reseach resource.

Contrary to lots of advice you'll see from financial and legal folk, becoming freelance (self-employed in other words) is very easy in terms of legal and regulatory set-up. I refer to the UK. In some other nations it will be a little more difficult, in others even easier. In the UK you do not need a limited company. You do not need a VAT number. You simply need to inform your tax office, which actually is a good source of advice about starting up.

If you have plans of substantial scale then seek qualified legal and financial advice, but for many new training business start-ups a freelance/self-employed approach is perfectly adequate for the authorities and the market place, as well as being very quick and inexpensive for the freelancer.

Public liability insurance is advisable because without the protection of a limited company you have unlimited personal liability for any damages arising against you. Many customers and venues insist on trainers having public liability insurance anyway. It's not necessarily very expensive, and is different to professional liability insurance of the sort that lawyers and doctors and high-powered consultants typically need, when potential liabilities run to £millions rather than a few £thousands. That said, insurance is a personal matter for you to decide and resolve as you think reasonable. I merely offer general pointers.

When starting a new business, especially from a marketing/advertising viewpoint, it's usually more effective to focus on a small number of strong unique specialisms - or even just a single very powerful offering - than to offer a one-stop shop or wide catch-all range. A good specialist will usually beat a widely-spread generalist in any single area.

Networking is a useful marketing method for new businesses - look at the processes for effective business networking.

Task Management Skills

- ☆ Identify aims and vision for the group, purpose, and direction - define the activity (the task)
- ☆ Identify resources, people, processes, systems and tools (inc. financials, communications, IT)
- ☆ Create the plan to achieve the task - deliverables, measures, timescales, strategy and tactics
- ☆ Establish responsibilities, objectives, accountabilities and measures, by agreement and delegation
- ☆ Set standards, quality, time and reporting parameters
- ☆ Control and maintain activities against parameters

- ☆ Monitor and maintain overall performance against plan
- ☆ Report on progress towards the group's aim
- ☆ Review, re-assess, adjust plan, methods and targets as necessary

Your Responsibilities as a Manager for the Group are

- ☆ Establish, agree and communicate standards of performance and behaviour
- ☆ Establish style, culture, approach of the group - soft skill elements
- ☆ Monitor and maintain discipline, ethics, integrity and focus on objectives
- ☆ Anticipate and resolve group conflict, struggles or disagreements
- ☆ Assess and change as necessary the balance and composition of the group
- ☆ Develop team-working, cooperation, morale and team-spirit
- ☆ Develop the collective maturity and capability of the group - progressively increase group freedom and authority
- ☆ Encourage the team towards objectives and aims - motivate the group and provide a collective sense of purpose
- ☆ Identify, develop and agree team- and project-leadership roles within group
- ☆ Enable, facilitate and ensure effective internal and external group communications
- ☆ Identify and meet group training needs
- ☆ Give feedback to the group on overall progress; consult with, and seek feedback and input from the group

Your Responsibilities as a Manager for each Individual are

- ☆ Understand the team members as individuals - personality, skills, strengths, needs, aims and fears
- ☆ Assist and support individuals - plans, problems, challenges, highs and lows
- ☆ Identify and agree appropriate individual responsibilities and objectives
- ☆ Give recognition and praise to individuals - acknowledge effort and good work
- ☆ Where appropriate reward individuals with extra responsibility, advancement and status
- ☆ Identify, develop and utilise each individual's capabilities and strengths
- ☆ Train and develop individual team members
- ☆ Develop individual freedom and authority

Action Centred Leadership and John Adair

John Adair, born 1934, British, developed his Action Centred Leadership model while lecturing at Sandhurst Royal Military Academy and as assistant director and head of leadership department at The Industrial Society. This would have been during the 1960s and 70s, so in terms of management theories, Adair's works is relatively recent.

His work certainly encompasses and endorses much of the previous thinking on human needs and motivation by Maslow, Herzberg and Fayol, and his theory adds an elegant and simple additional organisational dimension to these earlier works. Very importantly, Adair was probably the first to demonstrate that leadership is a trainable, transferable skill, rather than it being an exclusively inborn ability.

He helped change perception of management to encompass leadership, to include associated abilities of decision-making, communication and time-management. As well as developing the Action Centred Leadership model, Adair wrote over 40 books on management and leadership, including Effective Leadership, Not Bosses but Leaders, and Great Leaders.

Adair is now a management consultant and also has his own publishing company in Surrey, England. He also maintains links with the University of Surrey, where he was the first UK chair of leadership studies, 1979-83. The John Adair Leadership Foundation appointed Dr David Farady as its director in January 2006.

Carol Kennedy's excellent book 'Guide to the Management Gurus' supports the view that John Adair's ideas are fundamental and very significant in the development of management and leadership thinking.

Leadership is different to management. All leaders are not necessarily great managers, but the best leaders will possess good management skills. One skill-set does not automatically imply the other will be present.

Adair used the original word meanings to emphasise this: Leadership is an ancient ability about deciding direction, from an Anglo-Saxon word meaning the road or path ahead; knowing the next step and then taking others with you to it. Managing is a later concept, from Latin 'manus', meaning hand, and more associated with handling a system or machine of some kind. The original concept of managing began in the 19th century when engineers and accountants started to become entrepreneurs.

There are valuable elements of management not necessarily found in leadership, *e.g.* administration and managing resources. Leadership on the other hand contains elements not necessarily found in management, eg, inspiring others through the leaders own enthusiasm and commitment.

The Action Centred Leadership model is Adair's best known work, in which the three elements - Achieving the Task, Developing the Team and Developing Individuals - are mutually dependent, as well as being separately essential to the overall leadership role.

Importantly as well, Adair set out these core functions of leadership and says they are vital to the Action Centered Leadership model:

- ✰ Planning - seeking information, defining tasks, setting aims
- ✰ Initiating - briefing, task allocation, setting standards
- ✰ Controlling - maintaining standards, ensuring progress, ongoing decision-making
- ✰ Supporting - individuals' contributions, encouraging, team spirit, reconciling, morale
- ✰ Informing - clarifying tasks and plans, updating, receiving feedback and interpreting
- ✰ Evaluating - feasibility of ideas, performance, enabling self assessment

The Action Centred Leadership model therefore does not stand alone; it must be part of an integrated approach to managing and leading, and also which should include a strong emphasis on applying these principles through training.

Adair also promotes a '50:50 rule' which he applies to various situations involving two possible influencers, *e.g.* the view that 50 per cent of motivation lies with the individual and 50 per cent comes from external factors, among them leadership from another. This contradicts most of the motivation gurus who assert that most motivation is from within the individual. He also suggests that 50 per cent of team building success comes from the team and 50 per cent from the leader.

Adair is an example of how management thinking changes and becomes more sophisticated over time, and in response to the development of previous management thinking. Personally I have great respect for Adair's work - it's far more accessible and relevant than much of the traditional previous gurus' thinking - it's holistic as well - you can see how it works easily in a multi-dimensional way, and above all believe it gets right to the heart of the leadership role, which explains very clearly why some succeed and others do not.

Part VI

Communication and Diffusion of Innovations

28

Concept and Element of Communication

Introduction

The beginning and booming of human civilization, its growth and stability, its decline and demolition have all been attributed to the process of communication. Once the helpless and armless human being started scrambling across the meandering path of progress to reach the destination, the initiation of the process of communication also kept on moving on and on. Without symbol he remained culturally mute and with symbol he started creating his culture to communication. The logical transformation of forest life into an agrarian life is the resultant of a revolutionary technological problems spearheaded by the need of survival. But both for the nomadic life and agrarian life, communication went on contributing comprehensively. The transformation of blunt stones into sharp stone itself had been the most signification development communication in the annals of human civilization. For the first time the human need had been translated through a primitive version of technology, the development communication.

Communication

The word 'communication' comes from the Latin word communis, meaning common. 'Communication is a process by which two or more people exchange ideas, facts, feelings, or impressions in way that each gains a common understanding of meaning, intent and use of message.

Levels of Communication

According to Thayer (1968) there are four level of human communication. The levels of communication are the following.

1) Intrapersonal Communication

It is language use or thought internal to the communicator. Intrapersonal communication is the active internal involvement of the individual in symbolic processing of messages. The individual becomes his or her own sender and receiver, providing feedback to him or herself in an ongoing internal process. It can be useful to envision intrapersonal communication occurring in the mind of the individual in a model which contains a sender, receiver, and feedback loop. Although successful communication is generally defined as being between two or more individuals, issues concerning the useful nature of communicating with oneself and problems concerning communication with non-sentient entities such as computers have made some argue that this definition is too narrow. Intrapersonal communication can encompass: Day-dreaming Nocturnal dreaming, including and especially lucid dreaming Speaking aloud (talking to oneself), reading aloud, repeating what one hears; the additional activities of speaking and hearing (in the third case of hearing again) what one thinks, reads or hears may increase concentration and retention.

2) Interpersonal Communication

"A spoken word flies, you won't catch it". A Russian proverb.

"I will pay more for the ability to deal with people than for any other ability under the sun". John Rockefeller.

"Trust is the single most important factor in both personal and professional relationships".

Interpersonal communication skills are much more important than technical skills if you want to become a senior manager. Good communication is a very important aspect of interpersonal relationships. Listening is one part of communication which sounds fine in theory but is rare to find in practice. Empathy will help you to understand people and respond to their needs more effectively. Understanding the basic communication model should help to improve your communication skills. In the basic communication model there is a sender, a message and a receiver. In other words, it's a speaker, words and listeners.

Sender

First the sender conceives an idea which must be translated into words. The effectiveness with which the sender conveys the message will depend on choice and appropriateness of words, vocabulary, tone and gestures. All these factors help to project the message with precision and clarity. The message may be distorted, ignored or misunderstood, if any of these is mishandled by the sender.

Message

The message consists of words and non-verbal cues, what are words? Words are symbols which represent concrete objects and concepts. It is important to remember that words can create understanding and inspire people to great deeds or they can create misunderstanding, distract, annoyance and bad feeling. So pick your words with care and discretion. Words are not necessarily the precision tools of description that they are made out to be.

For example:

- ✰ Words mean different things in different contexts. In the context of a training programme, 'course' means curriculum, but in the context of a horse or a greyhound race it means route. Therefore meaning of words lies in the people and not in the words.
- ✰ Words mean different things in different cultures.
- ✰ Words mean different things to different people. A bull to a farmer means livestock. On the other hand, to a stockbroker, it means a person who buys shares.

Receiver

The third ingredient in the communication model is the receiver. The involvement of the receiver can be considered in three ways. Firstly, the receiver recognizes what the speaker is saying and tunes in. secondly, the receiver interprets the meaning of what is being said by mentally summarizing and reorganizing the idea. Thirdly, the receiver understands the message, which has been assimilated, associates and consolidates with exiting knowledge and only then responds to what has been said. You can now see just how complicated communication is. Mishaps can occur at any stage and interfere with the message. This is the communication cycle.

Verbal and Non-verbal Communication

The experts tell us that only thirty five (35) per cent of what we mean is in the verbal message. The other sixty five (65) per cent of the meaning is contained in the non-verbal communication. Facial expressions, eyes and other body movements must be considered when interpreting meaning. In any social interaction with other people you must be aware of the non-verbal cues and signals. Psychologists have been paying increasing attention to these non-verbal aspects of communication. The following are some of their findings.

Boredom

Boredom is "an unpleasant, transient affective state in which the individual feels pervasive lack of interest and difficulty concentrating on the current activity. Or in other way we can say "an affective experience associated with cognitive attentional processes". In positive psychology, boredom is described as a response to a moderate challenge for which the subject has more than enough skill. There are three types of boredom, all of which involve problems of engagement of attention. These include times when we are prevented from engaging in wanted activity, when we are forced to engage in unwanted activity, or when we are simply unable for no apparent reason to maintain engagement in any activity or spectacle. Boredom proneness is a tendency to experience boredom of all types.

Arms Folded

Folded arms usually indicate resistance; the listener is forming a protective barrier against you and your ideas. But beware, it can also mean other things.

Frown

The frown may suggest disagreement, lack of understanding or annoyance. Several responses are possible. You can back up and restate your last point. You can pause and ask for questions, or ask you can clarify your point.

Upward Glance

Watch people's eyes. The eyes are the minor to the soul. The sky ward glance may mean "I have heard that one before". Eye contact is essential for good communication.

Memory

The average person has a poorly trained memory. Both the long-term and short-term memories are important factors in communication. Research has shown that we forget 50 per cent of what we read immediately unless we review. We forget 80 per cent of what we hear within 24 hours. We forget 90 per cent within 48 hours. Don't rely on your over worked memory that we tend to remember what we like and forget what we dislike.

Encouraging People to like you

There are some practical steps we can take to encourage other people to like us better. We can put the following tips into practice straight away:

Positive Thinking

Many books have been written about the power of positive thinking. The problem with negative thinking is that it becomes a self-fulfilling prophecy. The good news is that positive thinking is likewise a self-fulfilling prophecy. Shakespeare said: there is nothing good or bad except that thinking makes it so. Believe in yourself. Think positive experiences. Think about your successes rather than your failures. Develop a mindset for success. Positive expectations are more likely to lead to positive outcomes. Positive thoughts proceed positive actions and positive actions produce positive results. Negative expectations are more likely to lead to negative outcomes. Positive thinking is about developing a favorable image of yourself.

Show Interest

Take a genuine interest in the people around you and in the work environment. To be interesting, be interested. Ask questions that the other person will enjoy answering. Encourage them to talk about themselves and their achievements. Disraeli said: talk to a man about himself and he will listen for hours. Develop a friendly and helpful attitude. Always try and be sincere. If you are artificial, people will eventually see through you. Be positive. If you don't know something, say: 'I don't know, but I'll certainly find out and let you know', rather than a curt: 'Don't know'. Develop a reputation for reliability. If you say you're going to do something, do it!

Remember to Smile

A smile is like a piece of sunshine. You'll be surprised at the friendly way people will react. Smiling facilitates positive thinking by giving you a psychological lift. You can't smile and entertain negative thoughts at the same time. Develop a cheerful disposition.

- ☆ **Name:** People like to hear the sound of their own names. One of the best ways of crating goodwill is to remember names and use them. To remember names, relate them to the person's features and repeat the name frequently during the conversation.
- ☆ **Manners:** Good manners are a sign of good breeding. Manners help make living pleasant. Like smiling, good manners cost nothing but reap many benefits.
- ☆ **Learning is a lifelong continuous process:** Seek out opportunities for self-development, not only in your work life but also in your personal life. Action cures fear. Don't procrastinate. Undertake that educational programme non-whether formal, such as certificate, diploma or degree programme, or informal, such as challenging recreational pursuits. It will make you a more interesting person.
- ☆ **Practice:** Practice the previous advice. It's like driving a car, if you practice these, they will become automatic confident responses. Bernard Sha once remarked: 'If you teach a man anything, he will never learn'. Learning is an active process. We learn by doing. Only knowledge that is used sticks in your mind.

Handling People

The following guidelines will help you to handle people.

Tact

Don't argue: The best way to win an argument is to avoid it. Let the other person do most of the talking. Use the feedback technique. Ask questions in a friendly voice. Use the open-ended questioning technique – Why? What? When? Where? How? And Who? Arguing encourages emotional responses such as anger. Anger leads to irrational behavior and anything that encourages anger should be avoided. If, despite everything, a person gets angry, don't get angry back. 'Anger blows out the light of reason, but, after a roaring hurricane expends its fury, the full of calm sets in'. Let the angry person blow themselves out.

Respect

Show respect for the other person's feelings and opinions. Tact and diplomacy are the watchwords here. Never say directly, or suggest to a person, 'You're wrong'. You could say you may be right, but let's look at the facts'. Nobody likes being told directly that they're wrong especially in front of a third party.

Empathize

See things from the other person's point of view. How would you feel if you were in their shoes? That is what empathy means. Think ahead and try and anticipate reactions.

Simple

Speak in simple language. Remember the KISS technique-Keep if Short and Simple. Avoid jargon or technical terms. Don't try to impress others by using big words. Always use the simple work in preference to the long work. Keep your sentences short. Big words and long-winded sentences amount to woolly thinking, confusing others as well as yourself.

Praise

Most people are very reluctant to praise others, even when it is due. We spend most of our time knocking others. Treat people as winners and they will live up to your expectations. A sincere compliment for work well done is a boost to morale and an incentive for sustained excellence in the future.

Appeal to Senses

Dramatic your ideas and involve the other person. Appeal to as many of the senses as possible – hearing, sight, taste, smell and touch. When making a presentation to a group use visuals-a picture speaks a thousand words. Remember the old proverb: 'I hear, I forget; I see, I member; I do, I understand'.

Synchronize with your Listener

Watch for body language and non-verbal communication. Try and synchronize your words with the speed of receptivity of the listener (you can gauge this from the actions and expressions of the listener). Mirror your body language to that of the other person.

Sell

Make other person feel the idea is theirs. This is the best way to implement your ideas. Use the open-ended questioning technique to obtain viewpoints. Concentrate and develop those on which there is common ground and agreement. Show what's in it for them. Most people are motivated by self-interest.

Making an impression on others

The critical points about making an impression on other people are:

- ☆ Posture
- ☆ Look, listen and learn
- ☆ Expression
- ☆ Appearance
- ☆ Speech
- ☆ Eagerness to help

Posture

Your posture can indicate clearly whether or not you are going to be friendly and helpful. Indifference, nerves, restlessness and lack of confidence can all be indicated by posture.

Look, Listen and Learn

For the total message listen for the words and observe the accompanying body language. Pay attention to what they say and show that you are really interested in them. People prefer to talk about themselves, rather than listen to you. Don't interrupt the speaker or change the subject or you will show lack of interest and may create resentment.

Expression

Most people look at your face, and particularly your eyes, at some point during your conversation. Shakespeare said that the eyes are a minor of the soul. Remember, your face and eyes reveal your feelings. They show the other person not only how your yourself feel – tired, interested, uninterested – but also how you feel about the other person. It's polite and a sign of attentiveness to look at the person who is talking to you (but avoid staring).

Appearance

When somebody comes up to you, both you and your work area are on view. Do you give the impression that you are friendly, neat, well-groomed, smart and organized? Remember, if you look untidy, your work may be untidy too.

Speech

Your first words create an impression, which colour the other person's reply. If they are friendly and positive, they invite a smile and a thank you from the other person. Nothing is more annoying than unhelpful, negative remarks. Remember to use the person/s name. There is nothing sweeter to a person's ear than the sound of their own name. Use the name frequently during the conversation.

Eagerness to Help

Adopt a positive attitude and helpful manner in your dealings with people. Show in a positive way that you are enthusiastic, eager and willing to help solve problems. If you want to make friends you must be seen to be obliging, unselfish and thoughtful.

Consequences of Poor Personal Inter-Communication Skills

- Stress
- Lack of communication
- Irritation
- Close mindedness
- No team spirit
- Lack of credibility
- Poor self-esteem
- Suspicion
- Loss of productivity

- ☆ Isolation
- ☆ Poor health
- ☆ Distrust
- ☆ Anger
- ☆ Prejudice
- ☆ Un co-operative behaviour
- ☆ Conflict
- ☆ Frustration
- ☆ Unhappiness

Interpersonal communication skills are much more important than technical skills if you want to become a senior manager. Good communication is a very important aspect of interpersonal relationships. Listening is one part of communication, which sounds fine in theory but is rare to find in practice. Empathy will help you understand people and respond to their needs more effectively. Understanding the basic communication model should help to improve your communication skills. Experts maintain that as much as 65 per cent of the message communication is non-verbal or body language.

Organizational Communication

Communication is the most vital ingredient of an organization. In fact, an organization cannot be conceived of without communication. An organization is a group of persons constituted to achieve certain specific objectives. The achievement of these objectives largely depends upon a proper co-ordination and integration of human effort in an organization. The people working in an organization are interrelated; their activities are also interrelated because all activities are performed only to achieve the organizational objectives. Communication is the nervous system of an organization. It keeps the member of the organization informed about the internal and external happenings relevant to a task and of interest to the organization. It coordinates the efforts of the member to achieving organizational objectives. It is the process of influencing the action person or a group. It is a process of meaningful interaction among human beings to initiate, execute, accomplish, or prevent certain actions. Communication is, thus, the life food of an organization. Without communication, an organization is lifeless and its very existence is in danger.

Types of Organizational Communication

The Organizational Communication may be grouped on the basis shown in Figure given on next page.

According to Organizational Structure

Formal Communications

Such communications are those communications, which are associated with the

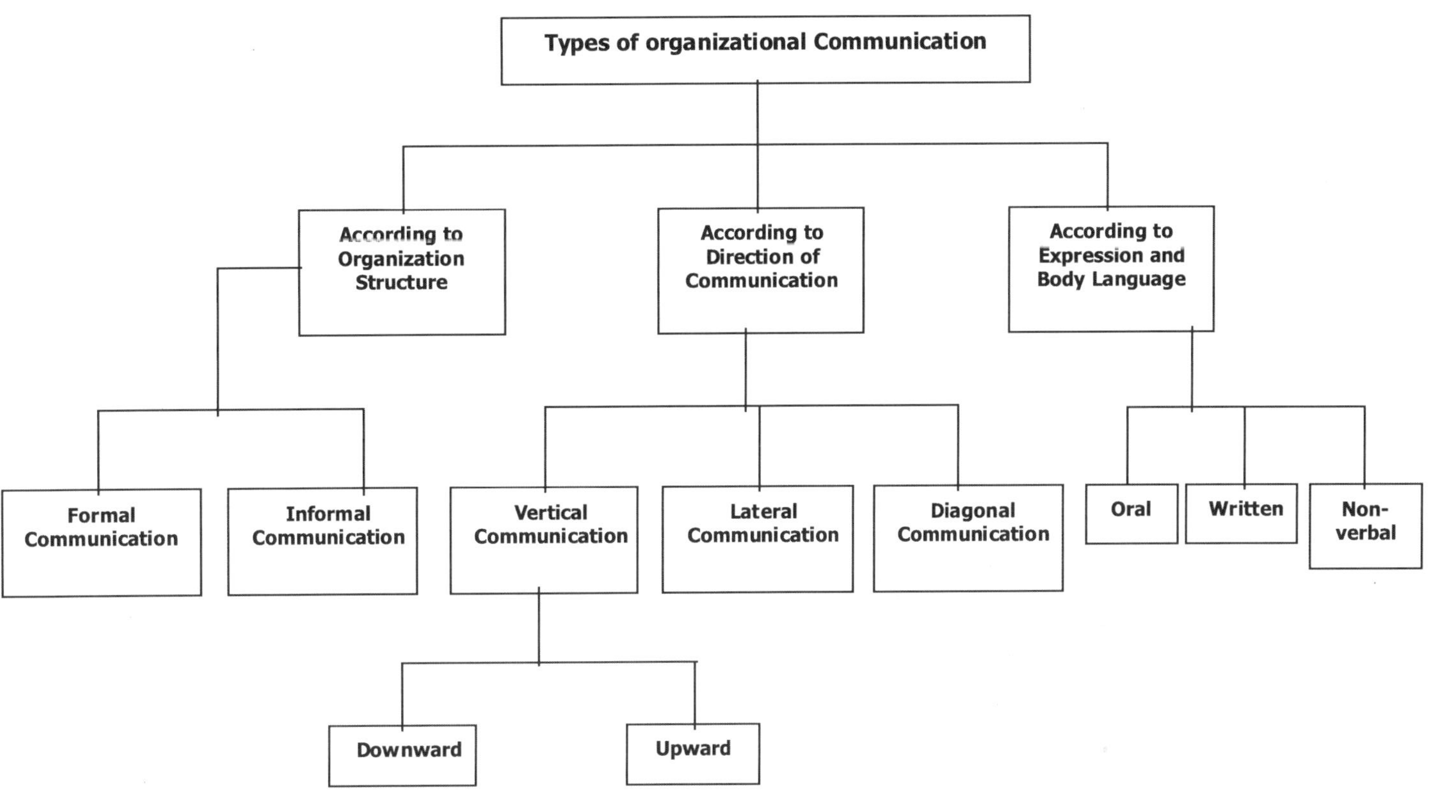
Types of organizational Communication
According to Organization Structure
According to Direction of Communication
According to Expression and Body Language
Formal Communication
Informal Communication
Vertical Communication
Lateral Communication
Diagonal Communication
Oral
Written
Non-verbal
Downward
Upward

formal organization structure. The travel through the formal channels – officially recognized positions in the organization chart. They are established mainly by the organization structure. Formal communications are mostly in black and white. We generally hear the phrase "through proper channel".

Informal Communications

Informal Communications are also known as "Grapevine Communications". They are free from all sorts of formalities, because they are based on the informal relationship between the parties, such as friendship, membership of the same club or association or origin from the same place. Such communication includes comments, suggestions of any other informal reaction also.

According to Direction of Communication

Downward Communication

Communications which flow from the superiors to subordinates are known as downward Communications. They include orders, rules instructions and policy directives *etc.* Their nature is directive. It would be impossible to manage an enterprise without downward Communications.

Upward Communications

Upward Communications are just reverse of the downward Communications. It flows from the sub-ordinates to their superiors. Such communications include reactions and suggestions from workers, their grievance *etc.* Contents of the upward Communication are reports, reactions, suggestions, statements and proposals prepared for the submission to the boss *etc.* Upward Communication is considered to be a main source of motivation in employees.

Exhibit : Vertical Communication

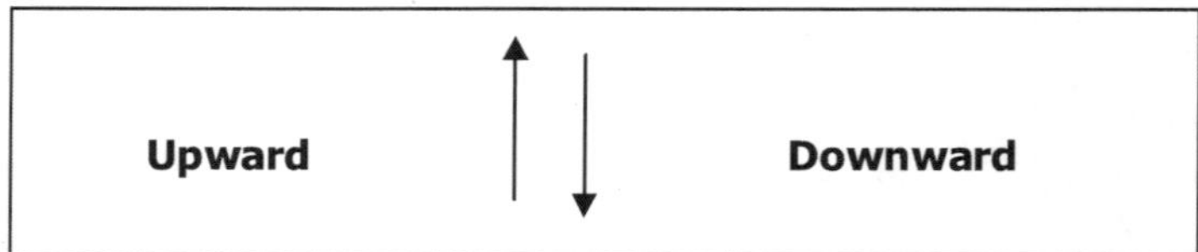

Lateral/Horizontal Communication

This type of communication refers to communication between various departments or units representing the same level, or people within the same or different departments, without having a superior – sub-ordinate relationship *viz.*, peer groups, friends and trainees. It flow between persons at the same hierarchical level. The main object of this type of communication is to co-ordinate the efforts of different departments performing different but related activities. Such communications may be oral or written.

Exhibit: Lateral Communication

Diagonal Communication

In an organization, communication does not necessarily traverse along a stipulated path. While vertical and lateral forms for transmission of messages are important, there is yet another mode to be taken cognizance of, that is, the diagonal.

Inter-organizational Communication

This is a system developed by each organization to communicate with another organization. Department or organizations do not intercommunicate as such rather people do. Communication always occurs between people.

The Communication Process

The dynamics of communication has been both explicit and implicit through the process. It starts from a source and reaches the ultimate receiver through a channel. While it is travelling and traversing along and across the channels, the message keeps moving in its original form or in a changed version. The source is in human or non human entity and so also is the receiver. The accepted definition of communication is, it is the process by which a Source is sending a Message through a Channel to some Receiver for some Effect. It is popularly known as SMCRE approach in defining communication process.

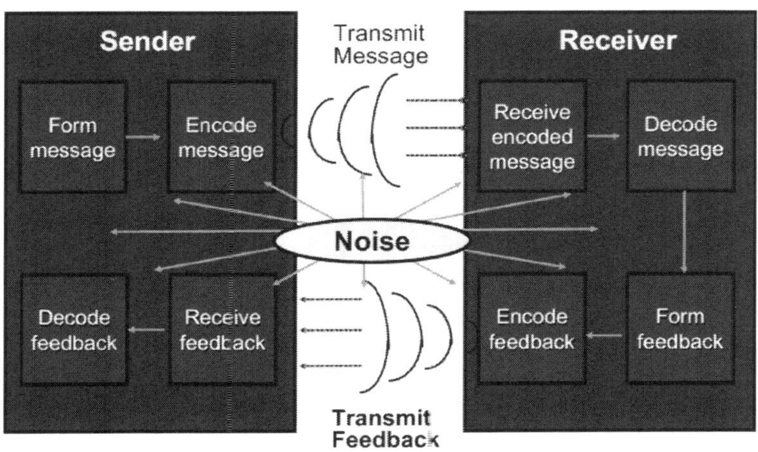

29

Models of Communication

Models are symbolic representations of structures, objects or operations. They are useful theoretical constructs that are frequently used in social sciences for explanatory purposes. They may be used to show the size, shape or relationship of various parts or components of an object or process. A model may also be useful in explaining the working of a system.

Concepts of Model

a) It is representation of the process of communication in the same sense that a blueprint is a representation of house.

b) Models are symbolic representations of structures, objects or operations.

Purpose of Models

Communication models serve three main purposes

a) They describe the process of communication.

b) They visually show relationship among the variables involved in communication and

c) They aid in finding and in correcting communication problems

Aristotle's Model

According to Aristotle, communication has three ingredients

1. Speaker – the person who speaks
2. Speech – the speech that the individual produces
3. Audience – the person who listens

Aristotle's Model

The Osgood and Schramm circular model emphasizes the circular nature of communication.

The participants swap between the roles of source/encoder and receiver/decoder.

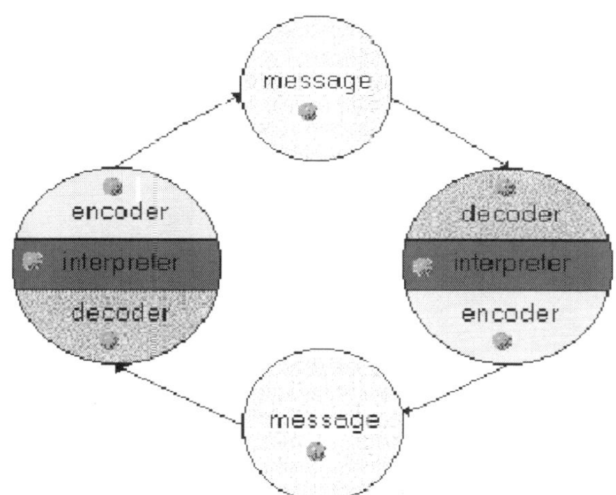

The Osgood & Schramm Circular Model
Please click a button for further information
(Message, encoder and decoder will take you to Shannon-Weaver)

***Source*: CCMS-Infobase at http://www.cultsock.ndirect.co.uk/**

Gerbner's General Model

Gerbner's General Model also emphasizes the dynamic nature of human communication.

Shannon-Weaver's model

The Shannon-weaver (1949) model is consistent with Aristotle's proposition. According to them, the ingredients of communication are :

1. Source
2. Transmitter
3. Signal
4. Receiver
5. Destination

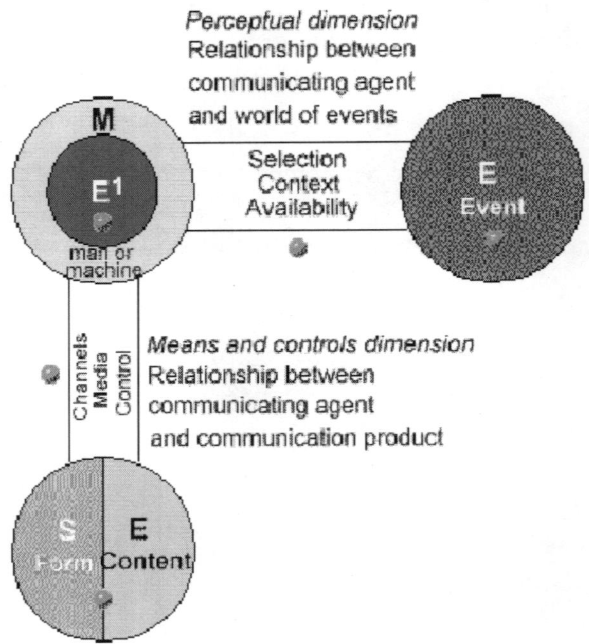

Gerbner's General Model

Please click a button for further information

***Source*: CCMS-Infobase at http://www.cultsock.ndirect.co.uk/**

Shannon and Weaver's Model

Compared with the Aristotelian model, the source is the speaker, the signal is the speech and the destination is the audience, plus two added ingredients, it transmits which sends out the source's message and a receiver which catches the message for the destination.

Berlo's Model

According to Berlo (1960) the model of communication consists of:

a) Source
b) Encoder
c) Message
d) Channel
e) Decoder
f) Receiver

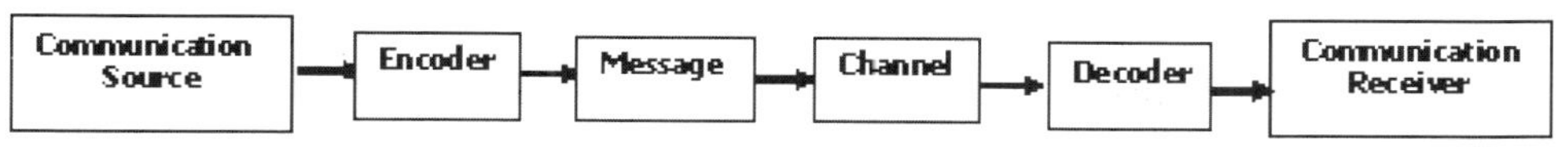

Berlo's Model

Constructionist Models

There is an additional working definition of communication to consider that authors like Richard A. Lanham (2003) and as far back as Erving Goffman (1959) have highlighted. This is a progression from Lasswell's attempt to define human communication through to this century and revolutionized into the constructionist model. Constructionists believe that the process of communication is in itself the only messages that exist.

Linear Model

This is a one way model to communicate with others. It consists of the sender encoding a message and channelling it to the receiver in the presence of noise. Draw backs – the linear model assumes that there is a clear cut beginning and end to communication. It also displays no feedback from the receiver.

☆ For example; a letter, email, text message, lecture.

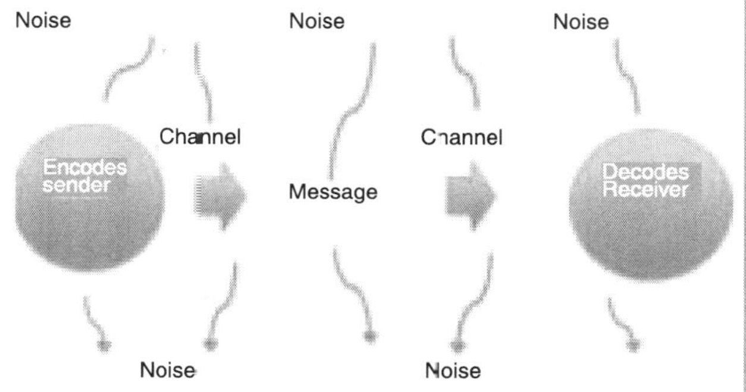

The Linear Model

Interactive Model

The two linear models are stacked on top of each other. The sender channels a message to the receiver and the receiver then becomes the sender and channels a message to the original sender. This model has added feedback, indicates that communication is not a one way but a two way process.

Transactional Model

It assumes that people are connected through communication; they engage in transaction. Firstly, it recognizes that each of us is a sender-receiver, not merely a sender or a receiver. Secondly, it recognizes that communication affects all parties involved.

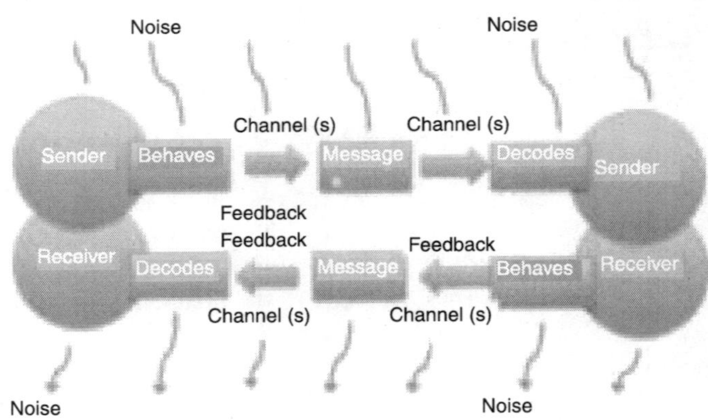

The Interactive Model

☆ For example – talking/listening to friends. While your friend is talking you are constantly giving them feedback on what you think through your facial expression verbal feedback without necessarily stopping your friend from talking.

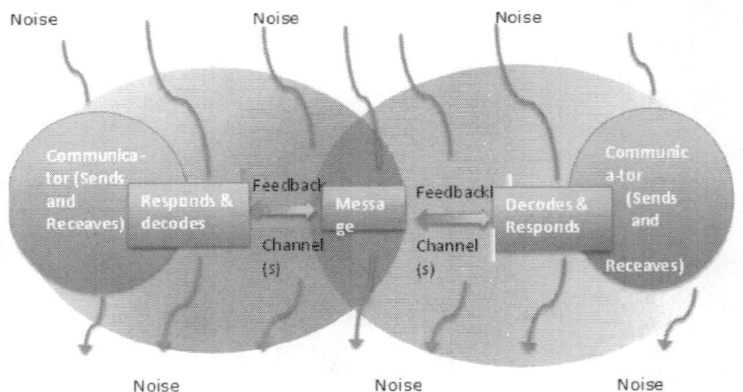

The Transactional Model

Schramm's Model

According to Schramm (1961), the communication process involves:

1. Source
2. Encoder
3. Signal
4. Decoder
5. Destination

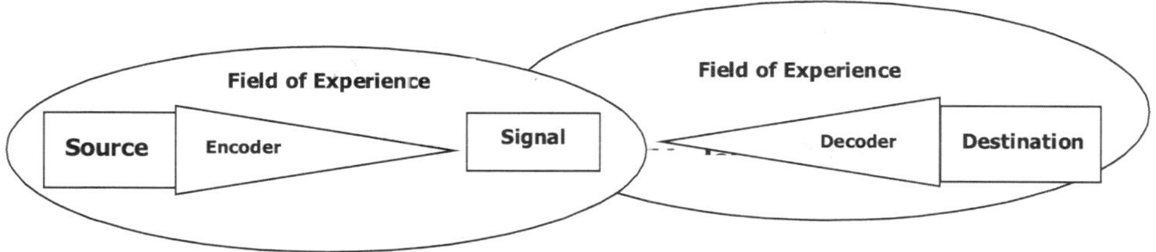

Schramm's Model

This model of communication is preferably relevant for the mass media. In human communication it is most important whether people can properly encode or decode the signal (message), and how they interpret in their own situations.

Jean Paul Leagan's Model

The communication model forwarded by Leagans (1963) has the following elements:

a) Communicator
b) Message
c) Channel
d) Treatment
e) Audience
f) Response

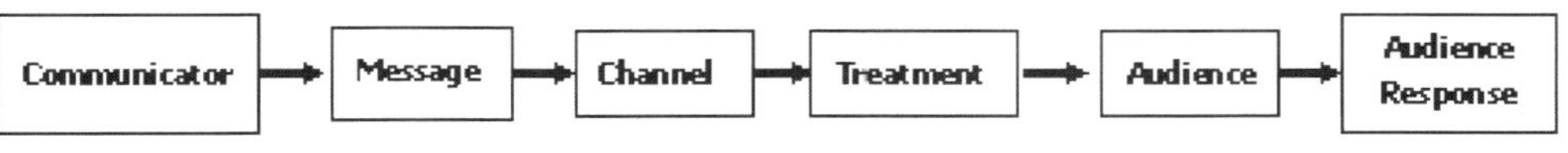

J.P. Leagans Model

Rogers and Shoemaker's Model

Rogers and shoemaker (1971) thought of the communication process in terms of the S-M-C-R-E model, the components of which are:

a) Source
b) Message
c) Channel
d) Receiver
e) Effects

Rogers and Shoemaker's Model

According to them a source **(S)** and message **(M)** via certain channels **(C)** to the receiving individual **(R),** which causes some effects **(E)** *i.e.,* changing the existing behaviour pattern of the receiver.

Westley and Maclean's Model

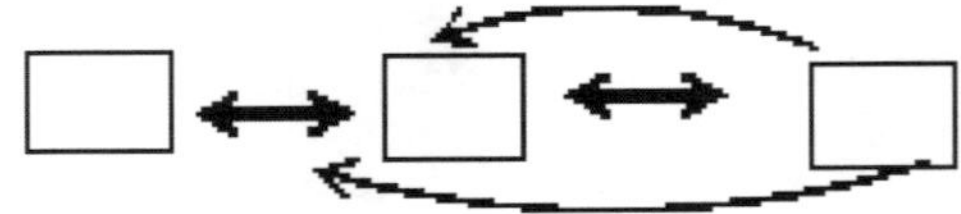

'A' is source (*E.g.,* ANGRAU), 'B' is receiver or public, 'C' is the mass media channel (*e.g.,* E-TV Annadata), 'A' gets feedback from the public refines the message and 'C' also gets feedback and refines and the total refinements are carried out by the interpreting source that is 'C' and then refined messages are transmitted to the users. This is typical in a TV channel broadcast. *e.g.,* Annadata of E-TV does similar to explained above.

Model on Extension Communication System

Communication in extension may also be thought of as two-way stimulus – response (S-R) situation in which the necessary stimulus is provided by the communicator, the extension agent, in the form of a message which produces certain response on the audience, the farmers and vice-versa. A favourable response by the audience reinforces learning. A diagrammatic representation of the extension communication system on the basis of the model suggested by Leagans (1963) is presented in the below figure.

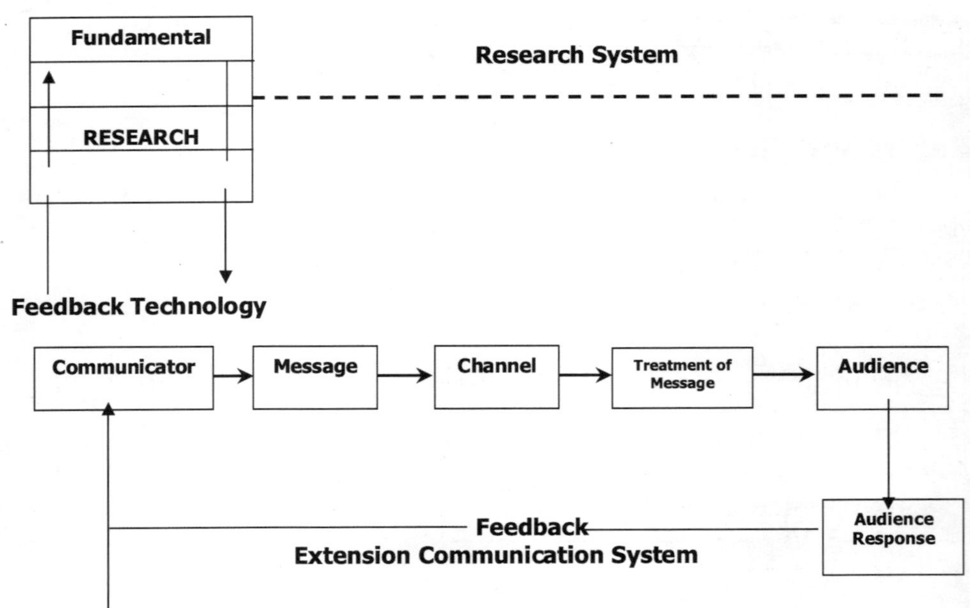

30

Principles of Effective Communication

Principles of Communication

Communication is a two-way process of giving and receiving information through any number of channels. Whether one is speaking informally to a colleague, addressing a conference or meeting, writing a newsletter article or formal report, the following basic principles apply:

- ☆ Know your audience.
- ☆ Know your purpose.
- ☆ Know your topic.
- ☆ Anticipate objections.
- ☆ Present a rounded picture.
- ☆ Achieve credibility with your audience.
- ☆ Follow through on what you say.
- ☆ Communicate a little at a time.
- ☆ Present information in several ways.
- ☆ Develop a practical, useful way to get feedback.
- ☆ Use multiple communication techniques.

Communication is complex. When listening to or reading someone else's message, we often filter what's being said through a screen of our own opinions. One of the major barriers to communication is our own ideas and opinions.

There's an old communications game, telegraph, that's played in a circle. A message is whispered around from person to person. What the exercise usually proves is how profoundly the message changes as it passes through the distortion of each person's inner "filter."

Environmental Factors

Communication can be influenced by environmental factors that have nothing to do with the content of the message. Some of these factors are:

- ☆ The nature of the room, how warm it is, smoke, comfort of the chair, *etc.*
- ☆ Outside distractions, what is going on in the area.
- ☆ The reputation/credibility of the speaker/writer.
- ☆ The appearance, style or authority of the speaker.
- ☆ Listener's education, knowledge of the topic, *etc.*
- ☆ The language, page layout, design of the message.

People Remember

- ☆ 10 per cent of what they read
- ☆ 20 per cent of what they hear
- ☆ 30 per cent of what they see
- ☆ 40 per cent of what they hear and see

Communication with Decision Makers

Innovation and change often depends upon persuading potential users of the benefits of an innovation. To deal persuasively with decision makers, it is necessary to know and understand their interests and opinions. The following questions are helpful in organizing technology transfer efforts:

a) Who are the key people to persuade?

b) Who will make the decisions about innovation and change?

c) What are these decision makers' past experiences with innovation and change?

d) What are the decision makers' current attitudes toward innovation and change? Are they neutral, friendly, hostile or apathetic?

e) What is the most appropriate way to approach the decision maker?

f) What are the work styles of the decision makers? Are they highly formal people who want everything in writing and all appointments scheduled in advance? Or are they more flexible, responding favorably to personal telephone calls and informal meetings?

g) What networks or groups is the decision maker a part of?

h) What programs or services will the new innovation improve?

i) What programs or services will the new innovation cause problems with?
j) How will the innovation or change benefit the decision maker?

Principles of Effective Persuasion

Whether making a formal presentation at a meeting or writing a report or fact sheet, the following principles hold.

- Do not oversell or overstate your case. Make effective use of understatement.
- Outline the topic you are trying to cover into two parts. The first part should give broad background information, while the second part provides a detailed summary.
- Persuasion depends on clarity and simplicity. Avoid the use of jargon and buzz words.
- Be prepared to back up claims or facts immediately.
- Incorporate major anticipated objections into your program or presentation.
- Address all relevant aspects of a topic, especially those that may affect the functioning of an organization.
- Use graphics and audiovisuals appropriately.
- Consider ways to get meaningful input from people. Find out what they think about the innovation or change.

Selecting the Best Communication Method

In communicating with decision makers, use the most appropriate communications method. One way to do this is to ask yourself the following questions.

- What is the purpose of your message? Do you plan to tell them something new? Inform? Do you plan to change their view? Persuade?
- What facts must be presented to achieve your desired effect?
- What action, if any, do you expect decision makers to take?
- What general ideas, opinions and conclusions must be stressed?
- Are you thoroughly familiar with all the important information on the innovation?
- What resources and constraints affect adoption of the innovation? How much time is available? How much money is available
- Which method, or combination of methods, will work most effectively for this situation? Personal contact—requires scheduling, time and interpersonal skills.

31

Theories of Communication

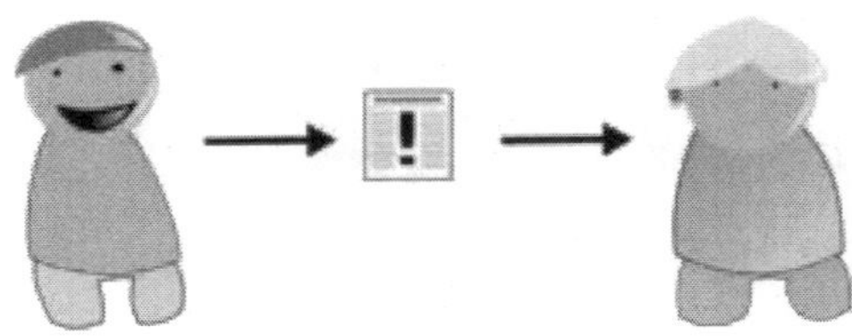

A simple communication model with a sender transferring a message containing information to a receiver

The theories of communication can be classified into three groups namely:

Theories about any phenomenon in general will highlight the key concepts involved in it. These theories of communication will be useful for a person who is trying to understand a given communication event by way of providing a framework which suggests what he should look for. These theories should be viewed based on the basic assumptions that communications is a transitive, multidimensional, multipurpose process.

Theories of Mass Communication (Message Flow Models)

Several models have been conceptualized to explain the flow of message from the communicator to the mass audience. Based on the number of ways involved in the flow of message between sender and receiver theories like hypodermic needle, one step flow, two step flow, and multistep flow are conceived.

Theories of Communication Distortion

The first one states that larger the volume of communication, larger would be the communication distortion. Increasing the volume of communication increases more chances for misunderstanding, misinterpretation, less concentration and low priorities.

Theory of Communication Distortion in Transit

The second one emphasizes that messages pass through many layers in an organization. Therefore, repetition and reproduction of the same message is multiplied into many which reduces the efficiency of communication. It can be also viewed from the point view of individual gathering information as he passes through his environment.

Communication Process and the Key Communicator

Key communicators or opinion leaders are persons in any social systems that are sought out for information and advice on general or specific topics. Rogers and Kincoid (1981) define opinion leadership as the degree to which an individual is able to informally influence other's knowledge, attitudes, or overt behaviour in a desired way with relative frequency. It is assumed that such persons are respected persons in each social system to whom some people look for advice and information and who, through such consultations, influence their behaviour and actions.

Identification Methods of Key Informants

Three methods are in common use for identifying key-informants. These methods are (i) Socio-metric method, (ii) self-designation technique and (iii) information ratings.

Socio-metric Method

In this method, members of a social system are asked to name the person(s) to whom they go for advice and information on a particular topic. It has several advantages over other methods in so far as socio-metric questions are easy to administer and are adaptable to different situations. This method has also relatively higher validity. However, the use of this method requires a large number of respondents and complex statistical analysis. This is concerned primarily with obtaining choices in interpersonal relations, such as with whom one would like to work, play, *etc.* or to whom one would go for advice on farming or other problems.

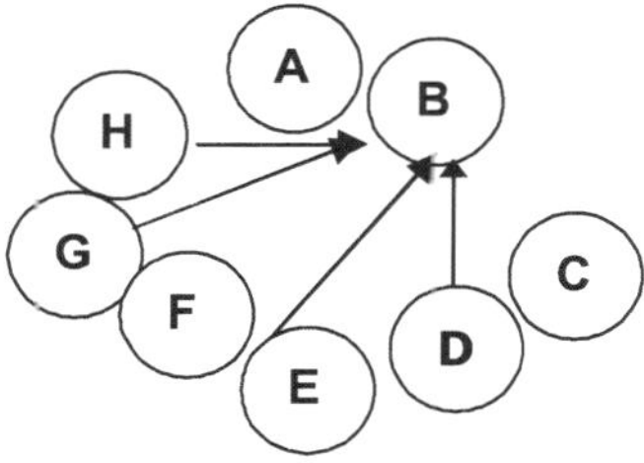

Socio-gram

Information Ratings

In this method, some members of a social system are purposively or randomly selected and asked to designate persons giving advice and information on a general or particular topic. This method is economical and quick as it saves cost and time.

Self-Designating Method

In this method, each selected person is asked a series of questions designed to determine the degree to which he perceives himself to be a key-communicator. This method has one important advantage in so far as it also measures the individual's perception of his being a key-communicator, which in turn influences his behaviour.

Characteristics

Key-communicators have been found to have distinctive characteristics. On the basis of a review of a large number of research studies, Rogers and Shoemarker (1971) drew the following conclusions regarding characteristics of key-informants.

(i) External Communication

a) Key-communicators have greater exposure to mass media than their followers

b) Key-communicators are more cosmopolitan than their followers.

c) Key-communicators have greater change agent contact than their followers.

(ii) Accessibility

Key-communicators exercise relatively greater social participation than their followers in a social system.

(iii) Social Status

Key communicators enjoy a relatively higher social status than their followers as far as conditions in that social system exists.

(iv) Innovativeness

a) Key-communicators are more innovative than their followers.

b) When the social system's norms favors change, key-communicators are more innovative but otherwise not especially so.

c) When the norms of a system are more modern, key-communicators are more monographic. Monography is the tendency of a key-communicator to act as such for only one topic.

(v) Relationship to the Social Systems

According to Shankaraiah (1969), the key-communicators under Indian conditions are:

a) Perceived as the best farmers when the social system is actively modern.

b) Social stars when the social system is relatively traditional.

On the basis of a review made by Sandhu (1970) of a large number of research conducted under Indian conditions, no specific conclusions could be drawn regarding age, education, farm size, *etc.*, although, key-communicators tend to belong to the middle-age group *i.e.* 30 to 50 years, be somewhat better educated and own relatively larger farms than their followers.

Role of Key Communicators

1. Communication of the new technology to other villages.
2. Keeping in touch with the scientists, other institutions, media and sources of information.
3. Coordinating the functions of the village organizations and the institutions and channelizing them towards the adoption of technologies for greater production.
4. Assisting the villagers in securing the supplies and services required by them.
5. Guiding and helping people in getting assistance from government.
6. Helping the fellow villagers in the preparation of the action plans.
7. Giving constant guidance and acting as the local consultants.
8. Focusing on the problems of villager and helping the extension workers and
9. Serving as the demonstrator in the case of agricultural innovations.

Functions of Communication

Communication is defined as the interaction, giving and taking of information, sending and receiving of messages through verbal and non verbal means. Functions of communication vary as one must determine the function of the communication. Known as the primary function and in certain circumstances the situation or position may have one, two or three other secondary functions.

Inform Functions

An organization needs a vast amount of information to function and operate a business. The top management would require timely and accurate information for the various departments to make effective decisions.

Control Functions

The management of any organization will always have plans with long, medium or long term objectives for the months and years ahead. To achieve these objectives, the daily and monthly activities must proceed as planned in order to achieve the objectives for the period.

Persuading Functions

Persuasion usually involves the 'selling' of an idea, product or services to a person or group. Persuasive words or actions, there is a better chance of the person or group accepting it.

Coordinating Functions

Every employee plays a part in controlling inputs, outputs, job scope and head count. In Any form of communication, there is a sender and receiver of the message. The question of whether the message is sent and how the message is received is of vital importance in communication. Communication is successful only when the receiver receives the intended message of the sender.

SENDER ⟶ ENDCODING ⟶ MESSAGE ⟶ DECODE ⟶
RECEIVER ⟶ FEEDBACK

The above is the communication process which has to be repeated until both parties finish expressing themselves. First, the sender has an idea then he/she transforms the idea into a message. After which he/she transmits the message and he receiver gets the message. The receiver then interprets the message (decoding) which in turn reacts and sends feedback to the sender. In conclusion, there are many functions and characteristics of communication and the above listed are just some of them.

i) Information Function

The basic requirement of adapting and adjusting oneself to the environment is information. There must be some information about what is going on in the environment which concerns the people. The receiving or giving of information underlines all communication functions, either directly or indirectly.

ii) Command or Instructive Function

Those who are hierarchically superior in the family, society or organization, often initiate communication either for the purpose of informing their subordinates or for the purpose of telling them, what to do, how to do when to do *etc.* The command and instructive functions of communication are more observable in formal organizations than in informal organizations.

iii) Influence or Persuasive Function

According to Berlo (1960), the sole purpose of communication is to influence people. Persuasive function of communication *i.e.* to induce people is extremely important for extension in changing their behaviour in the desirable direction.

iv) Integrative Function

A major function of communication is integration or of continuously offsetting any disintegration at the interpersonal or at the organizational level. This helps in maintaining individual, societal or organizational stability and identity.

The Communication Process

Before communication can take place, a purpose, expressed as a message to be conveyed is needed. It passes between a sender and a receiver. The message is encoded (converted to a symbolic form) and passed by way of some medium (channel) to the receiver, who retranslates (decodes) the message initiated by the sender. The result is transference of meaning from one person to another. The key parts of this model are: (1) the sender, (2) encoding, (3) the message, (4) the channel, (5) decoding, (6) the receiver, (7) noise, and (8) feedback.

Application of Communication Technology in Rural Development

Developing country governments are increasingly aware that they have a major responsibility for rural development and food security, but lack the capacity and solutions to meet the challenge. In 1996, the world's heads of state meeting in Rome committed their countries to eradicating hunger and reducing the number of undernourished people by 50 per cent by the year 2015 (FAO 1998). While on the one hand the agricultural yields in developing countries continue to decline, despite technological innovations, their populations continue to expand beyond food production capacities.

Potential of New ICTs for Rural Development

While the term 'ICTs' can be interpreted as including a wide range of media, 'new ICTs' is used to denote "the use of computers and communication systems between computers". These new ICTs are becoming more accessible, and users can obtain information from various sources, and one computer could meet the needs of a large rural community. Electronic mail is the most commonly used new ICT and has caused a cultural revolution in the way individuals and organisations interact, in terms of time, cost and distance. The second most significant use of new ICTs is the World Wide Web, which enables people to access information on millions of other computers. Although the Internet is not a panacea for food security and rural development problems, it can open new communication channels that bring new knowledge and information resources to rural communities (Bie 1996).

Decision Making Process

Sound decision making is dependent upon availability of comprehensive, timely and up-to-date information. Food security problems facing developing countries demonstrate the need for informed researchers, planners, policy makers, development workers and farmers.

Empowering Rural Communities

ICTs can empower rural communities and give them "a voice" that permits them to contribute to the development process. With new ICTs, rural communities can acquire the capacity to improve their living conditions and become motivated through training and dialogue with others to a level where they make decisions for their own development (Balit 1998). Giving rural people a voice means giving them a seat at the table to express their views and opinions and become part of the decision making process. The approach should be participatory and could lead to improved policy formation and execution. Farmers can also interact with other farmers, their families, neighbours, suppliers, customers and intermediaries and this is a way of educating rural communities. The Internet can also enable the remotest village to access regular and reliable information from a global library (the web). Different media combinations may, however, be best in different cases - through radio, television, video cassettes, audio cassettes, video conferencing, computer programmes, print, CD-ROM or the Internet. Rural areas also get greater visibility by having the opportunity to disseminate information about their community to the whole world.

Targeting Marginalized Groups

Most rural poor people lack the power to access information. ICTs could benefit all stakeholders including the civil society, in particular youth and women (UNDP 2000). Other disadvantaged groups that could be targeted include the disabled and subsistence peasants.

Creating Employment

Through the establishment of rural information centres, ICTs can create employment opportunities in rural areas by engaging telecentre managers, subject matter specialists, information managers, translators and information technology technicians.

Innovative Application of ICTs in Rural Areas

There are examples of initiatives in several developing countries (*e.g.* Bangladesh, Chile, India, Mauritius, Mexico, Peru, South Africa and Uganda) supported by FAO, ITU, IDRC, IFAD, UNESCO, DFID, British Council, GTZ and others.

ICTs for Improving Linkages

Weak linkages between researchers, frontline workers and farmers have been a major constraint that has resulted in research findings not being applied by poor rural farmers. ICTs can improve and strengthen these linkages and ensure knowledge and information, which are essential for improving food security are communicated to all stakeholders.

- ☆ FAO's Research, Extension and Training Division and the World Agricultural Information Centre (WAICENT) recently developed a prototype network - the Virtual Extension Research Communication Network (VERCON) concept that aims at improving linkages between agricultural research and extension institutions. The Internet tool allows network members to capture and develop local content, share, store, retrieve and disseminate information and connect geographically dispersed people from research and extension institutions, faculties of agricultural education, NGO workers and agricultural producers. The tool can also facilitate communication, sharing information and supporting improved agricultural production and can further broaden and strengthen collaboration through facilitating co-ordination of rural, local, national and regional development programmes.

- In Latin America, FAO has applied ICTs in a project to establish farmer information networks FARMNets - involving agricultural producers and farmer associations, extension services and NGOs in Chile and Mexico. Essential information on inputs, prices, markets, weather and credit are exchanged through the electronic network (via the Internet) to farmer organisations, co-operatives and local government. The project also provided training on how to analyse, retrieve and disseminate information of local relevance using ICTs (Van Crowder and Fortier 2000).
- An evaluation of the project indicated that the Internet was found appropriate for transmitting information across the network. Farmers and their associations were able to sell their produce at much higher prices than they could fetch in the local market and strategize on what quantities and when to produce. Users of the centres found the Internet to be a cheaper, timely and appropriate communication option (Balit 1998, Masias 1996). Knowing the future prices for cereals and oilseeds also enabled the association to better plan the quantities to plant. In addition vegetable producers reported that the information on weather conditions informed them of climatic conditions faced by competitors in other states and countries.
- The Farmers' Information and Technology Service (FITS - Techno-Pinoy) is a web-based information service initiative in the Philippines. FITS aims to contribute to the empowerment of farmers, processors, entrepreneurs and traders and provides information and technology services that facilitate decision-making by rural communities. This leads to improved production, processing, trading and marketing. The initiative intends to link with organisations, networks and technology services and existing resources into a centre near the farmers and uses the Internet, traditional media, and face-to-face information delivery and access at the local level.
- The Mango Information Network (MIN) is a web-based information service that provides information on market outlook, pest management, directory of players in the mango industry, a virtual meeting place and extension/research. MIN is relevant to farmers, farmer organizations, co-operatives, frontline agents and entrepreneurs and offers a question and answer service.

Inadequate Human Resources

To ensure more meaningful participation in rural development, and to pave way for the creation of a critical mass of people that effectively harness ICTs in developing countries, training and capacity building must be an integral part of all ICT projects. Most staff managing ICT-based projects lack adequate training that would enable them takes full advantage of the new technologies. There is need to invest in training and advisory services for information intermediaries, telecentre staff, frontline workers and women's groups. People in rural communities of developed countries are already reaping the benefits of electronic distance education and developing countries could offer similar services through selected centres (Richardson 1999). Focus should be on such skills as how to use ICTs through practical and participatory approaches. The 'Self Employed Women's Association' in India, for example, has trained rural women in the production and use of video to generate income, disseminate new skills and to advocate for changes in policy (Balit 1999).

Tele-centres

There is evidence that Tele-centres have played a major role in mobilizing communities to address their development problems. Tele-centres can be used as information hubs that capture, repackage and disseminate information to rural communities (FAO 1998). Bie (1996) recommends "an Internet and development strategy focused on rural and agricultural communities and the intermediary agencies that serve those communities with advice, project support, research, extension, and training".

Capacity Building and Training

One major constraint for delivery of food security initiatives in rural areas is weak institutional capacity and insufficient co-ordination. Non-governmental organizations and the private sector in particular possess a vast but often untapped potential. FAO could seek partners to fund capacity building activities. The partnership could also assist with building the required human and institutional capacities at national and regional level to provide training and education to rural communities on how to manage local knowledge and information, using ICTs. Training materials produced could be availed as electronic archive of training resources, and could be repackaged in preferred media such as the successful video-based model used in Latin America. The resources could then be translated into major languages to ensure that most developing countries benefit.

32

Barriers to Communication

There are many reasons why interpersonal communications may fail. In many communications, the message (what is said) may not be received exactly the way the sender intended. It is, therefore, important that the communicator seeks feedback to check that their message is clearly understood. The skills of Active Listening, Clarification and Reflection may help but the skilled communicator also needs to be aware of the barriers to effective communication and how to avoid or overcome them.

A. Active Listening

Active listening is a skill that can be acquired and developed with practice. However, active listening can be difficult to master and will, therefore, take time and patience. 'Active listening' means, as its name suggests, actively listening. That is fully concentrating on what is being said rather than just passively 'hearing' the message of the speaker. Active listening involves listening with all senses. Active listening not only means focusing fully on the speaker but also actively showing verbal and non-verbal signs of listening. Generally speakers want listeners to demonstrate 'active listening' by responding appropriately to what they are saying. Appropriate responses to listening can be both verbal and non-verbal:

Science of Active Listening

This is a generic list of non-verbal signs of listening, in other words people who are listening are more likely to display at least some of these signs. However these signs may not be appropriate in all situations and across all cultures.

Smile

Small smiles can be used to show that the listener is paying attention to what is being said or as a way of agreeing or being happy about the messages being received.

Eye Contact

It is normal and usually encouraging for the listener to look at the speaker. Eye contact can however be intimidating, especially for more shy speakers – gauge how much eye contact is appropriate for any given situation. Combine eye contact with smiles and other non-verbal messages to encourage the speaker.

Posture

Posture can tell a lot about the sender and receiver in interpersonal interactions. The attentive listener tends to lean slightly forward or sideways whilst sitting. Other signs of active listening may include a slight slant of the head or resting the head on one hand.

Mirroring

Automatic reflection/mirroring of any facial expressions used by the speaker can be a sign of attentive listening. These reflective expressions can help to show sympathy and empathy in more emotional situations. Attempting to consciously mimic facial expressions (*i.e.* not automatic reflection of expressions) can be a sign of inattention.

Distraction

The active listener will not be distracted and therefore will refrain from fidgeting, looking at a clock or watch, doodling, playing with their hair or picking their fingernails. Although a strong signal of attentiveness, caution should be used when using positive verbal reinforcement..

Remembering

The human mind is notoriously bad at remembering details, especially for any length of time. However, remembering a few key points, or even the name of the speaker, can help to reinforce that the messages sent have been received and understood *i.e.* listening has been successful.

Questioning

The listener can demonstrate that they have been paying attention by asking relevant questions and/or making statements that build or help to clarify what the speaker has said. By asking relevant questions the listener also helps to reinforce that they have an interest in what the speaker has been saying.

Reflection

Reflecting is closely repeating or paraphrasing what the speaker has said in order to show comprehension. Reflection is a powerful skill that can reinforce the message of the speaker and demonstrate understanding.

B. Clarification

Clarifying involves asking questions of the speaker to ensure that the correct message has been received. Clarification usually involves the use of open questions which enables the speaker to expand on certain points as necessary.

Summarisation

Repeating a summary of what has been said back to the speaker is a technique used by the listener to repeat what has been said in their own words. Summarising involves taking the main points of the received message and reiterating them in a logical and clear way, giving the speaker chance to correct if necessary. In communication, clarification involves offering back to the speaker the essential meaning, as understood by the listener, of what they have just said.

The purpose of clarification is to:

- Ensure that the listener's understanding of what the speaker has said is correct.
- Reassure the speaker that the listener is genuinely interested in them and is attempting to understand what they are saying.
- As an extension of reflecting, clarifying reassures the speaker that the listener is attempting to understand the messages they are expressing. Clarifying can involve asking questions or occasionally summarising what the speaker has said.

Some examples of non-directive clarification-seeking questions are:

- "I'm not quite sure I understand what you are saying."
- "I don't feel clear about the main issue here."
- "When you said. what did you mean?"
- "Could you repeat.?"

Clarifying involves:

- Non-judgemental questioning.
- Summarising and seeking feedback as to its accuracy.

Clarification Questions

When you are the listener in a sensitive environment, the right sort of non-directive questioning can enable the speaker to describe their viewpoint more fully. Asking the right question at the right time can be crucial and comes with practice. The best questions are open-ended as they give the speaker choice in how to respond, whereas closed questions allow only very limited responses.

Open Questions

If your role is to assist a speaker to talk about an issue, often the most effective questioning starts with 'when', 'where', 'how' or 'why'. These questions encourage speakers to be open and expand on their thoughts. For example:

"When did you first start feeling like this?"

"Why do you feel this way?"

Closed Questions

Closed questions usually elicit a 'yes' or 'no' response and do not encourage speakers to be open and expand on their thoughts. Such questions often begin with 'did you?' or 'were you?' For example:

"Did you always feel like this?"

"Were you aware of feeling this way?"

Guidelines for Clarifying

Clarification is the skill we use to ensure that we have understood the message of the speaker in an interpersonal exchange. When using clarification follow these guidelines to help aid communication and understanding.

- ☆ Admit if you are unsure about what the speaker means.
- ☆ Ask for repetition.
- ☆ State what the speaker has said as you understand it, and check whether this is what they really said.
- ☆ Ask for specific examples.
- ☆ Use open, non-directive questions - if appropriate.
- ☆ Ask if you have got it right and be prepared to be corrected.

Summarising

As a further extension to clarification a summary involves reviewing what has taken place during the whole conversation. It is important to keep only to the essential components of the conversation, and it must be given from the speaker's frame of reference, not an interpretation from the listener's viewpoint. The aim of a summary is to review understanding, not to give explanation, to judge, to interpret or provide solutions.

Summary of Clarification

In reflecting, clarifying and summarising, speakers must be allowed to disagree with, and correct, what the listener says. They should be encouraged to express themselves again, if necessary, giving the listener another chance at understanding, and to check understanding until agreement is reached.

C. Reflection

Reflecting is the process of paraphrasing and restating both the feelings and words of the speaker. The purposes of reflecting are:

- ☆ To allow the speaker to 'hear' their own thoughts and to focus on what they say and feel.
- ☆ To show the speaker that you are trying to perceive the world as they see it and that you are doing your best to understand their messages.
- ☆ To encourage them to continue talking.

Two Main Techniques of Reflecting

Mirroring

Mirroring is a simple form of reflecting and involves repeating almost exactly what the speaker says. Mirroring should be short and simple. It is usually enough to just repeat key words or the last few words spoken. This shows you are trying to understand the speakers' terms of reference and acts as a prompt for him or her to continue. Be aware not to over mirror as this can become irritating and therefore a distraction from the message.

Paraphrasing

Paraphrasing involves using other words to reflect what the speaker has said. Paraphrasing shows not only that you are listening, but that you are attempting to understand what the speaker is saying. Your responses should be non-directive and non-judgemental. It is very difficult to resist the temptation to ask questions and when this technique is first used, reflecting can seem very stilted and unnatural. You need to practice this skill in order to feel comfortable.

Reflecting Content, Feeling and Meaning

The most immediate part of a speaker's message is the content, in other words those aspects dealing with information, actions, events and experience, as verbalised by him or her. Reflecting content helps to give focus to the situation but, at the same time, it is also essential to reflect the feelings and emotions expressed in order to bring them into sharper focus. This helps the speaker to own and accept their own feelings, for quite often a speaker may talk about them as though they belong to someone else, for example using "you feel guilty" rather than "I feel guilty."

Reflecting needs to combine content and feeling to truly reflect the meaning of what the speaker has said. For example:

Speaker

"I just don't understand my boss. One minute he says one thing and the next minute he says the opposite."

Listener

"You feel very confused by him?"

Reflecting meaning allows the listener to reflect the speaker's experiences and emotional response to those experiences. It links the content and feeling components of what the speaker has said.

Guidelines for Reflecting

- Be natural.
- Listen for the basic message - consider the content, feeling and meaning expressed by the speaker.
- Restate what you have been told in simple terms.

- ☆ When restating, look for non-verbal as well as verbal cues that confirm or deny the accuracy of your paraphrasing.
- ☆ Do not question the speaker unnecessarily.
- ☆ Do not add to the speaker's meaning.
- ☆ Do not take the speaker's topic in a new direction.
- ☆ Always be non-directive and non-judgemental.

Common Barriers to Listening

There are many things that get in the way of listening and you should be aware of these barriers, many of which are bad habits, in order to become a more effective listener. Barriers and bad habits to effective listening can include:

- ☆ **Trying to listen to more than one conversation at a time**, this includes having the television or radio on while attempting to listen to somebody talk; being on the phone to one person and talking to another person in the same room and also being distracted by some dominant noise in the immediate environment.
- ☆ **You find the communicator attractive/unattractive** and you pay more attention to how you feel about the communicator and their physical appearance than to what they are saying. Perhaps you simply don't like the speaker - you may mentally argue with the speaker and be fast to criticise, either verbally or in your head.
- ☆ **You are not interested** in the topic/issue being discussed and become bored.
- ☆ **Not focusing** and being easily distracted, fiddling with your hair, fingers, a pen *etc.* or gazing out of the window or focusing on objects other than the speaker.
- ☆ **Feeling unwell or tired**, hungry, thirsty or needing to use the toilet.
- ☆ **Identifying rather than empathising** - understanding what you are hearing but not putting yourself in the shoes of the speaker. As most of us have a lot of internal self-dialogue we spend a lot of time listening to our own thoughts and feelings - it can be difficult to switch the focus from 'I' or 'me' to 'them' or 'you'. Effective listening involves opening your mind to the views of others and attempting to feel empathetic.
- ☆ **Sympathising rather than empathising** - sympathy is not the same as empathy, you sympathise when you feel sorry for the experiences of another, to empathise is to put yourself in the position of the other person.
- ☆ **You are prejudiced or biased** by race, gender, age, religion, accent, and/ or past experiences.
- ☆ **You have preconceived ideas or bias** - effective listening includes being open-minded to the ideas and opinions of others, this does not mean you have to agree but should listen and attempt to understand.

- **You make judgements**, thinking, for example that a person is not very bright or is under-qualified so there is no point listening to what they have to say.
- **Previous experiences** – we are all influenced by previous experiences in life. We respond to people based on personal appearances, how initial introductions or welcomes were received and/or previous interpersonal encounters. If we stereotype a person we become less objective and therefore less likely to listen effectively.
- **Preoccupation** - when we have a lot on our minds we can fail to listen to what is being said as we're too busy concentrating on what we're thinking about. This is particularly true when we feel stressed or worried about issues.
- **Having a Closed Mind** - we all have ideals and values that we believe to be correct and it can be difficult to listen to the views of others that contradict our own opinions. The key to effective listening and interpersonal skills more generally is the ability to have a truly open mind - to understand why others think about things differently to you and use this information to gain a better understanding of the speaker.

Non-Verbal Signs of Ineffective Listening

Although with all non-verbal signals a certain amount of error has to be expected, generally signs of inattention while listening include:

- **Lack of eye contact with the speaker** – listeners who are engaged with the speaker tend to give eye contact. Lack of eye contact can, however, also be a sign of shyness.
- **An inappropriate posture** - slouched, leaning back or 'swinging' on a chair, leaning forward onto a desk or table and/or a constantly shifting posture. People who are paying attention tend to lean slightly towards the speaker.
- **Being distracted** - fidgeting, doodling, looking at a watch, yawning.
- **Inappropriate expressions and lack of head nods** - often when a listener is engaged with a speaker they nod their head, this is usually an almost subconscious way of encouraging the speaker and showing attention. Lack of head nods can mean the opposite – listening is not happening. The same can be true of facial expressions, attentive listeners use smiles as feedback mechanisms and to show attention.

Barriers to Organizational Communication

When a communication is sent to the receiver, it must be effective and it is effective only when it has been understood by the receiver in the same sense in which the sender intended it. If it is not properly understood by the receiver, the very purpose of communication is lost and the communication proves ineffective. There may be so many obstructions in the way of its being effective.

Types of Barriers

Barriers, obstructions, and interruptions in communication may broadly be categorized into the following groups:

- ☆ External barriers
- ☆ Organizational barriers
- ☆ Personal barriers

However, this classification does not suggest that these categories are mutually exclusive.

External Barriers

External barriers are those caused by factors other than organizational and personal factors. Such external barriers may be (a) semantic barriers, (b) emotional or psychological barriers.

Organizational Barriers

An organization is a deliberate creation of management for the attainment of certain specific objectives. The day-to-day functioning of the organization is regulated in such a way as to contribute to the attainment of these objectives in the most effective manner.

Personal Barriers

As communication is basically an interpersonal process, many personal factors inherent in the two parties to communication, the sender and the receiver, influence the flow of communication and present many hurdles in the way of effective communication.

A Categorisation of Barriers to Communication

Language Barriers

Clearly, language and linguistic ability may act as a barrier to communication. However, even when communicating in the same language, the terminology used in a message may act as a barrier if it is not fully understood by the receiver(s).

Psychological Barriers

The psychological state of the receiver will influence how the message is received. For example, if someone has personal worries and is stressed, they may be preoccupied by personal concerns and not as receptive to the message as if they were not stressed. Stress management is an important personal skill that affects our interpersonal relationships.

Physiological Barriers

Physiological barriers may result from the receiver's physical state: for example, a receiver with reduced hearing may not grasp to entirety of a spoken conversation especially if there is significant background noise.

Physical Barriers

These barriers are those that separate people from each other and mark territories. This type of barrier can often be seen in the workplace where offices and closed doors stop communication an example of a physical barrier to communication is geographic distance between the sender and receiver(s). Communication is generally easier over shorter distances as more communication channels are available and less technology is required.

Systematic Barriers

Systematic barriers to communication may exist in structures and organisations where there are inefficient or inappropriate information systems and communication channels, or where there is a lack of understanding of the roles and responsibilities for communication. In such organisations, individuals may be unclear of their role in the communication process and therefore not know what is expected of them.

Attitudinal Barriers

Attitudinal barriers are behaviours or perceptions that prevent people from communicating effectively. Attitudinal barriers to communication may result from personality conflicts, poor management, resistance to change or a lack of motivation. Effective receivers of messages should attempt to overcome their own attitudinal barriers to facilitate effective communication.

Language Barriers

Not using words another can understand will certainly stop your message from being conveyed. This not only applies to actual languages, but that of expressions, buzz words, and other jargon. If one is not familiar with your language, misinterpretation will occur.

Gender Barriers

Variation exists among masculine and feminine styles of communication. While women often emphasize politeness, empathy, and rapport building, male communication is often more direct. Meshing these two styles without awareness could be becomes a barrier.

Interpersonal Barriers

These are barriers are created to distance themselves from others. These can be done through withdrawal, meaningless rituals which keep one devoid of real contact, superficial activities through pastimes, and more.

Perceptual Barriers

Different world views can create misunderstanding. Without thinking, one might only view a message from their mindset rather than looking to see it from another viewpoint.

Cultural Barriers

Ethnic, religious, and social differences can often create misunderstandings when trying to communicate. These differences can also affect perceptual factors, as mentioned above.

Emotional Barriers

Trouble listening can occur if one is consumed with emotion. Hostility, anger, fear, and other emotions make it hard to hear outside of one's self. Overcoming these barriers to communication is no easy task. It takes great awareness and a willingness to adapt and look at communication from new perspectives. But, if you begin to focus on how these communication barriers are affecting your everyday conversation, you will be well on your way to becoming an effective communicator.

33

Some Relevant Terminologies in Communication

Opinion Leaders/Change Agents

The opinions of people in a group are influenced by what they hear from "opinion leaders". An individual who is a member of a group manifests certain characteristics in his thinking and behaviour that contribute to the formation of "public opinion". The opinion of the leader is based on rational thinking due to education and experience. They weigh the pros and cons of the information they receive and then give their judgment on it.

Encoder

In the process of communication, the sender or source of the message is referred to as the encoder.

Decoder

The person receiving the message and decodes it is referred to as the decoder.

Sender

The initiator and encoder of a message

Receiver

The one that receives the message (the listener) and the decoder of a message.

Channel

The medium through which the message travels such as through oral communication (radio, television, phone, in person) or written communication (letters, email, text messages)

Message

The verbal and nonverbal components of language, that is sent to the receiver by the sender, which conveys an idea.

Feedback

Feedback, term form cybernetics, is the study of messages. It refers to an inquiry, response or experiment. Feedback can be positive (when the required result is achieved) or negative; instantaneous (when the response is immediate) or delayed. Feedback is used to gauge the effectiveness of a particular message put forth or situation that has taken place the receivers verbal and nonverbal responses to a message such as a nod for understanding (nonverbal), a raised eyebrow for being confused (nonverbal), or asking a question to clarify the message (verbal).

Positive Feedback

This feedback shows the communicator that his message was understood.

Negative Feedback

This feedback shows the communicator that his message was not understood. It does not imply disagreement just a lack of understanding.

Evaluation

Measuring the effectiveness of the message.

Noise

In all communication, there is a sender, a message/communication and a receiver. The meaning of a message is greatly dependent on the culture in which it is transmitted. The sender encodes a message, the receiver decodes it. Between the sender, the message and receiver, noise gets in the way and complicates the process. A noiseless communication does not exist. There always is some kind of noise entering the communication. Noise can be physical noise for example static or psychological *i.e.* when culture, taboos or values come into play to disrupt the normal transmission process of communication. Misunderstanding of a particular message *i.e.* distortion of meaning is a form of noise, example, the game of Chinese Whisper"a person starts off with a particular message and the original message may be distorted by the time it comes to the final player.

Communication Fidelity

Faithful performance of communication process by all its elements *i.e.* communicator, message, channel and receiver.

Problem

There is something that stand between you and person with whom you are communicating which interferes the complete transfer of information.

Interaction and Communication

The concept of interdependence it itself complex and can be illustrated by defining the possible relationship between any two concepts, such A and B. A and B are independent if and only if neither affects the others. For example, the colour of a person's here (A) and his left-handedness are independent. They don't affect each other. Interdependence can be defined as reciprocal or mutual dependence. If A affects B and B affects A, then A and B are interdependent. For example, in this country the farmer and the grocer are interdependent.

Levels of Communicative Interdependence

a) *Definitional-physical interdependence*

The functions of the source and receiver are physically interdependent, although the functions may be performed at different points in time and space. When two people are communicating, they rely of the physical existence of the other for the production or reception of messages.

b) *Action-reaction interdependence*

Communication often involves action-reaction interdependence. The action of the source affects the reaction of the receiver, the reaction of a receiver affect the subsequent reaction of the source, *etc.* The source or the receiver can make use of the reaction of the other. Reactions serve as feedback. They allow the source or receiver to check up on himself, to determine how well he is doing in accomplishing his purpose. Feedback also affects subsequent behaviour, if the source and receiver are sensitive to it.

c) *Interdependence of expectations: empathy*

Every communicator carries around with him an image of his receiver. He takes his receiver into account when he produces his message. He anticipates the possible responses of his receiver and tries to predict them ahead of time. These images affect his message behaviour.

Empathy

Empathy is "I feel how you feel".

Theories of Empathy

a) *Inference theory of empathy*

An inference theory of empathy is psychologically oriented. It argues that man can observed his own physical behaviour directly, and can relate his behaviour symbolically to his own internal psychological states-his feelings, thoughts, emotions, *etc.*

b) *Role taking theory of empathy*

Role taking theories argue that the new born infant cannot distinguish himself from other people, can't tell one person from another. In order to develop the concept of self, the infant must first look on himself as an object-must act toward himself as he acts toward other objects, other people.

Interaction: The Goal of Human Communication

One necessary condition for human communication is an interdependent relationship between the source and the receiver. Each affects the others. At one level of analysis, communication involves only a physical interdependence; i.e source and receiver are dyadic concepts, each requires the other for its very definition, each requires the other for its existence. At a second level of complexity, interdependence can be analyzed as an action-reaction sequence. At a third level of complexity, communication analysis is concerned with empathic skills, the interdependence produced by expectations about how others will respond to a message.

Types of Communication

The components of a communication process are: a sender who encodes and sends the message, the channel through which the message is sent, and the receiver who decodes the message, processes the information and sends an appropriate reply via the same channel back to the sender. Communication can occur via various processes and methods, and depending on the channel used and the style, there can be various types.

Communication Channels

Based on the channels used for communicating, the process of communication can be broadly classified as verbal and non-verbal communication. Verbal communication includes written and oral communication whereas non-verbal communication includes body language, facial expressions and visuals diagrams or pictures.

Verbal it is further divided into oral and written communication.

Oral Communication

Oral communication refers to the spoken words in the communication process. It can either be face-to-face or a conversation over the phone or on the voice chats over the Internet. Spoken conversations or dialogs are influenced by voice modulation, pitch, volume and even the speed and clarity of speaking.

Written Communication

The other type of verbal communication is written communication. It can be either via snail mail, or email. The effectiveness of written communication depends on the style of writing, vocabulary used, grammar, clarity and precision of language.

Nonverbal

Non-verbal communication includes the overall body language of the person who is speaking, which will include the body posture, the hand gestures, and the overall body movements. Facial expressions also play a major role in communication. Even expressions on a person's face say a lot about his/her mood. Also, gestures like a handshake, a smile or a hug can independently convey emotions and do not need to be accompanied by words. Non verbal communication can also be in the form of pictorial representations, signboards, or even photographs, sketches and paintings.

Categories of Non-Verbal Communication

Elements such as physique, height, weight, hair, skin color, gender, odors, and clothing too send non-verbal messages during interaction. The various sources and categories of Non-verbal communication have been described below:

Arbitrariness

While much Non-verbal communication is based on arbitrary symbols, which differ from culture to culture, a large proportion is also to some extent iconic and may be universally understood. Paul Ekman's influential 1960s studies of facial expression determined that expressions of anger, disgust, fear, joy, sadness and surprise are universal.

Clothing and Bodily Characteristics

Clothes have both a functional and a communicative purpose. A villager's clothes alone can identify him as a well-to-do farmer having a respectable position in the village community. Or the attire of a village headman will differ from an ordinary farmer. The Sarpanch will in all probability have a turban on in most part of the country.

Physical Environment

Environmental factors such as furniture, architectural style, interior decorating, lighting conditions, colors, temperature, noise, and music affect the behavior of communicators during interaction.

Proxemics: Physical Space in Communication

Proxemics is the study of how people use and perceive the physical space around them. The space between the sender and the receiver of a message influences the way the message is interpreted. The perception and use of space varies significantly across cultures and different settings within cultures. Space in non-verbal communication may be divided into four main categories: intimate, social, personal, and public space.

Chronemics: Time in Communication

Chronemics is the study of the use of time in Non-verbal communication. The way we perceive time, structure our time and react to time is a powerful communication tool, and helps set the stage for communication. Time perceptions include punctuality and willingness to wait, the speed of speech and how long people are willing to listen.

Monochronic Time

A monochronic time system means that things are done one at a time and time is segmented into precise, small units. Under this system time is scheduled, arranged and managed. Monochronic cultures place a paramount value on schedules, tasks and "getting the job done". These cultures are committed to regimented schedules and may view those who do not subscribe to the same perception of time as disrespectful.

Polychronic Time

A polychronic time system is a system where several things can be done at once, and a more fluid approach is taken to scheduling time. These cultures are much less focused on the preciseness of accounting for each and every moment. As Raymond Cohen notes, polychronic cultures are deeply steeped in tradition rather than in tasks – a clear difference from their monochromic counterparts. "Traditional societies have all the time in the world. The arbitrary divisions of the clock face have little saliency in cultures grounded in the cycle of the seasons, the invariant pattern of rural life, and the calendar of religious festivities". Instead, their culture is more focused on relationships, rather than watching the clock. They have no problem being "late" for an event if they are with family or friends, because the relationship is what really matters.

Movement and Body Position

In 1952, Ray Birdwhistell, an anthropologist studied how people communicate through posture, gesture, stance, and movement. Part of Birdwhistell's work involved making film of people in social situations and analyzing them to show different levels of communication not clearly seen otherwise. The study was joined by several other anthropologists, including Margaret Mead and Gregory Bateson.

Posture

Posture can be used to determine a participant's degree of attention or involvement, the difference in status between communicators, and the level of fondness a person has for the other communicator. Studies investigating the impact of posture on interpersonal relationships suggest that mirror-image congruent postures, where one person's left side is parallel to the other's right side, leads to favorable perception of communicators and positive speech; a person who displays a forward lean or a decrease in a backwards lean also signify positive sentiment during communication. Posture is understood through such indicators as direction of lean, body orientation, arm position, and body openness.

Gesture

A gesture is a non-vocal bodily movement intended to express meaning. They may be articulated with the hands, arms or body, and also include movements of the head, face and eyes, such as winking, nodding, or rolling ones' eyes. The boundary between language and gesture, or verbal and non-verbal communication, can be hard to identify.

Haptics: Touching in Communication

Haptics is the study of touching as Non-verbal communication. Touches that can be defined as communication include handshakes, holding hands, back slapping, a pat on the shoulder, and brushing an arm. Touching of oneself may include licking one's lips, picking, holding, and scratching. These behaviors are referred to as "adaptor" and may send messages that reveal the intentions or feelings of a communicator. The meaning conveyed from touch is highly dependent upon the context of the situation, the relationship between communicators, and the manner of touch.

Oculesics Eye Gaze

The study of the role of eyes in Non-verbal communication is sometimes referred to as "oculesics". Eye contact can indicate interest, attention, and involvement. Gaze comprises the actions of looking while talking, looking while listening, amount of gaze, and frequency of glances, patterns of fixation, pupil dilation, and blink rate.

Paralanguage: Vocalic Non-verbal Cues of the Voice

Paralanguage (sometimes called vocalic) is the study of Non-verbal cues of the voice. Various acoustic properties of speech such as tone, pitch and accent, collectively known as prosody, can all give off Non-verbal cues. Paralanguage may change the meaning of words.

G.W. Porter divides non-verbal communication into four broad categories

a) *Physical*

This is the personal type of communication. It includes facial expressions, tone of voice, sense of touch, sense of smell, and body motions.

b) *Aesthetic*

This is the type of communication that takes place through creative expressions: playing instrumental music, dancing, painting and sculpturing.

c) *Sings*

This is the mechanical type of communication, which includes the use of signal flags, the 21-gun salute, horns, and sirens.

d) *Symbolic*

This is the type of communication that makes use of religious, status, or ego-building symbols.

The relation between verbal and non-verbal communication and their relative importance:

In person-to-person communications our messages are sent on two levels simultaneously. If the Non-verbalcues and the spoken message are incongruous, the flow of communication is hindered. Right or wrong, the receiver of the communication tends to base the intentions of the sender on the non-verbal cues he receives.

Verbal vs Oral Communication in the Context of Non-verbal Communication

Scholars in this field usually use a strict sense of the term "verbal", meaning "of or concerned with words", and do not use "verbal communication" as a synonym for oral or spoken communication. Thus, vocal sounds that are not considered to be words, such as a grunt, or singing a wordless note, are non-verbal. Sign languages and writing are generally understood as forms of verbal communications, as both make use of words – although like speech, both may contain paralinguistic elements and often occur alongside non-verbal messages. Non-verbal communication can occur through any sensory channel – sight, sound, smell, touch or taste.

Interaction of Verbal and Non-verbal Communication

When communicating, non-verbal messages can interact with verbal messages in six ways: repeating, conflicting, complementing, substituting, regulating and accenting/moderating.

i) *Repeating*

Repeating consists of using gestures to strengthen a verbal message, such as pointing to the object of discussion.

ii) Conflicting

Verbal and non-verbal messages within the same interaction can sometimes send opposing or conflicting messages. A person verbally expressing a statement of truth while simultaneously fidgeting or avoiding eye contact may convey a mixed message to the receiver in the interaction.

iii) Complementing

Accurate interpretation of messages is made easier when non-verbal and verbal communication complements each other. Non-verbal cues can be used to elaborate on verbal messages to reinforce the information sent when trying to achieve communicative goals; messages have been shown to be remembered well when non-verbal signals affirm the verbal exchange.

iv) Substituting

Non-verbal behavior is sometimes used as the sole channel for communication of a message. People learn to identify facial expressions, body movements, and body positioning as corresponding with specific feelings and intentions.

v) Regulating

Non-verbal behavior also regulates our conversations. For example, touching someone's arm can signal that you want to talk next or interrupt.

vi) Accenting/Moderating

Non-verbal 1signals are used to alter the interpretation of verbal messages. Touch, voice pitch, and gestures are some of the tools people use to accent or amplify the message that is sent; non-verbal behavior can also be used to moderate or tone down aspects of verbal messages as well. For example, a person who is verbally expressing anger may accent the verbal message by shaking a fist.

Dance and Drama Non-verbal Communication

Dance and drama is a form of non-verbal communication that requires the same underlying faculty in the brain for conceptualization, creativity and memory, as doe's verbal language in speaking and writing. Means of self-expression, both forms have vocabulary (steps and gestures in dance), grammar (rules for putting the vocabulary together) and meaning. Dance, however, assembles (choreographs) these elements in a manner that more often resembles poetry with its ambiguity and multiple, symbolic and elusive meanings.

The Need and Ways to Improve Non-verbal Communication Skills in Extension Workers

Development experiences of the last decades have shown that human resources development is essential for food security and market integration. This focuses on human resources calls for increased knowledge and information sharing about agricultural production, as well as on appropriate communication methodologies, channels and tools. New agricultural technologies are generated by research institutes, universities, private companies, and by the farmers themselves.

The role of extension workers is to give highly accurate, specific and unbiased technical and management information and advice in direct response to the needs of the farmers. The adoption of new agricultural technologies by farmers is often very slow. In many countries low agricultural production has been attributed, among other factors, to poor linkages between Research-Advisory Service-Farmers and to ineffective technology delivery systems, including poor information packaging, inadequate communication systems and poor communication skills of the extension personnel. A lack of appropriate communication structures, methodologies and tools results in poor identification of farmers' needs and priorities, inappropriate research programs, poor or irrelevant extension information and technologies and finally, low farmers' take-up of technology innovations.

Rural communication is an interactive process in which information, knowledge and skills, relevant for development are exchanged between farmers, extension/advisory services, information providers and research either personally or through media such as radio, print and more recently the new "Information and Communication Technologies" (ICTs). It is here that it is not only what an extension worker says in a training/meeting that is important, but its how he/she says it that can make the difference to participating farmers in finally adopting a particular practice that can bring about significant change in their behaviour for a better livelihood and in the present context, it is rightly said:

"It's not what you say; it's how you say it"

Non-verbal messages are an essential component of communication in the teaching process. Extension workers should be aware of non-verbal behavior in a farmers' training/meeting for three major reasons:

- An awareness of non-verbal behavior will allow one to become better receivers of farmers' messages.
- One will become a better sender of signals that reinforce learning.
- This mode of communication increases the degree of the perceived psychological closeness between the extension worker and farmer.

Extension workers who make eye contact with the farmers open the flow of communication and convey interest, concern, warmth and credibility. Smiling is a powerful cue that transmits:

- Happiness
- Friendliness
- Warmth
- Liking
- Affiliation

Thus, if the trainer smiles frequently he/she will be perceived as more likable, friendly, warm and approachable. Smiling is often contagious and farmers will react favorably and learn more.

Cultural norms dictate a comfortable distance for interaction with farmers. The extension worker should look for signals of discomfort caused by invading farmers' space. Some of these are:

- ☆ Rocking
- ☆ Leg swinging
- ☆ Tapping
- ☆ Gaze aversion

Typically, in large gatherings space invasion is not a problem. In fact, there is usually too much distance. Increasing proximity enables one to make better eye contact and increases the opportunities for farmers to speak.

The vocal elements of non-verbal communication include:

- ☆ Tone
- ☆ Pitch
- ☆ Rhythm
- ☆ Timbre
- ☆ Loudness
- ☆ Inflection

Some more Tips for Successful Non-verbal Communication

- ☆ Take a time out if you're feeling overwhelmed by stress. Stress compromises your ability to communicate. When you're stressed out, you're more likely to misread other people, send off confusing or off-putting non-verbal signals, and lapse into unhealthy knee-jerk patterns of behavior. Take a moment to calm down before you jump back into the conversation. Once you've regained your emotional equilibrium, you'll be better equipped to deal with the situation in a positive way.
- ☆ Pay attention to inconsistencies. Non-verbal communication should reinforce what is being said. If you get the feeling that someone isn't being honest or that something is "off", you may be picking up on a mismatch between verbal and non-verbal cues. Is the person is saying one thing, and their body language something else? For example, are they telling you "yes" while shaking their head no?
- ☆ Look at non-verbal communication signals as a group. Don't read too much into a single gesture or non-verbal cue. Consider all of the non-verbal signals you are sending and receiving, from eye contact to tone of voice and body language. Are your non-verbal cues consistent – or inconsistent – with what you are trying to communicate?

Communication Based on Style and Purpose

Based on the style of communication, there can be two broad categories: formal and informal communication that has their own set of characteristic features.

Formal

It includes all the instances where communication has to occur in a set formal format. Typically this can include all sorts of business or corporate communication. The style of communication in this form is very formal and official. Official conferences, meetings and written memos and corporate letters are used for this form of communication. It can also occur between two strangers when they meet for the first time. Hence, formal communication is straightforward, official and always precise and has a stringent and rigid tone to it.

Informal Communication

The informal channel of communication is also known as 'Grapevine'. It is the result not of any official action, but of the operation of social forces at work-place. The term 'grapevine' arose during the days of U.S. Civil War. At that time, intelligence telephone lines were strung loosely from tree to tree in the manner of a grapevine, and the message thereon was often distorted; hence any rumour was said to be from the grapevine. Today the term applies to all informal communication. While formal communication exists to meet the utilitarian needs of the organization, informal communication is the method by which people carry on social, non-programmed activities within the formal boundaries of the system. It, thus, exists outside the official network. Though continuously interacting with it, this informal channel is generally multiple in nature; same person having social relationships with a number of people working in the same organization.

Informal communication includes instances of free and unrestrained conversation between people who share a casual rapport with each other. It requires two people to have a similar wavelength and hence occurs between friends and family. It does not have any rigid rules and guidelines. The conversations need not necessarily have boundaries of time, place or even subjects for that matter since we all know that friendly chats with our loved ones can simply go on and on.

Even though the whole process of communication may seem so simple, the effectiveness of each type depends to a great extent on certain internal and external environmental factors and also the communicator's ability to send, receive, decode and send a response. This is why the field of communication has been and continues to be widely studied and is given a place of importance in any walk of life.

34

Agricultural Communication and Journalism

Use of Television for Communication of Information

The electronic media- more so radio and also television are considered a potential vehicle for disseminating agricultural technology information. The Indian had it beginning in 1959. The objectives for assessing the value of different education of TV programmers are suitable for groups viewing in rural and urban communities. Television was first demonstrated in India in 1965 at an industrial exhibition. At the conclusion of the exhibition a part of the apparatus was few programmers designed for community viewing such as "responsibility of citizenships which include traffic and road sense, danger to community health *etc.* After the success of this experiment, it was decided to extend the scope of the community viewing experiment to school or educational television, which was launched in Oct, 23, 1961 in Delhi schools. From April 1965, television general service was increased to one hour, four days a week. But from August 15, that year it becomes a daily transmission krishi Darshan Programme for better farming precautions was introduced from January 26. 1967. The rural service primarily meant to familiarize the rural viewers with the technical and scientific know how about farming, weather forecasts. The countries second television center came up at Bombay on October 2, 1972. It was quickly followed by Shringar, Amritsar both 1973 Calcutta, Madras and Luck now (all three in 1975) but from August 1, 1975 Doordarshan undertook an historic and a unique steps in its march forward.

The satellite instructional television experiment popularly known as SITE, the country's most exciting and ambitious project was started by Doordarshan. In the beginning Doordarshan formed a part of all India Radio outfit. But on April 1, 1976,

Doordarshan was made an independent department of the ministry of information and Broadcasting. Unit 1982 television in India has not gone colour. From August 15, 1982 colour was introduced. Between 1980 and 1984 several new high power transmitters and low power transmitter's area and are assessable to the Annual Report of information and Broadcasting Ministry.

Doordarshan now has several channels. DD1 is the National channel available all over the country on certain fix hours. Dd2 and DD3 are also available. There are regional channels for putting out programmers in the 13 regional languages. Krishi Darshan is the oldest well established and the best known program for the rural area and farmers. The foresight and the pioneering sprit of the visionary space scientist Dr. Vikram Sarabhia and the initiative of Dr. M. S. Swaminathan. The basic objective of the program is to familiarize the rural audience with the latest technical and scientific know how about the farming practices, rural development programs, to acquaint the viewers with the importance of health, family planning, sanitation, *etc.*

Preparing of Television Program

Television has become a good medium to cover the message to the farmers. It has several advantages over others mass media by providing words with picture and sound effect. The effectiveness of television programmers can be increased by changing the place and time with the help of camera and giving sound effect to it. It reaches large number of people at short test possible time. The people learn through eye and ear remember thing better.

The TV Programme can be Prepared in Various Formats:

a) Interview of subject expert.
b) Discussion with subject expert.
c) Documentations.
d) Dramatized formats.
e) Musical formats *e.g.* Powada, Bharud and other folk from.
f) Features which include both dramatized, documentation and experts.

Out of these formats, most popular formats is the interview of subject expert and discussion, but these formats are used occasionally as they are more expensive, time consuming and tedious however these formats entertainment value and are poplar. But argil, programmers are informative in nature their success depends on the preparation by expert and his performance.

Procedure for Developing Television Programme

a) Selection of subject: The subject and matter selected for telecast should be on the felt needs of the audience and it should be timely. The subject should be location specific and within the resources of the people.
b) Collect the detail information of the subject by referring books and by discussing with experts farmers.

c) Determine the main points to be covered, concentrate on few important aspects of the subject. Never try to give maximum technical information by covering many points.

d) Divide the programme into important steps. It is better to list out the steps in logical consult the resources persons for checking accuracy of the information.

e) Decide the audience to whom the program is telecasted.

f) Choose the formats or method of presentations *i.e.* demonstration interview, dramatic presentation locals, forum *etc.*

g) Decide the visuals *i.e.* sample model, chart, and graph act to bus in the programme. Fix the time required place and method of presenting visual.

h) Prepare the put line of the programmed. Divide the sheet in columns. Writer the things, which are to be shown on, left while in the right hand column put the thing that are to be talked. Title the left column video and right column 'Audio '. The thing to be shown and thing to be said should be written opposite to each other on the page. Make the beginning attractive and interesting.

i) If possible give supporting information *e.g.* information of a variety give the experience of the farmer with actual photography of the crops At. The extreme left of the page indicate time required in minutes and seconds for each step and also showing visual.

j) Prepare the script by discussing with producer, experts and time available.

Think on the audience profile *i.e.* these who never see.

k) The programme, feel that not necessary to see the programme, known the information, for some time see the programme and the stop, see the programmed but do of think on it and see the programme carefully. Concentrate on those who see the programme but do not think on it and see the programme for time and then stop.

Points for Presenting the Television Program

a) Be yourself do not try to change sound, dressing mode of taking *etc.*

b) Dressing make up should be normal.

c) Be fresh and there should be ups and downs in the talk and smile on the face.

Remove the mechanical fear from the mind.

d) Arrange all the material f Visuals, equipment to be used in the programme.

Add or subtracts material to fit the script in to the desired time.

e) Speed of talk should be normal.

f) The language should be simple with familiar words and small sentence.

Keep the eye contact with camera.

g) Summarize the programme by giving stress on important points at the end.

Outline of Radio Talk

Radio is the on of the most effective and popular electronics mass media in the world today. In India radio we first standard by a private company in Bombay (presently Mumbai). In Madras a private organization known as the madras Presidency Radio club was functioning from 31 st July 1924, But due to financial difficulties the transmits was suspended within a short period.

In March 1935 B.C was approached to help properly organic the service. On June, 1936, the Indian Stale broadcasting Service was renamed as "All India Radio (AIR), In April 1937 a radio station was started in Peshawar. After the partition of the country three AIR Stations went to Pakistan – Lahore, Peshawar and Decca (now in Bangladesh). After independence several radio stations were setup. The All Indian Radio (AIR) was renamed as Akashvani on October 3, 1957.

All Indian radio has three tier system of broadcasting *viz.* national, Regional and local. Local Radio is a comparatively new concept of broadcasting in our country. Each of the station serving a small area provider utility services and reaches right into the heart of the community. The programming on local radio is area specific these are fixable and spontaneous enough to enable the station function as the mouth piece of the local community. The first local station was setup at nagercoil in kanyakumari district in Tamil Nadu in October 1984. It encourages the procedures to experiment an innovate a new format of broadcasting.

Radio acts as a catalytic agent between the sates a people. It provides a platform to both *i.e.* for discussion and dissemination of information. Radio and TV broadcasting essentially a part of mass media communication process shares the characteristic and limitation of such as system. In such a system, the amount of feedback readily available is low; the message flow tends to be one way and can be used to increase knowledge.

Radio is most widely accessible mass used to communicate information and news to the people. It is a good source of communicating agriculture information. It is a oral presentation heard by the audience. Hence the broadcaster has to attack and hold the attention of the audience otherwise the message is lost. The broadcast may be straight talks, interviews or forums. In all the types for broadcast is necessary. It is an outline of the broadcast.

Preparation of Radio Scripts

Radio script should be prepared in such a way that listener should feel that somebody is talking with him in a friendly manner and with informal style. For this purpose:

a) Collect the material from your experience.

b) Keep the purpose of the script and audience in mind.

c) Decide the method of presentation of information.

d) Considering the time available for the talk, the script should be prepared. Generally we talk two words per seconds and 8 to 10 lines are required per minute having 5-6 words per sentences.

e) Time for talk is always limited hence select one idea at a time never include maximum technical information in the script.

f) Arrange the information in the sequences.

g) Include supporting and illustrated facts

h) See that audience can understand your talk, for that uses short a simple sentence, familiar words and personal words like I, we, you.

i) Never use statistical table in a dialogue. In the interview do not frame such question of which the audience known the answer.

j) Repeat where ever necessary to emphasis the important points.

k) At the end summaries the main point.

Points for Presenting Radio Talk

a) Be yourself feel free and easy in front microphone. Remove the mechanical fear so that you will feel and it will help in building confidence.

b) Just talk to the people and not read, speak naturally, do not spend any time or worrying about your voice down to the room level as if conserving with any of your friend.

c) Observe milk manner. Turn aside from mike if you have to cough or clear your throats try to avoid noisy breathing.

d) Speed should be normal. There should be up and down in the voice never speak in same tone. It will lose the interest of the audience.

e) Report the sentence to stress the important point.

f) Be enthusiastic put smile in your voice and take with pleasant enthusiasm.

g) Give personal touch, reference human interest. It the end summarizes the while programme in time.

h) Complete the while programme in times

How to Use Radio Media More Efficiency for Communicating Agriculture Information

a) Encourage local people to hear the radio collectively.

b) Encourage them to listen the rural programme.

c) Encourage to corresponding with the broadcasting station to express their needs and opinions.

d) You should also maintain the contacts with broad casting center.

e) Specially announce introducing programme well in advance.

f) Have a list of talented persons in the area and arrange their participations in broadcasting programme.

Farm Advertisement in News Paper

An advertisement is a form of non personal presentation and promotion of idea, goods, and service by an identified sponsor. When product or inputs required in farming or any farm innovation are advertised, it is called farm advertisement in the newspapers, which considered as a most credible source of the information to evoke desired response from readers.

Components of Advertisement

1. **Heading:** Caption of advertisement
2. **Illustration:** Supporting picture or photograph or diagram
3. **Text:** information of product presented through words.
4. **Layout:** Attractive combination of heading, illustration and text.

Objective of Advertisement

1. To enhance the corporate image.
2. To reach new product
3. Persuasive
4. Mutative consumer to acquire product and create demand.
5. Educate consumer about correct use of the product.
6. Informing consumer about availability of product.
7. To establish product image.
8. To increase a sales and profit.

Advertisement Posses the following Characteristics

1. You pay for it.
2. You control what is said
3. You control who it is said
4. You control whom it is said
5. You control where it is put a publication or in the AIR.
6. You control the frequency of its use.

Type of Advertisement

1. Heading dominant
2. Illustration dominate
3. Combination of 1 and 2
4. Balance

Criteria for Good Advertisement

1. Advertisement should have proper border.
2. Advertisement should give details of important feature of the product.
3. it should have effective impact on appropriate audience.
4. Advertisement should aim at potential clan of people
5. Never underestimate the customers. (respect farmers)
6. Never make derogatory statement against competitors.
7. Honesty pays.

Advertisement is lifeline for newspapers. Newspaper is only product, first newspaper was published in world in 1622 and first advertisement in newspaper appeared in 1650.

Farm Advertisement of Radio and Television

The advertisements are means from selling the product by paying fees. The first advertisement on all India radio appeared in November 1967 from Vivid Bharti Bombay. The advertisement on radio and television are commercial in nature, while public service broadcasts (PSB) pay to the person involves.

Advertisement

1. Interest
2. Inform
3. Involve
4. Motivate
5. Direct

Two Types Radio Advertisements:

1. Spot (Prose/Form)
2. Jingle (Musical Form)

Criteria for Good Radio Advertisement

a) Advertisement should be fast and catchy (100 words per minutes)
b) Choose target audience and plan the advertisement accordingly.
c) Effective writing with appropriate slogans which will be remembered.
d) Background music and special sound effect, to draw the attention.
e) Advance copy should be in same language (local). Copy of advertisement is its heart.

   Constricting voices should be used, with few and appropriate words.
f) Good sound effect adds topics in advertisement.

g) Use signature tunes as per choice, without voice there will be void on radio.

h) Voice pronunciations should be clear is should create memorable visual image.

i) Avoid anything obscene or defamatory statement which can create law and order problem.

j) Humour in advertisement cab be very effective. Avoid order type language.

k) The advertisement agencies provide useful back up for producing advertisement on radio and television. However these agencies lack in agricultural knowledge. There is a vast scope for experts for advertising in farming sector.

There are various time categories and programme categories available for advertisement charge on radio and television, form 5 second onwards.

The television had added advantages of video presentation, which makes it a very effective mode of advertisement. For evoking desired response, timing of advertisement is necessary, to avoid waste, selection of short, light, colour combination. Voice and personality of presenters play important role in effectiveness of advertisement on television.

Types of Shots

1. BCU: Big close up shot
2. CU: Close up.
3. MCU: Medium shot.
4. MS: Medium close up.
5. MLS: medium long shot.
6. Ls: Long shot.

Types of Light

1. **Key Light:** Main source of light 9 suns).
2. **Full light:** Addition light to remove shadow.
3. **Black light:** For separation of background.
4. **Background light:** Illustration of background.

Writing and Evaluating Extension Literature

Literature is a basis of any teaching programmed. The written material is often supported by appropriate illustration and used as extension tool. Leaflet, pamphlet and bulletin play in important role in extension teaching. These are the simple and brief publication for givens information to the farmers.

How to Write

a) Select the subject of your interest and knowledge.

b) Approach the farmer (Probable reader 0 to know their interest. The reader's interest should be kept in mind rather than authors interest.

c) Review the relevant literature and discuss with experts.

d) After collecting the information, decide the information into three parts *i.e.* essential desirable and can be committed.

e) Arrange the information in the sequence.

Writing the Script

a) Introduction should be direct. In the beginning the subject should be clear.

b) Always start from simple to complex or familiar to unfamiliar, treat the reader with respect. Do not use many unfamiliar or complex highs-sounding. it will help to understand the subject.

c) Sentences should be simple and clear.

d) Only one idea in one sentence. The sentences should be understandable. The complex sentences should be avoided.

e) Be accurate, specific and brief while writing.

f) Have short paragraph.

g) Write the information in the sequence.

h) The cover should be appealing and attractive. The illustration should be appropriate to the subject.

i) The title should be short and appealing.

j) Symbols and complicated diagrams should be avoided.

k) Use suitable type of lettering. Bold and clear letters are better. Positioning and spacing of letters is important in increasing efficiency.

l) Prepare the complete script and after few hours or next day review it or show it to others.

m) Printing of extension literature should be on good quality proper with good type.

n) Printing of extension literature should be done on good quality paper with good type.

Leaflet

Leaflet is a single sheet of paper folded to make four page piece of print matter. It is usually given accurate or specific information on a particular topic. Leaflets are mostly for giving a "how to do or process story, a set of improved practice to be followed in solving a home or farm operation "Illustration is place important role in leaflet.

Pamphlet

Pamphlet is a small – unbound booklet on some optic. It has 12 – 24 pages and deals with one problem in details.

Bulletin

Bulletin is a bulkier publication with 24 to 50 pages and convey large amount of information for applying in local situation. It covers more technical information.

News Stories

What is News?

News is any timely information, which is collected from North, East, West and South direction, which interest a number of persons. The best news is that which has greatest interest for the number of people.

News Story

A news story is an account of events in sequence. It is used mainly to get information to many people quickly.

Sources of News

Some of the sources of news material are results of demonstration, review of research publication, accomplishment of farmers, account meeting *etc.*

Kinds of News

Extension news stories tend to group themselves into one or more of the following categories.

1. Advance event articles
2. Follow- up event articles
3. Information articles
4. Feature articles
5. Experience and success stories.
6. New developments
7. Predictions
8. Subject matter

Writing the Story

While writing follow these rules.

a) Use five 'W' and one 'H' as the guide *i.e.* to answer who, what, why, when, Where and how.

b) Easy to understand.

c) In the language of the village people who are reading the story.

d) Accurate in all details.
e) In short sentences and paragraph.
f) Write the lead sentence.
g) Avoid using your personal opinion.
h) Use A.B.C of new stories *i.e.* accuracy, brevity and clarity for writing.

News Structure

a) Head

It is capsule opening center at the top of news. Heading must summaries the entire news, which arouses interest. It should be brief, clear and stimulating. Effective words should be used to make it meaningful and catchy.

b) Lead

The lead is the opening part of news story or the introduction of the story. It is condensed news which gives abstract of entire information. Most of the news leads falls into following two major forms.

1. *Inverted Pyramid (Summary) Lead*

The name is self explanatory because a lead written in this forms summarizes its story. A summary lead is expended to answer as many as possible of the six questions: Who? When? Where? Why? and How?

2. *Upright Pyramid (Suspended interest) Lead*

Placing the real climax or features of a story somewhere other than in the lead is effective and exciting in what is known as the suspended interest story. Often the climax in a story is in the final paragraph.

3. *Body*

It gibes details information about the event. It should be made simple, clear and easy to read.

Guidelines for Making a News Copy

a) Always use white standard size paper (8 ½ X 11 inches)
b) Always type in double space.
c) Leave 3 to 4 inch margin at the top of your first page write your name and address at the upper leaf hand corner. Leave margin approximately one inch wide at the sides and bottom of the page.
d) If you include a headline, type it at the top copy on your first page, but still leave the three inch top margin.
e) Type in bold letter 'News for publication.'
f) if your story requires more than one page, write: more: or "continued" at the bottom of your first sheet and bottom of all following pages.

g) Use standard proof reading symbols for showing correction.

h) Write the editor's address.

i) It is best to deliver or mail copy flat; because that is the way it must be handled. But newspaper will not object too much if you mail your copy folded in thirds from bottom to top. Folding your copy more than this will make it too difficult to handle when unfolded.

Typing, Printing and Proof Reading

a) Typing instruction is given to the typists for obtaining a good copy of the material. Following precautions should be taken to obtain a good copy which will enable the printer to print properly an accurately.

b) White thick un-ruled paper should be used for manuscript. However for typing good quality should be used and typing left= 4 cm, Right= 3 cm/top and bottom = 3 cm. The first line of the paragraph should be five spaces from the merging.

c) The general text should be typed in double space. However tables long quotations foot notes figure caption should be single spaced.

d) Correcting should be done by use of erasers or with black ink.

e) From and spacing of the heading should be consistent thoroughly the script. Print instructions are given for obtaining a good print copy of the material. There are many types of printing processes. The important among them are letterpress, lithography, offset. Teleprint *etc.* the choice of printing process depends on the type and size number of colour involved, time and money available after writing the manuscript the author gives instruction to the typist to obtain a copy for giving to printing. While printing and proof reading symbols are used by the proofreader instruct the printer for correcting the proof.

f) Proof reading is an important part of editing and even editor must have proficiency in proof reading. It is an art of making correcting in the proofs pulled from a composed matter. This job is completed with the help of sings and symbols, which are marked by a proof reader in a particular manner with the assistance of copyholder. These marks or proof correction symbols are easily understood by a composter too.

A publication full mistake is nothing but only a leap of waste papers and the reader would not show any interest in it. Therefore proof reading is must for publication, if it has been a faultless works. Some of the important rule for the proof correcting is as below:

☆ All correction should be made in the margin only such marks should be made in the text which and required to indicate the place to which the correction refers.

If any changes are required in a letter on word, it should be struck through and the letter or matter to be substituted should be written in the margin.

Special attention should be paid to the checking of scientific manes and numerical data.

Words breaking or division at the end of the line should be checked and made to agree with the dictionary or the style manual of a publishing house.

- ☆ Running heads should be read carefully and the continuation from the last of the each page to the head of the next page be properly checked.
- ☆ In the pages carrying illustration the position of the blocks and the accompanying captions should be carefully checked. It is rare that a block in journal has been published upside down or with a wrong caption. Carefully comparisons with text will avoid such ugly situation.
- ☆ In proof reading certain instruction symbols are used by the proof reader. Usually the printer sends galley proofs (also called as the first proofs), page proofs and engraver's proofs whenever photography or other illustration are involved, if required revised proofs any stage are sent.
- ☆ In case of high quality of work, the printer provides a dummy, which is also called machine proof or press proof. The galley proof is not divided into pages. Galley is short name of galley proof.
- ☆ It is a single column set in the line width of text and sometimes, as long as 20 to 24 inches. The second's stage is to be carried out. The other mark is in the margin of page proof, given an idea of what each page will contain and look like.
- ☆ Dummy is the final stage, which represents the printed vision before copes are run off. Proofs have to be read letter by letter than words by words. Correction marks are indicated at two places. One mark is within the line and the other mark is in against the corresponding line giving the instructions.

Corrected Specimen

A large ear is a man; they say is a sign of luck. But, large or small the ear deserves all the care you can give it. Any neglect may mean trouble doctors warm us.

A large bar in a man they say is a sign of luck but large or small the ear deserve s all the care you can give it any neglect may mean trouble doctors warm us.

Proof Reading Symbols

Instruction in the Margin	*Marks in the Text*	*Explanation*
Bf	The institute	Change in bold face type
Bf ital	The institute	Set bold face in italic
UC	The institute	Set in capital letters
LC	The institute	Set in small letters
IF	The institute	Set in light Face
Rom	The institute	Set in roman type

Instruction in the Margin	*Marks in the Text*	*Explanation*
Ital	The institute	Change in italic
Us/lc	The institute	Set in main letters in caps and rest in small letters
'	The institute s	Insert apostrophe
" " ,	Indian Institute of Management Ahmadabad	Insert double inverted comma and comma
;	The institute	Insert Semicolon
:	The institute activities are	Insert semicolon
.	These are the institutes activities	Insert full stop
	The institute	Close up
-/	Semi annual	Insert Hyphen
/--/	The two institute at Ahmadabad and Calcutta	Insert dash
(	Institute	Move to the left
)	Institute	Move to the right
?	What are the activities of the institute	Insert question mark

Use of Electronic Media in Agricultural Journalism

Mass media sources like radio, television and type recorders are owned by large numbers of farmers, radio and television enjoy high level of credibility and farmers prider to seek information through radio and television, than other mass media. The electronic media is going to play more important role in dissemination of information. Thus there is a need to give more emphasis on electronic media in future.

Apart from radio and television, audio and video cassettes, satellite communication, interactive Visio disc and computer Assisted Instruction (CAI) and Internet have transformed the field of mass communication. The use of electronic media in journalism has made the world smaller through instant access to reality. It can effectively fulfil the criteria of five W's and one 'H' of journalism. Television provides viewers with realistic experience, which capture their attention and motivate them in proper direction.

Radio and television can easily break the literacy barrier. They are more accessible to rural population as compared to print media. Success stores, demonstration, question- answers, features. On improved from technique and latest information from research laboratories can easily made available to farmers. The market news, weather forecasting, information about GOVT. schemes *etc.* is made available to farmers through electronic media. The advent of privet channels on television has opened the nesters of journalism through electronic media and ides of farm rural journalism through media is spreading fast.

Reading Behavior of Rural People and Rural Libraries

The rural population include and neo-literates. There is undoubtedly some different between reading habit of rural and urban masses, which include greater percentage of highly literate and literate individuals. The rural people read literature published mostly in local language the time spent on reading, source of reading material, consistency in reading habit and the filed of liking. The study of reading behavior of rural people can give insight in using the print media as a means for establishing communication link with them.

The traditional meaning of library is collection of books for reading, catering the needs of various selection of the society.

However in rural areas the libraries can effectively perform the role of center for disseminating knowledge not only through reading material but also through other media of communication, which can enhance the knowledge of rural people in the fields of their interest. The process of continuous education of neo-literates and providing platform for non formal education are the important function of rural libraries.

35

Diffusion of Innovations

Diffusion of innovations is a theory that seeks to explain how, why, and at what rate new ideas and technology spread.

Everett Rogers, a professor, communication studies, popularized the theory in his book *Diffusion of Innovations*; the book was first published in 1962, and is now in its fifth edition (2003). Rogers argues that diffusion is the process by which an innovation is communicated over time among the participants in a social system. The origins of the diffusion of innovations theory are varied and span multiple disciplines. Rogers proposes that four main elements influence the spread of a new idea: the innovation itself, communication channels, time, and a social system. This process relies heavily on human capital. The innovation must be widely adopted in order to self-sustain. Within the rate of adoption, there is a point at which an innovation reaches critical mass. The categories of adopters are innovators, early adopters, early majority, late majority, and laggards Diffusion manifests itself in different ways and is highly subject to the type of adopters and innovation-decision process. The criterion for the adopter categorization is innovativeness, defined as the degree to which an individual adopts a new idea.

The concept of diffusion was first studied by the French sociologist Gabriel Tarde in late 19th century and by German and Austrian anthropologists and geographers such as Friedrich Ratzel and Leo Frobenius. The study of diffusion of innovations took off in the subfield of rural sociology in the mid-western United States in the 1920s and 1930s. Agriculture technology was advancing rapidly, and researchers started to examine how independent farmers were adopting hybrid seeds, equipment, and techniques. A study of the adoption of hybrid corn seed in Iowa by Ryan and Gross (1943) solidified the prior work on diffusion into a distinct paradigm that would be cited consistently in the future. Since its start in rural sociology, Diffusion of Innovations has been applied to numerous contexts, including medical sociology, communications, marketing, development studies, health promotion, organizational studies, knowledge management, and complexity studies, with a particularly large impact on the use of medicines, medical techniques, and health communications. In organizational studies, its basic epidemiological or internal-influence form was formulated by H. Earl Pemberton, who provided examples of institutional diffusion such as postage stamps and standardized school ethics codes.

In 1962, Everett Rogers, a professor of rural sociology, published his seminal work: *Diffusion of Innovations*. Rogers synthesized research from over 508 diffusion studies across the fields that initially influenced the theory: anthropology, early sociology, rural sociology, education, industrial sociology and medical sociology. Using his synthesis, Rogers produced a theory of the adoption of innovations among individuals and organizations. *Diffusion of Innovations* and Rogers' later books are among the most often cited in diffusion research. His methodologies are closely followed in recent diffusion research, even as the field has expanded into, and been influenced by, other methodological disciplines such as social network analysis and communication.

Elements

The key elements in diffusion research are:

Element	*Definition*
Innovation	Innovations are a broad category, relative to the current knowledge of the analyzed unit. Any idea, practice, or object that is perceived as new by an individual or other unit of adoption could be considered an innovation available for study.
Adopters	Adopters are the minimal unit of analysis. In most studies, adopters are individuals, but can also be organizations (businesses, schools, hospitals, *etc.*), clusters within social networks, or countries.
Communication channels	Diffusion, by definition, takes place among people or organizations. Communication channels allow the transfer of information from one unit to the other. Communication patterns or capabilities must be established between parties as a minimum for diffusion to occur.
Time	The passage of time is necessary for innovations to be adopted; they are rarely adopted instantaneously. In fact, in the Ryan and Gross (1943) study on hybrid corn adoption, adoption occurred over more than ten years, and most farmers only dedicated a fraction on their fields to the new corn in the first years after adoption.
Social system	The social system is the combination of external influences (mass media, organizational or governmental mandates) and internal influences (strong and weak social relationships, distance from opinion leaders). There are many roles in a social system, and their combination represents the total influences on a potential adopter.

Characteristics of Innovations

Studies have explored many characteristics of innovations. Meta-reviews have identified several characteristics that are common among most studies. These are in line with the characteristics that Rogers initially cited in his reviews. Potential adopters evaluate an innovation on its relative advantage (the perceived efficiencies gained by the innovation relative to current tools or procedures), its compatibility with the pre-existing system, its complexity or difficulty to learn, its trialability or testability, its potential for reinvention (using the tool for initially unintended purposes), and its observed effects. These qualities interact and are judged as a whole. For example, an innovation might be extremely complex; reducing its likelihood to be adopted and diffused, but it might be very compatible with a large advantage relative to current tools. Even with this high learning curve, potential adopters might adopt the innovation anyway.

Studies also identify other characteristics of innovations, but these are not as common as the ones that Rogers lists above. The fuzziness of the boundaries of the innovation can impact its adoption. Specifically, innovations with a small core and large periphery are easier to adopt. Innovations that are less risky are easier to adopt as the potential loss from failed integration is lower. Innovations that are disruptive to routine tasks, even when they bring a large relative advantage, might not be adopted because of added instability. Likewise, innovations that make tasks easier are likely to be adopted. Closely related to relative complexity, knowledge requirements are the ability barrier to use presented by the difficulty to use the innovation. Even when there are high knowledge requirements, support from prior adopters or other sources can increase the chances for adoption.

Characteristics of Individual Adopters

Like innovations, adopters have been determined to have traits that affect their likelihood to adopt an innovation. A bevy of individual personality traits have been explored for their impacts on adoption, but with little agreement. Ability and motivation, which vary on situation unlike personality traits, have a large impact on a potential adopter's likelihood to adopt an innovation. Unsurprisingly, potential adopters who are motivated to adopt an innovation are likely to make the adjustments needed to adopt it. Motivation can be impacted by the meaning that an innovation holds; innovations can have symbolic value that encourage (or discourage) adoption. First proposed by Ryan and Gross (1943), the overall connectedness of a potential adopter to the broad community represented by a city. Potential adopters who frequent metropolitan areas are more likely to adopt an innovation. Finally, potential adopters who have the power or agency to create change, particularly in organizations, are more likely to adopt an innovation than someone with less power over his choices.

Characteristics of Organizations

Organizations face more complex adoption possibilities because organizations are both the aggregate of its individuals and its own system with a set of procedures and norms. Three organizational characteristics match well with the individual characteristics above: tension for change (motivation and ability), innovation-system fit (compatibility), and assessment of implications (observability). Organizations can feel pressured by a tension for change. If the organization's situation is untenable, it will be motivated to adopt an innovation to change its fortunes. This tension often plays out among its individual members. Innovations that match the organization's pre-existing system require fewer coincidental changes and are easy to assess are more likely to be adopted. The wider environment of the organization, often an industry, community, or economy, exerts pressures on the organization, too. Where an innovation is diffusing through the organization's environment for any reason, the organization is more likely to adopt it. Innovations that are intentionally spread, including by political mandate or directive, are also likely to diffuse quickly.

Process

Diffusion occurs through a five step decision-making process. It occurs through a series of communication channels over a period of time among the members of a similar social system. Ryan and Gross first identified adoption as a process in 1943. Rogers' five stages (steps): awareness, interest, evaluation, trial, and adoption are integral to this theory. An individual might reject an innovation at any time during or after the adoption process. Abrahamson examined this process critically by posing questions such as: How do technically inefficient innovations diffuse and what impedes technically efficient innovations from catching on? Abrahamson makes suggestions for how organizational scientists can more comprehensively evaluate the spread of innovations. In later editions of *Diffusion of Innovation*, Rogers changes his terminology of the five stages to: knowledge, persuasion, decision, implementation, and confirmation. However, the descriptions of the categories have remained similar throughout the editions.

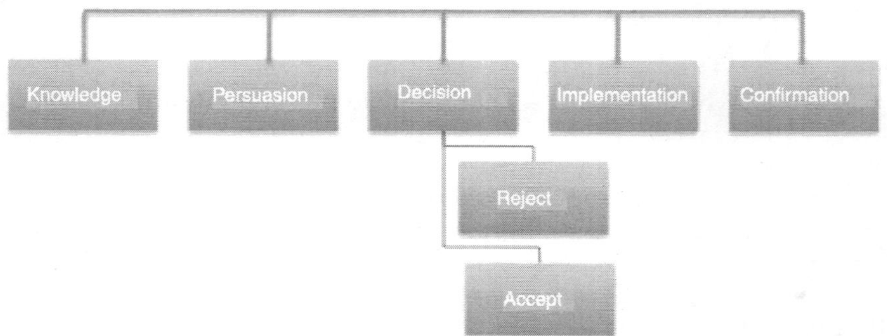

Five Stages of the Adoption Process

Stage	*Definition*
Knowledge	The individual is first exposed to an innovation, but lacks information about the innovation. During this stage the individual has not yet been inspired to find out more information about the innovation.
Persuasion	The individual is interested in the innovation and actively seeks related information/ details.
Decision	The individual takes the concept of the change and weighs the advantages/ disadvantages of using the innovation and decides whether to adopt or reject the innovation. Due to the individualistic nature of this stage, Rogers notes that it is the most difficult stage on which to acquire empirical evidence.[11]

Stage	Definition
Implementation	The individual employs the innovation to a varying degree depending on the situation. During this stage the individual also determines the usefulness of the innovation and may search for further information about it.
Confirmation	The individual finalizes his/her decision to continue using the innovation. This stage is both intrapersonal (may cause cognitive dissonance) and interpersonal, confirmation the group has made the right decision.

Decisions

Two factors determine what type a particular decision is:

- Whether the decision is made freely and implemented voluntarily
- Who makes the decision?

Based on these considerations, three types of innovation-decisions have been identified.

Type	Definition
Optional Innovation-Decision	Made by an individual who is in some way distinguished from others.
Collective Innovation-Decision	Made collectively by all participants.
Authority Innovation-Decision	Made for the entire social system by individuals in positions of influence or power.

Rate of Adoption

The rate of adoption is defined as the relative speed at which participants adopt an innovation. Rate is usually measured by the length of time required for a certain percentage of the members of a social system to adopt an innovation. The rates of adoption for innovations are determined by an individual's adopter category. In general, individuals who first adopt an innovation require a shorter adoption period (adoption process) when compared to late adopters.

Within the adoption curve at some point the innovation reaches critical mass. This is when the number of individual adopters ensures that the innovation is self-sustaining.

Adoption Strategies

Rogers outlines several strategies in order to help an innovation reach this stage, including when an innovation adopted by a highly respected individual within a social network and creating an instinctive desire for a specific innovation. Another strategy includes injecting an innovation into a group of individuals who would readily use said technology, as well as providing positive reactions and benefits for early adopters.

Diffusion vs Adoption

Adoption is an individual process detailing the series of stages one undergoes from first hearing about a product to finally adopting it. Diffusion signifies a group phenomenon, which suggests how an innovation spreads.

Adopter Categories

Rogers defines an adopter category as a classification of individuals within a social system on the basis of innovativeness. In the book *Diffusion of Innovations,* Rogers suggests a total of five categories of adopters in order to standardize the usage of adopter categories in diffusion research. The adoption of an innovation follows an S curve when plotted over a length of time. The categories of adopters are: innovators, early adopters, early majority, late majority and laggards. In addition to the gatekeepers and opinion leaders who exist within a given community, change agents may come from outside the community. Change agents bring innovations to new communities– ûrst through the gatekeepers, then through the opinion leaders, and so on through the community.

Adopter Category

	Definition
Innovators	Innovators are willing to take risks, have the highest social status, have financial liquidity, are social and have closest contact to scientific sources and interaction with other innovators. Their risk tolerance allows them to adopt technologies that may ultimately fail. Financial resources help absorb these failures.
Early adopters	These individuals have the highest degree of opinion leadership among the adopter categories. Early adopters have a higher social status, financial liquidity, advanced education and are more socially forward than late adopters. They are more discreet in adoption choices than innovators. They use judicious choice of adoption to help them maintain a central communication position.[41]
Early Majority	They adopt an innovation after a varying degree of time that is significantly longer than the innovators and early adopters. Early Majority have above average social status, contact with early adopters and seldom hold positions of opinion leadership in a system (Rogers 1962, p. 283)
Late Majority	They adopt an innovation after the average participant. These individuals approach an innovation with a high degree of skepticism and after the majority of society has adopted the innovation. Late Majority are typically skeptical about an innovation, have below average social status and little financial liquidity, in contact with others in late majority and early majority and little opinion leadership.
Laggards	They are the last to adopt an innovation. Unlike some of the previous categories, individuals in this category show little to no opinion leadership. These individuals typically have an aversion to change-agents. Laggards typically tend to be focused on "traditions", lowest social status, lowest financial liquidity, oldest among adopters, and in contact with only family and close friends.

Failed Diffusion

Failed diffusion does not mean that the technology was adopted by no one. Rather, failed diffusion often refers to diffusion that does not reach or approach 100 per cent adoption due to its own weaknesses, competition from other innovations, or simply a lack of awareness. From a social networks perspective, a failed diffusion might be widely adopted within certain clusters but fail to make an impact on more distantly related people. Networks that are over-connected might suffer from a rigidity that prevents the changes an innovation might bring, as well. Sometimes, some innovations also fail as a result of lack of local involvement and community participation.

For example, Rogers discussed a situation in Peru involving the implementation of boiling drinking water to improve health and wellness levels in the village of Los Molinas. The residents had no knowledge of the link between sanitation and illness. The campaign worked with the villagers to try to teach them to boil water, burn their garbage, install latrines and report cases of illness to local health agencies. In Los Molinas, a stigma was linked to boiled water as something that only the "unwell" consumed, and thus, the idea of healthy residents boiling water prior to consumption was frowned upon. The two-year educational campaign was considered to be largely unsuccessful. This failure exemplified the importance of the roles of the communication channels that are involved in such a campaign for social change. An examination of diffusion in El Salvador determined that there can be more than one social network at play as innovations are communicated. One network carries information and the other carries influence. While people might hear of an innovation's uses, in Rogers' Los Molinas sanitation case, a network of influence and status prevented adoption.

Heterophily and Communication Channels

Lazarsfeld and Merton first called attention to the principles of homophily and it's opposite, heterophily. Using their definition, Rogers defines homophily as "the degree to which pairs of individuals who interact are similar in certain attributes, such as beliefs, education, social status, and the like". When given the choice, individuals usually choose to interact with someone similar to themselves. Homophilous individuals engage in more effective communication because their similarities lead to greater knowledge gain as well as attitude or behavior change. As a result, homophilous people tend to promote diffusion among each other. However, diffusion requires a certain degree of heterophily to introduce new ideas into a relationship; if two individuals are identical, no diffusion occurs because there is no new information to exchange. Therefore, an ideal situation would involve potential adopters who are homophilous in every way, except in knowledge of the innovation.

Promotion of healthy behavior provides an example of the balance required of homophily and heterophily. People tend to be close to others of similar health status. As a result, people with unhealthy behaviors like smoking and obesity are less likely to encounter information and behaviors that encourage good health. This presents a critical challenge for health communications, as ties between heterophilous people are relatively weaker, harder to create, and harder to maintain. Developing heterophilous ties to unhealthy communities can increase the effectiveness of the diffusion of good health behaviors. Once one previously homophilous tie adopts the behavior or innovation, the other members of that group are more likely to adopt it, too.

The Role of Social Systems

Opinion Leaders

Not all individuals exert an equal amount of influence over others. In this sense opinion leaders are influential in spreading either positive or negative information about an innovation. Rogers relies on the ideas of Katz and Lazarsfeld and the two-step flow theory in developing his ideas on the influence of opinion leaders.

Opinion leaders have the most influence during the evaluation stage of the innovation-decision process and on late adopters. In addition opinion leaders typically have greater exposure to the mass media, more cosmopolitan, greater contact with change agents, more social experience and exposure, higher socioeconomic status, and are more innovative than others.

Research was done in the early 1950s at the University of Chicago attempting to assess the cost-effectiveness of broadcast advertising on the diffusion of new products and services. The findings were that opinion leadership tended to be organized into a hierarchy within a society, with each level in the hierarchy having most influence over other members in the same level, and on those in the next level below it. The lowest levels were generally larger in numbers and tended to coincide with various demographic attributes that might be targeted by mass advertising. However, it found that direct word of mouth and example were far more influential than broadcast messages, which were only effective if they reinforced the direct influences. This led to the conclusion that advertising was best targeted, if possible, on those next in line to adopt, and not on those not yet reached by the chain of influence.

Research on actor-network theory (ANT) also identifies a significant overlap between the ANT concepts and the diffusion of innovation which examine the characteristics of innovation and its context among various interested parties within a social system to assemble a network or system which implements innovation.

Other research relating the concept to public choice theory finds that the hierarchy of influence for innovations need not, and likely does not, coincide with hierarchies of official, political, or economic status. Elites are often not innovators, and innovations may have to be introduced by outsiders and propagated up a hierarchy to the top decision makers.

Electronic Communication Social Networks

Prior to the introduction of the Internet, it was argued that social networks had a crucial role in the diffusion of innovation particularly tacit knowledge in the book *The IRG Solution – hierarchical incompetence and how to overcome it*. The book argued that the widespread adoption of computer networks of individuals would lead to much better diffusion of innovations, with greater understanding of their possible shortcomings and the identification of needed innovations that would not have otherwise occurred. The social model proposed by Ryan and Gross is expanded by Valente who uses social networks as a basis for adopter categorization instead of solely relying on the system-level analysis used by Ryan and Gross. Valente also looks at an individual's personal network, which is a different application than the organizational perspective espoused by many other scholars.

Recent research by Wear shows, that particularly in regional and rural areas, significantly more innovation takes place in communities which have stronger inter-personal networks.

Organizations

Innovations are often adopted by organizations through two types of innovation-decisions: collective innovation decisions and authority innovation decisions. The collective decision occurs when adoption is by consensus. The authority decision occurs by adoption among very few individuals with high positions of power within an organization. Unlike the optional innovation decision process, these decision processes only occur within an organization or hierarchical group. Within an organization certain individuals are termed "champions" who stand behind an innovation and break through opposition. The champion plays a very similar role as the champion used within the efficiency business model Six Sigma. The process contains five stages that are slightly similar to the innovation-decision process that individuals undertake. These stages are: agenda-setting, matching, redefining/ restructuring, clarifying and routinizing.

Extensions of the Theory

Policy

Diffusion of Innovations has been applied beyond its original domains. In the case of political science and administration, policy diffusion focuses on how institutional innovations are adopted by other institutions, at the local, state, or country level. An alternative term is 'policy transfer' where the focus is more on the agents of diffusion and the diffusion of policy knowledge, such as in the work of Diane Stone. Specifically, policy transfer can be defined as "knowledge about how policies administrative arrangements, institutions, and ideas in one political setting (past or present) is used in the development of policies, administrative arrangements, institutions, and ideas in another political setting".

The first interests with regards to policy diffusion were focused in time variation or state lottery adoption, but more recently interest has shifted towards mechanisms (emulation, learning and coercion) or in channels of diffusion where researchers find that regulatory agency creation is transmitted by country and sector channels. At the local level, examining popular city-level policies make it easy to find patterns in diffusion through measuring public awareness. At the international level, economic policies have been thought to transfer among countries according to local politicians' learning of successes and failures elsewhere and outside mandates made by global financial organizations. As a group of countries succeed with a set of policies, others follow, as exemplified by the deregulation and liberalization across the developing world after the successes of the Asian Tigers. The reintroduction of regulations in the early 2000s also shows this learning process, which would fit under the stages of knowledge and decision, can be seen as lessons learned by following China's successful growth.

Technology

Peres, Muller and Mahajan suggested that diffusion is "the process of the market penetration of new products and services that is driven by social inûuences, which include all interdependencies among consumers that affect various market players with or without their explicit knowledge".

Eveland evaluated diffusion from a phenomenological view, stating, "Technology is information, and exists only to the degree that people can put it into practice and use it to achieve values".

Diffusion of existing technologies has been measured using "S curves". These technologies include radio, television, VCR, cable, flush toilet, clothes washer, refrigerator, home ownership, air conditioning, dishwasher, electrified households, telephone, cordless phone, cellular phone, per capita airline miles, personal computer and the Internet. These data can act as a predictor for future innovations.

Diffusion curves for infrastructure reveal contrasts in the diffusion process of personal technologies versus infrastructure.

Consequences of Adoption

Both positive and negative outcomes are possible when an individual or organization chooses to adôpt a particular innovation. Rogers states that this area needs further research because of the biased positive attitude that is associated with innovation. Rogers lists three categories for consequences: desirable vs. undesirable, direct vs. indirect, and anticipated vs. unanticipated. In contrast Wejnert details two categories: public vs. private and benefits vs. costs.

Public versus Private

Public consequences comprise the impact of an innovation on those other than the actor, while private consequences refer to the impact on the actor. Public consequences usually involve collective actors, such as countries, states, organizations or social movements. The results are usually concerned with issues of societal well-being. Private consequences usually involve individuals or small collective entities, such as a community. The innovations are usually concerned with the improvement of quality of life or the reform of organizational or social structures.

Benefits versus Costs

Benefits of an innovation obviously are the positive consequences, while the costs are the negative. Costs may be monetary or nonmonetary, direct or indirect. Direct costs are usually related to financial uncertainty and the economic state of the actor. Indirect costs are more difficult to identify. An example would be the need to buy a new kind of pesticide to use innovative seeds. Indirect costs may also be social, such as social conflict caused by innovation. Marketers are particularly interested in the diffusion process as it determines the success or failure of a new product. It is quite important for a marketer to understand the diffusion process so as to ensure proper management of the spread of a new product or service.

Mathematical Treatment

Main Article: Logistic Function

The diffusion of an innovation typically follows an S shaped curve which often resembles a logistic function. Mathematical programming models such as the S-D model apply the diffusion of innovations theory to real data problems. In addition to that, agent-based models follow a more intuitive process by designing individual-level rules to model diffusion of ideas and innovations.

Complex Systems Models

Complex network models can also be used to investigate the spread of innovations among individuals connected to each other by a network of peer-to-peer influences, such as in a physical community or neighbourhood.

Such models represent a system of individuals as *nodes* in a network (or graph). The interactions that link these individuals are represented by the edges of the network and can be based on the probability or strength of social connections. In the dynamics of such models, each node is assigned a current state, indicating whether or not the individual has adopted the innovation, and model equations describe the evolution of these states over time.

In threshold models, the uptake of technologies is determined by the balance of two factors: the (perceived) usefulness (sometimes called utility) of the innovation to the individual as well as barriers to adoption, such as cost. The multiple parameters that influence decisions to adopt, both individual and socially motivated, can be represented by such models as a series of nodes and connections that represent real relationships. Borrowing from social network analysis, each node is an innovator, an adopter, or a potential adopter. Potential adopters have a threshold, which is a fraction of his neighbors who adopt the innovation that must be reached before he will adopt. Over time, each potential adopter views his neighbors and decides whether he should adopt based on the technologies they are using. When the effect of each individual node is analyzed along with its influence over the entire network, the expected level of adoption was seen to depend on the number of initial adopters and the network's structure and properties. Two factors emerge as important to successful spread of the innovation: the number of connections of nodes with their neighbors and the presence of a high degree of common connections in the network (quantified by the clustering coefficient). These models are particularly good at showing the impact of opinion leaders relative to others. Computer models are often used to investigate this balance between the social aspects of diffusion and perceived intrinsic benefit to the individuals.

Criticism

Because there are more than four thousand articles across many disciplines published on Diffusion of Innovations, with a vast majority written after Rogers created a systematic theory, there have been few widely adopted changes to the theory. Although each study applies the theory in slightly different ways, this lack of cohesion has left the theory stagnant and difficult to apply with consistency to new problems.

Diffusion is difficult to quantify because humans and human networks are complex. It is extremely difficult, if not impossible, to measure what exactly causes adoption of an innovation. This is important, particularly in healthcare. Those encouraging adoption of health behaviors or new medical technologies need to be aware of the many forces acting on an individual and his or her decision to adopt a new behavior or technology. Diffusion theories can never account for all variables, and therefore might miss critical predictors of adoption. This variety of variables has also led to inconsistent results in research, reducing heuristic value.

Rogers placed the contributions and criticisms of diffusion research into four categories: pro-innovation bias, individual-blame bias, recall problem, and issues of equality. The pro-innovation bias, in particular, implies that all innovation is positive and that all innovations should be adopted. Cultural traditions and beliefs can be consumed by another culture's through diffusion, which can impose significant costs on a group of people. The one-way information flow, from sender to receiver, is another weakness of this theory. The message sender has a goal to persuade the receiver, and there is little to no reverse flow. The person implementing the change controls the direction and outcome of the campaign. In some cases, this is the best approach, but other cases require a more participatory approach. In complex environments where the adopter is receiving information from many sources and is returning feedback to the sender, a one-way model is insufficient and multiple communication flows need to be examined.

36
Innovation

Collaborative Innovation

Collaborative innovation is a process in which multiple players (within and outside an organization) contribute towards creating and developing new products, services, processes and business solutions. It might include the involvement of customers, suppliers and multiple stakeholders such as agencies and consultants.

Usually, firms that promote open forms of collaboration benefit from having access to different capabilities and knowledge, enhancing their competitiveness and accelerating their innovation process. On one hand, it enables small companies such as start-ups to partner with other players, complementing each other and taking advantage of different perspectives and resources. On the other hand, it helps large companies to speed-up their innovation process and time-to-market, overcoming bureaucracy and inflexible procedures.

Collaboration can occur in all aspects of the business cycle, depending on the context:

- ✰ Procurement and supplier collaboration
- ✰ Research and development of new products, services and technologies
- ✰ Marketing, distribution and commercialization

Collaborative Innovation Network (CoIN)

Collaborative innovation network is a type of collaborative innovation practice that makes use of the internet platforms such as email, chat, social networks, blogs, and Wikis to promote communication and innovation within self-organizing virtual teams. The difference is that people that collaborate in Collaborative innovation network are so intrinsically motivated that might not be paid nor get any advantage.

Thus, a Collaborative innovation network is a social construct with a huge potential for innovation. It has been defined by the originator of the term, Peter Gloor from MIT Sloan's Center for Collective Intelligence, as "a cyber team of self-motivated people with a collective vision, enabled by the Web to collaborate in achieving a common goal by sharing ideas, information, and work".

Indeed, Collaborative innovation network is a type of open collaboration that helps organizations to become more creative, productive and efficient. By adopting CoIN as part of their culture, these companies accelerate innovation, uncover hidden business opportunities, reduce costs and enhance synergies. They not only can engage employees from every level of hierarchy towards a common project (discovering new talents and promoting direct relation between employees) but also partner with external parties.

Similar is the concept of the "Self-Organizing Innovation Network", it has been described by author, Robert Rycroft of the Elliott School of International Affairs of George Washington University.

Overview

Collaborative innovation networks feature internal transparency and direct communication. Members of a Collaborative innovation network collaborate and share knowledge directly with each other, rather than through hierarchies. They come together with a shared vision because they are intrinsically motivated to do so and seek to collaborate in some way to advance an idea.

Collaborative innovation networks work across hierarchies and boundaries in which members can directly and openly exchange ideas and information. This collaborative and transparent environment fosters innovation. Gloor describes phenomenon as "swarm creativity". According to him, "COINs are the best engines to drive innovation".

Collaborative innovation networks existed well before the advent of modern communication technology. However, internet and instant communication improved productivity and enabled the reach of a global scale. Today, they rely on Internet, e-mail, and other communications vehicles for information sharing.

According to Peter Gloo, Collaborative innovation networks have 5 main characteristics:

- **Dispersed Membership**: technology allows members to be spread over the world. Regardless of the location, members share a common goal and are convinced of their common cause.
- **Interdependent Membership**: cooperation between members is key to achieve common goal. The work of one member is affected and interdependent on the others´ work.
- **No simple chain of command**: there is no superior command. It is a decentralized and self-organized system. Conflicts are solved without the need of a hierarchy or authority.

- **Work towards a common goal**: members are willing to contribute, work and share freely. They are intrinsically motivated to donate their work, create and share knowledge in favor of a common goal.
- **Dependence on trust**: cooperative behavior and mutual trust is needed to work efficiently within the network. Members act accordingly to an ethical code that states the rules and principles to be followed by all members. Usually, ethical codes include principles related to respect, consistency, reciprocity and rationality.

There also are five essential elements of collaborative innovation networks (what Gloor calls as "**genetic code**"):

1. **They are learning networks**: they set an informal and flexible environment which facilitates and stimulates collaboration and the exchange of ideas, information and knowledge.
2. **They need an ethical code**: they agree on an ethical code that guides the conduct and behavior of the members.
3. **They are based on trust and self-organization**: members trust each other without the need of a centralized management. They are brought together by mutual respect and strong sense of shared beliefs.
4. **They make knowledge accessible to everyone**: Collaborative innovation network nurture communication to an extent that information shared with everyone. Nowadays, with internet and social Medias, their ideas and concepts achieve a global level.
5. **They operate in internal honesty and transparency**: they create a system based on reciprocal trust and mutually established principles.

Examples of Collaborative Innovation Network

Collaborative innovation networks have been developing many disruptive innovations such as the Internet, Linux, the Web and Wikipedia. These inventions were created in universities or labs by a group of enthusiastic students with little or no budget. They were not focused on the money but on the sense of accomplishment.

The **Web** is the early version of Internet. It was driven by a Collaborative innovation network of intrinsically motivated people that wanted to improve technical development and launch a disruptive solution. Their goal was to link mainframes and allows multiple users simultaneously.

Another important contribution of Collaborative innovation network was **Linux**, an operating system for personal computing that directly competes with Microsoft. It was initially developed by a student called Linus Torvalds and later became open source software. Today, the code is publicly available and anyone can contribute or enhance it. The success of Linux is the constant and continuous updating which is done at a much lower cost than closed source software.

Last but not least, **Wikipedia** gathers thousands of volunteers that constantly write and update content. Although it does not have a hierarchy nor a central authority, the entries are mostly accurate and complete. Volunteers share a strong feeling of community and willingness to contribute towards knowledge without being paid for it.

Faced with these creations, large companies such as IBM and Intel have learnt to use the principles of open innovation to enhance their research learning curve. They increased or established collaborations with universities, agencies and small companies to accelerate their processes and launch new services faster.

Current Challenges

Although it exists for long time, Collaborative Innovation still needs to be empowered. A more collaborative approach involving stakeholders such as governments, corporations, entrepreneurs and scholars is key to tackle the main challenges facing today.

First of all, it is still important to raise the awareness of Collaborative innovation network and its benefits among companies and major economic fields. Policy makers and corporate leaders could support the development of programs, strategies and educational plans to stimulate Collaborative innovation networks in specific sectors, benefiting the whole economy.

Second, the overall legal and regulatory framework still needs to evolve to foster cross-firm collaboration. Fiscal and intellectual property regimes should be reviewed to provide the necessary infrastructure to nourish Collaborative innovation networks. A further stimulus is important to encourage the creation of start-ups and the development of a network of partners across companies.

Finally, financial aid should be granted to support collaborative projects related to technology, research and innovation. Collaborative innovation networks have an enormous potential to deliver innovation and drive significant gains in competitiveness. However, they need resources in order to fully operate and reach their maximum potential.

Future

As Collaborative innovation networks become increasingly popular among governments and corporations, the ethical, financial, economic, and cognitive issues which drive incentives will inevitably face challenges. Over time potential innovators may be unwilling to participate in projects merely on the basis of implied financial gain. As globalization begins to impact traditional models of planned social progress, the broader political context in which participants cooperate has become more relevant lately. This suggests an increased need for independent parties to collaborate on the basis of agreed upon principles and objectives, ultimately this could encompass the interests of humanity and the emergence of a global culture.

Critical Mass (Socio-dynamics)

In social dynamics, critical mass is a sufficient number of adopters of an innovation in a social system so that the rate of adoption becomes self-sustaining and creates further growth. The term is borrowed from nuclear physics and in that field it refers to the amount of a substance needed to sustain a chain reaction.

Social factors influencing critical mass may involve the size, interrelatedness and level of communication in a society or one of its subcultures. Another is social stigma, or the possibility of public advocacy due to such a factor.

Critical mass may be closer to majority consensus in political circles, where the most effective position is more often that held by the majority of people in society. In this sense, small changes in public consensus can bring about swift changes in political consensus, due to the majority-dependent effectiveness of certain ideas as tools of political debate.

Critical mass is a concept used in a variety of contexts, including physics, group dynamics, politics, public opinion, and technology.

History

The concept of critical mass was originally created by game theorist Thomas Schelling and sociologist Mark Granovetter to explain the actions and behaviors of a wide range of people and phenomenon. The concept was first established (although not explicitly named) in Schelling's essay about racial segregation in neighbourhoods, published in 1971 in the Journal of Mathematical Sociology, and later refined in his book, *Micro-motives and Macro-behavior*, published in 1978. He did use the term "critical density" with regard to pollution in his "On the Ecology of Micro-motives". Mark Granovetter, in his essay "Threshold models of collective behavior", published in the American Journal of Sociology in 1978 worked to solidify the theory. Everett Rogers later cites them both in his important work *Diffusion of Innovations*, in which critical mass plays an important role.

Predecessors

The concept of critical mass had existed before it entered a sociology context. It was an established concept in medicine, specifically epidemiology, since the 1920s, as it helped to explain the spread of illnesses.

It had also been a present, if not solidified, idea in the study of consumer habits and economics, especially in General Equilibrium Theory. In his papers, Schelling quotes the well-known "The Market for Lemons: Quality Uncertainty and the Market Mechanism" paper written in 1970 by George Akerlof. Similarly, Granovetter cited the Nash Equilibrium game in his papers.

Finally, Herbert A. Simon's essay, "Bandwagon and underdog effects and the possibility of election predictions", published in 1954 in Public Opinion Quarterly, has been cited as a predecessor to the concept we now know as critical mass.

Logic of Collective Action and Common Good

Critical mass and the theories behind it help us to understand aspects of humans as they act and interact in a larger social setting. Certain theories, such as Mancur Olson's Logic of Collective Action or Garrett Hardin's Tragedy of the Commons, work to help us understand why humans do or adopt certain things which are beneficial to them, or, more importantly, why they do not. Much of this reasoning has to do with individual interests trumping that which is best for the collective whole, which may not be obvious at the time.

Oliver, Marwell, and Teixeira tackle this subject in relation to critical theory in a 1985 article published in the American Journal of Sociology. In their essay, they define that action in service of a public good as "collective action". "Collective Action" is beneficial to all, regardless of individual contribution. By their definition, then, "critical mass" is the small segment of a societal system that does the work or action required to achieve the common good. The "Production Function" is the correlation between resources, or what individuals give in an effort to achieve public good, and the achievement of that good. Such function can be decelerating, where there is less utility per unit of resource, and in such a case, resource can taper off. On the other hand, the function can be accelerating, where the more resources that are used the bigger the payback. "Heterogeneity" is also important to the achievement of a common good. Variations (heterogeneity) in the value individuals put on a common good or the effort and resources people give is beneficial, because if certain people stand to gain more, they are willing to give or pay more.

Gender Politics

Critical mass theory in gender politics and collective political action is defined as the critical number of personnel needed to affect policy and make a change not as the token but as an influential body. This number has been placed at 30 per cent, before women are able to make a substantial difference in politics. However, other research suggests lower numbers of women working together in legislature can also affect political change. Kathleen Bratton goes so far as to say that women, in legislatures where they make up less than 15 per cent of the membership, may actually be encouraged to develop legislative agendas that are distinct from those of their male colleagues. Others argue that we should look more closely at parliamentary and electoral systems instead of critical mass.

Interactive Media

While critical mass can be applied to many different aspects of socio-dynamics, it becomes increasingly applicable to innovations in interactive media such as the telephone, fax, or email. With other non-interactive innovations, the dependence on other users was generally sequential, meaning that the early adopters influenced the later adopters to use the innovation. However, with interactive media, the interdependence was reciprocal, meaning both users influenced each other. This is due to the fact that interactive media have high network effect, where in the value and utility of a good or service increases the more users it has. Thus, the increase of adopters and quickness to reach critical mass can therefore be faster and more intense with interactive media, as can the rate at which previous users discontinue their use. The more people that use it, the more beneficial it will be, thus creating a type of snowball effect, and conversely, if users begin to stop using the innovation, the innovation loses utility, thus pushing more users to discontinue their use.

Markus Essay

In M. Lynne Markus' essay in *Communication Research* entitled "Toward a 'Critical Mass' Theory of Interactive Media", several propositions are made that try to predict under what circumstances interactive media is most likely to achieve critical mass and reach universal access, a "common good" using Oliver, *et al.*'s terminology. One proposition states that such media's existence is all or nothing, where in if universal access is not achieved, then, eventually, use will discontinue. Another proposition suggests that a media's ease of use and inexpensiveness, as well as its utilization of an "active notification capability" will help it achieve universal access. The third proposition states that the heterogeneity, as discussed by Oliver, *et al.*, is beneficial, especially if users are dispersed over a larger area, thus necessitating interactivity via media. Fourth, it is very helpful to have highly sought-after individuals to act as early adopters, as their use acts as incentive for later users. Finally, Markus posits that interventions, both monetarily and otherwise, by governments, businesses, or groups of individuals will help a media reach its critical mass and achieve universal access.

Fax Machine Example

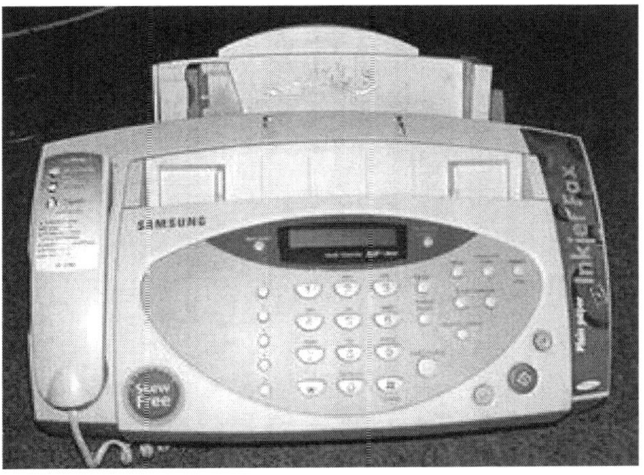

A Fax Machine

An example put forth by Rogers in *Diffusion of Innovations* was that of the fax machine, which had been around for almost 150 years before it became popular and widely used. It had existed in various forms and for various uses, but with more advancements in the technology of faxes, including the use of existing phone lines to transmit information, coupled with falling prices in both machines and cost per fax, the fax machine reached a critical mass in 1987, when "Americans began to assume that 'everybody else' had a fax machine".

Delphi Method

The Delphi method is a structured communication technique or method, originally developed as a systematic, interactive forecasting method which relies on a panel of experts. The experts answer questionnaires in two or more rounds. After each round, a facilitator or change agent provides an anonymised summary of the experts' forecasts from the previous round as well as the reasons they provided for their judgments. Thus, experts are encouraged to revise their earlier answers in light of the replies of other members of their panel. It is believed that during this process the range of the answers will decrease and the group will converge towards the "correct" answer. Finally, the process is stopped after a predefined stop criterion (*e.g.* number of rounds, achievement of consensus, stability of results) and the mean or median scores of the final rounds determine the results.

Delphi is based on the principle that forecasts (or decisions) from a structured group of individuals are more accurate than those from unstructured groups. The technique can also be adapted for use in face-to-face meetings, and is then called mini-Delphi or Estimate-Talk-Estimate (ETE). Delphi has been widely used for business forecasting and has certain advantages over another structured forecasting approach, prediction markets.

History

The name "Delphi" derives from the Oracle of Delphi, although the authors of the method were unhappy with the oracular connotation of the name, "smacking a little of the occult". The Delphi method is based on the assumption that group judgments are more valid than individual judgments.

The Delphi method was developed at the beginning of the Cold War to forecast the impact of technology on warfare. In 1944, General Henry H. Arnold ordered the creation of the report for the U.S. Army Air Corps on the future technological capabilities that might be used by the military. Different approaches were tried, but the shortcomings of traditional forecasting methods, such as theoretical approach, quantitative models or trend extrapolation, quickly became apparent in areas where precise scientific laws have not been established yet. To combat these shortcomings, the Delphi method was developed by Project RAND during the 1950-1960s (1959) by Olaf Helmer, Norman Dalkey, and Nicholas Rescher. It has been used ever since, together with various modifications and reformulations, such as the Imen-Delphi procedure.

Experts were asked to give their opinion on the probability, frequency, and intensity of possible enemy attacks. Other experts could anonymously give feedback. This process was repeated several times until a consensus emerged.

Key Characteristics

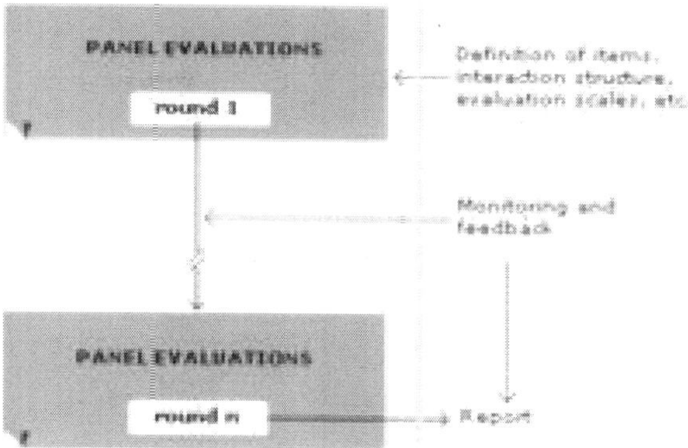

The Delphi Method Communication Structure

The following key characteristics of the Delphi method help the participants to focus on the issues at hand and separate Delphi from other methodologies: in this technique a panel of experts is drawn from both inside and outside the organisation. The panel consist of experts having knowledge of the area requiring decision making. Each expert is asked to make anonymous predictions.

Anonymity of the Participants

Usually all participants remain anonymous. Their identity is not revealed, even after the completion of the final report. This prevents the authority, personality, or reputation of some participants from dominating others in the process. Arguably, it also frees participants (to some extent) from their personal biases, minimizes the "bandwagon effect" or "halo effect", allows free expression of opinions, encourages open critique, and facilitates admission of errors when revising earlier judgments.

Structuring of Information Flow

The initial contributions from the experts are collected in the form of answers to questionnaires and their comments to these answers. The panel director controls the interactions among the participants by processing the information and filtering out irrelevant content. This avoids the negative effects of face-to-face panel discussions and solves the usual problems of group dynamics.

Regular Feedback

Participants comment on their own forecasts, the responses of others and on the progress of the panel as a whole. At any moment they can revise their earlier statements. While in regular group meetings participants tend to stick to previously stated opinions and often conform too much to the group leader; the Delphi method prevents it.

Role of the Facilitator

The person coordinating the Delphi method is usually known as a *facilitator* or Leader, and facilitates the responses of their *panel of experts*, who are selected for a reason, usually that they hold knowledge on an opinion or view. The facilitator sends out questionnaires, surveys *etc.* and if the panel of experts accept, they follow instructions and present their views. Responses are collected and analyzed, then common and conflicting viewpoints are identified. If consensus is not reached, the process continues through thesis and antithesis, to gradually work towards synthesis, and building consensus.

Applications

Use in Forecasting

First applications of the Delphi method were in the field of science and technology forecasting. The objective of the method was to combine expert opinions on likelihood and expected development time, of the particular technology, in a single indicator. One of the first such reports, prepared in 1964 by Gordon and Helmer, assessed the direction of long-term trends in science and technology development, covering such topics as scientific breakthroughs, population control, automation, space progress, war prevention and weapon systems. Other forecasts of technology were dealing with vehicle-highway systems, industrial robots, intelligent internet, broadband connections, and technology in education.

Later the Delphi method was applied in other places, especially those related to public policy issues, such as economic trends, health and education. It was also applied successfully and with high accuracy in business forecasting. For example, in one case reported by Basu and Schroeder (1977), the Delphi method predicted the sales of a new product during the first two years with inaccuracy of 3-4 per cent compared with actual sales. Quantitative methods produced errors of 10-15 per cent, and traditional unstructured forecast methods had errors of about 20 per cent. (This is only one example; the overall accuracy of the technique is mixed.)

The Delphi method has also been used as a tool to implement multi-stakeholder approaches for participative policy-making in developing countries. The governments of Latin America and the Caribbean have successfully used the Delphi method as an open-ended public-private sector approach to identify the most urgent challenges for their regional ICT-for-development eLAC Action Plans. As a result, governments have widely acknowledged the value of collective intelligence from civil society, academic and private sector participants of the Delphi, especially in a field of rapid change, such as technology policies.

Use in Policy-making

From the 1970s, the use of the Delphi technique in public policy-making introduces a number of methodological innovations. In particular:

- The need to examine several types of items (not only *forecasting* items but, typically, *issue* items, *goal* items, and *option* items) leads to introducing different evaluation scales which are not used in the standard Delphi.

These often include *desirability*, *feasibility*(technical and political) and *probability*, which the analysts can use to outline different scenarios: the *desired* scenario (from desirability), the *potential* scenario (from feasibility) and the *expected* scenario (from probability);

- ✰ The complexity of the issues posed in public policy-making leads to give more importance to the arguments supporting the evaluations of the panelists; so these are often invited to list arguments for and against each option item, and sometimes they are given the possibility to suggest new items to be submitted to the panel;
- ✰ For the same reason, the scaling methods, which are used to measure panel evaluations, often include more sophisticated approaches such as multi-dimensional scaling.

Further innovations come from the use of computer-based (and later web-based) Delphi conferences. According to Turoff and Hiltz, in computer-based Delphis:

- ✰ The iteration structure used in the paper Delphis, which is divided into three or more discrete rounds, can be replaced by a process of continuous (round less) interaction, enabling panellists to change their evaluations at any time;
- ✰ The statistical group response can be updated in real-time, and shown whenever a panelist provides a new evaluation.

According to Bolognini, web-based Delphis offer two further possibilities, relevant in the context of interactive policy-making and e-democracy. These are:

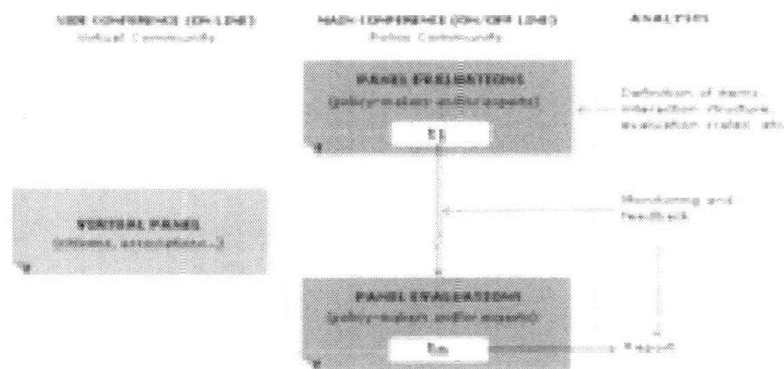

A Web-based Communication Structure (Hyperdelphi)

- ✰ The involvement of a large number of participants,
- ✰ The use of two or more panels representing different groups (such as policy-makers, experts, citizens), which the administrator can give tasks reflecting their diverse roles and expertise, and make them to interact within ad hoc communication structures. For example, the *policy community* members (policy-makers and experts) may interact as part of the *main conference* panel, while they receive inputs from a *virtual community* (citizens, associations *etc.*) involved in a *side conference*. These web-based variable communication structures, which he

calls *Hyperdelphi* (HD), are designed to make Delphi conferences "more fluid and adapted to the hypertextual and interactive nature of digital communication".

One successful example of a (partially) web-based policy Delphi is the five-round Delphi exercise (with 1,454 contributions) for the creation of the eLAC Action Plans in Latin America. It is believed to be the most extensive online participatory policy-making foresight exercise in the history of intergovernmental processes in the developing world at this time. In addition to the specific policy guidance provided, the authors list the following lessons learned include:

"(1) The potential of Policy Delphi methods to introduce transparency and accountability into public decision-making, especially in developing countries;

(2) The utility of foresight exercises to foster multi-agency networking in the development community;

(3) The usefulness of embedding foresight exercises into established mechanisms of representative democracy and international multilateralism, such as the United Nations;

(4) The potential of online tools to facilitate participation in resource-scarce developing countries; and

(5) The resource-efficiency stemming from the scale of international foresight exercises, and therefore its adequacy for resource-scarce regions."

Online Delphi Systems

Main Article: Real-time Delphi

A number of Delphi forecasts are conducted using web sites that allow the process to be conducted in real-time. For instance, the TechCast Project uses a panel of 100 experts worldwide to forecast breakthroughs in all fields of science and technology. Another example is the Horizon Project, where educational futurists collaborate online using the Delphi method to come up with the technological advancements to look out for in education for the next few years.

Variations

Traditionally the Delphi method has aimed at a consensus of the most probable future by iteration. Other versions, such as the Policy Delphi, are instead a decision support method aiming at structuring and discussing the diverse views of the preferred future. In Europe, more recent web-based experiments have used the Delphi method as a communication technique for interactive decision-making and e-democracy. The Argument Delphi, was developed by Osmo Kuusi, focuses on ongoing discussion and finding relevant arguments rather than focusing on the output. The dis-aggregative Policy Delphi, developed by Petri Tapio, uses cluster analysis as a systematic tool to construct various scenarios of the future in the latest Delphi round. The respondent's view on the probable and the preferable future are dealt with as separate cases. The computerization of Argument Delphi is relatively difficult because of several problems like argument resolution, argument aggregation and argument evaluation. The computerization of Argument Delphi, was developed by Sadi Evren Seker, proposes solutions to such problems.

Accuracy

Overall the track record of the Delphi method is mixed. There have been many cases when the method produced poor results. Still, some authors attribute this to poor application of the method and not to the weaknesses of the method itself. It must also be realized that in areas such as science and technology forecasting, the degree of uncertainty is so great that exact and always correct predictions are impossible, so a high degree of error is to be expected.

Another particular weakness of the Delphi method is that future developments are not always predicted correctly by consensus of experts. The issue of ignorance is important. If panellists are misinformed about a topic, the use of Delphi may only add confidence to their ignorance.

One of the initial problems of the method was its inability to make complex forecasts with multiple factors. Potential future outcomes were usually considered as if they had no effect on each other. Later on, several extensions to the Delphi method were developed to address this problem, such as cross impact analysis that takes into consideration the possibility that the occurrence of one event may change probabilities of other events covered in the survey. Still the Delphi method can be used most successfully in forecasting single scalar indicators.

Despite these shortcomings, today the Delphi method is a widely accepted forecasting tool and has been used successfully for thousands of studies in areas varying from technology forecasting to drug abuse.

Delphi vs Prediction Markets

Delphi has characteristics similar to prediction markets as both are structured approaches that aggregate diverse opinions from groups. Yet, there are differences that may be decisive for their relative applicability for different problems.

Some advantages of prediction markets derive from the possibility to provide incentives for participation.

1. They can motivate people to participate over a long period of time and to reveal their true beliefs.
2. They aggregate information automatically and instantly incorporate new information in the forecast.
3. Participants do not have to be selected and recruited manually by a facilitator. They themselves decide whether to participate if they think their private information is not yet incorporated in the forecast.

Delphi seems to have these advantages over prediction markets:

a) Participants reveal their reasoning

b) It is easier to maintain confidentiality

c) Potentially quicker forecasts if experts are readily available.

d) Delphi is applicable in situations where the bets involved might affect the value of the currency used in bets (*e.g.* a bet on the collapse of the dollar made in dollars might have distorted odds).

Hierarchical Organization

A hierarchical organization is an organizational structure where every entity in the organization, except one, is subordinates to a single other entity. This arrangement is a form of a hierarchy. In an organization, the hierarchy usually consists of a singular/group of power at the top with subsequent levels of power beneath them. This is the dominant mode of organization among large organizations; most corporations, governments, and organized religions are hierarchical organizations with different levels of management, power or authority. For example, the broad, top-level overview of the general organization of the Catholic Church consists of the Pope, then the Cardinals, then the Archbishops, and so on. Members of hierarchical organizational structures chiefly communicate with their immediate superior and with their immediate subordinates. Structuring organizations in this way is useful partly because it can reduce the communication overhead by limiting information flow; this is also its major limitation.

Visualization

A hierarchy is typically visualized as a pyramid, where the height of the ranking or person depicts their power status and the width of that level represents how many people or business divisions are at that level relative to the whole the highest-ranking people are at the apex, and there are very few of them; the base may include thousands of people who have no subordinates. These hierarchies are typically depicted with a tree or triangle diagram, creating an organizational chart or organogram. Those nearest the top have more power than those nearest the bottom, and there being fewer people at the top than at the bottom. As a result, superiors in a hierarchy generally have higher status and command greater rewards than their subordinates.

Common Models

All governments and most companies have similar structures. Traditionally, the monarch was the pinnacle of the state.

In many countries, feudalism and manorialism provided a formal social structure that established hierarchical links at every level of society, with the monarch at the top.

In modern post-feudal states the nominal top of the hierarchy still remains the head of state, which may be a president or a constitutional monarch, although in many modern states the powers of the head of state are delegated among different bodies. Below the head, there is commonly a senate, parliament or congress, which in turn often delegates the day-to-day running of the country to a prime minister. In many democracies, the people are considered to be the notional top of the hierarchy, over the head of state; in reality, the people's power is restricted to voting in elections.

In business, the business owner traditionally occupied the pinnacle of the organization. In most modern large companies, there is now no longer a single dominant shareholder, and the collective power of the business owners is for most purposes delegated to a board of directors, which in turn delegates the day-to-day running of the company to a managing director or CEO. Again, although the shareholders of the company are the nominal top of the hierarchy, in reality many companies are run at least in part as personal fiefdoms by their management; corporate governance rules are an attempt to mitigate this tendency.

Studies

The organizational development theorist Elliott Jacques identified a special role for hierarchy in his concept of requisite organization. The iron law of oligarchy, introduced by Robert Michels, describes the inevitable tendency of hierarchical organizations to become oligarchic in their decision making. The Peter Principle is a term coined by Laurence J. Peter in which the selection of a candidate for a position in an hierarchical organization is based on the candidate's performance in their current role, rather than on abilities relevant to the intended role. Thus, employees only stop being promoted once they can no longer perform effectively, and managers in an hierarchical organization "rise to the level of their incompetence."

Hierarchiology is another term coined by Laurence J. Peter, described in his humorous book of the same name, to refer to the study of hierarchical organizations and the behavior of their members. Having formulated the Principle, I discovered that I had inadvertently founded a new science, hierarchiology, the study of hierarchies. The term hierarchy was originally used to describe the system of church government by priests graded into ranks. The contemporary meaning includes any organization whose members or employees are arranged in order of rank, grade or class. Hierarchiology, although a relatively recent discipline, appears to have great applicability to the fields of public and private administration. The IRG Solution – hierarchical incompetence and how to overcome it argued that hierarchies were inherently incompetent, and were only able to function due to large amounts of informal lateral communication fostered by private informal networks.

Narcissists

Narcissists may like hierarchical organisations because they might think they will rise to high ranks and reap status and power. Narcissists may be less interested in hierarchies where there is little opportunity for upward mobility.

Criticism and Alternatives

In the work of diverse theorists such as William James (1842–1910), Michel Foucault (1926–1984) and Hayden White, important critiques of hierarchical epistemology are advanced. James famously asserts in his work "Radical Empiricism" that clear distinctions of type and category are a constant but unwritten goal of scientific reasoning, so that when they are discovered, success is declared. Hierarchies and hierarchical thinking has been criticized by many people, including Susan McClary and one political philosophy which is vehemently opposed to hierarchical organization: anarchism is generally opposed to hierarchical organization in any form of human relations. Heterarchy is the most commonly proposed alternative to hierarchy and this has been combined with responsible autonomy by Gerard Fairtlough in his work on Triarchy theory.

Amidst constant innovation in information and communication technologies, hierarchical authority structures are giving way to greater decision-making latitude for individuals and more flexible definitions of job activities and this new style of work presents a challenge to existing organizational forms, with some research studies contrasting traditional organizational forms against groups that operate as online communities that are characterized by personal motivation and the satisfaction of making one's own decisions. With all levels of an organization having access to information and communication via digital means, power structures align more as a hierarchy, enabling the flow of power and authority to be based not on hierarchical levels, but on information, trust, credibility, and a focus on results.

Information Revolution

The term information revolution describes current economic, social and technological trends beyond the Industrial Revolution. Many competing terms have been proposed that focus on different aspects of this societal development. The British polymath crystallographer J. D. Bernal introduced the term *"scientific and technical revolution"* in his 1939 book *The Social Function of Science* to describe the new role that science and technology are coming to play within society. He asserted that science is becoming a "productive force", using the Marxist Theory of Productive Forces. After some controversy, the term was taken up by authors and institutions of the then-Soviet Bloc. Their aim was to show that socialism was a safe home for the scientific and technical ("technological" for some authors) revolution, referred to by the acronym STR. The book *Civilization at the Crossroads*, edited by the Czech philosopher Radovan Richta(1969), became a standard reference for this topic. Daniel Bell (1980) challenged this theory and advocated post-industrial society, which would lead to a service economy rather than socialism. Many other authors presented their views, including Zbigniew Brzezinski (1976) with his "Technetronic Society".

Information in Social and Economic Activities

The main feature of the information revolution is the growing economic, social and technological role of information. Information-related activities did not come up with the Information Revolution. They existed, in one form or the other, in all human societies, and eventually developed into institutions, such as the Platonic Academy, Aristotle's Peripatetic school in the Lyceum, the Musaeum and the Library of Alexandria, or the schools of Babylonian astronomy. The Agricultural Revolution and the Industrial Revolution came up when new informational inputs were produced by individual innovators, or by scientific and technical institutions. During the Information Revolution all these activities are experiencing continuous growth, while other information-oriented activities are emerging. Information is the central theme of several new sciences, which emerged in the 1940s, including Shannon's (1949) *Information Theory* and Wiener's (1948) *Cybernetics*. Wiener stated: "information is information not matter or energy". This aphorism suggests that information should be considered along with matter and energy as the third constituent part of the Universe; information is carried by matter or by energy. By the 1990s some writers believed that changes implied by the Information revolution will lead to not only a fiscal crisis for governments but also the disintegration of all "large structures".

The Theory of Information Revolution

The term *information revolution* may relate to, or contrast with, such widely used terms as Industrial Revolution and Agricultural Revolution. Note, however, that you may prefer mentalist to materialist paradigm. The following fundamental aspects of the theory of information revolution can be given:

1. The object of economic activities can be conceptualized according to the fundamental distinction between matter, energy, and information. These apply both to the object of each economic activity, as well as within each economic activity or enterprise. For instance, an industry may process matter (*e.g.* iron) using energy and information (production and process technologies, management, *etc.*).
2. Information is a factor of production (along with capital, labor, land (economics)), as well as a product sold in the market, that is, a commodity. As such, it acquires use value and exchange value, and therefore a price.
3. All products have use value, exchange value, and informational value. The latter can be measured by the information content of the product, in terms of innovation, design, *etc.*
4. Industries develop information-generating activities, the so-called Research and Development (R and D) functions.
5. Enterprises, and society at large, develop the information control and processing functions, in the form of management structures; these are also called "white-collar workers", "bureaucracy", "managerial functions", *etc.*
6. Labor can be classified according to the object of labor, into information labor and non-information labor.
7. Information activities constitute a large, new economic sector, the information sector along with the traditional primary sector, secondary sector, and tertiary sector, according to the three-sector hypothesis. These should be restated because they are based on the ambiguous definitions made by Colin Clark (1940), who included in the tertiary sector all activities that have not been included in the primary (agriculture, forestry, *etc.*) and secondary (manufacturing) sectors. The quaternary sector and the quinary sector of the economy attempt to classify these new activities, but their definitions are not based on a clear conceptual scheme, although the latter is considered by some as equivalent with the information sector.
8. From a strategic point of view, sectors can be defined as information sector, means of production, means of consumption, thus extending the classical Ricardo-Marx model of the Capitalist mode of production (see Influences on Karl Marx). Marx stressed in many occasions

the role of the "intellectual element" in production, but failed to find a place for it into his model.

9. Innovations are the result of the production of new information, as new products, new methods of production, patents, *etc.* Diffusion of innovations manifests saturation effects (related term: market saturation), following certain cyclical patterns and creating "economic waves", also referred to as "business cycles". There are various types of waves, such as Kondratiev wave (54 years), Kuznets swing (18 years), Juglar cycle (9 years) and Kitchin (about 4 years, see also Joseph Schumpeter) distinguished by their nature, duration, and, thus, economic impact.
10. Diffusion of innovations causes structural-sectoral shifts in the economy, which can be smooth or can create crisis and renewal, a process which Joseph Schumpeter called vividly "creative destruction".

From a different perspective, Irving E. Fang (1997) identified six 'Information Revolutions': writing, printing, mass media, entertainment, the 'tool shed' (which we call 'home' now), and the information highway. In this work the term 'information revolution' is used in a narrow sense, to describe trends in communication media.

Measuring and Modeling the Information Revolution

Porat (1976) measured the information sector in the US using the input-output analysis; OECD has included statistics on the information sector in the economic reports of its member countries. Veneris (1984, 1990) explored the theoretical, economic and regional aspects of the informational revolution and developed a systems dynamics simulation computer model. These works can be seen as following the path originated with the work of Fritz Machlup who in his (1962) book "The Production and Distribution of Knowledge in the United States", claimed that the "knowledge industry represented 29 per cent of the US gross national product", which he saw as evidence that the Information Age had begun. He defines knowledge as a commodity and attempts to measure the magnitude of the production and distribution of this commodity within a modern economy. Machlup divided information use into three classes: instrumental, intellectual, and pastime knowledge. He identified also five types of knowledge: practical knowledge; intellectual knowledge, that is, general culture and the satisfying of intellectual curiosity; pastime knowledge, that is, knowledge satisfying non-intellectual curiosity or the desire for light entertainment and emotional stimulation; spiritual or religious knowledge; unwanted knowledge, accidentally acquired and aimlessly retained.

More recent estimates have reached the following results:

- ✰ The world's technological capacity to receive information through one-way broadcast networks grew at a sustained compound annual growth rate of 7 per cent between 1986 and 2007;
- ✰ The world's technological capacity to store information grew at a sustained compound annual growth rate of 25 per cent between 1986 and 2007;

- ✰ The world's effective capacity to exchange information through two-way telecommunication networks grew at a sustained compound annual growth rate of 30 per cent during the same two decades;
- ✰ the world's technological capacity to compute information with the help of humanly guided general-purpose computers grew at a sustained compound annual growth rate of 61 per cent during the same period.

Lateral Communication

In organizations and organisms, lateral communication works in contrast to traditional top-down, bottom-up or hierarchic communication and involves the spreading of messages from individuals across the base of a pyramid.

Lateral communication in organism or animals can give rise to collective intelligence, or the appearance of collective intelligence.

Examples of lateral communication in organisms include:

- ✰ A coordinated flock of birds or a shoal of fish all maintains their relative positions, or alters direction simultaneously due to lateral communication amongst members; this is achieved due to tiny pressure variations.
- ✰ Ants, termites, bees nest is not coordinated by messages sent by the queen ant/bee/termite but by the lateral communication, mediated by scent trails of the ants. Its physical structure is an emergent property of the individual entities.
- ✰ Bacterial colonies communicate with each other, coordinating for example an attack, or the production of slime using lateral communication based on chemical messengers so that as a group they can detect how many colleagues there are, and if they are likely to overwhelm a target.
- ✰ The pacemaker cells in the heart; cardiac pacemaker is a very small group of cells, where lateral communications sweeps through the cells, much like a Mexican wave as a three-dimensional circulating wave, which relays contraction signals to the whole heart.
- ✰ With slime mold millions of individual amoeba like creatures can spread out and graze the surface of a leaf. When conditions change, the amoeba all concentrate and form a slug like creature which can actually move off somewhere else before forming a spore body and releasing millions of individual spores.
- ✰ The positions of human cells, and which type of cell is mediated by lateral communication.

In Organizations and Communities

It is argued that communities communicate and store collective knowledge through lateral communication, and that it is an essential ingredient to make hierarchies work, by compensating for errors in hierarchies' information flows.

Lateral communication is communication between different individuals and, departments, or organisms on the same organizational level.

Lateral/Horizontal Communication

The term lateral communication can be used interchangeably as horizontal communication. In his text entitled *Organizational Communication,* Michael J. Papa defines horizontal communication as "the flow of messages across functional areas at a given level of an organization" (Papa and Daniels 55). With this system people at the same level are permitted "to communicate directly without going through several levels of organization" (Papa and Daniels 55). Given this elasticity, members within an organization have an easier time with "problem solving, information sharing across different work groups, and task coordination between departments or project teams" (Papa and Daniels 56). The use of lateral or horizontal communication in the workplace "can also enhance morale and afford a means for resolving conflicts (Koehler *et al.,* 1981) (Papa and Daniels 56).

According to research done by John E. Spillan, Mary Mino, and M. Susan Rowles, "lateral communication involves not only the movement of information from the upper levels to the lower levels of the organizational hierarchy but also is defined primarily as the quality of information sharing among peers at similar levels (McClelland and Wilmont, 1990). Specifically, lateral communication occurs among co-workers, during staff meetings and informational presentations, throughout shift changes, and among employees regardless of peer types.

"Communicating effectively laterally involves the exchange of information between and among all organizational members. While we may perceive that organizational information flows vertically or from top to bottom, in reality, information moves laterally. In other words, as information directives are communicated from an upper to a lower position on the hierarchy, peers at each hierarchical level should quickly interpret and communicate these directives between and among peers at similar hierarchical levels.

Quality: Structure and Culture

Structure

Mechanistic Structure

"A mechanistic or hierarchical organizational structure emphasizes specialization in position. Examples include healthcare and governmental organizations where information is communicated based on chain of command. This organizational structure type is not conducive to lateral communication and, in fact, discourages it. Since direction and coordination is achieved through upper hierarchical levels, peer information sharing is limited. Overall, a mechanistic structure promotes vertical communication or top down communication with strict alignment and unity of command within the organization".

Organic Organizational Structure

"An organic organizational structure is built upon an entrepreneurial concept. Here, the decisions made are decentralized and coordinated by mutual adjustment rather than command and control. Examples include a small business or a manufacturing facility where communication is promoted at all levels of the organization. This organizational structure allows for greater autonomy, promotes individual initiative, and allows employees to be involved in the decision-making process thus enabling employee decision-making to contribute to or detract from organizational goals. In short, an organic organizational structure can encourage and facilitate lateral communication".

Organizational Culture/Climate

"Organizational culture refers to the organization's shared visions, values, beliefs, goals, and practices (Gilsdorf, 1998). Deals and Kennedy (1982) have connected culture with effective communication. Strong cultures and effective communication result in employees who are more productive because they know exactly what is expected of them in organizational settings. Gilsdorf (1998) has implied that the more employees perceive a positive or strong organizational culture, the more productive they will be. Poole (1985) has contended that an organization's quality of communication is summed-up in its culture. In many cases, a strong positive organizational culture or climate can encourage employees to communicate effectively laterally when sharing achievements and disappointments. Thus, as Comer (1991) has asserted, managers should actively encourage employees to communicate effectively laterally with their peers.

Problems with Horizontal Communication

Although this system of communication can be effective, problems can often ensue within organizations. According to Papa's book *Organizational Communication*, "horizontal communication problems occur because of territoriality, rivalry, specialization, and simple lack of motivation." In addition to these problems and in general, "organizations that traditionally have functioned under rigid authority structures with fixed lines of communication may find that the values and expectations that members have acquired under such systems inhibit attempts at horizontal communication." Other problems with this form of communication can happen between multinational corporations. "Horizontal communication between subsidiaries of the same multinational corporation (MNC) is a problem faced by staff as the demands for communicating across borders are pushed downwards in the organizational hierarchy."

Territoriality

Territoriality often occurs when members of an organization "control task-related activity within a defined and fixed jurisdictional area" and as a result "regard others' involvement in that area as territorial encroachment." "Departments value their turf and strive to protect it. This problem may be compounded through interdepartmental rivalries that arise from win/lose competition for rewards and resources".

Rivalry

Rivalry within organizations occurs for example when the different levels of an organization fail "to cooperate with one another". For example, Papa gives an example of "corporate executives in a national department store chain" who "encountered territorial rivalry when they discovered that local stores within each of the company's major sales districts refused to cooperate with one another on sales promotions". Rivalries such as the example stated in this paragraph inhibit the effectiveness of horizontal communication.

Specialization

Specialization is a problem that often happens when organizations do not have uniformity within departments, causing communication difficulties. Specialization can occur with procedures or vocabulary used by different departments. For example, when "different specialties use the same terms in different ways," this can create confusion and miscommunication. When this occurs organizations have trouble functioning properly and do not run smoothly.

Lack of Motivation

"Horizontal communication often fails simply because organization members are unwilling to expend the additional effort that it requires." "Horizontal communication may require contact with people in units that are well removed from our own. The channels and rules of interaction may be unclear. We do not really know these people

Sociological Theory of Diffusion

The sociological theory of diffusion is the study of the diffusion of innovations throughout social groups and organizations. The topic has seen rapid growth since the 1990s, reflecting curiosity about the process of social change and "fuelled by interest in institutional arguments and in network and dynamic analysis." The theory uses a case study of the growth of business computing to explain different mechanisms of diffusion.

The Concept of Diffusion

In the 1962 book, *Diffusion of Innovations*, Everett Rogers defines sociological diffusion of innovation as a process in a social system where an innovative idea or concept is spread by members of the social group through certain channels. He identifies four elements that influence how and how quickly a new idea spreads:

- ✰ The innovation itself
- ✰ The types of communication channels used
- ✰ The amount of time the social group is exposed to the innovation
- ✰ The nature of the social group

In a study by Surry and Farquhar, researchers explain that the theory of diffusion is used in occupations ranging from marketing to agriculture in order to ensure that new products, ideas, and techniques are well adopted by the social group. The concept of diffusion is of particular interest in the marketing field, as this concept affects the success or failure of new ads or products. Understanding this theory helps marketers influence the way the public will perceive each innovation. The speed at which an innovation spreads through a mass of people depends on how favorably an idea is perceived by the audience. Innovations that are ill matched with existing techniques are not as well accepted and diffused through the group. Social structures are naturally designed in a hierarchy; thus, different ideas follow different routes or courses in the hierarchy, depending on the type and source of an innovation.

The study of the diffusion of innovations has led to advancements in awareness of three important aspects of social change: the qualities of an innovation which lead to successful diffusion, the effect of peer networking and conversations when it comes to spreading ideas, and the importance of various "user segments" (Robinson). The theory of diffusion of innovations differs from other theories about the processes of change since most changes are improvements, or "reinventions", of a previously existing product or technique. These changes are generally favorably perceived by the members of the group because they usually are more in line with the values and needs of the group. There are five important qualities that factor into the success or failure of innovations. First, the relative advantage is that, whether the new innovation surpasses similar existing ideas in terms of satisfaction and convenience secondly, the compatibility of the new idea with the needs and practices of the group members. Third, the simplicity of the innovation: usually, the simpler the innovation, the more quickly the concept is adopted. Fourth, the "trialability" of an innovation; that is, whether it can be tested without commitment for a period of time. An innovation that has an available trial period provides less uncertainty to the group member who will be trying it. Lastly, whether there are observable results with use of the innovation. The more positive and visible results, the higher the likelihood it gets adopted as a permanent idea for the group.

Why Diffusion Happens

Sociological diffusion occurs when a social group or organization develops an innovation: a new idea or behavior. Diffusion, in the context of corporations and businesses, is a way for an idea to be fleshed out. The diffusion of innovations provides insights into the process of social change: one can observe the qualities that make an innovation successfully spread and the importance of communication and networks. According to Rogers, a new idea is diffused through a decision-making process with five steps:

- **Knowledge** - An individual first becomes aware of the new innovation, but lacks information and inspiration
- **Persuasion** - The individual's interest in the innovation spikes, and he or she begins research
- **Decision** - The individual weighs the positive and negative results of changing to the new idea

- **Implementation** - The individual adds the innovation into the system. At this stage, he or she also begins to determine the innovation's usefulness.
- **Confirmation** - The individual decides to continue with the new innovation.

The key part of the five stages is the decision; this is the main reason why diffusion exists. The decision to either adopt or reject the idea is vitally important. Those responsible for evaluating innovations either determine that the new concept is likely to provide future success, and adopt it, or determine that it is likely to be a failure, and continue to move forward in search of other ideas. It is counterproductive for an organization to invest time, energy, and in most cases money, into a poorly developed or bad idea. An important aspect of the diffusion and decision process is communication. As an idea further develops and spreads, it flows and moves through an organization by communication. Communication is a necessary condition for an idea to take hold. The innovation depends on a communication network within the organization in order to take root. In Emanuel Rosen's book *The Anatomy of Buzz*, Rosen points out the importance of communication networks in the spread and development of an idea within an organizational system. (Dobson). Studies of the diffusion of innovation have shown that new ideas must fit with already established system in order for changes not only to occur, but also to occur easily. (Pinard) An innovation faced with structural or ideological barriers cannot diffuse. On the other hand, if a new idea or innovation has few obstacles and acknowledges places where change is logical, movement to it will occur. (Freeman)

Networks and Environment

A firm's interaction with other players, along with its environment and organizational culture, are keys in the social theory of diffusion.

The Use of Networks

The effects of networks and institutional environment on adoption of innovations can be explained using a social network theory model. In such a model, nodes represent agents (*e.g.* companies or organizations) and ties represent a connection between two entities (*e.g.* a company-client relationship or competitive relationship). Diffusion occurs when a novel idea, product, or process is implemented by an agent and permeates through these ties to others.

Internal and External Diffusion

Diffusion of information and ideas has been categorized into two modes:

Internal diffusion is the spread of information and innovations within a network, flowing within a single adopting population – a given industry or geographical network. Internal diffusion dynamics require that innovative and early adopter firms introduce new ideas into a network, which are then picked up by the majority of firms and laggard firms. DiMaggio and Powell (1983) argue that firms search for the best ideas and practices and mimic new ideas that prove to work. This phenomenon is known as mimetic isomorphism, and ironically may lead to clustering of firm structure and practices. Additionally, firms are often forced to adopt new ideas as they are constantly competing with other firms; that is, firms want to seem modernized and seek legitimacy in implementing innovative practices.

External diffusion refers to the introduction of ideas to a network from outside actors: firms or other agents on the edge of the network. Outside actors include the mass media and "change agents." Mass media can amplify trends and movements that occur in the marketplace, introducing new innovations to network members, exposing "best-practice" ideas, and conveying new principles. Change agents are usually business professionals (such as lawyers, consultants, bankers, or politicians) who spread new practices or aid in promoting new ideas. These individuals often introduce business models, legal strategies, or investment techniques that are picked up by several entities within a network and continue to diffuse. Often, such external diffusion leads to conformity of a set of corporate strategies or structures, a phenomenon DiMaggio and Powell called "normative isomorphism".

Environmental and Cultural Factors of Diffusion

An agent's environmental and cultural makeup influences the decision to adopt an idea diffusing through a network. Some of the major characteristics of firms that influence their decision to innovate are *clustering*, *weak ties*, and firm size. *Clustering', the existence of a group of tightly connected agents, is a frequent concept in network theory. It includes, for example, similar firms locating themselves in close proximity to each other (Silicon Valley for technology firms; New York for banking services). Such clustering and close proximity increases the diffusion rate of ideas for firms within a cluster, as other firms are more likely to adopt an idea if another firm has adopted it within its cluster.* An agent with *weak ties* has a connection to two or more clusters. These agents are integral in connecting groups, as they provide communication between large clusters. Firms with weak ties can be isolated firms, firms with business in two or more spaces, or those which are external change agents. Firms with weak ties introduce clusters to new, proven methods.Firm size has been shown to have an influence on the rate of diffusion. Strang and Soule (1998) have shown that large, technical, and specialized organizations with informal cultures tend to innovate much faster than other firms. Smaller and more rigid firms attempt to mimic these "early adopters" in attempt to keep up with competition.

Mathematical Treatment

Mathematical models can be used to study the spread of technological innovations among individuals connected to each other by a network of peer-to-peer influences, such as in a physical community or neighborhood. Complex system (particularly complex network) models can be used to represent a system of individuals as *nodes* in a network (or Graph (discrete mathematics)). The interactions that link these individuals are represented by the edges of the network and can be based on the probability or strength of social connections. In the dynamics of such models, each node is assigned a current state, indicating whether or not the individual has adopted the innovation, and model equations are used to describe the evolution of these states over time. In threshold models the uptake of technologies is determined by the balance of two factors: the (perceived) usefulness (sometimes called utility) of the innovation to the individual as well as barriers to adoption, such as cost. The multiple parameters that influence decisions to adopt, both individual and socially motivated, can be represented by such mathematical models. Computer models have been developed to investigate the balance between the social aspects of diffusion and perceived intrinsic benefit to the individuals. When the effect of each individual node was analyzed along with its influence over the entire network, the expected level of adoption was seen to depend on the number of initial adopters and the structure and properties of the network. Two factors in particular emerged as important to successful spread of the innovation: The number of connections of nodes with their neighbors, and the presence of a high degree of common connections in the network (quantified by the clustering coefficient).

Case Study: Diffusion of Business Computing in Organizations

To illustrate how different diffusion mechanisms can have varying effects in individual cases, consider the example of business computing. The 1980s and 1990s saw a rapid paradigm shift in the way many organizations operated; specifically, the rise of computers and related technologies saw organizations adopt these innovations to help run their business (Attewell 1992:1). Thus, the diffusion of business computing through organizations during this time period provides an informative case study through which to examine different mechanisms of diffusion and their respective roles.

Networks

The roles of communication networks, as described by traditional theories of diffusion, have been to facilitate information flow about a new innovation and thus remove one of the major barriers to adoption. In this model, those closest to the initial champions of a new innovation are quicker to respond and adopt, while those farther away will take more time to respond (Rogers 1983; Strang and Soule 1998:272). This theory about the roles of networks in diffusion, while widely applicable, requires modification in this particular case, among others. Attewell (1992) argues that in this case, knowledge of the existence of computers and their business applications far preceded their eventual adoption. The main barrier to adoption was not awareness, but technical knowledge: knowledge of how to effectively integrate computing into the workplace. Thus, the most relevant networks to the diffusion of business computing were those networks that transmitted the technical knowledge required to utilize the innovation, not those that simply transmitted awareness of the idea behind the innovation.

Institutions

New institutions, in particular those which acted as educators or consultants, also played an important role in the diffusion of business computing. In order to adapt to evolving trends in business computing, organizations first needed to gain the technical knowledge necessary to operate the technology (Attewell 1992:3-6). The "knowledge barrier" could be reduced or partially circumvented, however, by the formation of new institutions. The new institutions that formed during this time period – such as service bureaus, consultants, and companies creating simplifications of the technology – lowered the knowledge barrier and allow for more rapid diffusion of the ideas and technology behind business computing. This explains the phenomenon in which, at first, many organizations obtained business computing as an out-sourced service. However, after these service institutions effectively lowered the barrier to adoption, many organizations became capable of bringing business computing in-house (Attewell 1992:7-8).

Innovation Decisions

Rogers (1983) notes two important ways in which innovations are adopted by organizations: collective innovation decisions, and authority innovation decisions. "Collective innovation decisions" are best defined as a decision that occurs as the result of a broad consensus for change within an organization. "Authority innovation decisions", on the other hand, need only the consensus of a few individuals with large amounts of power within the organization. In the case of organizations adopting business computing, authority decisions were largely impossible. As J.D. Eveland and L. Tornatzky (1990) explain, when dealing with advanced technical systems such as those involved with business computing, "decisions are often many (and reversed), and technologies are often too big and complex to be grasped by a single person's cognitive power – or usually, to be acquired or deployed within the discretionary authority of any single organizational participant." Therefore, a much broader consensus within an organization was required to reach the critical mass of technical knowledge and authority necessary to adapt to business computing. This provided an opportunity for collective innovation decisions within the organization.

Diffusion of Innovation Theory

Diffusion research examines how ideas are spread among groups of people. Diffusion goes beyond the two-step flow theory, centering on the conditions that increase or decrease the likelihood that an innovation, a new idea, product or practice, will be adopted by members of a given culture. In multi-step diffusion, the opinion leader still exerts a large influence on the behavior of individuals, called adopters, but there are also other intermediaries between the media and the audience's decision-making. One intermediary is the change agent, someone who encourages an opinion leader to adopt or reject an innovation (Infante, Rancer, and Womack, 1997).

Innovations are not adopted by all individuals in a social system at the same time. Instead, they tend to adopt in a time sequence, and can be classified into adopter categories based upon how long it takes for them to begin using the new idea. Practically speaking, it's very useful for a change agent to be able to identify which category certain individuals belong to, since the short-term goal of most change agents is to facilitate the adoption of an innovation. Adoption of a new idea is caused by human interaction through interpersonal networks. If the initial adopter of an innovation discusses it with two members of a given social system, and these two become adopters who pass the innovation along to two peers, and so on, the resulting distribution follows a binomial expansion. Expect adopter distributions to follow a bell-shaped curve over time (Rogers, 1971).

Adopter Categorization

The criterion for adopter categorization is innovativeness. This is defined as the degree to which an individual is relatively early in adopting a new idea then other members of a social system. Innovativeness is considered "relative" in that an individual has either more or less of it than others in a social system (Rogers, 1971).

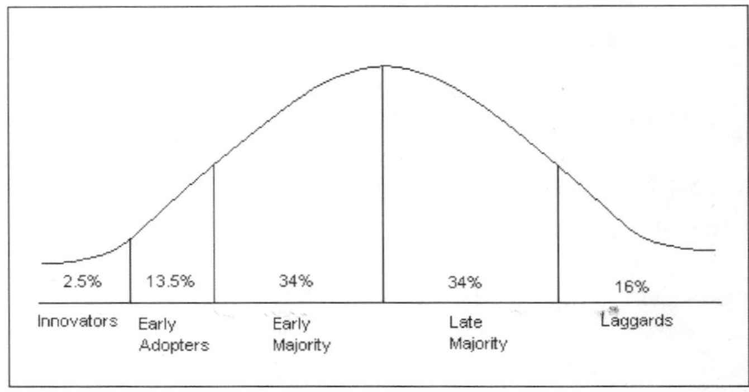

Adopter Categorization on the Basis of Innovativeness.

Adopter distributions closely approach normality. The above figure shows the normal frequency distributions divided into five categories: innovators, early adopters, early majority, late majority and laggards. Innovators are the first 2.5 per cent of a group to adopt a new idea. The next 13.5 per cent to adopt an innovation are labeled early adopters. The next 34 per cent of the adopters are called the early majority. The 34 per cent of the group to the right of the mean are the late majority, and the last 16 per cent are considered laggards (Rogers, 1971). The above method of classifying adopters is not symmetrical, nor is it necessary for it to be so. There are three categories to the left of the mean and only two to the right. While it is possible to break the laggard group into early and late laggards, research shows this single group to be fairly homogenous. While innovators and early adopters could be combined, research shows these two groups as having distinctly different characteristics. The categories are 1) exhaustive, in that they include all units of study, 2) mutually exclusive, excluding from any other category a unit of study already appearing in a category, and 3) derived from one classificatory principle. This method of adopter categorization is presently the most widely used in diffusion research (Rogers, 1971).

Adopter Categories

Innovators are eager to try new ideas, to the point where their venturesomeness almost becomes an obsession. Innovators' interest in new ideas leads them out of a local circle of peers and into social relationships more cosmopolite than normal. Usually, innovators have substantial financial resources, and the ability to understand and apply complex technical knowledge. While others may consider the innovator to be rash or daring, it is the hazardous risk-taking that is of salient value to this type of individual. The innovator is also willing to accept the occasional setback when new ideas prove unsuccessful (Rogers, 1971).

Early adopters tend to be integrated into the local social system more than innovators. The early adopters are considered to be localites, versus the cosmopolite innovators. People in the early adopter category seem to have the greatest degree of opinion leadership in most social systems. They provide advice and information sought by other adopters about an innovation. Change agents will seek out early adopters to help speed the diffusion process. The early adopter is usually respected by his or her peers and has a reputation for successful and discrete use of new ideas (Rogers, 1971).

Members of the early majority category will adopt new ideas just before the average member of a social system. They interact frequently with peers, but are not often found holding leadership positions. As the link between very early adopters and people late to adopt, early majority adopters play an important part in the diffusion process. Their innovation-decision time is relatively longer than innovators and early adopters, since they deliberate some time before completely adopting a new idea. Seldom leading, early majority adopters willingly follow in adopting innovations (Rogers, 1971).

The late majority are a skeptical group, adopting new ideas just after the average member of a social system. Their adoption may be borne out of economic necessity and in response to increasing social pressure. They are cautious about innovations, and are reluctant to adopt until most others in their social system do so first. An innovation must definitely have the weight of system norms behind it to convince the late majority. While they may be persuaded about the utility of an innovation, there must be strong pressure from peers to adopt (Rogers, 1971).

Laggards are traditionalists and the last to adopt an innovation. Possessing almost no opinion leadership, laggards are localite to the point of being isolates compared to the other adopter categories. They are fixated on the past, and all decisions must be made in terms of previous generations. Individual laggards mainly interact with other traditionalists. An innovation finally adopted by a laggard may already be rendered obsolete by more recent ideas already in use by innovators. Laggards are likely to be suspicious not only of innovations, but of innovators and change agents as well (Rogers, 1971).

Uses and Gratification

Uses and gratification is more a concept of research than a self-contained theory. Even contributors in this field of research find problems with the scope of the research and call uses and gratification an umbrella concept in which several theories reside (Infante *et al.*, 1997). Researchers in this field argue that scholars have tried to do too much and should limit the scope and take a cultural-empirical approach to how people choose from the abundance of cultural products available.

Critics claim the theory pays too much attention to the individual and does not look at the social context and the role the media plays in that social context. Rubin (1985), as cited in Littlejohn (1996), suggests that audience motive research based on uses and gratification research has been too compartmentalized within certain cultures and demographic groups, leading to the assumption this has thwarted synthesis and integration of research results, which are two key ingredients in theory building.

The uses and gratification theory is a basic extension of the definition of an attitude, which is a non-linear cluster of beliefs, evaluations, and perceptions. These beliefs, evaluations, and perceptions give individuals latitude over how they employ media in their lives; in other words, how individuals filter, interpret, and convey to others the information received from a medium. Basically, a person's attitude toward a segment of the media is determined by beliefs about and evaluations of the media. A key to this research is that the consumer, or audience member, is the focal point instead of the message. The research views the members of an audience as actively utilizing media contents, rather than being passively acted upon by the media, according to Katz, Blumer, and Gurevitch (1971) as cited in Littlejohn (1996). When audience members, not the media, are the action takers, the variations taken from the messages received are the intervening variables.

A core assumption of uses and gratification research is the assumption that individual needs are satisfied by audience members actively seeking out the mass media (Infante *et al.*, 1997). Rubin (1983), as cited in Littlejohn (1996), designed a study to explore adult viewers' motivations, behaviors, attitudes and patterns of interaction to see if behavioral and attitudinal consequences of the viewer could be predicted. In 1984, the researcher identified two types of television viewers. The first type is the habitual viewer who watches television for a diversion, has a high regard for the medium, and is a frequent user. The second type is the non-habitual viewer who is selective, likes a particular program or type of programs and uses the medium primarily for information. The non-habitual viewer is more goals oriented when watching television and does not necessarily feel that television is important. Rubin (1983) argues that habitual viewers use the medium as a companion and that non-habitual viewers are more actively involved in the viewing experience (Littlejohn, 1996).

Expectancy-Value Theory

Another theory to consider under this umbrella of uses and gratification research is expectancy-value theory from information-integration theorist Martin Fishbein (Littlejohn, 1996). The researcher proposes there are two kinds of belief; belief in something and belief about something. The example used by Fishbein is the person who believes in marijuana as a recreational drug or the person who believes that using marijuana will move on to other drugs and serious crimes in order to continue the habit. In Fishbein's theory development, attitudes are different from beliefs in that they are evaluative and are correlated with beliefs and predispose a person to behave a certain way toward the attitude object. The two beliefs about marijuana mentioned above would change dramatically if more serious drugs and crime were evaluated as bad. Also cited in Littlejohn is Philip Palm green, an early uses and gratification researcher, who claims that gratifications are sought in terms of a person's beliefs about what a medium can provide and that person's evaluation of the medium's content (Littlejohn, 1996).

Part VII

Audio-Visual Aids

37

Meaning of Audio-Visual Aids

Audio-Visual Aids are defined as any device used to aid in the communication of an idea. From this definition, virtually anything can be used as an aid, providing it successfully communicates the idea or information for which it is designed. In this Chapter we not only use the term "Audio-Visual Aids" but "Instructional Aids", "Teaching Aids" and "Visual Aids" as well. An audiovisual product is any audiovisual (AV) item such as still photography, motion picture, audio or video tape, slide or filmstrip, that is prepared singly or in combination to communicate information or to elicit a desired audience response. Even though early aids, such as maps and drawing are still in use, advances in the audio visual field have opened up new methods of presenting these aids such as videotapes and multimedia equipment which allow more professional and entertaining presentations to be presented.

'Audio-visual aids are any device which can be used to make the learning experience more concrete, more realistic, more dynamic.'

Kinder S James (1959) 'All materials used in the classroom or other teaching situations to facilitate the understanding of the written and spoken words are audio-visual aids.'

Dale (1965) an audio-visual aid is an instructional device that can be heard and seen.

Haas and packer (1964) Audio visual aids are those sensory objects or images which initiate or stimulate and reinforce learning.

Burton (1967) Audio visual aids are those aids which help in completing the triangular process of learning that is motivation, classification and stimulation.

Carter V Good (1963) Audio visual aids are anything by means of which learning process may be encouraged or carried on through the sense of hearing or sense of sight.

Good's dictionary of education: Audio visual aids are supplementary devices by which the teacher, through the utilization of more than one sensory channel is able to clarify, establish and correlate concepts, interpretations and appreciations.

McKean and Roberts: An audio visual aid is an instructional device in which the message can be heard as well as seen.

K P Neeraja: Literally audio refers to sound waves that can normally be heard by the human ear.The term visual implies relating to the sense of seeing. The term audio-visual aid refers to anything that an extension agent uses to help to convey the message when communicating with farmers. The spoken word is the agent's main communication tool, but, whether the agent is speaking to a large village meeting or discussing a problem in a field with a group of farmers, its impact and effectiveness can be greatly increased by the use of suitable audio-visual aids. When selected and used properly, audio-visual aids can help in the following ways:

- The interest of the audience can be maintained if the agent varies the mode of presentation. It is difficult to concentrate for long on what someone is saying; but if the agent refers to a wall chart, or illustrates a point with some slides, his audience's attention can be maintained.
- When information is presented to more than one sense (sight and touch, for example, as well as hearing), more is taken in and it is better understood and remembered.
- Processes and concepts that are difficult to express in words alone can be explained. The procedure for applying for a loan, for example, may sound confusing, but a simple chart or diagram can make the process clearer. Again, the life cycle of a crop pest can be explained by showing a series of slides or drawings.
- The effects of decisions and actions that farmers might take can be shown. Photographs of a cattle dip or a model of a cooperative store can give farmers a clear idea of just what it is they might be considering.
- Pictures can have a more immediate impact on our emotions than words. Photographs of a heavy crop, for example, are likely to arouse interest more effectively than details of yields read out by an extension agent

Values of Audio Visual Aids

- Best motivators
- Antidote to the disease of verbal instruction
- Clear images
- Vicarious experiences
- Variety
- Freedom
- Opportunities to handle and manipulate
- Retentively

- ☆ Based on maximum of teachers
- ☆ Helpful in attracting attention
- ☆ Helpful in fixing up new learning
- ☆ Saving of energy and time
- ☆ Realism
- ☆ Vividness
- ☆ Meeting individual difference Encouragement to healthy class room interaction
- ☆ Spread on education on a mass scale
- ☆ Promotion of scientific temper
- ☆ Development of higher faculties
- ☆ Reinforcement to learners
- ☆ Positive transfer of learning and training
- ☆ Positive environment for creative discipline

Characteristics

1. They should be meaningful and purposeful
2. They should be accurate in every aspect
3. They should be simple
4. They should be cheap
5. They should be up to date
6. They should be easily portable
7. They should be according to the mental level of students
8. They should be motivate the learners
9. They should be improvised.
10. They should be large enough to be properly seen by the students for whom they are mean.

Principles in the Use of Audio-Visual Aids

Principle of Selection: 1. the age level; 2. other personality angles; 3. they should have specific educational values; 4. they should help in the realization of learning desired objectives.

Principle of Preparation: 1. As far as possible, the local material should be used in the preparation of aids. 2. The teacher also must receive training in the preparation of aids. 3. The teacher him/herself can prepare some aids or can take help of students also.

Principle of Physical Control: This is concerned with the arrangement of keeping aids safely and also to facilitate to their lending to the teachers for se.

Principle of Proper Presentation: 1. Teacher should carefully visualize the use of teaching aids before their actual presentation; 2. They should be well acquainted themselves with the use and manipulation of the aids to be shown in the class room; 3. The aids should be displayed properly. So that, all the students can see it, observe it, and can derive maximum benefits from it.

Principle of Response: This is the important principle. This tells the teacher guide the students to respond actively to the audio visual stimulus so that they derive the maximum benefits in learning.

Principle of Evaluation: This Principle stipulates that there should be continuous evaluation of Audio Visual Aids materials and accompanying techniques in the light of desired objective

Advantages of Audio-Visual Aids

a) **Best Motivators:** They are the best motivators. Students work with more interest and zeal. They are more attentive.

b) **Fundamental to Verbal Instructions:** They help to reduce verbalism which is a major weakness of our schools. They convey the same meaning as words mean. They give clear concepts and thus help to bring accuracy in learning.

c) **Clear Images:** Clear images are formed when we see, hear, touch, taste and smell as our experiences are direct, concrete and more or less permanent. Learning through the senses becomes the most natural and consequently the easiest.

d) **Vicarious Experience:** Everyone agrees to the fact that the first hand experience is the best type of educative experience but such an experience cannot always be provided to the pupils an so in some situations certain substitutes have to be provided. For this we find a large number of inaccessible objects and phenomenon. For example all the students in India cannot possibly be shown Taj Mahal *etc.* In all such cases audio-visual aids provide us the best substitutes.

e) **Variety:** Audio-Visual aids provide variety and provide different tool s in the hands of the teacher.

f) **Freedom:** The use of audio-visual aids provide various occasions for the pupil to move about, talk, laugh and comment upon. Under such an atmosphere the students work because they want to work and not because the teacher wants them to work.

g) **Opportunities to Handle and Manipulate:** The use of audio-visual aids provides immense opportunities to the pupils to see handle and manipulate things.

The Range of Audio-Visual Aids

Extension agents often use sophisticated audio-visual aids which require electricity and complex machinery such as projectors or television sets. But there are many simple aids that the agent can make locally, and these have several advantages. They do not require a power source or heavy equipment, they do not cost much to produce and they can be made to suit the precise needs of the agent. Between these two extremes lies a wide range of more or less sophisticated aids. The distinction between a mass medium and an audio-visual aid lies only in the way it is used. All the mass media described earlier can be used as audio-visual aids. A film is a mass medium, in that it is shown to large audiences in many different places; but for an individual extension agent who uses it to increase the impact of a talk, it is an audio-visual aid. Many of the principles of media use discussed earlier also apply to audio-visual aids. The audio-visual aids available to the extension agent can now be examined.

Principles to be Followed for the Effective Use of Audio-Visual Aids

a) Audio visual materials should function as an integral part of the educational program.
b) Audio visual aids should be centralized, under specialized direction and leadership in educational programs.
c) An advisory committee consisting of representative from all areas of curriculum should be appointed to assist in selection and coordination of Audio visual materials.
d) An education program should be flexible.
e) Audio visual materials should be carefully located to eliminate duplication, easy accessibility and convenient use.
f) Audio visual materials should be available whenever and wherever they needed for effective utilization as an integral part of curriculum.
g) Budget appropriations should be made regularly for Audio visual education programs.
h) Periodic evaluation to be done to assess the function of, utilization and expenditure of the program.

Some Criteria for Selection and Use of Audio Visual Aids

What role the media play in people's lives:

How frequently and extensively is emotional or artistic expression? This would indicate the Vitality of the tradition in which they reset.

What is the usual size and composition of audience attracted in terms of age, sex, educational, economic and social groupings?

What is the importance of the functions performed in a framework of prevailing social needs and values?

a) Concern for plasticity of the medium, or capacity of a particular form to be "loaded" with specific innovative messages. Examination of religious or social functions attributed to the medium should be made so as not to distort its special role. In some, cases, examination may reveal such use to be unacceptable. However, it has been found that religious themes can be interpreted in modern context, providing impetus for social change.

b) It would be determine whether the medium has an age or individuality (in which case positive identification of audience with actor can be the agent of change), or whether it is entirely non-individual. It has sometimes been found that certain messages can be best expressed without the personalities of, actors in evidence during performance, as in puppet show, mask presentations and giant figures.

c) It should be noted, whether a particular medium or any aspect of its traditions and features could be used by or extended to, mass media presentation. If so, is this desirable in relation to both a medium and the message?

d) Are the required resources, talent, materials, texts, available for extensive utilization of the medium?

e) Does the selected medium have entertainment value and/or artistic appeal?

f) Is it unique enough to prove a valuable communication channel?

g) Is the form of the proposed medium flexible enough to incorporate a sufficiently broad range of content materials, such as family planning and other developmental messages?

h) Is the medium sufficiently versatile to reach varied audiences. To properly assess this criterion, the country must be in a position to assess the range of traditional media and their suitability for audiences according to local and provincial make-up.

i) Is the selected medium, relevant to the intended audience?

38

Classification of Audio Visual Aids

Various classifications are given for Audio visual aids according to the type of projection by various authors.

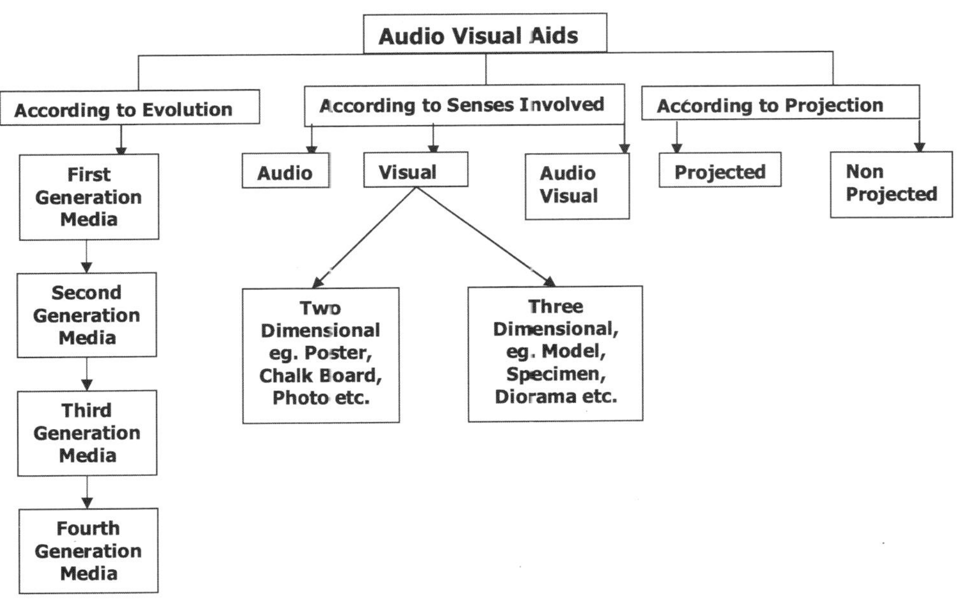

1. According to Evolution

a) First Generation Media

E.g. Hand made chart, graph, exhibits, model, hand writing materials *etc.*

b) Second Generation Media

E.g. Printed or illustrated text, printed graph, work books.

c) Third Generation Media

E.g. Photographs, slides, filmstrips, radio, films, recordings, telecast *etc.*

d) Fourth Generation Media

E.g. Television, programmed instruction, language laboratories, electronic digital computers *etc.*

While using the foregoing methods the extension worker uses these methods independently or takes the help of certain audio-visual aids. The word audio-visual comprises three words namely, Audio refers to sense of hearing; Visual refers sense of seeing and Aid for instructional devices.

2. According to Senses Involved

Audio	*Visual*	*Audio-Visual*
Tape recorder, Public address system, Telephone, Radio *etc.*	Flash cards Black board Pictures *etc.*	Cinema projector, telecast, Video recording, Television, Sound-synchronized slides, Drama *etc.*

Audio Aids

Tape Recorder

A tape recorder is a portable electronic gadget to record, reproduce, erase and re record sound on a magnetic tape. This device can be used without much fuss by anybody by operating the following press buttons attached to the recorder, viz, stop, play, wind, rewind, record, pause, and eject.

Uses: it is used to learn foreign languages, rhymes, and songs with clarity. GRAME PHONES Like radio gramophones are also important teaching devices. It helps to listen to famous speeches.

Using a Record Player for Teaching

a) A record player can be used in the following ways in the actual class room situation

b) A record player can be used to supplement a lesson.

c) A record player can be used for an appreciation lesson in music.

d) A record player can be used for an appreciation lesson in literature.

e) A record player can be used for students to acquire the singing ability, deliver a speech properly, and recite a poem in the right way.

f) The player can be used to end or conclude a lesson;

g) Introduce a lesson and review a lesson.

h) A record player can be used for physical exercises accompanied with music

Public Address System (PA system)

A public address system (PA system) is an electronic sound amplification and distribution system with a microphone, amplifier and loudspeakers, used to allow a person to address a large public, for example for announcements of movements at large and noisy air and rail terminals.

The term is also used for systems which may additionally have a mixing console, and amplifiers and loudspeakers suitable for music as well as speech, used to reinforce a sound source, such as recorded music or a person giving a speech or distributing the sound throughout a venue or building. Simple PA systems are often used in small venues such as school auditoriums, churches, and small bars. PA systems with many speakers are widely used to make announcements in public, institutional and commercial buildings and locations. Intercom systems, installed in many buildings, have microphones in many rooms allowing the occupants to respond to announcements. Sound reinforcement systems and PA systems may use some similar components, but with differing application, although the distinction between the two is not clear-cut. Sound reinforcement systems are for live music or performance, whereas PA systems are primarily for reproduction of speech. In Britain any PA system is sometimes colloquially referred to as a Tannoy, after the company of that name now owned by TC Electronic Group, which supplied a great many of the PA systems used in the past.

Telephone Paging Systems

Some analog or IP private branch exchange (PBX) telephone systems use a paging facility that acts as a liaison between the telephone and a PA amplifier. In other systems, paging equipment is not built into the telephone system. Instead the system includes a separate paging controller connected to a trunk port of the telephone system. The paging controller is accessed as either a designated directory number or central office line. In many modern systems, the paging function is integrated into the telephone system, and allows announcements to be played over the phone speakers. Many retailers and offices choose to use the telephone system as the sole access point for the paging system, because the features are integrated. Many schools and other larger institutions are no longer using the large, bulky microphone PA systems and have switched to telephone system paging, as it can be accessed from many different points in the school.

Radio

Radio is the wireless transmission of signals through free space by electromagnetic radiation of a frequency significantly below that of visible light, in the radio frequency range, from a few Hz to 300 GHz. These waves are called radio waves. Electromagnetic radiation travels by means of oscillating electromagnetic fields that pass through the air and the vacuum of space. Information, such as sound, is carried by systematically changing (modulating) some property of the radiated waves, such as their amplitude, frequency, phase, or pulse width. When radio waves strike an electrical conductor, the oscillating fields induce an alternating current in the conductor. The information in the waves can be extracted and transformed back into its original form

Etymology

The etymology of "radio" or "radiotelegraphy" reveals that it was called "wireless telegraphy", which was shortened to "wireless" in Britain. The prefix radio in the sense of wireless transmission was first recorded in the word radio conductor, a description provided by the French physicist Édouard Branly in 1897. It is based on the verb to radiate (in Latin "radius" means "spoke of a wheel, beam of light, ray").

The word "radio" also appears in a 1907 article by Lee De Forest. It was adopted by the United States Navy in 1912, to distinguish radio from several other wireless communication technologies, such as the photo-phone. The term became common by the time of the first commercial broadcasts in the United States in the 1920s. (The noun "broadcasting" itself came from an agricultural term, meaning "scattering seeds widely.") The term was adopted by other languages in Europe and Asia. British Commonwealth countries continued to commonly use the term "wireless" until the mid-20th century, though the magazine of the BBC in the UK has been called Radio Times ever since it was first published in the early 1920s.

In recent years the more general term "wireless" has gained renewed popularity through the rapid growth of short-range computer networking, *e.g.*, Wireless Local Area Network (WLAN), Wi-Fi, and Bluetooth, as well as mobile telephony, *e.g.*, GSM and UMTS. Today, the term "radio" specifies the actual type of transceiver device or chip, whereas "wireless" refers to the lack of physical connections; one talk about radio transceivers, but another talk about wireless devices and wireless sensor networks.

Radio Communication System

A radio communication system sends signals by radio. Types of radio communication systems depend on technology, standards, regulations, radio spectrum allocation, user requirements, service positioning, and investment. The radio equipment involved in communication systems includes a transmitter and a receiver, each having an antenna and appropriate terminal equipment such as microphone at the transmitter and a loudspeaker at the receiver in the case of a voice-communication system. The power consumed in a transmitting station varies depending on the distance of communication and the transmission conditions. The power received at the receiving station is usually only a tiny fraction of the transmitter's output, since communication depends on receiving the information, not the energy that was transmitted.

A radio communication system may send information only one way. For example, in broadcasting a single transmitter sends signals to many receivers. Two stations may take turns sending and receiving, using a single radio frequency; this is called "simplex." By using two radio frequencies, two stations may continuously and concurrently send and receive signals; this is called "duplex" operation.

Telephone

A telephone, or phone, is a telecommunications device that permits two or more users to conduct a conversation when they are not in the same vicinity of each other to be heard directly. A telephone converts sound, typically and most efficiently the human voice, into electronic signals suitable for transmission via cables or other transmission media over long distances, and replays such signals simultaneously in audible form to its user. The word telephone has been adapted into the vocabulary of many languages. First patented in 1876 by Alexander Graham Bell and further developed by many others, the telephone was the first device in history that enabled people to talk directly with each other across large distances. Telephones rapidly became indispensable to businesses, government, and households, and are today some of the most widely used small appliances.

The essential elements of a telephone are a microphone (transmitter) to speak into and an earphone (receiver) which reproduces the voice of the distant person. In addition, most telephones contain a ringer which produces a sound to announce an incoming telephone call, and a dial used to enter a telephone number when initiating a call to another telephone. A landline telephone is connected by a pair of wires to the telephone network, while a mobile phone, such as a cellular phone, is portable and communicates with the telephone network by radio transmissions. A cordless telephone has a portable handset which communicates by radio transmission with the handset base station which is connected by wire to the telephone network.

Visual Aids

Visual communication is communication through visual aid and is described as the conveyance of ideas and information in forms that can be read or looked upon. Visual communication in part or whole relies on vision, and is primarily presented or expressed with two dimensional images, it includes: signs, typography, drawing, illustration, Industrial design, graphic design, advertising, animation colour and electronic resources. It also explores the idea that a visual message accompanying text has a greater power to inform, educate, or persuade a person or audience.

Visual aids are two types

1. Two-dimensional visuals (Poster, Chalk Board, Photo *etc.*)
2. Three dimensional visuals (Model, Specimen, Diorama *etc.*)

Audio-Visual Aids

Movie Projector

Movie projector is an opto-mechanical device for displaying motion picture film by projecting it onto a screen. Most of the optical and mechanical elements, except for the illumination and sound devices, are present in movie cameras.

Physiology

According to the theory of persistence of vision, the perceptual processes of the brain and the retina of the human eye retain an image for a brief moment of time. This theory is said to account for the illusion of motion which results when a series of film images is displayed in quick succession, rather than the perception of the individual frames in the series.

Persistence of vision should be compared with the related phenomena of beta movement and phi movement. A critical part of understanding these visual perception phenomena is that the eye is not a camera, *i.e.*: there is no "frame rate" or "scan rate" in the eye. Instead, the eye/brain system has a combination of motion detectors, detail detectors and pattern detectors, the outputs of all of which are combined to create the visual experience.

The frequency at which flicker becomes invisible is called the flicker fusion threshold, and is dependent on the level of illumination. Generally, the frame rate of 16 frames per second (frame/s) is regarded as the lowest frequency at which continuous motion is perceived by humans. (Interestingly this threshold varies across different species; a higher proportion of rod cells in the retina will create a higher threshold level). It is possible to view the black space between frames and the passing of the shutter by the following technique:

Close your eyelids, then periodically rapidly blink open and closed. If it done fast enough, you will be able to randomly "trap" the image between frames, or during shutter motion. This will not work with television due to the persistence of the phosphors nor with LCD or DLP light projectors due to the continuity of image; although certain color artifacts may appear with some digital projection technologies.

Silent films usually were not projected at constant speeds but rather were varied throughout the show at the discretion of the projectionist, often with some notes provided by the distributor. Speeds ranged from about 18 numbers frame/s on up - sometimes even faster than modern sound film speed (24 numbers frame/s).

Contrary to received opinion, 16 numbers of frame/s though sometimes used as a camera shooting speed - was dangerously inadvisable for projection, due to the high risk of the nitrate-base prints catching fire in the projector. (A dramatic rendition of a nitrate print fire and its potentially devastating effects is famously found in Nuovo Cinema Paradiso, which revolves around the goings-on of a projectionist.)

Since the birth of sound film, virtually all film projectors in commercial movie theaters project at a constant speed of 24 numbers frame/s. This speed was chosen for both financial and technical reasons. There are some specialist formats (*e.g.* Showscan and Maxivision) which project at higher rates 60 numbers frames/sec for Showscan and 48 for Maxivision. The Hobbit was shot at 48 numbers frames/sec and projected at the higher frame rate at specially equipped theaters.

Tele-Text (or Broadcast Tele-text)

It is a television information retrieval service created in the United Kingdom in the early 1970s by the Philips Lead Designer for VDUs, John Adams. Teletext is a means of sending text and diagrams to a properly equipped television screen by use of one of the vertical blanking interval lines that together form the dark band dividing pictures horizontally on the television screen. It offers a range of text-based information, typically including news, weather and TV schedules. Subtitle (or closed captioning) information is also transmitted within the television signal. The first test transmissions were made by the BBC in 1973, known as Ceefax. After adoption in the UK the standards became international as the European Teletext standards and as the World System Teletext (WST). The World Wide Web began to take over some of the functions of teletext from the late 1990s, and many broadcasters have ceased broadcast of teletext CNN in 2006 and the BBC in 2012.

Electronic Video Recording

Electronic Video Recording, or EVR, was a film-based video recording format developed by Hungarian-born engineer Peter Carl Goldmark at CBS Laboratories in the 1960s. CBS announced EVR in October 1967. The 750-foot film was stored on a 7-inch-diameter (180 mm) spool in a plastic cartridge. It used a twin-track 8.75 mm film onto which video signals was transferred by electron beam recording, one monochrome track in each direction of travel.

Electronic Video Recorder

Television

Television, or TV for short, (from French télévision, meaning "television"; from Ancient Greek τῆλε (tèle), meaning "far", and Latin visio, meaning "sight") is a telecommunication medium typically used in the 2000s for transmitting and receiving moving color images and sound. In the broader sense, television can also refer to images that can be monochrome (black-and-white) or with or without accompanying sound. "Television" may also refer specifically to a television set, television program, or television transmission.

Commercially available since the late 1920s (but in extremely limited amounts during those years), the television set has since become commonplace in homes, businesses and institutions, particularly as a vehicle for advertising, a source of entertainment, and news. Since the 1950s, television has been the main medium for molding public opinion. Since the mid-1960s, color televisions have been widely available. Since the 1970s, the availability of media such as video cassettes (mid 1970s), laserdiscs (1978), DVDs (1997) and high-definition Blu-ray Discs (2006) has enabled viewers to use the television set to watch recorded material, such as movies, as well as broadcast material. In recent years, Internet television has seen the rise of television available via the Internet through services such as iPlayer and Hulu.

In 2009, 78 per cent of the world's households owned at least one television set, an increase of 5 per cent over 2003. One major change in television technology was the replacement of the long - dominant CRT technology with various flat-panel TVs using LCD, plasma or LEDs. In 2013, 87 per cent of TVs sold were color LCD screen TVs. Although other forms such as closed-circuit television (CCTV) are in use, the most common usage of the medium is for broadcast television, which was modeled on the existing radio broadcasting systems developed in the 1920s, and uses high powered radio frequency transmitters to broadcast the television signal to individual TV receivers.

The broadcast television system is typically disseminated via radio transmissions on designated channels in the 54-890 MHz frequency band. Signals are now often transmitted with stereo or surround sound in many countries. Until the 2000s, broadcast TV programs were generally transmitted as an analog television signal but during the decade several countries went almost exclusively in digital. In addition to over-the-air transmission, television signals are also distributed by cable and satellite systems. A standard television set comprises multiple internal electronic circuits, including those for receiving and decoding broadcast signals. A visual display device which lacks a tuner is properly called a video monitor, rather than a television. A television system may use different technical standards such as digital television (DTV) and high definition television (HDTV). Television systems are also used for surveillance, industrial process control, and guiding of weapons, in places where direct observation is difficult or dangerous. Some studies have found a link between infancy exposure to television and ADHD.

Drama

Drama is the specific mode of fiction represented in performance. The term comes from a Greek word meaning "action" (Classical Greek: δρᾶμα, drama), which is derived from the verb meaning "to do" or "to act" (Classical Greek: δράω, draō). The enactment of drama in theatre, performed by actors on a stage before an audience, presupposes collaborative modes of production and a collective form of reception. The structure of dramatic texts, unlike other forms of literature, is directly influenced by this collaborative production and collective reception.. The early modern tragedy Hamlet (1601) by Shakespeare and the classical Athenian tragedy Oedipus the King (c. 429 BC) by Sophocles are among the masterpieces of the art of drama. A modern example is Long Day's Journey into Night by Eugene O'Neill (1956).

The two masks associated with drama represent the traditional generic division between comedy and tragedy. They are symbols of the ancient Greek Muses, Thalia and Melpomene. Thalia was the Muse of comedy (the laughing face), while Melpomene was the Muse of tragedy (the weeping face). Considered as a genre of poetry in general, the dramatic mode has been contrasted with the epic and the lyrical modes ever since Aristotle's Poetics (c. 335 BC) the earliest work of dramatic theory.

According to Projection Audio-Visual Aids can be Classified as Projected and Non Projected Aids

Projected Audio Visual Aids

a) Magic Lantern
b) Slide projector
c) Over Head Projector
d) Motion pictures
e) Over Head Transparencies
f) The Opaque Projector
g) Filmstrip
h) LCD Projector
i) Home theatre Projector

a) Magic Lantern

The Magic Lantern was the forerunner of the modern slide projector. However, researchers do not know who invented the first one. In 1676, a type of Magic Lantern called the Sturm Lantern was invented and it may have been one of the first.

The practice of projecting images from glass plates began centuries before the invention of photography. However, in the 1840s, Philadelphia daguerreotypists, William and Frederick Langenheim, began experimenting with The Magic Lantern as an apparatus for displaying their photographic images. Because the opaque nature of the Because the opaque nature of the daguerreotype disallowed its projection, the brothers looked for a medium that could create a transparent image.

b) Slide Projector

A slide is a small piece of transparent material on which a single pictorial image or scene or graphic image has been photographed or reproduced otherwise.

Slides are a form of projected media that are easy to prepare. They are still pictures on positive film which you can process and mount individually yourself or send to a film laboratory. The standard size of the slides is 2 "X 2 "any 35mm camera will make satisfactory slides.

Types of Slides

1. Photographic Slides

a) Black and white 2" X 2"
b) Colored 3" X 4"

2. Handmade Slides

Can be made with

a) Acetate sheet
b) Cellophane
c) Etched glass
d) Plain glass

Slides can be made from photographs and pictures by teachers and pupils taking photographs and snapshots when they go on fieldtrips for historical, geographical, literacy or scientific excursions. The arrangement of slides in proper sequence, according to the topic discussed, is an important aspect of teaching with them.

Advantages

1. Requires only filming, processing and mounting by self or laboratory.
2. Results in colorful, realistic, reproduction original subject.
3. Preparation with any 35mm camera for most uses.
4. Easy to revise and up-date.
5. Easily handled, stored and re-arranged for various uses.
6. Can be combined with tape narration or can control time for discussion.
7. May be adapted to group or individual use.

c) Over Head Projector (OHP)

The over head projector is the most used in all Audio visual aids. It projects transparencies with brilliant screen images suitable for use in a lighted room. The teacher can write or draw diagrams on the transparency while he teaches; these are projected simultaneously on the screen by the OHP.

During Presentation

a) Keep the screen above the heads of the participants.
b) Keep the screen in full view of participants
c) Make sure you are not blocking any ones view when presenting.
d) Darken the room appropriately by blocking out sunshine and dimming nearby.
e) Turn the screen off between slides if you are going to talk for more than two.
f) Talk to the audience, not to the screen

Purposes

a) To develop concepts and sequences in a subject matter area;
b) To make marginal notes on the transparencies for the use of the teacher that can carry without exposing them to the class;
c) To test students performances, while other classmates observe;
d) To show relationships by means of transparent overlays in contrasting color;
e) To give the illusion of motion in the transparency.

Advantages

a) It permits the teacher to stand in front of the class while using the projector, thus enabling her to point out features appearing on the screen by pointing to the materials at the projector itself and at the same time, to observe the students reactions to her discussion

b) Gains attention of the student

d) Motion Picture

A film, also called a movie or motion picture, is a series of still images which, when shown on a screen, creates the illusion of moving images due to phi phenomenon. A film is created by photographing actual scenes with a motion picture camera; by photographing drawings or miniature models using traditional animation techniques; by means of CGI and computer animation; or by a combination of some or all of these techniques and other visual effects. Contemporary definition of cinema is the art of simulating experiences, which communicate ideas, stories, perceptions, feelings, beauty or atmosphere by the means of recorded or programmed moving images along with other sensory stimulations.

The process of filmmaking is both an art and an industry. Films were originally recorded onto plastic film which was shown through a movie projector onto a large screen; more modern techniques may use wholly digital filming and storage, such as the Red One camera which records onto hard-disk or flash cards. Films usually include an optical soundtrack, which is a graphic recording of the spoken words, music and other sounds that are to accompany the images. It runs along a portion of the film exclusively reserved for it and is not projected.

Films are cultural artefacts created by specific cultures. They reflect those cultures, and, in turn, affect them. Film is considered to be an important art form, a source of popular entertainment, and a powerful medium for educating-or indoctrinating-citizens. The visual basis of film gives it a universal power of communication. Some films have become popular worldwide attractions by using dubbing or subtitles to translate the dialog into the language of the viewer. Some have criticized the film industry's glorification of violence.

Advantages

1. Attracts audience and arouses their interests.
2. Presents facts in a realistic and interesting way.
3. Exposes the audience to a new practice in a short time.
4. Provides experiences normally difficult to get.
5. Helps in changing attitudes.
6. Can reach illiterate people also.
7. May be shown according to convenience over and over origin.

e) Over Head Transparencies

Transparencies are popular instructional medium. They are simple to prepare and easy to prepare and easy to operate with the over head projector which is light weight. A 10x10 inches sheet with printed, written or drawn material is placed on the platform of the projector (OHP) and a large image is projected on a screen behind you. The projector is used from near to the front of the room with the teacher standing or sitting beside, facing the student.

Guidelines for making Effective Transparencies

a) Have one main idea an each transparency.
b) Include only related figures and diagrams.
c) Use simple lettering style in writing.
d) Use diagrams in proposition to its lettering.
e) Keep the message clear and simple.
f) Emphasize the key messages.
g) Use color and lettering with discretion.

Advantages

a) Permits face to face interaction with the students.
b) Can be used in daylight conditions.
c) Can present information in systemic developmental sequences.
d) Requires limited planning and can be prepared in variety of inexpensive methods.
e) Easily available.

f) The Opaque Projector

Opaque projector is the only projector on which you can project a variety of materials ex: book pages, objects, coins, postcards, or any other similar flat material that is non-transparent.

The opaque projector will project and simultaneously enlarge, directly from the originals, printed matter, all kinds of written or pictorial matter in any sequence derived by the teacher. It requires a dark room, as projector is large and not reality movables.

Advantages

a) Stimulates attention and arouses interest;
b) Can project a wide range of materials like stamps, coins, specimen, when one copy is available;
c) Can be used for enlarging drawings, pictures and maps;
d) Does not require any written or typed materials, hand-written material can be used;
e) Helps students to retain knowledge for longer period;
f) Review instructional problems;

g) Test knowledge and ability;
h) Simple operation.

Disadvantages

a) Costly equipment;
b) Needs to use it with care;
c) Needs a dark room for projector

g) Filmstrip

Film strips are sequence of transparent still pictures with individual frames on 35mm film. A tap recorded narration can be synchronized with film strip. Each strip contains from 12 to 18 or more pictures. It is a fixed sequence of related stills on a roll of 35mm film or 8mm film.

Principles

a) Preview filmstrips before using them and selected carefully to meet the needs of the topic to be taught.
b) Show again any part of the filmstrip needing more specific study.
c) Use filmstrip to stimulate emotions, build attitudes and to point up problems.
d) It should be introduced appropriately and its relationship to the topic of the study brought out.
e) Use a pointer to direct attention, to specific details on the screen.

Types of Filmstrip

a) **Discussion filmstrip**: it is continuous strip of film consisting of individual frames arranged in sequence usually with explanatory titles.
b) **Sound slide film**: it is similar to filmstrip but instead of explanatory titles or spoken discussion recorded explanation is audible, which is synchronized with the pictures.

Advantages

a) Are compact, easily handled and always in proper sequence?
b) Can be supplemented with recordings;
c) Are inexpensive when quantity reproduction is required;
d) Are useful for group or individual study at projection rate are controlled by instructor or user;
e) Are projected with simple light weight equipment.

h) LCD Projector

A liquid-crystal display (LCD) is a flat panel display, electronic visual display, or video display that uses the light modulating properties of liquid crystals. Liquid crystals do not emit light directly. LCDs are available to display arbitrary images (as in a general-purpose computer display) or fixed images which can be displayed or hidden, such as preset words, digits, and 7-segment displays as in a digital clock. They use the same basic technology, except that arbitrary images are made up of a large number of small pixels, while other displays have larger elements.

LCDs are used in a wide range of applications including computer monitors, televisions, instrument panels, aircraft cockpit displays, and signage. They are common in consumer devices such as video players, gaming devices, clocks, watches, calculators, and telephones, and have replaced cathode ray tube (CRT) displays in most applications. They are available in a wider range of screen sizes than CRT and plasma displays, and since they do not use phosphors, they do not suffer image. LCDs are, however, susceptible to image persistence. The LCD screen is more energy efficient and can be disposed of more safely than a CRT. Its low electrical power consumption enables it to be used in battery-powered electronic equipment. It is an electronically modulated optical device made up of any number of segments filled with liquid crystals and arrayed in front of a light source (backlight) or reflector to produce images in color or monochrome. Liquid crystals were first discovered in 1888. By 2008, annual sales of televisions with LCD screens exceeded sales of CRT units worldwide, and the CRT became obsolete for most purposes. Each pixel of an LCD typically consists of a layer of molecules aligned between two transparent electrodes, and two polarizing filters (parallel and perpendicular), the axes of transmission of which are (in most of the cases) perpendicular to each other. Without the liquid crystal between the polarizing filters, light passing through the first filter would be blocked by the second (crossed) polarizer. Before an electric field is applied, the orientation of the liquid-crystal molecules is determined by the alignment at the surfaces of electrodes.

LCDs have many characteristics that have caused their demand to increase over the past several years. They are lightweight, aesthetically appealing, energy efficient and long-term cost effective. Some of the disadvantages are that they are not suitable for dimly lit applications, can require frequent adjustments throughout the day, have limited viewing angles, and their purchase cost is relatively expensive.

LCD Advantages

a) **Brightness**: Produces very bright images due to high peak intensity. Very suitable for environments that are brightly lit.

b) **Emissions**: Produce considerably lower electric, magnetic and electromagnetic fields than CRTs.

c) **Geometric Distortion**: No geometric distortion at the native resolution. Minor distortion can occur for other resolutions.

d) **Power Consumption**: Energy efficient. Consume less than 1/3 the power of a comparable CRT. Consume less electricity than a CRT and produce little heat.

e) **Physical Aspects**: Take up about 40 per cent less desk space. LCDs are thin and compact.

f) **Screen Shape**: Completely flat screen.

g) **Sharpness**: At the native resolution, the image is perfectly sharp. Adjustments are required at all other resolutions which can result in measurable degradation to the image.

LCD Disadvantages

a) **Aspect Ratio**: The aspect ratio and resolution are fixed.

b) **Black-Level**: Not proficient at producing black and very dark grays. In a "standard" configuration, not appropriate for use in dimly lit and dark conditions.

c) **Contrast**: Lower contrast than CRTs due to a poor black-level.

d) **Color and Gray-Scale Accuracy**: Color saturation is reduced at low intensity levels due to a poor black-level. Images are satisfactory, but not accurate due to problems with black-level, gray-scale and Gamma.

e) **Cost**: Considerably more expensive purchase price than comparable CRTs. (Cheaper lifetime cost: lasts about 13,000 - 15,000 more hours than a typical CRT.)

f) **Gray-Scale**: Have an irregular intensity scale and typically produce fewer than 256 discrete intensity levels. For some LCDs portions of the gray-scale may be dithered.

g) **Motion Artifacts**: Slow response times and scan rate conversion result in severe

i) 3D Home Theatre Projector

At one time, the slide projector was the presentation method of choice when visual media needed to be presented to a large group. Travellers often took photographs on slide film for the express purpose of holding slide shows of their trips when they returned, and slides could also be prepared for business and educational presentations. In an art history course, for example, the teacher might use slides of famous works of art to display examples for discussion.

Slides are actually a great presentation tool, because of the very high image quality. However, slide projectors can be difficult to work with, especially in the case of carousel projectors, which can become recalcitrant with time. It is also difficult to remember to insert the slides properly so that the images are not reversed or inverted, and to organize the slides in order. As many people who have made presentations with a slide projector know, one of the worst things which can happen is dropping the carousel on the way to the presentation podium, causing all the slides to fall out.

Because slides can be difficult to work with, other projection methods such as projectors designed to link to computers or read presentations from CDs and DVDs have become more commonplace. These projectors may be of lower quality, but they are simple enough for almost anyone to operate.

Non Projected Audio-Visual Aids

Graphic Aids

It is a combination of graphic and pictorial material designed for the orderly and logical visualizing of relationships between key facts and ideas example: comparisons, relative amounts developments, processes, classification or organization. It includes the following:

a) Charts

Introduction

These visual symbols used for summarizing, comparing, contrasting or performing other services in explaining subject matter. A chart is a combination of pictorial, graphic, numerical or vertical material, which presents a clear summary.

Definition

Chart is defined as a visual aid which depicts pictorial and written key information in systematic way to summarize, compare, example: anatomical charts and figure, diagrams *etc.*

Purposes

a) To visualize an item, it is otherwise difficult to explain only in words.
b) To highlight important points.
c) To provide outline for materials covered in presentation.
d) To show continuity in process.
e) For creating problems and stimulating thinking.
f) For showing development of structure.

Types of Charts

a) **Narrative chart**: Arrangement of facts and ideas for expressing the events in the process or development of a significant issue to its point of resolution or we can show an improvement over a period of years;
b) **The cause and effective chart:** Arrangement of facts and ideas for expressing the relationship between rights and responsibilities or between a complex of conditions and change or conflict;
c) **The chain chart**: arrangement of facts and ideas for expressing transitions or cycles;
d) **The evolution chart**: facts and ideas for expressing changes in specific items from beginning data and its projections in to future;

e) **Strip tease chart:** it enables speaker to present the information step by step. It increases the interest and imagination of the audience. The information on the chart is covered with thin paper strips to which it has been applied either by wax, tape or sticky substance or pins. As the speaker wishes to visually reinforce a point with words or symbols, he removes the appropriate strip or paper. It produces interest.

f) **Pull chart:** it consists of written messages which are hidden by strips of thick paper. The message can be shown to the viewer, one after another by pulling out the concealing strips;

g) **Flow chart:** diagrams used to show organizational elements or administrative or functional relationships. In this chart lines, rectangles, circles, are connected by lines showing the directional flow;

h) **Tabulation chart**: it shows the schedule of an activity or of an individual ex: time-table of a class. These are very valuable aid in the teaching situation where breakdown of a fact or a statement is to be listed. Also it is a useful aid for showing points of comparison, distinction, and contrasts between two or more things. While making the table charts the following points must be kept in the mind.

   The chart should be 50 X 75 cm or more in size.

   The chart should be captioned in bold letters.

   The vertical columns should be filled in short phrases rather than complete sentences;

i) **Flip chart:** a set of charts related to specific topic have been tagged together and hang on a supporting stand. The individual charts will carry a series of related materials or messages in sequence. The silent points of specific topic will be presented;

j) **Pie chart**: a circle will be drawn and divisions will be made into different sections, each section will be coded differently and code key will be given at right corner of the chart as legend. The circumference is divided into suitable sections. It is relevant for showing the component part.

b) Flash Cards

Definition

"Flash cards are a set of pictured paper cards of varying sizes that are flashed one by one in a logical sequence."

"Flash cards can be self made or commercially prepared and are made up of chart or drawing paper, plain paper using colors or ink on them for drawings."

Purposes

a) To teach the students.

b) To give health education.

c) Useful for small group.

d) Used in group discussions.

Principles

a) The messages can be brief, simple line drawing or photographs, cartoons and the content will be written in few lines at the back of the each card.

b) 10″ X 12″ or 22″ X 28″ is commonly used size.

c) 10-12 cards for one talk can be used. It should not be less than 3 and more than 20.

d) Prepare a picture for each idea which will give visual impact to the idea.

e) The height of writing on the flash card is to be approximately 5cm for better visualization.

Using the Flashcards

For class room instruction, the flash card s is to be properly used. The following steps are used while displaying flash cards.

a) Give brief introduction about the lesson to students;

b) Give instructions to students about their actions while you flash the cards;

c) Flash the card in front of the class by holding it high with both your hands so that all the students can see it;

d) Let the student respond as per instructions already given;

e) Review the lesson by selectively using flash cards.

Advantages

a) Flash cards can be used to introduce and present topics;

b) It can be used to apply information already gained by students to new situations;

c) It can be used to review a topic;

d) Can be used for drill and practice in elementary classes;

e) To develop the cognitive abilities of recognition and recall of students;

f) It can work as a useful supplementary aid and can be effectively used with other material.

Disadvantages

a) Cannot be used for a large group

b) Prone to get spoiled soon

c) Preparation is time consuming.

c) Posters

Definition

"Posters are the graphic aids with short quick and typical messages with attention capturing paintings."

Purposes

a) To provide general motivation;

b) To create an esthetic or atmospheric effect;

c) To communicate a more general idea. To thrust the message for leading to action;

d) For the class room and community.

Preparation and Rules

a) To do a special job;

b) To promote one point;

c) To support local demonstration;

d) Planned for specified people;

e) Tell the message at single glance;

f) Use bold letters;

g) Use pleasing colors....;

h) It should place, where people pass or gather.

Features of a Good Poster

a) Brevity: message should be concise

b) Simplicity: message should be easily understandable

c) Idea: should base on single idea and it should be relevant.

d) Color: suitable color and combination should be used to make the poster attractive and eye catching.

e) Display: while displaying one should be sure to find a place where there is adequate light and where the larger population will see it.

Advantages

a) It attracts attention;

b) It conveys the message very quickly;

c) It does not require a detailed study;

d) Good poster leads to action with good motivation;

e) It can stand alone and is self explanatory.

Disadvantages

a) Poster does not always give enough information;

b) When a poster is seen for longer time it may not attractive;

c) So it should be dynamic.

d) Graphs

Definition

Graphs are the visual teaching aids for presenting statistical data and contrasting the trends or changes of certain attributes.

Method of Preparation

a) Before making the bar chart makes a rough sketch of it in a note book;

b) For drawing the bar graph use the chart paper of 50x 72 cm size;

c) Use two different color shades for the two contrasting groups;

d) The bars should be equi-spaced;

e) Write the key to the bar graph in a box on the right hand side corner of the chart paper;

f) Numbers specifying the magnitude of the bars should be on the top on the bars.

Types

Pie Graph

These are called as circle diagram. The data are presented thorough the sections of portions of a circle.

a) In determining the circumference of a circle we have to take in to consideration a quantity known as pie;

b) The surface area of a circle is to cover 360 degree;

c) The total frequencies or value us equated to 360 degree and then the angles corresponding to component parts are calculated;

d) After determining their angle, the required sectors in the circle are drawn.

Bar Graph

The graphic presentation extends the scale horizontally along the length of bars. Each bar must be of the same width, height of the bar over a period represents the corresponding time of the variable. Graphs are available in 2 forms that is vertical and horizontal

Line Graph

To show the trends and relationships ex: single line shows the relation and the variation in the quantity. Quantitative data are plotted or when the data is continuous. The concepts are represented with the help of lines drawn either horizontally or vertically. The plotted points are connected to one another, instead of the base thus producing the curve.

Pictorial Graph

It is an outstanding method of graphic representation. Pictures are used for the expression of ideal; they are more attractive and easily understood. Vivid pictures will be used to create rapid association with the graphic message; each visual symbol may be used to indicate quantity.

e) Map

Definition

A map is a graphic aid representing the proportionately as a diagram, the surface of the earth, world or parts thereof. It conveys the message by lines, symbols, words and colors.

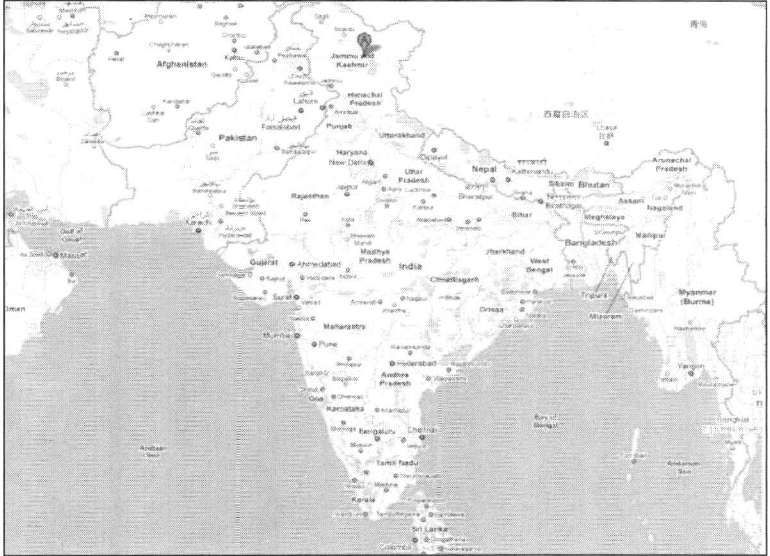

Map of India

Types of Maps

- ☆ **Political maps**: these maps show political divisions of the world, a continent, a nation.
- ☆ **Physical maps**: shows the physical contour of a place, area, and region.
- ☆ **Relief maps**: it shows the actual elevations and depressions in a place, area, and region.
- ☆ **Weather maps**: shows the amount of rains, temperature extremes, humidity in an area, region country.

- ☆ **Population maps**: shows the distribution of population in various parts of region, country.
- ☆ **Picture or tourist maps**: shows historical spots monumental sites.etc.
- ☆ **Road maps**: shows the roads of a region connecting various parts and points together.
- ☆ **Railway maps**: shows the railway links between various points.
- ☆ **Air maps**: shows the air routes between various points.
- ☆ **Sea root maps:** shows the sea routes between various sea ports

f) Cartoons

The word cartoon has various meanings, based on several very different forms of visual art and illustration. The term has evolved over time. The original meaning was in fine art, and there cartoon meant a preparatory drawing for a piece of art such as a painting.

Definition

A cartoon is humorous caricature which gives a subtle message. In a cartoon the features of objects and people are exaggerated along with generally recognized symbols.

Principles

a) The quality of the drawing should be high primarily for visual effectiveness.
b) The symbols used should be familiar and represent a concept or idea to which students can react intellectually.

Advantages

A cartoon can be effectively used to initiate certain lesson. It can be used for making a lesson lively and interesting.

a) Fantasy
b) Satire
c) Exaggeration.

g) News Papers

It can furnish health messages in local languages which cancan reach to the public easily. The information will be available in low cost, easy to read and understand simple language.the people may learn to read and interpret the contents along with pictures to enhance easy grasping.

Advantages

a) Best method to reach a large group;
b) Pictures will help in easy understanding;

c) Attractive and easy to understand;

d) Lot of information can be obtained in various fields.

Disadvantages

a) Useful for literates only;

b) Detailed information cannot be procuces.

h) Comic Strips

Definition

A comic strip is the graphic depiction in a series of pictures or sketches of some character and events full of action. This medium of communication is found very interesting and exciting by children.

Uses

a) Comic strips fire the imagination of children;

b) It boosts the courage of children and builds up the spirit of adventure;

c) It communication detailed and vivid;

d) It stimulates reality and involvement.

Limitations

a) Comic strips misguide children by depicting characters with supernatural powers divorced from the hard realities of life;

b) Comic strips hamper the development of language of children;

c) Classics brought out in the form of comics develop the tendency in children to ignore or by pas s the original work;

d) Comics can soon become an obsession with young children and they tend to avoid serious studies.

i) Display Board

i) Chalk Board

Definition

A **chalkboard** or **blackboard** is a reusable writing surface on which text or drawings are made with chalk or other erasable markers. Blackboards were originally made of smooth, thin sheets of black or dark grey slate stone. Modern versions are often green or brown and are thus sometimes called a green board or brown board instead.

A blackboard can simply be a piece of board painted with matte dark paint (usually black or dark green). A more modern variation consists of a coiled sheet of plastic drawn across two parallel rollers, which can be scrolled to create additional writing space while saving what has been written. The highest grade chalkboards are made of rougher version porcelain enameled steel (black, green, blue or sometimes other colours). Porcelain is very hard wearing and chalkboards made of porcelain usually last 10-20 years in intensive use.

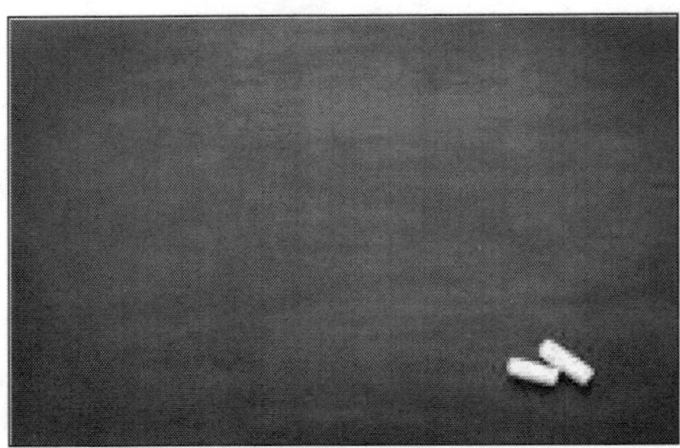

Disadvantages

They produce a fair amount of dust, depending on the quality of chalk used. Some people find this uncomfortable or may be allergic to it, and there has been speculation about links between chalk dust and respiratory problems.

The dust also precludes the use of chalk in areas shared with dust-sensitive equipment such as computers. However, these alternative methods of displaying information have drawbacks of their own. The scratching of fingernails on a blackboard is a sound that is well-known for being extremely irritating. Blackboards are also used in many establishments (typically public houses) as a form of advertising often for upcoming events and menus - as well as to keep the score in darts matches.

(ii) Flannel Graph

Flannel graph is a storytelling system that uses a board covered with flannel fabric, usually resting on an easel. It is very similar to Fuzzy felt, although its primary use is as a story telling medium, rather than as a toy. This teaching tool is called by different names as Visual Board, Frick Board, Slap Board, Felt Board, Cohero graph, Video graph.

How to Use

The principle involved is the inter1ooking of fibers of two rough or bairy surfaces, so that the pieces pressed on to a background which is hard and vertical will stay. It can be illustrated on a larger scale by pressing two tooth brushes or hair brushes together, so the bristle inter-1ook. In case of flannel graph similar principle of friction helps an object to cling to the surface of the board. The flannel board is usually painted to depict a background scene appropriate to the story being told. Paper cutouts of characters and objects in the story are then place on the board, and moved around, as the story unfolds. These cutouts are backed, either with flannel, or with some other substance that adheres lightly to the flannel background, such as coarse sandpaper.

Advantages

1. Permits numerous and varied arrangements of visua1 materials;
2. Permits the use of either chart or small pieces of material Materials can be packed and transported complete notes;
3. Permits the development of a complete story;
4. Promotes conscientious planning, which must precede the development of the material in the first place;
5. Challenges one to develop symbols to portray such things as abstractions;
6. Easier to construct materials for flannel board than to make slides or movies.

Disadvantages

1. Transportation and storing of boards and materials is a problem. Suitable tables to support boards must be available.
2. Time and cost of making material for presentation present a problem.
3. Cost of boards themselves can't be overlooked.
4. Presentation is limited a new idea involves a lapse of time before the new material can be added

5. Might tend to deter one from using other more effective methods and techniques when it is evident that other methods might be more appropriate.
6. To tell a complete story it often takes either too much board space or smaller designs and materials some of which cannot be seen well.

(iii) Bulletin Board

Definition

It is a soft board which will hold pins or tags almost suitable. Simple device placed either indoor or outdoor. Items generally displayed are photographs, publications, posters, news paper cut outs.

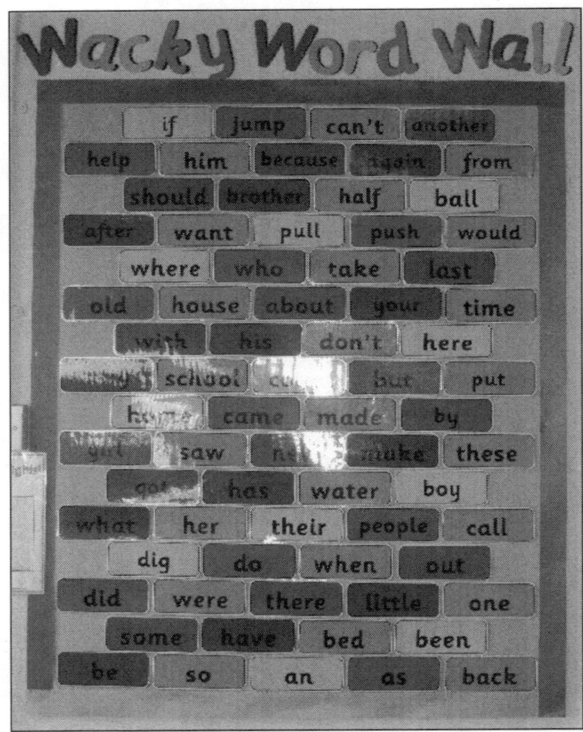

Advantages

1. Explains important events
2. Reports special activities

Disadvantages

1. Not effective for illiterate group.
2. Takes lot of preplanning and preparation

A bulletin board (pin board, pin board or notice board in British English) is a place where people can leave public messages, for example, to advertise things to buy or sell, announce events or provide information. Dormitory corridors, well-trafficked hallways, lobbies, and freestanding kiosks often have cork boards attached to facilitate the posting of notices. At some universities, lampposts, bollards, trees, and walls often become impromptu postering sites in areas where official boards are sparse in number.

(iv) Peg Board

It is a type of board which contains small holes to fix certain letters into the holes which is used especially in the offices to display certain items, name of the personal or faculty member.

(v) Magnetic Board

It is a framed iron sheet carrying porcelain coating in some dark color generally black or green. It can be used to display pictures, cutouts and light objects with disc magnets or magnetic holders.

Advantages

1. Movement of visual material is easy.

j) 3-Dimentional Aids

Models

Definition: a model is a recognizable representation of a real thing three dimensionally, that is height, width, and depth is felt as reality.

Types of Models

a) **Solid models**: it is the replica of an original thing made with some suitable material like clay, plaster of Paris, wood, iron *etc.* to show the external parts of the things. Example: globe, clay model of human and animal;

b) **Cutaway and x-ray models**: are the replicas of the original things to show internal parts of a thing. Cross sectional models are difficult to make in the class room or institutions as they require expertise to construct them. Example: cross sectional model of human body;

c) **Working models**: these models are either actual working things or their miniature replicas. For illustrating an operation. Example: a motor, a generator;

d) **Sand models**: made by using sand, clay, saw dust, example a tribal village, a forest area.

Advantages

a) Models heighten reality of things and make learning direct and meaningful as they are three dimensional;

b) Models illustrate the application side of certain principles and laws;

c) Models explain the complex and intricate operations in a simplified way and thus make comprehension easier;

d) Models are lasting and ultimately work out to be cheaper teaching aids;

e) Still models are easy to make with the help of discarded materials like empty boxes, pins, clips, nails, and clay;

f) Models are to reasonable size and convenient to handle;

g) Models involve the use of all the five senses and thus make learning effective.

Limitations

a) It requires expertise to make;

b) Time consuming;

c) Some of the models may be very expensive.

k) Objects and Specimen

Definition

A collection of real things for instructional use refers to objects.

A specimen is a sample of the real object or a material.

Using objects and specimens: while using the specimen and objects as teaching aids, a teacher must keep the following points in her mind.

a) Plan your teaching with certain simple and direct observations of the object or specimen being referred to;
b) Ask questions from the students to elicit more details of the features of the object or specimen under observation;
c) Clarify and emphasize important structural details of the object or specimen under observation;
d) Provide review and practice to make learning permanent.

Sources of Objects and Specimens

a) Local markets;
b) Manufacturers and factories;
c) Discarded material from the houses;
d) Specimen found in the nature can be collected by students from field trips and nature hunt;
e) Plasters casts can be purchased;
f) Wild flowers, leaves shells, stones butterflies moths, insects can also be procured.

Mounting the Objects and Specimens

Objects and specimens should be mounted in shallow boxes in an artistic way and the boxes should be covered with cellophane paper. Also label each object or specimen using self adhesive paper.

Advantages of Objects and Specimens

a) Collection of objects and specimens by students requires interaction with others leading to development of social skills and values;
b) Students when collect and display objects and specimens derive satisfaction of contributing to the school and teacher something worthwhile;
c) Student's power of observation and first hand experiences is enhanced by collection of objects and specimens;
d) Student's personal collection of objects and specimens can be good source of doing investigatory projects;
e) Collection of objects and specimens become an interesting educational pursuit of the teacher and students alike;
f) It arouse some interest among students in learning;
g) Objects and specimens involve all the five senses in the process of learning;

h) It heighten the reality in the class room;

i) It makes teaching lively.

l) Exhibition

Many times in the school, a department of the school or a class put up their work for showing it to the people outside the school, and such a show called exhibitions. The pieces of work done by the students for an exhibition are called exhibits.

Requisites for Exhibition

a) The exhibition should have a central theme with a few sub themes to focus attention to a particular concept;

b) The exhibits should be clean, labeled properly;

c) The concepts of contrast in color and size should be used for lying out the exhibitions;

d) The exhibits should be so placed so the most visitors, can see them;

e) The place and exhibits should be well lighted;

f) To capture attention and interest of visitors, both motion and sound should be utilized;

g) The exhibition should have some exhibits with operative mechanism such as switches, handles, to be operated by the visitors to observe some happenings;

h) The exhibition should include lot of demonstrations as they involve deeply the students and the visitors;

i) The exhibition should be able to relate various subjects' areas to provide integrated learning.

Advantages

a) Exhibitions inspire the students to learn by doing things themselves and they get a sense of involvement;

b) Exhibitions give students a sense of accomplishment and achievement

c) Exhibitions develop social skills of communication, cooperation, coordination;

d) Exhibitions foster better school community relations and make community members conscious about the school;

e) Exhibitions couple information with pleasure;

f) Exhibitions foster creativity among students.

Disadvantages

a) Requires thorough preparation

b) Time consuming

c) Require funds or budget.

m) Museums

Definition

A museum is a building displaying a collection of historical relics, antiques, curiosities, works of arts, works of science, literature and other artifacts of general interest. Museums can be useful both for public education and specific class room instructions.

Setting up School Museum

a) School should have enough space

b) Take the help of students, collect old and new objects and articles

c) Accept donations from various organizations who donates the articles

d) Students can be guided to prepare the exhibits

e) All the collected and prepared articles should be displayed and labeled

f) A detailed report book should be maintained giving a brief description of each museum pieces

g) The museum rooms should be well lighted

h) It should be cleaned and maintained timely.

n) Dioramas

Definitions

A diorama is a three dimensional arrangement of related objects, models, and cut outs to illustrate a central theme or concept. The objects and models are generally placed in a big box or show case with a glass covering and background printed with a shade or a scene. Ex: a harvest scene, a planting scene *etc.*

Advantages

a) Provide a good opportunity to learn

b) It gives the appearance of actual things which can not be brought to the class room

c) Interesting and enhance creativity

d) Live things also can shown in diorama ex: aquarium

e) Provides students to do project works

Disadvantages

a) Needs expatriation for the preparation;

b) Sometimes cost effective

c) Require budget;

d) Sometimes it may misguide the student if is not the replica of actual thing.

o) Mock Ups

It emphasizes the functional relationship between the device reality and its workability. Certain element of the original reality is emphasized to make it more meaningful for the purpose of instruction. In common usage, a mockup is a scale model of a structure or device, usually used for teaching, demonstration, testing a design, *etc*. Mockups are also used in the Consumer goods industry, as part of the product development process, when the size, impression and/or artworks have to be tested and approved. Mockup is also a frequently used term when talking about an early layout or sketch of a Web site or GUI program.

Examples: An artificial kidney to demonstrate dialysis.

p) Fiups Moulage

Moulage (French: casting/moulding) is the art of applying mock injuries for the purpose of training Emergency Response Teams and other medical and military personnel. Moulage may be as simple as applying pre-made rubber or latex "wounds" to a healthy "patient's" limbs, chest, head, *etc.*, or as complex as using makeup and theatre techniques to provide elements of realism (such as blood, vomits, open fractures, *etc.*) to the training simulation. The practice dates to at least the Renaissance, when wax figures were used for this purpose. Mould can be made up of plastic material to stimulate some life in objects. Example: body which shows evidence of trauma, infection, disease, surgical intervention.

Image Perspectives Moulage Process

The basic material we use to create our soft tissue injuries is very inexpensive; it costs only pennies per simulation. If someone walks off with a simulation it can be recreated in a matter of minutes! You are not limited to mass-produced latex or plastic "one-size-fits-all injuries". You determine the type, scope and size of the injuries and create as many as you wish, in whatever size you need. Our simulations can be handled and bandaged, with care, as you would a real injury. The simulations will not shift and/or be damaged. When bandages are removed, the injury simulation will remain intact.

The simulations can be made liquid-proof, and with care - reusable!

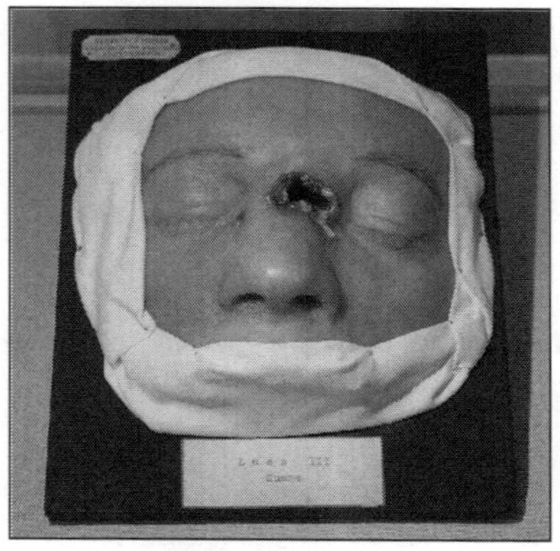

q) Folk Media

Reasons for using Folk Media

No communication strategy would be complete unless it included the traditional media, which must rank alongside the more commonly accepted channels of communication.

Status and Function of Folk Media

a) As an intrinsically valuable form of popular entertainment or artistic expression it is worth preserving and developing for its own sake;

b) It is a means of changing values, attitudes and norms in order to provide a proper climate for social and economic progress;

c) It is a method of promoting certain behaviour acts or patterns. The aim is to get people to perform specific acts to achieve objectives of national policy (*e.g.* visiting clinics, investing in bonds, using fertilizer);

d) It is a channel for conveying information about available techniques and facilities which people may use to solve problems.

Classification of Traditional Forms

Traditional forms of expression were identified. The classification covering the performing arts and the visual arts included music, song, drama, skits, puppet shows, poetry, speech, sounds, gesture, gossip, jokes, proverbs, painting, certain printed literature, sculpture, handicrafts, costuming, patterning and colouring of material and use of head-gear. Symbolic meanings may often be an important consideration in use of any one or combination of these forms of expression and therefore the thought-processes behind selection of forms assumed significance.

Advantages

Traditional media come from the people and antecede the mass media; their appeal has historically been both functional and aesthetic. Whether of roughly hewn folk origins or of classifically rigid moulds, they have always served to entertain, educate to reinforce existing ideas or ideologies or to change existing values and attitudes.

Another advantage of the traditional media is that they attract people who might not attend an educational meeting. With skill, new content might be added to the old forms which are already familiar and dear to the people. Finally, unlike mass media programmes, produced for large and often diverse audience, the folk forms can use familiar dialects for the most intimate and local communication at the village level.

Limitations, on the other hand, included the dangers in "overloading" certain traditional channels with too many "messages" of an instrumental nature. Channels could only be loaded according, to their capacity and with respect, in particular to cultural, religious, or other sensitively integral elements. Sociologists and anthropologists could help planners define these sensitive factors affecting such limitations.

r) Puppets

One of the old and popular arts in Indian villages is puppetry. Puppetry is an education cum entertaining aid in which puppets manipulated by the performer is a person termed as a characters in a story to be depicted.

Definition

A puppet is a manipulative doll dressed as a character and the performer is a person termed as a puppeteer. A good puppeteer has to blend his art with dramatization to produce the desired effect. It is used as an effective teaching aid for languages and social sciences.

Types of Puppets

1. **String or marionettes puppets:** Marionettes consist of puppets with hinged body parts which are controlled by nine strings produces required movements in the puppet. These puppets are mainly manipulated by professional puppeteers.

2. **Stick puppets:** Stick puppet are the painted cutouts attached by sticks. The actions of these puppets are manipulated by the teacher and students

by hiding behind a screen so that only puppets are visible to the audience or the class.

3. **Shadow puppets**: Shadow puppets are silhouettes of cardboard which produce shadows on white screen. The motion of these silhouettes is manipulated by the teacher and students.
4. **Finger of hand puppet:** Hand puppets are round balls painted as heads with overflowing colorful costumes. These are worn on fingers which operate their movements. These are operated from below the stage.

Selection

In writing or selecting a puppet play, the age, background and tastes of the students should be taken in to consideration. A short puppet play is always preferable.

Advantages

1. Creates interest;
2. Gives the knowledge in a brief period;
3. Puppet is an effective method in teaching;
4. Motivate students;
5. Easy to carry and operate.

Disadvantages

1. Needs group cooperation and coordination;
2. Requires skills in preparation and supply;
3. Skills needed in presentation.

s) Activity Aids

There are certain learning situations in which student participation through direct experiences can be easily incorporated, these are called activity aids. The activity teaching aids are really of great value as they put students in a role of active seekers of knowledge. There are five important activity teaching aids, which are listed below:

a) Field trips
b) Demonstrations
c) Experiments
d) Dramatizations

(i) Field Trips

Definitions

According to Hedger ken Field trip may be defined as "an educational procedure by which the student studies firsthand objects and materials in their natural environment."

Types of Field Trips

Depending on the place of visit and its duration, field trips are mainly of the following four types, namely:

a) Local school trips
b) Community trip
c) Educational trips
d) The natural hunt

Advantages of Field Trip

Field trip provides learning experience in the real life situation by direct contact with objects, process, and systems and thus has many advantages which are enumerated as follows:

1. It provides accurate information objects, process, and systems in their real life setting.
2. It provides meaningful direct experience and hence results in lasting learning.
3. The students learning can be easily diverted towards effective learning.
4. Field trips are valuable aids to what students are curious about the natural and man-made process and objects.
5. Field trips can effectively supplement the classroom learning through application and reviewing the experiences of student.

Limitations of Field Trip

a) A field trip may be occasional activity which at best supplement some learning segments of the syllabus.
b) They can be expensive and out of reach for many disadvantaged and poor students.
c) Field trips require proper and detailed planning to make them meaningful otherwise the trip leads to confusion, and fails to fulfill the requirement

(ii) Demonstrations

Demonstration method is a concrete visual aid, because of its wide use in the teaching of nurses. In nursing education, it is used for this purpose and also for clinics, conferences, laboratory classes, symposia, autopsies, and teaching of health to patients. The demonstration method teaches by explanation and exhibition. In short, it is a performance to show a process or activity to others. When a teacher demonstrates, students observe and imitate to learn.

Farmers are keen on seeing how a new idea works and how it might affect their crop production. Both of these can be done by a demonstration. An appropriate and practical demonstration is an important method in extension, especially among farmers who cannot read. This is because they have an opportunity to watch the differences between the new idea and the old one. The demonstration should be simple and it should illustrate concrete results to the farmers. In extension work, the agents use two different kinds of demonstrations. These are called method demonstration and result demonstration. Both of these require a great deal of thought, planning and competent implementation.

Advantages of Demonstration

The following are the advantages of demonstration method.

a) It activates several senses. This increases learning, because it gives a better opportunity for observational learning.

b) It clarifies the underlying principles by demonstrating the 'why' or 'how' of the procedure.

c) It provokes interest by use of concrete illustrations.

d) It correlates theory with practice engages student's attention and concentration.

e) It encourages student's participation in learning through questions and answers as the teacher performs.

(iii) Experiment

An experiment is a learning activity in which students collect and interpret observations using measuring instruments to reach some conclusions. In science subjects experiments are used invariably used as instructional aid as they encourage learning by doing. While giving a lesson on an experiment, the teacher should organize the instruction so as to make the students aware of the following steps of the experiments:

a) Objectives of the experiments

b) Apparatus required

c) Procedure or methodology

d) Observations of data

e) Computation (totaling) of the observations made.

f) Results or conclusion

g) Precautions

h) Ideas for future work

The student performs the experiment and writes a report on it. Showing the cause and effect relationship.

National Demonstration

A nationwide programme of demonstrations known as National Demonstration on major food crops was launched in 1964-1965 with a view to obtain maximum return per unit area per unit time on the cultivators fields. It was to provide opportunity to scientists to demonstrate validity of their findings on farmer's field. NDP was implemented in selected district through State Agricultural Universities with the help of a team of scientists, training as well as demonstration and other extension activities. The rationale behind the scheme was that unless the scientists could demonstrate what they advocated, their advice might not be heeded by the farmers.It was a nationwide project with a uniform design and pattern. The scientists were expected to conduct different types of demonstrations as given below:

- Multiple cropping demonstrations
- Rain fed cropping demonstration
- Special problem soil demonstration
- Entire farming system demonstration

It differed from other demonstration programme in the following four major aspects:

- There was a specific yield target and there was no separate control plot near the demonstration. The idea behind this principle was that the entire living memory of the farmers about the yield potential of the crop, as well as the entire block in which the demonstration had been laid out, would serve as the control.
- The area of the demonstration plot was about one hectare it could be one acre if bigger plots were not available. So that the feasibility of raising a good crop could be strikingly and unquestionably demonstrated.
- The farmers in whose plots the demonstrations were laid out were the actual cultivators with small holdings, so that the high yields obtaine4d were not attributed to the effects of affluence.
- The agricultural scientists conducted these demonstrations in association with local extension agencies/workers.

Method Demonstration

Method demonstration, basically, show how to do something to the farmers. The farmers are shown step by step how to do something, for example how to sow fodder seeds or operate a specific machine. The extension agent is most likely dealing with farmers, who are already familiar with the practice being illustrated. However, now they want to learn how to do it by themselves. Major advantage of this method is that it allows the agent to communicate with the farmers and give details about simple farming methods to numerous people. In other words, the agent is growing the impact of his extension work. In addition, when the farmers have a chance to try it by themselves, the possibility of the demonstration helping them increases rather than in a situation where the farmers are given a lecture about the same subject. The key disadvantage of the method is if the demonstration is being observed by a large number of farmers, only some of the farmers get to see, hear and do.

Result Demonstration

In result demonstration, the agent shows the farmers that the new idea that is being demonstrated can work in local conditions. The difference is important in result demonstration, whether it is the difference between poor seed and selected seed, or using a fertilizer or not using it. The farmers tend not to believe the agent's words if he only says it to the farmers, that is why the old saying "seeing is believing" applies here. After the farmer sees the results, the agent builds confidence among the farmers and can encourage the farmers to try it by themselves. The major disadvantage of the method is that it takes a long time to mature; hence it is a costly method. If a demonstration fails, due to for example lack of rain, it can have devastating consequences.

Front-Line Demonstration

It is the new concept of field demonstration evolved by the Indian Council of Agricultural Research with the inception of the Technology Mission on Oilseed Crops during mid-eighties. The field demonstrations conducted under the close supervision of scientists of the National Agriculture Research System are called front-line demonstrations because the technologies are demonstrated for the first time by the scientists themselves before being fed into the main extension system of the State Department of Agriculture. The main objective of Front-Line Demonstrations is to demonstrate newly released crop production and protection technologies and its management practices in the farmers' field under different agro-climatic regions and farming situations. While demonstrating the technologies in the farmers' field, the scientist are required to study the factors contributing higher crop production, field constrains of production and thereby generate production data and feedback information. Front-Line Demonstrations are conducted in a block of two or four hectares land in order to have better impact of the demonstrated technologies on the farmers and field level extension functionaries.

Special Features

The Front Line Demonstrations are different than the normal demonstrations conducted by the extension functionaries. The special features of Front Line Demonstrations are:

- ☆ Front Line Demonstrations are conducted under the close supervision of the scientists of the National Agriculture Research System comprising of ICAR Institute, National Research Centers, Project Directorates, Krishi Vigyan Kendras, and the State Agricultural Universities and its regional Research Stations.
- ☆ Only newly released technologies or those likely to be released in near future are selected for the Front Line Demonstrations.

- ☆ Front Line Demonstrations are organized in a block of two to four hectares involving all those farmers whose plots fall in the identified demonstration block.
- ☆ Only critical inputs and training are provided from the scheme budget, remaining inputs are supplied by the farmers themselves.
- ☆ Training of the farmers associated with the Front Line Demonstrations is a prerequisite for conducting such demonstrations.
- ☆ The target audience of the Front Line Demonstration is both farmers and the extension officers. The purpose is to be convince extension functionaries and farmers together about the potentialities of the technologies for further wide scale diffusion; and
- ☆ Front Line Demonstration is used as a source of generating data on factors contributing higher crop yields and constraints of production under various farming situations.

Field Demonstration

Field demonstration is a long term educational activity conducted in a systematic manner in farmers' fields to show worth of a new practice/technology. "Seeing is believing" is the basic philosophy of field demonstrations. Only proven technologies are therefore selected for field demonstrations. Field demonstrations educate farmers through results obtained in terms of varieties resistant to disease and pest, quality of the grains and overall higher yields. In addition, it also educates the farmers in term of input-output ratio and economic gains in terms of net returns. Basically, there are two types of field demonstrations:

(i) Single practice demonstration and

(ii) Composite demonstration.

Single Practice Demonstration

Single Practice Demonstration aims at proving the worth of a single practice such as effect of balance fertilizers in rice crops, higher yields from newly released varieties of hybrid maize, effect of irrigation at crown root initiation stage of wheat, effect of new pesticide on fruit borer in gram *etc.*

Composite Demonstration

Composite Demonstration is a combination of field based result demonstrations and a chain of skill oriented method demonstrations. A long term sequential method demonstrations (Composite Demonstration), aim at demonstrating the superiority of a package of practices in growing a field crops. Here the effect of one practice in harnessing the effect of other practices is also demonstrated and studied. For example, combined effect of irrigation and fertilizer application on grain yield and quality of a newly released mustard variety may be demonstrated as a composite demonstration. Many times, there is an over emphasis on field demonstrations in extension work. Good field demonstrations lead to higher adoption of demonstrated practices by the farmers as they developed the confidence amongst them in the practices demonstrated. Field demonstrations provide an effective learning situation as farmers "See the crops themselves", "interact with the scientists and extension workers on the fields", and "get doubts clarified then and there itself. It is, therefore, essential that whatever demonstrations are conducted, they should be well planned and execute giving no chance to fail. One bad demonstration can vanish the impact of many good demonstrations. Scientists, therefore, have to be very careful in planning and conducting the field demonstrations. Without specific purpose, demonstrations should not be organized. Demonstration should be conducted only when situation demands.

Steps in Conducting Field Demonstrations

Since field demonstration is often used as an extension method, it is sometimes laid out in a routine manner. A well conducted demonstration should help the scientists to give finishing touch to changing attitude of farmers and extension workers and improve their knowledge, understanding and skills. The following steps need to be followed in conducting field demonstration.

A. Planning Phase

1. Know the Vicinity

The scientists need to develop an understanding of the farmers, their farming systems, resources and establish rapport with them. It is essential to gather information on cropping system, present level of use of inputs and productivity of major crops of the area. There are different ways of knowing vicinity. Some are formal and some are informal. A few areas under

a) Visiting villages and farmers.
b) Collection of information using PRA tools.
c) Meeting people individually and in groups.
d) Meeting opinion leaders.
e) Exchanging information with local extension workers; and
f) Consulting office records of population and basic agriculture.

2. Select Technologies

Select only proven technologies which have higher potentialities interms of yield, disease resistance, quality, and can fit in the existing farming systems and situations of the area/farmers. Technology should be frontier ones *i.e.* recently released technologies or such which are at advance stage of release. Be sure that the technology selected for demonstration is much superior to the technology being already in use. At this stage scientists responsible for demonstration must have a lot of consultations with the research scientists who are responsible for released of the technology. They should ask a number of questions from the research scientists and be satisfied about the superiority of the technology.

3. Select Demonstration Site

Avoid isolated farm. Demonstration site should be easily accessible for the farmers and extension workers. As far as possible, block of demonstration site should have a good number of farmers of all categories of land holding and status. Never conduct block demonstration in a single farmer's plot. Pay also attention to farm size, layout of the field, soil type, fertility status, irrigation facilities and drainage system.

4. Select Demonstration Farmers

A group of farmers land holdings in the selected demonstration block and who are willing to cooperate in the conduct of demonstration should be selected. Demonstration farmers should be selected finally by holding a meeting in the village where the purpose of demonstration should be clearly stated and suggestion sought from the farmers. Any difference in opinion may be sorted out tactfully otherwise there is a chance of having non cooperation from those who are not selected as demonstrating farmers which may ultimately jeopardize the very purpose of demonstration.

5. Finalize Package of Practices

This is an important step in planning the field demonstrations. Collect the new technologies from the ICAR Institute/SAU's and ensure these technologies are frontier ones showing substantial increase in yields. Involve as many scientists of the research station in the discussion as possible. This will help in working out minute details of sequences of method demonstration required, identification of important tasks/practices in which presence of scientists should be necessary and critical input for demonstration. Knowledge about farming conditions will be useful at this stage. Involve also demonstrating farmers in finalizing package of practices. This will help in understanding the level of farmer's practices, resource base to sustain the technologies and their perspectives.

6. Prepare for Demonstration

Arrange critical inputs for the demonstration. Critical inputs are those agricultural inputs which are vital to help the selected technologies to exhibit its production potentialities on farmer's field and not earlier being used by the farmers. Arrange such inputs *viz.* seeds, fertilizers, farm equipments and other inputs in time. Only critical inputs need to be supplied by the scientists. Other inputs should be arranged by the farmers themselves. Ensure that the inputs which are to be given by the farmers are available with them. The farmers should never be given an impression that the demonstration is a means of receiving free inputs. Rather they should be educated to understand the educational value of such demonstration.

B. Conducting Phase

1. Layout of Demonstration

Guide and assist the farmers in laying out the field. Special training programme may be arranged for all farmers in whose plots demonstrations are to be laid. Keep the control plot if needed; otherwise treat all other neighboring/surrounding plots as control plots. In case of block demonstration, one acre plot as a control is adequate. Sometimes a control plot is not necessary as the "entire memory of the farmer" is taken as a "control" or the neighboring plots would serve the purpose of control plots. Put a publicity board on the fringe of the demonstration plot. Mention the details of the demonstration on the board.

2. *Crucial Farm Operations*

Ensure your presence at the time of important operations like, seeding, fertilizer application, weeding, irrigation, plant protection measures, harvesting, threshing and weighing of produce. Two things are important at this stage: (i) using demonstration for farmers' training, and (ii) record keeping in demonstration. Each operation should be used as input of training of farmers. Encourage questions from the farmers at each of these operations. This will help in better understanding of the task/operations.

3. *Field Day*

Arrange a field day to project the new technologies demonstrated in front of a large manageable group of interested farmers. It is an intensive educational activity in which farm experts, extension workers and farmers are involved and learn from each other. Plan the field day when the crop is fully matured yet green. Ask the demonstration farmers to explain the story of demonstration one by one to the assembled group of farmers and extension workers. Arrange a few method demonstrations on operation of farm machines and equipments, operation of seed drill, seed treatment, fertilizer application, plant protection, *etc.* Farmer-scientist-extension workers discussion should be an important feature of the Field Day.

4. *Harvesting*

Make an eye estimate of the field. Arrange harvesting in the presence of identified groups of farmers. Ask the farmers to estimate the yield and to say in what way the demonstrated technologies are superior to the earlier ones. Are they satisfied with the performance of the technologies? What lessons they have learned from the demonstration? Will they advice other fellow farmers to adopt this practice? Will they exchange the seed materials of new variety with other fellow farmers? What are the expected profits? Will it be more than what they used to get from their own practices? What are the difficulties in following the demonstrated practices? Idea is to ascertain as to what extent farmers are satisfied with the demonstrated technology and what is the possibility of their continued adoption.

C. Follow up Phase

Some farmers may revert to old practices in the absence of follow-up. They need information reinforcement, timely supply of inputs or on the spot guidance. Group approach in follow-up will give better results. It is better to link your follow-up programmes with the local institutions like Farmers Club, Farmers Cooperative Society, Village Panchayat *etc.*

D. Record Keeping

There are two types of records which one should maintain for each block demonstration.

1. Information Card

This card contains basic information about the demonstration site *viz.* previous crops and varieties grown, fertility status of the plots, present productivity of crops, size of holdings of each farmers in the demonstration block, extent of use of inputs *etc.* It should also contain detail information of demonstration like size of block, variety of crop, seed rate, sowing date, inputs applied, irrigation schedule followed, intercultural operation performed, plant protection measures taken, date of maturity, date of harvesting, incidence of disease and pests, average numbers of tillers, yields of crop *etc.* The card should remain with the demonstration farmers and is filled up by them or by an educated person in their family or by the KVK staff who visit the demonstration site from time to time.

2. Technical Report

The technical report should contain information on soil analysis, variety of crop, germination, plant population, pest and diseases, irrigation, fertilizer application, harvesting, final yield, extension activities undertaken *etc.* It should also contain information on cost-benefit ratio of the demonstration. This will help to work out the economic returns. A favorable cost-benefit ratio will fully convince the extension officers and the farmers about the profitability of the technologies demonstrated. Submission of technical reports to Zonal Coordinator/Council in time is as important as concluding good field demonstrations. Late submission of reports may constrain the Zonal coordinator/Council not to include them in the Zonal report/national report. Remember, demonstration report helps in planning agricultural development and farm research programme of your area, state and country. Role of the scientists in writing of feature story is very important. Such story becomes the important learning materials in the Krishi Vigyan Kendras.

(iv) Dramatization

Dramatization is a very potent method of keeping the class room instruction lively and interesting. When a teacher dramatizes a lesson, the students become both the spectators and participants. This makes learning easy and permanents.

Types of dramatizations suitable for class room instruction:

a) Role-play
b) Lets Play
c) Pageant
d) Pantomime

Role Playing

Role-playing refers to the changing of one's behaviour to assume a role, either unconsciously to fill a social role, or consciously to act out an adopted role. While the Oxford English Dictionary offers a definition of role-playing as "the changing of one's behaviour to fulfil a social role", in the field of psychology, the term is used more loosely in four senses.

Let's Play

Play Let is a series of screenshots or a recorded video documenting a play through of a video game, usually including commentary by the gamer. A Lets Play differs from a walkthrough or strategy guide by focusing on an individual's subjective experience with the game, often with humorous, irreverent, or even critical commentary from the gamer, rather than being an objective source of information on how to progress through the game.

Pantomime

Pantomime (informally Panto), not to be confused with the theatrical medium of mime, is a type of musical comedy stage production, designed for family entertainment. It was developed in the United Kingdom and is generally performed during the Christmas and New Year season. Modern pantomime includes songs, slapstick comedy and dancing, employs gender-crossing actors, and combines topical humour with a story loosely based on a well-known fairy tale. It is a participatory form of theatre, in which the audience is expected to sing along with certain parts of the music and shout out phrases to the performers. An important part of the pantomime, until the late 19th century, was the harlequinade. The pantomime is performed today throughout Britain and, to a lesser extent, in other English-speaking countries.

- ✰ To refer to the playing of roles generally such as in a theatre, or educational setting;
- ✰ To refer to taking a role of an existing character or person and acting it out with a partner taking someone else's role, often involving different genres of practice;
- ✰ To refer to a wide range of games including role-playing video game, play-by-mail games and more;
- ✰ To refer specifically to role-playing games.

Advantages of Dramatization

a) Dramatization gives an added advantage of students working as both observers (spectators) and doers (participants) unlike in experiment where there are just doers and in demonstration where there are just observers;

b) Dramatization makes learning a pleasure children love to act and show off;

c) Dramatization involves students totally and they appreciate the lessons remember it better;

d) Dramatization develops the social skills required for them such as cooperation, co-ordination, punctuality, and human relations *etc.*;

e) Dramatization makes students creative, sensitive, and alert.

The Disadvantages of Audio Visual Aids in Teaching

Audiovisual aids reinforce the information being given in a presentation.

Audio-visual aids can be a great tool to use in classrooms. Often they can spice up otherwise dull topics, keep students more engaged in their lesson and enliven their imaginations. Video presentations, slide shows, PowerPoint and other media have revolutionized the way teachers reach their students. But these aids are not without their disadvantages. It is important to weigh both the pros and the cons when considering how you will use audiovisual aids in your classroom.

Technical Problems

Regardless of your equipment, there are many things that could go wrong that may derail your presentation. A bulb might burn out on your projector or it might be blurry or hard to focus. When using a PowerPoint presentation, font and colors may show up differently on screen or the music and sound might not play. CDs can skip. DVDs may be scratched or not compatible with your player.

Student Distractions

Visual aids are more of a distraction if used throughout the entire presentation versus during key points. Special effects and light from the projector can draw attention away from the speaker. Too much information can also be distracting, because the students will be trying to read and/or take notes while the speaker has moved on to the next subject. There also may be too many slides for the students to interpret or the audio and visual may not match up. An entertaining video can take the focus off the speaker and the presentation, while a dry video and low lighting can put students to sleep.

Expenses

Professionally put together audiovisual aids can be expensive. Slides, videos, and films can cost money unless borrowed from a library, and the necessary equipment can be cost-prohibitive. Due to budget constraints, some schools may not have enough audiovisual equipment for every classroom, and availability could be limited.

Time

Audiovisual presentations can take a considerable amount of time to prepare. It takes time to produce your own video or slides. Films may be difficult to obtain, or it can take time to get permission to use them. Posters and transparencies may require extensive preparation. Creating a PowerPoint presentation can also be time consuming. Time spent in preparing these materials could take away from the time it takes to prepare a clear, well-organized lesson plan.

Space

If you choose to use an audiovisual aid, the size of the room should be taken into consideration. It is critical that all students are able to see or hear your presentation. If the room is too large for everyone to see the visual aid, or if part of your audience is forced to view the presentation at odd angles, some students will struggle to keep up with your lesson.

Convenience

Certain audiovisual aids such as VCRs for older video recordings can be bulky and difficult to transport.

Part VIII

Research, Teaching and Extension

39

History of Agricultural Research in India

Agriculture, as the backbone of Indian economy, plays the most crucial role in the socio-economic sphere of the country. Indian agriculture is a diverse and extensive sec-tor involving a large number of stakeholders. It has been one of the remarkable success stories of the post-independence era through the association of Green Revolution technologies. The Green Revolution contributed to the Indian economy by providing food self-sufficiency and improved rural welfare. Moreover, since independence, India has observed considerable increase in the production of oilseeds (through the yellow revolution), milk (through the white revolution), fish (through the blue revolution), and fruits and vegetable (through the golden revolution). The role of the National Agricultural Research System (the NARS) was imperative in the con-text of all these momentous and successful upheavals. The main events in the history of agricultural research in India can be grouped into the following seven categories:

a) Establishment of agriculture departments and agriculture colleges.

b) Establishment of the Imperial Council of Agricultural Research.

c) Initiation of commodity committees.

d) Project for intensification of regional research on cotton, oilseeds and millets.

e) Initiation of all-India coordinated crop improvement projects.

f) Reorganization of the Indian Council of Agricultural Research (ICAR).

g) The development of agricultural universities.

Among these, the first three signify the development of agriculture in the colonial India, whereas the next four were prominent in the post-colonial era. India has one of the largest and institutionally most complex agricultural research systems in the world. Historically, the Indian agricultural research system is the contribution of a process which started in the 19th century and which resulted in the establishment of the Imperial (now Indian) Council of Agricultural Research on the recommendation of a Royal Commission on Agriculture in 1929. Since then, there has been an evolution of agricultural research in India.

Agricultural Research in Colonial India

The early development of agricultural research in India was associated with the occurrence of famines. The recur-rence of famines and starvation death of peasants in the second half of the 19th century opened the eyes of the imperial rulers. During this period 24 famines, big and small, took an estimated toll of more than 28 million souls. However, at that time, little precedence was accorded to agricultural research and development in the country.

Genesis and Development of Agricultural Science Universities in India

Introduction

Chung emphasizes on Agricultural Education. According to him,".ingredient for agricultural development is educational institutions, including provision of training, extension and research. It may be said that the relative standard of education among different people, farmers in particular is a key factor in increasing agricultural productivity and in achieving higher standards of living. Accordingly, universal education for all people of all ages is essential for attainment of these ends".

India ranks as one of the top 12 industrial countries in the world in terms of output, but it remains a predominantly agricultural country with 70 per cent of the population dependent on, and 35 per cent of its annual value of production from, agriculture. Agriculture provides a number of industries with raw materials and is important for the industrial development of the country. Growth and development of agricultural economy depend upon a number of factors. Some of the environmental factors are climate, land fertility, nature of the soils, and extent of rainfall. Besides natural environmental factors, the development of agriculture depends upon a number of institutional factors like irrigation facilities, availability of transport, quality of seeds, fertilizers, crop diseases, pesticides and so on.

To get optimum productivity from agriculture, farmers must have knowledge about these factors. For instance, it is essential to know about the different types of soils and the crops that will be cultivated on these soils, various pests and their control, weather forecasting to cultivate the land and so on. In India majority of the farmers were illiterate and following traditional (conventional) farming.

In traditional farming method, productivity of land and labourers are found lower than that of modem farming methods. Technique so f production in the agriculture sectors was quite simple and primitive. Wooden ploughs, simple implement sand bullocks were used for cultivation. The rain was the main source of water. The energy used in the sector was low and in the form of animal power. Manure was used as fertilizers. Storage Method of agricultural products was defective. Consequently, productivity of the land is also low.

Hence the need was arisen to improve and develop modern methods in agriculture. In this way, Research and developmental activities become necessary in Agriculture and related sciences. Further, education of the farmers felt necessary so as to adopt modern methods and techniques in Agriculture. It was only through the education; the output of the research will reach the farmers and indirectly improve the agricultural methods and techniques. To provide education and training for the farmers and undertaking research in agriculture and related sciences, the various Agricultural Science Universities, Colleges and Research Institutes were established in India.

Records reveal that agricultural education in India began with the Taksasila University. The University had a curriculum which included eighteen arts, that includes Medicine, Surgery, Astronomy, Agriculture, Accountancy, Archery, Astrology and Snake Charming. This University flourished as an educational centre till the fifth century A.D., while Nalanda University was destroyed towards the end of the twelfth century.

Later in the medieval period, College and Universities established by Mohammeden rulers were not provided agricultural education. The curriculum of these institutes included Arabic and Persian Literature, Grammer, Rhetoric, Logic, Law, Geometry, Astronomy, Natural Philosophy, Metaphysics and Theology. Under British rule, several schools and colleges were founded but they are concentrated on education on law, literature and religion. At the end of 1856, the Government of India approved the general plan for the establishment of three Universities sat Calcutta (Kolkata) Madras (Chennai) and Bombay (Mumbai) from 1857. But they are aimed to cover various subject courses. The application of chemistry to soils in 1840 and the establishment of the Roth Amsted Research Station in 1843, followed by the opening of the first agricultural college in England at Cirencester in 1845 are known land marks of development of agricultural science and education in the Report of the Royal Commission on Agriculture in India. The first proposal for a Special Department of Agriculture in India originated from the Commission in Bengal and Orissa in 1866 and later in Uttar Pradesh. Reviewing the conditions prevailing in India during 1920s, the Royal Commission on Agriculture in India concluded in 1928 that" however efficient an organization might be built up for demonstration and propaganda, unless it was based on the solid foundations provided by research, it was merely a house built on sand. A direct outcome of the work of the Royal Commission of Agriculture was the establishment of the Imperial Council of Agricultural Research in 1929 (Now known as Indian Council of Agricultural Research). In 1936 and 1937, Sir John Russell, Director of Rothamsted Experimental Station, visited India to review the condition of Indian agriculture and here ported that" in general. the men who actually till the soil are scarcely touched by the national programme of agricultural education".

By 1947, the development of education and research facilities in agriculture and related fields advanced rapidly as India had 17 college so far agriculture with an annual enrolment of about 1500 students. Agricultural research was conducted largely in the fields where industry directly infringed on agriculture, such as in cotton through the activities of the Indian Central Cotton Committee and in jute, sugar, coffee, tea, rubber and tobacco. Particularly in sugar, the progress was spectacular and no less than 80 per cent of the sugarcane grown in India by 1947 was çlaimed to be under improved varieties, for which sugarcane research could claim credit.

By 1947, the major institutions functioning in agricultural science education and research are Indian (formerly Imperial) Council of Agricultural Research, Indian Institute of Fruit Technology, Lyalpur (1945), the Central Agricultural Marketing Department (1934), Indian Veterinary Research Institute (1890), the Indian Dairy Research Institute (1923), the Central Rice Research Institute (1946) and various research centres like the Pusa Institute and the Sugarcane Breeding Institute, Coimbatore.

Among the colleges of Agriculture functioning by 1947, in Coimbatore (1868), near Madras (1878), College of Agriculture, Pune (1879) formed as a branch of College of Science. Later the Allahabad Agricultural Institute was founded in 1910 and was sponsored co-operatively by a number of churches and missions and currently prepares students in agriculture, agricultural engineering and dairying. Then a few colleges included courses in agronomy and biochemistry.

After the independence, the importance attached by the new nation to educationist reflected in the Government of India resolution of 4th November 1948, setting up of the Indian University Education Commission under the Chairmanship of Dr. S. Radhakrishnan. Among the major recommendations, the Commission emphasized Professional Education in Agriculture. 'The study of Agriculture in primary, secondary and higher education be given high priority in national economic planning. So far as is feasible, agricultural education is given in a rural setting.

40

Establishment of the Imperial Council of Agricultural Research (The Present-Day ICAR)

The Royal Commission on Agriculture (Linlithgow Com-mission), which was appointed in the year 1926, authoritatively reviewed the position of agriculture in India and reported the same in 1928. Having duly recognized the work done by the agriculture departments, the Commission stressed upon the enormous possibilities for future work and made comprehensive recommendations regarding the different problems of agriculture. The Commission proposed that an Imperial Council of Agricultural Research should be set up to endorse, direct and organize agricultural, including veterinary, research all over India. The Council was supposed to guide the research activities of central and provincial departments of agriculture. According to the proposal of the Royal Commission on Agriculture, the GoI, Department of Education, Health and Lands set-up the Imperial Council of Agricultural Research (now ICAR) on 16 July 1929.

The developments during the half century (1900-1949) in Agricultural Science Education led to the establishment of 21 institutions for higher education in agriculture. The number of enrolments to 17 agricultural colleges rose steadily from 30 in 1924 to 1448 in 1948. In a resolution 0f 24th November 1954, the Indian Ministry of Food and Agriculture as per the advice of Dr Frank W Parker, set up a First Joint Indo American Study Team to provide some useful guide lines for strengthening agricultural colleges in India and for promoting Centre-State co-operation and co ordination in the field of agricultural research. Accordingly, that resolution contained the decision of the Government of India to constitute a joint team of Indian and American specialists to make a comparative study of the institutions in the USA and India and make the recommendations. The team was headed by Mr. K. R. Damle, the then Vice-President of ICAR. There port submitted by the team in September 1955, contained 118 recommendations, which together laid the true foundation for all subsequent developments in India leading to the establishment of Agricultural Universities and enhancing the value of research work in agricultural sciences in India.

The Second Joint Indo-American Study Team on Agricultural Education, Research and Extension was constituted following a resolution dated 12th September 1959, under the chairmanship of Dr. M. S. Randhawa, the then Vice-President of ICAR. This team reviewed the progress of the work done in the preceding five years and made supplementary recommendations designed to strengthen agricultural education, research and extension. The summary of the team's recommendations submitted on 11th July included 67 items. On Agricultural University the team Recorded that there was wide spread demand from many states for the establishment of agricultural universities and without any awareness of the needs of such university. The team there for recommended that assistance to establish an agricultural university should not be granted unless there is adherence to basic principles such as:

a) Autonomous status
b) Location of Agricultural, Veterinary/Animal Husbandry, Home Science, Technological and Science Colleges on the same campus
c) Integration of teaching by offering courses in any of these institutions to provide a composite course and
d) Integration of Education, Research and Extension. Two more Committees that is High level Committee and Agricultural Research Review Team were constituted in 1963, but they were not contributed much towards the agricultural education and research.

By the beginning of 1961, as a sequel to the recommendation soft he First and Second Joint Indo-American Study Teams, actively stimulated by the sustained efforts of many American Advisors, interest in establishing Agricultural Universities grew to become a movement in various states. The Government of India, therefore felt the need to appoint a Committee to examine the proposals received from the State Governments to establish Agricultural Universities from the point of view of prerequisites for the integration of teaching, research and extension and the evolution of a workable relationship between existing institutions and departments. This Committee was headed by Dr. Ralph. W. Cummings, Field Director, the Rockfeller Foundation, who was to play for many years a dominant role in shaping the course development of many agricultural universities in India.

After discussing with the Planning Commission, the Government of India decided, however, that during the Second Five Year Plan only one Agricultural

University be established (in Uttar Pradesh) as an experimental measure, but in view of the wide spread demand from many states for such universities, the Government of India accepted in 1961 the need for setting up a few more such universities during the Third Plan period.

In 1966, seven institutions were attempting to establish new Agricultural Universities in various states. The Government of India set up another body known as the Education Commission through are solution dated 14th July 1964, which was headed by Dr. D. S. Kothari, Chairman of the University Grants Commission. The Commission covered the entire field of education in India and devote done chapter of 20 pages to 'Education for Agriculture'. Specifically, the Commission supported the establishment of at least one agricultural university in each state. The Commission desired that these universities should have the following features:

1. Their concern with all aspects of increasing, disseminating and supplying knowledge related to agriculture, including basic and applied research;
2. Their primary emphasis on teaching and research directly and immediately related to the solution of the social and economic problems of the country side;
3. Their readiness to develop and teach the wide range of applied sciences and technologies needed to build up the rural economy;
4. Their readiness, not only to teach undergraduates, post-graduates and research students but also to give specialized technical training to young people who are not candidates for degrees;and
5. Their emphasis on adult and continuing education side by side with teaching regularly enrolled students.

The Commodity Committee

Several semi-autonomous Central Commodity Commit-tees were set up by the Ministry of Food and Agriculture. These were concerned with research and development activities related to specific crops. These committees were semi-autonomous bodies in the sense that they were financed partly by the Government and partly by the taxes collected on the export of the concerned commodities. The Indian Central Cotton Committee was the first one to be established in 1921 on the recommendation of the Indian Cotton Committee (1917–18). The chief function of the Central Cotton Committee was cotton improvement with special focus on the development of improved methods of growing and marketing cotton. The Committee's support led to the development of 70 im-proved varieties and considerably improved fibre quality. Most of the commodity committees related to specific crops had their own research stations or institutes which were located in the regions where the concerned crop was most widely grown. However, some other commodity committees financed research schemes conducted by the State Departments of Agriculture, *e.g.* Spices and Cashew nut Committee. The achievement by the Indian Central Cotton Committee led to the setting up of commodity committees on crops such as lac, jute, sugarcane, tobacco, coconut, oilseeds, spices, cashew nut and arecanut. The Vice-President of ICAR presided over all the commodity committees. However, there was a great necessity to conduct researches on various crops within the different agro climatic regions of the country as in the context of India, the soil and climate fluctuate to a great extent from one region to another. These apprehensions led to the formulation of the Project for Intensification of Regional Research on Cotton, Oilseeds and Millets (PIRRCOM), which was the first step in the country to-wards coordinated approach to agricultural research. The Central Commodity Committees were abolished in 1965 and the research institutes under their control were transferred to ICAR.

The National Academy of Agricultural Research Management is yet another exclusive institution under ICAR to conduct research and training in agricultural research management.

Other organizations involved in agricultural research are:

- About 27 traditional universities are carrying out agricultural research.
- Scientific organizations such as the Council of Scientific and Industrial Research (CSIR), and Bhabha Atomic Research Centre (BARC).
- Government departments such as the Department of Science and Technology and the Department of Bio-technology.
- Private and voluntary organizations.
- Scientific societies and institutions like the National Dairy Development Board which has been instrumental in transforming the dairy industry for rural development.

Department of Agricultural Research and Education setup in 1973, in Ministry of Agriculture, is responsible for coordinating research and educational activities in agriculture, animal husbandry and fisheries. Besides it helps to bring about inter-departmental and inter-institutional collaboration with national and international agencies engaged in the same and allied fields. The Department provides Government support, service and linkage to Indian Council of Agricultural Research

Plantation Research in Colonial India

Though the experimental farms were established in 1884, the provincial agricultural departments could seldom go beyond the collection of revenue data and famine relief operations. For conducting experiments, the Government looked more to agricultural societies than to its own agencies. The Agricultural and Horticultural Society of India (AHSI) was established with specific objectives such as:

- To obtain precise and trustworthy details as to the cost of cultivation and produce per acre of fibre-bearing plants of promising character, so that the Agriculture Department may be able to form a decisive conclusion as to the prospects of a profitable exploitation of the plants in question.
- Evaluate the machines and processes for the extraction of fibres.

The official experimental farms were obsessed with cot-ton. Mounting pressure from British cotton tycoons had forced the GoI to initiate a vigorous cotton improvement programme. However, earlier projects of the 1840s and 1860s were not successful mainly because of insufficient botanical knowledge or the necessary market research. Later, in 1890, the association of an expert botanist in cotton experiments was specifically called for. The cultivation and marketing of existing varieties gave a relatively stable and acceptable return to money-lenders and dealers. New and untested varieties involved different methods of cultivations and great labour input, without a higher level of output or profit, and with the risk of severe losses to each of these classes. The last quarter of the 19th century saw the closure of several experimental farms. But private farms patronized by cotton mills proved remunerative. For example, the Government's Mungeli farm at Bilaspur had to be abandoned, whereas the nearby Kyragarh and Nandgaon cotton farms owned by Bengal, Nagpur Cotton Mills Company produced 600,000 lb of cotton. The Government's fascination towards imported technologies and ideas proved not to be advantageous. Instead, the dependence on traditional knowledge proved to be beneficial for the mills. Sericulture remained a purely private concern for a long period of time. As for example, at the turn of the century Ms Tata and Sons successfully started a silk firm in Bangalore for the introduction of Japanese methods of sericulture.

Tea and indigo were the major items of export. In the 1870s, A. W. Blyth established a process by which it could be easily known whether the merest fragment of a plant belonged to the 'theine' class or not. However, no experiment was conducted to control the scourge of blight. Although the AHSI made an attempt in this direction, it failed because of lack of funds and skilled entomologists. In 1891, M. K. K. Bamber, a chemist appointed by the joint committee of the Indian Tea Association and the AHSI, performed analysis of the soil and tea leaves, gave advice on manures, drainage, insecticides, *etc.* and brought out a book called *The Chemistry and Agriculture of Tea*. The establishment of a permanent tea research station at Toclai, Assam, highlights the progress of tea research in colonial India. Almost the same thing happened in the case of indigo. Indian indigo remained unrivalled till the end of the 19th century when Germany perfected its synthetic counterpart. In early 1899, the GoI deputed its agricultural chemist J. W. Leather to tour the indigo districts of north Bihar. An indigo experimental farm was started at Dalsinghsarai in July the same year. The interests of the indigo lobby were at work behind the creation of the Pusa Agricultural Research Institute, Bihar. Pusa was selected as the site because of its proximity to the plantations of north Bihar.

Agricultural Research in Post-Colonial India

Since independence, there has been a substantial growth in the Indian agricultural research system. ICAR is the chief public body at the national level for directing and endorsing agricultural research and education in the country. Likewise, State Agricultural Universities are responsible for doing the same at the state level. After independence, agricultural research was given much emphasis which in turn led to increased agricultural production and near self-sufficiency in food grains in the country.

The greater emphasis given to agricultural research could be established by the fact that the system guided by ICAR now has 64 ICAR institutes, 15 national research centres, 6 bureaus, 4 Deemed University, 13 Project Directorates, 59 All-India Coordinated Research Projects (AICRPs) and the 20 All India Network Projects (AINPs) and 10 other projects, 690 Krishi Vigyan Kendras (KVK), 64 State Agricultural Universities, 3 Central Agricultural University, 4 Central Universities having Faculty of Agriculture.

Integrated Functioning of Teaching, Research and Extension in ICAR and SAUs Systems

Functioning of ICAR

The Indian Council of Agricultural Research (ICAR) is an autonomous apex body responsible for the organization and management of research and education in all discipline of agricultural and allied sciences. Agriculture must continually move forward for which adoption of newer methods of farming become imperative. The ICAR has a division of Agricultural Extension that helps in the assessment and refinement of technologies in all discipline of research conducted in the system. The specific extension roles of ICAR has been given below:

- Technology assessment and refinement for identifying the location specificity of technology under various farming systems.
- Vocational training of farmers.
- Training of extension personnel to update their knowledge in frontier areas of technology.
- Providing single window delivery system of technology products, services and information through Agricultural Technology Information Centres (ATIC).
- Development of gender specific technologies.

Achievements of ICAR

- Large network of front-line ICAR-SAU extension system consisting of 690 Krishi Vigyan Kendras (KVKs), 10 Trainers' Training Centres (TTC), 42 IVLP centres and 44 ATICs.
- Training of 4.33 lakh farmers and farm women every year in agriculture and allied fields such as crop production, plant protection, livestock production and management, soil and water management, farm machinery and tools, and home science.
- Vocational training covering 66,000 rural youths in the areas of poultry, dairying, piggery, beekeeping, fisheries, fruit and vegetable preservation, maintenance and repairing of farm machinery and tools, and hybrid seed production.
- In-service training programme to upgrade the knowledge and skills in transfer of technologies in agriculture and allied areas for 40,000 extension functionaries in KVKs and TTCs.
- Front-line Demonstration (FLD) on oilseed and pulse crops benefiting 23,000 farmers. FLDs are conducted to demonstrate the production

potential of newly released production technologies at the farmers' fields in a given farming situation.

- Training of trainers in frontier areas of technology such as dryland agriculture, animal production, horticulture, freshwater acquaculture, marine fisheries, hill agriculture, agriculture engineering and women in agriculture in TTCs.
- Multi locational advance varietal of 17 selected crops in 60 contres under irrigated and rainfed conditions under different cropping sequences in a joint collaborative Technology Evaluation and Impact Assessment Project with Crop Science Division.
- Technology Assessment and Refinement through Institute-Village Linkage Programme (IVLP) in 42 centres located at the ICAR research institutes and SAUs.
- Identification, assessment and development of gender-specific technologies at the National Research Centre for Women in Agriculture, Bhubaneshwar (Orissa). Training of farm women in farm implements and tools at the sub-centre of NRCWA at the Central Institute of Agricultural Engineering, Bhopal (Madhya Pradesh).
- Establishment of 44 Agricultural Technology Information Centres (ATICs) in ICAR Institutes and SAUs. These centres will provide a 'Single Window' delivery system for technology products, services and information available in the institutions to the farmers.
- Networking of 108 Krishi Vigyan Kendras under Agricultural Research Information System (ARIS).

ICAR acts as a repository of information and provides consultancy on agriculture, horticulture, resource management, animal sciences, agricultural engineering, fisheries, agricultural extension, agricultural education, home science and agricultural communication. It has the mandates to co-ordinate agricultural research and development programmes and develop linkages at national and international level with related organisations to enhance the quality of life of the farming community.

ICAR has established various research centres in order to meet the agricultural research and education needs of the country. It is actively pursuing human resource development in the field of agricultural sciences by setting up numerous agricultural universities spanning the entire country. The Technology Intervention Programmes also form an integral part of ICAR's agenda which establishes Krishi Vigyan Kendras (KVKs) responsible for training, research and demonstration of improved technologies.

Research Centres

The Central Research Institutes have been established to meet agricultural research and education needs of the country in terms of pursuit of basic and strategic research in the concerned disciplines in focused areas. They develop, evaluate and refine technologies for packaging them in a form ready to be transferred from Lab to land. They also help in human resource development for agricultural sector. The Institutes/Directorates/Bureau have a mandate for working on single or selected group of crops, animals or commodities with each major discipline having status of the division/section. Some of the Institutes have regional stations to cover diverse agro-ecological areas for developing area-specific technologies.

The National Research Centres (NRCs) have been established for concentrated attention with a mission approach by a team of scientists under a single leader on selected topics with relevance to resolving national problems in a particular animal, crop or commodity. There is no divisional set up and rarely and regional station for the NRCs.

The creation of All India Co-ordinated Research Project (AICRP) under the ICAR system is a landmark in the history of agricultural research in India. The AICRP is a mechanism in building nation-wide co-operative, inter-disciplinary research network linking ICAR Institutes with the SAUs to focus attention on commodities and species of national importance. The AICRPs have succeeded in mobilizing country's scarce resources through inter-institutional and inter-disciplinary interaction and joint evaluation of new technologies to arrive at collective recommendation. The AICRPs also strengthen research base in each agricultural university. A few projects are elevated to the level of Project Directorates with additional research responsibilities for important areas *e.g.* wheat, rice, oilseed, pulses and vegetables.

In addition, the ICAR supports a number of short-term ad-hoc research schemes, implemented by scientists in various colleges, universities and institutes. The programmes of ad-hoc research schemes are need-based and are formulated and executed in accordance with the recommendations of the scientific panels related to different disciplines.

Human Resource Development

A total of 64 Agricultural Universities have been established in the different parts of the country, including a Central Agricultural University for the north-eastern region. In addition the Indian Agricultural Research Institute, New Delhi, the Indian Veterinary Research Institute, Izatnagar (Bareilly), the National Dairy Research Institute, Karnal and the Central Institute of Fisheries Education, Mumbai have the status of deemed university.

The agricultural education system in the country offers degree programmes in 11 specific disciplines *viz.* agriculture, veterinary science, agricultural engineering, forestry, home science, dairy technology, fisheries, sericulture, marketing, banking and co-operation, horticulture and food science with a total intake of about 11,000 students. It also offers post graduate programmes in more than 55 fields of specialization with a total intake capacity of about 5,000 students. Under the human resource development programme the council offers about 1200 scholarships and fellowships from the undergraduate to post-doctoral levels. Special fellowships are also offered for socially and economically weaker groups.

A World Bank assisted project on the Agricultural Human Resource Development (AHRD) has been launched to improve he educational system in agriculture and meet its future challenges. In its initial stage the participating states are Andhra Pradesh, Haryana and Tamil Nadu involving the agricultural universities of these states. During the 5 years of the project period steps will be taken for establishment of accreditation board, syllabus reform, faculty quality improvement, upgrading and modernization of the laboratory facilities, farms, libraries, hostels, communication systems and other ancillary facilities. A National Agricultural Technology Project (NATP) is also contemplated to bridge critical gaps of technology generation, assessment, refinement and transfer and to enhance ICAR's institutional capability to meet future challenges on research and development fronts.

Technology Intervention Programmes

The Technology Intervention system of the ICAR is devoted to the first-line extension activities. The KVKs are responsible for vocational training, on-farm research and demonstration of the improved technologies. In majority of the cases, the technologies developed were found appropriate only in the area where they were generated. Even slight variations in conditions turn the technology irrelevant. Keeping this in view a new approach is being experimented for technology assessment and refinement through Institute-Village Linkage Programme (IVLP). The IVLP has been started initially at 42 centres through various ICAR institutes and State Agricultural Universities.

In addition to the new initiative of the IVLP, the technology intervention programmes are going on in eleven zones throughout the country. Zone I has 36 KVKs and one TTC serving the farming communities. The Zone II is constituted to serve the farming communities with 26 KVKs. There are 12 KVKs and one TTC in the Zone III. The Zone IV has 29 KVKs. Andhra Pradesh and Maharashtra are being served by 26 KVKs and one TTC of the Zone V. The State of Gujarat, Rajasthan and Dadar Nagar-Haveli have been included in Zone VI functioning through 42 KVKs. Zone VII consists 3 KVKs and two TTC. The states of Karnataka, Kerala, Tamil Nadu, Pondicherry and Goa are being served by 22 KVKs and 3 Trainers Training Centres in Zone VIII. The farmers' fairs, farmers' days, Kisan Goshthies (farmers' meetings) and farmers' visits were organised in all the zones.

41

Agricultural Universities

Agricultural Universities or 'AUs' are mostly public universities in India that are engaged in teaching, research and extension in agriculture and related disciplines. In India, agricultural education has evolved into a large and distinct domain, often separately from other areas of higher education. Many of these universities are member of a registered society, the Indian Agricultural Universities Association. Indian Council of Agricultural Research is the main regulatory authority of agricultural education in India, while the disciplines of veterinary medicine and forestry are regulated by the Veterinary Council of India and Indian Council of Forestry Research and Education respectively. Based on the statutes establishing these universities, agricultural universities in India can be grouped into various types.

State Agricultural Universities

State Agricultural Universities (SAUs) are the predominant class of agricultural universities in India. A SAU is usually a university established by an act of state legislature with a dedicated mandate of teaching, research and extension in agriculture and related disciplines.

History of SAUs

After Independence, one of the greatest challenge before India was agriculture and rural development. This necessitated the availability of trained human resources in these areas. The first Education Commission of India (1949), headed by Sarvepalli Radhakrishnan, recommended setting up rural universities in India on the American land-grant model. In the 1950s, Indian Parliament was already setting up specialized engineering universities - Indian Institutes of Technology - as Institutes of National Importance. However, the Parliament had limited mandate in establishing such specialized institutions for agricultural education, despite the urgent need, as the Constitution of India had conferred the power to make laws in subjects of agriculture (including agricultural research and education) exclusively in the domain of the states.

The state of Uttar Pradesh (UP) took the first step for establishing an agricultural university in 1954, when it invited an Indo American team headed by Dr K R Damle, the Vice-President of Indian Council of Agricultural Research (ICAR), to consider an area around Tarai State farm in Nainital district as a possible site for a rural university. Encouraged by the favourable view of the Damle team, senior UP government officials H S Sandhu and A N Jha visited the United States to look for collaborations with US universities. In consultation with University of Illinois dean Dr H W Hannah, the state of UP presented a proposal to the central government in 1956 for establishing a land-grant style university. Thereafter, a contract was signed between the Government of India, the Technical Cooperation Mission and a few US land-grant universities, to promote agricultural education in India. The US universities included the University of Tennessee, Ohio State University, Kansas State University, the University of Illinois, Pennsylvania State University and the University of Missouri.

The UP Act XI-V of 1958 - the founding legislative act for establishing an agricultural university was promulgated by state legislature of UP. The task of mentoring the proposed university in UP was assigned to the University of Illinois, which signed a contract in 1959. Thus, the first state agricultural university of India, G. B. Pant University of Agriculture and Technology, Pantnagar, was established in UP. It was inaugurated by the first Prime Minister of India Jawaharlal Nehru on 17 November 1960. The model of establishing Pantnagar University paved the way for establishing state government supported dedicated universities for agricultural education, *i.e.* SAUs, in India. There are 54 SAUs in India.

Features of SAUs

SAUs are characterized by their triple mandate of teaching, research and agricultural extension. Since the initial SAUs were established with technical cooperation from United States land-grant universities, their academic programmes are strongly influenced by the American system and present a departure from the academic system prevailing in other general universities in India. The undergraduate degree programs of SAUs are of minimum four years' duration. Since their inception, SAUs have followed a trimester or a semester system with credit-based, continuous evaluation on a 5-point or 10-point GPA scale. This is different from general universities, which usually follow a yearly term and percentage based end-session evaluation. As envisioned in ICAR's Model Act for Agricultural Universities, most SAUs are non-affiliating universities/However, SAUs in states like Maharashtra, Tamil Nadu, Chhattisgarh *etc.* have affiliated private agricultural colleges. SAUs also have territorial jurisdiction.

Deemed Universities

Deemed universities are not established by an act of independent legislation, but declared to function as universities by Government of India under Section 3 of the University Grants Commission Act 1956. As per this section, "The Central Government may, on the advice of the Commission, declare by notification in the Official Gazette, that any institution for higher education, other than a University, shall be deemed to be a University for the purposes of this Act, and on such a declaration being made, all the provisions of this Act shall apply to such institution as if it were a University within the meaning of clause (f) of section 2". Thus, the provision of deemed universities enables the central government to incorporate an agricultural university without the need of Parliamentary legislation, thus circumventing the complexities of federal division for legislative powers, which has put agriculture in the state list.

Central Agricultural Universities

Notwithstanding the limited power of the Indian Parliament for legislating on subjects of state list such as agriculture (including agricultural research and education), there are certain provisions in the Constitution of India which allows it to legislate on these matters in certain conditions, such as when the Rajya Sabha passes a resolution to this effect by two-thirds majority (Article 249) or when legislatures of two or more states pass a resolution to do so (Article 252) or to implement decisions taken in international conferences (Article 253).

The Central Agricultural University at imphal, Manipur was incorporated by an act of Parliament (No.4 of 1992) and caters to many states in Northeast India. There are proposals and demands to establish more CAUs. The Rani Lakshmi Bai Central Agricultural University Bill was introduced in the Rajya Sabha on 22 May 2012; it aims to incorporate a CAU at Jhansi catering to the Bundelkhand region spanning the two states of Uttar Pradesh and Madhya Pradesh. The Parliament passed the Finance Bill 2012-2013 for budgetary allocation for establishing a CAU in Bihar. Chief Minister of Orissa has demanded setting up a CAU in Kalahandi Balangir Koraput Region. The West Bengal government placed a proposal before the Centre to set up a CAU at Goaltore in West Midnapore district on a 1,000-acre (4.0 km^2) land parcel.

SAUs work under the control of state administration. They are funded up to 90 per cent by the ICAR and rest of their budget is met by State Government. Since agricultural research, education and extension is the primary responsibility of States, the growth of SAUs has been faster.

On Organization and Coordination of agricultural science universities the National, Commission on Agriculture recommended".agricultural education programme could best be ensured by an apex body consisting of representatives of the !ICAR, Directorate of Extension, NCERT, Ministry of Education and Social Welfare should, after careful study of the pros and cons, develop programmes form idle level training".

Science University in India is headed by the Vice-Chancellor. The term of Vice-Chancellor varies from 3 to 5 years in different agricultural universities. The highest executive body is commonly known as the Executive Council Committee (This is also called as the Board of Management, Board of Regents and Governing Body in different Universities).The membership of the Council ranges from 12 to 38. In some universities there is a body, called Court of General Council, which delegates' administrative power to the Executive Council/Committee. The Executive Council meets as often as necessary, as it has to give guidance and directive to the academic and other bodies entrusted with the management of the university. The Executive Councils generally presided over by the Vice-Chancellor.

The Deans, Directors and Heads of Departments are specifically entrusted, somewhat differently indifferent universities, with the administration of academic programmes. They receive directions from the Academic Council, which is a subordinate body of the Executive Council. Each of the officials of a university has his/her own areas of administration but coordinated on top by the Deans. Academic Councils of different universities mostly consists of Deans of various faculties and Post Graduate Studies, Directors of Research, Extension, Instruction and Student Affairs, Heads of departments and in some cases all the teachers up to the rank of Associate Professor or Reader. The Vice-Chancellor is the Chairman of the Academic Council. It works through Committees in all facets of the academic life of the institution. The faculty of the various departments and colleges receives directives from the Academic Council and the department/college is administered by the Dean as its Chairman.

Functioning of SAUs

SAUs have an important role in imparting training for extension personnel at the middle rung. They run national courses in agricultural technology or Extension and Communication methods in chosen areas with the help and support from Directorate of Extension, Govt. of India. SAUs were set up in India on the pattern of Land Grant University in USA with integrated teaching, research and extension functions. Each SAU was expected to perform some extension roles in the area under its jurisdiction. However, extension role of SAU was supposed to be complementary to the major extension efforts made by State Department of Agriculture. The specific extension roles of SAUs are as given below:

- ☆ To prepare personnel to perform various extension roles and provide dynamic and analytic leadership.
- ☆ To collect, process and disseminate latest research findings to extension clientele through extension personnel.
- ☆ To identify field problems to which scientific solutions are not yet available and communicate these to subject matter specialist.
- ☆ To produce information materials and teaching aids for extension personnel, trainers and farmers.
- ☆ To plan, organize and conduct production and problem oriented training of various types.

- To provide authentic and systematic guidance through an upto date and specialized "Farm and Home Advisory Service" to all categories of farmers.
- To undertake researches on various facets of extension education.
- To undertake adaptive and operational research projects for testing the applicability of findings.

42

Krishi Vigyan Kendra (Farm Science Centre)

The National Agricultural Research System (NARS) in India has made efforts for propelling agricultural growth and achieving sustainable production in agriculture and its allied sectors through its Research and Extension education programmes. This led to ushering in green, white, blue and yellow revolutions in agriculture in the country and significantly increased the food grains, oilseeds, milk, fish and meat production. However, ever increasing population and changes in food habits, climate change, natural resources base degradation necessitates new strategies and innovations for accelerating agricultural growth. Increasing agricultural production and making farming remunerative assumes importance and overriding priority because of globalization of agriculture. Agriculture sector is witnessing radical changes and challenges national and globally. The demand for agriculture commodities are changing in view of increased income and changing food choices. Globalization has also opened uncommon opportunities for our farmers to derive benefit. Therefore, efforts need to be focused to make agriculture globally competitive. To meet these new challenges it is necessary to have paradigm shift in the way agriculture enterprise is practiced by our farmers. So far our achievements in respect of agricultural growth and food production have been spectacular and we have achieved food grain self-sufficiency. One of the pillars for the success has been the development of sound technology dissemination institutional system in the country.

To address the above in agriculture, the **Krishi Vigyan Kendra (KVK) – Agricultural Science Centre**, a grass root level plan scheme designed and nurtured by Indian Council of Agricultural Research (ICAR) for the past four decades. The Krishi Vigyan Kendras (KVKs) have played a very important and crucial role in technology refinement and dissemination. Right from 1974 when the first KVK was established, these institutions have proven and met stakeholders' expectations. As a result of which there have been demands for opening more KVKs. Across the country so far ICAR has established 630 KVKs under different host organizations like State Agricultural Universities, ICAR institutes, Central institutes/Deemed Universities, State Government, Public Undertakings and Non Governmental Organizations.

The KVKs have played role in

- ✰ Showcasing the frontier technologies
- ✰ Capacity building among stakeholders
- ✰ Front runner in technological application, information and inputs
- ✰ Participatory approaches in planning, implementing, executing and evaluation
- ✰ Refinement of technologies to suit different agro-climatic condition

However for meeting new challenges it is important to focus on conservation of natural resources, achieving higher productivity keeping sustainability of agriculture, enhancing farmers income through bringing in farming to market value chain efficiency.

The new agriculture will have to be

1. **Remunerative agriculture:** Increasing productivity and production, management of post harvest losses, value chain for increasing farmer's income and ultimately system sustainability.
2. **Climate smart agriculture:** Conservation of soil and water, bio-conservation, reducing environmental pollution, climate resilience, and increasing productivity and production of green house/protected cultivation.
3. **Food and dietary diversification agriculture:** Removing malnutrition, nutritional disorders, year round house hold food security, stable and sustainable food supplies, improving livelihood at grass root level through convergence of all schemes for eliminating hunger and poverty.

KVKs as an institution having played important role in propelling agricultural transmission, will have to play more important role in economic empowerment and accelerating agricultural growth in the years to come. In order that the KVKs to become more efficient, vibrant and meet stakeholders expectations, the High Powered Committee has developed KVK Management Guidelines which will help ushering in economic prosperity for the farming community.

Genesis and Growth

The Education Commission (1964-66) recommended that a vigorous effort be made to establish specialized institutions to provide vocational education in agriculture and allied fields at the pre and post-matriculate levels to cater the training needs of a large number of boys and girls coming from rural areas. The Commission, further, suggested that such institutions be named as 'Agricultural Polytechnics'. The recommendation of the Commission was thoroughly discussed: during 1966-72 by the Ministry of Education, Ministry of Agriculture, Planning Commission, Indian Council of Agricultural Research (ICAR) and other allied institutions. Finally, the ICAR mooted the idea of establishing **Krishi Vigyan Kendras (Agricultural Science Centres)** as innovative institutions for imparting vocational training to the practicing farmers, school dropouts and field level extension functionaries.

The ICAR Standing Committee on Agricultural Education, in its meeting held in August, 1973, observed that since the establishment of Krishi Vigyan Kendras (KVKs) was of national importance which would help in accelerating the agricultural production as also in improving the socio-economic conditions of the farming community, the assistance of all related institutions should be taken in implementing this scheme. The ICAR, therefore, constituted a committee in 1973 headed by Dr. Mohan Singh Mehta of Seva Mandir, Udaipur (Rajasthan), for working out a detailed plan for implementing this scheme. The Committee submitted its report in 1974.

The first KVK, on a pilot basis, was established in 1974 at Puducherry (Pondicherry) under the administrative control of the Tamil Nadu Agricultural University, Coimbatore. In 1976-77, the Planning Commission approved the proposal of the ICAR to establish 18 KVKs during the Fifth Five Year Plan. With the growing demand for more such Kendras, the Governing Body (GB) of the Council approved 12 more KVKs in 1979 and they were established in the same year from Agricultural Produce Cess Fund (AP Cess Fund). Pending clearance of the Sixth Five Year Plan scheme on KVK by the Planning Commission, the GB of the Council again approved 14 KVKs in 1981, which were established during 1982-83 from AP Cess Fund. A High Level Evaluation Committee on KVK was constituted by the ICAR in 1984, after thorough review of the programme, strongly recommended for the establishment of more KVVKs in the country. Keeping this in view the Planning Commission approved to establish 44 new KVKs during the Sixth Plan. Thus by the end of Sixth Plan, 89 KVKs had started functioning in the country.

During the Seventh Plan, 20 new KVKs were established. The success of KVKs at many locations created a great demand for establishment of more KVKs in the remaining districts of the country. Accordingly, the Planning Commission further approved 74 new KVKs to be established during the period 1992 -93. Again in the Eighth Plan (1992-97), 78 new KVKs were approved and the same were established in the country, making total number of functional KVKs by the end of the Eighth Plan to The number of KVKs increased to 290 during Ninth Plan with the establishment of 29 KVKS. On the occasion of the Independence Day Speech on 15th August, 2005 the Hon'ble Prime Minister of India announced that by the end of 2007 there should be one KVK in each of the rural districts of the country. This has taken the total number of KVKs to 551 at the end of Tenth Plan.

At present there are 690 KVKs, which include 466 under State Agricultural Universities (SAU) and Central Agricultural University (CAU)/Deemed Universities, 64 under ICAR Institute, 105 under NGOs, 36 under State Governments, Public Sector undertaking 3 and the remaining 16 under other educational institutions. All KVKs are working towards reducing the time lag between generation of technology at the research institution and its application to the location specific farmer fields for increasing production, productivity and net farm income on a sustained basis.

Vision, Mission and Mandate

The Mohan Singh Mehta Committee (1974) enunciated the following three basic concepts of Krishi Vigyan Kendra:

- ☆ The Kendra will impart learning through work-experience and hence will be concerned with technical literacy, the acquisition of which does not necessarily require as a pre-condition for the ability to read and write.
- ☆ The Kendra will impart training to only those extension workers who are already employed or too the practicing farmers and fishermen. In other words, the Kendra will cater to the needs of those who are already employed or those who wish to be self-employed.
- ☆ There will be no uniform syllabus for a Kendra. The syllabus and programme of each Kendra will be tailored to the felt needs, natural resources and the potential for agricultural growth in that particular area.

The Committee further suggested that the success of Kendras would depend upon adherence to the following three basic principles.

- ☆ Accelerating agricultural and allied productions in the operational area of the Kendra should be the prime goal;
- ☆ Experiential learning *i.e.*, "teaching by doing" and "learning by doing" should be the principal methods of imparting skill training' and
- ☆ Training efforts should not be made to make economically good people better but the poor ones good so as to raise the living conditions of the poorest of the poor.

The main idea was to influence the production system with social justice by creating a favourable condition for the have-nots.

Keeping the above in view, the Vision, Mission and Mandate of KVK are tailored as follows:

Vision

Take a stock of appropriate technologies in agro-ecosystems perspective in a continuum for overall improvement in the social and economic well being of the farming community.

Mission

Bringing desirable behavioural changes associated with awareness, knowledge, skill and attitude among the farmers and other stakeholders for technology-led growth in farming and allied enterprises.

Mandate

The mandate of KVK is application of technology/products through assessment, refinement and demonstration for adoption.

To implement the mandate effectively, the following activities are envisaged for each KVK:

1. On-farm testing to identify the location specificity of agricultural technologies under various farming systems.
2. Frontline demonstrations to establish production potentials of technologies on the farmers' fields.
3. Training of farmers and extension personnel to update their knowledge and skills in modern agricultural technologies.
4. Work as resource and knowledge centre of agricultural technologies for supporting initiatives of public, private and voluntary sector for improving the agricultural economy of the district.
5. Provide farm advisories to farmers on marketing of agricultural produce.
6. Identify, document and validate selected farm innovations
7. Select suitable ICT for reaching the unreached.
8. Produce and make available technological products like seed, planting material, bio agents, young ones of livestock *etc.* to the farmers as per their resources.
9. Organize extension activities to create awareness about improved agricultural technologies.

In addition, KVKs would also implement sponsored schemes/programmes, which are within the mandate of KVK.

Coordination and Convergence

The biggest challenge that the policy makers and development agencies are facing today is how best a synergy can be achieved among the related schemes and programmes so that the national resources are made best use of, without any duplication. As the frontline extension system of ICAR, KVKs have the onerous job of strengthening the technology adaptation mechanisms among farmers. Having been recognized as the only scientific institution at the district level in agriculture and allied sectors, every KVK has to work with many organizations for achieving its goals and objectives, which necessitate functional linkages with the research, extension, development departments and the farmers.

Following are the important areas that require convergence:

State Level Interface

On the lines of Kharif and Rabi conferences organized by Department of Agriculture Cooperation, Govt. of India, interface meetings at the state level involving all heads of State Department of Agriculture and other line departments would be desirable. These meetings may be organized by SAUs in collaboration with the Department of Agriculture under the Co-Chair of Vice Chancellor and Principal Secretary (Agriculture) with the participation of Zonal Project Director, State Nodal Officer of ATMA, Director of SAMETI, Directors of ICAR Institutes,

Directors of Research, Directors of Extension, Heads of Regional Stations of ICAR Institutes and SAUs, Head of KVKs, Scientists from ICAR and SAUs.

Scientific Advisory Committee (SAC) Meetings

KVKs conduct SAC meetings twice a year, preferably during pre-kharif and pre-rabi seasons. The SAC meeting provides an opportunity for critical reviewing of the progress achieved and to plan for need-based and problem-solving activities in the ensuing season. Participation of heads of all the line departments is very much essential to ensure flow of feedback on the technologies/activities carried out by the KVK and to take on the spot decisions regarding dovetailing of schemes and to facilitate sharing of human and material resources for conducting the activities aimed at the overall development of agriculture in the district.

KVK Farm and Demonstration Units

All KVKs must have a fully developed farm with all the demonstration units. Available financial provisions of ICAR shall be supplemented by the host organizations to develop the KVK infrastructure in such way that the farm is a miniature of the agro-climatic situation of the district with all the crops and enterprises. KVKs must make efforts to seek funding available under various schemes (*e.g.* ATMA provides support one-time strengthening of District Training Centres)

Production of critical inputs (seeds, planting material, bio-products, young ones of animals and fish *etc.*) and providing them to needy farmers should be taken up on partnership mode. KVKs need to train rural youth on such activities and encourage them to be the entrepreneurs in potential areas of agribusiness. Vocational training programmes of sufficiently longer duration with emphasis on hands-on experiential methods have to be specially organized by the KVKs for this purpose. Links with banks would facilitate establishment of agri-entrepreneurs in rural areas.

On Farm Trials

This is a highly specialized and trademark activity of the KVKs. In order to plan for problem solving OFTs, support from ICAR Institutes and SAUs is needed for technological inputs. KVKs must coordinate and ensure the visit of the scientist concerned to OFT plots at critical stages to check if the treatments have been properly laid out and to get guided on observations being recorded. The extension system should play a key role in bringing the field problems to the attention of the KVK and support the KVK financially (as provided under ATMA cafeteria) to identify the location specific and problem-solving technologies. Scientists-extension functionaries-farmers interfaces also need to be organized in successful OFT plots to reaffirm the utility of results to the local situations and to further upscale such technologies through Front Line Demonstrations.

Frontline Demonstrations

Frontline demonstrations may be conducted by the KVKs on the latest and best technologies under the direct supervision of KVK scientists. The FLDs validate proven technologies in farmers' fields and educate and train the demonstrating farmers to reduce the risk factors faced in adopting new technologies. FLDs also provide valuable feedback to the Agricultural Research Institutes and Universities that stimulate to take up new research or for refining the existing technologies/ recommendations. In order to conduct successful demonstrations the preparatory activities are crucial (selection of farmers, identifying and arranging critical inputs, arranging pre-demonstration training *etc.*). This demands involvement of many agencies and organizations at various stages of FLDs. Officials of the development department concerned, panchayat representatives and farmers organizations/ groups play vital role in selection of farmers, arranging field activities like visits, training and field days. Press and media need to be fully involved at crucial stages to provide publicity and propaganda to the successful technologies. Field days should be organized involving all the stakeholders for the successfully demonstrated technology.

Training for Farmers, Youth and Extension Functionaries

It is likely that more than one training institute are functioning in a district on agriculture and allied areas. It is necessary to have convergence with all the training institutes in the district not only to share the resources, but also to avoid duplication of training activities in terms of courses and participants. A common training calendar for the district should be developed by involving all the training institutes in the district. The KVKs shall maintain a database of organizations having training facilities in the district and strive to utilize these facilities for organizing collaborative and peripatetic training programmes. In the districts where other departments do not have district level training centres, district line departments may release funds to KVK so that the KVK can perform as district training centre for line departments.

Training need assessment and skill gap analysis of farmers and youth involved in demonstrations are to be conducted on a regular basis in collaboration with Panchayats, Development departments, NGOs and farmers organizations working in the district.

Extension Activities

Extension activities conducted by the KVKs are of varied nature, ranging from individual contact to mass contact. Most of the activities need to be through convergence with other stakeholders. The KVKs organize Technology Week, a mega event to showcase all the available technologies through live demonstrations, exhibition-cum-sale, and farmers-scientists interactions. This is an annual event coinciding with the bountiful crop conditions. This requires massive collaboration from all the stakeholders like SAUs, ICAR Institutes, development departments, private firms, input companies/agencies, banks, cooperatives, local administration, press and media, farmers groups/associations. Mobile exhibition-cum-diagnostic units provided to Directorate of Extension Education of SAU will provide necessary support in increasing the instant out reach of KVKs. This activity requires convergence with the local administration (panchayat), development departments, press and media, NGOs, Input agencies *etc.* The ICAR Institutes, SAUs, National Seed Corporation and State Seed Corporations may collaborate with KVKs to organize seed villages in their cluster villages in which seed production activities of those varieties, which are in great demand in the district, are taken up in participatory and partnership mode.

Group approach to agricultural extension has been the order of the day. Many organizations are supporting promotion of groups in rural areas. KVKs have been involved in capacity building of members of the Farmers' Clubs supported by NABARD. This has the potential to be implemented in all the districts. ATMA also has separate funds for mobilization of groups, their capacity building and to provide initial seed money. There could be many more similar schemes in other departments as well. A mechanism is needed to institutionalize the process of capacity building of these groups in agriculture and allied areas. KVKs may take up these activities for all the schemes in their cluster villages.

Action Points for Convergence and Coordination of KVK with ATMA

- ☆ Quarterly meetings may be organized by Project Director (PD), ATMA at District level for strategy, implementation and dissemination of information to farmers.
- ☆ KVK will undertake capacity building of Block Technology Team, Block Technology Managers and Subject Matter Specialists of ATMA on emerging technologies. The Development Departments may communicate additional training needed, if any, so that KVKs arrange need-based capacity building activities to extension functionaries.
- ☆ SAUs/ICAR will identify a senior expert for their KVK district for coordinating flow of technology information and advisories to district development officials through KVK. This is subject to transfer of earmarked funds under ATMA to University/ICAR Institutes in the beginning of the year itself.
- ☆ KVKs also have to be associated in prioritizing district technological interventions along with SAU scientists.
- ☆ KVKs implement the mandated activities in a "cluster villages" approach. The KVKs will provide the list of villages and the plan of activities in these cluster villages to the Development Departments in the SAC Meetings, which are conducted twice a year.
- ☆ The PD ATMA and his team should visit these cluster villages along with the Chief Scientist of KVK for gaining first-hand knowledge of the technologies demonstrated in these villages so that activities could be initiated under ATMA for large-scale dissemination of successful technologies. PD ATMA shall keep the Collector/Commissioner/ Secretary (Agriculture) informed about the spread of new technologies in the district and impact of such technologies on production and productivity.
- ☆ The senior expert mentioned in para 3 above, would interact with ATMA and KVK in respective district.

- The identified expert for each district will coordinate between development departments and research system. However, apart from KVK adopted cluster villages, the problems or issues noticed by PD ATMA and his staff in other villages could be brought to the notice of Director Extension of SAU who will ensure necessary follow-up.
- The Vice Chancellor of SAU and Principal Secretary (Agriculture) will jointly organize pre-kharif and pre-rabi interfaces in which developmental department secretaries, senior officers of university, KVK CSs, ATMA PDs and Collectors take part and chalk out strategies for the ensuing season. This may be scheduled immediately after pre-kharif and pre-rabi conferences organized by the DAC in Delhi.
- ATMA may sponsor the technology week to be conducted by KVK at its farm besides brining all stakeholders for the participation.
- KVKs may also provide an Agricultural Technology Update (ATU) on half yearly basis *i.e.* before the start of Kharif and Rabi crop seasons to the ATMA for its wider dissemination among the farmers of the district.
- Funds earmarked should be transferred electronically to the respective heads of host organisations of the KVKs, who in turn transfers to KVKs within a week. KVKs will undertake the assigned activities only after receipt of funds.
- SREPs need to be revisited in most districts. The revisited SREPs should be made available to all Agricultural Research Stations/ICAR Research Institutes. As most ZARS are under-staffed, the State Governments should permit SAUs to fill up the vacant posts of ZARS on priority basis so that the researchable issues identified in the SREPs can be addressed as suggested.
- KVK will undertake capacity building of PD, Dy.PDs, BTT, BTM and SMS who will be able to implement programmes at District and Block level.
- KVKs will also provide technology backstopping for different programmes of the district. The PD ATMA should regularly liaise with KVK for the purpose.
- The KVK will extend required assistance wherever the connectivity is ensured and the calls are routed to KVK in respect of Kisan Calls.
- Successful and innovative models of convergence between KVKs, SAUs and Development Departments of the State in Agriculture and Allied Sectors need to be developed and replicated in all the States by ATMA.
- The respective Departments from their own budgetary allocations will meet additional requirement of TA/DA, *etc.* However, funds available under ATMA Cafeteria for Research-Extension-Farmer linkages may also be appropriately utilized. Separate budget line need to be provided for all activities to be undertaken. The source and the pattern of flow of funds need to be made very clear by ATMA.

Operational Terminology

KVK is an institutional innovation inspiring the World of the 21st Century for the farmers, which has frontline decentralized application of technology at district level on various aspects of agriculture and allied sectors. They impart the latest technical know-how and do-how to the different clientele. At this juncture, it is worth to know about the different terminologies related to KVK system so as to meaningful perspective. The KVKs have witnessed several changes in their functions over the years. Accordingly their functional definition also has radically got refined so as to meet the new challenges in agriculture. ***"KVKs are grass root level organizations meant for application of technology through assessment, refinement and demonstration of proven technologies under different 'micro farming' situations in a district"*** **(Das, 2007).** It should be clearly understood that transfer of technology is not a primary function of KVKs. The KVKs on the other hand will assess and refine (if needed) the newly released technologies, demonstrate the proven ones and train farmers and extension functionaries on the same. Hence it is a Frontline Extension System.

Technology Assessment and Refinement

Technology

Technology is any systematic knowledge applicable in a particular area in order to achieve some values. Technology involves application of science and knowledge to practical use.

Agricultural Technology

Agricultural Technology is a complex blend of materials, processes and knowledge. Agricultural technologies are of two types:

a) **Material Technology** (When knowledge is embodied into technological products such as tools, equipments, seed materials *etc.*)

b) **Knowledge-based technology** (Such as the technical knowledge, methodologies, managerial skills, motivational techniques *etc.* that farmers and rural people need for better production in their enterprises.)

Technology Management

Technology management can be defined as the integrated planning, designing, optimization, operation and control of technological products, processes and services, for human advantage.

Technological Gap

Technological Gap is the gap between the level of recommendation and the extent of adoption (against recommendations). The technological gaps in respect of different aspects of the technology are computed in percentage by the following formula:

$$\text{Technological Gap (TG)} = \frac{R - A}{R} \times 100$$

Where, R= Maximum possible adoption score that a respondent could be awarded in respect of a given component of the technology.

Technology Assessment and Refinement

Technology Assessment and Refinement (TAR) refers to a set of procedures whose purpose is to develop recommendations for a particular agro-climatic situation/location through assessment and refinement of recently released technology through participatory approach. It refers to the process or a set of activities before taking up new scientific information for its dissemination in a new production system. *OFTs conducted by KVKs are based on this concept and thus distinguish it from agronomic and research trials.* TAR has three components. They are technology testing, technology adaptation and technology integration.

On Farm Trials (OFT)

An On-Farm Trial aims at testing a new technology or an idea in farmer's fields, under farmers' conditions and management, by using farmer's own practice as control. It should help to develop innovations consistent with farmer's circumstances, compatible with the actual farming system and corresponding to farmer's goals and preferences. On-farm-trial is not identical to a demonstration plot, which aims at showing farmers a technology of which researchers and extension agents are sure that it works in the area.

Stakeholders of On-farm Trials

There are various stakeholders in an on-farm trial.

Understanding them and their roles can help KVKs to develop better OFTs.

The stakeholders are:

- The farmers who are the clients for the out-coming results,
- The SMS who should help the farmers to overcome their problems and improve their economical situation.
- The Scientist who needs to apply promising on-station results under farmers' conditions before releasing the technology to the extension service,
- The extension system and government itself, who is interested in seeing an efficient and participatory technology development model evolving, since most top-down approaches have failed miserably.

Typology of On Farm Trials

There are three types of OFTs according to the stakeholder who is going to take the lead role:

Type 1 - Research driven: Research system designed and managed (with the assistance of extension)

Type 2 - **Extension driven:** Extension System or KVK system designed and managed by farmers

Type 3- **Farmer driven:** Farmers designed and managed, with the assistance of Extension system/KVK system.

Adaptive Research

It is essentially research done for solving a local problem and uses research information generated by applied research to tailor appropriate technology for specific areas and defined groups of farmers. This research is primarily conducted to test the performance of the technology under farmers' condition, which is different from the condition of the research stations.

Agro-ecosystem

It refers to grouping of elements in relation to agriculture and ecology within a boundary and analysis of functional relationship of these elements to each other as an individual or as a group to assess the potential of the system as a whole.

Participatory Technology Development (PTD)

It is the process of combining local farmers' knowledge and skills with those of external agents to develop site specific and socio-economically adopted farming technologies.

Demonstration

Frontline Demonstration

Frontline Demonstration is the concept of field demonstration evolved by the Indian Council of Agricultural Research during mid-eighties. The field demonstrations conducted under the close supervision of scientists of the National Agriculture Research System is called front-line demonstrations because the technologies are demonstrated for the first time by the scientists themselves before being fed into the main extension system of the State Department of Agriculture.

Method Demonstration

It is a relatively short-time demonstration given before a group of people to show the skill involved in an entirely new practice or an old practice in a better way. Demonstrator shows the step-by-step procedure in the operation, explaining each succeeding step as he proceeds. The learners watch the process, listen to the oral explanation and ask questions during the demonstration. Then as many as possible members of the group repeat the demonstration in the presence of the demonstrator and other group members for effective exposure. This helps to fix the process in the minds of the learners and increases confidence in their ability to master the technique.

Result Demonstration

It is a way of showing people the value or worth of an improved practice, whose success has already been established on the research station, followed by district trials/minikit trials. In this method, new practice is compared with the old practice so that the people see and judge the results by themselves. It involves careful planning, a substantial period of time, adequate records and comparison of results. Result demonstration is otherwise called as field demonstration where a farmer in his field conducts the demonstration in a systematic manner under the direct supervision of extension personnel to prove the advantages of a recommended practice or combination of practices. ***"Seeing is believing"*** is the basic philosophy of field demonstration and it will teach other farmers in addition to the farmer who conducts the demonstration.

Field demonstrations are further classified as simple or single or comparative and composite demonstrations.

a) **Simple or single or comparative demonstration** - It shows the difference in the control plot and the plot treated with only one practice like local seed vs. improved seed. Thus, it deals with a single practice at a time.

b) **Composite demonstration** - It includes many improved practices like improved seed, fertilizer, irrigation *etc.* are combined together to show the maximum results. Thus, it is concerned with a series of related practices or entire package of practices at a time.

Training

Training is a systematic instructional means to develop knowledge, skills and attitude for adults in order to keep pace with the changes in life. It is the process by which individuals are helped to acquire certain specific skills related to a given set of operations in certain specified contexts only.

Training Needs Assessment (TNA)

Assessment/evaluation of training can be defined as a process by which the desired behavioural changes (in knowledge, skills and attitudes) due to training are measured against set of objectives/standards. The assessment can be conducted by training institution, participants (trainees) and participating institution (sponsoring agency). The literature of training needs assessment has traditionally focused on the development of better techniques and method for improving the quality of information collected about organizational tasks and person analyses.

Training for Extension Personnel

It can be defined as a planned and systematic effort to increase Knowledge, improve Skill, inculcate appropriate Attitudes and develop other attributes in extension personnel to enable them to better serve their clients-the farmers.

Capacity Building

Capacity building is the ability to perform appropriate tasks effectively and sustainably. UNDP (1997) defined capacity building as the ability of an individual or organisation to perform functions effectively, efficiently and sustainably.

Human Resource Development (HRD)

It refers to planned approach of developing human capabilities, latent talents and contributes to full growth of all aspects of people in terms of skills, maturity, competence, self-awareness, continuous adjustment with changing environment and self-confidence.

Vocational Training

It refers to long -term vocation-based and skill oriented training for rural youth, farm women, school drop-outs *etc.* in agriculture and allied activities for income generation and self-employment.

Training Effectiveness

Training Effectiveness can be defined as the degree to which a training course helps to transfer to make effective performance in one's job through application of knowledge gained, skills acquired and changed attitudes.

Training Impact Assessment/Evaluation

An Impact Evaluation is a method of assessing changes in on-the-job behaviour (*i.e.* improved performance) as a result of training effort. It is concerned with the overall effect of training on farmers or the rural community served by the trainees. Under this evaluation, the impact is usually measured in terms of productivity (crop yield per hectare) production (total crop production in an area) and per capita income of farmers.

KVK as Knowledge and Resource Centre

Data

Data usually means raw, unevaluated facts, figures, symbols, objects, events *etc.* They are the collection of natural phenomena descriptors including the results of experience, observation or experiment, or a set of premises. This may consist of numbers, words, or images, particularly as measurements or observations of a set of variables. A data base is a structured collection of records or data. A computer database relies upon software to organize the storage of data. The software models the database structure in what are known as database models.

Database Management System (DBMS)

A Database Management System (DBMS) is computer software designed for the purpose of managing databases based on a variety of data models. A DBMS is a complex set of software programs that controls the organization, storage, management, and retrieval of data in a database.

Information

When data are processed conveying some meaning, it is called information. Information – usually consisting of facts and figures collected at the initial stages of a project – that provides a basis for measuring progress in achieving project objectives and outputs. Davis and Olson (1985) provides a general definition of information which reads as "information is data that has been processed into a form that is meaningful to the recipient and is real or perceived value in current or prospective actions or decisions". It bears a diversity of meanings, from everyday usage to technical settings. Generally speaking, the concept of information is closely related to notions of constraint, communication, control, data, form, instruction, knowledge, meaning, mental stimulus, pattern, perception, and representation.

Knowledge

Knowledge is the retained information concerning facts, concepts and relationship. Oxford English Dictionary defined Knowledge as (i) expertise, and skills acquired by a person through experience or education; the theoretical or practical understanding of a subject, (ii) what is known in a particular field or in total; facts and information or (iii) awareness or familiarity gained by experience of a fact or situation.

Knowledge Management

Knowledge management is a management theory, which emerged in the 1990s. It seeks to understand the way in which knowledge is created, used and shared within organizations. A significant part of Knowledge Management theory is the **DIKW** model, which places *Data, Information, Knowledge and Wisdom* into an increasingly useful pyramid.

Technology Management

Technology management can be defined as the integrated planning, designing, optimization, operation and control of technological products, processes and services, for human advantage.

Technology Backstopping

Backstopping refers to any precaution taken against an emergency condition. Accordingly, technology backstopping can be defined as any technology precaution taken to combat technology fatigue. In simple terms, technology backstopping is the process of making available ready to use technologies for farm families through assessment, refinement and demonstration processes in order to combat the existing/forecasted technology fatigue.

Innovation

An innovation is an idea, practice or object that is perceived as new by an individual or other unit of adoption. The technologies/practices developed through research are innovations, which may be new varieties of crops and plants, new breeds of livestock, new chemicals and medicines, new technique of doing things *etc.* Farmers themselves may develop some new practices, which are also innovations. Irrespective of the time period the idea or practice was originally developed, when a person first becomes aware of it, it is an innovation to that person.

Extension Programmes

Group Discussion

The lecture method may be made more participative, if at the end the audience are allowed to discuss the topic in presence of the speaker and elicit the latter's comments and clarification on the points raised. This shall lead to a better understanding of the topic, as the participants are more actively involved, have the opportunity to express their views and get their doubts clarified. Learning is reinforced through the interaction of the audience with the speaker. Limitations of traditional group discussion are that group members may pursue an idea to the exclusion of other alternatives, and pressures to confirm can discourage the expression of deviant opinions.

Field Day

A field day is a day in which an area containing successful farming or other practices is open for people to visit. This way, a group of farmers could witness the performance or results of certain practice (s) under local conditions. Filed days could be planned to deal with one specific topic at a time or a number of items of interest to the farmers. The purpose of the field day is to permit extension clientele to observe personally, ask about successful and locally applicable practices, and to create a situation in which informal contacts and learning can take place. Field days are normally held once or twice a year, usually in each crop season. They are held on farmers' fields, regional research stations, agricultural universities and government farms to demonstrate successful farming techniques or research.

Farmers' Field School (FFS)

The farmers' field school (FFS) is basically "a school without wall". FFS consists of a group of people with a common interest who get together on a regular basis to study the "how or why" of a particular topic. The topic covered can vary considerably from IPM, organic agriculture, and husbandry *etc.*, to income generating activities such as handicrafts.

Exhibition

Exhibition is a planned and systematic display of real objects, specimens, models, charts and posters presented to public view for instruction, judging a competition, advertising or entertainment. It helps to create awareness, develop interests and stimulate action amongst the public. It accommodates both exhibits and display. Displays use two dimensional or flat materials like pictures, photograph, chart, posters *etc.*, while exhibits tend to use more three dimensional materials like real objects, specimens and models.

Field visit:

Visit to successful farmers' fields in addition to KVK farms to reinforce the relevant technologies imparted in the training programme. The principle of "seeing is believing" can be fulfilled. Field visit gives participants exposure to the real life situation and also provide an idea about the possible solutions to the problems.

Monitoring and Evaluation

Stakeholders

An agency, organisation, group or individual who has a direct or indirect interest in the project/programme, or who affects or is affected positively or negatively by the implementation and outcome of it. Primary stakeholders are the term used for the main intended beneficiaries of a project.

Accountability

Obligation of Government, public services or funding agencies, to demonstrate the citizens that entrusted work has been conducted in compliance with agreed rules and standards or to report fairly and accurately on performance results vis-à-vis mandated roles and/or plans. This may require a careful, even legally defensible, demonstration that the work is consistent with the contract terms. Projects commonly focus on upward accountability to the funding agency, while downward accountability involves making accounts and plans transparent to the primary stakeholders. Ensuring accountability is one part of the function of monitoring and evaluation (learning and management are the other two).

Input

Something that is put into a process/program *e.g.* an amount put in an investment, a fertilizer application for increasing the crop yield. It is a component of production such as land, water, labour, seed, package of practices, plant protection measures, energy *etc.*

Output

The tangible (easily measurable, practical), immediate and intended results to be produced through sound management of the agreed inputs. Examples of outputs include goods, services or infrastructure produced by a project. These may also include changes, resulting from the intervention, that are needed to achieve the outcomes at the purpose level.

Outcome

The results achieved at the level of "purpose" in the objective hierarchy. Outcome is part of impact (result at purpose and goal level).

Impact

It is the change that has taken place across the system over a period of time due to a development initiative. It is something that follows as a result or consequence.

Success Stories

Experience of successful farmers as success stories carry special significance both for trainees as well as the KVKs and boost the technology dissemination process. For example, the success stories reported by KVK West Tripura on True Potato Seeds (TPS) have stimulated farmers in the surrounding villages. Other mass media like Radio and Television can broadcast and telecast success stories to create awareness and motivate farmers about successful technologies that could be practiced by them as well.

Case Study

In this method, the trainer gives the trainees information about a situation and directs them to come to a decision or solve a problem concerning the situation. For this, a written case or a problem situation is presented to the participants in a programme for careful study and examination from all facets, so as to enable them to exercise their analyzing, synthesizing and decision making powers. As there is no single infallible solution to a problem, different solutions will be evolved and, in course of evaluation of these solutions, participants are enabling to appreciate the viewpoints of others and also see the lacunae in their own thinking and analysis. The interrelatedness of various factors operating in the situation is also highlighted.

General

Coordination

Coordination means establishing harmonious relationship between the efforts of individuals and groups for accomplishment of enterprise objectives. For smooth running of an enterprise coordination is necessary within the organization, as well as with outside organisations. Some sort of basic coordination is essential within an organisation throughout its lifetime for its survival and for unified action.

Farmer

The term "farmers" refers to both men and women, and include landless agricultural labourers, sharecroppers, tenants, small, marginal and sub- marginal cultivators, farmers with larger holdings, fishers, livestock and poultry rearers, pastoralists, small plantation farmers, as well as rural and tribal families engaged in a wide variety of farming related occupations such as apiculture, sericulture and vermiculture. The KVKs should also include tribal families engaged in shifting cultivation and in the collection and use of non- timber forest products. Farm and Home Science Graduates earning their livelihoods from crop and animal husbandry, fisheries and agro-forestry will also have their rightful place in the world of farmers and farming.

Rural Youth

The Ministry of Human Resource Development (1985) considers 'youth group' in India as persons in the age group 15 to 35 years. Youth forms nearly one-third of the population of India. Rural Youth constitute over two-and half times of the size of urban youth.

Self Help Group (SHG)

SHG is a people's institution, which provides the poor with the space and support necessary to take effective steps towards greater control of their lives in private and in society. SHG is not a static institution. It grows on resources and management skills of its members and their increasing confidence to get involved in issues and programmes that require their involvement in the public and private sectors.

Sustainability

Sustainability refers to the successful management of resources to satisfy changing human needs, while maintaining or enhancing the quality of environment and conserving natural resources. The KVK staff while formulating their programs should focus on maintaining the sustainability of the micro farming system in their districts.

Social System

It is a set of interrelated units that are engaged in joint problem solving to accomplish a common goal.

Rapid Rural Appraisal (RRA)

It is a systematic, semi-structured activity conducted on-site by a multidisciplinary team with the aim of quickly and efficiently acquiring new information and hypotheses about rural life and rural resources.

Participatory Rural Appraisal (PRA)

It is a method by which research team can quickly and systematically collect information for general analysis of specific topic, question or problem, need assessment, identifying and prioritizing projects, project or programme evaluation *etc.* It is a way of enabling local people to analyze their living conditions, to share the outcomes and to plan their activities. Other words, it is handing over a stick to the insider in methods and actions wherein outsider's role is, that of a catalyzer, a facilitator and convener of processes within a community, which is prepared to alter their situation.

43

Agricultural Technology Management Agency (ATMA)

Agricultural Technology Management Agency **(ATMA)** is a society of key stakeholders involved in agricultural and allied activities for sustainable agricultural and allied development in the district. It is a focal point for integrating Research and Extension activities and decentralizing day-to-day management of the public Agricultural Technology System **(ATS).** It is a registered society responsible for technology dissemination for all line departments at the district level. As a society, it would be able to receive and expend project funds, entering into contracts and agreements and maintaining revolving accounts that can be used to collect fees and thereby recovering operating cost.

The ATMA at district level would be increasingly responsible for all the technology dissemination activities at the district level. It would have linkage with all the line departments, research organizations, non-governmental organizations and agencies associated with agricultural development in the district. Research and Extension units within the project districts such as **ZRS** or substations, **KVKs** and the key line Departments of Agriculture, Animal Husbandry, Horticulture and Fisheries *etc.* would become constituent members of ATMA. Each Research-Extension **(R-E)** unit would retain its institutional identity and affiliation but programmes and procedures concerning district-wise R-E activities would be determined by ATMA Governing Board to be implemented by its Management Committee **(AMC).**

ATMA is supported by Governing Board **(GB)** and Management Committee (AMC). The Governing Board is a policy making body and provide guidance as well as review the progress and functioning of the ATMA. The Management Committee would be responsible for planning and executing the day-to-day activities of ATMA.

ATMA Governing Board (GB)

Composition:		
Sl.No	*Members*	*Status*
1	District Magistrate/Collector	Chairman
2	Chief Executive officer (CEO) Chief Development Officer (CDO)	Vice – Chairman
3	Joint Director/Deputy Director (Agri)	Member
4	A representative from ZRS/Krishi Vigyan Kendra	"
5	One Farmer representative	"
6	One Livestock Producer	"
7	One Horticulture Farmer	"
8	Representative of Women Farmers interest group	"
9	One SC/ST farmer representative	"
10	A representative of NGO	"
11	Lead Bank Officer of the District	"
12	A representative of District Industrial Center	"
13	Representative of Agriculture Marketing Board	"
14	Representative of input supplying Associations	"
15	One Fisheries/Sericulture representative	"
16	District Panchayati Raj Institution members associated with Agriculture	"
17	Project Director* ATMA	Member-Secretary cum Treasurer (Ex-officio)
Appointment/Nomination/Term of Members:		
Non – official members of GB will be appointed for a period of 2 years by Chairman of GB		
i. Some initial appointments would be staggered to ensure that about two-thirds of the members would be carry over for an additional year on the GB. Thirty per cent of the farmer representatives on the GB would be reserved for women farmers to ensure their interests is fully represented. On the basis of local requirement other members may be nominated		
Principal Agriculture Officer in West Bengal will act as Project Director Sub-Divisional Officers (Agriculture) are as nominated members.		

ATMA Management Committee (AMC)

1	Project Director of ATMA	**Chairman**
2	District Head of Dept. Agri.	Member
3	District Head of Dept. Horticulture	"
4	District Head of Dept. Animal Husbandry	"
5	District Head of Dept. Fisheries	"
6	District Head of Dept. Sericulture	"

7	District Head of other appropriate line Departments, that may be important within a district	"
8	Head, Krishi Vigyan Kendra	"
9	Head, Zonal research Station	"
10	Sub Divisional Agricultural Officers	"
11	One representative of NGO' in-charge of farmers organization	"
12	Two representatives of Farmer's organizations (one year rotation basis)	"

In pursuance of the number of mechanisms built into the project design and to ensure effective coordination among the departments like agriculture, animal husbandry, fisheries, horticulture, soil conservation *etc.* it is proposed to constitute a state level inter departmental working group under the chairmanship of the Agriculture Production Commissioner/Secretary Agriculture with the following composition:

In departments like horticulture, soil conservation *etc.*, where separate secretaries do not exist, director of the concerned departments may act as a member on the interdepartmental group.

A **State Nodal Cell (SNC)** would be established in each States with the office of the Director of Agriculture. This SNC would monitor project activities being carried out in each pilot district and ensure that project funds released to the States are included within state budget.

The State Agricultural Extension Management and Training Institutes **(SAMETI)** would be strengthened by way of providing training managerial and equipments, communication support, programme cost SAMETIs are proposed to be autonomous institutes with greater flexibility in structural and operational aspects. The Key functions of SAMETI are:

- ✰ To provide capacity building support in Extension Management related areas to the extension functionaries both from public and private sector.
- ✰ To provide consultancy in the areas like project planning, appraisal, implementation *etc.*:
- ✰ Develop and promote the application of management tools for improving the effectiveness of Agricultural Extension services through better management of human and material resources.
- ✰ Organize need based training programmes for middle level and grass-root level agricultural extension functionaries: and
- ✰ Develop modules on Management, Communication, Participatory Methodologies *etc.*, as a sequel to the feedback from training programmes.

Farm information and advisory centre (FIAC) consists of two bodies namely Farmer Advisory Committee **(FAC)** and Block Technology Team **(BTT).** The FAC is a body of farmer representatives (11-17) members representing various Enterprises and socio economic strata. The BTT on the other hand is a group of technical advisors operating at block level representing agriculture and allied sectors. FAC and BTT, taken together, act as planning and operational arm of ATMA. It would be created at the block level.

It would be the common meeting point for line departments to prepare detailed extension programmes and coordinate their implementation. It would also be the level where farmer input could be more effectively mobilized through a Farmer Advisory Committee (FAC). Such a mechanism, including representatives of all major stakeholders, would help set extension priorities across each program area and allocate resources.

The FIAC team would be responsible for operationalising the **SREP** in each block and moving toward a single window extension system. The FIAC team would prepare Block Action Plans **(BAPs)** that would detail extension activities to be undertaken. This plan should be approved by the FAC before it could be forwarded to the ATMA. The ATMA Management committee (AMC) would ensure that these plans were technically and administratively feasible, and consistent with the SREP, before being forwarded to the ATMA Governing Board (GB) for approval. The district – level line departments and research units would also prepare annual WPs to (1) maintain diagnostic and support services (*e.g.* soil testing laboratories), (2) organize in service training and technical support activities for FIAC and field level extension staff, (3) carry out research programs and (4) periodically up-date the district SREP.

Block technology team (BTT) is an Inter Departmental Team of Agriculture and Line Departments operating at block level. An indicative composition of BTT is given below; however, the composition would change from place to place depending on the critical areas pertaining to different states.

The Farmer Advisory Committee consists of 11-17 members covering different categories of farmers covering under the given block, with due representation to women farmers and weaker sections of the society. Composition given below may suitably modified in composition as per their agro-ecological situation.

Farmer's organisations would be encouraged at village level and village level groups would, in turn, evolve into Commodity Associations **(CAs),** Marketing Cooperatives and other types of **FOs** at the block and district level. At village level Farmer Interest Groups **(FIGs)** and Farmer Associations **(FAs)** will be effectively involved in the preparation of block action plans. These organizations will coordinate in organizing demonstrations, on-farm and adaptive trails and give feedback to the extension and research. Their representatives would be directly involved in the block-level **FACs** and also at Governing Board of ATMA. The GB of ATMA would select and utilize project funds to support one or more NGOs to assist different types of farmers in becoming organized into different types of FOs within the district.

ATMA will have operational flexibility in use of project funding. They will be expected to adapt plan activities at the district level in consultation with the participating entries as necessary in response to unfolding events. The ATMA Management Committee will be authorized to release project funds onwards to the public/private partners in the agreed activities included in the framework of the district extension plan and will maintain separate accounts/sub accounts partner-wise and activity-wise. The accounts (audited by Chartered Accountants) and reimbursement claims will then be routed through the Nodal Cell for onward transmission to the GOI.

The FIAC team would prepare block action plans **(BAPs)** and budgets that would outline extension and farmer training activities to be undertaken during the coming season. These coordinated plans must address key constraints and opportunities outlined within the SREP if they are to be funded by ATMA. In addition, the Convenor, FIAC would be responsible for coordinating these proposed block-level extension activities and for submitting these proposals to the FAC for review. After the FAC has approved these proposals, then they would be submitted to the ATMA. The Convenor, BTT and the Chairman, FAC from each block would jointly present these extension plans to the ATMA Management Committee prior to their submission to the ATMA GB for approval. In case of programmatic disagreements between the AMC and the FIAC, then these issues would be resolved by the GB.

Once a Block Action Plans (BAPs) have been approved by the GB, then the ATMA Project Director would forward a check to the Convenor, FIAC in each block to cover the budgeted cost of approved extension programs. The FIAC would maintain a bank account and funds would be allocated to each FIAC member in implementing their approved program of extension activities. The Convenor, FIAC and chairman of the FAC would sign all disbursement checks. The Convenor, BTT would be responsible for maintaining complete financial records, including expenditure receipts, for approved extension activities. Also, the Convenor, BTT would periodically submit detailed financial records to ATMA. The flow of funds to individual blocks would be suspended if financial and performance records are not submitted to ATMA in accordance with agreed upon procedures.

All BTT team members would continue to be employed by their respective line departments, but they would function as a multi-disciplinary technology team or working group that would address the four main programmatic thrusts within the SREP in designing and implementing an integrated extension program. Village extension workers (like KPS/AEO/Livestock Assistant/Fishery Agent/ Fishery Assistant *etc.*) would have prime responsibility for day-to-day program implementation; with FIAC team members assisting with demonstration plots installation, teaching farmer-training courses, and conducting farm field days, other group activities, *etc.* The goal of this proposed new arrangement is to create an integrated or ***single window*** extension system.

To the extent possible, developmental activities financed under central and state government schemes would be utilized to demonstrate and support extension and technology transfer activities within the district and block. In the long-term, the goal would be for more of these central, states, and district funds to be directly transferred to the ATMA in support of SREP and BAPs implementation. In the short run, however, the FIAC, in consultation with the FAC, would determine where these development activities (especially for agriculture and horticulture) could be most effectively utilized in support of on-going block-level extension programs.

The organogram on the preceding page outlines the proposed structure for implementing block and field extension activities under this new arrangement where ATMA would serve as a decentralized mechanism to finance extension activities in the district. The key to effective program implementation would rest with the establishment of a FIAC team that would *ground* or fine-tune the extension program for specific agro-ecological zones **(AEZs)** within the block. It is also at the block level that farmer participation can be most effectively mobilized in the development of block action plans. It is also at this level where representatives of Self Help Groups (**SHGs**), Farmer Interest Groups **(FIGs),** and block-level Farmer Associations **(FAs)** can be directly involved in the block-level FAC.

To describe in more detail how this proposed organizational structure would be expected to operate, a set of job description and qualifications have been presented below:

The Convenor, Farm Information and Advisory Centre (FIAC) have to play the roles like:

- ✰ Provide technical coordination and administrative supervision for all block-level extension programs carried out under the auspices of the FIAC and funded by the ATMA.
- ✰ Prepare and submit block action plans (BAPs) to the ATMA. These BAPs would be compiled from proposals submitted by Block Technology Center team. Then these combined BAPs would be reviewed and approved by a Farmer Advisory Committee (FAC) before it was forwarded to the ATMA for approval and financing.
- ✰ Maintain a bank account in which ATMA funds would be deposited. These funds would then disburse to FIAC staff so that they can carry out those extension activities that were approved and funded by the ATMA.
- ✰ Maintain complete and accurate financial accounts indicating how ATMA funds were spent; these financial records would be submitted periodically to the ATMA.
- ✰ Establish and maintain an active Farmer Advisory Committee (FAC) to help set extension priorities, to review and approve block action plans, and to evaluate extension programs within the block. The composition of the FAC should reflect the major groups of stakeholders within the block.

The Convenor, FIAC would be under the administrative supervision of his/her respective line department head. In addition, the Convenor, FIAC would receive technical and programmatic guidance from the ATMA Project Director and the District Management Committee.

Agriculture Development Officer, Horticulture Development Officer, and Block Livestock Development Officer, Fishery Extension Officer, Sericulture Extension Officer *etc.* posted at block level have to play the roles like using participatory rural appraisal **(PRA)** methods, the AEO, HEO and/or LEO would identify and prioritize important technical, management, and resource constraints faced by different socio-economic groups of farmers within their block or mandal. Particular attention would be given to the needs of resource poor farmers and other disadvantaged groups, including those from scheduled castes and tribal groups. In addition, the **AEO, HEO** and **LEO** would identify important natural resource management problems that pose a threat to the different cropping systems and the natural resource base within the block.

Working with the **OIC** and district-level research and extension specialists, this technology team would develop a annual action plan for the block, including attention to farming system innovations, farmer organizations, technology gaps within important crop and livestock production systems, and serious natural resource management problems within the block or mandal.

Working in collaboration with village extension workers **(VEWs),** and with the technical guidance of subject matter specialists **(SMSs)** and researchers, this technology team would implement their respective block action plans **(BAPs).** These **BAPs** would include an appropriate mix of farmer exposure visits to see farming system innovations or *success stories*, demonstrations, farmer training, and, possibly, the use of the mass media. In addition, where farmer interest groups (FIGs) and farmer associations (FAs) exist, technology team members would utilize these groups in disseminating technology and providing training to key target groups within the block or mandal.

Working with both NGOs and VEWs, this technology team would help establish and then provide technical support to different types of FIGs and FAs–farmers who are producing different types of high value crops or livestock enterprises.

The FIAC members would need training in (a) participatory rural appraisal, (b) extension methodology and farmer training procedures, and (c) developing and working with farmer organizations. Finally, the FIAC members should receive periodic short-term technical trainings organized by district-level SMSs, in collaboration with ZRS researchers and KVK trainers. These monthly training sessions would be carried out at the district (or sub-divisional) headquarters, with the content being tailored to reflect the major crops, livestock, and farming systems within the different agro-ecological zones of the district.

The FIAC members would be under the administrative control of their respective line department heads. However, the Convener for the FIAC would provide technical and programmatic supervision.

The officers of sub divisional level will monitor the activity of BTT/FAC and guide for implementation as when necessary. They will also provide support for technical matters whenever it is needed. Village Extension Worker (like KPS/AEO/ Livestock Assistant/Fishery Agent/Fishery Assistant/*etc.*) working at below the block level have played the roles like using PRA methods, the VEW like KPS/AEO/ Livestock Assistant/Fishery Agent/Fishery Assistant/*etc.*) would inventory the predominant crop, livestock, and farming systems in his/her service area (circle) and then determine the primary technology gaps that limit the productivity of field crop, horticultural, fishery, livestock *etc.* production systems. Included in this analysis would be attention to identifying those more homogeneous socio-economic groups of farmers (*e.g.* scheduled caste, tribal and women farmers) who carry out particular farming systems. These similar groups of socio-economic farm households could be targeted for specific farming system innovations (success stories within the district or block) and for organization into FIGs.

Working with under the technical and operational guidance of the Convener and the BTT, the VEW (like KPS/AEO/Livestock Assistant/Fishery Agent/ Fishery Assistant/*etc.*) would carry out a combination of on-farm visits, meetings, demonstrations, farmer field days, and by encouraging farmers to participate in training activities being offered at the village, block, district *etc.* level. Emphasis would be given to assisting resource poor farmers and other disadvantaged groups, including women farmers.

In addition, the VEW (like KPS/AEO/Livestock Assistant/Fishery Agent/ Fishery Assistant/*etc.*) would encourage and help farmers gain access to other technical services, such as soil testing and artificial insemination (AI), and resources (credit).

Working in collaboration with NGOs, the VEW (like KPS/AEO/Livestock Assistant/Fishery Agent/Fishery Assistant/*etc.*) would work with similar socio-economic farm households in each village who have expressed an interest in producing a particular high value commodity (dairy, chilli, *etc.*) and who might be organized into an *informal* farmer interest group (FIG). Once these groups are established, then the VEW (like KPS/AEO/Livestock Assistant/Fishery Agent/ Fishery Assistant/*etc.*) would assist these groups of farmers by making available the necessary information, technology, and training that would enable them to successfully produce these commodities, and by helping them gain access to inputs, credit and marketing services.

Finally, working with the FIAC team, the VEW would encourage newly established FIGs to join similar types of FIGs within the Block or Gram Panchayat, to form a FA that could begin to provide a broader range of commodity specific services to its members.

VEWs would work under the administrative control of the block-level Officers, and under the technical direction of FIAC team.

Project Director (ATMA) is Playing the Roles like

The Project Director (PD) for the Agricultural Technology Management Agency (ATMA) will be expected to perform the following duties:

Provide overall leadership for agricultural extension and related technology transfer activities within the district.

Work closely with the ATMA Governing Board (GB); serve as Member Secretary of the GB; schedule GB meetings; and forward meeting notices, agenda and necessary reports to GB members in advance of scheduled meetings.

Prepare and submit Annual Action Plans to the ATMA Governing Board (GB) for their review, possible modification, and approval. Prepare and submit other types of proposals, as authorized by the GB that will enhance the work and financial sustainability of the ATMA and its programs.

Develop and maintain a close working relationship with those line departments that participate in ATMA activities, as well as with other government offices in the district.

Develop and maintain a positive working relationship with NGOs, cooperatives, and private sector firms, including input supply, agro-processing, and marketing.

Promote the development of farmer organizations throughout the district.

Promote and help establish new marketing and agro-processing opportunities for farmers within the district.

Establish, equip, and maintain a suitable office for the ATMA to carry out its assigned functions, activities and tasks.

Hire and supervise ATMA staff in carrying out ATMA business.

Establish and maintain ATMA bank accounts; receive and disburse funds as directed by the GB.

Maintain accurate and up-to-date financial accounts that are periodically audited by Chartered Accountant. Submit financial reports to the TDU as required by under Government of India and World Bank agreements.

Collect project performance data and compile information needed to monitor project progress; submit performance reports to the TDU as specified in project agreements.

Carry out other duties and tasks necessary to ensure an active program of agricultural technology dissemination within the district.

The PD reports directly to the District Collector as the Chairman of the ATMA Governing Board.

Subject Matter Specialists are Playing the Roles like

At the district or sub-division level, there are subject matter specialists (SMSs). In addition, some of the other line departments, such as the Departments of Animal Husbandry (DAH) and Horticulture (DOH) have technical specialists assigned at the district level who could provide technical support to the FIAC team.

Working in collaboration with the KVK trainers and ZRS research staff, the district and/or division-level subject matter specialists (SMSs) would:

Plan, coordinate, and help provide specialized technical training and support to the block and field level extension staff.

Assist the FIAC team in planning their seasonal or annual work plans. Then, help the field staff to implement specialized extension activities, including demonstrations, farming training, and farmer field days.

Monitor and carry out periodic assessments of soil fertility levels, water management practices, pesticide use, and other factors that may effect the sustainability of different farming systems within different agro-ecological zones of the district.

Technical specialists would be administratively responsible to the district or sub-division heads of their respective line department. In addition, they would coordinate their technical program activities with ZRS researchers and KVK trainers.

ATMA Finance

Source of Funds

The resources required for the scheme shall be shared between centre and the state in the ratio of 90:10. The 10 per cent state's share shall consist of cash contribution of the State, beneficiary contribution or the contribution of other non-governmental organizations (The actual contribution made by other than the State Government in a year may be counted for the next year's State's contribution.)

Coverage of Districts

The scheme would cover 252 districts during the Xth Plan period, which will include 1/3rd of the districts taken from all the States except North-eastern States and Jammu and Kashmir wherefrom 50 per cent of the districts shall be covered. Any district receiving external assistance for the reform activities covered under this scheme shall be excluded for support. The funds available under the Scheme shall be allocated to the States in proportion to the number of Blocks being covered. It is estimated that, on an average, about Rs. 3.3 lakhs (90 per cent central share plus 10 per cent state share) would be available per block (excluding the allocation for state level activities and innovative activities to be approved at Government of India level) during the each of two year period 2005-06 to 2006-07, Rs. 8.33 lakhs per blocks (2007-08) for undertaking extension activities under the Scheme. The actual release of the resources to the States shall be on the basis of Work Plans submitted by them, keeping in mind Utilization Certificates received in respect of funds released earlier. The Work Plan provides enough flexibility to the States to propose need-based allocations for different districts within the State.

Phasing of Activities

The resources for preparation of SREP and SEWP documents shall be released in the first year along with resources needed for capacity building and infrastructure support at district and State levels. Resources meant for infrastructure support shall be released keeping in mind the infrastructure available with the concerned institutions.

Schedule of Release of Funds

Funds will normally be released in two installments in a year. Release of second installment would depend upon furnishing of UCs that have become due and release of corresponding State's share against funds released earlier by the Center. The permissible carry over of unspent balance would be 25 per cent of the annual allocation. This permissible carry over would be adjusted in the second installment. Funds will be released to States preferably through an autonomous Institution identified by the States concerned, which is closely linked to State Agriculture Programmes or through the State Agricultural Management and Extension Training Institute (SAMETI).

Monitoring and Evaluation of SEWPs

Project activities at **district** level shall be monitored by the ATMA Governing Board at periodic intervals. At **State** level, the project shall be monitored through a mechanism similar to those existing in ATMA, *i.e.*, an Inter Departmental Working Group (IDWG) functioning under Chairmanship of APC or Secretary (Agriculture) of the state. The monitoring mechanisms shall include quarterly reports, field inspections, workshops, *etc.* The Cafeteria specifically provides for third party Monitoring and Evaluation to be organized as a state level activity. The DAC will also organize concurrent Monitoring and Evaluation, including impact evaluation, as needed as part of its scheme "Extension Support to Central Institutions".

Cafeteria of Activities

- ✰ A Cafeteria of activities, which would be supported under the scheme, includes cost norms and ceilings applicable for each activity. Under the Cafeteria, activities to be undertaken at State and District level are categorized separately. The state level activities include support for upgrading state level training institutions such as SAMETI, HRD of extension functionaries, organization of state level Agri-Exhibitions and Monitoring and Evaluation of the Scheme.
- ✰ The district level activities are further categorized in four groups namely, farmer oriented activities, farm information dissemination, research-extension-farmer (R-E-F) linkages and administrative expenses. **Farmer oriented activities** include development of SREP, mobilization of farmer groups, training/exposure visit of farmers, field demonstrations, all aimed at empowering the farmers and improving their participation in technology dissemination process. Under the category **farm information dissemination**, local level agricultural exhibitions, information dissemination through printed materials and development of technology packages in electronic form are covered. The **R-E-F linkages** based activities include organization of Farmer-Scientist Interaction at local level, organization of Field days and Kisan Goshties and support for local level researchable issues emanated from the SREP. The **administrative expenses** under district level activities provide support for running ATMA like Institutions and a few block level Farm Information and Advisory Centres.
- ✰ It has been specified in the Cafeteria that funds for 'Farm-Information Dissemination' category can't exceed 20 per cent of total resources available for district level activities under the Scheme. These percentage for 'REF Linkage' and 'Administrative Expenses' categories are 15 per cent

and 20 per cent respectively. Any savings in any of these three categories could be utilized for 'Farmer-Oriented Activities'.

- The cost norms and ceilings may be exceeded by a maximum of 10 per cent of the specified amount in exceptional circumstances by a local decision. All such cases shall have to be reported in the next year's Work Plan and shall be subject to ex-post guidance. Any other deviation from the prescribed norms/ceilings or any new activity not specified in the Cafeteria could be taken up by States with prior approval of the DAC.
- The sum total of all activities specified in the cafeteria is more than the funds likely to be available for State/district level activities. States will have to prioritize their proposed activities based on gaps identified in the SREP.

Policy Parameters Governing the Cafeteria

In order to ensure that key reforms under the scheme are adequately addressed, notes given at the bottom of the Cafeteria specify the following policy parameters within which the Cafeteria is to be used:

- **Multi-agency extension strategies**: In order to ensure promotion of multi-agency extension strategies, minimum 10 per cent of allocation on recurring activities at district level is to be used through non-governmental sector *viz.* NGOs, Farmers Organization (FOs), Panchayati Raj Institutions (PRIs), Para-Extension Workers, Agripreneurs, Input Suppliers, Corporate Sector *etc.*
- **Farming system approach:** The activities specified in the Cafeteria are broad enough to ensure extension delivery consistent with farming system approach and extension needs emerging through SREPs.
- **Farmer Centric Extension Services**: The Cafeteria provides for group-based extension as it has necessary allocation for activities related to organizing and supporting farmer groups. In order to supplement these efforts, a provision for rewards and incentives to the best-organized farmer groups has also been provided in the cafeteria.
- **Convergence:** The SREP will also be a mechanism for ensuring convergence of all activities for extension. At present, resources for extension activities are being provided under different schemes of Center/State governments. However, under the scheme, it is being mandated that the Work Plan to be submitted by the State Governments for funding under the scheme shall explicitly specify the activities to be supported from the resources of other schemes as well as from the proposed scheme.
- **Mainstreaming Gender Concern:** The gender concerns are being mainstreamed by specifying in the Cafeteria that minimum 30 per cent of resources on programmes and activities are utilized for women farmers. Similarly, 30 per cent of resources meant for extension workers are proposed to be spent for women functionaries.

- **Sustainability of Extension Services:** With a view to ensure sustainability of extension services, it is being mandated that minimum 10 per cent contribution should be realized from beneficiaries with respect to beneficiary oriented activities.

The scheme has no provision for creating new posts. The manpower required for its implementation has to be provided by the concerned State governments from out of its existing staff strength on full time basis. Similarly, no infrastructure, civil works, vehicle or subsidy is permissible except to the extent specified in the Cafeteria.

Any deviation from the prescribed norms will be treated as ineligible expense and adjusted from the next year's allocation of the State.

Limited funds are being provided under the scheme for purchase of equipments for ATMA at district level and for State level activities.

Strategic Research and Extension Plan (SREP)

One of the first tasks of ATMA like institutions is to facilitate the preparation of Strategic Research and Extension Plan (SREP) of the district. The SREP is prepared through participatory methodologies such as Participatory Rural Appraisal (PRA) involving all the stakeholders and farmers. The SREP contains detailed analysis of all the information on existing farming systems in the district and research – extension gaps required to be filled-up. It also prioritizes the research – extension strategies within the district. It becomes the basis for development of work plans at block/district level.

State Extension Work Plan (SEWP)

Based on the research-extension strategies given in the SREPs, block/district level plans shall be developed by each ATMA like institutions. The State Extension Work Plan developed at state level shall contain a consolidated activity-wise plan incorporating all the District Agriculture Action Plans (DAAPs) in the state and state level activities to be carried out with activity-wise budgetary requirement as per the norms prescribed in the cafeteria. It will also indicate all other extension activities that may be undertaken from out of resources provided under any other scheme of the Centre/State Governments. The DAAPs developed under the scheme should be processed consistent with Article 243 ZD of the Constitution.

Important Additional Guidelines for Implementation of the Scheme

- There should be no duplication between the activities of KVKs and extension activities to be taken up under the scheme. ICAR should ensure that research priorities of KVK are fully aligned with and sensitive to SREPs at the District level. KVK should be an active partner to operationalize the ATMA model.
- Macro Management programmes and Centre/State sponsored schemes for agriculture and allied sectors should be integrated with the activities

of ATMA by creating desired linkages for effective convergence and delivery.

- ☆ Base line surveys should be undertaken in each ATMA District against which the scheme impact should be evaluated. Baseline survey should be made an integral part of the Strategic Research and Extension Plan (SREP) process. Adequate funds would be provided to ATMAs for baseline surveys.
- ☆ The States may consider associating Panchayati Raj Institutions at all levels of institutional framework under the Scheme.
- ☆ The Secretary of Department of Panchayati Raj of the State Government should be included as a member of IDWG.
- ☆ ATMA should propagate only economically viable and proven technologies since it aims to help resource poor families.
- ☆ ATMA should be part of an integrated extension system consisting of farm schools in the field of farmers, KVKs, agri-clinics and agri-business centres.
- ☆ The ultimate aim of ATMA should be to make small farmers credit worthy (without collateral).

44

Different Extension Programmes in India

Attracting and Retaining of Youth in Agriculture (ARYA)

In order to create interest and confidence among rural youth in agriculture, there is needed to make agriculture more profitable. Retaining youth in agriculture and making agriculture more profitable are thus, big challenges. There is a continuous increase in migration of rural youth to urban areas. On the other hand, small holdings are on the rise which posses challenge to food security for increasing population. Thus, it was felt to bring a comprehensive model for the development of rural youth in general and agricultural youth in particular. Thus, realising the importance of rural youth in agricultural development especially from the point of view of food security of the country, ICAR has initiated a programme on "Attracting and Retaining of Youth in Agriculture (ARYA)".

Objective of ARYA Project

1. To attract and empower the Youth in Rural Areas to take up various Agriculture, allied and service sector enterprises for sustainable income and gainful employment in selected districts.
2. To enable the Farm Youth to establish network groups to take up resource and capital intensive activities like processing, value addition and marketing.
3. To demonstrate functional linkage with different institutions and stakeholders for convergence of opportunities available under various schemes/program for sustainable development of youth.

ARYA project will be implemented in 25 States through KVKs, one district from each State. In one district, 200-300 rural youths will be identified for their skill development in entrepreneurial activities and establishment of related micro-enterprise units. KVKs will involve the Agricultural Universities and ICAR Institutes as Technology Partners. At KVKs also one or two enterprise units will be established so that they serve as entrepreneurial training units for farmers. The purpose is to establish economic models for youth in the villages so that youths get attracted in agriculture and overall rural situation is improved. Skill development of rural youths will help in improving their confidence levels and encourage them to pursue farming as profession, generate additional employment opportunities to absorb under employed and unemployed rural youth in secondary agriculture and service related activities in rural areas. The concurrent monitoring, evaluation and mid-term correction will be an integral part of project implementation.

Farmer FIRST Programme (FFP)

The Farmer FIRST Programme (FFP) is an ICAR initiative to move beyond the production and productivity, to privilege the smallholder agriculture and complex, diverse and risk prone realities of majority of the farmers through enhancing farmers-scientists interface. There are concepts and domains that are new in emphasis like resource management, climate resilient agriculture, production management including storage, market, supply chains, value chains, innovation systems, information systems, *etc.* The Farmer FIRST as a concept of ICAR is developed as farmer in a centric role for research problem identification, prioritization and conduct of experiments and its management in farmers' conditions. The focus is on farmer's Farm, Innovations, Resources, Science and Technology (**FIRST**). Two terms 'enriching knowledge' and 'integrating technology' qualify the meaning of Farmer FIRST in Indian context. Enriching knowledge signifies the need for the research system as well as farmers to learn from each other in context to existing farm environment, perception of each other and interactions with the sub-systems established around. Technology integration is looked from the perspective that the scientific outputs coming out from the research institutions, many times do not fit as such in the farmers' conditions and thus, certain alterations and adaptations are required at field level for their acceptance, adoption and success.

Indian agriculture embraces diverse actors in its endeavour to feed 1.21 billion people. The small, marginal and landless farmers are extremely vital for food security as land holdings are shrinking day by day. The contribution of women farmers is also immense particularly in on farm operations, harvesting, post harvest management, savings and in other activities especially in horticulture and animal husbandry sector.

In today's scenario innovation in agriculture is key to address growing challenges. The innovations in agriculture from scientists of NARS to Farmer Innovators need to be validated, integrated and scaled up. The innovation development process consists of all the decisions, activities, and their impacts that occur from recognition of a need or problem, through research, development, and commercialization of an innovation, through diffusion and adoption of the innovation by users.

There are many agencies other than public organizations and various kinds of public-private farmer's partnerships. Earlier farming was an adapting performance. Participatory approaches and processes were central. Relationships and interactions are seen more important. Partnerships like public-private, multi-stakeholder partnership *etc.*, and collaborations like co-management, co-breeding, co-creation, co-evolution, co-development, *etc.* have emerged in present day context. Learning of different kinds has been evolved- interactive, experiential, alternative, *etc.* Now concern is not only with capacity building but with mind sets, perceptions, priorities and realities.

About the Project

There is growing perception tha1t the emerging demand of the farmers about the recent technological and institutional needs is not adequately addressed. Also many times, research system is not getting adequate feedback to plan and conduct demand driven research thereby, a huge gap exists in the quality of research output required at the farm level and that being developed. It is also being perceived that the research system should play a pro-active role in reaching to the farmers for getting first hand information, farmers' perceptions, feedback on generated technologies and develop new more appropriate processes, methodologies and technologies for diverse farm environments.

In the past ICAR led in responding the contemporary challenges and other issues of Indian agriculture whether it was Green Revolution (complimented with National Demonstration, 1964), upliftment of small and marginal farmers (complimented with Lab to Land Programme, 1979), watershed development, soil improvement, crop protection, focus on weaker section (complimented with Operational Research Project, 1974), technology assessment and refinement (complimented with IVLP, 1995), *etc.*, its response was timely and need based. The Initiatives of NAIP have extended the efforts towards improving rural livelihood of farmers living in less favoured, marginal or more complex environment.

The new proposed project – 'Farmer FIRST' is an ICAR initiative to move beyond the production and productivity and to privilege the complex, diverse and risk prone realities of majority of the farmers through enhancing farmers-scientists contact with multi stake holders-participation. Many aspects are multiple or multi; multiple stakeholders, multiple perspectives, multiple realities, multi-functional agriculture, multi-method approaches. There are concepts and domains that are new or new in emphasis like food systems, trade, market chains, value chains, innovation pathways and most of all innovation systems.

'Farmer FIRST' aims at enriching farmers-scientists interface for technology development and application. It will be achieved with focus on innovations; feedback; multiple stakeholders participation, multiple realities, multi method approaches, vulnerability and livelihood interventions.

The project is conceptualized to deal with focus on:

i) Enriching Farmers–Scientist interface

Enabling involvement of researchers for continuous interaction with farm conditions, problem orientation, exchange of knowledge between farmers and other stakeholders, prioritization of problems and setting up of research agenda.

ii) Technology Assemblage, Application and feedback

Integrating components of technology for application in different agro ecosystems with focus on innovations and feedback.

iii) Partnership and Institutional Building

Building partnerships involving different stakeholders; development of rural based institutions; agro-ecosystem and stakeholders analysis and impact studies.

iv) Content Mobilization

Using the platform of the project having commodity institutions as partners to develop commodity specific contents for e-enabled knowledge sharing.

Objectives

The specific objectives are:

1. To take up technology development based on feedback with the participation of farmers and landless for enhancing production, productivity, income and equitability of the households.
2. To build a network of linkages with different entities around the farm house holds *viz.*, technology institution, input support system, market, *etc.* For facilitating access of information, technology and marketability of produce for higher returns.
3. To find out the technical, socio-economic and environmental impact of the project to develop a database on performance of NARS technologies; perception of the farmers on technologies; agriculture as a profession in the rural settings; migration, *etc.*
4. To identify and integrate economically viable and socially acceptable entrepreneurial activities as models of enhanced earnings to the farmers in different agro-ecological situations.
5. To initiate special modules for farm women to enhance their participation in agricultural activities for higher earnings and livelihood security.
6. To utilize the strength of the technology institution (partners) to develop commodity specific contents for knowledge sharing among various stakeholders specially farmers.

Mera Gaon Mera Gaurav

Mera Gaon Mera Gaurav is a path breaking program introduced by Hon'ble Prime Minister of India for upliftment of farmers of the country. The programme has been initiated in the year 2015-16.

Second Green Revolution

As we all know current state of agriculture in India is result of green revolution which is in place since late 1960's, which was heavily backed by government. It has delivered India food security and sufficiency which was critical at that time. This progress and security had its own costs in terms of environment and economic viability. Green revolution rampantly used fertilizers and other chemicals, which made food and water, toxic to some extent. It supported indiscriminate use of water as rice was main target of revolution. India's 80 per cent of fresh water is consumed by agriculture, more particularly by rice farming. Further, in past GR constantly increased yields per hectare of various food crops. Originally yield per hectare for rice was merely 1 ton and now it stands over 3 tons. In case of wheat it's around 2.4 tons. This increase was incremental over all these years, as more and more cultivated area kept coming under irrigation, high yield varieties and fertilizers. But in recent past, these measures are showing fatigue and yield growths are flattening up. We have seen in last articles that increasing costs of fertilizers are pushing agro costs upwards and turning food unaffordable for masses. Also, due to indiscriminate use of fertilizers, nutrients in the soil are depleting rapidly.

In short, it is being realized that original means adopted by green revolution are to some extent exhausted. As yields on Indian lands are still far below world average, our policy makers will have to look somewhere else, to bring it at par with world average. Current state of agriculture is not sustainable at all. Consequently, new agriculture policy of India aims at sustainable agriculture, which is popularly called 'second green revolution' or 'Evergreen Revolution'.

Almost every current policy can be covered under category of second green revolution. It aims at information technology revolution, bio technology revolution, R and D and targeting of specific crops *etc.* Information Technology will create awareness in farmers, introduce them with more cost efficient technologies and enable them to choose among different alternatives. It is, in fact a support service like customer care, where information deficit on part of farmer will be addressed. We tried to cover this at length in previous article.

Bio technology revolution aims at improving genetic traits of crops, by making them draught, pest, weed, climate resistant. This is to be done by genetically engineered crops, by continuously improving them. However use of GMO crops and its benefits are controversial and this issue was also covered in previous article. Another aspect of bio tech revolution is use of bio pesticides and bio fertilizers. This is actually part of bigger strategy targeted towards organic farming.

Organic Farming in India

Organic farming system in India is not new and is being followed from ancient time. It is a method of farming which is primarily aimed at cultivating the land and raising crops in such a way, as to keep the soil alive and in good health by use of organic wastes (crop, animal and farm wastes, aquatic wastes) and other biological materials along with beneficial microbes (bio fertilizers) to release nutrients to crops for increased sustainable production in an ecofriendly pollution free environment.

To promote organic farming a central scheme named 'National Project on Organic Farming' is in place. Apart from general things (creating awareness, promoting organic fertilizers, training, capacity building *etc.*) scheme provides financial assistance through Capital Investment Subsidy Scheme for agro-waste compost production units, bio-fertilizers/bio-pesticides production units, development and implementation of quality control regime, human resource development, *etc.* It is worth mentioning that, north eastern India has highest potential for organic farming. It has distinct agro- climatic zone and its soil is immensely rich in organic matter. Further, it was aloof of any green revolution and consequently farmers there still use traditional methods. All this obviously is perfect opportunity for promoting organic farming rigorously in the north east. There is 'Horticulture Mission for North East and Himalayan States' under which financial assistance is provided for adopting organic farming and setting up of vermin compost unit. Note that scheme is also available in other Himalayan states. Also, there is 'Indo Israel Agricultural cooperation action plan' in which Israel shares its expertise (mainly in organic farming and related technologies) with India. This cooperation includes structured interactions between governments, experts and most important – farmers. This comprises of bilateral sharing of knowledge and technologies, establishment of Centers of Excellence in various agriculture sectors across India, exchange visits of experts and farmers and Post-doctoral scholarships for Indian agriculture researchers in Israel. Two Centers of Excellence have already been made fully operational – one for vegetables in Gharaunda (Karnal) and the other for fruits in Mangiana (Sirsa) — in Haryana while work is in progress to establish three such centers in Maharashtra- a citrus centre in Nagpur, a pomegranate centre in Rahuri and a Kaisar mango centre in Aurangabad.

Mission for Integrated Development of Horticulture (MIDH)

It is a Centrally Sponsored Scheme for the holistic growth of the horticulture sector covering fruits, vegetables, root and tuber crops, mushrooms, spices, flowers, aromatic plants, coconut, cashew, cocoa and bamboo. Scheme took off in 2014. While Government of India (GOI) contributes 85 per cent of total outlay for developmental programs in all the states except the states in North East and Himalayas, 15 per cent share is contributed by State Governments. In the case of North Eastern States and Himalayan States, GOI contribution is 100 per cent. National Horticulture and Bamboo mission are integrated under this mission. These both come under National Horticulture Board.

National Horticulture Mission

National Horticulture Mission, a centrally sponsored scheme is in place since 2005-06. The objective of this scheme is to provide holistic growth of horticulture sector in India and to enhance horticulture production. All the States and Union Territories are covered under the Mission except the eight North Eastern States including Sikkim and the States of Jammu and Kashmir, Himachal Pradesh and Uttrakhand, which are covered under previously mentioned separate mission. The assistance from Government of India will be 85 per cent with 15 per cent contribution by the State Government. All crops except coconut and medicinal plants are covered under NHM.

National Bamboo Mission

Mission envisages promoting holistic growth of bamboo sector by adopting area based, regionally differentiated strategy and to increase the area under bamboo cultivation and marketing. Steps have been taken to increase availability of quality planting material by supporting setting up of new nurseries and strengthening of existing ones. It addresses forward integration by taking steps toward marketing of bamboo products, particularly of handicraft items. Here also biggest potential lies in north eastern India.

Rashtriya Krishi Vikas Yojna

Introduced in 2007 as a new 'Additional Central Assistance scheme' to incentivize States to draw up plans for their agriculture sector more comprehensively, taking agro-climatic conditions, natural resource issues and technology into account, and integrating livestock, poultry and fisheries more fully. This involves new schemes for Additional Central Assistance to State Plans, administered by the Union Ministry of Agriculture over and above its existing Centrally Sponsored schemes, to supplement the State-specific strategies including special schemes for beneficiaries of land reforms. There is no. of sub schemes under RKVY, some of them are:

a) Bringing Green Revolution to Eastern India

Specifically targets improvement in rice productivity in 7 eastern states *viz.*

Bengal, Bihar, Orissa, Assam, Chhattisgarh, Jharkhand and eastern Uttar Pradesh as original green revolution brought a little impact here. It includes some components of green revolution, such as irrigation and also focuses on water harvesting/ conservation *etc.*

b) National Saffron Mission

Kashmir valley's Karewa soils are perfect for growth of saffron which is considered as one of the most expensive cash crop. However area under saffron cultivation and yield per hectare are on constant decline. To arrest this decline National Saffron Mission was initiated in 2010. The objective (as in other missions) of the Saffron National Mission is to improve the overall production of saffron, enhancing quality of saffron, enhancement of research and extension capability and develop appropriate system for organized marketing for the growers.

c) Nutri-cereals

To promote production of millets viz. Jowar, Bajra, Ragi *etc.* through better technologies and creating awareness about their health benefits. These crops are more nutrient and protein rich then rice and wheat, but less remunerative to farmers.

d) Accelerated Fodder Development Program

Targets fodder development through technologies. It is quite important for Animal Rearing sector as it is getting short of fodder

e) Promotion of Oil Palm

f) National Mission for Protein Supplements

To promote animal based protein production through livestock development, dairy farming, piggery, goat rearing *etc.*

g) Rainfed Area Development Program

Improve quality of life of small and marginal farmers in rainfed areas.

h) Crop Diversification in Original Green Revolution States

States of Punjab, Haryana and Uttar Pradesh are involved in 'Intensive farming' of rice and wheat which is causing depletion of soil nutrients. To stop this it is very important to diversify cultivation towards horticulture, which is precisely aimed by the scheme.

i) National Mission for Sustainable Agriculture

It is one of the eight Missions under the National Action Plan on Climate Change and seeks to address issues regarding 'Sustainable Agriculture' in the context of risks associated with climate change by devising appropriate adaptation and mitigation strategies for:

1. Ensuring food security,
2. Equitable access to food resources,

3. Enhancing livelihood opportunities and
4. Contributing to economic stability at the national level.

It seeks to transform Indian agriculture into a climate resilient production system in the domain of crops and animal husbandry. These interventions would be embedded in research and development activities, absorption of improved technology and best practices, creation of physical and financial infrastructure and institutional framework, facilitating access to information and promoting capacity building. Promotion of dryland agriculture would receive prime importance by way of developing suitable drought and pest resistant crop varieties. The Mission would also expand its coverage to rainfed areas for integrating farming systems with management of livestock and fisheries, so that agricultural production continues to grow in a sustainable manner. Under integrated farming system, agriculture, horticulture or animal husbandry is integrated, to develop a system which derives synergies from each other. An example, Poultry waste can be fed to fishes hence saving costs and it can also be used as effective manure for horticulture.

j) National Food Security Mission

Food Security Mission was launched in 2007, which comprised rice, wheat and pulses to increase the production of rice by 10 million tons, wheat by 8 million tons and pulses by 2 million tons by the end of the Eleventh Plan. The Mission is continued during 12th Five Year Plan with new targets of additional production of food grains of 25 million tons of food grains comprising of 10 million tons rice, 8 million tons of wheat, 4 million tons of pulses and 3 million tons of coarse cereals by the end of 12th Five Year Plan. Mission also aims at improving soil fertility and improving employment opportunities. Mission during 12th FYP has five components:

1. NFSM- Rice
2. NFSM-Wheat
3. NFSM-Coarse cereals
4. NFSM-Commercial Crops
5. NFSM-Pulses: This mission is further supplemented by Accelerated Pulses Production Program (A3P). India is biggest producer of pulses at 25 per cent of world's production and still it has to import pulses. This is mainly because of low yields. A3P aims at Integrated Nutrient Management and Plant protection to enhance yields and productivity.

k) Integrated Pest Management

The term IPM refers to – "the careful consideration of all available pest control techniques and subsequent integration of appropriate measures that discourage the development of pest populations and keep pesticides and other interventions to levels that are economically justified and reduce or minimize risks to human health and the environment. IPM emphasizes the growth of a healthy crop with the least possible disruption to agro-ecosystems and encourages natural pest control mechanisms"

l) Integrated Scheme for Oilseed, Pulses, Oil Palm and Maize (ISOPOM)

Four erstwhile schemes of Oilseeds Production Program, Oil palm Development Program, National Pulses Development Program and Accelerated Maize Development Program – have been merged into one Centrally Sponsored 'Integrated Scheme of Oilseeds, Pulses, Oil palm and Maize' (ISOPOM) being implemented from 01.04.2004. Different components (i.e. pulses, oil palm *etc.*) of scheme are in force in different states as per dominant cropping pattern of the state. For e.g. for oil palm it is applicable in 9 major oil palm producing states. It aims to promote crop diversification and provides focused approach by giving financial assistance in purchase of various agriculture inputs such as seeds, agro chemicals, technology, expertise, infra *etc.* Subsidy is funded by Central and state government in 75:25 ratio.

m) Technology Mission on Cotton

The Govt. of India launched Technology Mission on Cotton in February 2000. The objective of TMC was as under:

1. To improve the yield and quality of cotton, particularly in respect of staple length, micronnaire, strength, *etc.* through development of better cotton varieties as well as through improved seeds, and integrated water, nutrient and pest management technologies.
2. To increase the income of the cotton growers by reducing the cost of cultivation as well as by increasing the yield per hectare through proper transfer of technology to the growers.
3. To improve the quality of processing of cotton, particularly in respect of trash, contamination, *etc.* by improving the infrastructure in the market yards for cotton and by modernizing the existing ginning and pressing factories and setting up new units.

Right now India is at second position behind china in cotton production and it is expected to overtake china this year. After this mission there has been constant increase in cotton production on back of high yield variety and GMO seeds.

Initiatives in Budget 2014

1. Price Stabilization Fund

A Rs. 500 crore fund was announced to compensate farmers when there is fall in agricultural prices. The Price Stabilization Fund Scheme for the Plantation crops (Tea, Coffee and Rubber) was already there since 2003 with the objective of providing financial relief to growers on account of fall in the prices of commodities below a specified level. A Price Stabilization Fund for agricultural commodities has been announced in the Union Budget 2014-15 with a view to mitigate volatility in the prices of agricultural produce.

2. Pradhan Mantri Krishi Sinchayee Yojana

The Government has proposed to initiate the scheme "Pradhan Mantri Krishi Sinchayee Yojana". It is proposed to meet the need to provide assured irrigation to mitigate risk to the farmer since bulk of the farm lands are rainfed and depend on monsoon. This scheme would facilitate access to irrigation. A sum of Rs.1000 crores is being set aside for this scheme.

3. Soil Health Card

One of the most serious problems of Indian agriculture is bad targeting of inputs, be it water, fertilizer or choice of crops to be grown. Soil of different farmlands differ in their nutrient composition and can support (or not) different specific suppose. If farmer is aware which crop to grow and how much and which nutrient to supply, then he will save enormous costs and at same time get better yields. For this Scheme of Soil health card is initiated on pilot basis. Under this mobile soil testing laboratories will be set up which will do soil certification and farmers will be issued a soil health card.

State Governments have adopted innovative practices like involvement of agricultural students, NGOs and private sector in soil testing, determining average soil health of villages, *etc.*, to issue Soil Health Cards.

This is prerequisite for effective 'integrated nutrient management'

The scheme 'Soil Health Card' has been approved by the Ministry of Agriculture and Farmers Welfare and for implementation during the remaining period of 12th Plan with an outlay of Rs. 568.54 crore. Under the Scheme, Soil Health Card will be provided to all the farmers in the country at an interval of 3 years so as to enable the farmers to apply appropriate recommended dosages of nutrients to realize improved and sustainable soil health and fertility status and also to ensure higher per unit yields.

Given the problems of chemical based agriculture, second green revolution is quite important that prefers organic farming. It is true that heavy current dependence upon chemicals and water intensive techniques, gives little room for any remarkable change in near future. But as time passes, conventional means will get even more expensive and this coupled with growing concern for water security will force farmers to adopt sustainable methods. Of late farmers are showing their interest in organic farming. Its advantages are many – environment friendly, nontoxic, its value is significantly higher than normal crops. Interestingly, this sector is attracting investments from non-farmers too. Further, R&D is focusing on developing water efficient technologies. One such technology developed is 'system of rice intensification 'which not only saves water significantly, but also increases yield of rice.

References

Aarnink, N., and Kingma, K. (1991). *Female farmers and male extension workers: Women and agriculture in Tanzania II.* Leiden: Women and Autonomy Centre, Leiden University.

ACC (Administrative Committee on Coordination), United Nations (1984). *Guiding principles for the design and of monitoring and evaluation in rural development projects and programmes.* Rome: International Fund for Agricultural Development (IFAD).

Ackoff, R.L. (1999) *Ackoff's Best: His Classic Writings on Management.* Wiley, NY.

Adams, M. E. (1982). *Agricultural extension in developing countries.* Burnt Mill, Harlow, Essex: Longman

Adhikarya, R. (1978). Guideline proposal for a communication support component in transmigration projects. A consultancy report prepared for FAO/United Nations (project 6/INS/01/T). Rome: FAO.

Adhikarya, R. (1985). Planning and development of rat control campaign objectives and strategies for the state of Penang, Malaysia. A summary report of a workshop on multimedia campaign planning for integrated pest control, organized by FAO/United Nations (projects GCP/RAS/101/NET and GCP/RAS/092/AUL).

Adhikarya, R. (1994). *Strategic extension campaign: A participatory-oriented method of agricultural extension.* Rome: FAO/United Nations.

Adhikarya, R., and Middleton, J. (1979). *Communication planning at the institutional level: A selected annotated bibliography.* Honolulu, Hawaii: East-West Communication Institute, the East-West Center.

Adhikarya, R., and Posamentier, H. (1987). *Motivating farmers for action: How strategic multi-media campaigns can help*. Eschborn, Frankfurt, Germany: GTZ (Deutsche Gesellschaft für Technische Zusammenarbeit).

Adnan, S., Barren, A., Nurul Alam, S. M., and Brustinow, A. (1992). *People's participation: NGOs and the flood action plan*. Dhaka, Bangladesh: Research and Advisory Services.

Agricultural Extension Service. (1978). *Agricultural extension in Japan*. Tokyo: Japan Agricultural Extension and Development Association.

Agricultural Research and Extension Network Paper No. 32. London: Overseas Development Institute.

Aguirre, P., and Namdar-Irani, M. (1992). Complementarities and tensions in NGO-state relations in agricultural development: The trajectory of AGRARIA (Chile).

Ahituv, N., Neumann, S., and Riley, H. N. (1994). *Principles of information systems for management* (4th ed.). Dubuque, IA: Wm. C. Brown Communications.

Ahmed, A. (1982). *The role of the information system in development*. Studies Series, No. 314. Baghdad, Iraq: Ministry of Culture and Information.

Ahmed, SA (2008) "Community Led Total Sanitation in Bangladesh:Chronicles of a People's Movement" IDS Conference paper Accessed 2015-02-27

Aidoo, A. A. (1988). Women and food security: The opportunity for Africa. *Development*, 2(3), 56-61.

Akor, R. (1990). *The role of women in agriculture and constraints to their effective participation in agricultural development in Nigeria*. Paper presented at UNDP/ILO/DFRRI Training Workshop on Monitoring and Evaluation of Rural Women in Productive Skills Project.

Albrecht, H., Bergman, H., Diederich, G., Grober, E., Hofman, V., Keller, P., Payr, G., and Sülzer, R. (1990). *Agricultural extension: Examples and background material* (Vol. 2). Eschborn: BMZ/GTZ/CTA.

Albrecht, H., Bergmann, H., Diederich, G., Grober, E., Hoffmann, V., Keller, P., Payr, G., Sülzer, R. (1989). *Agricultural extension*. Vol. 1: Basic concepts and methods. GTZ Eschborn.

Albrecht, H., Bergmann, H., Diederich, G., Grosser, E., Hoffmann, V., Keller, P., Payr, G., and Sülzer, R. (1990). *Agricultural extension: Vol. 2. Examples and background material*. Eschborn: Deutsche Gesellschaft für Technische Zusammenarbeit (GTZ).

Alexandratos, N. (1995). *World agriculture: Towards 2010, an FAO study*. West Sussex, England: FAO and John Wiley and Sons.

Allen, W.J. (2001) Working Together for Environmental Management: The Role of Information Sharing and Collaborative Learning. PhD Thesis, Massey University, Auckland, NZ.

Altieri, M. A., and Yurjevic, A. (1992). Changing the agenda of the universities. *ILEIA Newsletter, 2*, 39.

Amado, G. (1993) "La résonance psychosociale". PhD Thesis, Université Paris-VII, Paris.

Amanor, K., and Farrington, J. (1991). NGOs and agricultural technology development. In W. M. Rivera and D. J. Gustafson (Eds.), *Agricultural extension: Worldwide institutional evolution and forces for change.* Amsterdam: Elsevier.

An Open Letter in response to the World Development Report 2015. Retrieved 23 March 2015.

Andrew Bartlett. (2005) Farmer Field Schools to promote Integrated Pest Management in Asia: the FAO experience, *Workshop on Scaling Up Case Studies in Agriculture*. IRRI.

Antholt, C. H. (1992). *Relevancy, responsiveness and cost-effectiveness: Issues/or agricultural extension in the 21st century.* Washington, DC: Asia Region, Technical Department, World Bank.

Antholt, C. H. (1994). *Getting ready for the twenty-first century: Technical change and institutional modernisation in agriculture.* World Bank Technical Paper 217. Asia Technical Department Series. Washington, DC: World Bank.

Anzieu, D. and Martin, J.-Y. (1966) *La dynamique des groupes restreints*. PUF, Paris.

Apps, J. W. (1991). *Mastering the teaching of adults.* Malabar, FL: Krieger.

Arasingham, S. R. (1981). Agricultural extension in Sri Lanka. *Tropical Agriculturist,* 137, 41-54.

Archer, T., and Layman, J. (n.d.). Focus group interview. *Evaluation Guide Sheet.* Columbus: The Ohio Cooperative Extension Service, The Ohio State University.

Argyris, C. (1993) *Knowledge for Action: A Guide to Overcoming Barriers to Organizational Change*. Jossey-Bass, CA.

Argyris, C. and Schön, D.A. (1989) "Participatory Action research and Action Science Compared." *American Behavioral Scientist* 32: pp. 612–23.

Argyris, C., Putnam, R. and Smith, D. (1985) *Action Science: Concepts, Methods and Skills for Research and Intervention*. Jossey-Bass, CA.

Arnstein, S. (1969). The ladder of citizen participation. *Journal of American Institute of Planners, 35,* 221.

Asian Productivity Organization (APO). (1994). *Agricultural extension systems in Asia and the Pacific.* Tokyo: Asian Productivity Organization.

Attah-Krah, A. N., and Francis, P. A. (1987). The role of on-farm trials in the evaluation of composite technologies: The case of alley farming in southern Nigeria. *Agricultural Systems, 23,* 133-152.

Avault, J.W., Jr., 1984. Plan for research: Inland Fisheries Department of the Central Agricultural Research Institute, Liberia. Baton Rouge, Louisiana (USA), Louisiana State University

Awad, E. M., and Gotterer, M. H. (1992). *Database management*. Danvers, MA: Boyd and Fraser.

Axinn, G. H. (1988). T&V (tragic and vain) extension. *INTERPAKS, INTERCHANGE, International Agriculture, 5* (3). Urbana: College of Agriculture, University of Illinois at Urbana-Champaign.

Axinn, G. H., and Thorat, S. (1972). *Modernising world agriculture: A comparative study of agricultural extension education systems.* New Delhi: Oxford and IBH Publishing.

Axinn, G.H. (1988). Guide on alternative extension approaches. Agricultural Education and Extension Service, Human Resources Institutions and Agrarian Reform Division. Rome, FAO, M/S8971E/1/12.88/1000

Ayers, A. J. (1992). Conflicts or complementarities? The state and NGOs in the colonization zones of San Julian and Berlin, eastern Bolivia. *Agricultural Research and Extension Network Paper No. 37.* London: Overseas Development Institute.

Bagadion, B. U., and Korten, F. F. (1991). Developing irrigators' organizations: A learning process approach. In M. M. Cernea (Ed.), *Putting people first* (2nd ed.). Oxford: Oxford University Press.

Bagadion, B. U., and Korten, F. F. (1991). Developing irrigators' organizations: A learning process approach. In M. M. Cernea (Ed.), *Putting people first* (2nd ed.). Oxford: Oxford University Press.

Bahal, R., Swanson, B. E., and Farner, B. J. (1992). Human resources in agricultural extension: A worldwide analysis. *Indian Journal of Extension Education, 28* (3, 4), 1-9.

Bailey, C. (1988). The social consequences of tropical shrimp mariculture development. Ocean and Shoreline Management, 11:31-44

Balint, M. (1954) "Training General Practitioners in Psychotherapy". *British Medical Journal*, vol I, pp. II5-20.

Banerjee, U. K., and Sachdeva, R. K. (1995). *Management information system: A new frame work.* New Delhi: Vikas Publishing House.

Bannister, M. E., and Nair, P. K. R. (1990). Alley cropping as a sustainable agricultural technology for the hillsides of Haiti: Experience of an agroforestry outreach project. *American Alternative Agriculture, 5* (2), 51-59.

Barr, N. P., and Cary, J. W. (1992). *Greening a brown land: The Australian search for sustainable land use.* Melbourne: Macmillan.

Bartram, J; Charles,K; Evans, B; O'Hanlon, L; Pedley, S (2012). "Commentary on community-led total sanitation and human rights: Should the right to community-wide health be won at the cost of individual rights?". Journal of Water and Health. ***10*** *(4): 499–503. doi:10.2166/wh.2012.205.*

Bartunek, J.M. and Schein, E.H. (2011) "Organization Development Scholar Practitioners: Between Scholarship and Practice", in S.A. Mohrman and E.E. Lawler (eds), *Useful Research: Advancing Theory and Practice*, Berrett Koehler, CA, pp. 233–250.

Barus-Michel, J. (1987) *Le sujet social: étude de psychologie sociale clinique*. Dunod, Paris.

Barus-Michel, J., Enriquez, E. and Lévy, A. (2006) *Vocabulaire de psychosociologie: Positions et références*. Erès, Paris.

Bass, B. M., and Vaughan, J. A. (1966). *Training in industry: The management of learning*. Belmont, CA: Wadsworth Publishing.

Bathgate, F. (Ed.). (1956). *Experiment in extension-The Goon Sathi.* London: Oxford University Press.

Bawden, R. (1992). Creating learning systems: A metaphor for institutional reform for development. Paper presented at the joint conference of IIED/IDS. IDS, University of Sussex: IIED, London.

Bawden, R. (1994). *A learning approach to sustainable agriculture and rural development: Reflections from Hawkesbury (mimeo)*. Australia: Hawkesbury College.

Baxter, M. (1990). Agricultural extension investment. In D. C. Misra (Ed.), *Training and visit system of agricultural extension in India in action.* New Delhi: Ministry of Agriculture, Government of India.

Bebbington, A. J., and Riddell, R. (1994). *New agendas and old problems: Issues, options and challenges in direct funding of southern NGOs.* Draft, mimeo. London: Overseas Development Institute.

Bebbington, A. J., and Thiele, G. (Eds.). (1993). *NGOs and the state in Latin America: Rethinking roles in sustainable agricultural development.* London: Routledge.

Beck, R. (1986). Nimba County rural development project. Aquaculture promotion programme: review and recommendations. Monrovia (Liberia), Nimba County Rural Development Project

Beebe, J. (1987). Rapid appraisal: The evolution of the concept and the definition of issues (p. 47-68). In *Proceedings of the 1985 international conference on rapid rural appraisal.* Khon Kaen, Thailand: Khon Kaen University.

Belenky, M.F., Clinchy, B.M., Goldberger, N.R. and Tarule, J.M. (1986). *Women's Ways of Knowing*. Basic Books, NY.

Bennet, C. (1990). *Cooperative extension roles and relationships for a new era: Summary.* Washington, DC: Extension Service, U.S. Department of Agriculture.

Benor, D. (1987). Training and visit extension: Back to basics. In Rivera and Schram, *Agricultural Extension Worldwide* (p. 137-149). New York: Croom Helm.

Benor, D., and Harrison, J. Q. (1977). *Agricultural extension: The training and visit system.* Washington, DC: The World Bank.

Benor, D., Harrison, J. Q., and Baxter, M. (1984). *Agricultural extension: The training and visit system.* Washington, DC: World Bank.

Bentley, I. W., Rodriguez, G., and Gonzalez, A. (1993). Science and the people: Honduran compressions and natural pest control inventions. In D. Buckles (Ed.), *Gorras y sombreros: Caminos hacia la colaboración entre técnicos y campesinos.* El Zamarano, Honduras: Department of Crop Protection.

Bentley, J. (1994) "Facts, Fantasies, and Failures of Farmer Participatory Research", *Agriculture and Human Values*, vol 11, no 2-3, pp. 140–150.

Ben-Yami, M. (1987). Aquaculture: the importance of knowing its limitations. Rome, FAO, Ceres, 112:15-19

Bergdall, T. D. (1993). *Methods for active participation: Experiences in rural development from East and Central Africa.* Nairobi: Oxford University Press.

Berger, M., DeLancey, V., and Mellencamp, A. (1984). *Bridging the gender gap in agricultural extension.* Washington, DC: ICRW.

Berger, Peter L. (1967). The Scared Canopy: Elements of a Sociological Theory of Religion. Garden City, NYC: Doubleday and Company, Inc. p. 3.

Bessette, G. (2004) *Involving the Community: A Guide to Participatory Development Communication.* IDRC, Ottawa.

Bessette, J. (1994) *The Mild Voice of Reason: Deliberative Democracy and American National Government.* University of Chicago Press, Chicago.

'Beyond Subsidies - Triggering a Revolution in Rural Sanitation' Institute of Development Studies (IDS) In Focus Policy Brief 10 July 2009.

Bhasin, K. (1976). *Participatory training for development.* FFHC/AD. Bangkok: FAO.

Bhatia, C. R., Review of the book '*Revitalizing Higher Agicul-tural Education in India: Journey Towards Excellence*' by P. M. Tamboli and Y. L. Nene, Published by Asian Agri-History Foun-dation. *Curr. Sci.*, 2011, **101**, 1378.

Bhatnagar, B., and Williams, A. (1992). *Participatory development and the World Bank: Potential directions for change.* Washington, DC: World Bank.

Bijlmakers, H. (2005). Farmer Field Schools for IPM - Refresh your Memory, IPM DANIDA

Bindlish, V., and Evenson, R. E. (1993). *Evaluation of the performance of T&V extension in Kenya.* World Bank Agricultural and Rural Development Series No. 7. Washington, DC: World Bank.

Bindlish, V., Gbetibouo, M., and Evenson, R. (1993). *Evaluation of T&V extension in Burkina Faso.* World Bank Technical Paper No. 226. Washington DC: African Technical Department.

Bion, W.R. (1961) *Experiences in Groups and Other Papers.* Tavistock, London.

Birkhaeuser, D., Evenson, R. E., and Feder, G. (1991). The economic impact of agricultural extension: A review. *Economic Development and Cultural Change, 39*, 607-650.

Birkhauser, D., Evenson, R. E., and Feder, G. (1988). *The economic impact of agricultural extension: A review.* Washington, DC: The World Bank.

Black, A. W. (1976). *Organizational genesis and development: A study of Australian agricultural colleges.* St. Lucia: University of Queensland Press.

Black, M. and B. Fawcett (2008) The Last Taboo: Opening the Door on the Global Sanitation Crisis. London: Earthscan

Blackburn, D. J., and Flaherty, J. (1994). Transitions and directions in extension. In D. J. Blackburn (Ed.), *Extension handbook: Processes and practices* (p. 8-17). Toronto: Thompson Educational Publishing.

Blackburn, D. J., and Vist, D. L. (1984). Historical roots and philosophy of extension. In D. J. Blackburn (Ed.), *Extension handbook* (1st ed.). Guelph, Ontario: University of Guelph.

Blanckenburg, P. V. (1984). *Agricultural extension systems in some African and Asian countries.* FAO Economic and Social Development Paper 46. Rome: FAO.

Blangy, S. (2010) "Co-construire le tourisme autochtone par la recherche action participative et les Technologies de l'Information et de la Communication: Une nouvelle approche de la gestion des ressources et des territoires". PhD thesis, Université Paul Valéry - Montpellier III.

Bloom, C. (1980). The managerial role (p. 21-46). In L. J. Goodman and R. N. Love (Eds.), *Project planning and management: An integrated approach.* New York: Pergamon Press.

Blum, A. (1987). The Israeli experience in agricultural extension and its application to developing countries. In W. M. Rivera and S. G. Schram (Eds.), *Agricultural extension worldwide - Issues, practices and emerging priorities.* London: Croom Helm.

BMZ (1990). *Querschnittsanalyse von Projekten der Landlichen Regionalentwicklung.* Bonn: Bundesministerium für Wirtschaftliche Zusammenarbeit.

Bne Saad, M. H. A. Al-h. (1990). *An analysis of the needs and problems of Iraqi farm women: Implications for agricultural extension services.* Unpublished doctoral thesis, University College, Dublin.

Bongartz, Petra *et al.* (eds) (2010) "Tales of shit: Community-Led Total Sanitation in Africa. Vol. 61. IIED, 2010.

Bonoma, T, and Slevin, D. P. (1978). *Executive survival manual.* Boston: CBI Publishing Company.

Bos, J. T. M., Proost, M. D. C., and Kuiper, D. (1991). Reorganizing the Dutch agricultural extension service: The IKC in focus. In D. Kuiper and N. G. Röling (Eds.), *Proceedings of the European seminar on knowledge management and information technology.* Wageningen, The Netherlands: Agricultural University, Department of Extension Service.

Bossard, P. (1987). Small project review: evaluation of projects. Kinshasha (Zaïre)

Boulet, M. (n.d.). Un type original d'enseignant: Le professeur departmental d'agriculture. Unpublished paper.

Bourassa, M., Bélair, L. and Chevalier, J.M. (eds.) (2007) "Les outils de la recherche participative". *Éducation et Francophonie,* vol XXXV, no 2.

Boyle, P. G. (1981). *Planning better programs.* New York: McGraw-Hill.

Braun, A.R. and Hocdé, H. (2000) "Farmer Participatory Research in Latin America: Four Cases", in W.W. Stur, P.M. Horne, J.B. Hacker and P.C. Kerridge (eds.) *Working with Farmers: The Key to Adoption of Forage Technologies*. ACIAR Publication, PR095, pp. 32–53.

Bray, P. (1984). *Joseph Needham's science and civilization in China: Vol. 6. Biology and biological technology. Pt. II: Agriculture.* Cambridge: Cambridge University Press.

Brinkman, R. (1994). Recent developments in land use planning. Keynote address at the 75-year Anniversary Conference of the Wageningen Agricultural University. In L. 0. Fresco *et al.* (Eds.), *Future of the land: Mobilising and integrating knowledge for land use options.* Chichester: John Wiley and Sons.

Brock, K. and Pettit, J. (2007) *Springs of Participation: Creating and Evolving Methods for Participatory Development*. Practical Action, Warwickshire, UK.

Brouwers, J., and Röling, N. (in press). Living local knowledge for sustainable development. Accepted for publication in D. M. Warren, S. Fujisaka, and G. Prain (Eds.), *Biological and cultural diversity: The role of indigenous experimentation in development.* London: Intermediate Technology Publications.

Brown, D. (2004) "Participation in Poverty Reduction Strategies: Democracy Strengthened or Democracy Undermined?", in S. Hickey and G. Mohan (eds) *Participation - From Tyranny to Transformation? Exploring New Approaches to Participation in Development*. Zed, London, pp. 278–283.

Brown, L.D. (1993) "Social Change through Collective Reflection with Asian Nongovernmental Development Organizations", *Human Relations*, vol 46, no 2, pp. 249–274.

Brown, L.D., and Gaventa, J. (2010) "Constructing Transnational Action Research Networks: Reflections on the Citizenship Development Research Centre", *Action Research*, vol 8, no 1, pp. 5–28.

Brown, L.D., and Tandon, R. (1983) "Ideology and Political Economy in Inquiry: Action Research and Participatory Research", *Journal of Applied Behavioral Science*, vol 1, no 2, pp. 277–294.

Brulin, G. (1998) "The New Task of Swedish Universities". *Concepts and Transformation*, vol 3, no 1/2, pp. 113–127.

Bruneau, Hilarion. (1994) *Managing Financial Resources-Principles and Recommendations for Agricultural Research Managers.* The Hague: International Service for National Agricultural Research (ISNAR).

Brunner, E. and Hsin Pao Yang, E. (1949) Rural America and the Extension Service, Columbia University

Brunner, I., and Guzman, A. (1989). Participatory evaluation: A tool to assess projects and empower people. In R. F. Conner and M. Hendricks (Eds.), *International innovations in evaluation methodology* (p. 9-18). New Directions for Program Evaluation, No. 42. San Francisco: Jossey-Bass.

Bryant, C., and White, L. G. (1982). *Managing development in the third world.* Boulder, CO: Westview Press.

Brydon-Miller, M. (2003) "Why Action Research?". *Action Research,* vol 1, no 1, pp. 9–28.

Buckland, J., and Graham, P. (1990). The Mennonite Central Committee's experience in agricultural research and extension in Bangladesh. *Agricultural Research and Extension Network Paper No. 17.* London: Overseas Development Institute.

Buckles, D.J. and Khedkar, R. (with B. Ghevde and D. Patil) (2012) *Fighting Eviction: Katkari Land Rights and Research-in-Action.* Cambridge University Press India, New Delhi.

Buford, J., Jr. and Bedeian, A. G. (1988). *Management in Extension* (2nd ed.). Alabama: Alabama Cooperative Extension Service, Auburn University.

Building the Knowledge Society. Report to Government, December (2002). Information Society Commission, IrelandArchived 21 November 2007 at the Wayback Machine. Retrieved 20 October 2009.

Bunch, R. (1991). People-centred agricultural improvement. In B. Haverkort, van der Kamp, and A. Waters-Bayer (Eds.). *Joining farmers' experiments: Experiences in participatory development.* London: IT Publications.

Bunch, R. (1991). *Low input soil restoration in Honduras: The Cantarranas farmer-to-farmer extension programme.* Sustainable Agriculture Programme Gatekeeper Series SA23. London: IIED.

Bunney, P. M. G., and Bawcutt, D. E. (1991). Making a business of an extension service. *Agricultural Progress, 66,* 36-43.

Burkey, S. (1993). *People first: A guide to self-reliant participatory development.* London: Zed Books.

Burns, D. (2007) *Systemic Action Research: A Strategy for Whole System Change.* Policy, Bristol.

Buttel, F. (1991). The restructuring of the American public agricultural research and technology transfer system: Implications for agricultural extension. In W. M. Rivera and D. J. Gustafson (Eds.), *Agricultural extension: Worldwide institutional evolution and forces for change.* Amsterdam: Elsevier.

CAAP (1991, December). Generación y transferencia de tecnología agropecuaria: Sistematización de experiencias en el CAAP. Paper presented at *Taller regional para América del Sur: Generación y transferencia de tecnología agropecuaria; el papel de las ONGs y el sector publico,* Santa Cruz, Bolivia. (Summarized in English as "El centro Andino de acción popular" in A. J. Bebbington and G. Thiele [Eds.], *NGOs and the state in Latin America: Rethinking roles in sustainable agricultural development.* London: Routledge, 1993.)

Caffarella, R. S. (1982). Identifying client needs. *Journal of Extension, 20* (July -August), 5-11.

Cameron, K.S. and Quinn, R.E. (2011) *Diagnosing and Changing Organizational Culture: Based on the Competing Values Framework*. Josey-Bass, CA.

Camic, C. and Joas, H. (2003) *The Dialogical Turn: New Roles for Sociology in the Postdisciplinary Age*. Rowman and Littlefield, Maryland.

Campbell, A. (1994). *Landcare: Communities shaping the land and the future*. Sydney: Alien and Unwin.

Campbell, D. (1982). *L'occidentalisation des campagnes francaises: Le cas du plateau ardechois*. Unpublished doctoral thesis, Université Paul Valery, Montpellier, France.

Carman, K., and Keith, K. (1994). *Community consultation techniques: Purposes, processes and pitfalls - A guide for planners and facilitators*. Brisbane: Department of Primary Industries, Queensland.

Carr, W. and Kemmis, S. (1986). *Becoming Critical: Education, Knowledge and Action Research*. Falmer, London.

Carroll, T. (1992). *Intermediary NGOs: The supporting link in grassroots development*. West Hartford, CT: Kumarian Press.

Cary, J. W. (1983). Social dilemmas in catchment management for pollution control. In *Non-point sources of pollution in Australia*. Australian Water Resources Council, Canberra: Australian Government Publishing Service.

Cary, J. W. (1993a). Changing foundations for government support of agricultural extension in economically developed countries. *Sociologia Ruralis, 33* (3/4), 334-345.

Cary, J. W. (1993b). Three eras of extension in Australia: From public folly to private good. *Australia-Pacific extension conference proceedings: Vol. 1*. Brisbane: Department of Primary Industries.

Cary, J. W., and Wilkinson, R. L. (1992). *The provision of government extension services to the Victorian farming community*. Parkville: University of Melbourne, School of Agriculture and Forestry.

Cary, J. W., and Wilkinson, R. L. (1994). *An evaluation of farm management 500: A programme promoting better business management*. Parkville: University of Melbourne, School of Agriculture and Forestry.

Case, D'A.D. (1990) *The Community's Toolbox: The Idea, Methods and Tools for Participatory Assessment, Monitoring and Evaluation in Community Forestry*. FAO, Rome.

Casley, D. J., and Kumar, K. (1987). *Project monitoring and evaluation in agriculture*. Baltimore and London: Johns Hopkins University Press (published for the World Bank).

Casley, D. J., and Lury, D. A. (1982). *Monitoring and evaluation of agriculture and rural development projects*. Baltimore and London: Johns Hopkins University Press (published for the World Bank).

Casley, D. J., and Lury, D. A. (1987). *Monitoring and evaluation of agriculture and rural development projects.* Baltimore and London: Johns Hopkins University Press (published for the World Bank).

Castelloe, Paul; Gamble, Dorothy N. (2005). *"Participatory methods in community practice: popular education and participatory rural appraisal". In Weil, Marie; Reisch, Michael. Handbook of community practice. Thousand Oaks, CA: Sage Publications. pp. 261–275.*

Catley, A., Burns, J., Abebe, D. and Suji, O. (2009) *Participatory Impact Assessment: A Guide for Practitioners.* Feinstein International Centre, Tufts University, Medford.

Cavill, S. with Chambers, R. and Vernon, N. (2015) 'Sustainability and CLTS: Taking Stock', Frontiers of CLTS: Innovations and Insights Issue 4, Brighton: IDS, p. 18.

Centro de Desarrollo Pesquero (1989). Programa nacional de pesca y acuicultura, 1989-1993. La Paz (Bolivia), Ministerio de Asuntos Campesinos y Agropecuarios

Cernea, M. (1992). The building blocks of participation: Testing a social methodology. In B. Bhatnagar and A. C. Williams (Eds.), *Participatory planning and the World Bank: Potential directions for change* (p. 96-108). Washington, DC: World Bank.

Cernea, M. M. (1991). *Putting people first* (2nd ed.). Oxford: Oxford University Press.

Cernea, M. M., and Topping, B. J. (1977). *A system for monitoring and evaluating agricultural extension projects.* World Bank Staff Working Paper No. 272. Washington, DC: World Bank.

Cervero, R. M., and Wilson, A. L. (1994). *Planning responsible for adult education.* San Francisco: Jossey-Bass.

CGIAR Secretariat. (1988). *Financial Guidelines series No. 1 - Financial management.* Washington, D.C.

CGIAR Secretariat. (1993). *Financial Guidelines series No. 2 - Accounting policies and reporting practices manual.* Washington, D.C.

Chaguma, A., and Gumbo, D. (1993). ENDA-Zimbabwe and community research. In K. Wellard and J. G. Copestake (Eds.), *NGOs and the state in Africa: Rethinking roles in sustainable agricultural development.* London: Routledge.

Chakroff, M.S., 1982. Aquaculture extension services review. Rome, FAO, Fish. Circ. No. 747

Chamala, S. (1990). Establishing a group: A participative action model. In S. Chamala and P. D. Mortiss, *Working together for Landcare: Group management skills and strategies,* p. 13-38. Brisbane: Australian Academic Press.

Chamala, S. (1991). Establishing a group: A model for participative action management. In P. D. Mortiss and S. Chamala, *Group management skills/or Landcare: A trainers' guide,* p. 33-60. Brisbane: Australian Academic Press.

Chamala, S. (1995a). Overview of participative action approaches in Australian land and water management. In S. Chamala and K. Keith (Eds.), *Participative approaches to Landcare: Perspectives, policies and programs.* Brisbane: Australian Academic Press.

Chamala, S. (1995b). Group effectiveness: From group extension methods to participative community Landcare groups. In S. Chamala and K. Keith (Eds.), *Participative approaches to Landcare: Perspectives, policies and programs.* Brisbane: Australian Academic Press.

Chamala, S., and Mortiss, P. D. (1990). *Working together for Landcare: Group management skills and strategies,* p. 368. Brisbane: Australian Academic Press.

Chambers, R. (1974). *Managing rural development.* Uppsala: Scandinavian Institute of African Studies.

Chambers, R. (1983) *Rural Development: Putting the Last First*. Longman, London.

Chambers, R. (1991, June). *Farmers' practices, professionals and participation: Challenges for soil and water management.* Paper presented at the 1991 workshop on Farmers' Practices and Soil and Water Conservation Programmes, ICRISAT.

Chambers, R. (1992a). The self-deceiving state: Psychosis and therapy. *IDS Bulletin, 23* (4), 31-42.

Chambers, R. (1992b, September). *Methods for analysis by farmers: The professional challenge.* Paper presented at the 12th Annual Symposium of the Association for FSR/E, Michigan State University, East Lansing, MI.

Chambers, R. (1992c). *Rural appraisal: Rapid, relaxed and participatory.* IDS Discussion Paper 311. Brighton: IDS.

Chambers, R. (1993). *Challenging the profession - Frontiers for rural development.* London: Intermediate Technology Publications, p. 67.

Chambers, R. (2008) "PRA, PLA and Pluralism: Practice and Theory", in *The Sage Handbook of Action Research: Participative Inquiry and Practice*. Reason, P. and H. Bradbury (eds). Sage, pp. 297–318.

Chambers, R., Pacey, A., and Thrupp, L. A. (Eds.). (1989). *Farmer first: Farmer innovation and agricultural research.* London: Intermediate Technology Publications.

Chambers, Robert (1984) [1983]. *Rural development: putting the last first.* London; New York: Longman.

Chambers, Robert (1993). *Challenging the professions: frontiers for rural development.* London: Intermediate Technology Publications.

Chambers, Robert (1997). *Whose reality counts?: putting the first last.* London: Intermediate Technology Publications. p. 106.

Chambers, Robert (April 1981). *"Rapid rural appraisal: rationale and repertoire" (PDF). Public Administration and Development.* 1 (2): 95–106.

Chambers, Robert (July 1994). "The origins and practice of participatory rural appraisal" (PDF). *World Development.* 22 (7): 953–969. doi:10.1016/0305-750X(94)90141-4.

Chandy, K. T., Agricultural Research and Education in India. Booklet No. 435, Agricultural Extension Education: AEES-4, 2002; Available at www.inseda. org/. per cent 20Agriculture per cent 20and- per cent 20Environment per cent 20Edu. (accessed on 29 August 2017).

Chattopadhyay, S., and Pareek, U. (1982). *Managing organizational change.* New Delhi: Oxford and IBH Publishing Company.

Checkland, P. B. (1981). *Systems thinking, systems practice.* Chichester: John Wiley.

Checkland, P., and Scholes, J. (1990). *Soft systems methodology in action.* Chichester: John Wiley and Sons.

Chen, M., Mitra, M., Athreya, G., Dholakia, A., Law, P., and Rao, A. (1986). *Indian women: A study of their role in the dairy movement.* New Delhi: Shakti Books.

Cherrington, D. J. (1989). *Organizational behavior: The management of individual and organizational performance.* Boston: Allyn and Bacon.

Chevalier, J.M. and Buckles, D.J. (2008) *SAS2: A Guide to Collaborative Inquiry and Social Engagement.* Sage India and IDRC, Ottawa and New Delhi.

Chevalier, J.M. and Buckles, D.J. (2013) *Participatory Action Research: Theory and Methods for Engaged Inquiry,* Routledge UK.

Choudhary, A., and Tandon, R. (n.d.). *Participatory evaluation: Issues and concerns.* New Delhi, India: Society for Participatory Research in Asia.

Chung, Yong-Bok, and Dong, Youl-Mo. (1984). *A study of the agricultural extension experiences in the Republic of Korea.* Rome: FAO.

Chung, Yong-Bok, *et al.* (1985). *Agricultural extension services in Korea.* Seodun-dong, Suwon, Republic of Korea: Rural Development Administration.

Claar, J. B., and Bentz, R. P. (1984). Organizational design and extension administration. In B. E. Swanson (Ed.), *Agricultural extension: A reference manual.* Rome: FAO.

Clark, R.C. (1973). *Basic concepts and theories of administration and supervision* (revised edition). Madison:

Clayton, E. (1983). Role, characteristics, and operational features of agricultural monitoring systems (p. 1-9). In E. Clayton and F. Petry (Eds.), *Monitoring systems for agricultural and rural development projects* (vol. 1). Rome: FAO.

CLVS (1845). Wiesenbewasserung. *Wochenblatt für Land-und Hauswirthschaft, Gewerbe und Handel* (Centralstelle des Landwirthschaftlichen Vereins zu Stuttgart), 24,133.

Coghlan, D. and Brannick, T. (2007) *Doing Action Research in Your Own Organization.* Sage, Thousand Oaks, CA.

Cohen, J. (1989) "Deliberative Democracy and Democratic Legitimacy", in A.P. Hamlin and P.N. Pettit (eds) *The Good Polity: Normative Analysis of the State.* Blackwell, Oxford, pp. 17–34.

Coletti, F. ([1900] 1985). Le associazioni agrarie in Italia della meta del secolo XVIII al XIX. Reprinted in *Il Dottore in Science Agrarie e Forestall*, 15(1985, October), 17-30.

Collinson, M.P. (ed.) (2000) *History of Farming Systems Research*. FAO, Rome.

Compton, J. L. (1984). Extension programme development. In B. E. Swanson (Ed.), *Agricultural extension: A reference manual* (p. 108-119). Rome: FAO.

Conway, G. R. (1987). The properties of agroecosystems. *Agricultural Systems, 24*, 95-117.

Cook, T, and Campbell, D. (1979). *Quasi-experimentation: Design and analysis issues for field settings*. Chicago: Rand McNally.

Cooke, B. and Kothari, U. (eds) (2001) *Participation: The New Tyranny?*. Zed, London.

Coombs, P. H., and Ahmed, M. (1974). *Attacking rural poverty: How nonformal education can help*. Baltimore and London: Johns Hopkins University Press.

Cooper, C.B., Dickinson, J., Phillips, T. and Bonney, R. (2007) "Citizen Science as a Tool for Conservation in Residential Ecosystems", *Ecology and Society*, vol 12, no 2.

Coumans, C., Moodie, S. and Sumi, L. (2009) *Mining and Health: A Community-Centred Health Assessment Toolkit*. Canary Research Institute for Mining, Environment and Health, Ottawa.

Coutts, J. (1994). *Process, paper policy and practice: A case study of the introduction of a formal extension policy in Queensland, Australia, 1987-1994*. Den Haag: Koninklijke Bibliotheek.

Covey, S. R. (1992). *The 7 habits of highly effective people*. Toronto: Simon and Schuster, p. 29.

Crézé, F. and Liu, M. (coord.) (2006) *La recherche-action et les transformations sociales*. Paris, L'Harmattan.

Cristóvão, A. (1990). Research in extension: Popular participation as a major issue. In M. Andersson and U. Nitsch (Eds.), *Proceedings of the 9th European seminar on extension education* (p. 204-229). Uppsala: Swedish University of Agricultural Sciences, Department of Extension Science, Rapport 2.

Cristóvão, A., Botelho, C., and Born Jesus, A. (1994). *Diagnóstico rápido participado*. Plancas I. S. Tome e Principe: CENFOPA, Centre de Formação Profissional Agrária.

Cromwell, E., and Wiggins, S. (1993). *Sowing beyond the state: NGOs and seed supply in developing countries*. London: Overseas Development Institute.

Cronbach, L. J., and Associates. (1981). *Toward reform of program evaluation: Aims, methods, and institutional arrangements*. San Francisco: Jossey-Bass.

Crozier, M. (2000) *À quoi sert la sociologie des organisations?* 2 vol, SeIi Arslan, Paris.

Cummings, T.G. (ed.) (2008) *Handbook of Organization Development*, Sage, London.

Dahama, O. P. (1979). *Extension and rural welfare.* New Delhi: Ram Parsad and Sons.

Davis, G.B., and Olson, M. H. (1985). *Management information systems: Conceptual foundations, structure, and development.* New York: McGraw-Hill.

Davis. L. E. and Taylor, C. (Eds.). (1979), *Design of jobs.* Santa Monica, CA: Good Year.

de Failly, D. (1970). Histoire de l'enseignement agricole au Congo (Pt. 1). *Cahiers Congolais, 13* (1), 100-133.

de Janvry, A., Marsh, R., Runsten, D., Sadoulet, E., and Zabin, C. (1989). Impacto de la crisis en la economia campesina de América Latina y el Caribe. In F. Jordan (Ed.), *La economía campesina: Crisis, reactivación, políticas* (p. 91-205). San Jose, Costa Rica: Institute Interamericano de Cooperación para la Agricultura.

De Koning, K. and Martin, M. (eds) (1996) *Participatory Research in Health: Issues and Experiences.* Zed Books, London.

De Wilde, J. C. (1967). *Experiences with agricultural development in tropical Africa: Vol. 2. The case studies.* Baltimore: Johns Hopkins University Press.

Deardon, K.M. (1998) "Participatory Action Research as Service Learning", *New Directions for Teaching and Learning,* vol 1998, no 73, pp. 57–64.

Deepak Kumar, *Science and the Raj,* Oxford University Press, 2000.

Dejours, C. (1988) *Plaisir et souffrance dans le travail.* AOCIP, Paris.

Delbecq, A. (1975). *Group techniques for program planning: A guide to nominal group technique and Delphi process.* Glenview, IL: Scott Foresman.

Delman, J. (1991). *Agricultural extension in Renshou County, China: A case-study of bureaucratic intervention for agricultural innovation and change.* Aarhus, Denmark: Institute of East Asian Studies.

Delp, P., Thesen, A., Motiwalla, J., and Seshadri, N. (1977). *System tools for project planning.* Bloomington, IN: PASITAM.

Denhardt, K. G. (1989). *The ethics of public service: resolving moral dilemmas in public organisations.* Greenwood Press.

Department of Agricultural and Extension Education, University of Wisconsin at Madison.

Desgagné, S., Bednarz, N., Couture, C., Poirier, L. and Lebuis, P. (2001) "L'approche collaborative de recherche en éducation un nouveau rapport à établir entre recherche et formation". *Revue des sciences de l'éducation,* vol 27, no 1, pp. 33–64.

Deshler, D. (1990). Concept mapping. In J. Mezirow and Associates (Eds.), *Fostering critical reflection in adulthood: A guide to transformative learning.* San Francisco: Jossey-Bass.

Dick, B. and Dalmau, T. (1991) *Values in Action: Applying the Ideas of Argyris and Schon,* Chapel Hill, Australia.

Dillman, D. A. (1978). *Mail and telephone surveys: The total design method.* New York: John Wiley and Sons.

Dillman, D. A. (1986). Cooperative extension at the beginning of the 21st century. *The Rural Sociologist, 6* (2), 102-119.

Dobbs, T. L., Becker, D. L., and Taylor. (1991). Sustainable agriculture policy analyses: South Dakota on-farm case studies. *Journal of Farming Systems Research-Extension,* 2(2), 109-124.

Doss, C. R. (1991). African professional women in agriculture: An analysis of two roundtable discussions. Development Studies Paper Series. Morrilton, AR: Winrock International Institute for Agricultural Development.

Drewes, E. (1980). Role of women in small-scale fisheries of the Bay of Bengal. Madras, India, Development of Small-Scale Fisheries in the Bay of Bengal, FAO/SIDA, (GCP/RAS/040/SWE), BOBP/REP/4, 41 p

Drewes, E. (1985). Income-earning activities for women from fishing communities in Sri Lanka. Madras, India, Development of Small-Scale Fisheries in the Bay of Bengal, FAO/SIDA (GCP/RAS/040/SWE), BOBP/REP/21, 27 p

Drucker, P. F. (1977). *People and performance.* New York: Harper's College Press.

Drucker, P. F. (1992). *Managing the non-profit organization: practices and principles.* New York: Harper.

Drucker, P. R (1966). *The Effective executive.* New York: Harper and Row.

Dube, S. C. (1958). *India's changing villages.* London: Routledge.

Dubet, F. (1991) *Les lycéens*. Seuil, Paris.

Dubet, F. (2001) *Plaidoyer pour l'intervention sociologique*, in O. Kuty and D. Vrancken (eds) *La sociologie et l'intervention: enjeux et perspectives.* De Boeke, Bruxelles, pp. 89–110.

Dubost, J. (1987) *L'intervention psychosociologique*. PUF, Paris. pp. 84-88.

Dubost, J. (1987) *L'intervention psychosociologique*. PUF, Paris. pp. 287-291.

Dubost, J. (1987) *L'intervention psychosociologique*. PUF, Paris. pp. 90-101.

Dusseldorp, D. van, and Zijderveld, K. (1991). *Preparation and implementation of development projects in the third world.* Wageningen: Wageningen Agricultural University.

Eckbom, A. (1992). *Economic impact assessment of implementation strategies for soil conservation: A comparative analysis of the on-farm and catchment approach in Trans Nwia, Kenya.* Sweden: Unit for Environmental Economics, Department of Economics, Gothenburg University.

Ecological Basis for Low-toxicity Integrated Pest Management (IPM) in Rice and Vegetables, Kevin Gallagher *et al.*, In, Jules Pretty, *The Pesticide Detox*. 2005. Earthscan [2]

Economic Commission for Africa (ECA) (1973). *Women and national development in African countries.* Position paper prepared by Human Resources Development Division, ECA. *ed. by James Wicken; et al. (2008). Beyond construction : use by all - a collection of case studies from sanitation and hygiene promotion practitioners in South Asia. Delft: IRC International Water and Sanitation Centre and WaterAid. p. 181.*

Effland, R. 1998. The Cultural Evolution of CivilizationsArchived 15 May 2016 at the Portuguese Web Archive.

Eisenberg, E.M., Baglia, J. and Pynes, J.E. (2006) "Transforming Emergency Medicine Through Narrative: Qualitative Action Research at a Community Hospital". *Health Communication*, vol 19, no 3, pp. 197–208.

Eisner, E. (1983). Educational connoisseurship and criticism: Their form and function in educational evaluation. In G. F. Madaus, M. Scriven, and D. Stufflebeam (Eds.), *Evaluation models.* Boston: Kluwer-Nijhoff.

El Titi, A., and Landes, H. (1990). Integrated farming system of Lautenbach: A practical contribution toward sustainable agriculture. In C. A. Edwards, R. Lal, P. Madden, R. H. Miller, and G. House (Eds.), *Sustainable agricultural systems.* Ankeny: Soil and Water Conservation Society.

Elam, S. (1971). *Performance based teacher education: What is the state of the art.* Washington, DC: AACTE.

Elkins, A. (1980). *Management structures, functions, and practices.* Massachusetts: Addison-Wesley, p. 539.

Elliott, H. (1994). Applying the agricultural technology systems approach: Lessons from ISNAR's experience. In R. M. A. Loyns and F. Datepa-Mupondwa (Eds.), *Management for researchers.* Department of Agricultural Economics and Farm Management, University of Manitoba. Winnipeg, Manitoba: Friesen Printers.

Elvin, M. (1973). *The pattern of the Chinese past.* London: Methuen.

Encyclopedia of public relations Archived April 16, 2009, at the Wayback Machine., by Robert L. Heath, 2005. Retrieved October 7, 2009.

Engel, P. (1995). *Facilitating innovation: An action-oriented approach and participatory methodology to improve innovative social practice in agriculture.* Published doctoral dissertation. Agricultural University, Wageningen.

Engel, S; Susilo, A (2014). "Shaming and Sanitation in Indonesia: A Return to Colonial Public Health Practices?". *Development and Change.* 45 (1): 157–178. doi:10.1111/dech.12075.

Engle, C.R. (1985). Segundo informe trimestral. Panama, Consultoria No. 16, Convenio de Prestamo No. 98-IC-PN B1/BNP/MIDA

Engle, C.R. (1986). Criteria for the evaluation of subsistence aquaculture projects. Santiago (Chile), Food and Agriculture Organization of the United Nations, FAO/RLAC/85/44-PES-8

Engle, C.R. (1987a). Analisis economico de la produccion comercial de la tilapia, colossoma, y Macrobrachium rosenbergii en monocultivo y policultivo en Panama. Revista Latinoamericana de la Acuicultura 33:7-25

Engle, C.R. (1987b). Women in training and extension services in aquaculture. In: C.E. Nash, C.R. Engle, and D. Crossetti (eds). Women in Aquaculture. Proceedings of the ADCP/NORAD Workshop on Women in Aquaculture, Rome, Italy, 13-16 April, 1987. Rome, FAO, ADCP/REP/88/28

Ennals, R. (2004) "Europe as a Knowledge Society: Integrating Universities, Corporations, and Government", *Systemic Practice and Action Research*, vol 17, no 3, pp. 237–248.

Enriquez, E. (1992) *L'organisation en analyse*. PUF, Paris.

Eponou, T. (1993). *Partners in agricultural technology.* ISNAR Research Report No. 1. The Hague: ISNAR.

Epstein, K.K., Lynch, K.M. and Allen-Taylor, J.D. (2012). *Organizing to Change a City*. Peter Lang, NY.

Esman, M. J., and Uphoff, N. T. (1984). *Local organisations: Intermediaries in rural development.* Ithaca, NY: Cornell University Press.

Evans, J. P., and Dahl, D. T. (1984). Organizing for extension communication. In B. E. Swanson (Ed.), *Agricultural extension: A reference manual.* Rome: FAO.

Evenson, R. E. (1986). The economics of extension. In G. Jones (Ed.), *Investing in rural extension: Strategies and goals* (p. 65-91). Amsterdam: Elsevier Applied Sciences Publishers.

Evenson, R. E. (1987). *Economic issues in agricultural extension policy.* New Haven, CT: Yale University, Economic Growth Center.

Evenson, R. E. (1988). Estimated economic consequences of PIDAP I and PIDAP II programs for crop production. Unpublished manuscript. New Haven, CT, Yale University, Economic Growth Center.

Evenson, R. E., and Westphal, L. E. (1994). *Technological change and technology strategy.* UNU/INTECH Working Paper No. 12.

Extension Committee on Organization and Policy (ECOP), Futures Task Force (1987). *Extension in transition: Bridging the gap between vision and reality.* Washington, DC:NASULGC.

Faeth, P. (Ed.). (1993). *Agricultural policy and sustainability: Case studies from India, Chile, the Philippines and the United States.* Washington, DC: WRI.

Fals Borda, O. (1995) "Research for social justice: Some North-South convergences", Plenary Address at the Southern Sociological Society Meeting, Atlanta, April 8.

Fals Borda, O. and Rahman, M.A. (1991) *Action and Knowledge*. Rowman and Littlefield, Lanham, Maryland.

FAO (1982). *Guidelines: Women in land and water development.* Rome: FAO, Land and Water Division.

FAO (1985). *Report of the expert consultation on linkages of agricultural extension with research and agricultural education.* Bangkok: RAPA/FAO.

FAO (1988). *Socio-economic indicators for monitoring and evaluating agrarian reform and rural development.* WCARRD: Ten years of follow-up (1979-1989). Rome: Author.

FAO (1990). *Participation in practice: Lessons learned from the FAO People's Participation Programme.* Rome: Food and Agriculture Organization, People's Participation Programme.

FAO (1990). *Report of the global consultation on agriculture extension.* Rome: Author.

FAO (1992). *Sociological analysis in agricultural investment project design.* FAO Investment Centre Technical Paper 9. Rome: Author.

FAO (1993). *Agricultural extension and farm women in the 1980s.* Rome: FAO.

FAO (1993). *The potentials of microcomputers in support of agricultural extension, education and training.* Rome: FAO.

FAO and Ministry of Agriculture, the Netherlands. (1991). FAO/Netherlands Conference on Agriculture and the Environment: The den Bosch Declaration and Agenda for Action on Sustainable Agriculture and Rural Development - Report of the Conference. Rome: FAO and the Ministry of Agriculture, Nature Management and Fisheries of the Netherlands.

FAO (1971). A comparative analysis of agricultural extension systems of eight East African countries. Addis Ababa, Economic Commission for Africa/Food and Agriculture Organization Joint Agriculture Division. Rome, FAO

FAO (1981). The peasants' charter. The Declaration of Principles and Programme of Action of the World Conference on Agrarian Reform and Rural Development. Rome, FAO

FAO (1990). *Women in agricultural development: FAO's plan of action.* Rome: FAO.

FAO (1991). *International directory of agricultural extension organizations.* Rome: FAO.

Farnworth, E. G. (1991, June). *The Inter-American Development Bank's interactions with non-governmental environmental organizations.* Paper presented at the Third Consultative Meeting on the Environment, Caracas, Venezuela.

Farrington, J., and Bebbington, A. J. (1993). *Reluctant partners? Non-governmental organisations, the state, and sustainable agricultural development.* London: Routledge.

Farrington, J., and Lewis, D. (Eds.). (1993). *NGOs and the state in Asia: Rethinking roles in sustainable agricultural development.* London: Routledge.

Farrington, J., and Martin, A. (1988). Farmer participation in agricultural research: A review of concepts and practices. *Occasional Paper No. 11.* London: Overseas Development Institute.

Farrington, J., and Martin, A. (1993). *Farmer participation in agricultural research: A review of concepts and practices.* London: ODI.

Feder, G, Murgai, R and Quizon, J.B. (2004). Sending Farmers Back to School: The impact of Farmer Field Schools in Indonesia, *Review of Agricultural Economics,* 26(1), 45-62 [7]

Feder, G., and Slade, R. (1986). The impact of agricultural extension: The training and visit system in India. *The World Bank Research Observer, 1,* 139-161.

Feder, G., Slade, R., and Sundaram, A. (1986). The training and visit extension system: An analysis of operations and effects. *Agricultural Administration, 21,* 48.

Federal Extension Service (FES). (1974). *Extension international exchange.* Washington, DC: Federal Extension Service, U.S. Dept. of Agriculture.

Federal Office of Statistics (FOS) (1966). *Rural economic survey of Nigeria: Household enquiry - 1963/64 and 1964/65.* Lagos: FOS.

Feldstein, H. S. (1994). Training in gender analysis in agricultural research: The use of case studies (chapter 32). In H. S. Feldstein and J. Jiggins (Eds.), *Tools for the field.*West Hartford, CT: Kumarian Press.

Feldstein, H. S., and Jiggins, J. (Eds.). (1994). *Tools for the field: Methodologies handbook for gender analysis in agriculture.* West Hartford, CT: Kumarian Press.

Feldstein, H. S., and Poats, S. V. (1989). *Working together: Gender analysis in agriculture.* West Hartford, CT: Kumarian Press.

Fernandez, A. (1993a). NGOs and government: A love-hate relationship. In J. Farrington and D. Lewis (Eds.), *NGOs and the state in Asia: Rethinking roles in sustainable agricultural development.* London: Routledge.

Fernandez, A. P. (1993b). *The Myrada experience: The interventions of a voluntary agency in the emergence and growth of people's institutions for sustained and equitable management of micro-watersheds.* Bangalore, India: Myrada.

Fernandez, A. P., and Mascarenhas, J. (1993). MYRADA-participatory rural appraisal and participatory learning methods. In J. Farrington and D. Lewis (Eds.), *NGOs and the state in Asia: Rethinking roles in sustainable agricultural development.* London: Routledge.

Fine, M. and Torre, M.E. (2008) "Theorizing Audience, Products and Provocation", in P. Reason and H. Bradbury (eds) *The SAGE Handbook of Action Research.* Sage, London, pp. 407–419.

Finsterbusch, K., and van Wicklen, W. A. (1989). Beneficiary participation in development projects: Empirical tests of popular theories. *Economic Development and Cultural Change, 37* (3), 573-593.

Fishkin, J.S. (2009) *When the People Speak: Deliberative Democracy and Public Consultation.* Oxford University Press, Oxford.

Flemming, P. (1989). *Strategic planning: Fact sheet.* Ontario: Ministry of Agriculture and Food, Rural Organizations and Services Branch.

Flippo, E. B. (1961). *Principles of personnel management.* New York: McGraw Hill.

Flower, Charlotte; Mincher, Paul; Rimkus, Susan (*2000*). *"Overview participatory processes in the North". PLA Notes.* ***38****: 14–18.*

Forest, L. B., and Baker, H. R. (1994). In D. J. Blackburn, (Ed.), *Extension handbook: Processes and practices* (p. 86-99). Toronto: Canada: Thompson Educational Publishing.

Forester, J. (1999) *The Deliberative Practitioner: Encouraging Participatory Planning Processes.* MIT, MA.

Fowler, F. J., and Mangione, T. W. (1990). *Standardized survey interviewing: Minimizing interviewer-related error.* Applied Social Research Methods Series Volume 18. Newbury Park, CA: Sage Publications.

Fox, J., Suryanata, K. and Herschock, P.D. (eds) (2005) *Mapping Communities: Ethics, Values, Practice.* East-West Center, Honolulu.

Freire, P. (1969) *Extension and Communication*, translated by Louise Bigwood and Margaret Marshall and re-printed in 'Education: The Practice of Freedom' (1976), Writers and Readers Publishing Cooperative

Freire, P. (1970) *Pedagogy of the Oppressed.* Herder and Herder, New York.

Freire, P. (1982) "Creating alternative research methods. Learning to do it by doing it", in Hall, B., Gillette, A. and R. Tandon (eds.) *Creating Knowledge: A Monopoly.* Society for Participatory Research in Asia, New Delhi, pp. 29–37.

French, W. L. (1982). *The personnel management process* (5th ed.). Boston: Houghton Miffin.

Fresco, L. O., Stroosnijder, L., Bouma, J., and van Keulen, H. (Eds.). (1994). *The future of the land.* Chichester: John Wiley and Sons.

Freudenberger, K. S. (1994). *Tree and land tenure: Rapid appraisal tools.* Rome: FAO.

Friedlander, F., and Brown, L.D. (1974) "Organization Development", *Annual Review of Psychology*, vol 25, pp. 313–341.

Fujisaka, S. (1991). Thirteen reasons why farmers do not adopt innovations intended to improve the sustainability of agriculture. In *Evaluation for sustainable land management in the developing world: Volume 2.* Technical Papers. IBSRAM Proceedings, 72(2), 509-522. Thailand: IBSRAM.

Funtowicz, S. O., and Ravetz, J. R. (1993). Science for the post-normal age. *Futures*, 25(7, September), 739-755.

Galvin, M (2015). "Talking shit: is Community-Led Total Sanitation a radical and revolutionary approach to sanitation?". Wiley Interdisciplinary Reviews: Water. ***2****: 9–20.*

Garforth, C. (1986). Mass media and communications technology. In G. E. Jones (Ed.), *Investing in rural extension: Strategies and goals* (p. 185-192). London and New York: Elsevier Applied Science Publishers.

Garforth, C. (1993). Rural people's organizations and extension communication in northern Thailand. *Journal of Extension System*, 9(2), 33-64.

Garforth, C. (1993). Seeing the people for the trees: Training for social forestry in Karnataka, India. *Rural Extension Bulletin, 2,* 33-39.

Garkovich, L. E. (1989). Local organizations and leadership in community development. In J. A. Christenson and J. W. J. R. Robinson (Eds.), *Community development in perspective,* p. 196-218. Ames: Iowa State University Press.

Gaujelac (de), V. (1997) "Introduction", in N. Aubert, V. de Gaujelac and K. Navridis (eds) *L'aventure psychosociologique*. Desclée de Brouwer, Paris.

Gaventa, J. and Barrett, G. (2010) "So What Difference Does it Make? Mapping the Outcomes of Citizen Engagement", *IDS Working Paper* 347, IDS, Brighton.

Gaventa, J., and Tandon, R. (2010) *Globalizing Citizens: New Dynamics of Inclusion and Exclusion*. Zed, London.

Gergen, K.J. (2009) *Relational Being: Beyond Self and Community*. Oxford University Press, New York.

Gha, D., Hopkins, M., and McGranahan, D. (1988). *Some reflections on human and social indicators for development.* Discussion Paper No. 6. Geneva: United Nations Research Institute for Social Development (UNRISD).

Gilbert, E. (1990). Non-governmental organisations and agricultural research: The experience of the Gambia. *Agricultural Research and Extension Network Paper No. 12.* London: Overseas Development Institute.

Gill, G. (1991). But how does it compare with the "real" data? *RRA Notes, 14,* 5-13.

Gill, G. (1993). OK, the data's lousy, but it's all we've got (being a critique of conventional methods). Sustainable Agriculture Programme. *Gatekeeper Series* SA38. London: IIED.

Giordano, E., Satish, S., and Farrington, J. (1993). Greenwork at Auroville: From survival to inter-institutional collaboration. In J. Farrington and D. Lewis (Eds.), *NGOs and the state in Asia: Rethinking roles in sustainable agricultural development.* London: Routledge.

Gittinger, J. P., Chernick, S., Horenstein, N. R., and Saito, K. (1990). *Household food security and the role of women.* Washington, DC: World Bank.

Giust-Desprairies, F. (1989) *L'enfant rêvé: Significations imaginaires d'une école nouvelle.* Armand Colin, Paris.

Goe, W. R., and Kenney, M. (1988). The political economy of the privatization of agricultural information: The case of the United States. *Agricultural Administration and Extension, 28* (2), 81-99.

Gold, N. (1988). Stakeholders and program evaluation: Characterizations and reflections. In A. S. Bryk (Ed.), *Stakeholder-based evaluation.* New Directions for Program Evaluation, No. 17 (p. 62-63). San Francisco: Jossey-Bass.

Goldstein, I. L. (1986). Evaluation procedures. In I. L. Goldstein, *Training in organizations: Needs assessment, development and evaluation* (2nd ed.). Monterey, CA: Brooks/Cole.

Gonsalves, J., and Miclat-Teves, A. (1993). The International Institute for Rural Reconstruction (IIRR): Developing an agroforestry kit. In J. Farrington and D. Lewis (Eds.), *NGOs and the state in Asia: Rethinking roles in sustainable agricultural development.* London: Routledge.

Gonsalves, J., Becker, T., Braun, A., Campilan, D., Chavez, H. de, Fajber, E., Kapiriri, M., Rivaca-Caminade, J. and Vernooy, R. (2005) *Participatory Research and Development for Sustainable Agriculture and Natural Resource Management: A Sourcebook,* 3 vols. IDRC, Ottawa.

Goodell, G. E. (1983). Improving administrators' feedback concerning extension, training and research relevance at the local level: New approaches and findings from South-east Asia. *Agricultural Administration, 13,* 39-55.

Gray, E. M. (1952). *The educational work of Philipp Emanuel von Fellenberg* (*1771-1844*). Unpublished master's thesis, The Queen's University, Belfast.

Greene, J. C. (1988). Stakeholder participation and utilization in program evaluation. *Evaluation Review, 12,* 91-116.

Greene, J. C., Caracelli, B. J., and Graham, W. F. (1989). Toward a conceptual framework for mixed-method evaluation designs. *Educational Evaluation and Policy Analysis, 11,* 255-274.

Greenwood, D.J. and González Santos, J. (1991) *Industrial democracy as process: participatory action research in the Fagor Cooperative Group of Mondragón.* Van Gorcum Arbetslivscentrum, Assen/Maastricht-Stockholm.

Greenwood, D.J. and Levin, M. (2006) *Introduction to Action Research: Social Research for Social Change.* Sage, CA.

Grobel, M. (1933). *The Society for the Diffusion of Useful Knowledge, 1826-1846, and its relation to adult education in the first half of the nineteenth century.* Unpublished master's thesis. University of London.

GTZ (1992). *The spark has jumped the gap.* Eschborn: Deutsche Gessellschaft für Technische Zusammenarbeit (GTZ).

Guba, E. G. (1981). Criteria for assessing the trustworthiness of naturalistic inquiries. *Educational Communication Technology Journal, 29,* 75-92.

Guba, E. G., and Lincoln, Y. S. (1989). *Fourth-generation evaluation.* Newbury Park, CA: Sage.

Guggisberg, K. (1953). *Philipp Emanuel von Fellenberg und sein Erziehungsstaat* (2 vols.). Bern, Switzerland: Verlag Herbert Lang.

Guijt, I. (1991). *Perspectives on participation: An inventory of institutions in Africa.* London: IIED.

Gulick, L., and Urwick, L. (1959). *Papers on the science of administration in extension.* Madison: National Agricultural Extension Center for Advanced Study, University of Wisconsin at Madison.

Gupta, A. K. (1981). Small farmers' credit constraints: A *paradigm. Agricultural System,* 7(4), 157-161.

Gupta, A. K. (1985). On organizing equity: Are solutions really the problem? *Journal of Social and Economic Studies, 4,295-312.*

Gupta, A.K. (2006) "From Sink to Source: The Honey Bee Network Documents Indigenous Knowledge and Innovations in India", *Innovations: Technology, Governance, Globalization*, vol 1, no 3, pp. 49–66.

Gupta, T. S. (1963). *Characteristics of effective village level workers.* Unpublished master's thesis, New Delhi, Division of Agricultural Extension, Indian Agricultural Research Institute.

Gupta, V. K. (1989). *Guide to monitoring and evaluation of small farmers' cooperatives in Asia,* p. 121. Rome: FAO.

Gupta, V. K., and Gaikwad, V. R. (1986). *Guide to management of small farmers' cooperatives,* p. 172. New Delhi: Concept Publishing Company.

Gustafson, D. J. (1990). Developing extension within a complex institutional arena. *Journal of Extension Systems, 8* (1),87-99.

Gustafson, D. J. (1994). Developing sustainable institutions: Lessons from cross-case analysis of 24 agricultural extension programmes. *Public Administration and Development, 14 ft.*

Gustavsen, B. (1985) "Workplace Reform and Democratic Dialogue", *Economic and Industrial Democracy*, vol 6, no 4, pp. 461–479.

Gustavsen, B. (2008) "Action Research, Practical Challenges and the Formation of Theory", *Action Research*, vol 6, no 4, pp. 421–437.

Gutierrez Alva, F.W. (1986). Informe anual del proyecto piloto "promocion y extension de la produccion piscicola". Departamento de Ucayali-Programacion 1985. Lima (Peru), Instituto Veterinario de Investigaciones Tropicales y de Altura

Halim, A., and Ali, M. M. (1988). Administration and management of training programmes. *Bangladesh Journal of Training and Development, 1* (2), 1-19.

Hall, B.L. (1975) "Participatory research: an approach for change", *Convergence*, vol 8, no 2, pp. 24–32.

Hall, B.L. (1981) "Participatory Research, Popular Knowledge, and Power: A Personal Reflection", *Convergence*, vol 14, no 3, pp. 6–17.

Hall, B.L. (1992) "Form Margins to Center? The Development and Purpose of Participatory Research", *The American Sociologist*, vol 23, no 4, pp. 15–28.

Hall, B.L. (2005) "In from the Cold: Reflections on Participatory Action Research From 1970-2005"], *Convergence*, vol 38, no 1, pp. 5–24.

Hambers, R. (1994) "Participatory Rural Appraisal (PRA): Analysis of Experience". *World Development*, vol 22, no 9, pp. 1253–1268.

Hanson, J. C., Johnson, D. M., Peters, S. E., and Janke, R. R. (1990). The profitability of sustainable agriculture on a representative grain farm in the mid-Atlantic region, 1981-1989. *Northeastern Journal of Agriculture and Resource Economics, 19* (2), 90-98.

Harkavy, I., Puckett, J. and Romer, D. (2000) "Action Research: Bridging Service and Research". *Michigan Journal of Community Service Learning*, Special issue, pp. 113–118.

Harter, D. (1992). Commercialization in Britain. *Interpaks Interchange*, 9(1), 5-6.

Hatch, L.U. and C.R. Engle, 1987. An economic analysis of integrated livestock-fish culture: some data from Panama. The Journal of Aquaculture in the Tropics, 2:93-105

Haverkort, B., van der Kamp, and Waters-Bayer, A. (Eds.). (1991). *Joining farmers' experiments: Experiences in participatory development*. London: IT Publications.

Hayami, Y., and Ruttan, V. W. (1985). *Agricultural development: An international perspective*. Baltimore, MD: Johns Hopkins University Press.

Hayward, J. A. (1990). Agricultural extension: The World Bank's experiences and approaches. In B. E. Swanson (Ed.), *Report of the global consultation on agricultural extension* (p. 115-134). Rome: FAO.

Hedges, L. E., and Rawls, W. (1988). *The status of agricultural education and extension in Uganda and the establishment of an agricultural education and extension department a Makerere University, Faculty of Agriculture and Forestry*. Kampala, Uganda: Makerere University, Faculty of Agriculture and Forestry.

Heikkinen, H., Kakkori, L. and Huttunen, R. (2001) "This is my truth, tell me yours: some aspects of action research quality in the light of truth theories", *Educational Action Research*, vol 9, no 1, pp. 9–24.

Hemp, P., Kaczor, C., and Zwilling, A. (1983). *Course of study: Citizen's advisory councils in education*. Champaign, IL: University of Illinois, Department of Vocational and Technical Education.

Henderson, P. A., and Singh, R. (1990). NGO-government links in seed production: Case studies from the Gambia and Ethiopia. *Agricultural Research and Extension Network Paper No. 14*. London: Overseas Development Institute.

Henk van den Berg (2004). IPM Farmer Field School: a synthesis of 25 Impact Evaluations, Wageningen University.

Hercus, J. M. (1991). The commercialization of government agricultural extension services in New Zealand. In W. M. Rivera and D. J. Gustafson (Eds.), *Agricultural extension: Worldwide institutional evolution and forces for change*. Amsterdam: Elsevier.

Heron, J. (1995) *Cooperative Inquiry: Research into the Human Condition*. Sage, London.

Herr, K.G. and Anderson, G.L. (2005) *The Action Research Dissertation: A Guide for Students and Faculty*, Sage, CA.

Hershkowitz, M. (1973). A regional ETV network: Community needs and system structure. *Report for regional education service agency of Appalachian Maryland*. Silver Spring, MD: Operations Research Inc.

Herzberg, F. (1966). *Work and nature of work*. Cleveland: World Publishing.

Heyzer, N. (1992). Gender, economic growth and poverty. *Development, 1,* 50-53.

Hickey, S. and Mohan, G. (2005) *Participation: From Tyranny to Transformation? Exploring New Approaches to Participation in Development.* Zed, London.

Hildebrand, P. E. (1981). Combining disciplines in rapid appraisal: The sondeo approach. *Agricultural Administration, 8,* 423-432.

Hilderbrand, P. E. (1986). *Perspectives on farming systems research and extension.* Boulder, CO: Lynne Rienner.

Hills, M., Mullett, J., and Carroll, S. (2007) "Community-Based Participatory Research: Transforming Multi-disciplinary Practice in Primary Health Care". *Pan American Journal of Public Health,* vol 21, no 2/3, pp.

Hinchcliffe, F., Thompson, J., Pretty, J. N., Guijt, I. and Shah, P. (eds) (1999) *Fertile Ground: The Impacts of Participatory Watershed Management.* Intermediate Technology, London.

Honadle, G. (1982). Supervising agricultural extension:

Horenstein, N. R. (1989). *Women and food security in Kenya.* Washington, DC: World Bank.

Hortan, D., Peterson, W., and Ballantyne, P. (1993). M and E principles and concepts (p. 5-19). In D. Hortan *et al.* (Eds.), *Monitoring and evaluating agricultural research: A sourcebook.* Oxon, UK: CAB International.

Horton, M. and Freire, P. (1990) *We Make the Road by Walking: Conversations on Education and Social Change.* Temple University Press, Philadelphia.

Hortona, D. and Ronald, M., Using evaluation to enhance institu-tional learning and change: recent experiences with agricultural research and development. *Agric. Syst.,* 2005, **78**, 127–142.

Hourihan, J.J. (1986). Women in development: guidelines for the fisheries sector. Report to the Asian Development Bank, Manila, the Philippines

Howell, J. (1985). *Recurrent costs and agricultural development.* London: Overseas Development Institute.

Howell, J. (Ed.). (1988). *Training and visit extension in practice.* London: Overseas Development Institute.

Hruska, A. J. (1993). Care policy promotes IPM. *International Ag-Sieve,* 5(6), 4-5.

http://planningcommission.nic.in/plans/planrel/fiveyr/welcome. html (accessed on 19 August 2012).

http://www.iari.res.in/(accessed on 12 August 2017).

http://www.icar.org.in/(accessed on 4 August 2017).

http://www.icar.org.in/en/mandate.htm

http://www.nddb.org/English/Pages/default.aspx (accessed on 7 August 2017).

https://www.theguardian.com/global-development/poverty-matters/2011/may/30/mdg-sanitation-offtrack-but-community-led-approach-is-working by Robert Chambers Accessed 2017-02-27

Huang Xiushen (1995). *Implement projects, learn skills and get benefits - Discussion about the new way of rural women's technical training.* Paper presented at international seminar on women in agriculture, Beijing, CECAT.

Huang, R. Q. (1992). *The level of cooperation among agricultural extension organizations in the greenhouse vegetable sector in the Westland area in the Netherlands.* Paper presented at the eighth World Congress for Rural Sociology, University Park, Pennsylvania State University.

Hudson, K. (1972). *Patriotism with profit: British agricultural societies in the eighteenth and nineteenth centuries.* London: Hugh Evelyn.

Hussain, S. S., Byerlee, D., and Heisey, P. W. (1994). Impacts of the training and visit extension system on farmers' knowledge and adoption of technology: Evidence from Pakistan. *Agricultural Economics, 10,* 39-47.

IBRD (1987). *World Bank experience with rural development, 1965-1986.* Washington, DC: The World Bank.

ICAR (1988). *International conference on appropriate agricultural technologies for farm women: Proceedings and recommendations.* New Delhi, ICAR.

ICRISAT. (1993). Will the pod-borer become the farmer's pall bearer? *SAT News,* p. 7-10.

Idachaba, F. (1987). Agricultural research in Nigeria: Organization and policy. In V. W. Ruttan and C. Pray (Eds.), *Policy for agricultural research.* Boulder, CO: Westview Press.

IDS/IIED (1994). *PRA and PM and E annotated bibliography.* Sussex: IDS; and London: IIED.

IFAD (1985). *Rural women in agricultural investment projects 1977-1984: A report of findings based on an assessment study of IFAD experience relating to rural women.* Nairobi: IFAD.

IIRR (International Institute of Rural Reconstruction), in collaboration with IDRC and CIDA (1998), *Participatory Methods in Community-Based Coastal Resource Management,* 3 vols, Cavites, Philippines.

Imboden, N. (1980). *Managing information for rural development projects.* Paris: Organization for Economic Co-operation and Development.

Information Society Policies at a Glance. From Europa.eu.Archived 24 March 2010 at the Wayback Machine. Retrieved 20 October 2017.

International Finance Corporation (2013). Working with Smallholders: A Handbook for Firms Building Sustainable Supply Chains. *"Archived copy". Archived from the original on 2013-12-03. Retrieved 2017-12-18.*

IRRI. (1990). *Training and technology transfer course performance objectives manual.* Manila: International Rice Research Institute.

Ison, R. (1990). *Teaching threatens sustainable agriculture.* Gatekeeper Series SA21. London: IIED.

Israel, A. (1987). *Institutional development: Incentives to performance.* Baltimore and London: Johns Hopkins University Press (published for the World Bank).

Ivanevich, J. M., Donnelly, J. H., Jr., and Gibson, J. L. (1980). *Managing for performance.* Georgetown, Ontario: Irwin Dorsey.

Jack Goody. "The Labyrinth of Kinship". New Left Review. Archived from the original on 27 September 2007. Retrieved 24 July *2007.*

Jacobs, L. C., and Chase, C. I. (1992). *Developing and using tests effectively: A guide for faculty.* San Francisco: Jossey-Bass.

Jamal, S. (1994). *Women in dairy development.* New Delhi: Concept Publishing.

James, E.A., Milenkiewicz, M.T. and Bucknam, A. (2007) *Participatory Action Research for Educational Leadership: Using Data-Driven Decision Making to Improve Schools.* Sage, CA.

James, E.A., Slater, T. and Bucknam, A. (2011) *Action Research for Business, Nonprofit, and Public Administration: A Tool for Complex Times.* Sage, CA.

Janelid, I. (1975). *The role of women in Nigerian agriculture.* Rome: FAO.

Jaques, E. (1951) *The Changing Culture of a Factory,* Tavistock, London.

Jenkins, H. M. (1884). *Report on agricultural education in north Germany, France, Denmark, Belgium, Holland, and the United Kingdom* (C.3981-1). London: Eyre and Spottiswoode.

Jhamtani, A., and Singh, Y. P. (1989). Perceptual variations of individuals on organizational environment dimensions of a development department. *Indian Journal of Extension Education, 25,* 53-58.

Jhamtani, A., and Singh, Y. P. (1992). Inter-unit variation in organizational environment of a development department. *Manpower Journal, 28,* 17-30.

Jiggins, J. (1989). Agricultural technology: Impact, issues and action (chapter 1, part I). In R. S. Gallin, M. Aronoff, and A. Ferguson (Eds.), *The women and international development annual: Vol. I.* Boulder, CO: West view Press.

Jiggins, J. (1989). How poor women earn income in sub-Saharan Africa and what works against them. *World Development, 17* (7), 953-963.

Jiggins, J. (1994). *Changing the boundaries. Women-centred perspectives on population and the environment.* Washington, DC: Island Press.

Jiggins, J., and de Zeeuw, H. (1992). Participatory technology development in practice: Process and methods. In C. Reijntjes, B. Haverkort, and A. Waters-Bayer (Eds.), *Farming for the Future.* Netherlands: Macmillan and ILEIA.

Joerges, B. (1967). Animation Rurale in Africa - Die Method-en der IRAM. *Zeitschrift für ausländische Landwirtschaft, 6,* 293-309.

Johannessen, K.S. (1996) "Action Research and Epistemology: Some Remarks Concerning the Activity-Relatedness and Contextuality of Human Language", *Concepts and Transformation*, vol. 1, no 2-3, pp. 281–297.

John P McKay, Bennett D Hill, John Buckler, Clare Haru Crowston and Merry E Wiesner-Hanks: *Western Society: A Brief History*. Palgrave Macmillan, 2009. Archived 1 January 2016 at the Wayback Machine.

Jones, G. (1986). *Investing in rural extension: Strategies and goals.* London: Elsevier Applied Science Publishers.

Jones, G. E. (1981). The origins of agricultural advisory services in the nineteenth century. *Social Biology and Human Affairs, 46* (2), 89-106.

Jones, G. E. (1982). The Clarendon letter. In G. E. Jones and M. J. Rolls (Eds.), *Progress in rural extension and community development: Vol. 1. Extension and relative advantage in rural development* (p. 11-19). Chichester: John Wiley and Sons.

Jones, G. E. (1994). Agricultural advisory work in England and Wales: The beginnings. *Agricultural Progress, 69,* 55-69*Journal d'Agriculture Pratique* (1874). 38th year, vol. 2.

Jones, G.E. and Garforth, C. (1997) *The history, development, and future of agricultural extension* in Swanson, B. "Improving Agricultural Extension: A Reference Manual (3rd Edition)" FAO

Jordan, F. (Ed.). (1989). *La economía campesina: Crisis, reactivación, políticas.* San Jose, Costa Rica: Institute Interamericano de Cooperación para la Agricultura.

Jordan, V. W. L., Hutcheon, J. A., and Glen, D. M. (1993). *Studies in technology transfer of integrated farming systems: Considerations and principles for development.* Bristol: AFRC Institute of Arable Crops Research, Long Ashton Research Station.

Jucious, M. J. (1963). *Personnel management* (5th ed.). Homewood, IL: Richard D. Irwin.

Judd, A., Boyce, J., and Evenson, R. E. (1986). Investing in agricultural supply: The determinants of agricultural research and extension investment. *Economic Development and Cultural Change, 35,* 77-113.

Kaimowitz, D., Snyder, M., and Engel, P. (1990). A conceptual framework for studying the links between agricultural research and technology transfer in developing countries. In D. Kaimowitz (Ed.), *Making the link: Agricultural research and technology transfer in developing countries.* London: Westview Press.

Kal, K and Chambers, R (2008) Handbook on Community-led Total Sanitation, Plan UK Accessed 2015-2-26

Kamp, K., Gregory, R., and Chowhan, G. (1993). Fish cutting pesticide use. *ILEIA Newsletter, 2,* 22-23.

Kang, B. T., Wilson, G. F., and Lawson, T. L. (1984). *Alley cropping: A stable alternative to shifting agriculture.* Ibadan: IITA.

Kasl, E. and Yorks, L. (2002) "An Extended Epistemology for Transformative Learning Theory and Its Application Through Collaborative Inquiry", Lyle Yorks, *Teachers College Record.*

Kath Pasteur and Preetha Prabhakaran (2014) Lessons in Urban Community Led Total Sanitation from Nakuru, Kenya, implemented by Practical Action and Umande Trust in collaboration with County Government of Nakuru, health services department, CLTS Foundation

Kaufman, R. A. (1982). *Identifying and solving problems: A systems approach* (3rd ed.). San Diego, CA: University Associates.

Keen, P. G. W., and Morton, M. S. S. (1978). *Decision support systems.* Reading, MA: Addison-Wesley.

Keller, B. (1986). Development for rural Zambian women. In 1985: *Training for agriculture and rural development.* Rome: FAO.

Kelsey, L., and Hearne, C. (1963). *Cooperative extension work.* Ithaca, NY: Comstock Publishing Associates of Cornell University Press.

Kemmis, S. and McTaggart, R. (1982) *The Action Research Planner*. Deakin University Press, Victoria, Australia.

Kemmis, S. and McTaggart, R. (2000) "Participatory action research", in N.K. Denzin and Y.S. Lincoln (eds) *Handbook of Qualitative Research* (2nd ed.). Sage, CA, pp. 567–605.

Kenmore, P. (1991). *How rice farmers clean up the environment, conserve biodiversity, raise more food, make higher profits: Indonesia's IPM - A model for Asia.* Manila, Philippines: FAO.

Kerkhof, P. (1990). *Agroforestry in Africa: A survey of project experience.* London: Panos Institute.

Kerr, J. (1994). *How subsidies distort incentives and undermine watershed development projects in India.* Paper presented at the IIED Conference, New Horizons: The Social, Economic and Environmental Impacts of Participatory Watershed Development, Bangalore, India. IIED, London.

Kesby, M. (2007) "Spatialising participatory approaches: the contribution of geography to a mature debate", *Environment and Planning A*, vol 39, no 12, pp. 2813–2831.

Kile, O. M. (1921). *The farm bureau movement.* New York: Macmillan.

Kindon, S.L., Pain, R. and Kesby, M. (eds) (2007) *Participatory Action Research Approaches and Methods: Connecting People, Participation and Places*. Routledge UK.

Kingsley, M. A., and Musante, P. (1994). *Activities for developing linkages and cooperative exchange among farmers' organisations, NGOs, GOs and researchers:Case study of IPM in Indonesia.* Paper presented at IIED symposium, Brighton.

Kirk, V., and Miller, M. L. 1986. *Reliability and validity in qualitative research.* Qualitative Research Series 1. Beverly Hills, CA: Sage Publications.

Kirkpatrick, D. (1976). Evaluation of training. In R. L. Craig (Ed.), *Training and development handbook.* New York: McGraw Hill.

Kiss, A., and Meerman, F. (1991). *Integrated pest management in African agriculture.* World Bank Technical Paper 142. African Technical Department Series. Washington, DC: World Bank.

KKU (1987). *Rapid rural appraisal: Proceedings of an international conference.* Thailand: Rural Systems Research Project, Khon Kaen University.

Knowles, M. S. (1977). *The modern practice of adult education: Andragogy versus pedagogy.* New York: Association Press.

Koehnen, T. L., Portela, J., and Cristóvão, A. (1993). Bilateral curriculum development for assistance in food security in Lusophone Africa. In A. Bonano (Ed.), *The agricultural and food sector in the new global era* (p. 195-212). New Delhi: Concept Publishing.

Kohl, B. (1991). Protected horticultural systems in the Bolivian Andes: A case study of NGOs and inappropriate technology. *Agricultural Research and Extension Network Paper No. 29.* London: Overseas Development Institute.

Koontz, H., and O'Donnel, C. (1978). *Essentials of management.* New Delhi: Tata McGraw-Hill, p. 104-105.

Korten, D. (1987). Third generation NGO strategies: A key to people-centred development. *World Development,* 75 (Supplement), 145-159.

Korten, D. C. (1980). Community organization and rural development: A learning process approach. *Public Administration Review,* 40(Sept.-Oct.), 480-512.

Korten, D. C. (1981). Rural development programming: The learning process approach. *Rural Development Participation Review,* 2(2), 1-8.

Korten, D. C. (1989). *Getting to the 21st century: Voluntary action and the global agenda.* Manila: Bookmark.

Korton, D. C. (1980). Community organizations and rural development: A learning process approach. *Public Administration Review, 40* (5), 480-511.

Krishna, A. (1993). *Watershed development in Rajasthan: The new approach.* Jaipur: Watershed Development, Government of Rajasthan.

Krueger, R. A. (1994). *Focus groups: A practical guide for applied research* (2nd ed.). Thousand Oaks, CA: Sage Publications.

Kumar, K. (Ed.). (1993). An overview of rapid appraisal methods (p. 1-22). In K. Kumar (Ed.), *Rapid appraisal methods.* Washington, DC: World Bank.

Lal, R. (1989). Agroforestry systems and soil surface management of a tropical alfisol. I: Soil moisture and crop yields. *Agroforestry Systems, 8,* 7-29.

Lapassade, G. and Lourau, R. (1971) *Clés pour la sociologie*. Segher, Paris.

Le Gouis, M. (1991). Alternative financing of agricultural extension: Recent trends and implications for the future. In W. M. Rivera and D. J. Gustafson (Eds.), *Agricultural extension: Worldwide institutional evolution and forces for change.* Amsterdam: Elsevier.

Leal Filho, W. (ed.) (2011) *Experiences of Climate Change Adaptation in Africa*. Springer, Berlin.

Lee, J. W. (1990). *More than accountability: Evaluating agricultural extension programmes.* TRDC Publication No. 172. Amidale, Australia: University of New England, Rural Development Centre.

Leeuwis, C. (1993). *Of computers, myths and modelling: The social construction of diversity, knowledge, information, and communication technologies in Dutch horticulture and agricultural extension.* Wageningen Studies in Sociology, 36. Wageningen: Agricultural University.

Leeuwis, C. and van den Ban, A. *Communication for Rural Innovation: Rethinking Agricultural Extension* (3rd Edition), Blackwell Publishing

Lenski, G. (1974). *Human Societies: An Introduction to Macrosociology.*

Leonard, D. K. (1977). *Reaching the peasant farmer: Organizational theory and practice in Kenya.* Chicago: University of Chicago Press.

Leonard, D. K. (1985). African practice and the theory of user fees. In J. Howell (Ed.), *Recurrent costs and agricultural development* (p. 130-145). London: Overseas Development Institute.

Leurs, Robert (1997). "Critical reflections on rapid and participatory rural appraisal". *Development in Practice*. 7 (3): 290–293. JSTOR 4029070.

Lévy, A. (2001) with G. Amado, *La recherche-action: perspectives internationales*. Eska, Paris.

Lévy, A. (2010) *Penser l'événement: pour une psychosociologie critique*. Parangon/Vs, Lyon.

Lewin, K. (1946) "Action Research and Minority Problems", *Journal of Social Issues*, vol 2, no 4, pp. 34–46.

Lewin, K. (1948) *Resolving Social Conflicts: Selected Papers on Group Dynamics*. Harper and Row, NY. pp. 82, 202-6.

Lewin, K. (1951) "Problems of Research in Social Psychology", in D. Cartwright (ed.) *Field Theory in Social Science; Selected Theoretical Papers*, Harper and Row, New York

Lewis, H. and Gaventa, J. (1988) *The Jellico handbook: A teacher's guide to community-based economics*. Highlander Center, Knoxville, Tenn.

Liebhardt, W, Andrews, R. W, Culik, M. N., Harwood, R. R., Janke, R. R., Radke, J. K., and Rieger-Schwartz, S. L. (1989). Crop production during conversion from conventional to low-input methods. *Agronomy Journal, 81* (2), 150-159.

Lincoln, Y. S. (1990). The making of a constructivist: A remembrance of transformations past. In E. G. Guba (Ed.), *The paradigm dialog.* Newbury Park, CA: Sage.

Lincoln, Y. S., and Guba, E. (1985). *Naturalistic inquiry.* Newbury Park, CA: Sage Publications.

Lionberger, H. F. (1968). *Adoption of new ideas and practices.* Ames, IA: Iowa State University Press.

Lionberger, H., and Chang, H. C. (1970). *Farm information/or modernizing agriculture: The Taiwan system* (p. 282-283). New York: Praeger.

Liu, M. (1997) *Fondements et pratiques de la recherche-action.* L'Harmattan, Paris.

Lobo, C., and Kochendorfer-Lucius, G. (1992). *The rain decided to help us: An experience in participatory watershed development in Maharashtra State, India.*Ahmednagar: Social Centre.

Logan, G. N. (1984). Man and the land in Queensland education 1874-1905: Agricultural education as a solution to man-physiography dissonance in colonist Queensland. Unpublished M.Ed. thesis, University of Queensland.

Long, N., and Long, A. (Eds.). (1992). *Battlefields of knowledge: The interlocking of theory and practice in social research and development.* London: Routledge.

Lord Clarendon's practical instructors in mid-nineteenth century Ireland. In *Proceedings of the 4th international seminar on extension education.* Dublin: University College, Department of Agricultural Extension.

Loudon, David L.; Bitta, Albert J. Della (1993). *Consumer behavior: concepts and applications.* McGraw-Hill Series in Marketing (4th ed.). New York: McGraw-Hill.

Lourau, R. (1970) *L'analyse institutionnelle.* Minuit, Paris.

Lourau, R. (1996) *Interventions socianalytiques: les analyseurs de l'Église.* Anthropos, Paris.

Lovshin, L.L. and R. Pretto Malca, 1983. A strategy for the use of tilapias in rural Latin America: the Panamanian integrated approach. Proceedings of the Symposium on Tilapia, Nir-David, Israel

Lovshin, L.L., N.B. Schwartz, V.G. de Castillo, C.R. Engle, and L.U. Hatch (1986). Cooperatively managed rural Panamanian fish ponds: the integrated approach. Auburn University (USA), Alabama Agricultural Experiment Station, Research and Development Series No. 33

Lucas, C. P. (1913). *A historical geography of the British colonies: Vol. Ill, West Africa* (3rd ed.). Oxford: Clarendon Press.

Lucas, H. C., Jr. (1990). *Information systems concepts for management.* New York: McGraw-Hill.

Lukenya Notes Taking Community Led Total Sanitation to Scale with Quality Outputs from a workshop in Nairobi, Kenya, 24th-27th July 2011. *(PDF). wash-liberia.org/. September 2011. pp. 3-4.* Retrieved 14 March 2015.

Lynton, R. P., and Pareek, U. (1990). *Training for development.* West Hartford, CT: Kumarian Press.

Maalouf, W. D. (1987). In W. M. Rivera (Ed.), *Planning adult learning: Issues, practices and directions* (p. 106-126). London: Croom Helm.

Maalouf, W. D., and Contado, T. E. (1983). Basic and in-service training for agricultural extension. *Training for Agriculture and Rural Development, 31,* 39-47.

Macload, R. and Kumar, D., *Technology and the Raj*, Sage Publi-cation India Pvt Ltd, 1995.

Maguire, P. (1987) *Doing Participatory Research: Feminist Approach.* University of Massachusetts, Amherst.

Maier-Bode, F. (1910). *Die Organization und die Erfolge des landwirtschaftlichen Wanderunterrichts im Königreich Bayern.* Landsberg am Lech: Georg Verza.

Majumdar, R. C., *An Advanced History of India*, Macmillan and Co Ltd, 1958, p. 944.

MALDM (1988-1994). Reports of catchment approach planning and rapid catchment analyses. Nairobi: Soil and Water Conservation Branch, Ministry of Agriculture, Livestock Development and Marketing.

Malone, V. M. (1984). Inservice training and staff development. In B. E. Swanson (Ed.), *Agricultural extension: A reference manual.* Rome: FAO.

Manalili, A. G. (1990). *Community organizing for people's empowerment.* Manila: Kapativan-Kaunlaran Foundation.

Mara, D; Lane, J; Scott, BA; Trouba, D (2010). "Sanitation and Health". *PLOS Medicine*. 7 (11): e1000363. doi:10.1371/journal.pmed.1000363.

Maredia, M. K. and Raitzer, D. A. (2012).Review and analysis of docu-mented patterns of agricultural research impacts in Southeast Asia. *Agric. Syst.*, **106**, 46–58.

Marshall, C. (1990). Goodness criteria: Are they objective or judgment calls? In E. G. Guba (Ed.), *The paradigm dialog.* Newbury Park, CA: Sage Publications.

Marshall, J., Coleman, G. and Reason, P. (2011) *Leadership for Sustainability: An Action Research Approach.* Greenleaf UK.

Marshall, P. (1992). Introduction to the management process. In *Managing people at work.* Guelph: University of Guelph Press.

Martin, J. (1990). *Telecommunications and the computer* (3rd ed.). Englewood Cliffs, NJ: Prentice-Hall.

Mascarenhas, J., Shah, P., Joseph, S., Jayakaran, R., Devararam, J., Ramachandran, V., Fernandez, A., Chambers, R., and Pretty, J. N. (1991). Participatory rural appraisal. *RRA Notes, 13.* IIED, London.

Masefield, G. B. (1950). *A short history of agriculture in the British colonies.* Oxford: Clarendon Press.

Mason, R. D., and Swanson, B. E. (1981). *Measurements for management decision.* Reading, MA: Addison-Wesley.

Massie, J. L. (1979). *Essentials of management.* New Delhi: Prentice-Hall.

Masters, J. (1995) "The History of Action Research", in I. Hughes (ed.) *Action Research Electronic Reader*. University of Sydney, Australia.

Mathur, P. N., and Sinha, B. P. (1991). Extension and communication strategies for rainfed agriculture - Indian experience. In C. Prasad and P. Das (Eds.), *Extension strategies for rainfed agriculture.* New Delhi: Indian Society of Extension Education.

Mattee, A. Z., and Lassalle, T. (1994). Diverse and linked: Farmers' organizations in Tanzania (p. 19-29). In *Linking farmers' organizations and researchers: Four case studies.* Network Paper 50. London: Overseas Development Institute.

Matteson, P. C. (1992). "Farmer First" for establishing IPM. *Bulletin of Entomological Research, 82,* 293-296.

Matteson, P. C., Gallagher, K. D., and Kenmore, P. E. (1992). Extension of integrated pest management for planthoppers in Asian irrigated rice: Empowering the user. In R. F. Denno and T. J. Perfect (Eds.), *Ecology and Management of Planthoppers.* London: Chapman and Hall.

Maunder, A.H. (1973). Agricultural extension: a reference manual (abridged edition). Rome, FAO

Mazhar, F, Buckles, D., Satheesh, P.V. and Akhter, F. (2007) *Food Sovereignty and Uncultivated Biodiversity in South Asia: Essays on the poverty of food policy and the wealth of the social landscape.* Academic Foundation/IDRC, New Delhi/Ottawa.

McConkey, D. D. (1983). *How to manage by results* (4th ed.). New York: American Management Association.

McGhee, W., and Thayer, P. W. (1961). *Training in business and industry.* New York: John Wiley and Sons.

McGregor, D. (1960). *The human side of enterprise.* New York: McGraw-Hill.

McIntyre, A. (2008) *Participatory Action Research.* Sage, CA.

McLeod, R., Jr. (1995). *Management information systems: A study of computer-based information systems* (6th ed.). New Delhi: Prentice Hall of India.

McNeil, A., and Clemmer, J. (1988). *The V.I.P. strategy-leadership skills for exceptional performance.* Toronto: Key Porter Books Ltd.

McNiff, J. (2010) *Action Research for Professional Development: Concise Advice for New (and Experienced) Action Researchers.* September Books, Dorset.

McNiff, J. and Whitehead, J. (2006) *All You Need to Know About Action Research.* Sage, London.

McNiff, J. and Whitehead, J. (2009) *Doing and Writing Action Research.* Sage, London.

McTaggart, R. (1997) "16 Tenets of Participatory Action Research", in Y. Wadsworth, *Everyday Evaluation on the Run,* Allen and Unwin.

Means, K. and Josayma, C., with E. Nielsen and V. Viriyasakultorn (2002) *Community-Based Forest Resource Conflict Management: A Training Package,* Volume 1 and Volume 2. FAO, Rome.

Mendel, G. (1980) "La sociopsychanalyse institutionnelle, *in J. Ardoino, J. Dubost, A. Levy, E, Guattari, G. Lapassade, R. Lourau and G. Mendel,* L'intervention institutionnelle. *Payot, Paris, pp. 237–301.*

Mendel, G. and Prades, J-L. (2002) *Les méthodes de l'intervention psychosociologique.* La Découverte, Paris.

Merrill-Sands, D. (1992). *Managing links with technology users: A training module.* The Hague: International Service for National Agricultural Research.

Merrill-Sands, D., and Kaimowitz, D. (1989). *The technology triangle.* The Hague: International Service for National Agricultural Research.

Mesnier, P.-M. and Vandernotte, C. (eds) (2012) *En quête d'une intelligence de l'agir,* 2 vols. L'Harmattan, Paris.

Michelot, C. (2008) "Le discours de la méthode de Guy Palmade", *Nouvelle revue de psychosociologie,* vol 1, no 5, pp. 97–104.

Middleton, J., and Hsu Lin, Y. (1975). Planning communication for family planning. Honolulu, Hawaii: East-West Communication Institute, the East West Center.

Miles, M. B., and Huberman, M. A. (1994). *Qualitative data analysis: An expanded sourcebook.* Thousand Oaks, CA: Sage Publications.

Miller, E. L., Burack, E. H., and Albrecht, M. H. (1980). *Management of human resources.* Englewood Cliffs, NJ: Prentice-Hall.

Miller, L. C., and Hustedde, R. J. (1987). Group approaches. In D. E. Johnson, L. R. Meiller, L. C. Miller, and G. F. Summers (Eds.), *Needs assessment: Theory and methods*(p. 91-125). Ames, IA: Iowa State University Press.

Ministère de l'Agriculture (1882). *Bulletin,* 1st year, vol. 1.

Ministry of Agriculture (1993). *Outline of cooperative agricultural extension service in Japan.* Tokyo: Ministry of Agriculture, Forestry and Fisheries (Extension and Education Division).

Ministry of Agriculture, Liberia, 1979. Proposal for the development of inland fisheries in the rural areas of Liberia. Monrovia, Ministry of Agriculture

Minkler, M. and Wallerstein, N. (eds) (2008) *Community-Based Participatory Research for Health: From Process to Outcomes.* Jossey-Bass, CA.

Mintzberg, H. (1988). *Mintzberg on management.* New York: Free Press.

Mintzberg. H. (1973). *The nature of managerial work.* New York: Harper and Row.

Misra, D. C. (1994). *Agricultural extension effectiveness in India (mimeo).* University of Oxford, Queen Elizabeth House, International Development Centre.

Moehl, J.F. and K.L. Veverika, (1988). Final report: Rwanda National Fish Culture Project. Auburn University (USA)

Moely, B., Billig, S.H. and Holland, B.A. (2009) *Creating Our Identities in Service-Learning and Community Engagement.* Information Age, Charlotte, NC.

Molnar, J.J. and B.L. Duncan, (1989). Monitoring and evaluating aquaculture projects. In: Impacts of Small-Scale Fishery Projects: Aspects of Monitoring and Evaluation. University of Rhode Island (USA), International Center for Marine Resource Development (in press)

Mook, B. T. (1982). *The world of the Indian field administrator.* New Delhi: Vikas Publishing House.

Moris, J. (1990). *Extension alternatives in Africa.* London: ODI.

Moris, J. (1991). *Extension alternatives in tropical Africa.* London: Overseas Development Institute, p. 184.

Morize, J., Dutilleul, J. P., and Beaulier, A. (1983). *Guide de l'agent du développement rural: Milieu physique, humain et agricole.* Presses Universitaires de France.

Morrison, J.R. (1988). Evaluation of the Peace Corps fish culture program in Zaïre (Projet Pisciculture Familiale au Zaïre). Washington D.C. (USA), U.S. Peace Corps

Morss, E.R. and D.D. Gow, (1985). Implementing rural development projects: lessons from AID and World Bank experiences. Boulder, CO (USA), Westview Press

Morss, E.R., J.K. Hatch, D.R. Mickelwait and F.C. Sweet, (1976). Strategies for small farmer development. Boulder, CO (USA), Westview Press

Mortiss, P. D., and Chamala, S. (1991). *Group management skills for Landcare: A trainers' guide,* p. 407. Brisbane: Australian Academic Press.

Mruthyunjaya and Ranjitha, P. (1998). The Indian agricultural research system: structure, current policy issues, and future orientation. *World Develop.,* **26,** 1089–1101.

Murphy, J. (1993). Good enough, soon enough. *The Rural Extension Bulletin, 1,* 27-33.

Murphy, J., and Marchant, T. (1988). *Monitoring and evaluation in extension agencies.* Washington, DC: The International Bank for Reconstruction and Development/The World Bank.

Murphy, J., and Marchant, T. J. (1988). *Monitoring and evaluation in extension agencies.* World Bank Technical Paper No. 79. Washington, DC: World Bank.

Musembi and Musyoki (2016). "CLTS and the Right to Sanitation, Frontiers of CLTS issue 8" (PDF). Brighton: IDS.

Mustafa, S., Rahman, S., and Sattar, G. (1993). Bangladesh Rural Advancement Committee (BRAC): Backyard poultry and landless irrigators programmes. In J. Farrington and D. Lewis (Eds.), *NGOs and the state in Asia: Rethinking roles in sustainable agricultural development.* London: Routledge.

NAFES (2005) *Consolidating Extension in the Lao PDR,* National Agricultural and Forestry Extension Service, Vientiane

Nagel, U. J. (1980). *Institutionalization of knowledge flows: An analysis of the extension role of two agricultural universities in India.* Frankfurt (Main): DLG-Verlag.

Nagel, U. J., Baas, S., Chiyanika, P., Eckert, S., Edsen, J., Geiger, M., Laue, R., Lubke, G., and Marbach, H. (1992). *Developing a participatory extension approach: A design for Siavonga district, Zambia.* Berlin: Centre for Advanced Training in Agricultural Development.

Narayan, D. (1993). *Focus on participation: Evidence from 121 rural water supply projects.* UNDP-World Bank Water Supply and Sanitation Program. Washington, DC: World Bank.

Narayanasamy, N. (2009). Participatory rural appraisal: principles, methods and application. Los Angeles: Sage Publications. OCLC 265732154.

Neilson, D. (1993). *Institutional reform and privatization in research and extension.* Paper presented at the Latin American Agricultural Retreat, Harpers Ferry, WV. Washington, DC, The World Bank.

Netherlands Ministry of Foreign Affairs (1988). *Women in agriculture.* Sector Papers No. 1. The Hague: Directorate General for Foreign Affairs.

Nichols, P. (1991). *Social survey methods: Afield guide for development workers.* Development Guidelines No. 6. Oxford: Oxfam.

Nielsen, K.A. and Svensson, L. (eds) (2006), *Action Research and Interactive Research: Beyond Practice and Theory*. Shaker Verlag, Hamburg.

Noel, Hayden (2009). *Consumer behaviour*. Lausanne, Switzerland La Vergne, TN: AVA Academia Distributed in the USA by Ingram Publisher Services.

Noffke, S.E. and Somekh, B. (2009) *The SAGE Handbook of Educational Action Research.* Sage, London.

Noppakun, M., and Chamala, S. (1992). Adoption of livestock practices by successful and less successful extension groups of small farmers. In P. W. Daniels, S. Holden, E. Lewin, and S. Dadi (Eds.), *Livestock services for smallholders: A critical evaluation.* Proceedings of a seminar held in Yogyakarta, Indonesia, November 10-15, 1992. Jakarta: Directorate General of Livestock Services, Indonesia.

NRC (1989). *Alternative agriculture.* Washington, DC: National Research Council; National Academy Press.

Oakley, P., and Marsden, M. (1984). *Approaches to participation in rural development.* Geneva: International Labor Office.

Obermaier, D. (1990). Promotion of women remains controversial. *D + C: Development and Cooperation, I,* 7-8.

OECD (Organization of Economic Cooperation and Development) (1989). *Survey on effects and consequences of different forms of funding agricultural services.* Paris: OECD doc. AGR/REE 89, 7.

Ogilvy, J.A. (2002) *Creating Better Futures: Scenario Planning as a Tool for a Better Tomorrow*. Oxford University Press, Oxford.

Okali, C., Sumberg, J., and Farrington, J. (1994). *Farmer participatory Research: Rhetoric and reality.* London: IT Publications.

Olawoye, J. E. (1985). Rural women's role in agricultural production: An occupational survey of women from six selected rural communities in Oyo State, Nigeria. *Nigerian Journal of Rural Sociology,* 2(1-2), 34-37.

Olawoye, J. E. (19890). Women in extension (chapter 14). In W. M. Rivera and T. O. Ogunfiditimi (Eds.), *Extension for development: Concepts and issues.* Ibadan: Lidato Press.

Olawoye, J. E. (1989a). Difficulties for rural women in securing resources for agricultural production: Two case studies from Oyo State, Nigeria. *Rural Development in Nigeria, 3* (2), 77'-81.

Olawoye, J. E. (1991). Harmonizing development efforts for rural women in Nigeria. In T. A. Olowu and J. A. Akinwumi (Eds.), *Development directions for 21st century Nigeria.* Ibadan: Ibadan Socio-Economic Group.

Olawoye, J. E. (1994a). *Women and forestry in Nigeria.* Technical Report to the Tropical Forest Action Programme, Ibadan.

Olawoye, J.E. (1994b, February-March). *Gender priorities and issues in agricultural extension delivery.* Paper presented at the inaugural conference of the Society for Agricultural Extension of Nigeria at ARMTI, Ilorin, Nigeria.

Olofin, Olutola Olayinka (OIL) (1988). *Occupational health hazards in traditional methods of processing gari and palm oil: A study of rural women in selected areas of Oyo State.* Unpublished bachelor's project. University of Ibadan.

Olukossi, J. O. (1993). Participatory research methodology: A few hints. *Newsletter National Farming Systems Research Network,* Zaria, Nigeria.

Onyango, C. A. (1987). Making extension effective in Kenya: The district focus for rural development. In W. M. Rivera and S. G. Schram (Eds.). *Agricultural extension worldwide* (p. 149-162). London: Croom Helm.

Oram, P. A., and Bindlish, V. (1984). *Investment in agricultural research in developing countries: Progress, problems, and the determination of priorities.* Washington, DC: IFPRI.

Ornes, F. (1988). Community training centres for organic agriculture and appropriate technology. Dominion Republic. In C. Conroy and M. Litvinoff (Eds.), *The greening of aid.* London: Earthscan Publications.

Osborne, D. and Gaebler, T. (1993) *Reinventing government-How the entrepreneurial spirit is transforming the public sector.* Reading, Mas: Addison-Wesley P. C. Inc.

Ospina, S., Dodge, J., Foldy, *e.g.* and Hofmann, A. (2008) "Taking the Action Turn: Lessons From Bringing Participation to Qualitative Research", in P. Reason and H. Bradbury, *The Sage Handbook of Action Research: Participative Inquiry and Practice.* Sage, CA, pp. 420–433.

Ottaway, R. N., and Terjeson, L. (1986). A proposed change evaluation scaling method. *Journal of Managerial Psychology, 1* (2), ii.

Pareek, U. (1993). *Making organizational roles effective.* New Delhi: Tata McGraw-Hill.

Pareek, U., and Rao, T. V. (1992). *Designing and managing human resource systems.* New Delhi: Oxford and IBH Publishing Company.

Paris, T, and Frio, A. L. (1994). Training in integrating a gender perspective into farming systems research (chapter 33). In H. S. Feldstein and J. Jiggins (Eds.), *Tools for the field.* West Hartford, CT: Kumarian Press.

Park, P., Brydon-Miller, M. Hall, B. and Jackson, T. (1993) *Voices of Change: Participatory Research in the United States and Canada.* Bergin and Garvey, Wesport, Conn.

Patil, Sumeet; Arnold, Benjamin; Salvatore, Alicia; Briceno, Bertha; Ganguly, Sandipan; Colford Jr., John; Gertler, Paul (2014). "The Effect of India's Total Sanitation Campaign on Defecation Behaviors and Child Health in Rural Madhya Pradesh: A Cluster Randomized Controlled Trial". PLOS Medicine. **11**: e1001709. doi:10.1371/journal.pmed.1001709. PMC 4144850/ . PMID 25157929.

Patton, M. Q. (1991). Beyond evaluation myths. *Adult Learning* (October), 9-10, 28.

Perkins, D. H. (1969). *Agricultural development in China, 1368-1968.* Chicago: Aldine Publishing; Edinburgh: Edinburgh University Press.

Perraton, H., Jamison, D., and Orival, F. (1983). Mass media for agricultural extension in Malawi (p. 147-202). In *Basic education and agricultural extension: Costs, effects and alternatives.* Staff Working Paper No. 564. Washington, DC: World Bank.

Perumal, G. (1975). *A study of job performance, psychological characteristics, communication behaviour and training aspects of the agricultural extension officers of Tamil Nadu.* Unpublished doctoral thesis. New Delhi, Division of Agricultural Extension, Indian Agricultural Research Institute.

Peters, S. (2004) "Educating the Civic Professional: Reconfigurations and Resistances", *Michigan Journal of Community Service-Learning,* vol 11, no 1, pp. 47–58.

Pétry, F. (1983). Designing a monitoring and evaluation system (p. 20-49). In E. Clayton and F. Pétry (Eds.), *Monitoring systems for agricultural and rural development projects* (vol. 1). Rome: FAO.

Philip Vincent Otieno - Defecation mapping in progress CLTS FIRE IGNITED IN DRCAccessed 2015-02-16

Phillips, L.J. and Kristiansen, M. (2012) "Characteristics and Challenges of Collaborative Research: Further Perspective on Reflexive Strategies", in *Knowledge and Power in Collaborative Research: A Reflexive Approach.* Routledge UK, ch. 13.

Pickering, Amy J; Djebbari, Habiba; Lopez, Carolina; Coulibaly, Massa; Alzua, Maria Laura (2016). Effect of a community-led sanitation intervention on child diarrhoea and child growth in rural Mali: a cluster-randomised controlled trial". *The Lancet Global Health.* 3 (11): e701–e711.

Pimbert, M. (2011) *Participatory Research and On-Farm Management of Agricultural Biodiversity in Europe.* IIED, London.

Pimbert, M., and Pretty, J. N. (1994). *Parks, people, and professionals: Putting "participation" into protected areas management.* Geneva: UNRISD and WWF; London: IIED.

Pine, G.J. (2009). *Teacher Action Research: Building Knowledge Democracies.* Sage, CA.

Plan USA and The Water Institute at UNC (2014) Testing CLTS Approaches for Scalability Accessed 2015-03-04

Plunkett, H. (1901-02). Agricultural education for Ireland. *Department of Agriculture and Technical Instruction for Ireland, Journal,* 2(1), 18-41

Poats, S. (1991). The role of gender in agricultural development. Issues in Agriculture No. 3. Washington, DC: CGIAR

Pollnac, R.B. (1985). Social and cultural characteristics in small-scale fishery development. In: M. Cernea (ed.). Putting People First. New York, Oxford University Press, Chapter 7

Pontius, J, Dilts, R, Bartlett, A. (2002). Ten Years of IPM Training in Asia - From Farmer Field School to Community IPM, FAO [4]

Portela, J. F. (1983). *Avaliação de projectos de desenvolvimento: Alguns problemas e conceitos.* Vila Real: UTAD.

Portela, J. F. (1990). Farmer-managed irrigation systems in the Trás-os-Montes Region of Portugal. In R. Yoder and J. Thurston (Eds.), *Design issues in farmer-managed irrigation systems: Proceedings of an international workshop of the farmer-managed irrigation systems network* (p. 295-299). Colombo, Sri Lanka: IRMI.

Pound, B., Snapp, S., McDougall, C. and Braun, A. (eds) (2003) *Managing Natural Resources for Sustainable Livelihoods: Uniting Science and Participation.* Earthscan/ IDRC, Ottawa.

Practices and procedures for improving field performance. *Agricultural Administration, 9,* 29-45.

Pratt, B., and Loizos, P. (1992). *Choosing research methods: Data collection for development workers.* Oxford: Oxfam.

Pretto Malca, R. (1985). La acuicultura y el desarrollo rural: capacitacion y promocion. Santiago (Chile), Organizacion de las Naciones Unidas para la Agricultura y la Alimentacion, PES-6

Pretty, J. N. (1994). Alternative systems of inquiry for sustainable agriculture. *IDS Bulletin,* 25(2), 37-48. University of Sussex: IDS.

Pretty, J. N. (1994a). Policies that work for a more sustainable agriculture. London: IIED.

Pretty, J. N. (1994b). Alternative systems of inquiry for sustainable agriculture. *IDS Bulletin,* 25(2), 37-48. University of Sussex: IDS.

Pretty, J. N. (1995). *Regenerating agriculture: Policies and practice/or sustainability and self-reliance.* London: Earthscan Publications; and Washington DC: National Academy Press.

Pretty, J. N., and Chambers, R. (1993). *Towards a learning paradigm: New professionalism and institutions for sustainable agriculture.* IDS Discussion Paper DP 334. Brighton: IDS.

Pretty, J. N., and Howes, R. (1993). *Sustainable agriculture in Britain: Recent achievements and new policy challenges.* London: IIED.

Pretty, J. N., and Sandbrook, R. (1991, October). *Operationalizing sustainable development at the community level:*

Pretty, J. N., and Shah, P. (1994). *Soil and water conservation in the 20th century: A history of coercion and control.* Rural History Centre Research Series No. 1. Reading: University of Reading.

Pretty, J. N., Guijt, I., Scoones, I., and Thompson, J. (1995). *A trainers' guide to participatory learning and action.* IIED Training Materials Series No. 1. London: IIED.

Pretty, J. N., Thompson, J., and Kiara, J. K. (1994). Agricultural regeneration in Kenya: The catchment approach to soil and water conservation. *Ambio.*

Pretty, J., Guijt, I., Thompson, J. and Scones, I. (1995) *Participatory Learning and Action: A Trainer's Guide*. IIED, London.

Pretty, Jules N.; Vodouhê, Simplice D. (1997) [1984]. "Using rapid or participatory rural appraisal". In Swanson, Burton E.; Bentz, Robert P.; Sofranko, Andrew J. Improving agricultural extension: a reference manual (2nd ed.). Rome: Food and Agriculture Organization of the United Nations. OCLC 456586976.

Primary environmental care. Paper presented to the DAC Working Party on Development Assistance and the Environment, OECD, Paris.

Proceedings of the 1985 International Conference on Rapid Rural Appraisal: Khon Kaen University. Khon Kaen, Thailand: Published by Rural Systems Research Project and Farming Systems Research Project for Khon Kaen University. 1987. OCLC 19025969.

Prokopenko, J. (1987). *Productivity management: A practical handbook.* Geneva: International Labour Office (ILO).

Proost, J., and Röling, N. (1991). "Going Dutch" in extension. *Interpaks Interchange, 9* (1), 3-4.

Province of Ontario (1900). *Report of the superintendent of farmers' institutes of the Province of Ontario, 1899-1900.* Toronto: L. K. Cameron.

Provus, M. M. (1971). *Discrepancy evaluation/or education program improvement and assessment.* Berkeley, CA: McCutchan.

Quarry, W. and Ramirez, R. (2009) *Communications for Another Development: Listening Before Telling*. Zed, London.

Quigley, B. (2000) "The practitioner-research: a research revolution in literacy?", *Adult Learning*, vol 11, no 3, pp. 6–8.

Raab, R. T., Swanson, B. E., Wentling, T. L., and dark, C. D. (Eds.). (1987). *A trainer's guide to evaluation.* Rome: FAO.

Raheja, S. K., and Jai Krishna (1991). *Manual for monitoring and evaluation of T&V agricultural extension system.* New Delhi: Centre for Agricultural and Rural Development Studies.

Rahman, Md. A. (2008) "Some Trends in the Praxis of Participatory Action Research", in P. Reason and H. Bradbury (eds) *The SAGE Handbook of Action Research*. Sage, London, pp. 49–62.

Rahman, Md. A. (2011) *Tagore: Social and Environmental Thinking of Rabindranath Tagore in the Light of Post-Tagore World Development*. Bangla Academy, Dhaka.

Rahnema, M. (1992). Participation. In W. Sachs (Ed.), *The development dictionary*. London: Zed Books.

Rajeswari, S. (1995). Agricultural research effort: conceptual clarity and measurement. *World Develop.*, **23**, 617–635.

Rama, B. R., Etling, A. W. W., and Bowen, B. E. (1993). Training of farmers and extension personnel. In R. K. Samanta (Ed.), *Extension strategy for agricultural development in 21st century*. New Delhi: Mittal Publications.

Ramesh Babu, A., and Singh, Y. P. (1987). Management information system in an agricultural extension organization. In *Proceedings of the national seminar on management of information system in management of agricultural extension* (p. 1-15). Hyderabad: NIRD.

Ramesh Babu, A., and Singh, Y. P. (1990). Agricultural administration at block level: A case study. *Indian Journal of Extension Education, 26* (1 and 2), 88-90.

Rao, A., Appel, M., Moore, K., and Feldstein, H. S. (Eds.). (1994). *Reflections and learnings: Gender trainers workshop report*. Amsterdam: Royal Tropical Institute; and New York: Population Council.

Rao, C. S. S. (1985). Agricultural extension management system in India: Past, present and modalities in future. *Indian Journal of Extension Education, 21* (1 and 2), 32-35.

Rauch, T. (1993). *Regional rural development RRD update: Elements of a strategy for implementing the RRD concept in a changed operational context*. Eschborn: Deutsche Gesellschaft für Technische Zusammenarbeit (GTZ).

Reason, P. (ed.) (1995) *Participation in Human Inquiry*. Sage, London.

Reason, P. and Bradbury, H. (2008) (eds) *The Sage Handbook of Action Research: Participative Inquiry and Practice*. Sage, CA.

Reij, C. (1988). The agroforestry project in Burkina Faso: An analysis of popular participation in soil and water conservation. In C. Conroy and M. Litvinoff (Eds.), *The greening of aid* (p. 74-77). London: Earthscan Publications.

Reijntjes, C., Haverkort, B., and Waters-Bayer, A. (1992). *Farming for the future: An introduction to low input and sustainable agriculture*. London: Macmillan.

Reijntjes, C., Haverkort, B., and Waters-Bayer, A. (1992). *Farming for the future: An introduction to low-external-input and sustainable agriculture*. The Information Centre for Low-External-Input and Sustainable Agriculture (ILEIA). London: Macmillan.

Report of the Recess Committee on the establishment of a Department of Agriculture and Industries for Ireland (1896). Dublin: Browne and Nolan.

Research projects on CLTS. *Projects. Sustainable Sanitation Alliance*. Retrieved 18 June 2015.

Rhoades, R. (1990). *The coming revolution in methods for rural development research.* Mimeo. Manila, Philippines: User's Perspective Network International Potato Center.

Rice, A.K. (2003) *Learning for Leadership*. Routledge, London.

Rivera, W, and Gustafson, D. (1991). *Agricultural extension: Worldwide institutional evolution and forces for change.* Amsterdam: Elsevier.

Rivera, W. M. (1989). An overview of agricultural extension systems. In A. M. Kesseba (Ed.), *Technology systems for small farmers: Issues and options.* Boulder, CO: Westview Press, with the International Fund for Agricultural Development.

Rivera, W. M. (1993). Impacts of extension privatization. *Journal of Extension,* Fall, 28-29.

Rivera, W. M. (1994). *Rapport de la deuxième mission du consultant coordinateur.* Projet PNUD/FAO-T.

Rivera, W. M., and Gustafson, D. J. (1991). New roles and responsibilities for public sector agricultural extension: The impact of multi-institutional activities. In W. M. Rivera and D. J. Gustafson (Eds.), *Agricultural extension: Worldwide institutional evolution and forces for change.* Amsterdam: Elsevier.

Rivera, W. M., and Gustafson, D. J. (1991). *Agricultural extension: Worldwide institutional evolution and forces for change.* Amsterdam: Elsevier.

Rivera, W. M., and Schram, S. G. (1987). *Agricultural extension worldwide: Issues, practices and emerging priorities.* New York: Croom Helm.

Roberts, N. (Ed.). (1989). *Agricultural extension in Africa.* World Bank Symposium. Washington, DC: World Bank.

Robinson-Pant, Anna (1995). "PRA: a new literacy?". PLA Notes. **24**: 78–82.

Roche, C. (1992). *Operationality in turbulence: The need for change.* London: ACORD.

Rocheleau, D.E. (1994) "Participatory Research and the Race to Save the Planet: Questions, Critique, and Lessons from the Field", *Agriculture and Human Values,* vol 11, no 2, pp. 4–25.

Rogers, E. M. (1962 and 1983). *Diffusion of innovations.* New York: Free Press.

Rogers, E. M. (1995). *Diffusion of innovations (4th ed.).* New York: Free Press.

Rogers, Everett M. (1962). *Diffusion of innovations* (1st ed.). New York: Free Press of Glencoe. OCLC 254636.

Rogers, Everett M. (1983). *Diffusion of innovations* (3rd ed.). New York: Free Press of Glencoe.

Rogers, F. E., and Olmsted, A. G. (1957). *Supervision in the cooperative extension service.* Madison, WI: National Agricultural Extension Center for Advanced Study.

Roling, N. (1988). *Extension science: Information systems in agricultural development.* Cambridge: Cambridge University Press.

Röling, N. (1991). Knowledge systems in their political context. In *Proceedings of the international workshop: Agricultural knowledge systems and the role of extension*. Bad Boll, May 21-24, 1991.

Röling, N. (1992, October). *Facilitating sustainable agriculture: Turning policy models upside down.* Paper presented at IIED/IDS Conference, Brighton.

Röling, N. (1993). Agricultural knowledge and environmental regulation: The crop protection plan and the Koekoekspolder. *Sociologia Ruralis, 33* (2), 212-231.

Roling, N. (1994). Agricultural knowledge and information systems. In D. Blackburn (Ed.), *Extension handbook: Processes and practices* (p. 56-67). Toronto: Thompson Educational Publishing.

Röling, N. (1994a). Platforms for decision making about ecosystems. In L. Fresco (Ed.), *The future of the land.* Chichester: John Wiley and Sons.

Röling, N. (1994b). Creating human platforms to manage natural resources: First results of a research programme. In *Proceedings of the International Symposium on Systems Oriented Research in Agriculture and Rural Development* (p. 391-395). Montpellier, France, November21.25.1994.

Roling, N. and Wagemakers, A. Editors.(1998), *Facilitating Sustainable Agriculture: Participatory learning and adaptive management in times of environmental uncertainty,* Cambridge University Press

Röling, N., and Engel, P. (1989). IKS and knowledge management: Utilizing indigenous knowledge in institutional knowledge systems. In D. Warren *et al.* (Eds.), *Indigenous knowledge systems: Implications for agriculture and international development.* Studies in Technology and Social Change, 11. Ames: Technology and Social Change Program, Iowa State University.

Röling, N., and Jiggins, J. (1993). Policy paradigm for sustainable farming. *European Journal of Agricultural Education and Extension,* 7(1), 23-44.

Röling, N., and Van de Fliert, E. (1994). Transforming extension for sustainable agriculture: The case of integrated pest management in rice in Indonesia. *Agriculture and Human Values, 11* (2, 3), 96-108.

Rondinelli, D. A. (1987). Administrative decentralization of agricultural and rural development programs in Asia: A comparative analysis. In W. Rivera and S. Schram (Eds.), *Agricultural extension worldwide* (p. 22-57). London: Croom Helm.

Rossi, P. H., and Freeman, H. E. (1982). *Evaluation: A systematic approach.* Beverly Hills, CA: Sage.

RRA Notes (1988-1995). Issues 1 to the present. London: Sustainable Agriculture Programme.

Ruddell, E. D. (1994). *A simplified methodology/or training peasant farmers how to conduct site-specific scientific field trials in deprived rural areas.* 1994 Conference Refereed Papers. Association for International Agricultural and Extension Education. Tenth Annual Conference, Arlington, VA.

Rushkoff, D. (2004) *Open Source Democracy*. Text. EBook-No. 10753.

Russ Dilts. (2001). From farmers field schools to community IPM: scaling up the IPM movement, LEISA magazine. 17/3 [1]

Russell, D. B., and Ison, R. L. (1991, April). *The research-development relationship in rangelands: An opportunity for contextual science.* Plenary paper presented at the 4th International Rangelands Congress, Montpellier, France.

Russell, E. J. (1966). *A history of agricultural science in Great Britain, 1620-1954.* London: George Alien and Unwin.

Russell, H. M. (1979). A review of management information systems for agriculture. In H. M. Russell (Ed.), *Information for agriculture: Proceedings of the national workshop on agricultural information* (p. 41-51). Melbourne: Department of Agriculture, Victoria.

Ruthenberg, I. (1985). *Innovation policy for small farmers in the tropics: The economics of technical innovations for agricultural development.* Oxford: Oxford University Press.

Sachdeva, R. K. (1990). *Management handbook of computer usage.* Oxford: NCC Blackwell.

Sadowske, P. S. (1991). *Towards futuristic program planning in adult education.* Unpublished doctoral dissertation, University of Wisconsin-Madison.

Saito, K., and Weidemann, C. J. (1990). *Agricultural extension for women farmers in Africa.* Washington, DC: World Bank.

Samanta, R. K. (1994). They reap less than they sow. *The Hindu* (April), No. 7. Madras: India.

Sanders, D. H. (1988). *Computers today* (3rd ed.). New York: McGraw-Hill.

Satish, S., and Vardhan, T. J. P. S. (1993). Action for world solidarity and the red-headed hairy caterpillar. In J. Farrington and D. Lewis (Eds.), *NGOs and the state in Asia: Rethinking roles in sustainable agricultural development.* London: Routledge.

Saville, A. H. (1965) Extension in Rural Communities: A Manual for Agricultural and Home Extension technician Workers. Oxford University Press.

Saville, A. H. (1965) Extension in Rural Communities: A Manual for Agricultural and Home Extension technician Workers. Oxford University Press.

Schön, D.A. (1983) *The Reflective Practitioner: How Professional Think in Action*. Basic Books, NY.

Schultz, T. W. (1964). *Transforming traditional agriculture.* New Haven, CT: Yale University Press.

Schulz, M. (1973). *Landwirtschaftliche Neuerungsverbreitung an der Elfenbeinkuste.* Saarbrucken: Verlag der SSIP-Schriften.

Scoones, I., and Thompson, J. (1994, October). *Beyond farmer first: Rural people's knowledge, agricultural research and extension practice: Towards a theoretical framework.* Paper presented at the Beyond Farmer First Workshop, Institute of Development Studies, University of Sussex, Brighton.

Scoones, I., and Thompson, J. (Eds.). (1994). *Beyond Farmer First.* London: IT Publications.

Scott, J.C., 1976. The moral economy of the peasant. New Haven, CT (USA), Yale University Press

Scriven, M. (1972). Pros and cons about goal-free evaluation. *Evaluation Comment, 3,* 1-5.

Search results for 'participatory rural appraisal' and 'participatory learning and action' on Google Ngram Viewer

Sebillotte, M. (2007) "Quand la recherche participative interpelle le chercheur", in M. Anadón (ed.) *La recherche participative: Multiples regards*. Presses de l'Université du Québec, Québec, pp. 49.84.

Second World Summit on the Knowledge Society. Retrieved 20 October 2009. Archived 30 October 2009 at the Wayback Machine.

Seetharaman, S. P., and Shingi, P. M. (1992). *Agribusiness cooperatives,* p. 210. ICA Domust Trust. New Delhi: Oxford and IBH Publishing Compan.

Selener, J.D. (1997) *Participatory Action Research and Social Change: Approaches and Critique*. Cornell University, NY.

Seligman, M.E.P. (2002) *Authentic Happiness: Using the New Positive Psychology to Realize Your Potential for Lasting Fulfillment*. Free Press, NY.

Sen, C. K. (1993). Nepal: Group extension. *Rural Extension Bulletin, 3,* 17-23. University of Reading, UK.

Senge, P. M. (1990). The leader's new work: Building learning organizations. *Sloan Management Review,* p. 7-22.

Senge, P. M. (1990). *The fifth discipline: The art and practice of the learning organization.* New York: Doubleday.

Senge, P.M. and Scharmer, C.O. (2001) "Community Action Research: Learning as a Community of Practitioners, Consultants and Researchers", in P. Reason and H. Bradbury (eds) "Handbook of Action Research: Participative Inquiry and Practice. *Sage, London, pp. 195-205.*

Shah, P. (1994). Participatory watershed management in India: The experience of the Aga Khan Rural Support Programme. In I. Scoones and J. Thompson (Eds.), *Beyond Farmer First.* London: IT Publications.

Shah, P., and Mane, P. M. (1993). Networking in participatory training and extension: Experiences of the Aga Khan Rural Support Programme in Gujarat State. In J. Farrington and D. Lewis (Eds.), *NGOs and the state in Asia: Rethinking roles in sustainable agricultural development.* London: Routledge.

Shaner, W. W., Phillipp, P. F., and Schmehl, W. R. (1982). *Farming systems research and development: Guidelines for developing countries.* Boulder, CO: Westview Press.

Sherman, F.T. and Torbert, W. (2000) *Transforming Social Inquiry, Transforming Social Action: New Paradigms for Crossing the Theory/Practice Divide in Universities and Communities*. Kluwer Academic Publishers, Boston.

Shingi, P. M., and Bluhm, L. H. (1987). Participation in irrigation projects: Changing patterns in northwestern India. In H. K. Schwarzweller (Ed.), *Research in Rural Sociology and Development. Volume 3*, p. 65-84. Greenwich: Jai Press

Shiva, V. (1991). *Most farmers in India are women.* New Delhi: FAO.

Shotter, J. (2012) "Situated Dialogic Action Research Disclosing 'Beginnings' for Innovative Change in Organizations". *Organizational Research Methods*, vol 13, no 2, pp. 268–85.

Sigler, R; Mahmoudi, L; Graham JP. (2015). "Analysis of behavioral change techniques in community-led total sanitation programs". *Health Promot Int*. 30 (1): 16–28.

Simon, H. A. (1977). The new science of management decision. New Jersey: Prentice-Hall.

Singh, B. D., Organisation for crop improvement in India. In *Plant Breeding: Principles and Methods*, Kalyani Publishers, Ludhiana, 2001, pp. 801–830.

Singh, Kamal (2001). "Handing over the stick: the global spread of participatory approaches to development". In Edwards, Michael; Gaventa, John. *Global citizen action*. Boulder, Colorado: Lynne Rienner Publishers. pp. 175–187.

Singh, Y. P., and Ramesh Babu, A. (1985). Basic management issues in extension. *Indian Journal of Extension Education*, 27 (1 and 2), 20-31.

Slade, R. H., and Feder, G. (1985). *Training and visit extension: A manual of instruction* (*mimeo*). Washington, DC: World Bank.

Smith, H. (1972). *The Society for the Diffusion of Useful Knowledge, 1826-1846*. London: Vine Press.

Smith, J. K. (1990). Alternative research paradigm and the problems of criteria. In E. Guba (Ed.), *The paradigm dialog*. Newbury Park, CA: Sage Publications.

Smith, L., Ronsenzweig, L. and Schmidt, M. (2010) "Best Practices in the Reporting of Participatory Action Research: Embracing Both the Forest and the Trees", *The Counseling Psychologist*, vol 38, no 8, pp. 1115–1138.

Smith, V. (1994, Summer). A learner centred approach to social skills for technical foresters. In *From the field*. Rural Development Forestry Network Paper 17e (p. 12-20). London: Overseas Development Institute.

Snyder, M. (1990). *Women: The key to ending hunger*. The Hunger Project Papers No. 8.

Society for the Diffusion of Useful Knowledge (1827). *Rules of the Society/or the Diffusion of Useful Knowledge*. London: William Clowes.

Sofranko, A., and Khan, A. (1988). It is not that simple. Improving needs assessment. *Journal of Extension, 26* (Winter), 14-16.

Sollows, J., Thongpan, N., and Leelapatra, W. (1993). NGO-governmental interaction in ricefish culture in NE Thailand. In J. Farrington and D. Lewis (Eds.), *NGOs and the state in Asia: Rethinking roles in sustainable agricultural development*. London: Routledge.

Sontheimer, Sally; Callens, Karel; Seiffert, Bernd (1999). "*PRA tool box*". Conducting a PRA training and modifying PRA tools to your needs: an example from a participatory household food security and nutrition project in Ethiopia. Rome: Food and Agriculture Organization of the United Nations.

Sotomayor, O. (1991). GIA and the new Chilean public sector: The dilemmas of successful NGO influence over the state. *Agricultural Research and Extension Network Paper No. 30.* London: Overseas Development Institute.

Spring, A. (1985). The women in agricultural development project in Malawi: Making gender free development work (p. 71-76). In R. S. Gallin and A. Spring (Eds.), *Women creating wealth: Transforming economic development.* Washington, DC: Association for Women in Development.

Spring, A. (1986). Reaching female farmers through male extension workers in Malawi. In *1985 - Training for agriculture and rural development.* Rome: FAO.

Sriskandarajah, N., Bawden, R. J., and Packham, R. G. (1991). Systems agriculture: A paradigm for sustainability. *Association for Farming Systems Research-Extension Newsletter*, 2(2), 1-5.

State Extension Leaders Network (SELN) (2006). 'Enabling Change in Rural and Regional Australia: The role of extension in achieving sustainable and productive futures', online resource: www.seln.org.au.

Staudt, K. (1975). Women farmers and inequities in agricultural services. *Rural Africana, 29,* 81-93.

Stevens, H. (1980). Problems of agricultural extension in Africa. *Training for agriculture and rural development, 21,* 23-37.

Stevenson, W. J. (1989). *Introduction to management science.* Boston: Irwin.

Stock, T. (1994). *The impact of integrated pest management farmer field schools in the Cordillera region of the Philippines.* Unpublished master's thesis. University of Reading.

Stocking, M. (Ed.). (1994). *Report of the Expert Consultation on Integrating Environmental and Sustainable Development Themes into Agricultural Education and Extension Programmes.* Rome: FAO.

Stoecker, R. (1999) "Are Academics Irrelevant? Roles for Scholars in Participatory Research", *American Behavioral Scientist* vol 42, no 5, pp. 840-54.

Stone, Diane (1999). "Learning lessons and transferring policy across time, space and disciplines". *Politics*. Wiley. 19 (1): 51–59.

Stone, Diane (2000). "Non-governmental policy transfer: the strategies of independent policy institutes". *Governance*. Wiley. 13 (1): 45–70.

Stone, Diane (2004). "Transfer agents and global networks in the 'transnationalization' of policy". *Journal of European Public Policy*. Taylor and Francis. 11 (3): 545–566.

Stoner, J. A. R, and Freeman. R. E. (1992). *Management* (5th ed.). New Delhi: Prentice-Hall of India.

Stringer, E.T. (2007) *Action Research in Education*. Prentice Hall.

Stufflebeam, D. L. (1971). *Educational evaluation and decision making.* Itasca, IL: F. E. Peacock.

Sülzer, R., and Payr, G. (1990). Socio-economic development approach: "Animation Rurale" in francophone Africa. In H. Albrecht *et al.* (Eds.), *Agricultural extension: Vol. 2. Examples and background material,* p. 31-35. Eschborn: Deutsche Gesellschaft fiir Technische Zusammenarbeit (GTZ).

Summers, G. F. (1987). Democratic governance. In D. E. Johnson, L. R. Meiller, L. C. Miller, and G. F. Summers (Eds.), *Needs assessment: Theory and methods* (p. 3-19). Ames, IA: Iowa State University Press.

Summers, J. C. (1977). *Dimensions of program effectiveness and accountability.* Morgantown: West Virginia University, Office of Research and Development.

Susianto, A, Purwadi, D, Pontius, J. (1998). Kaligondang: A Case History of an IPM Sub-district, FAO

Swanson, B. (Ed.). (1984). *Agricultural extension: A reference manual.* Rome: FAO.

Swanson, B. (Ed.). (1990). *Global Consultation on Agricultural Extension: A report.* Rome: FAO.

Swanson, B. E. (1993). *Identifying linkage problems using systems analysis: A training module.* Urbana, IL: International Program for Agricultural Knowledge Systems.

Swanson, B. E., and Claar, J. B. (1984). The history and development of agricultural extension. In *Agricultural extension: A reference manual.* Washington, DC: FAO.

Swanson, B. E., and Peterson, W. E. (1989). *Afield manual for analyzing agricultural technology development and transfer systems.* Urbana, IL: International Program for Agricultural Knowledge Systems.

Swanson, B. E., and Peterson, W. E. (1991). Strengthening research-extension linkages to address the needs of resource-poor farmers in rainfed agriculture. In C. Prasad and P. Das (Eds.), *Extension strategies for rainfed agriculture.* New Delhi: Indian Society of Extension Education.

Swanson, B. E., Farmer, B. J., and Bahal, R. (1990). The current status of agricultural extension worldwide. In B. E. Swanson (Ed.), *Report of the global consultation on agricultural extension.* Rome: FAO.

Swanson, B. E., Farner, B. J., and Bahal, R. (1990). The current status of agricultural extension worldwide. In FAO, *Report of the global consultation on agricultural extension,* p. 43-76. Rome: FAO.

Swanson, B. E., Farner, B. J., and Bahal, R. (1990). The current status of agricultural extension worldwide. In B. E. Swanson (Ed.), *Report of the Global Consultation on Agricultural Extension.* Rome: FAO.

Swanson, B. E., Farner, B. J., and Bahal, R. (1990). The current status of agricultural extension worldwide. In B. E. Swanson (Ed.), *Report of The Global Consultation on Agricultural Extension.* Rome: FAO.

Swanson, B. E., Roling, N., and Jiggins, J. (1984). Extension strategies for technology utilization. In B. E. Swanson (Ed.), *Agricultural extension: A reference manual.* Rome: FAO.

Swanson, B. E., Sands, C. M., and Peterson, W. E. (1990). Analyzing agricultural technology systems: Some methodological tools. In R, G. Echeverria (Ed.), *Methods for diagnosing research system constraints and assessing the impact of agricultural research* (*Vol. I*). The Hague: International Service for National Agricultural Research.

Swanson, B.E. (Ed.). (1984). *Agricultural extension: A reference manual.* Rome: FAO.

Swantz, M.L. (2008) "Participatory Action Research as Practice", in P. Reason and H. Bradbury, *The Sage Handbook of Action Research: Participative Inquiry and Practice*. Sage, CA, pp. 31–48.

SWCB (1994). The impact of the catchment approach to soil and water conservation: A study of six catchments in Western, Rift Valley and Central provinces of Kenya. Nairobi: Ministry of Agriculture, Livestock Development and Marketing.

Tajima, S. (1991). Processes of development of agricultural extension in Japan. In *Agricultural extension in Asia and the Pacific* (p. 19-36). Tokyo: Agricultural Productivity Organization.

Tandon, R. (ed.) (2002) *Participatory Research: Revisiting the Roots*, Mosaic, New Delhi.

TCPS 2 (2014) the latest edition of Tri-Council Policy Statement: Ethical Conduct for Research Involving Humans . Panel on Research Ethics.

The Addis Agreement: Using CLTS in urban and peri-urban areas *(PDF). 2016.*

The CLTS approach". *Community-Led Total Sanitation. 2011.* Retrieved 12 March 2015.

The Information Society. Indiana University. Archived 7 October 2009 at the Wayback Machine. Retrieved 20 October 2016.

The Intervention Design and Analysis Scorecard: A Planning. : Journal of Occupational and Environmental Medicine". Retrieved 2015-09-21.

Thiele, G., Davies, P., and Farrington, J. (1988). Strength in diversity: Innovation in agricultural technology development in eastern Bolivia. *Agricultural Research and Extension Network Paper No. 1.* London: Overseas Development Institute.

Thomas-Slayter, B., Esser, A. L., and Dale Shields, M. (1993). *Tools of gender analysis: A guide to field methods for bringing gender into sustainable resource management.* Worcester, MA: dark University.

Timmer, W. J. (1982). *The human side of agriculture: Theory and practice of agricultural extension.* New York: Vantage Press.

Todhunter, C. (2001) "Undertaking Action Research: Negotiating the Road Ahead", *Social Research Update*, Issue 34.

Tolman, D.L. and Brydon-Miller, M. (eds) (2001) *From Subjects to Subjectivities: A Handbook of Interpretive and Participatory Methods*. New York University Press, NY.

Torbert, W.B. (2004) *Action inquiry: The Secret of Timely and Transforming Leadership.* Berrett-Koehler, SF.

Torrans, L. (1988). Review of the U.S. Peace Corps' Assistance to the Liberian inland fisheries (fish culture) program. Pine Bluff, AR (USA), Cane Creek Farm

Torrans, L. (1989). A practical approach to tilapia culture in Africa. Pan-African Fisheries Conference. Washington D.C. (USA), U.S. Peace Corps, Information Collection and Exchange

Tosquelles, F. (1984) *Éducation et psychothérapie institutionnelle.* Hiatus, Nantes, France.

Tosquelles, F. (1992) *L'enseignement de la folie.* Dunod, Paris.

Touraine, A., Hegedus, Z., Dubet, F. and Wievorka, M. (1980) *La prophétie antinucléaire.* Seuil, Paris.

Trager, J. (1996) *The Food Chronology,* Aurum Press, London

Tripodi, T, Fellin, P., and Epstein, I. (1971). *Social program evaluation: Guidelines for health education and welfare administration.* Itasca, IL: F. E. Peacock.

Trist, E.L. and Bamforth, K.W. (1951) "Some Social and Psychological Consequences of the Longwall Method of Coal-Getting", *Human Relations,* vol 4, no 1, pp. 3–38.

Triulzi, L. (2001) "Empty and populated landscapes: the Bedouin of the Syrian Arab Republic between "development" and "state"," in *Land Reform, Land Settlement and Cooperatives,* 2, pp. 30-47.

True, A. C. (1895). Education and research in agriculture in the United States. In *Yearbook of the U.S. Department of Agriculture, 1894* (p. 81-116). Washington, DC: U.S. Government Printing Office.

True, A. C. (1900). Agricultural education in the United States. In *Yearbook of the U.S. Department of Agriculture, 1899* (p. 157-190). Washington, DC: U.S. Government Printing Office.

True, A. C. (1928). *A history of agricultural extension work in the United States, 1785-1923.* USDA, Miscellaneous Publication No. 15. Washington, DC: U.S. Government Printing Office.

Trujillo, E. (1991, December). *Investigación y extensión por El Ceibo en el Alto Beni.* Paper presented at the Regional Workshop for South America, "The generation and transfer of agricultural technology: The role of NGOs and of the public sector." Santa Cruz, Bolivia.

Tusser, T. ([1580] 1984). *Five hundred points of good husbandry.* Oxford: Oxford University Press.

Tyndale-Biscoe, P, Bond, M, Kidd, R (2013). ODF Sustainability Study, Plan Australia

U.S. Department of Agriculture (USDA) (1993). Agriculture Data. Washington, DC.

U.S. Office of Technology Assessment (OTA) (1986). *Technology, public policy and the changing structure of American agriculture.* Washington, DC.

Umali, D. L., and Schwartz, L. (1994). *Public and private agricultural extension: Beyond traditional frontiers.* Washington, DC: The World Bank.

UN/90/005, Assistance technique au développement de la vulgarisation agricole en Tunisie. Rome: FAO.

UNDP (1991). *Agricultural extension programme advisory note.* New York: United Nations Development Programme.

UNDP (1992). *The benefits of diversity: An incentive towards sustainable agriculture.* New York: UNDP.

UNICEF (2008). *"Field Notes: UNICEF Policy and Programming in Practice" (PDF).* UNICEF. Retrieved 12 March 2015.

UNICEF (2009) Field notes: UNICEF Policy and Programming in Practice - Community Approaches to Total Sanitation Accessed 2015-03-04

UNICEF (2010). CLTS Training manual for natural leaders - UNICEF and Sierra Leone Government, Freetown, Sierra Leone

UNICEF (2014) CLTS in Fragile and Insecure Contexts: Experience from Somalia and South Sudan, UNICEF WASH Field Note, Eastern and Southern Africa Sanitation and Hygiene Learning Series

UNIFEM and IWTC (1990). *Reaching rural women with information on technology.* New York: UNIFEM.

United Nations. (1991). *Government financial management in least developed countries.* New York: United Nations Department of Technical Cooperation for Development.

Uphoff, N. (1989). *Afield methodology for participatory self-evaluation of P.P.P. group and inter-group association performance.* Ithaca, NY: People's Participation Programme of the U.N. Food and Agriculture Organization.

Uphoff, N. (1992). *Learning from Gal Oya: Possibilities for participatory development and post-Newtonian science.* Ithaca, NY: Comell University Press.

Uphoff, N. (1992). *Participatory evaluation of rural development.* New York: IFAD Monitoring and Evaluation Division for the Monitoring and Evaluation Panel.

Urban, A.M., and Rojas, M. H. (1994). *Shifting boundaries: Gender, migration and community resources in the foothills of Choluteca, Honduras.* ECOGEN Case Studies Series. Worcester, MA: dark University.

US AID (1988). Project plan 1989-1994: fish culture expansion project. Kinshasha (Zaïre), United States Agency for International Development

US Embassy, Zaïre (1984). Transmittal to US AID Director (Zaïre). Kinshasha (Zaïre), United States Agency for International Development

US Peace Corps (1988). Trainee assignment criteria: freshwater fish culture. Washington D.C. (USA), U.S. Peace Corps

USAID (1987). *Women in development: A.I.D.'s experience, 1973-1985.* AID Program Evaluation Report No. 18. Washington, DC: Agency for International Development.

Van de Fliert, E. (1993). *Integrated pest management: Farmer field schools generate sustainable practices: A case study in central Java evaluating IPM training*. WU Papers 93-3. Published doctoral dissertation. Wageningen: Agricultural University.

Van de Fliert, E., van Elsen, K., and Nangsir Soenanto, F. (1993). Integrated rat management: A community activity. Results of a pilot programme in Indonesia. *FAO Plant Protection Bulletin, 41* (3).

Van Den Ban, A.W, and Hawkins, H. S. (1988). *Agricultural extension.* New York: Longman Scientific and Technical, p. 268-271.

Van der Poel, M. M., and Van Woerkum, C. J. M. (1994). *De positionering van communicatie in Beleid.* Deelrapport 1 (literatuuronderzoek) en Deel 2 (Weergave van bijeenkomsten, conclusies van het onderzoek en aan-bevelingen voor vervolgonderzoek). Wageningen: Agricultural University, Department of Communication and Innovation Studies, carried under a contract from the Government Information Office.

Van Dorsal, W. R. (1962). *The successful supervisor.* New York: Harper and Row.

Van Weeperen, W., Röling, N., Van Bon, K., and Mur, P. (1995). Introductie geïhtegreerde akkerbouw: Het veranderingsproces. Ervaringen van akkerbouwers bij het omschakelen naar een geintegreerde bedrijfsvoering. Leiystad: Informatie en Kenniscentrum Landbouw, afdeling AGV.

Vander Werf, E., and de Jager, A. (1992). Ecological agriculture in South India: An agro-economic comparison and study of transition. Den Haag: LEI and *etc.* Foundation, Leusden.

Vasimalai, M. (1993). Professional Assistance for Development Action (PRADAN): An NGO de-mystifies and scales down technology. In J. Farrington and D. Lewis (Eds.), *NGOs and the state in Asia: Rethinking roles in sustainable agricultural development.* London: Routledge.

Venkataramanan, V (2012) "Systematic Literature Review (Grey Literature) of Publications on Community-led Total Sanitation" The Water Institute at the University of North Carolina at Chapel Hill Accessed 2015-02-16

Venkataramanan, V; Rowe, R (2014). *"Research Summary of a Systematic Literature Review of Grey Literature Publications on Community-led Total Sanitation"*. The Water Institute at the University of North Carolina at Chapel Hill. Retrieved 2 January 2016.

Vereijken, P. (1992). A methodic way to more sustainable farming systems. *Netherlands Journal of Agricultural Science, 40,* 209-223.

Vereijken, P., Wijnands, F., Stol, W., and Visser, R. (1994). Progress reports of research network on integrated and ecological arable farming systems for EU and associated countries. Wageningen: DLO Institute for Agrobiology and Soil Fertility.

Verhagen, V. (1987). *Self-help promotion: A challenge to the NGO community.* Amsterdam: CEBEMO/Royal Tropical Institute.

Vernooy, R. (2003) *Seeds that Give: Participatory Plant Breeding*. IDRC, Ottawa.

Vickers, M.H. (2006) *Working and Caring For a Child with Chronic Illness: Disconnected and Doing it All*. Palgrave Macmillan.

Vijayaragavan, K. (1994). *Agricultural administration in India.* New Delhi: Concept Publishing Company.

Vijayaragavan, K., and Singh, Y. P. (1989). Job design and unit structural characteristics of department of agriculture. *Indian Journal of Extension Education,* 25, 1-12.

Vijayaragavan, K., and Singh, Y. P. (1991). Supervisory behaviour in agricultural departments. *Indian Journal of Extension Education, 27,* 16-22.

Vijayaragavan, K., and Singh, Y. P. (1992). Pay administration in agricultural departments. *Indian Journal of Extension Education, 28,* 60-64.

Voss, P. R., Tordella, S. J., and Brown, D. L. (1987). Role of secondary data. In D. E. Johnson, L. R. Meiller, L. C. Miller, and G. F. Summers (Eds.), *Needs assessment:Theory and methods* (p. 156-170). Ames, IA: Iowa State University Press.

Vroom, V.H., and Yetton, P. W. (1973). *Leadership in decision-making.* Pittsburgh: University of Pittsburgh Press.

Wakeford, T., Singh, J., Murtuja, B., Bryant, P. and Pimbert, M. (2007) "The Jury is Out: How Far Can Participatory Projects Go Towards Reclaiming Democracy?", in P. Reason and H. Bradbury (eds) *The SAGE Handbook of Action Research.* Sage, London, pp. 333–349.

Waldron, M.W. (1984). Management of adult and extension education. In D. Blackburn (Ed.), *Extension handbook.* Guelph: University of Guelph Press, p. 127-140.

Waldron, M.W. (1994a). Models for the future. In M. Brooke and M. Waldron (Eds.), *University continuing education in Canada: Current challenges and future opportunities.* Toronto: Thompson Educational Publishing.

Waldron, M.W. (1994b). Management and supervision. In D. Blackburn (Ed.), *Extension handbook: Processes and practices.* Toronto: Thompson Educational Publishing.

Walker, A. B. (1993). Recent New Zealand experience in agricultural extension. *Australia-Pacific extension conference proceedings: Vol. 1.* Brisbane: Department of Primary Industries.

Wang, Caroline C. (1999) "Photovoice: a participatory action research strategy applied to women's health", *Journal of Women's Health,* vol 8, no 2, 185-192.

Warren, D. (1991). The role of indigenous knowledge in facilitating a participatory approach to agricultural extension. Paper presented at the International Workshop on Agricultural Knowledge Systems, Bad Boll, Germany.

Warren, D.M., Slikkerveer, L.J. and Brokensha, D. (eds) (1995) *The Cultural Dimension of Development: Indigenous Knowledge Systems*. Intermediate Technology, London.

Water Aid (2011) Revitalising Community led Total Sanitation: A process guide. Accessed 2015-03-04

Watson, A. S., Hely, R., O'Keeffe, M., Cary, J. W, and dark, N. (1992). *Review of field-based services in the Victorian Department of Food and Agriculture.* Melbourne: Agmedia.

Watson, H. R., and Laquihon, W. A. (1993). The Mindanao Baptist Rural Life Center's sloping agricultural land technology (SALT) research and extension in the Philippines. In J. Farrington and D. Lewis (Eds.), *NGOs and the state in Asia: Rethinking roles in sustainable agricultural development.* London: Routledge.

Wejnert, Barbara (2002). "Integrating models of diffusion of innovations: a conceptual framework". *Annual Review of Sociology. Annual Reviews.* 28: 297–326.

Wellard, K., and Copestake, J. G. (Eds.). (1993). *NGOs and the state in Africa: Rethinking roles in sustainable agricultural development.* London: Routledge.

Wentling, T. L. (1992). *Planning for effective training: A guide to curriculum development.* Rome: FAO.

Wentling, T. L., and Wentling, R. M. (1993). *Introduction to microcomputer technologies. Rome: FAO.*

WEPA (2013) Community-based Sanitation lessons learned from Sanimas Programme in Indonesia Accessed 2015-03-04

Werner, F. and Totterdill, P. (eds) (2004) *Action Research in Workplace Innovation and Regional Development*. John Benjamins, Amsterdam.

Werther, W. B., Jr., and Davis, K. (1982). *Personnel management and human resources.* Tokyo: McGraw-Hill International Book Company.

Westfall, J.M., VanVorst, R.F., Main, D.S. and Herbert, C. (2006) "Community-Based Participatory Research in Practice-Based Research Networks", *Annals of Family Medicine*, vol 4, no 1, pp. 8–14.

White, K. D. (1970). *Roman farming.* London: Thames and Hudson.

White, K. D. (1977). *Country life in classical times.* London: Eiek Books.

White, L. G. (1989). Public management in a pluralistic arena. *Public Administration Review, 49 (6)*, 522-531.

Whitehead, J. (1993) *The Growth of Educational Knowledge: Creating Your Own Living Educational Theory*. Hyde, Bournemouth.

Whitehead, J. and McNiff, J. (2006) *Action Research Living Theory*. Sage, London.

WHO (World Health Organization) (1989). *Evaluation of programmes to ensure food safety.* Geneva: Author.

Whyte, W.F. (ed.) (1991) *Participatory Action Research*. Sage, CA.

Wiest, J, D., and Levy, F. K. (1982). *A management guide to PERT/CPM.* New Delhi: Prentice-Hall of India.

Wiggins, S. (1986). Agricultural policy and agricultural extension: The African experience. In G. E. Jones (Ed.), *Investing in rural extension: Strategies and goals* (p. 99-106). London: Elsevier Applied Service Publishers.

Williams, J. and Lykes, M.B. (2003) "Bridging Theory and Practice: Using Reflexive Cycles in Feminist Participatory Action Research", *Feminism and Psychology*, vol 13, no 3, pp. 287–294.

Willis, J. C. (1922). *Agriculture in the tropics: An elementary treatise* (3rd ed.). Cambridge: Cambridge University Press.

Wilson, M. (1991). Reducing the costs of public extension services: Initiatives in Latin America. In W. M. Rivera and D. J. Gustafson (Eds.), *Agricultural extension: Worldwide institutional evolution and forces for change.* Amsterdam: Elsevier.

Winarto, Y. (1993). Farmers' agro-ecological knowledge construction: The case of integrated pest management among rice farmers on the north-west coast of West Java. *Rural People's Knowledge, Agricultural Research and Extension Practice, 1* (3), London: IIED.

Wise, C. R. (1990). Public service configurations and public organizations: Public organization design in the post-privatization era. *Public Administrations Review,* 50(2), 141-155.

Worden, P. E., and Neumaier, P. (1987). *A source book for program evaluation and accountability.* Fort Collins: Colorado State University, State Cooperative Extension.

WORDOC (1988). Women in agriculture. *African Notes,* ^(Special Number). Institute of African Studies, Ibadan.

World Bank (1991a). *World Development Report 1991: The challenge of development.* Oxford: Oxford University Press.

World Bank (1991b). *Trends in developing countries 1991.* Washington, DC: World Bank.

World Bank (1992a). *Designing and implementing agricultural extension for women farmers.* Technical Note. Washington, DC: World Bank, Women and Development Division.

World Bank (1992b). *Towards a gender strategy for Nigeria: Integrating women's issues into the development agenda.* Washington, DC: World Bank.

World Bank (1994). *Agricultural extension: Lessons from completed projects.* Washington, DC: World Bank.

World Bank (1994). *Fiscal Systems for Agricultural Extension,* unpublished.

World Bank (1994, April). *The World Bank and participation.* Report of the Learning Group on Participatory Development. Washington, DC: World Bank.

World Bank. (1994). *Agricultural extension: lessons from completed projects, Operations Evaluation Department Report No. 13000.* Washington, D.C.: World Bank.

WSIS Implementation by Action Line. From ITU.int.Archived 26 March 2012 at the Wayback Machine. Retrieved 20 October 2009.

Young, A. (1989). *Agroforestry for soil conservation.* Wallingford: CAB International.

Yuki, G. (1994). Effective leadership in open systems. *Leadership in organizations* (3rd ed.). Englewood Cliffs, NJ: Prentice-Hall.

Yuryevic, A. (1994). *Community-based sustainable development in Latin America: The experience of CLADES.* Paper presented at IIED Symposium, Brighton.

Zangwill, B. (1977). *A compendium of laws and regulations requiring needs assessment.* Washington, DC: Office of the Assistant Secretary for Planning and Evaluation, U.S. Department of Health, Education, and Welfare.

Zijp, W. (1993). Being a good communicator doesn't solve all of extension's problems. *Development Communication Report,* 80, 20.

Zijp, W. (1994). *Improving the transfer and use of agricultural information: A guide to information technology.* World Bank Discussion Paper 247. Washington, DC: The World Bank.

Index

C

D

E

F

G

H

N

O

P

Q

R

S

T